Native Plants of the Sydney District

Native Plants

of the

Sydney District

An Identification Guide

Alan Fairley and Philip Moore

Kangaroo Press

in association with

The Society for Growing Australian Plants – NSW Ltd

To Marie and Chiew...

Foreword

Having lived in the Sydney area for most of my life, it was not until I had finished study at university that my awareness of the unique flora of the Sydney District was awakened. As a child my parents had taken me on picnics to many of the favourite parks and on holidays to the Central Coast where I spent many hours wandering through the bush in Bouddi. The bush was just scrub that prickled and needed to be avoided so that one did not get scratched.

In spring, though, it was a different story. The whole countryside burst into a profusion of colour, scent and flowers. I still recall being chastened for picking a bunch of beautiful pink flowers *(Eriostemon australasius)* which I thought were so incredible.

The fascination with the flora of the Hawkesbury sandstone and its related area of Wianamatta shales has been prsent since Sir Joseph Banks first recorded his finds at Botany Bay. In the 1900s people went to National Parks to pick wildflowers just to satisfy their interest. Not only did they collect them, but they drew, painted and wrote about them in many varied ways out of sheer curiosity. Such was the interest that Joseph Maiden and Ernst Betche in 1916 published the *Census of New South Wales Plants* which concentrated mainly on the Sydney District. Even at that stage it was noted that 'the Census will, as a matter of fact, be out of date the day after publication'.

We are fortunate that the Royal National Park was considered waste land between Sydney and the Illawarra and that this land was set aside as a recreational park. Its worth today as a resource of Sydney's unique flora is still being discovered. People always marvel when they discover for the first time the beauty of the waratah, boronia or the flannel flower which have grown and evolved on poor soils for millions of years. The Sydney native flora, however, is much more than this. It grows in a diverse range of habitats from heath rock communities to rainforests and includes orchids, grasses and ferns as well as many sclerophyll plants.

It is through this fascination and interest that Alan and Philip came together to record the flora. The Society for Growing Australian Plants in New South Wales had long recognised the need to have more than a botanical record of the flora. A descriptive catalogue of the flora including photographs was required which could be used by the bushwalker, the gardener and the enthusiast to learn more about the form and beauty of Sydney's unique native flora.

The Society learned of Alan's and Philip's interest to record the flora and encouraged them to proceed with the work. Many hours have been spent by the authors, in conjunction with members of the Society for Growing Australian Plants on many occasions to locate, to photograph and to record the habitat of the flora so that this book could be presented in such an extensive way. It will fill the gap in the understanding between the botanist and the keen observer. In this way it will serve to help all people to understand Sydney's native plants and the need to conserve them rather than ignore them and watch their gradual destruction.

Ray Page,
Vice President,
Past President,
The Society for Growing Australian Plants - NSW Ltd

Reprinted in 1995
First published in 1989 by Kangaroo Press Pty Ltd
3 Whitehall Road Kenthurst NSW 2156 Australia
PO Box 6125 Dural Delivery Centre NSW 2158
Printed in Hong Kong through Bookbuilders Ltd

ISBN 0 86417 261 3

Fairley, Alan.
Native plants of the Sydney district.

Bibliography.
Includes index.
ISBN 0 86417 261 3.

1. Botany - New South Wales - Sydney Region - Classification. 2. Plants - Identification. I. Moore, Philip, 1939- . II. Society for Growing Australian Plants - NSW. III. Title.

581.9944'1

Contents

Introduction

The Sydney district is one of the great wildflower regions of Australia. With over 2000 native plant species, it compares favourably with other recognised wildflower areas, such as the sand-plains of south-west Western Australia, the Grampians in Victoria, the wallum heathlands in Queensland and the arid Centre after rains. This book illustrates and describes nearly 1500 of these plants. It is an identification guide that includes all the common and conspicuous native species, many rare and restricted plants and most of the small and inconspicuous herbs and shrubs that are often overlooked. The amazing variety shown in the following pages is proof, if any is needed, that the Sydney district is the home of a significant and enduring part of the nation's floral wealth.

Botanists who collected in the early years of the settlement at Sydney were overwhelmed by the strangeness and diversity as much as by the staggering number of new plants they found. The impact which the plant life had on early Europeans is evident in the naming of Botany Bay by James Cook in April 1770. Accompanying Cook were the botanists Joseph Banks and Daniel Solander who collected enough plants in their short stay to require 2400 sheets of drying paper. Banks's collection aroused such interest in Europe that the great botanist, Linnaeus, suggested that the newly discovered southern land be called Banksia. Throughout the nineteenth century, Australia was a new world open to botanical enquiry, attracting many scientists and enthusiastis who worked with boundless energy exploring, collecting, drawing and classifying.

Today, over 200 years after the first European settlement, Sydney's flora still evokes a special fascination. Increasing numbers of people are visiting national parks and other natural areas specifically to see, photograph and study our native plants. There has also been a strong trend in both public and private landscaping towards the growing of native plants. Current attitudes are reflected in the healthy level of activity of such organisations as the Society for Growing Australian Plants, in the popularity of plant identification courses at tertiary colleges and in the large range of wildflower books available in bookshops.

Many people enjoy the special feeling of 'being in the bush', and are aware of its nuances and beauty without needing to know the names of particular plants. For others, a familiarity with plants and their names provides an added dimension. There is much pleasure to be had by the discovery of a 'new' plant or the rediscovery of an old favourite. We may become observant of the subtle interaction of plants with each other and with the environment. Often without noticing, we develop an insight into plant relationships and begin to see a degree of order in the apparent confusion.

Over the years numerous books have been published which assist in the identification of species in the Sydney district, but because of the richness of species in the area none has been completely satisfactory. Some cover large areas (e.g., New South Wales and Southern Queensland, South-eastern Australia) so that the Sydney flora is dealt with inadequately. Others, limited by size, illustrate or describe a tiny sample, perhaps only one or two species from common genera. The comprehensive text *Flora of the Sydney Region* by N.C.W. Beadle, O.D. Evans and R.C. Carolin, which classifies and describes all the native plants and many naturalised exotics, is the standard reference for botanists and students. It is, however, sparsely illustrated and its specialised botanical language is often a deterrent to the interested amateur. In addition, many monographs dealing with genera and families provide detailed information for the professional botanist.

This book seeks to fill the gap between the inadequate and the incomprehensible. It features a large number of photographs, descriptions in simple language and a limited area of coverage. It is not possible because of restrictions of space and cost, nor practicable, to include every native species in the area. Nevertheless, if it is a tree, a shrub or herbaceous plant that you wish to identify, you will probably find it illustrated in this book. Excluded are some of the sedges and grasses which are difficult to distinguish in photographs and which have limited popular appeal. Also excluded are introduced plants, some rarer ferns and orchids, species with doubtful records and a few plants occurring on the fringes of the area which are atypical of those in the Sydney district.

What the book contains

This is an identification guide, not just to flowering shrubs, but to the full range of native plants in the area. Its coverage extends beyond the flora of the sandstone areas to include plants of the rainforests, fresh water and saline swamps, the drier western mountains and the ancient eroded rocks of fringe areas. The introductory section includes a detailed account of the ecological and geographical features of the area, a geological outline and a discussion of plant communities. This is followed by the main section consisting of illustrations and descriptions of individual species. They are arranged in families, in a modern system of classification (see Classification and Arrangement, page 25). Family descriptions highlight similarities between genera in a family and the probable relationship of the family to other families. Generic and specific descriptions are designed to aid identification by pointing to those characteristics that help distinguish one genus from another and one species from another. Although botanical terms have been kept to a minimum, they cannot be avoided entirely, especially when describing ferns, eucalypts, sedges and grasses which need specific attention to details for which botanical terminology is the most efficient. Simple identification keys are provided, together with a glossary and an index of names. The keys are based on readily observable features and require no special botanical training or knowledge.

Photographs have been chosen on the basis of their assistance in identification, not necessarily because they are 'pretty pictures'. Photographs showing habit, leaves and fruit may be more useful than close-ups of flowers. It should be remembered that most photographs were taken *in situ*, often under difficult conditions. Plant names follow the latest nomenclature. As revision of classification is an ongoing process, some name changes will inevitably occur. For example, further revisions of important genera such as *Prostanthera* and *Leptospermum*, will result in additional name changes. Previous names of recently revised genera are cross-referenced in the index.

Each species description is followed by notes on habitat, distribution and flowering time. This information is based on records, research and personal observation over 20 years; however, plants seldom conform to strict categorisation and these records must be regarded as a guide only. Details of distribution beyond the Sydney district are also given for completeness and to make the book useful to those who live or travel beyond the area's boundaries.

A historical record

Two hundred years of European settlement, and growth of the Sydney district into the most populous region in Australia, have obviously had a deleterious effect on the flora. Much of the natural environment of the city area and its immediate surrounds has been destroyed. The degradation continues with widespread clearing for housing and agriculture, excessive burning and 'hazard reduction', logging of sensitive areas and pollution of streams. It is 15 years since we naively embarked on what appeared to be a simple project. In that time we have come to appreciate the complexity of the Sydney flora and to be aware of its vulnerability. Each year we have returned to familiar locations to look for specific plants only to find that the roadside has been cleared, a house has been built or fire has altered the habitat. More by luck than by design, there are still many beautiful areas of bushland close to the city where people can find the abundance and variety of plants that delighted earlier Australians. This book is, to some extent, a record of these plants in the late 1980s. It is also a plea for their protection, so that in another 200 years the people of Sydney might still derive the same inspiration and pleasure from them.

The Sydney Environment

The area covered by this book extends from Newcastle in the north to the Shoalhaven River in the south and from the coast in the east to the Great Dividing Range in the west. This area of about 23 000 square kilometres includes most of the geological unit known as the Sydney Basin and encompasses the catchment area of the Nepean-Hawkesbury Rivers (except the upper Wollondilly River). It also represents the botanical subdivisions of central coast and the central tablelands east of the Great Dividing Range. With some slight variation in the north and north-west, this is the same area covered by *Flora of the Sydney Region* by N.C.W. Beadle, O.D. Evans and R.C. Carolin, which has been the standard reference since it was first published as the *Handbook of the Vascular Plants of the Sydney District and Blue Mountains* in 1963.

Boundaries in the north and west are determined by dividing ranges. The Watagan Mountains and the complex of sandstone ridges and plateaux running westward through Bucketty along the Hunter Range, north of Howes Valley to Mount Coricudgy (1254 m) separate the north-flowing streams of the Hunter system from the south-flowing streams of the Hawkesbury system. This natural boundary has been extended in the north-east to include the southern portion of Newcastle and those small creeks and slow-moving rivulets which flow into Lake Macquarie.

From Mount Coricudgy to west of Wombeyan Caves, the Great Dividing Range forms a natural boundary between the Sydney district and the central west. The upper reaches of the Wollondilly River, including the area bordered by Marulan, Goulburn, Crookwell and Taralga, have been excluded, since this is part of the southern tablelands botanical subdivision. The Shoalhaven River forms a convenient southern boundary.

The Sydney district as defined above is large enough to contain a complex variety of physical features and a wide range of conditions which have led to an amazing floristic diversity. It also provides seemingly endless opportunities for exploration and discovery. It is still possible within a few hours travel of Australia's largest city to discover a new plant species, or a new form, or to record a rare species or a distribution not previously known. Although the area can logically be regarded as a floristic unit, the variety within the unit is impressive. Each zone has its own dominant species, its own special-interest plants and its own characteristic plant communities, and, within each zone, subtle differences occur, often within a matter of metres. The old myth that the Australian bush is 'all the same' ought to be buried forever.

Most parts of the Sydney district can easily be explored during a day trip, although more remote areas may require an overnight stay. The size of the area can be seen by some comparative distances. Sydney to Newcastle: 165 km; Sydney to Nowra, 155 km; Sydney to Lithgow: 140 km. Further afield are Howes Valley

(175 km from Sydney), Wombeyan Caves (180 km), Kanangra Walls (200 km), Glen Davis (210 km) and Yerranderie (300 km). The north coast zone and Watagan Mountains are within easy reach of the residents of Newcastle; Illawarra, lower Shoalhaven and the southern highlands are an easy drive from Wollongong.

The visitor to Sydney is often surprised that Australia's largest city, with a population of over three million, has so much natural bushland within easy access. Sydney and the Cumberland Plain are surrounded by a semi-circular belt of rugged sandstone plateaux, which are eroded into a complex of walled gorges, steep slopes and infertile ridges. This rough country effectively hemmed in early Sydney and, because it was unsuitable for agriculture or grazing, was mostly left uncleared. Today's road and rail links with the rest of the State have been built along interconnecting ridges or by massive construction feats cutting deep into sandstone ridges or infilling gullies. National parks and water catchment areas have been established over much of this sandstone belt, so that the plant species discovered by the first European explorers and botanists may still be found today. (Map 1.)

The Sydney district can be divided into a number of zones based on rock types and topography. Because the distribution and variety of plants are linked to the nature of the soils, topography and climate, these zones are also convenient botanical divisions. The zones are shown in Map 2.

Hornsby Plateau

The area north of Sydney, known as the Hornsby Plateau, rises gradually from Port Jackson like the rim of a giant saucer to an elevation of 640 m at Mount Warrawolong, the highest point in the Watagan Mountains on the Hunter-Hawkesbury divide. The general height of the plateau is 200 to 300 m and the major outcropping rock is Triassic Hawkesbury sandstone which has been eroded by numerous streams to form deep V-shaped valleys and a maze of rugged, narrow ridges and rough, tree-clad slopes. The Hawkesbury River carves deep valleys through the plateau before emptying into Broken Bay. Feeding the Hawkesbury River are streams such as the Macdonald River, Mangrove Creek and Mooney Mooney Creek which have cut through the Hawkesbury sandstone to expose the underlying strata of the Narrabeen group of rocks, also of Triassic age. More recent deposits of silt and sand have been built up along the major watercourses. In the lower reaches of the Hawkesbury, the valleys have been flooded by rising sea-levels, forming the 'drowned valleys' such as Cowan Waters, Berowra Creek and Mooney Mooney Creek, which are a distinctive feature of Broken Bay.

Hawkesbury sandstone is a glistening quartz-rich rock consisting of up to 60% sand grains, with small percentages of mica, feldspar and occasional lenses of shale and siltstone. Sandstone soils have a poor water-holding capacity and are very acidic and infertile. Shallow skeletal sands are common on the ridges, but elsewhere the sand may be metres deep and enriched by silt and decaying plant materials. Pockets of heath and low open woodland are commonly found on rocky platforms and ridgetops. Taller open forest occurs on the deeper plateau soils and on the slopes. This open forest usually has a varied understorey of medium-sized hard-leaved shrubs and small trees, with smaller monocots and herbs.

Soil removed from the slopes accumulates in the valleys where it may be relatively deep and contain a high percentage of humus. Where the Narrabeen group has been exposed in the deeper valleys along the western shores of Pittwater and along the eastern rim of the plateau, the eroded shales form deep clay-rich soils with a resultant change in the vegetation. In these deeper cool valleys, especially with southern or eastern aspects protected from drying winds and the direct rays of the sun, there is a lush growth of tall trees, rainforest shrubs, ferns and herbs. Here emergent trees may be up to 30 m tall and there is often a dense canopy cover. Trees such as *Eucalyptus deanei*, Turpentine *(Syncarpia glomulifera)* and Cedar Wattle *(Acacia elata)* are mixed with rainforest elements such as Lillypilly, Backhousia, Coachwood, Sassafras and Cabbage Tree Palms.

Volcanic soils are uncommon in the area, although there are a few small intrusions of Tertiary basalt such as at Mount Wareng (Howes Valley), Basalt Hill (Kulnura) and Dillons Crater (Brisbane Water National Park) which have weathered to dark clay soils rich in iron and magnesium. These areas of higher fertility were once islands of lushness surrounded by the vegetation of infertile sandy soils, but unfortunately they were often cleared, leaving little of the original vegetation.

A feature of the tidal inlets of Broken Bay, such as Patonga Creek, Mooney Mooney Creek, Mangrove Creek, Mullet Creek and Marramarra Creek, is the presence of mangrove communities backed by reedlands and saltmarsh flats. These are subject to frequent tidal inundation. On higher ground of the alluvial flats, the main trees are Swamp Oak *(Casuarina glauca)* and Prickly-leaved Paper-bark *(Melaleuca styphelioides).*

A good sample of the Hornsby Plateau is protected by national parks. Ku-ring-gai Chase, Marramarra, Muogamarra Nature Reserve, Dharug, Brisbane Water and Bouddi form a ring of national parks in the lower Hawkesbury and Broken Bay area, about 56 000 ha in total. Further north, the Watagan State Forests give a degree of protection to valuable natural areas. Here are plants not found elsewhere in the Sydney district, often species that are common further north. Of special interest are the patches of rainforest in the protected south-facing gullies, such as in Gap Creek Flora Reserve, where species such as *Alpinia caerulea* and *Embelia australiana* reach their southern limit.

North Coast

In the north-eastern corner, the Hornsby Plateau is eroded to expose the more fertile Narrabeen shales and sandstones. These extend from Brisbane Water northward and for the most part have been cleared of native vegetation. The southern section along the coastline of Bouddi National Park has a rugged cliff-line of Narrabeen Sandstone and extensive eroded rock platforms. Higher ridges are capped by Hawkesbury sandstone. Further north the landscape is dominated by a series of shallow lakes—Tuggerah Lake, Munmorah Lake and Lake Macquarie—around which are extensive sand and alluvial deposits. Sands transported by the currents along the shore have been deposited as beaches and tombolos, long strips of sands which almost entrap the waters of the lakes. In some areas these sands support unique examples of native vegetation, such as the *Angophora costata* forest at The Entrance North, magnificent coastal heaths in exposed sites and swampy gullies dominated by Cabbage Tree Palms and sedges.

Slow-moving streams, such as Ourimbah Creek, Wyong Creek and Dora Creek, meander across the coastal flats before empyting into the lakes. Deposits of unconsolidated sands and silts on the lower reaches of these streams are often fringed with remnant rainforest vegetation.

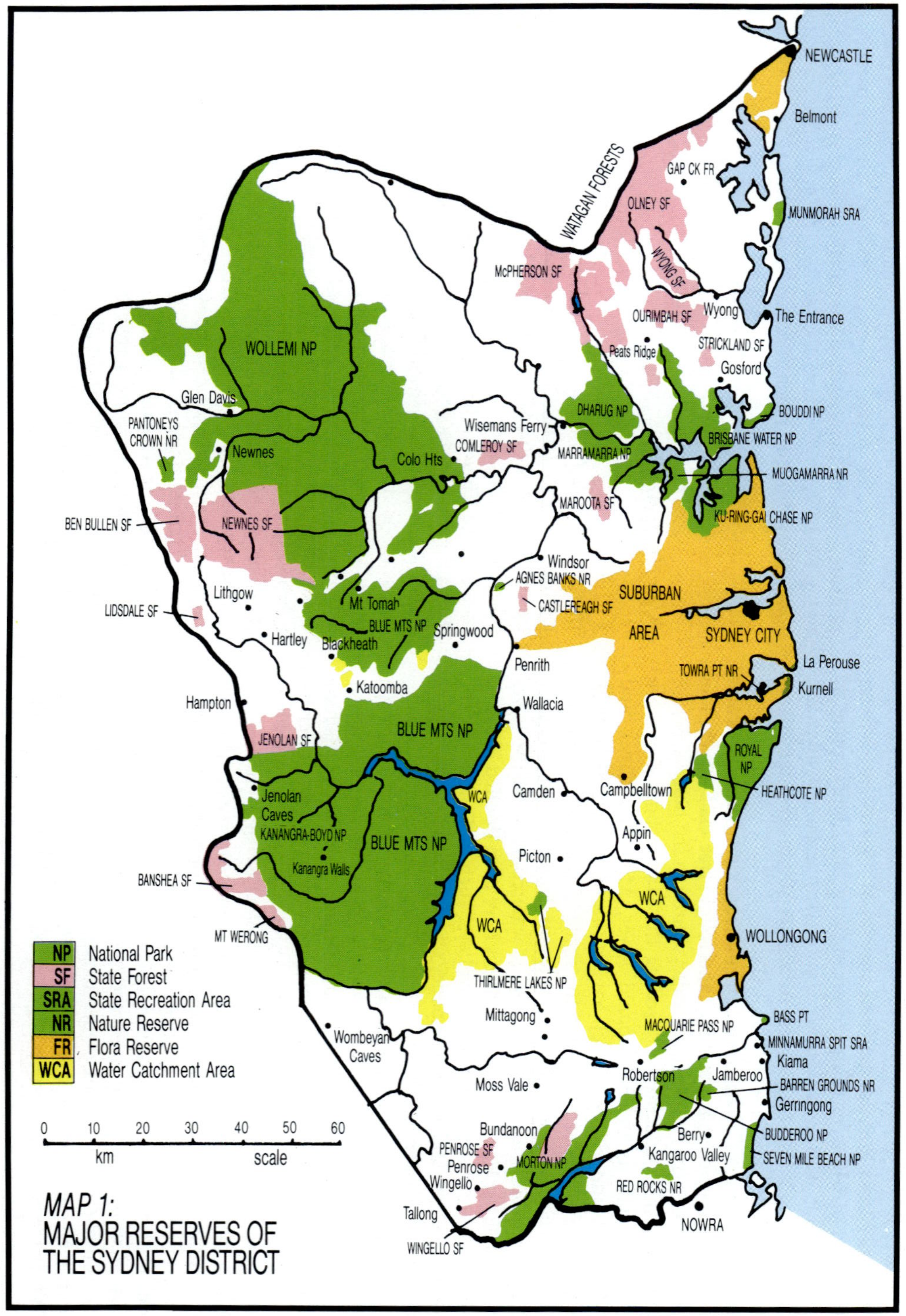

MAP 1:
MAJOR RESERVES OF
THE SYDNEY DISTRICT

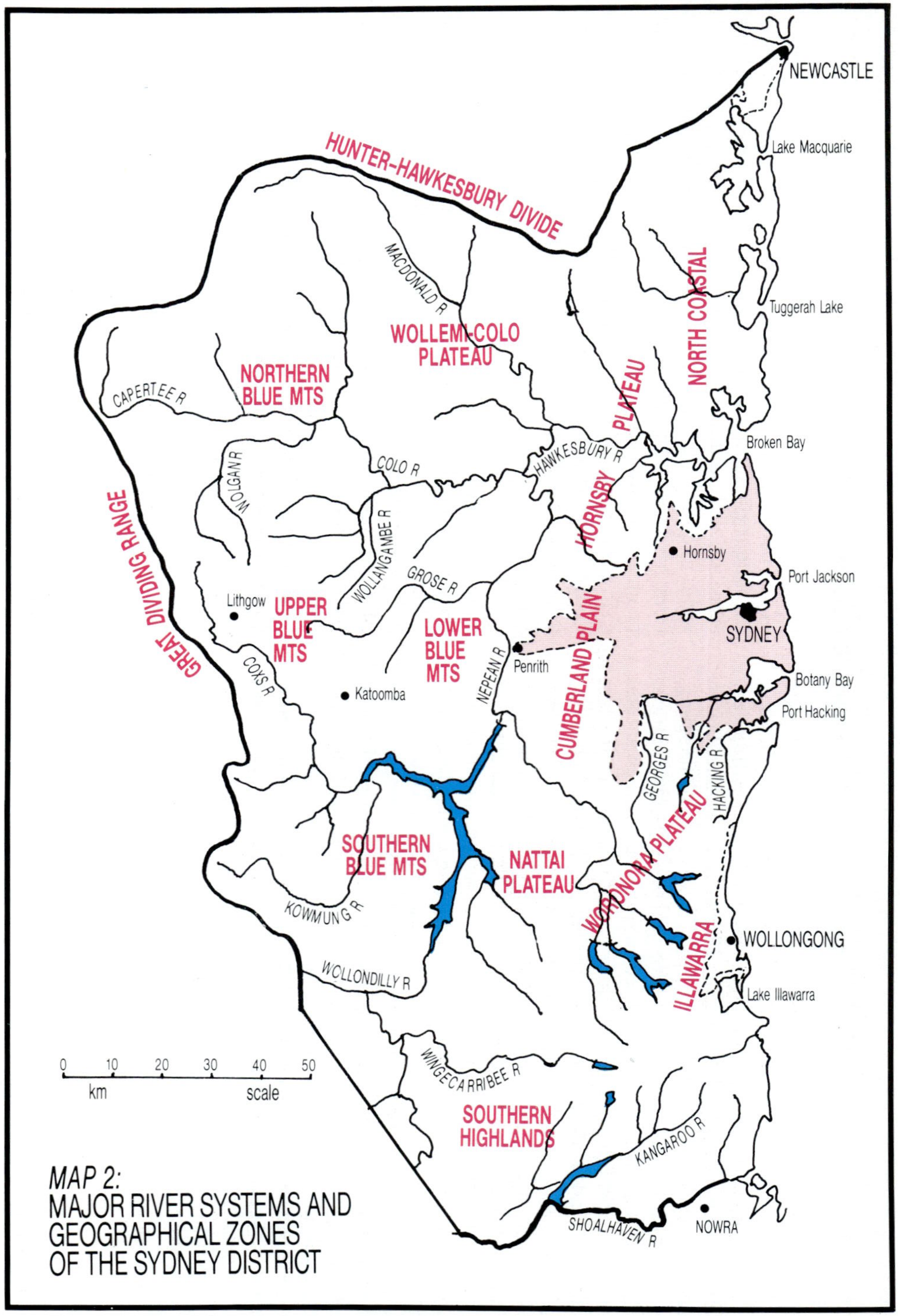

MAP 2:
MAJOR RIVER SYSTEMS AND
GEOGRAPHICAL ZONES
OF THE SYDNEY DISTRICT

A feature of the Gosford-Ourimbah area is the presence of numerous rainforest communities, some just narrow strips along creeks, others, such as Strickland, Ourimbah and Wyong State Forests and Katandra Reserve, containing significant stands with a diversity of species. These are Warm Temperate Rainforests which have developed on sheltered slopes and gullies on Narrabeen soils. An inventory of rainforests in the City of Gosford compiled in 1985 identified 25 rainforest sites containing 110 species of flowering plants. Although these communities are often small and badly weed-infested, they are part of a large biogeographical complex extending from Broken Bay northwards and are important natural museums contributing to the preservation of rainforest species in the Sydney district.

No significant coastal rainforest stands occur between Broken Bay and the Narrabeen outcroppings of the upper Hacking River (Royal National Park), so it is not surprising that a number of rainforest species do not extend southward beyond Broken Bay. Rose Walnut *(Endiandra discolor)* is one species that reaches its southern limit at Gosford. Other species, such as Whip Vine *(Flagellaria indica)* and *Cordyline stricta* are rare south of Broken Bay. The distribution of Brown Myrtle *(Choricarpia leptopetala)* is an interesting example. It is common in Olney and Wyong State Forests and occurs at Palm Grove near Gosford. South of Gosford, it grows only on Narrabeen shales along a creek at Stanwell Park and in Palm Jungle in the Royal National Park.

Colo Plateau–Wollemi

The Colo Plateau is a high area of Hawkesbury sandstone eroded in the east by the Macdonald River and in the west by the Colo River and Wollemi Creek. Each river has a dense network of smaller tributary creeks which are eating into the sandstone producing a maze of gorges and ridges. The main north-south ridge is known as Mellong Range and carries the Putty road; branching out from it like stubby fingers are Howes Range, Womerah Range and Culoul Range. The general height of the plateau is 200 to 300 m on the eastern edge, rising to 600 m on the western edge. The sandstone is up to 250 m thick.

Remnants of Wianamatta shale occur on the higher ridge-tops, especially along the Putty road north of Colo Heights. Because it provided reasonably good agricultural soil, most of the original vegetation has been cleared for small farms. A small patch of Wianamatta shale also occurs on the Culoul Range where the richer soils on the protected southern slopes support a tall open forest of Mountain Mahogany *(Eucalyptus notabilis)*, Grey Ironbark *(E. paniculata)* and Grey Gum *(E. punctata)* and a variable understorey of shrubs, herbs, grasses and twiners.

An unusual deposit of alluvial sands occurs along the Mellong Range. As it has a clayey zone just below the surface, water is retained near the surface and the ground is often swampy. These conditions produce a distinctive plant community dominated by Narrow-leaved Apple *(Angophora bakeri)*, Drooping Red Gum *(Eucalyptus parramattensis)* and Hard-leaved Scribbly Gum *(E. sclerophylla)*.

The Hawkesbury sandstone ends abruptly in great cliffs overlooking the Wollemi and Colo Gorges. West of the Colo Plateau, Narrabeen rocks dominate but it is often difficult to distinguish between the Narrabeen and Hawkesbury sandstone geology. Narrabeen sandstones are up to 800 m thick in parts and form the spectacular clifflines on the western rim of the Wollemi National Park where the Wolgan and Capertee Rivers cut deep gorges on their way eastward from Newnes and Glen Davis.

A dominant feature of the scenery of the Wollemi area is the Colo River deeply entrenched about 250 m below the plateau in a narrow gorge often only 100 m wide, with steep rocky cliffs towering above. Tributary streams, such as the lower Wollemi Creek, are also deeply entrenched. Shrubs cling tenaciously to rocky ledges on the cliffs; open forest and woodland occur on the small talus slopes at the foot of the cliffs. The floor of the gorge has scattered trees of River Oak *(Casuarina cunninghamiana)*, Water Gum *(Tristaniopsis laurina)* and an unnamed red-flowered tea-tree *(Leptospermum)*.

The skyline of Wollemi is broken by scattered humps of mountains. Many of these such as Monundilla and Uraterer retain a topping of basalt which flowed over the area 30 to 40 million years ago. This basalt breaks down to rich red soils capable of supporting tall forests. The basalt is 150 m thick on Mount Coricudgy. The vegetation of most of these volcanic outcrops has been altered by grazing and logging. Where native vegetation occurs, the southern and eastern slopes are most favourable. On the southern side of Mount Monundilla there is a rainforest dominated by Sassafras and Coachwood and thick with vines, mosses and ferns. The northern and western slopes, by contrast, are covered with open eucalypt forests. Another basalt patch near the western end of the Culoul Range carries a 30 m tall forest of Sydney Blue Gum *(Eucalyptus saligna)* with an understorey of *Acacia parramattensis* and many grasses and herbs.

The Wollemi is one of the last true wilderness areas in New South Wales. Almost the entire sandstone plateau and associated gorges between the Putty road and Glen Davis-Newnes are included in the Wollemi National Park, an area of over 480 000 ha extending from south of the Colo to the Hunter-Goulburn Valleys. A few fire trails lead from the Putty road to the edge of the Colo Gorge, but most of the area is untracked and unspoilt.

Northern Blue Mountains

This zone extends from Lithgow and Mt Wilson in the south northward across the Wollangambe Wilderness and Newnes Plateau to the old shale-mining settlements of Newnes and Glen Davis. The north-western corner is bounded by the Great Dividing Range sweeping in a broad arc to encircle the upper reaches of the Capertee River. This zone includes some of the highest parts of the Blue Mountains, rising from about 920 m above sea-level at Lithgow to over 1000 m on the Newnes Plateau (Birds Rock, 1181 m). The dissected plateau above Glen Davis is between 750 and 850 m and temperatures are often very low in winter. The average minimum temperature for July in Newnes State Forest is $-1°C$; in the hottest month, January, the average maximum is $23.5°C$.

The Newnes Plateau and Wollangambe Wilderness consists of Triassic Narrabeen sandstones and shales. Soils derived from these rocks are yellow-brown sands, often with ironstone gravel on the higher ridges. To the east, a complex of streams cut into the Narrabeen series forming long spurs, deep gullies, cliff-lined canyons and isolated rocky outcrops. A feature of the area is the sandstone formation known as 'pagodas', eroded flat-topped rocks standing conspicuously above the surrounding bushland. Another feature of the plateau is the presence of headwater swamps, formed in broad shallow valleys on a clay base which impedes drainage.

In the north-west, the Capertee and Wolgan Rivers have cut deep gorges through the Narrabeen rocks, exposing older Permian coal measures and siltstones, shales and sandstones of the Shoalhaven Group. Along Coco Creek and the upper Capertee, some of the most ancient rocks of the Sydney district dating from the Lower Devonian period have been exposed. Permian rocks have layers of softer shales, tuffs and coal seams which are more easily eroded, causing giant slabs of overlying Narrabeen sandstone to crack and fall, leaving sheer cliff-lines. These are a spectacular feature of the Glen Davis–Newnes area. The valley of Glen Davis is only 280 m above sea level, but the tops are 600 m or more. Erosion and undercutting have produced many prominent topographical features such as Baal Bone Point, Genowlar Mountain and isolated flat-topped mountains of Narrabeen sandstone, such as Mt Airly (1034 m) and Pantoney's Crown (1000 m). Beneath the cliff-lines are talus slopes consisting of sandstone blocks, Permian soils and scree material from the cliffs above. This is a very specialised habitat on which are found a number of interesting plants including the rare *Prostanthera cryptandroides* and endemic *Apatophyllum constablei.*

Tertiary basalt occurs as a capping on Mt Cameron, a peak which rises above the surrounding Newnes Plateau. Here on the richer soils there is a tall open forest of Ribbon Gum *(Eucalyptus viminalis)* and Blaxland's Stringybark *(E. blaxlandii).* The understorey is mostly grass, herbs and bracken. On the more protected southern slopes, rainforest elements, such as Sassafras *(Doryphora sassafras),* Native Mulberry *(Hedycarya angustifolia),* Tree Fern *(Cyathea australis)*, twiners and ferns are common.

Rainfall is very variable over this zone and has a marked effect on the plant communities. Whereas Mt Wilson to the south-east receives an average of 1228 mm of rain a year and Mt Tomah 1450 mm a year, this decreases to the north. Newnes State Forest receives 1047 mm and because of the rain-shadow effect of the surrounding plateau and ranges, the valleys of the Capertee and Wolgan Rivers receive less than 900 mm. A combination of low rainfall and infertile, skeletal soil produces woodland communities of trees such as *Eucalyptus albens, E. melliodora, E. blakelyi* and *Brachychiton populneus* which are common further to the west.

With such a variety of soil types, drainage and rainfall patterns, altitude, exposure and topography, it is not surprising that there is a complex of plant communities within the zone and a number of endemic and rare species. The cold higher plateau has woodland dominated by *Eucalyptus dives, E. radiata* and *E. sieberi*, with Snow Gum *(E. pauciflora)* common in some areas. Mallee eucalypts, *E. stricta* and *E. gregsoniana*, are common in heath on exposed shallow sandy soils. The high swamps contain disjunct populations of Snow Daisy *(Celmisia longifolia)* and the endemic *Boronia deanei.* At least two unnamed Mint-bushes *(Prostanthera)* have been found on or near pagoda rocks on Newnes Plateau. *Banksia conferta* var. *penicillata* is scattered in open forest on the plateau. The number of rare or uncommon plants found in the area, such as *Isopogon prostratus, Isopogon dawsonii, Acacia kybeanensis, Acacia dorothea, Gompholobium uncinatum, Haloragodendron lucasii* and *Prostanthera ovalifolia*, make it one of the finest wildflower areas in the Sydney district.

The majority of the Newnes Plateau is protected by the northern part of the Blue Mountains National Park and by Wollemi National Park. Significant areas of Newnes Plateau and the upper Coxs Valley are State Forests, some of which are cleared for exotic pine plantations. The Forestry Commission has established a small Forest Reserve at Birds Rock (447 ha). Pantoney's Crown is a Nature Reserve of 3230 ha controlled by the National Parks and Wildlife Service. Most of the Capertee and Wolgan valleys have been cleared for grazing and small-scale agriculture.

Blue Mountains

West of the Cumberland Plain and the Nepean River is a steep monocline marking the eastern edge of the Blue Mountains. The mountains rise gradually, almost in steps: 200 m above sea level at Glenbrook, 500 m near Linden, 850 m at Wentworth Falls, 1040 m at Katoomba. Mt Piddington (1094 m), near Mt Victoria, is the highest point on the main ridge which carries the Great Western Highway. The western rim of the Blue Mountains consists of a spectacular abrupt cliff-line running north-south. West and south of these cliffs are the valleys of Hartley, Megalong and Kanimbla. The highest parts of the area are in the south-west along the Great Dividing Range near Hampton.

The watercourses are part of the Hawkesbury-Nepean River System and have a major influence on the Blue Mountains' topography. Coxs River rises north of Lithgow, flows south through the western valleys and then turns east to join Lake Burragorang, a huge reservoir formed by the damming of the narrow gorge of the Warragamba River. Smaller streams such as Kedumba, Jamison and Bedford Creeks flow southward from the main east-west range, carving deep sheltered gullies where tall eucalypts or narrow bands of rainforest plants find moisture and protection. The Grose River, draining the Blue Mountains between the main ridge and Bells Line of Road, flows eastward cutting deep gorges through the sandstone and exposing the Permian coal measures beneath. The coal outcrops are mined at the head of the Grose River between Bell and Mt Victoria.

The lower Blue Mountains from Lapstone to about Woodford consist of Hawkesbury sandstone. This gives way to rocks of the Narrabeen group which form the major part of the central and upper Blue Mountains plateau and the sheer cliffs at Wentworth Falls, Katoomba and Blackheath. This sandstone is highly eroded and frequently giant pillar-like fragments, such as the Three Sisters and Orphan Rock at Katoomba, are left standing separated from the main plateau surface. The cliff-fringed remnant of Mt Solitary and the long fingers of Narrow Neck and Kings Tableland are evidence of the erosion processes affecting the Narrabeen group.

Residual Wianamatta shale caps occur along the Bells Line of Road from Kurrajong to Mt Tomah, but much of their native vegetation has been cleared for the growing of apples and oranges. Where there is native vegetation remaining, the well-drained clay soils support an open forest of Blue Mountain Mahogany *(Eucalyptus notabilis),* White Stringybark *(E. globoidea)* and Turpentine *(Syncarpia glomulifera)*, with occasional examples of Mountain Blue Gum *(E. deanei).* There is also a wedge of Wianamatta shale near Mt Portal, Glenbrook, trapped behind the Lapstone Monocline. It supports an open forest which contrasts with the surrounding shrubby sandstone community. A thin band of shale, about 45 m thick, underlies the basalt cap of Mt Tomah. Its exposure on the eastern slopes adds to the richness of the Tomah vegetation.

The Blue Mountains landscape is dotted with prominent peaks capped by basalt remnants. Mounts Wilson, Tomah, Irvine, Caley, Hay, Banks, Bell and Haystack all have a thin basalt topping. This is 80 m thick at Mt Tomah and covers an area of 240 ha. The rich volcanic soils of Mt Wilson encouraged wealthy Sydney residents to settle there in the 1880s and build large houses with spacious gardens planted with exotic deciduous trees. Due to a high rainfall, rainforest communities are present on the sheltered hillsides and gullies of Mt Wilson and Mt Tomah. Unusual species, such as Plume Bush *(Calomeria amaranthoides)* and *Prostanthera caerulea*, occur on these richer soils.

Eroded volcanic necks provide further evidence of past volcanic activity in the Blue Mountains. These necks are usually amphitheatre-shaped depressions supporting lush grassland, climbers and ferns and dominated by the tall straight trunks of *Eucalyptus deanei*. Murphy's Glen, Euroka, Machin Crater, St Helena and Toby's Glen are good examples of such communities.

Permian coal measures are exposed in narrow seams at the base of the western escarpment. Underlying these coal measures are Permian siltstones, sandstones and shales. The Bathurst batholith is exposed in the Hartley and upper Coxs Valleys, resulting in areas of undulating granite country. The Devonian Lambie group of ancient sediments outcrops west of Lithgow, south of Hartley, in the Kedumba Valley and extends southward to the rugged country between the Coxs and the Kowmung Rivers.

The western valleys are in a rain shadow receiving only 750 to 780 mm of rainfall a year. Due to their height, they are often very cold in winter with severe frost on about 20 days a year. Such climatic conditions, combined with acidic and relatively infertile soils, result in woodland communities. In sharp contrast is the environment in the gullies of the eastern escarpment, such as at Wheeny Creek or Green Scrub. Here the rainfall is about 1300 mm a year and the Narrabeen soils have been enriched by silts, shales and plant material. In these protected gullies are Red Cedar *(Toona australis)*, the palm-like lily *Cordyline stricta*, Cabbage Tree Palms *(Livistona australis)*, the rare Small-leaved Bleeding Heart *(Omalanthus stillingiifolius)* and numerous ferns and climbers.

The Blue Mountains are the refuge of many rare or threatened plants. *Acrophyllum australe* is known only from cliff-lines in the Woodford–Linden area. *Isopogon fletcheri* grows only on the cliffs at Govetts Leap near Blackheath. *Eucalyptus burgessiana* is restricted to ridges near Linden and Mt Tomah. *Hakea constablei, Darwinia fascicularis* ssp. *oligantha, Pultenaea incurvata* and *Acacia ptychoclada* are all endemic in the Blue Mountains. Some species have localised populations in the mountains separated from other occurrences. *Darwinia peduncularis* occurs on Kings Tableland and in the Hornsby–Hawkesbury area. *Eucalyptus bauerlenii* grows on south-facing cliff-lines at Wentworth Falls and near Braidwood. *Eucalyptus pulverulenta* is found along the Coxs River and at Bredbo in the southern tablelands. The Blue Mountains are clearly of great botanical significance in the interplay of factors which determine the geographical distribution of many plant species.

The Blue Mountains' population is confined to a chain of towns along the Great Western Highway and Railway Line. Smaller settlements are located along the Bells Line of Road, with a few farms in the western valleys and along the Great Divide. The remainder is rugged bushland, protected for the most part by the Blue Mountains National Park (215 955 ha) and by the Warragamba Catchment Area.

Southern Blue Mountains–Nattai Plateau

The area from Lake Burragorang to the Great Dividing Range and south of Jenolan Caves is probably the least visited part of the Sydney district. The rugged terrain and the restricted access to water catchment areas require visitors to make a long roundabout journey to reach scenic spots such as Kanangra Walls. Distance and difficult access make a trip to the historic silver-mining town of Yerranderie an exercise for the more intrepid. Wombeyan Caves in the south is reached by a narrow dirt road which winds down from the Nattai Plateau to the Wollondilly River and then twists and turns back up towards the Great Divide.

A dominant feature of the area is the Boyd Plateau, covering more than 13 000 ha. The highest parts of the plateau are over 1300 m above sea level, with Ben Lomand reaching 1330 m. The rocks of the plateau have withstood the erosive processes better than surrounding sedimentary rocks. The Boyd River and upper Kowmung River cut through major granite outcrops; other granite outcrops occur south of Jenolan Caves.

The forests of the Boyd Plateau were once part of a State Forest and have been extensively logged, mainly for *Eucalyptus fastigata. E. radiata, E. dalrympleana* and *E. dives* are common on the plateau, with *E. pauciflora* on the highest elevations. In poorly drained depressions *E. ovata* and *E. stellulata* occur with an understorey of sedges and ferns, including the Alpine Water-fern *(Blechnum penna-marina)*. Other species found in this high country and also in sub-alpine regions further south include *Oxylobium ellipticum, Pultenaea altissima* and *Grevillea australis.* A new species of *Hakea* has recently been discovered on hilltops of the Boyd Plateau.

Surrounding the plateau is an area of ancient sedimentary rocks of the Devonian Lambie group with an outcropping of Permian sandstone which form the vertical cliffs of Kanangra Walls. The horizontal strata of these sandstones contrast with the west-dipping Devonian sediments which form the Thurat Walls and a complex of eroded valleys and ridges of Kanangra–Boyd National Park. Here the rivers have cut deep gorges into the sediments. The Kanangra Gorge is 800 m deep; the gorge of the Kowmung River is equally impressive.

The geology south of the Kowmung is complex, with outcroppings of Ordovician and Silurian rocks overlain by Permian sandstones near Yerranderie and small intrusions of granite. Extensive outcrops of Devonian porphyry, dacite and tuff occurring along the Wollondilly River are responsible for the topography along the road between Bullio and Wombeyan Caves. Due to a rainfall of only about 800 mm a year, the hillsides, especially on the exposed northern and eastern slopes, are dry and open, supporting a woodland of Grey Box *(Eucalyptus moluccana)* and Yellow Box *(E. melliodora)*, with a low prickly understorey. Caves occur in limestone outcrops at Jenolan, Colong, Billys Creek and Wombeyan. At least two plants, the fern *Pleurosorus subglandulosus* from Jenolan and the wattle *Acacia chalkeri* from Wombeyan, are restricted to limestone habitats.

Of special interest in the Yerranderie–Kowmung area is the presence of a number of rare or uncommon plants. Unnamed species of *Kunzea, Leptospermum* and *Prostanthera* occur on Scotts Range. *Acacia clunies-rossii, Grevillea obtusiflora, Acacia flocktoniae* and *Seringia arborescens* are all recorded for the area.

East of the Wollondilly River the high plateau country forming the catchment of the Nattai, Wingecarribee and Little Rivers consists predominantly of Hawkesbury sandstone. The contrast between this and the older folded and eroded rocks to the west can be clearly seen along the Mittagong–Wombeyan Caves road. At the edge of the Hawkesbury sandstone, the cliff-line has been cut through by a road tunnel. Below this tunnel are the richer soils of the narrow band of Permian coal measures and then the cleared dry hillsides and open forests which result from the outcropping of Devonian rocks.

The geology of the Nattai Plateau is similar to that of the Hornsby and Woronora Plateaux. It consists of Hawkesbury sandstone about 700 m above sea level capped on the higher ridges with Wianamatta shale. Several peaks, such as Mt Jellore, Mt Wanganderry and Mt Flora are of Tertiary volcanic origin. The plateau is dissected by rivers which cut deep gorges into the Permian coal measures and Shoalhaven group. Plant communities follow the general pattern, with taller forests on richer soils, especially in the river valleys and protected slopes, woodlands on drier skeletal soils, heath on rocky outcrops and tea-tree scrubland on sandstone areas with poor drainage. Tall River Oaks *(Casuarina cunninghamiana)* line the banks of the Wingecarribee and Wollondilly Rivers.

Two national parks, Kanangra–Boyd and Blue Mountains, protect much of the southern Blue Mountains. There are significant areas of State Forest, some with pine plantations. Ridge tops of the Nattai Plateau have mostly been cleared for grazing. Most of the uncleared area of the Nattai is included in the Warragamba Catchment Area and has restricted access.

Cumberland Plain

The Cumberland Plain is a gently undulating basin dipping westward from Parramatta to the Nepean River and from Wilberforce and Richmond southward to Picton and Menangle. It consists of a deep, almost unbroken layer of Wianamatta shale, at altitudes ranging from 20 to 100 m above sea level. Poorly consolidated Tertiary alluvial deposits, mainly clays and silts, and often laterised, occupy the area from Penrith to Windsor and Riverstone. Recent sand and clay alluvium occurs along the banks of the Nepean–Hawkesbury River, along South and Eastern Creeks and around Liverpool.

The heavy clay soils of the Plain are often poorly drained and poorly aerated, hence swampy depressions are common. Summer temperatures are usually higher than in coastal zones which receive the sea breeze, and winter frosts may be severe. The open topography exposes the area to winds from the west and north-west. Rainfall is considerably less than along the coast and in some parts is about half that recorded for Sydney where the average is 1209 mm per year. The highest rainfall occurs on the eastern margin but the western margin is in a low rainfall zone (Richmond, 790 mm; Castlereagh State Forest, 635 mm; Camden, 790 mm). It is not surprising then that many plants that are adapted to the conditions of this zone are absent from the surrounding sandstone areas.

On the hills around Hoxton Park, Cecil Park and The Oaks, there are stands of Spotted Gum *(Eucalyptus maculata)* and Grey Box *(E. moluccana)* with undershrubs such as *Bursaria spinosa* and *Dillwynia juniperina*. Cabbage Gum *(E. amplifolia)* is common on alluvial flats of tributary creeks, and the presence of Swamp Oak *(Casuarina glauca)* seems to indicate a saline influence in the ground water. Forest Red Gum *(E. tereticornis)* occurs on the better-drained slopes. The clay flood plains of Eastern Creek near Doonside and South Creek near Shane Park support a number of uncommon herbs and shrubs, including *Murdannia graminea, Zornia dyctiocarpa, Opercularia diphylla, Phyllanthus filicaulis, Myoporum debile* and *Dichopogon fimbriatus.* On the western margin, on the river bank and flood-plain of the Nepean River at Bents Basin, there is a stand of *Eucalyptus benthamii*, a tree once much more widespread but now rare because of clearing and the flooding of part of its range by the Warragamba Dam.

Near the settlement of Agnes Banks is an unusual area of white sand of Pliocene or Pleistocene age. It originally covered an area of about 600 ha, but sand extraction has reduced it to less than half that size. The sand forms a low stable set of east-west dunes with an average height of 3 to 4 m and with swampy depressions between the main dunes. The area is botanically interesting because of its floristic and structural affinities with the sand dune vegetation of the coast. Here, for example, is *Banksia aemula*, a coastal species not found south of Botany Bay. Dominant trees of the sand deposit are *Eucalytpus sclerophylla, E. parramattensis* and *Angophora bakeri.* There is a rich understorey including *Dillwynia glaberrima, Acacia brownii* and *Ricinocarpos pinifolius.* A sedge, *Restio pallens*, common on the north coast of New South Wales, reaches its southern limit at Agnes Banks. On the clays and silts surrounding the sand, the larger trees are *Eucalyptus fibrosa* and *Melaleuca decora.* A small portion of this sand deposit is protected by a nature reserve.

The Castlereagh State Forest is also of interest since it contains the best remaining open forest community of *Eucalyptus fibrosa* and the most extensive area of undisturbed clay soil vegetation on the Cumberland Plain. here are found an unnamed species of *Allocasuarina*, the only stand of *Dodonaea falcata* in the Sydney district, a large number of rare or restricted ground orchids, and such clay-loving species as *Chorizema parviflorum, Grevillea juniperina, Prostanthera scutellarioides, Macrozamia spiralis, Logania pusilla* and *Boronia polygalifolia.*

Further south at Razorback Mountain, between Camden and Picton, a protected west-facing slope of Wianamatta shale supports a dry rainforest containing a number of unusual species. There is a disjunct occurrence of the woody climber *Deeringia amaranthoides*, the yellow-flowered *Abutilon oxycarpum*, the rare climber *Cissus opaca*, and the leafless shrub *Spartothamnella juncea*, and coastal rainforest trees *Alectryon subcinereus* and *Guioa semiglauca.*

Little of the original vegetation of the Cumberland Plain remains since the area was extensively cleared for grazing and small-scale farming in the early days of settlement. Today the rapid spread of suburban sub-division threatens the little that remains. A small but significant portion of the clay flood-plain of Eastern Creek is protected by Nurragingy Reserve at Doonside and reserves like Agnes Banks Nature Reserve, Castlereagh State Forest, Long Neck Lagoon and Bents Basin State Recreation Area, provide welcome, but inadequate protection against the suburban sprawl.

Woronora Plateau–Illawarra

The Woronora Plateau forms the southern part of the great saucer-like basin of sandstone around Sydney. Although often called a 'plateau', it is more like a ramp ascending gradually from Botany Bay and the Georges River to link up with the richer shale and basalt of the 700 m high Robertson Plateau in the southern highlands.

The eastern rim of the plateau ends abruptly in a line of sandstone cliffs which form the upper layer of the Illawarra Escarpment. West of this escarpment, the sandstone slopes gently towards the Wianamatta shales between Bargo and Campbelltown. The dip is about 2 to 3° in a north-westerly direction, so that streams such as the Woronora, Cataract, Cordeaux and Avon which have their headwaters very close to the coast on the top of the escarpment, flow away from the sea to link up with the Nepean River and Georges River systems.

The shallow acidic soils of the plateau are derived from Hawkesbury sandstone and support dry open forest and a rich variety of flowering shrubs. On the higher ridges at Menai and between Sutherland and Heathcote there are narrow cappings of Wianamatta shale. Elsewhere, the upper reaches of rivers have cut through the Hawkesbury sandstone into the Narrabeen group below. Here in the gullies, with shelter, increased humidity and enriched soil, communities of tall wet forest and rainforest are common. Along the upper Hacking River, for example, Blackbutt *(Eucalyptus pilularis)* and Turpentine *(Syncarpia glomulifera)* are common and there are many rainforest species. Sixteen species of rainforest climbers and over 50 species of rainforest trees and shrubs have been recorded from Bola Creek, a tributary of the Hacking River.

In the east, the Narrabeen Group of rocks is exposed in a narrow coastal strip from Garie to Coalcliff. The shales and mudstones have weathered to chocolate-brown soils which support a number of species absent from the drier sandy soils. On the seaward slopes near Garawarra in the Royal National Park is Palm Jungle, a unique remnant of littoral rainforest exposed to wind and sea spray. Here we find plants more at home in the rainforests of the north coast, such as Whip Vine *(Flagellaria indica)*, Bird-lime Tree *(Pisonia umbellifera), Psychotria loniceroides* and *Baloghia lucida.*

Sedgeland is common on the plateau where drainage of shallow upland valleys is poor and organic material builds up to create a swampy environment. Maddens Plains is a fine example of this type of community. Such areas are dominated by sedges and rushes, mainly belonging to the families Cyperaceae and Restionaceae. Shrubs, such as *Banksia robur, Melaleuca linariifolia* and *Leptospermum* species fringe the swamps, and herbs such as *Drosera spathulata, Goodenia dimorpha* and *Xyris gracilis* and fern-allies *Lycopodium* and *Selaginella* are common.

The Illawarra Escarpment runs NNE to SSW for about 85 km from the Royal National Park to the west of Gerringong. It is marked by a dramatic cliff-line of weathered Hawkesbury sandstone below which are exposed the softer Narrabeen rocks, Permian coal measures, Gerringong volcanics and the Shoalhaven group of silt-stones, shales and sandstones. These softer rocks form talus slopes, benches and foothills, and contribute to the alluvial soils in the valleys and in the Illawarra Coastal Plain. The coastal plain is very narrow in the north at Thirroul, but widens to about 25 km near Shellharbour where the scarp recedes inland encircling Macquarie Pass National Park. Around Lake Illawarra the landscape is gently undulating due to the erosion of the Shoalhaven group of rocks and the extensive deposits of alluvium on the floodplains and at the mouth of the lake.

South of Lake Illawarra, the coastal plain is dominated by the Gerringong Volcanics which form the spectacular Bass Point, seacliffs near Bombo and the Blow Hole at Kiama. Tertiary basalt occurs as a capping on Saddleback Mountain, south-west of Kiama, resulting in patches of rainforest with isolated occurrences of *Zieria granulata, Prostanthera lasianthos, Parsonsia brownii* and *Pimelia ligustrina* var. *hypericina.*

Sands and alluvial soils have been deposited along smaller streams and mangrove communities occur where there is tidal influence. At the mouth of the Minnamurra River, a low tombolo of stabilised sand, known as Minnamurra Spit, carries depauperate rainforest on the sheltered lagoon side and a sand dune succession of plants on the seaward side. Of special interest in this area are stands of *Celtis paniculata* at its southern limit, *Crinum pedunculatum, Myoporum insulare,* the rare sedge *Cyperus enervis* and large trees of *Endiandra sieberi.*

The Shoalhaven delta consists of a beach sand ridge system, backed by alluvial floodplains and terraces containing swamps and low-lying marshy ground. The zone behind Seven Mile Beach has a fine succession from foredunes covered with Spinifex *(S. sericeus)* to tall eucalypt forest. On the permanently vegetated hind-dunes are thickets of tea-tree, wattle, and scattered Coast Banksia *(B. integrifolia)*, Mock Olive *(Pittosporum undulatum)* and Cheese Tree *(Glochidion ferdinandi)*. Further from the beach are eucalypts, such as Bangalay *(E. botryoides)* and Blackbutts *(E. pilularis)* with dense patches of Burrawang *(Macrozamia communis)* in the understorey.

A mixture of rainforest types occurs in the Illawarra along the escarpment, with some isolated patches on the coastal plain. Warm-temperate rainforest is the most widespread, occurring in gullies, and along creeks on soils derived from the Illawarra coal measures. This type of rainforest also grows on the talus slopes above the benches of the escarpment. These rainforests, at altitudes of 150 to 450 m above sea level, receive about 1600 mm of rain a year and are frequently enshrouded by mist. The dominant trees are Coachwood *(Ceratopetalum apetalum),* Sassafras *(Doryphora sassafras)*, Brown Beech *(Cryptocarya glaucescens)* and Lillypilly *(Acmena smithii)*, with many broad-leaved shrubs, vines and ferns. Subtropical rainforest elements occur in gully bottoms where soils are enriched by latite from the Gerringong volcanics (such as at Minnamurra Falls and near Jamberoo). Although rainforest occurs almost continuously along the escarpment, some areas are more extensive and have easier access. Sites worth exploring are gullies at Stanwell Park, Mt Keira, Mt Kembla, Calderwood, Macquarie Pass and Minnamurra Falls.

Some of the best rainforest remaining on the coastal plain grows on private land at Berkeley and Flagstaff Hills, north of Lake Illawarra. This is on the northernmost outcrop of latite, most of which has been cleared for grazing and housing. Huge emergent figs *(Ficus macrophylla* and *Ficus superba* var. *henneana)* dominate these small patches. The lower canopy contains such interesting species as Whalebone Tree *(Streblus brunonianus)*, Brush Wilga *(Geijera salicifolia)*, Native Holly *(Coelebogyne ilicifolia)* and Scolopia *(S. braunii).*

A large part of the Woronora Plateau is reserved for catchment area and is therefore restricted to visitors. Further restricted areas include a military reserve occupying the plateau between Woronora River and upper Georges River and coal leases of the Illawarra Escarpment. People wishing to enjoy the bushland of the Woronora Plateau generally do so in the Royal National Park (15 014 ha) and the Heathcote National Park (2251 ha). The Macquarie Pass National Park (1064 ha) is the largest reserve on the escarpment. Smaller protected areas include the Illawarra Escarpment State Recreation Area (five small parcels of land totalling 1259 ha extending from Bulli Pass to Bong Bong Pass), and Minnamurra Falls Reserve. On the coast, Bass Point, Minnamurra Spit and Seven Mile Beach National Park (730 ha) are important conservation sites.

Southern Highlands

The southern highlands is a fertile plateau of more than 2600 km^2, with peaks rising to between 600 and 800 m above sea level. It covers a rough triangle between Mittagong, Robertson and Bundanoon, with fingers extending to Wingello and Barren Grounds. Its fertile nature is mainly a result of an extensive cover of Wianamatta shale which has weathered into a dark grey soil enriched from basalt hills scattered throughout the area. Basalt cappings occur at Robertson, Bowral, Exeter, Kangaloon, near Moss Vale and around Bundanoon. Most of the rich, dark-brown soils derived from these cappings have been cleared for grazing, potato growing, and orchards. An outcrop of volcanic trachyte rock at Mt Gibraltar (863 m) between Mittagong and Bowral retains much of its original vegetation.

The plateau of the southern highlands is besieged on all sides by streams which are eating away at the fertile capping. In the west beyond Berrima, the upper reaches of the Wingecarribee have cut deep gorges into the rolling landscape. In the north, the Nepean tributaries have exposed the Hawkesbury sandstone and in the south-east the Kangaroo River has eroded its way through the Triassic rocks and Illawarra coal measures to the Shoalhaven group below. However, it is in the south that the most spectacular change occurs, for it is here that the plateau ends abruptly in a series of sheer cliffs and deep V-shaped gorges. Streams such as Bundanoon Creek, Yarrunga Creek and Barrengarry Creek start in small swamps on the plateau, then plunge in spectacular waterfalls over the cliff-line before making their way in a series of falls and cascades down narrow valleys to the Shoalhaven River. These gorges and falls such as Fitzroy, Belmore, Carrington and Gerringong are a major scenic feature of the area.

In the lower reaches of these creeks, older siltstones, shales and sandstones of the Shoalhaven group are exposed and conglomerate is often a conspicuous component. Ancient Ordovician and Devonian rocks form the banks of the deep Shoalhaven Gorge between Bungonia and Burrier, although much of this land is now covered by the waters of Tallowa Dam.

Little remains of the original vegetation of the basaltic soils. There is a small patch at Robertson Nature Reserve (45 ha) where a mixed warm-temperate/cool-temperate rainforest occurs. Here the foliage is dense and there are many ferns and vines. Species include Sassafras *(Doryphora sassafras)*, Possum-wood *(Quintinia sieberi)*, Featherwood *(Polyosma cunninghamii)* and *Rapanea howittiana*. Rainforest vegetation also occurs in the deeper valleys to the south where Wianamatta Shale supports open forests of *Eucalyptus fastigata, Eucalyptus radiata* and *Eucalyptus obliqua* with trees 20 to 30 m tall.

Extensive swamps or closed sedgelands occur on the plateau, especially in Barren Grounds Nature Reserve, Budderoo National Park and near Carrington Falls. Such areas are dominated by sedges, with Grass Trees *(Xanthorrhoea resinosa), Pultenaea linophylla,* Sundews *(Drosera* species) and large stands of Christmas Bells *(Blandfordia nobilis)*. Heaths are also common on poorly drained soils derived from Hawkesbury sandstone; the Barren Grounds Nature Reserve was established primarily to protect the magnificent variety of heathland plants. A number of rare species occurs in these areas, including *Grevillea rivularis*, found only above Carrington Falls, *Boronia deanei*, a rare plant from the higher Blue Mountains but found along swampy margins near Fitzroy Falls, and the uncommon *Oxylobium arborescens* from the cliff edge at Fitzroy Falls.

Of special botanical interest is the area between Penrose and Tallong which has a complex geological base, providing a varied topography. It has a drier climate and the plant communities show relationships to those of the southern tablelands. A number of species found in this area are absent from, or uncommon in, the rest of the Sydney district. Such species include *Grevillea molyneuxii, Hibbertia sericea, Actinotus gibbonsii,* a new *Prostanthera* species, *Acacia jonesii* and *Asterolasia asteriscophora*.

The more rugged valleys and lower ridges to the south of the area are protected by the northern section of Morton National Park and Budderoo National Park (5700 ha), Barren Ground Nature Reserve and Red Rocks Nature Reserve. The State Forests at Penrose and Tallong are mostly devoted to exotic pine plantations, although there are some fringe areas of native bushland. Much of the central plateau has been cleared but there are small roadside reserves and natural areas where it is possible to sample some of the vegetation which was once more widespread.

Further Reading

Benson, D.H. (1981). Vegetation of the Agnes Banks Sand Deposit, Richmond, New South Wales. *Cunninghamia* 1 (1):35–57. Royal Botanic Gardens, Sydney.

Benson, J.S. and Fallding, H. (1981). Vegetation Survey of Brisbane Water National Park and Environs. *Cunninghamia* 1 (1):79–113. Royal Botanic Gardens, Sydney.

Fairley, Alan (1976). *A Field Guide to the Sydney Bushland.* Adelaide: Rigby.

Fallding, H. and Benson, J.S. (1985). Natural Vegetation and Settlement at Macquarie Pass, Illawarra Region, New South Wales. *Cunninghamia* 1 (3):285–311. Royal Botanic Gardens, Sydney.

Fuller, Leon (1980). *Wollongong's Native Trees.* Wollongong: Fuller and Badans.

Geological Outline of the Sydney District

Ordovician Period

(430 to 500 million years ago)

The most ancient exposed rocks are grey slate, greywacke and andesite. These occur in a few remote areas in the south-west such as at the base of the deep gorges carved by the Shoalhaven River below Point Lookout (Tallong) and at the junction of Paddys and Wollondilly Rivers. Small patches, outliers of larger areas further west, occur along the Great Divide north of Mt Werong and at Mt Shivering and the valley of Murruin Creek. They produce steep unstable slopes with infertile soils.

Silurian Period

(395 to 430 million years ago)

During this period, deep seas covered much of the east coast of Australia and deep layers of sediment, with material from intense volcanic activities, were deposited. Deep coral beds formed during this period are responsible for the limestone at Jenolan Caves, Hollanders Creek, Mt Colong and Wombeyan Caves. This limestone, 300 m thick at Jenolan Caves, is part of a sedimentary sequence containing siltstone, chert and quartzite interbedded with volcanic tuff and lava.

Devonian Period

(345 to 395 million years ago)

This was a time of extensive volcanic activity and a succession of lakes and shallow seas in which sediments were deposited from the west. In the north-west, the Capertee River and Coco Creek cut into a succession of shales, siltstones, quartzites, tuffs, limestones and cherts of Lower Devonian age. A band of acid volcanic rocks containing tuff, lavas, quartz felspar, porphyries and siltstones runs east of the Jenolan Caves and is the major sequence outcropping on the steep hillsides of the Wollondilly-Wombeyan Caves area. Rocks of Upper Devonian age outcrop over an extensive area south-west of Katoomba in the drainage basin of the Kowmung and Coxs Rivers. These are sedimentary rocks, consisting of massive quartzites and sandstones, with layers of shales and siltstones. Sandstones, conglomerates and siltstones of the Devonian period are also exposed in the valley of the Shoalhaven River between Tallowa Dam and Burrier. In general, Devonian rocks form soils with low permeability, low pH and low fertility.

Carboniferous Period

(280 to 345 million years ago)

Granite of Carboniferous age intrudes older sediments and volcanics in a belt along the western edge of the Sydney district. Granite is an igneous rock originating from molten magma forced up to the earth's crust. In many places pre-Carboniferous rocks have been metamorphosed by contact with granite intrusion. Granite country is characterised by rounded hills, exposed grey outcrops and tor formation; this can be seen along the Lett River at Hartley, at Boyd Crossing near Kanangra Walls and along the Coxs River in Megalong Valley. Granite produces moderately deep, permeable acidic soils of relatively low fertility.

Permian Period

(225 to 280 million years ago)

The Shoalhaven Group of Permian rocks was formed in swampy marine conditions, filled by sediments from the west. Layers of sand, silt and conglomerate were deposited. The Group is exposed in a band from Capertee in the north-west, along the valleys of the western Blue Mountains to the south and on the Illawarra Plains. As it underlies the Triassic rocks, it is exposed whenever the Hawkesbury sandstones and Narrabeen group have been worn away. The Shoalhaven group consists of a number of units, some with broad layers of sandstone and conglomerate, others with grey siltstone, shales and sandstones. The group is exposed in the valleys of the Blue Mountains (Jamison, Megalong, Kanimbla, Wolgan, Grose), forms the cliff-line of Kanangra Walls and peaks around Yerranderie and is exposed below the southern highlands escarpment in the Shoalhaven catchment and around Lake Illawarra. These deposits produce hard-setting sandy soils, which are moderately permeable, acidic and relatively infertile. Volcanic rocks of Permian age outcrop on the Illawarra Plain, producing richer soils and the more gentle topography seen around Kiama.

The Illawarra coal measures outcrop at the base of the Triassic cliff-line. Drilling confirms that these coals beds are a continuous sequence between the Illawarra and the western margin of the Sydney district. Coal is mined from these beds at Wollongong, West Berrima, West Mittagong, Burragorang Valley, Lithgow and upper Grose River; abandoned coal and oil shale mines occur at Bundanoon, Joadja, Hartley Vale, Katoomba, Newnes and Glen Davis. Layers of claystones, conglomerate and sandstone are part of the coal measures. The erosion of these coal measures helps enrich the talus slopes beneath the cliff-line. Coal beds also outcrop

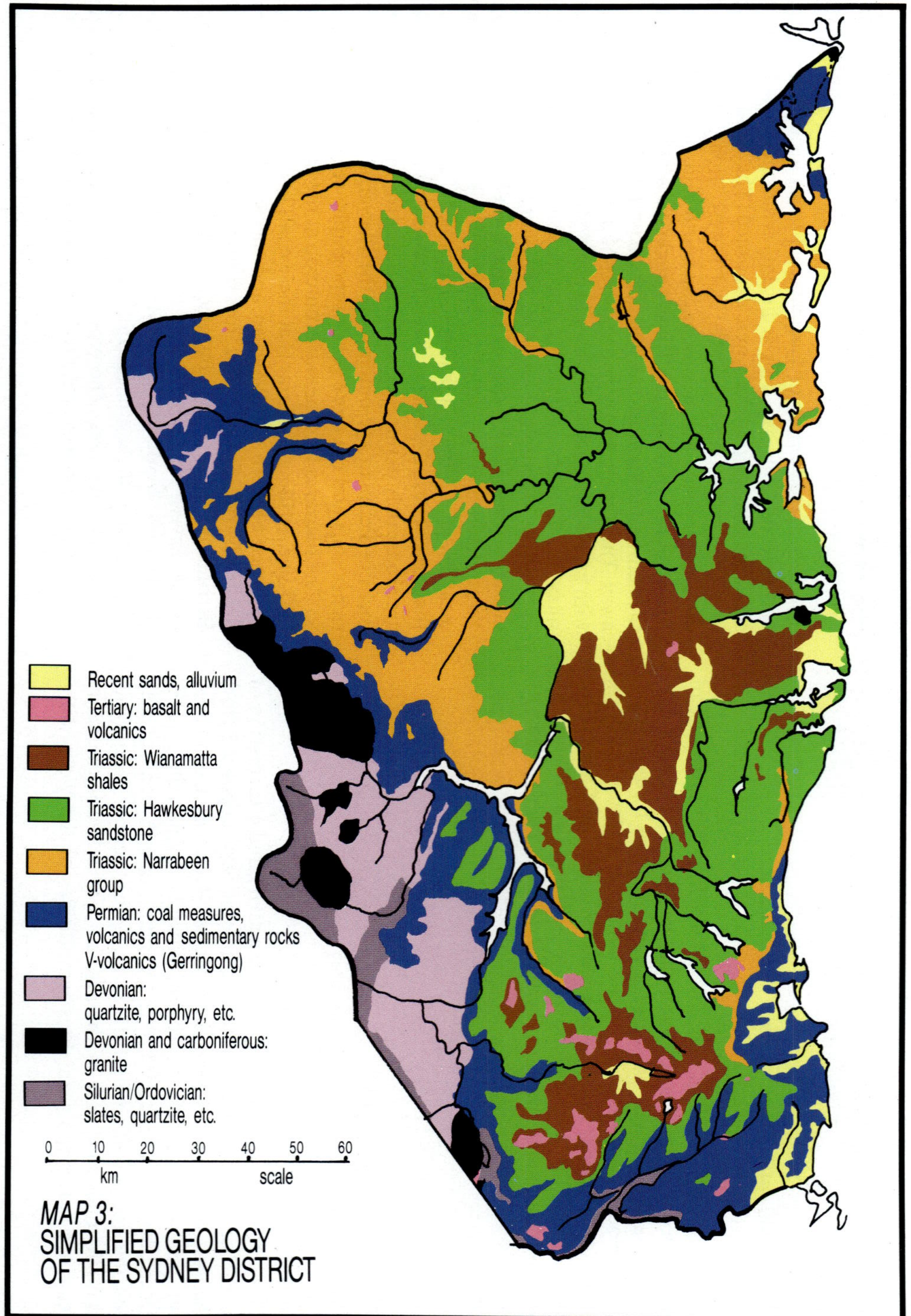

MAP 3:
SIMPLIFIED GEOLOGY OF THE SYDNEY DISTRICT

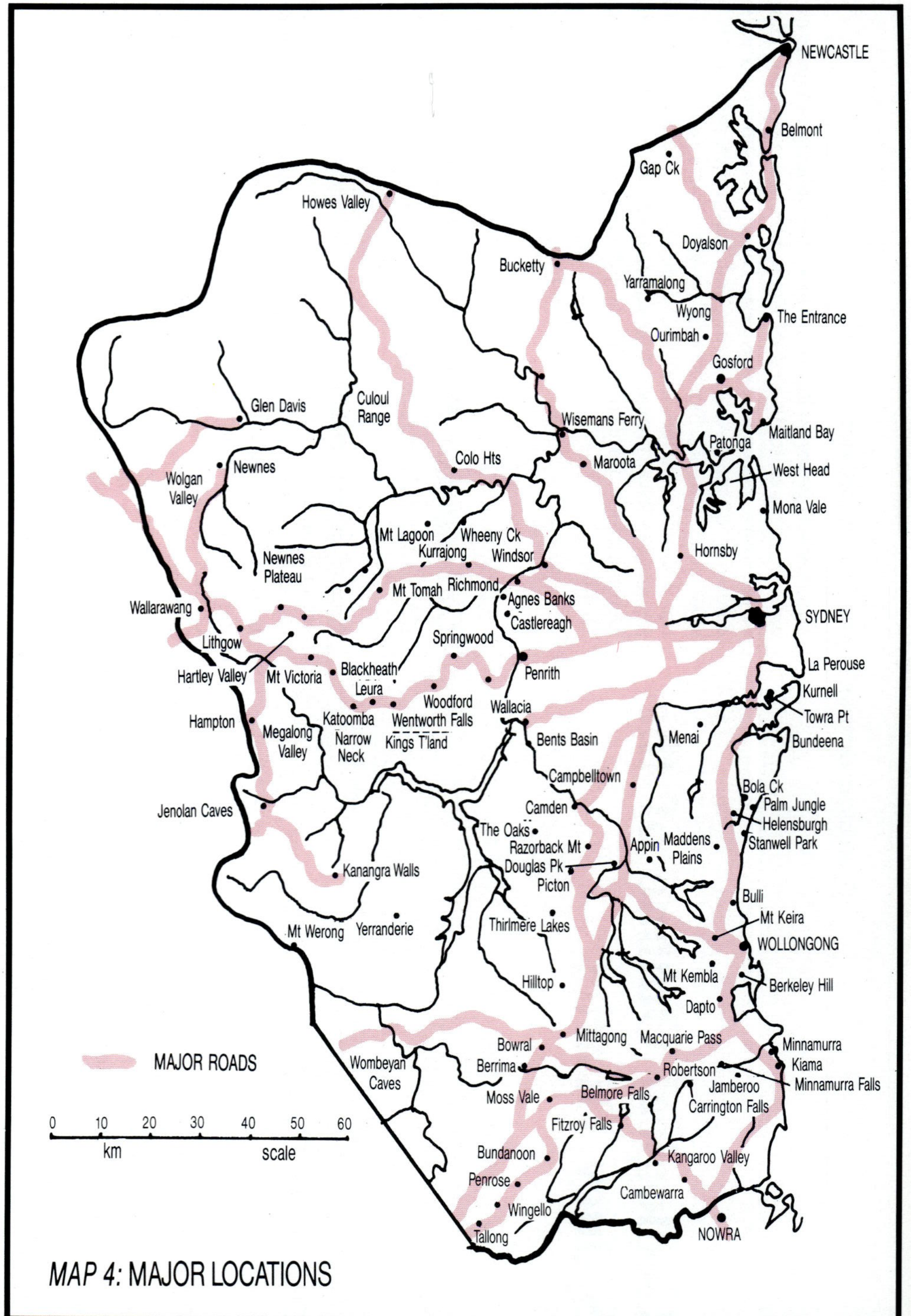

MAP 4: MAJOR LOCATIONS

around Newcastle and Lake Macquarie, and are probably the northern tip of the great saucer-shaped coal strata underlying the Triassic rocks and reaching a depth of 915 m below Sydney.

Triassic Period

(190 to 225 million years ago)

The unique topography of the Sydney district is largely the result of Triassic rocks and their characteristic erosion patterns. These rocks were formed in freshwater conditions when large streams from the west deposited huge amounts of sand, mud, silt and pebbles in river valleys, deltas and lagoons. First deposited was the Narrabeen group, followed by the Hawkesbury sandstone, then the Wianamatta group. Over millennia, the various layers of sediment were converted by pressure into sandstone, mudstone and shale.

The Narrabeen group consists of sandstone with regular bands of siltstone, claystone and other materials. On the coast between Pittwater and Terrigal, and from Garie Beach southward, these strata form a more rounded topography than Hawkesbury sandstone, their red-brown soils enriched by the eroded shale and mudstone layers. In the upper Blue Mountains, the massive sandstone layer (up to 240 m thick above the Grose Valley) forms spectacular sheer cliff-lines. Hawkesbury sandstone occupies about 12 900 km^2 of the Sydney geological basin and has a maximum thickness of 274 m near Sydney, thinning out at the extremities. It is a quartz-rich rock cemented by ferriferous clay and yields light-coloured sandy loams lacking in minerals, rapidly permeable and very acidic. Wianamatta shales have been eroded from most areas, but persist as remnant cappings along some ridges and in the basin of the Cumberland Plain. They produce somewhat richer clayey soils.

Jurassic Period

(136 to 190 million years ago)

Area relatively stable. No significant changes.

Cretaceous Period

(65 to 135 million years ago)

Area relatively stable. No significant changes.

Tertiary Period

(1.8 to 65 million years ago)

In the early Tertiary period, extensive lava flows covered parts of the area. These survive today as basalt remnants up to 60 m thick on the southern highlands and as cappings of olivine basalt on the higher peaks of the Blue Mountains. Many minor intrusions also occurred, leaving dykes and volcanic necks in the surrounding sandstones. Soils formed from the erosion of these volcanic rocks are red-brown, friable and of high fertility.

Deposits of Tertiary gravel, clay and silt cover a large area in the Windsor-Castlereagh district, with smaller patches at Wallacia and Maroota. They are thought to be Tertiary in age, perhaps deposited 12 million years ago during the Miocene Epoch, and subsequently weathered to leave reddish laterite gravels and a hard-pan near the surface.

During the latter part of the Tertiary period, the Sydney Basin was gradually uplifted. At this time, streams cut deeply into the uplifting plateau removing surface material and carrying it down to the plains and out to sea.

Quaternary Period

(Present to 1.8 million years ago)

In the last million years, rising sea levels have flooded the lower valleys of coastal streams, resulting in the picturesque inlets of Broken Bay, Port Jackson, Georges River and Port Hacking. With slower flowing rivers, sands and silts were deposited at deltas and along the flood plains. Coastal dunes at Munmorah, The Entrance North, La Perouse, Kurnell, Marley, Minnamurra Spit and Seven Mile Beach were formed mostly by wind-blown (aeolian) sand filling coastal depressions and covering the base rock from a coastline which was several kilometres to the east of today's coastline. Subsequent deposition of sand by coastal currents added to the wind-blown sand, building up beaches, linking former islands (e.g., Barrenjoey) and cutting off shallow lakes from the sea. Sand particles of eroded Hawkesbury sandstone were blown from the west and deposited near Agnes Banks in Pliocene/Pleistocene times, so that it lies over older Tertiary alluviums and adjoins the more recent Nepean River deposits. This process of shifting sands and river-borne depositions is a continuing one, as evidenced by the siltation of the Hacking River, the problems with sand covering the road at Kurnell and the regular blocking by sand of outlets of many coastal lagoons.

References

Bembrick, C. (1980). Geology of the Blue Mountains, Western Sydney Basin. In: *C. Herbert and R. Helby, eds A Guide to the Sydney Basin*. Geological Survey of NSW Bulletin 26. Department of Mineral Resources.

New South Wales Department of Mines (1969). *Sydney Basin. 1:500 000.* Geological Map.

Taylor, Griffith (1970). *Sydneyside Scenery.* Sydney: Angus and Robertson.

Major Vegetation Types

Heath

The heaths of Australia are among the richest plant communities in the world. Over 750 species have been recorded for the coastal heaths of Sydney and 1850 species in south-west Western Australia. Fossil records show that distinctive heath flora evolved in the southern hemisphere very early in the development of flowering plants. If defined according to structure, a heath community is one dominated by shrubs less than 2 m tall, having a foliage cover of 30 to 70% (open heath) or 70 to 100% (closed heath). It is characterised by a variety of hard-leaved plants whose size and toughness seem to be a response to soil depth and the low level of nutrients. Heaths are found on coastal sands exposed to onshore wind and sea spray, on cliff tops where there is little soil, in isolated depressions on rock platforms and on exposed sites with shallow sandy soils. Closed heaths are common in the Blue Mountains in shallow headwater valleys where poorly drained, sandy, acid peat soils have developed.

The floristic compositon of heaths varies considerably. On the sea-cliff tops, heaths may be dominated by *Westringia fruticosa, Baeckea imbricata, Leptospermum laevigatum, Correa alba* and *Hakea* and *Pultenaea* species. Away from the cliffs, common species are *Allocasuarina distyla, Darwinia fascicularis*, and many representatives of Proteaceae, Fabaceae and Epacridaceae. Wet heaths contain *Melaleuca* spp, *Leptospermum* spp, *Blandfordia* spp, *Banksia robur* and *Sprengelia incarnata*. Mallee eucalypts occur in many heaths, with *E. obtusiflora* most common along the coast and *E. stricta* in the Blue Mountains. Heaths on exposed sites with thin sandy soils in the Blue Mountains, such as along Narrow Neck and Kanangra Walls, are usually dominated by *Allocasuarina nana*. Heath species often form an understorey in open woodland where the two communities merge and many species dwarfed by the harsh heath conditions grow tall in nearby scrubs.

Scrub

Scrub communities are associated with sandy, poorly-drained soils on a wide range of topography. They are especially common on ridge tops and slopes and along drainage lines bordering swampy sedgelands. The community consists of shrubs 2 to 8 m tall, taller than shrubs in a heath, but lacking the tree cover of a woodland. Often these scrubs are very thick with prickly-leaved species such as *Hakea teretifolia* or thickets of *Banksia ericifolia*. Scrub-lined swamps are a feature of Blue Mountains plateau areas where shrub height and density are influenced by drainage, soil type and fire frequency. *Leptospermum* species are common in such areas. Large trees are absent, although small trees from neighbouring woodland may invade the more open areas of scrub.

Woodland

Woodlands are relatively open areas, dominated by trees (chiefly eucalypts), where the canopy foliage covers 10 to 30% of the ground area. The scattered nature of woodland trees is often a result of drier conditions; hence woodlands are common on the foothills of the western valleys, on drier plateau tops and slopes and in the low rainfall zone of Cumberland Plain. In the coastal zone, low open woodlands occupy less exposed positions than heaths and may grade into scrub-forests.

Angophora costata, Eucalyptus haemastoma, E. gummifera and *Banksia serrata* are the common trees in coastal woodlands. The understorey is rich in species. Woodlands of *Eucalyptus moluccana* and *E. tereticornis* were once common on the clay soils of the drier part of the Cumberland Plain but clearing has greatly reduced the extent of this community. *E. moluccana* prefers slightly higher sites than *E. tereticornis*; where drainage is impeded (such as along Eastern Creek at Nurragingy Reserve, Doonside), *E. tereticornis* is replaced by the closely-related *E. amplifolia*. Ground cover on these clays is generally low and open, with scattered shrubs and native grasses such as *Themeda australis* and *Aristida* spp. Dense patches of *Bursaria spinosa* are also common.

The structure and floristic composition of woodland varies considerably. The woodland of *Eucalyptus longifolia* and *Angophora floribunda* found on Permian sediments on the Illawarra Plain near Yallah bears little resemblance to the *Eucalyptus rossii* woodland on the dry steep slopes of Wolgan Valley. Different again is the woodland of *Eucalyptus sclerophylla* and *Angophora bakeri* occurring on the sandy Tertiary alluvium in Castlereagh State Forest. Along the banks and flats of some streams and rivers is a specialised woodland with tall trees of *Casuarina cunninghamiana* and *Angophora floribunda* dominant.

Open Forest

The predominant vegetation type over much of the Sydney district is open forest. It consists of trees to 30 m tall, with a canopy cover of 30 to 70% of the ground area, and a dry understorey of smaller trees, shrubs, herbs and monocots. The structure is varied and species composition depends on the particular environmental condition. In the coastal zone, open forest occurs on plateaux with deeper sands, in gullies and on sheltered hillsides, especially on southern and eastern aspects. The main trees are *Eucalyptus piperita, Angophora costata, E. gummifera, E. punctata, Banksia serrata* and *Allocasuarina littoralis*. The shrub layer is dominated by members of the families Proteaceae, Fabaceae and Myrtaceae. On plateaux, *Eucalyptus haemastoma* and *E. sieberi* may be locally common.

Eucalyptus fibrosa forests are found on alluvial soils in the western part of Cumberland Plain with other ironbarks. A small tree layer of *Melaleuca decora* is often conspicuous. Open *Eucalyptus sieberi/E. piperita* forest predominates on the sandstone plateaux and ridges of the upper Blue Mountains between Wentworth Falls and Bell; stringybarks may also be present.

Many of the most characteristic and beautiful of Sydney's wildflowers occur in the understorey of open forest. *Telopea speciosissima* may be locally common. *Doryanthes excelsa* and *Xanthorrhoea* spp. are often conspicuous. Wattles, boronias, grevilleas, pea-flowers, native irises, lilies and ground orchids all add their colour and magic to the understorey of the open forest.

Tall Open Forest

This community is found on the richer fertile soils of basalt caps and eroded necks, on enriched soils of sheltered slopes and in deep sheltered gullies. It consists of trees 30 m or more tall, usually with straight trunks, with a canopy cover of 30 to 70%. Rainforest elements are often present in the understorey. The slopes of basaltic peaks like Mt Wilson support tall trees of *Eucalyptus viminalis, E. cypellocarpa* and *E. fastigata. E. oreades* is common on more rocky slopes on the edge of the basalt. In the eroded dykes of the lower Blue Mountains there are beautiful trees of *E. deanei* and tall *Syncarpia glomulifera*. At Blue Gum Forest in the Grose Valley, *E. deanei* is associated with *Backhousia myrtifolia* and *Casuarina cunninghamiana*, with an open grassy understorey. In sheltered gullies of the coastal zone, *Eucalyptus saligna* may dominate. On the upper slopes of the Illawarra Escarpment on benches between rainforest or on foothill spurs, the main trees are *E. smithii, E. quadrangulata, E. muelleriana* and *Allocasuarina torulosa*.

The understorey of these taller, wetter eucalypt forests usually includes many climbers and scramblers, such as *Smilax australis, Cissus hypoglauca* and *Stephania japonica*, some tall broad-leaved shrubs and a groundcover of ferns and monocots.

Rainforest

Rainforest occurs in the Sydney district where there is a suitable combination of fertile soils, adequate rainfall, and shelter. Since soils are a reservoir of water and nutrients, they have a profound effect on the distribution and floristic compositon of rainforest.

The basaltic soils of Mt Wilson, Mt Tomah and the southern highlands support temperate rainforests as do the soils of the Gerringong volcanics and coal measures along the Illawarra Escarpment. In the Gosford–Wyong district to the north, Palm Jungle–upper Hacking in the south, and valleys of the Blue Mountains in the west, the eroded Narrabeen strata have produced nutrient-enriched soils which in protected sites also support rainforest vegetation. Aspect is an important limiting factor as it affects exposure to wind, humidity, light and temperature. Most rainforest stands in the area are confined to sheltered gullies (e.g., Wheeny Creek, Bola Creek) or slopes having a southerly aspect (e.g., Gap Creek Flora Reserve, Strickland State Forest).

Rainforest is defined as a tree-dominated community with a closed canopy (70 to 100% foliage cover) where the upper leaves are in close and intermingled contact. In such an environment, trees tend to be straight-stemmed, flowers and fruit are borne high in the canopy, leaves are dark and broad and there are plentiful vines and epiphytes. Little light penetrates the understorey, so only scattered shrubs and ferns are present. Important trees of this community include *Doryphora sassafras, Ceratopetalum apetalum, Cryptocarya glaucescens, Acmena smithii* and *Livistona australis*, but there are significant differences in species composition between the littoral rainforest of Palm Jungle and Minnamurra Spit, the mixed warm-temperate and subtropical rainforests of Macquarie Pass, the temperate rainforests on basalt on Mt Wilson and the shale-based rainforests of the Watagan Mountains.

Sedgeland Swamp

Swamplands dominated by members of the families Cyperaceae and Restionaceae occur regularly on plateau areas, especially in headwater valleys where impervious sandstone or clay layers create waterlogged conditions and a build-up of organic matter in the soil. As places of permanent moisture, they support their own specialised plants including the insectivorous sundews and bladderworts, yellow-flowered *Xyris*, various *Leptospermum* spp., *Epacris obtusifolia, Symphionema paludosum* and *Callistemon* spp. *Gahnia* thickets are frequent along drainage lines. The presence of alpine flora such as *Celmisia, Luzula* and Sphagnum moss in many higher mountain swamps indicates former links between these areas and the alpine bogs of Kosciusko.

Fresh Water

There are few freshwater lakes, lagoons or river overflows in the Sydney district. Thirlmere Lakes, Mountain Lagoon, Marley Lagoon, some open swamps on the Cumberland Plain, scattered farm dams and river backwaters provide a few suitable sites for freshwater plants but pollution and infilling have destroyed much of this habitat. A number of freshwater plants are rooted in the mud and have leaves which float on the surface. These include *Ottelia ovalifolia, Triglochin procera, Nymphoides geminata* and the rare water-lily *Brasenia schreberi*. Others like *Spirodela pusilla* are floating herbs; a few like *Vallisneria gigantea* are completely submerged. Lagoons and swamps are often bordered by rushes (*Juncus* spp.) and such wetland species as *Typha orientalis, Philydrum lanuginosum, Villarsia exaltata* and *Damosonium minus*.

Mangrove and Saltmarsh

Mangrove swamp

Estuarine vegetation occurs in protected coastal areas along the tidal zone or on alluvial flats subject to regular tidal inundation but not exposed to wave action. This complex consists of a sequence of zones determined by the frequency and duration of flooding and salinity levels. On the seaward edge of the mudflats is a zone of mangroves. Two species may be present, the taller Grey Mangrove *(Avicennia marina)* and the shorter River Mangrove *(Aegiceras corniculatum)*. River Mangrove is usually on the landward side of Grey Mangrove and often extends further up estuaries where the water is less salty. Behind the mangroves may be a zone of saltmarsh, dominated by low succulent-stemmed Samphire *(Sarcocornia quinqueflora)* and Seablite *(Suaeda australis)*. Saltmarsh areas range from small (e.g., Patonga Creek, Salt Pan Creek) to extensive (e.g., Towra Peninsula).

Behind the saltmarsh, in areas infrequently covered by salt water, there is usually a rushland or reedland zone. The rush *Juncus kraussii* dominates boggy saline areas and the reed *Phragmites australis* grows in less saline swampy ground. Here too may be found the small herbs *Wilsonia backhousii, Triglochin striata* and *Samolus repens*. Swamp Oak *(Casuarina glauca)* and Swamp Mahogany *(Eucalyptus robusta)* occupy slightly higher elevations on the landward side.

Saltmarsh

Coastal Dunes

The area between beach and hind-dune woodland is an inhospitable area for plants. Frontal dunes are usually very unstable and have poor water-holding capacity. On the seaward slope, the first plants to stablise the sand are the grasses, *Spinifex sericeus* and *Festuca littoralis*. On more stable areas *Lomandra longifolia* and *Acacia longifolia* var. *sophorae* occur, often with the exotic *Hydrocotyle bonariensis*. On the inner dune slope away from salt winds *Leptospermum laevigatum*, clumps of *Hibbertia scandens* and the climber *Stephania japonica* may occur. Shallow depressions which retain moisture often form between the dunes and some areas develop small lagoons immediately behind the dune area.

Further Reading

Clarke, L.D. and Hannon, N.J. (1967). The mangrove swamps and salt marsh communities of the Sydney district. I. Vegetation, soils and climate. *Journal of Ecology.* 55: 753–771.

Kartzoff, M. (1969). *Nature and a City.* Sydney: Edwards and Shaw.

National Parks and Wildlife Service (1977). *Rain Forests.* Sydney: NPWS.

National Parks and Wildlife Service (1981). *Heaths in New South Wales.* Sydney: National Parks and Wildlife Service.

Specht, R.L. (1970). Vegetation. In G.W. Leeper, ed. *The Australian Environment.* Melbourne: CSIRO-Melbourne University Press, 4: 44–67.

Where to See Wildflowers

Sydney, Newcastle and Wollongong are surrounded by extensive areas of Crown land and reserves where the natural vegetation is relatively undisturbed (see Map 1). However, it is not necessary to travel far to see a good range of native plants, as there are a number of wildflower gardens open to the public. Most have their plants labelled for instant identification. The best known are: Ku-ring-gai Wildflower Gardens (St Ives); Katandra Bushland Sanctuary (Mona Vale); Stony Range Flora Reserve (Dee Why); Bankstown Native Gardens (Sylvan Grove, Picnic Point); Joseph Banks Native Garden (Kareela); Glenbrook Native Gardens (lower Blue Mountains); Wirrimbirra Sanctuary (Bargo); Wollongong Botanic Gardens (near Wollongong University).

The Royal Botanic Gardens in Sydney has some Australian plant beds and the associated Herbarium has a collection of pressed and labelled plant specimens for public reference. The new botanic gardens at Mt Annan (between Campbelltown and Camden) and Mt Tomah (Bells Line of Road) will have excellent collections of native plants when fully established. Further afield, the Canberra Botanic Gardens has many Sydney species planted throughout its large area.

Many small urban bushland areas protect samples of native vegetation. There are many of these reserves, some large, some very small, and some more interesting than others. In the Sydney area, those recommended include Old Manly Reservoir, Davidson Park, Pennant Hills Park, Elouera Bushland, Cumberland State Forest (West Pennant Hills), Oatley Park, Lane Cove River Park, Wolli Creek, Jannali Park, Georges River Parkland, Nurragingy Reserve (Doonside), Castlereagh State Forest, Captain Cooks Landing Place (Kurnell) and the various headlands of Sydney Harbour National Park.

Membership of the Society for Growing Australian Plants is highly recommended. This organisation holds regular meetings in various urban centres and organises appropriate outings. It also provides a seed bank, information on the growing of native plants and a high quality magazine. Information may be obtained from the Honorary Secretary, 5 Ellesmere Road, Gymea Bay 2227.

The Naming and Classification of Plants

Naming and categorising are important and useful because nameless items cannot be easily dealt with or fully understood.

Even primitive cultures have names for plants that are useful or have some religious or magical significance. Simple rural communities generally have a working knowledge of the indigenous flora of their area handed on from one generation to the next. Usually there is little understanding of relationships between plants, and naming is homely and haphazard. Frequently names reflect the utilitarian value of the plant rather than its physical attributes. In more complex societies with broader horizons, such locally applied names are found to be inadequate for communication between different regions or countries. At some point it becomes necessary to devise a universal scientifically based means for classifying and naming all plants. To do this, a hierarchy of ranks or categories is created to take account of both general and specific features of the plants. For ease of reference, each rank must be named in a logical consistent way. From time to time minor or major modifications may be made to the classification as new information from various scientific disciplines is assimilated or increasing knowledge suggests alternative bases for classifying certain groups.

When reasonably developed, a systematic classification provides a two-way flow of information. New facts can be interpreted and old ideas re-examined in the context of relationships existing at various levels. Plant selection and breeding may be assisted, and the understanding and management of ecosystems improved. Comparisons and generalisations can be made. The system sometimes has predictive value, for example, in the search for new drugs or for plants with particular attributes. In comparing new information with what is already known, the classification becomes a convenient filing and information-retrieval system. The plant name is the key that provides access to that system.

Plant Names

Since the time of Linnaeus's *Species Plantarum* published in 1753, scientific plant names have consisted of two parts, the first with a capital initial letter signifying the genus (the generic name), the second with a small initial letter identifying the species (the specific epithet), the two together constituting the species name. To be complete these two must be followed by a third element called the authority. This is the name, in abbreviated form, of the person who formally described the plant and gave it its name. It is an important source of information to botanists, enabling them to trace name changes and decide priorities in nomenclature, but is frequently omitted in popular works. Rules governing the assignment and legitimacy of names are set out in the International Code of Botanical Nomenclature (ICBN). When botanists revise a family or a genus, it is frequently necessary to change a well-established name as new information highlights some past anomaly (perhaps the plant received different names in the past), or relationships become better understood. Sometimes a single genus or species is split into smaller units requiring new names, or possibly different genera or species are lumped together. The ICBN gives order to such changes.

Problems of Plant Classification

All systems of classification attempt to bring together individuals which by their form and structure show that they are related, and to arrange the groups so formed into progressively higher levels of classification based on broader, more generalised characteristics shared by them. Unlike early systems of classification which tended to rely on one or a few features and sometimes produced groupings of unrelated taxa (units of classification), modern systems take account of as many features as possible because the recognition of a number of shared attributes is more likely to lead to groupings in which there is an underlying relationship. Systems which attempt to reflect natural relationships are termed 'natural systems'. Originally these were intended to show God's plan of creation; the idea of lineage in the evolutionary sense was not at first envisaged.

Some difficulties arise because in attempting to make the universe intelligible we impose a degree of tidiness that does not actually exist. Nature does not segregate organisms into water-tight boxes. There are many instances where the boundaries between taxa are not clearly marked. Consequently concepts such as species and genus, although useful, and indeed real, cannot be precisely defined.

Even when similarities and differences are apparent, there is often considerable room for disagreement as to their significance and the level at which they should be recognised. Often, for example, distinctive groups of genera, or even a single genus, may be segregated into smaller families in one scheme, but are regarded as tribes or subfamilies of larger, more diverse families in another. Less often there is some difficulty in deciding into which higher group a taxon should be placed; those who regard a particular set of characteristics as predominant will place it in one group while those who put greater value on other characteristics may place it in another. Although there will always be some disagreement between 'splitters' and 'lumpers', the use of new and sophisticated methods of evaluating data may one day make consensus possible on the placement of certain taxa over which disagreement presently exists.

Classification and Arrangement

It is not necessary to have previous knowledge of plant taxonomy to understand the arrangement of families used in this book or to gain some appreciation of relationships which exist above and below the family level.

The plants are in five distinct groups corresponding to the category of division in the plant kingdom. These categories of classification in descending order (excluding sub-categories) are: kingdom; division; class; order; family; genus; and species. Because of the many differences between the various systems at higher levels of classification, we have deliberately avoided botanical nomenclature above the category of order.

The five groups are arranged from the most ancient and primitive (the fork ferns) to the most recent and advanced (the angiosperms). The angiosperms are by far the largest and commonest group. Each of the groups, but particularly the angiosperms, is arranged as far as possible so that families appearing earlier have a higher proportion of character states regarded as primitive while those appearing later have a higher proportion regarded as advanced or derived from more primitive states. Those appearing earlier are generally of greater antiquity. Families sharing a number of attributes are placed near each other in the arrangement and are considered to be more closely related than those that are apart. However, some taxa are relatively isolated, therefore close juxtaposition does not always imply a close relationship.

The monocots are treated as a group after the dicots because it is generally believed that they developed from the dicots at an early stage in angiosperm history. It does not follow that the present-day monocots are more advanced than present-day dicots; it is preferable to regard the two groups as concurrent lines of development in which individuals display both primitive and advanced features in differing proportions.

The classification and arrangement of families used in this book is based on the systems proposed by Takhtajan[1,2] and Cronquist[3]. Family delimitation and species names follow with few exceptions Jacobs and Pickard[4].

References

[1]Takhtajan, A. (1969). *Flowering Plants: Origin and Dispersal* translated by Jeffrey, C. Edinburgh: Oliver and Boyd.

[2]Takhtajan, A. (1980). Outline of the Classification of Flowering Plants *Botanical Review* 46: 225.

[3]Cronquist, A.J. (1981). *An Integrated System of Classification of Flowering Plants.* New York: Columbia University Press.

[4]Jacobs, S.W.L. and Pickard, J. (1981). *Plants of New South Wales.* Sydney: National Herbarium of New South Wales.

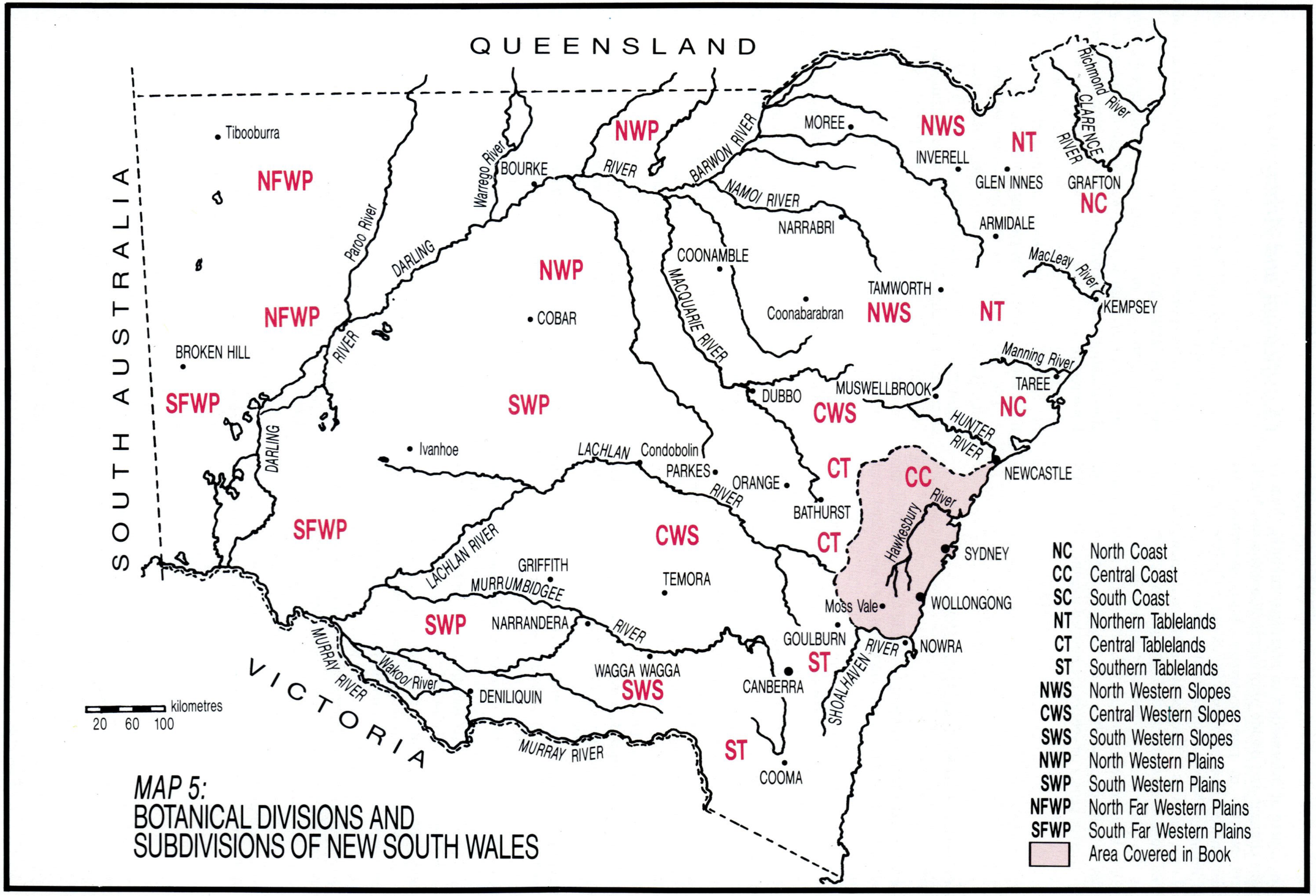

MAP 5:
BOTANICAL DIVISIONS AND SUBDIVISIONS OF NEW SOUTH WALES

1. *Psilotum nudum*

2. *Tmesipteris truncata*

1. Fork Ferns

Only two small genera, *Psilotum* and *Tmesipteris*, survive as descendants of a group of land plants which was widespread and abundant during the Devonian and Silurian periods (345–430 million years ago), possibly the earliest and most primitive of the vascular plants. Both the extinct and living members lack true roots and leaves, their functions being performed by the stems which provide photosynthesis in the upper part and anchor the plant and absorb water in the lower.

The fork ferns can be combined with the true ferns, the clubmosses and several other plants having similar characteristics and a similar life cycle into a group called the pteridophytes. Although all these plants represent different evolutionary lines, this classification allows us to distinguish them from the bryophytes (liverworts and mosses) on the one hand and the spermatophytes (cycads, conifers, flowering plants) on the other. The pteridophytes differ from the bryophytes in having a well-developed vascular system consisting of water-conducting xylem and food-conducting phloem arranged in vascular bundles. They differ from the spermatophytes in that they reproduce by means of spores rather than by seeds.

All plants have a life cycle consisting of alternating sexual and asexual generations, respectively a gamete-forming gametophyte and a seed or spore-forming sporophyte. In the pteridophytes the gametophyte is a separate free-living entity lacking true roots, stems or leaves, but possessing sexual organs called archegonia and antheridia which produce respectively eggs and sperms. The tiny gametophyte grows in moist places or under the soil litter where a film of moisture allows the motile sperm to swim to the neck of an archegonium and down the passage to the egg. This means of reproduction, more suited to an aquatic than a terrestrial existence, restricts the pteridophytes to certain moist habitats where they are often quite successful, occupying areas that are too damp or shady or in other ways unsuitable for most seed-bearing plants.

Psilotaceae

In *Psilotum*, the only genus in the family Psilotaceae, both the aerial and the underground stems fork repeatedly into 2 equal branches. Frequently one of the underground branches turns upward and develops into an aerial shoot. The aerial shoots are green and angular or flattened; the rhizomes are covered with brown rhizoids (hair-like structures which absorb moisture). The synangia (i.e. united sporangia, the spore-producing structures) are 3-lobed and appear to be in the branch forks, but are actually terminal on very short side branches which bear 2 tiny appendages called sporophylls. Spores are released by the splitting of the sporangia along longitudinal slits.

Unlike more advanced plants, *Psilotum* produces spores of one type only. These germinate to form a gametophyte which lives a saprophytic existence underground. This is the alternate sexual generation which, after fertilisation, produces a new sporophyte.

Psilotum

Psilotum is widespread in the tropics and subtropics. It has 3 species, of which 2 occur in Australia, 1 in the Sydney region.

Psilotum nudum

Skeleton Fork-fern **Plate 1**

In south-eastern Australia this small wiry plant is usually found in rock crevices, although in tropical areas it is more often epiphytic on other plants. The rhizome is short but much branched, producing erect or trailing rigid 3-ribbed yellowish stems up to 60 cm long which repeatedly fork in the upper part into equal branches. Tiny scale-like leaves occur repeatedly on the upper stems intermixed with 3-lobed spore-producing structures called synangia.

Habitat Moist crevices of sandstone cliffs in open forest, rarely epiphytic in rainforest.

Distribution Northern, eastern and central Australia, NZ, Lord Howe Island, pantropical.

Tmesipteridaceae

Tmesipteris, a small, hanging, fern-like epiphyte native to Australia, the Philippines, New Zealand and some Pacific islands, is the only genus in this family. Like *Psilotum*, to which it is related, it lacks true roots and leaves but possesses a vascular (i.e. fluid-conducting) system. Unlike *Psilotum*, it is commonest in cool-temperate regions.

The aerial stems of *Tmesipteris* branch only occasionally and have flattened leaf-like structures arranged pinnately along their length. These 'leaves' have a prominent central vein or nerve but no secondary veins. Synangia (united sporangia) are elongated and 2-lobed and attached just below the fork on branched leaf-like sporophylls which are borne irregularly along the stems. The sporangia produce spores of one type which germinate to form a small flat bisexual gametophyte.

Tmesipteris

Six of the seven species of *Tmesipteris* occur in Australia. Four species occur uncommonly in the Sydney area in cool moist places in rock crevices and on tree-fern trunks.

Tmesipteris truncata

Fork-fern **Plate 2**

The commonest of the 4 types of fork-ferns in the Sydney district, this species usually grows in rock crevices rather than on the trunks of tree-ferns which is the usual habitat of the others. The pendulous stems are up to 30 cm long with numerous linear curved leaf-like appendages about 20 mm long and 2 or 3 mm broad. The lower stems which bear few 'leaves' are ridged on all sides. The brown 2-lobed synangia are usually pointed at both ends.

Habitat Moist rock crevices in cool shady places.

Distribution NSW, south-eastern Qld and Lord Howe Island.

2. Club Mosses

The club and spike mosses are a much larger group than the fork or whisk ferns, and are quite distinct from them. Their ancestors appeared somewhat later, in the Devonian period, about 350 million years ago. It is not at all apparent from the lowly present-day members that these ancestors included many arborescent (tree-like) forms such as Lepidodendron which, with the horsetails, ferns, seed ferns and a few primitive conifers, dominated the Carboniferous vegetation, and today form the major black coal beds of the northern hemisphere.

Although still showing many primitive characters, they are more highly developed than the fork ferns. The stems branch dichotomously and bear tiny simple leaves with a mid-vein. The sporangia are borne singly on the upper surface of special leaves called sporophylls which form aggregate structures known as strobili. In the club mosses the strobili bear sporangia of 1 type and produce spores of 1 type, but in the spike mosses the strobili have 2 types of sporangia which produce either microspores or macrospores.

Lycopodiaceae

The Lycopodiaceae consists of 2 living genera, *Lycopodium* with about 200 mostly tropical species, and *Phylloglossum* with 1 species restricted to temperate Australia. *Lycopodium* is a heterogeneous assemblage in which 2 to 5 subgenera or genera are recognised. Although subdivision is based on difference in both the sporophyte (the plant as we know it) and the gametophyte (the alternate generation), 2 broad groups can be distinguished by the manner of branching: in one group the stem forks into 2 equal branches giving a sympodial structure, and in the other it branches unequally giving a monopodial structure with rooting along the stems.

Lycopodium

Club mosses *(Lycopodium)* are small terrestrial or epiphytic herbs with narrow small leaves which are either all alike and arranged spirally or of 2 types arranged in 1 plane. The sporangia (kidney-shaped spore-producing structures) at the ends of the branches or scattered along them are subtended by tiny crowded leaves called sporophylls, the whole

3. Lycopodium deuterodensum

forming a cone-like structure known as a strobilis. In some species the gametophyte develops into a small green plant reaching sexual maturity in less than a year, but in others it develops underground and may take 5 or more years to reach maturity. Sexual reproduction is the same as for other pteridophytes.

Sixteen species of *Lycopodium* occur in Australia with 3 widespread species in the Sydney district. The 3 Sydney species are terrestrial plants of the monopodial type.

Lycopodium deuterodensum

Bushy Clubmoss **Plate 3**

In this species and *L. cernuum* the strobili are always terminal on the branches. There is little chance of confusion, however, since the stems of this species are erect and much branched, giving the appearance of a miniature conifer, whereas those of *L. cernuum* are long and scrambling or hanging. The stems of *L. deuterodensum* rise at intervals from a long branched rhizome and have leaves of two types, sometimes with both on the same plant. The common leaf type is about 2 mm long, crowded, overlapping and appressed to the branches; the other is about 4 mm long, less crowded and more spreading. In protected conditions this clubmoss approaches a metre in height.

HABITAT In sheltered moist places in heath, open forest or rainforest margins, usually on sandstone soils.

DISTRIBUTION Qld to Vic, Tas, SA, NZ, New Caledonia, and Norfolk Island.

4. Lycopodium cernuum

Lycopodium cernuum **Plate 4**

This essentially tropical species has long, much-branched scrambling stems with fine curved leaves about 5 mm long ending in a long point. The strobili, borne at the ends of short branches, are about 5 mm long, pale in colour and more compact and crowded than the leafy stems. They are nodding at maturity, rather than erect as in *L. deuterodensum*.

HABITAT Moist cliff faces and clay banks.

DISTRIBUTION Qld, NSW, NT, WA, NZ, New Guinea and Polynesia.

5. Lycopidium laterale

Lycopodium laterale

Slender Clubmoss **Plate 5**

This small erect or straggling plant is readily identified by the sessile lateral strobili occurring at intervals along the stems. The stems are 10 to 40 cm long and either undivided or once or twice branched. Leaves are about 4 to 7 mm long, narrow linear and finely pointed. The sporophylls on the strobili have a broad base and a pointed apex with finely toothed margins.

HABITAT Bogs, swamps and damp areas of coastal heaths.

DISTRIBUTION Qld, NSW, Vic, Tas, SA, NZ and New Caledonia.

6. *Selaginella uliginosa*

Selaginellaceae

Although the Selaginellaceae and the Lycopodiaceae are related, the differences between them are sufficient for them to be regarded as separate orders as well as separate families. The single genus, *Selaginella*, consists of about 700 species of small trailing or upright terrestrial herbs occurring throughout the world, mostly in the moist tropics.

The vegetative or stem leaves of *Selaginella* are small and narrow and arranged in 4 rows. The leaves are all alike in some species, or of 2 kinds in others. They have an unbranched midvein and a tiny raised gland called a ligule on the upper surface near the base. In some cases this serves to distinguish *Selaginella* from *Lycopodium*. The fertile leaves (sporophylls) at the ends of the stems are smaller and more crowded than vegetative leaves.

Unlike *Lycopodium, Selaginella* is heterosporous (i.e., produces spores of 2 types—male and female). These are produced in 2 types of sporangia called microsporangia (male) and megasporangia (female), the former yellow or orange-brown, the latter pale green or whitish. The spores develop into separate male and female gametophytes, undergoing most of their development while still within their respective sporangia before they are shed. Final development and fertilisation of the female gametophyte by sperm from a male gametophyte occurs on the ground.

Selaginella

Nine species of *Selaginella* occur in Australia, mostly in the eastern states. One species occurs fairly commonly in damp sandy sites in the Sydney district.

Selaginella uliginosa

Swamp Selaginella **Plate 6**

With just one species of *Selaginella* in the area, the only possibility of confusion is with a clubmoss (*Lycopodium* sp). Swamp Selaginella is a smaller plant than local clubmosses with simple or pinnately branched erect stems bearing spreading leaves of one kind in 4 decussate rows. The leaves are 2–3 mm long, narrow lanceolate, acute at the apex and keeled on the back. They are light green, often turning bronze coloured in sunny locations. The sporophylls are similar to the leaves, though slightly smaller, more pointed at the apex and more crowded.

HABITAT Damp sandy soil in heath and open forest, and on the margins of swamps and ponds.

DISTRIBUTION All states except SA.

3. Ferns

The ferns are an ancient group of vascular plants with an ancestry stretching back nearly 350 million years to the Carboniferous and Devonian periods.

As with other pteridophytes, the fern life cycle alternates between 2 distinct generations represented by separate and dissimilar plants. The familiar fern plant with an underground rhizome, stem and leaves represents the sporophyte phase of the cycle. It produces spores from sporangia occurring on the underside of ordinary leaves or on specialised fertile fronds. The sporangia usually occur in small clusters called sori which often have a characteristic size, colour, shape and arrangement on the leaf. In some cases the sorus is protected by an outgrowth of tissue known as an indusium.

Germination of spores produces the alternate generation, the gametophyte. In ferns this is a small green heart-shaped disc less than a centimetre in diameter which is often referred to as the prothallus. The prothallus grows on wet surfaces or in soil litter in places where dew at least keeps it moist. On its underside the prothallus bears organs called antheridia and archegonia which produce respectively sperm and eggs. As with other pteridophytes, the male sperm swims in a film of water to fertilise the egg. The zygote formed by the union of the male and female gametes forms an embryo which eventually grows into a new fern plant.

Ferns differ from other pteridophytes in a number of respects. All pteridophytes except ferns have tiny simple leaves with a single unbranched midvein. Furthermore the vascular bundles in the leaves merge smoothly with those of the stems suggesting that the leaves are merely surface modifications of the stem. Most ferns, by contrast, have large leaves with a simple, pinnatifid or compound structure, a complex network of diverging veins and a vascular system which does not merge with that of the stem. Also, fern leaves or fronds are characteristically coiled in bud, and open by unrolling from the base to the tip.

There are possibly 10 000 species of ferns, which makes them by far the largest group among the pteridophytes. Only 350 species belonging to 100 different genera occur naturally in Australia, mostly on the fringes of the continent, particularly along the coast and ranges of eastern Australia. Many of the ferns in northern Australia have tropical affinities and are absent from south-eastern Australia where the fern flora is most closely related to that of New Zealand and the Antarctic islands. Since fern spores are readily dispersed over long distances by wind, fern species tend to have wide distributions governed mainly by ecological factors such as temperature, moisture and light.

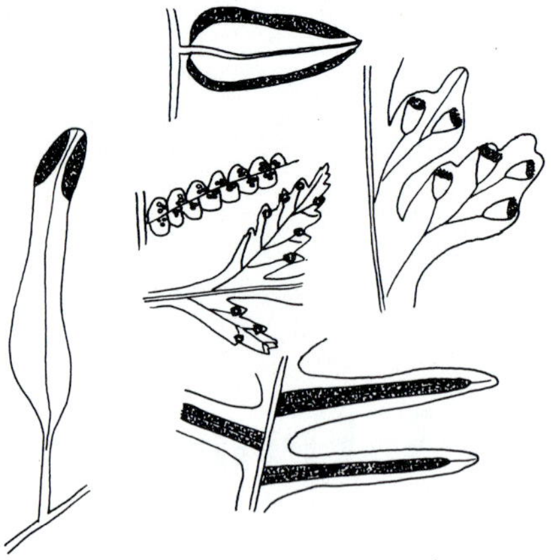

Some arrangements of sporangia

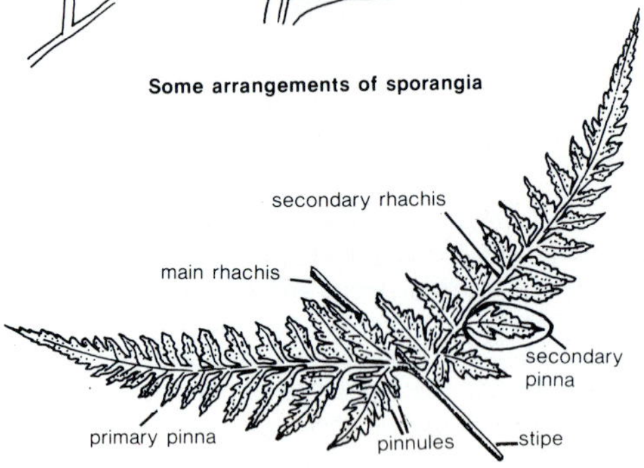

Section of a frond

Figure 1

Fern classification is based on attributes of both the gametophytic and sporophytic generations, particularly the latter. These include details of spore structure and ornamentation, the structure of sporangia, the size, shape and arrangement of sori, the presence or absence of an indusium, leaf shape and form, stem anatomy, the nature of stem hairs and scales, and base chromosome numbers. Often relationships between fern groups at the higher levels are more apparent through connecting species than by the common possession of particular character states. For some time there have been disagreements and differences in approach regarding fern classification, which are not yet entirely settled. This, added to the difficulty or impossibility of describing relationships in a way that is useful for identification purposes, has persuaded us to adopt the following treatment in which, although species are arranged in genera and families, specific rather than generic or family characteristics are emphasised.

Osmundaceae

Todea

Todea barbara

King Fern **Plate 7**

The King and Royal Ferns belong to a small family of only 3 genera and about 23 species considered to be among the most primitive of the modern ferns. The ferns in this family have a short hard erect rhizome which may become massive and trunk-like with age, bearing several large crowns of fronds. In *T. barbara* the fertile and sterile fronds are similar, 0.5 to 2 m long with shiny light green, bipinnate, leathery laminae. The main rhachis is smooth and pale, the secondary rhachises narrowly winged. Secondary pinnae (i.e. pinnules) are narrow with serrated margins. The spore-producing organs (sporangia) are borne on the underside of the lower pinnae. They are not clustered into sori as in most ferns, but form initially on the veins of the pinnules and eventually spread to cover most of the underside of the pinnules with a thick powdery velvet-brown mass. The spores are green, and are shed early in the development of the frond.

HABITAT Commonly found on shady creek banks and rocky creek beds, but also occurring in other wet places such as caves and rock crevices where growth may be stunted.

DISTRIBUTION Qld, NSW, Vic, Tas, SA, NZ and South Africa.

Leptopteris

Only 1 of the 7 *Leptopteris* species occurs in Australia.

Leptopteris fraseri

Crepe Fern **Plate 8**

An attractive rather delicate fern with a crown of large arching fronds rising from the ground or, in older specimens, from a short woody trunk. Fronds often exceed a metre in length and have a long, dark-green, thin, softly herbaceous narrow-elliptical lamina which is mostly bipinnate-pinnatifid. The stipes and main rhachises are clothed with soft woolly brown hairs. The light brown sporangia are clustered towards the centre of the ultimate segments and release their green spores at an early stage.

HABITAT Humid, cool shady places such as rainforest gullies, wet caves, under waterfalls and beside streams, mostly on the tablelands.

DISTRIBUTION Coast and tablelands of eastern Australian from about Sydney northwards. Locations include Neates Glen and Grand Canyon, Blackheath; Waterfall Gully, Mt Wilson.

7. Todea barbara

8. Leptopteris fraseri

Schizaeaceae

Schizaea

Small terrestrial ferns with erect simple or dichotomously branched fronds rising in tufts from a short horizontal rhizome, the lamina reduced to a narrow wing. Fertile segments pinnate at the apex, resembling a cock's comb. Sporangia brown, sessile in rows on the pinnae.

9. Schizaea rupestris

Schizaea rupestris Plate 9

This tiny fern has simple (i.e. unbranched) smooth flat fronds less than 15 cm tall and 1 to 2.5 mm broad. Fertile fronds are usually slightly longer and narrower than the sterile ones. In the absence of the comb-like fertile pinnae this species could at first be mistaken for *Grammitis* or *Dictymia* species, but it never has sporangia in sori on the back of the fronds as in those genera.

HABITAT On damp shady mossy rocks near waterfalls and in wet sandstone caves or shelves.

DISTRIBUTION Endemic in NSW.

10. Schizaea bifida

Schizaea bifida
Forked Comb-fern **Plate 10**

The narrow wiry fertile fronds of this species are 10 to 25 cm high, smooth or slightly rough and dichotomously divided once, or rarely twice. Each branch bears at its apex the fertile pinnae in a comb-like arrangement 1 to 2 cm long and 4 to 6 mm broad. The sterile fronds are much smaller and are undivided.

HABITAT On sandy soil in open forest, scrub anu heath.

DISTRIBUTION All states except WA.

Schizaea fistulosa
Narrow Comb-fern **Plate 11**

This species has characteristic slender unbranched fertile fronds 10 to 30 cm tall, mostly smooth but sometimes sandpapery. Each fertile frond bears a small comb of 5 to 9 pairs of fertile segments fringed with sharp teeth. Sterile fronds are undivided, terete or slightly flattened near the top and very narrow (less than 0.5 mm wide).

HABITAT Wet boggy soils.

DISTRIBUTION NSW, Vic, Tas, SA, WA and New Caledonia.

11. Schizaea fistulosa

Schizaea dichotoma
Branched Comb-fern **Plate 12**

S. dichotoma is the largest and most branched of the Australian comb-ferns. Fertile fronds have as many as 60 comb-like heads in a broad flat arrangement. The rough, slightly flattened sterile fronds are usually shorter than the fertile ones and are branched dichotomously many times above the middle.

HABITAT As for *S. bifida.*

DISTRIBUTION Qld, NSW, WA, NT and tropics.

12. Schizaea dichotoma

Gleicheniaceae

Gleichenia

A genus of 10 species, 5 in Australia, 3 in the Sydney district. Scrambling terrestrial ferns with a slender long-creeping scaly rhizome. Primary branches in pairs, often branched or forked 2 to 4 times, with a dormant apical bud in each fork. Pinnules linear, at right angles to the minor rhachises, the ultimate segments small, rounded or angular, flat or recurved. Sori solitary without indusia, consisting of 2 to 5 large sporangia at the base of the ultimate segments near the margin.

13. Gleichenia dicarpa

Gleichenia dicarpa
Pouched Coral-fern **Plate 13**

The commonest of the coral-ferns, this species forms large tangled masses up to 2 m high in sheltered and sunny moist situations. The long slender brown stipes are smooth and shiny, the minor rhachises densely clothed with stellate brown hairs. The primary branches fork about 3 times, with pinnules produced around the dormant buds in the forks. Pinnae are 3 to 5 cm long, narrow linear, with close-packed oblong ultimate segments 1 mm long with rounded ends. Ultimate segments are midgreen to yellowish green, leathery and slightly glossy with strongly recurved margins, at least on the fertile segments, forming a pouch on the lower surface. The lower surface has pale scurfy hairs and sometimes a bar of tissue across the base which partly obscures the two sporangia in the pouch. Although hardy, coral-ferns resent disturbances and will not grow on when transplanted.

HABITAT Occurs in a variety of situations, both shaded and sunny, always on damp to wet ground, often on stream banks.

DISTRIBUTION Qld, NSW, Vic, Tas, NZ, New Caledonia and South-east Asia.

14. Gleichenia microphylla

Gleichenia microphylla
Scrambling Coral-fern **Plate 14**

This species resembles the more common *G. dicarpa* in habit, but is easily distinguished upon closer inspection. The wiry brown stipes are smooth and shiny, the minor rhachises are clothed with brown stellate hairs. The pinnules are 4 to 8 cm long with flat or only slightly recurved fine oblique saw-tooth ultimate segments 1 to 1.5 mm long. Ultimate segments are dark glossy green above, pale green below. Fertile segments usually have 3 or 4 large sporangia at the base near the upper margin.

Habitat As for *G. dicarpa*, but more often in shaded rocky gullies near streams.

Distribution All states, NZ, New Caledonia and South-east Asia.

Gleichenia rupestris **Plate 15**

This relatively uncommon species is slightly smaller and does not form the dense tangled banks of the other two species. It has glossy brown hairless stipes and rhachises. The pinnae are 2.5 to 6 cm long with flat or slightly recurved round bead-like ultimate segments about 2 mm long. The ultimate segments are pale green to mid green, slightly glossy above and glaucous below. Fertile segments have 3 or 4 large sporangia at the base near the upper margin.

Habitat Usually on wet sandstone cliffs, clayey banks, rainforest margins, and as stunted forms in moist sandstone crevices.

Distribution NSW, Qld, New Caledonia.

Sticherus

A genus of about 100 species, 3 in Australia, 3 in the Sydney district. Erect terrestrial ferns with a long-creeping scaly rhizome. Fronds falsely dichotomous as in *Gleichenia*, forking 2 to 4 times with a dormant apical bud in each fork. Ultimate pinnules linear, entire or serrated. Sori numerous, in single rows on both sides of midvein, each with 2 to 5 (usually 4) sporangia.

Sticherus flabellatus

Shiny Fan-fern, Umbrella Fern **Plate 16**

A common species, often forming large colonies in shaded damp places, *S. flabellatus* grows to 1 m high with spreading umbrella-like fronds. The stipes are smooth green or brown, the rhachises smooth or slightly scaly. Ultimate pinnules have finely but distinctly serrated margins and are attached to the rhachises at about 45°. None of the pinnules is lobed, as in *S. lobatus*, nor attached to the primary rhachises (i.e. in the lowermost fork), as in the other 2 species. The pinnules are 2.5 to 3.5 cm long (rarely to 5.5 cm) and 2.0 to 2.5 mm wide, shiny light green above, slightly paler and usually glabrous beneath.

Habitat Moist gullies and creek banks in open forest.

Distribution Qld to eastern Vic, NZ and New Caledonia.

Sticherus lobatus

Spreading Fan-fern **Plate 17**

This species resembles the more common and widespread *S. flabellatus*, but is distinguished by the ultimate pinnules which have entire margins and are attached at right angles to the rhachises. Unlike the other species in the area, it has clustered lobed pinnules in, or on, the lowermost fork, and lobed or simple pinnules in higher forks. The pinnules are 2 to 4 cm long, dull light green above, paler and glabrous below. Generally, the pinnules widen at the base and are concurrent on the rhachises.

Habitat Hillsides in moist eucalypt forest and rainforest margins, favouring the tablelands.

Distribution Qld, NSW and Vic.

15. Gleichenia rupestris

16. Sticherus flabellatus

17. Sticherus lobatus

18. Sticherus tener

Sticherus tener

Silky Fan-fern **Plate 18**

More common in Victoria and Tasmania than in New South Wales, this species has ultimate pinnules attached at 45° to 70° to the rhachises. A few, usually simple (i.e. unlobed), pinnules are present in the lowermost fork as well as in the upper forks. The pinnules are 2 to 3 cm long, entire or with obscurely lobed margins, dull green above, paler and with fine stellate hairs below, the midveins usually bearing fine scales.

Habitat Moist sheltered gullies along creeks and in sandstone crevices.

Distribution NSW, Vic and Tas.

19. Hymenophyllum cupressiforme

Hymenophyllaceae

Hymenophyllum

Hymenophyllum cupressiforme

Common Filmy-fern **Plate 19**

The generic limits within this family are still the subject of debate, with some authorities recognising 2 or 3 large variable genera, others as many as 33. The group, consisting of about 300 species, is mainly tropical, although about 10 species extend well into temperate parts of eastern Australia. All these occur in rainforests and are small terrestrial or epiphytic ferns with simple or divided membranous laminae one cell thick. Sporangia are enclosed in a marginal 2-lipped or tubular indusium, known as an involucre, terminal on the vein of an ultimate segment. Common Filmy-fern can be recognised by the toothed margins on both the pinnate-pinnatifid fronds and the lips of the deeply slit involucres which are mostly carried on basal segments close to the main rhachis.

Habitat Forming mats on rocks in rainforest, or less often on rocks beside creeks in moist open forest.

Distribution Qld, NSW, Vic and Tas.

Cyatheaceae

Cyathea

A genus of 700 to 800 species, 11 in Australia, 3 in the Sydney district. Tree ferns with an erect trunk to 20 m high, usually less than 5 m. Stipe bases rough or prickly, persistent or deciduous, clothed with flat scales. Lamina large, lanceolate to ovate, bipinnate or tripinnate, pinnules toothed or lobed. Indusia present or absent. Sori in 1 to 3 rows on each side of pinnule midvein, in the forks of veinlets, away from the margin.

Cyathea australis

Rough Tree-fern **Plate 20**

Found in sheltered gullies and in rainforest, this is the commonest tree-fern in New South Wales. Its large rough trunk is covered on the upper part by the persistent bases of old broken fronds and is thickened at the base by wiry dark brown adventitious roots. Stipe bases are dark brown and rasp-like with many hard pointed tubercles. They bear flat, shiny-brown narrow scales which also cover the trunk at the crown. Fronds grow to 4 m and are tripinnate with finely toothed pinnules, pale to midgreen above, paler or slightly glaucous beneath. Sori are round, without indusia, and occur in a single row on each side of the midvein of the pinnules.

HABITAT Forested gullies, creek banks, and rainforests on the coast and tablelands.

DISTRIBUTION Qld, NSW, Vic and Tas.

Cyathea cooperi

Cooper's Tree-fern, Scaly Tree-fern **Plate 21**

This species resembles *C. australis* but can be distinguished by the trunk and the scales on the stipe bases and crown. The middle and upper trunk of *C. cooperi* is patterned with large oval leaf scars left by the clean detachment of old fronds. Scales are of 2 kinds; long silky and whitish mixed with shorter, narrower and red-brown. Older stipe bases are brown with pointed tubercles as in *C. australis*. The fronds are also similar. Sori have round indusia consisting of scales. This is a fast-growing attractive species which, for cultivation, is usually raised from spores and sold while young since *Cyathea* are less able to withstand transplanting than the Soft Tree-fern, *Dicksonia*.

HABITAT Sheltered gullies and rainforest on the coast and tablelands.

DISTRIBUTION North Qld to the Illawarra district of NSW.

21. Cyathea cooperi

Cyathea leichhardtiana

Prickly Tree-fern **Plate 22**

This species has a tall, slender, light trunk covered by persistent prickly or spiny purplish black stipe bases. The scales on the stipe bases and crown are pale straw coloured to almost white. Fronds are up to 3 m long, relatively broad, dark green and shiny above, paler beneath. They are tripinnate with more distinctly toothed pinnules than in the other 2 species. Sori are round, without indusia, in a single row on each side of the midvein and close to it. Although the fronds are attractive, this species has little horticultural merit.

HABITAT Slopes of rainforest gullies and gorges, usually on the tablelands although present along several creeks in the Royal National Park.

DISTRIBUTION North Qld to Gippsland, Vic.

20. Cyathea australis

22. *Cyathea leichhardtiana*

Dicksoniaceae

Dicksonia

Dicksonia is a small genus with 2 Australian representatives, one rare and confined to Queensland and northern New South Wales, the other, one of the commonest treeferns in eastern Australia, particularly in cool moist mountain areas.

Dicksonia antarctica

Soft Tree-fern **Plate 23**

Soft Tree-fern is closely related to *Cyathea*, but differs in several respects. Less tall than the common *Cyathea* species, the trunk is often heavy and stout due to a dense matted growth of fibrous roots which covers the lower part. Stipe bases are persistent on the upper trunk which is clothed in fine red-brown hairs, not flat scales as in *Cyathea*. Stipes are smooth, green to brown and clothed at the base in red-brown hairs. New fronds are often produced in flushes of 10 to 20 or more and grow to about 4 m. Mature fronds are dark glossy green above, paler beneath, rather harsh in texture and tripinnate with deeply lobed pinnules. Sori are large and globular on the lobes of the pinnules, protected on the outside by the recurved margin of the lobe, and on the inside by an indusium.

HABITAT Usually along creeks in moist gullies and rainforest, rarely, if ever, covering hillsides as in Vic.

DISTRIBUTION On coast and tablelands from Bunya Mountains (Qld) to Vic and Tas. Extinct in SA.

23. *Dicksonia antarctica*

Culcita

Culcita is another small genus with 2 Australian species, one rare in northern Queensland, the other very common throughout eastern Australia.

24. *Culcita dubia*

Culcita dubia

Common Ground-fern, Soft Bracken, False Bracken, Rainbow Fern **Plate 24**

This species is often present in the forest understorey, clothing the ground densely or sparsely. It also forms large fern banks in protected open areas where its soft yellow-green fronds contrast with the shiny dark green of bracken *(Pteridium esculentum)*. The rhizome is stout, long-creeping and densely covered with silky pale and darker brown hairs. Stipes are dark near the base with long brown hairs, reddish brown or yellowish brown higher up, shortly woolly or almost glabrous. The groove on the upper surface is continuous on the main and secondary rhachises which are sparsely and softly hairy beneath. Fronds grow to 1.5 m with a large, broadly triangular tripinnate-pinnatifid lamina. As in *Dicksonia*, the sori are near the margin, protected on one side by a folded-over pinnule lobe, and on the other by a small indusium.

HABITAT In a variety of sheltered habitats in eucalypt forest and rainforest.

DISTRIBUTION Qld, NSW, Vic and Tas.

25. *Dennstaedtia davallioides*

Dennstaedtiaceae

Dennstaedtia

Dennstaedtia davallioides

Lacy Ground-fern **Plate 25**

The only Australian species in a genus of about 70, this is a handsome terrestrial fern with large, delicate finely divided fronds a metre or more high. Its shiny reddish brown stipes rise at intervals along a creeping branched rhizome covered with reddish brown hairs. Stipes and rhachises are grooved on the upper surface. The lamina is dark green, soft, thin, very broadly triangular and tripinnate-pinnatifid. Ultimate pinnules are unequally sided and deeply divided into toothed lobes with the sori situated between the lower teeth on the upper margin of the lobes. The indusium is thin and cup-like with 2 lips opening outwards.

HABITAT Alluvial soil beside creeks in rainforest and moist eucalypt forest.

DISTRIBUTION Qld, NSW, Vic and Norfolk Island. Occurs in patches along the Hacking River in the Royal National Park; also Wheeny Creek.

Hypolepis

A genus of about 45 widely distributed species, 6 in Australia, 3 in the Sydney district. Rhizome slender, long-creeping, clothed with brown hairs. Stipes and rhachises usually hairy. Lamina tripinnate-pinnatifid (in local species), characterised by intermittent growth of the apex, often bearing bent septate hairs on midveins and veinlets. Sori terminal on veinlets, close to the margin. Indusium when present formed by the reflexed pinnule lobe.

26. *Hypolepis punctata*

Hypolepis punctata
Downy Ground-fern **Plate 26**

This is a coarse unattractive fern about 1 to 2 m high with sticky hairs on the stipes and rhachises. Stipes rise from a thick, long-creeping rhizome and are 7 to 10 mm broad with a deep groove on the upper surface. They are pubescent and dark brown at the base, green-brown to orange-brown or yellowish higher up, becoming slightly shiny and rough in older fronds. The lamina is large, triangular, soft, dull pale green with a generally sticky feel. Frequently the development of the apex is arrested as each pair of pinnae expands. Ultimate pinnules are shallowly lobed, the lobes with irregular margins. Sori are near the margin protected by a tiny reflexed pinnule lobe.

HABITAT In or near rainforest on hillsides and sandy creek banks, usually forming large colonies.

DISTRIBUTION Qld, NSW, Vic, NZ, New Guinea, Asia and Polynesia. Occurs locally in several places along the Hacking River in the Royal National Park.

27. *Hypolepis muelleri*

Hypolepis muelleri
Harsh Ground-fern **Plate 27**

Usually about 60 cm high, *H. muelleri* and *H. rugosula* are considerably smaller than *H. punctata*, and are not sticky-hairy. The rhizome of *H. muelleri* is slender, long-creeping, branched and unevenly clothed in brown hairs. The almost glabrous stipes are grooved and ridged, becoming irregularly 4-sided in the upper part and on the rhachises. The colour on drying is pale straw to yellowish green. Rhachises and veins are covered with bent septate hairs. The lamina is dull mid-green with a harsh texture, broadly triangular, tripinnate-pinnatifid in the lower part, less divided in the upper. Pinnules on fertile fronds are narrower than on sterile fronds and are covered with short stiff hairs on the upper surface. Sori are usually away from the margin and unprotected by false indusia.

HABITAT Sheltered alluvial flats in eucalypt forest near rainforest, e.g. Wattle Flat, Royal National Park.

DISTRIBUTION Qld, NSW, Vic and Tas.

Hypolepis rugulosa
Ruddy Ground-fern

In this species the stipes are fairly close together on a slender long-creeping rhizome. They are deeply grooved on the upper surface, reddish brown in colour, slightly glossy and finely raspy to the touch. Rhachises are angular, reddish and also finely raspy. Minor rhachises have bent septate hairs as in other *Hypolepis* species. The lamina is dull dark green above, paler beneath, tripinnate in the middle and upper parts. Sori are close to the margin and protected by a false indusium (a reflexed pinnule lobe) when young.

HABITAT In wet gullies near creeks or on swampy land.

DISTRIBUTION Qld, NSW, Vic, Tas and SA.

Histiopteris

Histiopteris incisa
Bat's-wing Fern **Plate 28**

The smooth stipes of this fern rise from a long-creeping rhizome clothed in glossy hairs. They are reddish brown at the base with a few hairs, becoming shiny green and glabrous higher up. They are more or less round, without a groove. The lamina is pale green, often glaucous, glabrous, soft and membranous. Fronds are bipinnate with the secondary pinnae opposite and lobed shallowly or almost to the midrib, the lobes large, opposite, blunt at the apex and broad at the base. Fertile fronds are frequently less divided than barren ones. Sori are continuous or interrupted around the margins, protected by a false indusium formed by the reflexed leaf margin.

HABITAT Mostly wet gullies, near creeks, at the base of wet sandstone cliffs, in dry caves, or stunted in rock crevices.

DISTRIBUTION Qld, NSW, Vic, Tas, SA, NT, pantropic.

28. *Histiopteris incisa*

Pteridium

Bracken ferns are found on every continent. Generally, 6 or 7 species are recognised, although sometimes they are regarded as different forms or subspecies of the one variable species. Three species occur in Australia, 2 in north-eastern Queensland, 1 widespread and common in rural areas. Bracken is poisonous to horses and cattle when eaten in large quantities.

29. *Pteridium esculentum*

Pteridium esculentum
Bracken, Common Bracken **Plate 29**

This fern is so well known that it hardly requires description. Fronds are leathery, glossy or slightly dull, medium to dark green above, whitish with fine appressed hairs beneath. The lamina is tripinnate to bipinnate, usually with a small lobe where each branch (pinna) joins the main rhachis. Sori are inconspicuous, continuous around the slightly recurved margins, protected on the inner side by a thin linear indusium. The spread of bracken is assisted by frequent burning and probably owes more to the vigorous widely creeping rhizome than to propagation by spores.

HABITAT Common in dry open forest, grassy clearings, roadsides, creek banks, and neglected pastures.

DISTRIBUTION All states, NT, NZ, Malesia, Polynesia.

30. Lindsaea linearis

Lindsaeaceae

Lindsaea

A genus of mostly small terrestrial ferns with slender wiry stipes. Fronds mostly pinnate to pinnately decompound. Ultimate segments thin, variously shaped, the lower edge thickened and decurrent on the rhachis. Sori on or near the upper or outer margins of the segments, the indusia opening outwards.

Lindsaea linearis

Screw Fern **Plate 30**

The fronds of this small fern rise no more than 20 cm in small rosettes from a short-creeping rhizome. They are pinnate with a dark shiny stipe and small green to yellowish green triangular pinnae, the barren fronds shorter and broader than the fertile ones. Sori are near the outer edge of the pinnae protected by a long outward-facing indusium. This fern is widespread and common but easily overlooked when in taller ground cover.

HABITAT Often forming small colonies in heath, scrub and open forest.

DISTRIBUTION All states, NZ, New Caledonia, Norfolk Island.

31. Lindsaea microphylla

Lindsaea microphylla

Lacy Wedge-fern **Plate 31**

This small fern, growing to about 30 cm tall, is fairly common in moist bushland around Sydney. Its slender, smooth shiny brown or yellowish brown flexuose stipes rise in a tuft from a short-creeping rhizome. The delicate lacy fronds are light yellowish green, thin and rather flaccid with the primary pinnae divided into little simple or compound segments with wedge-shaped lobes. Sori are marginal on the broad ends of the lobes. This fern does not survive transplanting.

HABITAT On sandy and clayey soils in moist bushland with eucalypt cover.

DISTRIBUTION South-east Qld and NSW. Rare in Vic.

32. Pteris tremula

Pteridaceae

Pteris

A large genus of about 250 species, mainly tropical and subtropical, 7 species in Australia, 5 in the Sydney district of which only 2 are common. Terrestrial ferns with a thick scaly rhizome. Stipe and main rhachis grooved on the upper surface. Lamina 1-pinnate to 4-pinnate, the lower pinnae sometimes branched. Sori linear, marginal, protected when young by an indusium formed by the reflexed modified margin, the indusium opening inwards.

Pteris tremula

Tender Brake **Plate 32**

In this species the smooth shiny brown grooved stipes rise together from an erect tufted rhizome. Fronds are up to 150 cm high, large, membranous, light green and glabrous on both sides. The lamina is triangular, tripinnate on the lower primary and secondary pinnae, mostly bipinnate with finely toothed pinnules in the upper pinnae (i.e. at the apex and ends of the branches). This species differs from the rare *P. comans* in having simple forked veins on the ultimate segments that do not form a network. Easily grown.

HABITAT Moist coastal gullies, creek banks, below rock faces, in or near rainforest.

DISTRIBUTION Qld, NSW, Vic, Tas, SA, NZ, Norfolk Island and Fiji.

33. Pteris umbrosa

Pteris umbrosa

Jungle Brake **Plate 33**

A large attractive species popular in cultivation, this fern grows in colonies in cool moist rainforest. Fronds are about a metre high rising from a short-creeping rhizome on grooved, slightly rough reddish or yellowish brown stipes. The lamina is dark green and mostly pinnate, with the middle and upper pinnae long, linear and decurrent on the main rhachis. The lower pinnae are slightly shorter and are pinnately divided, sometimes imperfectly, into long linear segments similar to those above. Margins are finely and irregularly toothed. Fertile and sterile fronds are similar.

HABITAT Rainforest, usually on rocky slopes.

DISTRIBUTION Qld, NSW and Vic. Occurs locally at Macquarie Pass and on basalt soils at Robertson and Mt Wilson.

Adiantaceae

Adiantum

A genus of about 200 species, mainly tropical, 8 or 9 in Australia, 5 in the Sydney district. Small ferns with lacy fronds borne on thin, dark, hard, shiny stipes. Lamina 1-pinnate to 4-pinnate. Ultimate pinnules membranous, variously shaped or lobed, midrib absent, the veins radiating from the stalk. Sori marginal, crescent or kidney shaped, protected by a false indusium formed by the leaf margin. The popular name 'maidenhair' comes from the specific name of the Mediterranean and European species, *A. capillus-veneris.*

34. Adiantum aethiopicum

35. Adiantum formosum

Adiantum aethiopicum

Common Maidenhair Fern **Plate 34**

Well known and popular, this hardy species is plentiful in moist open situations where it may form extensive patches. Fronds are smaller and simpler than in other local species, except *A. diaphanum*, rising in small tufts, 10 to 30 cm high, at intervals along wiry branched long-creeping rhizomes. Stipes are thin, reddish brown, smooth and shiny, the lamina tripinnate, divided into tiny membranous rounded pinnules attached centrally on the lower margin to a fine stalk. Fertile pinnules have 1 to 4 sori, each deeply embedded in the margin opposite the stalk.

HABITAT In a variety of moist situations with open tree cover, amongst grass, on clayey banks, at the base of rocks or in rock crevices, often in gullies near streams where night-time dews linger in the morning.

DISTRIBUTION All states, NT, tropics and subtropics.

Adiantum formosum

Giant Maidenhair Fern, Black Stem **Plate 35**

This handsome large fern grows in colonies in coastal rainforest where it is often the dominant fern species. Stipes are long, dark brown or black, shiny and slightly rough, rising at intervals along long-creeping branched rhizomes. The lamina is 2-pinnate to 4-pinnate, broadly triangular, reaching 90 cm wide at the base and 60 cm to the apex. The main rhachis is mostly smooth and zig-zagged from one primary branch to the next. Ultimate leaflets (pinnules) are attached asymmetrically at the base and vary greatly in shape. The sori are on the upper margins, on the ends of lobes rather than in the notches as in other species.

HABITAT Rainforests, often covering extensive areas of the forest floor.

DISTRIBUTION Qld, NSW, Vic and NZ.

Adiantum hispidulum

Rough Maidenhair Fern **Plate 36**

The fronds of this species are tufted on a short-creeping rhizome, forming a small clump in shaded places such as earth banks and rock crevices. Stipes are dark brown, shiny, angular and slightly rough. The lamina is characteristically divided into long and short pinnae along which the ultimate leaflets (pinnules) are pinnately arranged. The longer pinnae are central near the main forks, the shorter ones on the outside near the ends of the branches. Pinnules are obliquely 3- or 4-sided, broader than high, and attached to a short stalk at the bottom corner. They are sparsely covered with short white hairs, and are often a delicate pink when young, becoming mid-green or dark green and dull at maturity. Sori are attached to folds or notches in the upper and outer margins of fertile pinnules.

HABITAT Mostly in moist eucalypt forest, often on earth banks and in rock crevices.

DISTRIBUTION Qld, NSW, Vic, NZ, Africa, Tropical Asia and Polynesia.

36. Adiantum hispidulum

Pellaea

A genus of about 80 species, 2 in Australia, both in the Sydney district. Small terrestrial ferns with a short thick rhizome densely clothed with narrow dark scales. Stipes wiry, scaly or smooth and glossy. Lamina leathery, shiny, pinnate (in local species), the pinnae attached by short stalks or almost sessile. Sori marginal, at first separate at the ends of inconspicuous veins, later becoming confluent, protected when young by the reflexed leaf margin.

Pellaea falcata var. *falcata*

Sickle Fern **Plate 37**

This is a hardy attractive fern resembling in shape the common fishbone fern, *Nephrolepis*, but with smooth bright green fronds and shiny dark brown wiry stems clothed with long narrow scales. Fronds taper slightly towards the apex and have numerous pinnae in opposite or alternate pairs along the stem and a terminal pinna at the apex. In this variety the fronds grow to about 60 cm high, with the pinnae 25 to 60 mm long and 5 to 10 mm broad. The sori form a continuous line along the underside of the margins.

HABITAT In moist shady open forest and rainforest, often in rocky places.

DISTRIBUTION Qld, NSW, Vic, Tas, NZ, Malaysia and India.

37. Pellaea falcata var. *falcata*

Pellaea falcata var. *nana*

Small Sickle Fern **Plate 38**

This fern is similar to var. *falcata*, but smaller, with the pinnae less than 20 mm long and only 2 to 6 mm broad. Whereas var. *falcata* is often seen in moist open forest, this variety is restricted to rainforests.

HABITAT In rainforest, usually amongst rocks or growing on mossy rocks with other small ferns.

DISTRIBUTION Qld and NSW.

38. Pellaea falcata var. *nana*

Pellaea paradoxa **Plate 39**

This species is much less common than *P. falcata* and is readily distinguished by its sparse, short, semi-erect fronds, usually no more than 20 cm long, and its large, broad pinnae. These are in 2 to 11 alternate pairs along the stem, with an odd one terminal at the apex. They are shortly stalked, leathery, dull green, 2 to 9 cm long and up to 5 cm broad, ovate, rounded or truncate at the base, or sometimes heart-shaped on juvenile fronds. Sori are in wide bands along the underside of the margins. *P. paradoxa* is a hardy species, rewarding in cultivation.

HABITAT Usually in rainforest among, or growing on, rocks.

DISTRIBUTION Qld and NSW.

39. Pellaea paradoxa

Cheilanthes

A large genus of about 180 species, about 10 in Australia, 2 or 3 in the Sydney district. Rhizome wiry, short-creeping or tufted, clothed with scales. Stipes dark coloured, scaly, hairy or glabrous. Lamina generally small, narrow, erect. Sori marginal, terminal on veins, at first separate, later spreading, discontinuous. Indusium formed by modified leaf margin, discontinuous. Plants often shrivelled in appearance but reviving with rain.

Cheilanthes sieberi

Poison Rock Fern, Mulga Fern **Plate 40**

Widespread and variable, this species has often been confused with (or combined with) *C. tenuifolia*, and many instances of poisoning of cattle and sheep, apparently by the former, have been attributed to the latter. Whereas *C. sieberi* has a compact, narrow linear or oblong lamina, that of *C. tenuifolia* is open and triangular with longer primary and secondary pinnae. The two are otherwise very similar, and at times the separation is indistinct. The stipes of *C. sieberi* are wiry, slender, dark brown and shiny. Fronds are glabrous except for a few scales on the stipe, erect, usually 10 to 30 cm high and 2 to 4 cm across. They are 2- to 3-pinnate with the ultimate segments narrow, blunt and close together. Sori are on the lobe margins.

HABITAT Terrestrial in grassy ground cover or in shallow soil pockets, cracks and crevices on sandstone rock in dry eucalypt forest.

DISTRIBUTION All states.

40. Cheilanthes sieberi

Cheilanthes distans

Cloak-fern, Bristly Cloak-fern **Plate 41**

Less common than *C. sieberi*, this fern is easily recognised by its short erect hairy fronds. It usually consists of a few fronds rising close together from a scaly short-creeping branched rhizome, sometimes reaching 15 cm high, but often much less. The lamina is narrow and pinnate-pinnatifid with scales on the rhachises and undersurface of the pinnae, and dense hairs on the upper surface of the pinnae. Frequently the lower pinnae are reduced in size.

HABITAT On rocky slopes and in soil pockets on sandstone in dry eucalypt forest.

DISTRIBUTION All states except Tas.

41. Cheilanthes distans

Davalliaceae

Davallia

Davallia pyxidata

Hare's Foot Fern **Plate 42**

The common name of this fern derives from the furry brown tip of the thick long-creeping rhizome, often seen emerging from protected rock crevices. The straw-coloured stipes are in 2 irregular rows along the exposed bare rhizome. They are articulated to the rhizome and glabrous except at the base. The lamina is stiff and leathery, shiny light green when young, turning dark green when mature. It is broadly triangular in shape, tripinnate with irregularly lobed ultimate segments, more finely dissected on fertile fronds than on sterile ones. Sori are marginal at the ends of veinlets, protected in cup-shaped indusia which open outwards. Members of this genus are popular in horticulture, being particularly suitable for hanging baskets and large pots where the rhizomes can intertwine and overflow the rim.

HABITAT Rock crevices in moist eucalypt forest and rainforest.

DISTRIBUTION Qld, NSW and Vic (where rare and localised).

42. Davallia pyxidata

Rumohra

Rumohra adiantiformis

Shield Hare's Foot, Leathery Shield Fern **Plate 43**

The position of this wide-ranging species in Davalliaceae is somewhat controversial. Although typically Davallioid in its rhizome, rhachises and laminae, it differs from *Davallia* in chromosome number and shows links with the Aspidiaceae, particularly with *Arachniodes*, in its round non-marginal sori, shield-like indusia and anadromous venation. The ultimate segments are, however, bluntly lobed, never aristate as in *Arachniodes* and related genera. The long-creeping rhizome varies from thin to thick and is densely clothed with brown scales. The stipes are not articulated to the rhizome as in *Davallia*. The lamina is triangular, leathery, shiny dark green and usually tripinnate. Sori are large, more or less central on the lobes, and have a dark orbicular-peltate (i.e. round and centrally attached) indusium. The species is uncommon in the area.

HABITAT On rocks and tree trunks in or near rainforest.

DISTRIBUTION Southern Qld, NSW, Vic, Tas, NZ, South America and South Africa.

Oleandraceae

Arthropteris

Arthropteris is closely related to the fishbone fern *Nephrolepis* which includes such well-established cultivars as the Boston Fern and others with lacy fronds. The two genera differ principally in the rhizome which is long-creeping with distant articulated stipes in *Arthropteris*, erect with branched runners and tufted non-articulated stipes in *Nephrolepis*. In recent years the pair has been placed with various other genera into several families or subfamilies.

43. Rumohra adiantiformis

Arthropteris tenella — Plate 44

A common rainforest fern with a wiry rhizome densely clothed with pale to dark brown scales. Fronds grow to 60 cm long, but are usually much shorter. They have a slender, smooth rachis with 6 to 20 well-spaced alternate pairs of stalked pinnae, and often a large terminal one. Generally, the middle and upper pinnae are the largest, the others gradually diminish in size towards the base. Pinnae are bright shiny green, ovate to lanceolate, often falcate with an oblique base. Sori are large and globular without indusia, spaced around the margin with each sorus terminal on the upper branch of a forked veinlet.

HABITAT On trees and rocks in rainforest.

DISTRIBUTION Eastern Qld, coast, tablelands and western slopes of NSW, NZ, Lord Howe Island and Norfolk Island.

44. Arthropteris tenella

Arthropteris beckleri — Plate 45

Only 2 of the 4 Australian *Arthropteris* species extend southwards as far as Sydney. One of these, *A. beckleri* is both uncommon and small, and hence rarely seen. It forms small but dense patches on rocks and trees in the cooler, darker parts of the rainforest. The rhizome is threadlike, long-creeping, much branched and densely clothed with brown matted hairs and narrow scales. Fronds grow to 20 cm but in the Sydney district are usually only a few centimetres long. They have closely spaced stalked pinnae over most of their length, the larger ones about 4 to 8 mm long and situated near the middle, the others becoming progressively shorter towards the apex and base. The pinnae are oblong with a wavy upper margin, an obtuse or rounded apex and a very unequally sided oblique base. Sori are in a single row near the margin and, unlike those of *A. tenella*, have an indusium.

HABITAT On rocks and trees in rainforest.

DISTRIBUTION Qld and coast, tablelands and western slopes of NSW. Locations include Mt Keira, Belmore Falls and Fitzroy Falls.

45. Arthropteris beckleri

46. *Christella dentata*

47. *Cyclosorus interruptus*

Thelypteridaceae

Christella

Christella dentata — Plate 46

Previously known as *Cyclosorus nymphalis*, this fern is part of a species complex which has caused much difficulty and confusion in the past. The rhizome is short-creeping and densely covered with brown scales. The stipes have an indumentum of fine hairs and rise close together on the rhizome, giving the fern a tufted appearance. As is common in this family, the fronds are pinnate-pinnatifid. They are 60 to 100 cm long and 20 to 25 cm wide in the middle, becoming progressively smaller in the upper and lower pinnae, with the lowest 2 or 3 pinnae often quite reduced in size. Young fronds are light green maturing to a dull dark green with the upper surface sparsely covered with short stiff hairs. The primary and secondary rhachises are densely hairy on the grooved upper surface. Sori are very prominent on the pinnae lobes, the dark glossy sporangia contrasting with the pale-coloured round or kidney-shaped indusia.

Habitat Usually along streams in shady eucalypt forest or rainforest.

Distribution Northern Qld to Vic, SA, WA and NZ.

Cyclosorus

Cyclosorus interruptus — Plate 47

This is the only Australian species remaining after revision of the previously large genus *Cyclosorus*. It has a long-creeping rhizome sparsely clothed with scales. The fronds are well separated on the rhizome, rising to about a metre. They have a fairly long stipe, up to 30 cm, and a pale green, leathery pinnate-pinnatifid lamina with at least 10 pairs of opposite pinnae. The lower pinnae are not noticeably reduced in size, or are sometimes longer than the middle ones. Pinnae lobes are shallow and ovate-triangular in shape. On fertile fronds the sori are in crowded rows close to the lobe margins, often becoming confluent at maturity and forming a zig-zag or a series of Vs along the margins. Each sorus has a round or kidney-shaped indusium.

Habitat Usually in swampy ground near creeks, sometimes forming large but not dense colonies.

Distribution Qld, NSW and northern Australia. Rare in the Sydney district. Locations include Scarborough Park, Kogarah.

48. *Asplenium bulbiferum*

Aspleniaceae

Asplenium

A genus of about 650 species, 20 in Australia, 8 in the Sydney district. Terrestrial or epiphytic ferns. Rhizome short-creeping to erect, clothed with clathrate scales. Fronds simple to bipinnate. Veins mostly free and forked. Sori elongated, on the basiscopic side of the veinlets. Indusia tapering at each end, opening towards the rhachis or midvein.

Asplenium bulbiferum

Mother Spleenwort — Plate 48

This species is much more common in the fern gullies of Victoria and Tasmania than in the Sydney district. It has a tufted rhizome with pale brown scales and large soft herbaceous fronds which are divied 2 or 3 times into deeply and narrowly lobed pinnules. Fertile pinnules have 2 to 4 short elliptical sori near the margin on the inner basal part of the larger lobes. This species can usually be recognised by tiny plantlets which are produced towards the ends of the pinnae. By the time the frond withers these plantlets are capable of growing into new plants.

Habitat On rocks and trees in rainforest creeks and gullies.

Distribution Southern Qld, NSW, Vic, Tas, SA and NZ. Uncommon in the Sydney district, mostly on the tablelands (e.g. Mt Wilson).

Asplenium flaccidum

Weeping Spleenwort — Plate 49

A. flaccidum has finely divided fronds resembling some forms of *A. bulbiferum*, except that they are usually weeping and do not bear tiny plantlets.

49. *Asplenium flaccidum*

Fronds are thick and leathery, up to 70 cm long, and mostly bipinnate with large narrow incurved linear pinnules. Fertile pinnules have 1 linear sorus near the upper or inner margin (i.e. the one facing the rhachis). This is an attractive but very uncommon species found in cool humid places such as rainforest gorges and gullies.

HABITAT On rocks and trees in wet rainforest.

DISTRIBUTION Southern Qld, NSW, Vic, Tas, NZ, Norfolk Island and Fiji. Probably confined, in the Sydney district, to the tablelands (e.g. Bundanoon, Carrington Falls).

Asplenium falcatum

Willow Spleenwort **Plate 50**

Widely distributed but uncommon in temperate areas, this attractive epiphytic fern is perhaps better known in cultivation than in its natural state. It is similar to *A. polyodon* from northern Queensland, and the two may be merely different forms of the one species. It has a short rhizome densely clothed with narrow dark scales, and pendent fronds up to 80 cm long with a short dark shiny stipe and a long narrow oblong pinnately divided lamina. The irregularly toothed pinnae are 4 to 10 cm long, attached by a short winged stalk, oblique and broad at the base and gradually tapering to a drawn-out narrow apex. Sori are about 10 mm long, linear, numerous and parallel with the veinlets.

HABITAT On rocks, trees and rotting logs in rainforest, sometimes forming large clumps.

DISTRIBUTION Qld, NSW, Vic (rare or extinct), NZ and tropics.

Asplenium australasicum

Bird's-nest Fern **Plate 51**

Formerly known as *A. nidus* (a name now applied to a similar species endemic in northern Queensland), this large epiphytic fern is a familiar sight in the rainforests of eastern Australia. It has long been popular in cultivation owing to its impressive appearance and easy care coupled with its ready availability as a byproduct of forest logging and clearing. The fronds, which grow to 2 m long and 20 cm broad, are semi-erect or spreading from the crown of a short thick erect rhizome. They are simple, entire or rarely divided into segments at the base, bright green, glossy, thick and leathery. The glabrous midrib or rhachis is sharply keeled on the underside (rounded in *A. nidus*). The sori are very numerous on the undersurface, forming oblique lines parallel to the veins, 1 to 2 mm apart and 4 to 6 cm long, on both sides of the midrib.

HABITAT On trees and rocks in rainforest, eventually forming massive nest-like clumps which collect debris.

DISTRIBUTION Cape York (Qld) to south coast of NSW, Norfolk Island, New Caledonia and Polynesia.

51. Asplenium australasicum

50. *Asplenium falcatum*

Asplenium flabellifolium

Necklace Fern **Plate 52**

Necklace Fern grows widely and abundantly in a variety of situations, forming small sparse to dense colonies. These consist of a tangled mass of slender trailing fronds emerging in small tufts from short-creeping rhizomes and becoming very attenuated and devoid of pinnae towards the apex. Sometimes a few of the longer fronds are rooted at both ends due to growth of a proliferous bud at the apex. Fronds have smooth shiny dark brown stipes and numerous pairs of tiny thin pinnae along the rhachis. Pinnae vary greatly in shape, but are generally fan-shaped with a rounded or irregular margin opposite the fine point of attachment to the rhachis. Sori are pinnately arranged on the large pinnae, or radiate from the base on small pinnae.

HABITAT In open forest and rainforest in soil pockets, surface debris, cracks and crevices of rocks, in the ground amongst rocks, or occasionally on the trunks of trees and treeferns.

DISTRIBUTION All states and NZ.

52. Asplenium flabellifolium

Athyriaceae

Diplazium

Diplazium australe

Austral Lady-fern **Plate 53**

In recent years this fern has been known by the names *Athyrium australe* and *Allantodia australis* in addition to the one given above. It is an uncommon large fern favouring cool shady creek banks in rainforests. The rhizome is tufted, and in older specimens forms a short trunk covered with dull black scales. Fronds are up to 1.5 m high, bright green, fine, tender, very broad and bipinnate with the pinnules divided almost to the midvein, or often tripinnate in the lower part. Ultimate segments are oblong with a broad rounded apex and prominently toothed margins. Sori are in 2 rows on the ultimate segments, on veinlets on each side of and close to the midvein.

HABITAT In cool wet places near streams in rainforest.

DISTRIBUTION Qld, NSW, Vic, Tas and NZ.

53. *Diplazium australe*

Lunathyrium

Lunathyrium encompasses about 45 species, of which only 1 occurs in Australia.

Lunathyrium japonicum

Japanese Lady-fern

Assigned over the years to at least 5 different genera, this fern is better known by the now superseded name, *Athyrium japonicum*. It grows 30 to 80 cm high from a long-creeping rhizome with broad, pale brown scales. The stipes are 1 to 2 cm apart on the rhizome and are scaly near the base. The pale green herbaceous lamina is 15 to 30 cm long and pinnate-pinnatifid with the pinnae deeply cut into rounded shallowly toothed lobes. The lobes are pinnately veined with 4 to 6 pairs of simple or forked lateral veins on each lobe. On fertile fronds the sori are linear on the middle portion of each lateral vein, with a linear indusium opening towards the apex of the lobe.

HABITAT On stream banks in moist open forest and rainforest, sometimes forming colonies.

DISTRIBUTION Southern Qld, NSW, Vic (rare and localised), Polynesia, Malaysia, India, China and Japan.

54. *Lastreopsis acuminata*

Aspidiaceae

Lastreopsis

A genus of about 33 species, 11 in Australia, 4 in the Sydney district, very restricted in the Blue Mountains. Mainly tropical, but extending to cool temperate regions in eastern Australia and New Zealand. Terrestrial ferns with two ridges on the upper surface of the main rhachis continuous on secondary rhachises and lower margin of ultimate segments. Lamina finely divided, bipinnate-pinnatifid to quadripinnate, more or less triangular in outline. Sori small, round, about halfway between midvein and margin, protected by shield-like indusia.

55. *Lastreopsis decomposita*

Lastreopsis acuminata

Shiny Shield-fern **Plate 54**

L. acuminata differs from other shield-ferns in the area in its short, erect, tufted rhizome which is crowded with the broken bases of old stipes. The stipes are mostly 20 to 40 cm long, olive-brown, smooth and glossy except at the base which is slightly scabrous. The main rhachis and secondary rhachises are covered with short stiff white hairs. The lamina is narrowly triangular with shiny dark green segments. It is less divided than in the other species, being bipinnate-pinnatifid near the base, gradually becoming pinnate-pinnatifid higher up, reducing to small decurrent segments at the apex. The pinnae are stalked and do not have a basal segment attached to the main rhachis as in *L. decomposita* and *L. microsora*. The ultimate segments are broadly rounded with a few sharp teeth near the apex. Sori are slightly larger than in other species.

HABITAT Rainforest, often among rocks near creeks and waterfalls.

DISTRIBUTION Coast and rainforest gullies of the Blue Mountains. Range from south-eastern Qld to Tas and SA.

56. *Lastreopsis microsora*

Lastreopsis decomposita

Trim Shield-fern **Plate 55**

Although variable, *L. decomposita* can be distinguished by a combination of features. It has a thick short-creeping horizontal rhizome, 5 to 10 mm diameter with erect stipes crowded along it. Both the rhizome and the stipe bases are covered with glossy dark brown flat scales. The brownish or greenish stipes vary from slightly scabrous to dull and very hairy. The primary and secondary rhachises are sometimes so densely clothed with short appressed hairs as to obscure the ridges on the upper surface. The lamina is medium to dark green, usually dull, sometimes thin but firm. It is tripinnate-pinnatifid in the lower part, gradually becoming bipinnate-pinnatifid towards the middle, with the tertiary and ultimate segments smaller and narrower than in *L. acuminata* and *L. microsora*. *L. decomposita* is similar to *L. microsora* in having the lowest segment of the primary branches displaced downwards onto the main rhachis, but differs in also showing this feature on the secondary and tertiary rhachises. It also differs in the ultimate segments which are finely pointed at the apex.

HABITAT Often a dominant fern species in rainforest on the coast and tablelands.

DISTRIBUTION South-eastern Qld to just beyond the Vic border, also Norfolk Island.

57. *Polystichum australiense*

Lastreopsis microsora

Creeping Shield-fern **Plate 56**

This species has a long-creeping branched rhizome, hardly thicker than the stipes which are spaced along it up to 3 cm apart. The light brown stipes and the main rhachis are prominently ridged and slightly shiny on the upper surface (except in the channel between the ridges which is covered in matted hairs) and pubescent on the lower surface. The 2 ridges of the main rhachis are very obviously continuous on the secondary rhachises and the lower margin of the first ultimate segment. The main rhachis is narrowly winged towards the apex; the secondary rhachises are winged for most of their length. The lamina is often large, sometimes broader than long. It is usually shiny, pale green, rather thin and soft, tripinnate-pinnatifid at the base with the lower secondary pinnae enlarged and produced downwards. Frequently the lowest pinnule of the upper primary branches is displaced towards or onto the main rhachis. Ultimate segments are broadly rounded at the apex and have fine, sharply toothed margins. Despite the specific name, the sori are no smaller than in other species.

HABITAT Rainforest on the coast and tablelands, occasionally a dominant fern species.

DISTRIBUTION North-eastern Qld to east Gippsland, Vic.

Polystichum

A genus of about 200 species, 4 in Australia, 3 in the Sydney district. Terrestrial ferns with a thick tufted rhizome densely clothed with scales. Stipes and main rhachis scaly. Proliferous buds often present near apex of main rhachis. Lamina narrow lanceolate, leathery, anadromous, 1 to 4 pinnate. Ultimate segments unequal sided with aristate marginal teeth or apex. Sori large, rounded, protected by a centrally attached indusium.

Polystichum australiense **Plate 57**

This is the commonest of the 3 species in the area, particularly in coastal forests. In these forests it is readily recognised by its dark green, narrow triangular fronds carrying a scaly pale-brown proliferous bud on the main rhachis near the apex. When the arching fronds touch the ground, the buds may germinate, producing roots and fronds. The tiny plantlets eventually separate and grow into new plants. The rhizome, the lower part of the stipes and upper side of the main rhachis bear narrow, dull-brown scales. The lamina may reach 100 cm in length, but is usually about 40 cm long and 10 to 15 cm wide at the base. It is bipinnate with dark green, leathery pinnules 10 to 15 mm long. These have sharply pointed marginal serrations shorter than the spine-like apex. As in other *Polystichum* species, they are less developed on the lower side of the midvein, and have a large acroscopic basal lobe.

HABITAT Rainforest margins, usually on rocky or grassy hillsides.

DISTRIBUTION Coast and tablelands of NSW.

58. Polystichum proliferum

Polystichum proliferum

Mother Shield-fern **Plate 58**

Although coastal in parts of its range, this species appears to be restricted in the Sydney district to the higher parts of the tablelands. The thick erect rhizome often forms a short trunk in older plants. The rhizome and the lower part of the stipes are covered with glossy dark brown or black scales, usually with a pale border. The scales on the main rhachis are similar, except that towards the apex they are fawn to chestnut in colour. As in *P. australiense*, a scaly proliferous bud occurs near the apex of each main rhachis. The lamina is narrow lanceolate, sometimes reaching 80 cm in length, but usually much shorter. It is mostly bipinnate, with light green leathery pinnules on young fronds becoming darker when mature. Pinnules have serrated margins with acute or obtuse teeth and a mucronate or obtuse apex.

HABITAT Occurs locally in temperate rainforest on the highlands, mostly on volcanic basalt soils.

DISTRIBUTION Dorrigo (NSW) to Vic and Tas. Locations include Mt Wilson, Mt Irvine, Mt Tomah, Mt Victoria and Mt Werong.

Polypodiaceae

Platycerium

This remarkable genus of large bracket-epiphytes consists of 17 species distributed from tropical Africa through tropical Asia to eastern Australia where there are 4 species, with one extending to southern New South Wales.

Platycerium bifurcatum

Elkhorn **Plate 59**

Fronds are of 2 types; broad bowl-shaped sterile fronds which protect the rhizome and anchor the plant to the host or support, and large forked fertile fronds resembling the antlers of deer. Sterile fronds are initially green, turning brown and papery with age. Older fronds slowly decay as new ones are added to the outside, gradually building up a large humus-collecting receptacle. Fertile fronds are characteristically shaped and divided in each species. In the local one these are narrow and wedge-shaped at the base, becoming broader and forking 2 or 3 times in the upper half. The sporangia form brown velvety masses which cover most of the forked segments.

HABITAT Mostly on rock surfaces in rainforest, sometimes on trees, less often in moist eucalypt forest on large rocks, usually in inaccessible or remote places.

DISTRIBUTION Qld, NSW, New Caledonia and Lord Howe Island.

Microsorium

A genus of about 60 species, 6 in Australia, 2 in the Sydney district. Epiphytes with a long-creeping rhizome covered with deciduous or persistent scales. Lamina simple, entire to pinnatifid, thin textured or slightly leathery, the veins forming a complex network with free vein endings. Sori round or elongated, slightly sunken in the surface in local species.

59. Platycerium bifurcatum

60. *Microsorium scandens*

61. *Microsorium diversifolium*

62. *Pyrrosia rupestris*

Microsorium scandens

Fragrant Fern **Plate 60**

The two local species of *Microsorium* are very variable in leaf shape and size, and are not always easy to tell apart. The pair occupy similar habitats in rainforests, with *M. scandens* common on the coast and *M. diversifolium* favouring the cooler conditions of the tablelands. The somewhat flattened rhizome of *M. scandens* is usually 3 or 4 mm wide, green and densely covered with persistent shiny brown scales. Stipes are attached every 2 or 3 cm by small outgrowths known as phyllopodia. Fronds grow to 40 or 50 cm, but are usually much less. They are thin textured, drying to a dark green colour, and have an elusive musky smell (an unsatisfactory diagnostic feature), more noticeable on the dried fronds than when fresh. The lamina is entire or pinnatifid, with the lobes varying greatly in size and number (up to about 20 pairs). Sori are smaller than in *M. diversifolium* and are nearer the frond margin than the midrib.

HABITAT Commonly on rocks, less often on tree trunks, fallen tree ferns, logs, etc. in rainforest, often growing profusely.

DISTRIBUTION Qld to eastern Vic, NZ, Lord Howe Island and Norfolk Island. Occurs on the coast and tablelands in the Sydney district.

Microsorium diversifolium

Kangaroo Fern **Plate 61**

The rhizome of this species is long-creeping, up to 6 mm wide, fleshy, green or glaucous and clothed with appressed, deciduous shiny brown scales. Fronds are light shiny green, thin textured to slightly leathery and unscented. They grow to 40 or 50 cm, but are sometimes much smaller. The lamina is variable in shape from entire to pinnatifid, tending to have a few (up to about 8 pairs) large well-spaced lobes rather than many fine lobes as is common with *M. scandens*. The lamina tapers less gradually to the stipe than that of *M. scandens*, and the venation is much more complex. Sori are large (1.5 to 2.5 mm diameter) and are situated midway between the midrib and the frond margin, or slightly closer to the midrib. The common name is due to the resemblance of some leaf shapes to a kangaroo's 3-toed foot.

HABITAT On rocky creek banks, the trunks and limbs of trees, treeferns *(Dicksonia antarctica)*, fallen trees, etc., in rainforest, often covering large areas.

DISTRIBUTION South-east Qld, NSW, Vic, Tas, NZ and Norfolk Island. Probably restricted to the tablelands in the Sydney district. Locations include Mt Wilson, Megalong Valley and Kanangra Walls.

Pyrrosia

Small epiphytic ferns with simple erect fleshy fronds at intervals along slender long-creeping scaly rhizomes, the fronds dimorphic in some species. Stipes usually short, clothed with stellate hairs. Sori without indusia, in several rows in the upper part of the lamina, often confluent in older fronds. The genus contains approximately 100 species, 5 in Australia.

Pyrrosia rupestris

Rock Felt-fern **Plate 62**

In moist forests this fern is frequently seen covering the trunks and limbs of trees and large areas of rock surfaces. The soft fleshy fronds are covered with fine stellate hairs. They are markedly dimorphic, with the sterile fronds shorter and broader than the fertile ones. Sterile fronds are spoon shaped, usually 2 or 3 cm long and 5 to 15 mm broad. Fertile fronds are oblong with an obtusely pointed apex and large sori in several irregular rows on each side of the midrib, occupying most of the undersurface.

HABITAT On trees and rockfaces in moist open forest and rainforest.

DISTRIBUTION Qld, NSW, Vic and New Guinea. Widespread in the Sydney district.

Pyrrosia confluens

Plate 63

Common in north-eastern Australia, this species barely enters the region under consideration. Fronds are larger than in *P. rupestris*, mostly 3 to 10 cm long including the stipe, and 10 to 15 mm broad, the fertile fronds usually longer than the sterile ones. Sori are restricted to the margins at the apex, often becoming confluent and forming a thick brown band around the apex.

HABITAT On trees and rockfaces in rainforest and moist open forest.

DISTRIBUTION North Qld to the Hunter River district of NSW, New Caledonia, Lord Howe Island and Norfolk Island. Occurs at The Gap Flora Reserve and in other parts of the Watagan Mountains.

Dictymia

Dictymia brownii

Plate 64

The only Australian species in this small genus, *D. brownii* resembles some species of *Pyrrosia*, but differs in having glabrous instead of hairy fronds. Its rhizome is long-creeping and densely covered with spreading dark scales. Fronds vary from 5 to 60 cm in length and 5 to 20 mm in width. The lamina is dark green, thick and leathery, simple and entire or undulate on the margins with the base tapering

63. *Pyrrosia confluens*

64. Dictymia brownii

gradually into the stipe. The midrib is very prominent, the other veins sunken and indistinct. Sori are large, oblong to oval, on the upper part of the frond on both sides of the midrib halfway between midrib and margin. The species is uncommon in the area.

HABITAT On trees and rocks in rainforest, sometimes forming large clumps.

DISTRIBUTION Northern Qld to southern NSW.

Grammitidaceae

Grammitis

These are small epiphytic ferns with simple tufted fronds commonly growing on the trunks of rainforest trees and on damp mossy rocks in rainforest. The genus contains about 150 species, with 9 presently recognised in Australia. Some of these are difficult to distinguish owing to small differences between species and to a considerable dependence of size upon environmental conditions.

Grammitis billardieri

Finger Fern **Plate 65**

G. billardieri is distinguished from *G. meridionalis* by its densely hairy stipes and larger fronds. Fronds of the former range from about 8 to 16 cm in length (including the stipe) and 4 to 7 mm in width, and have finely wavy margins, compared to about 4 to 8 cm in length (including the sparsely hairy stipe) and 2 to 5 mm in width with entire margins for the latter. The sori of both species are in oblique rows on either side of the midrib, and close to it.

HABITAT On rocks and tree trunks in rainforest.

DISTRIBUTION Qld, NSW, Vic, Tas, NZ, Antarctic islands, South America and South Africa.

Blechnaceae

Blechnum

A genus of 200 species, 18 in Australia, 11 in the Sydney district. Rhizome clothed with dark shiny scales, short-creeping to erect, sometimes forming a short trunk. Fronds leathery entire, pinnate or pinnatifid, uniform or dimorphic with the fertile fronds much narrower than the sterile ones. Veins free, parallel, simple or forked. Sori linear, usually continuous, on each side of the midvein and close to it, often confluent with age and covering the whole of the lower surface of fertile pinnae. Indusia linear, opening towards the midrib.

Blechnum patersonii

Strap Water-fern, Strap Fern **Plate 66**

This species has an erect tufted rhizome covered by the old stipe bases. New fronds are semi-erect and sometimes pinkish; the older ones are leathery, dark green and glossy, and mostly lie on the ground or hang down on steep-sided creek banks. Fronds are usually 20 to 40 cm long and either strap-shaped or irregularly pinnatifid with up to about 10 linear segments of various shapes and sizes. Frequently both entire and divided fronds occur on the one plant. Sterile fronds are winged on both sides of the rhachises, the wings 5 to 15 mm wide. Fertile fronds are fewer and much narrower, the wings scarcely wider than the linear sori.

HABITAT In rainforest, usually on steep slopes near streams.

DISTRIBUTION Qld, NSW, Vic and Tas.

65. Grammitis billardieri

66. Blechnum patersonii

Blechnum cartilagineum

Gristle Fern **Plate 67**

This common fern is readily distinguished from other *Blechnum* species by its light yellowish green, leathery or gristly lamina in which the pinnae are attached to the rhachis by their broad bases. The lamina is triangular in shape with the upper pinnae alternate, confluent at their bases and directed towards the apex, the lower one usually bending downwards, and the lowest pair distant, sub-opposite and lobed at the base. Unlike other species with broadly attached pinnae, the lower pinnae are the longest, often exceeding 15 cm in length and 15 mm in width. All pinnae taper to an acute apex and have fine, sharply serrated margins. Fertile pinnae are long and narrow with linear sori close and parallel to the midvein, but are otherwise similar to the sterile ones.

HABITAT In open forest and rainforest in a wide variety of situations.

DISTRIBUTION Qld, NSW, Vic and Tas.

67. Blechnum cartilagineum

Blechnum nudum

Fishbone Water-fern, Fishbone Fern **Plate 68**

This common fern has an erect tufted rhizome which, in older plants, grows into a short black trunk. Fronds are erect, rather stiff, glabrous, up to 80 cm long and about 10 cm wide in the middle. They have a fairly short, smooth, shiny black stipe and a long lanceolate lamina pinnately divided into narrow straight lobes with broad decurrent bases. The dark fertile fronds are shorter and narrower with very narrow pinnae which are often incurved and bent back exposing the dark sori. This fern should not be confused with the rampant *Nephrolepis cordifolia*, also known as Fishbone Fern, which is not native to the Sydney district.

HABITAT Rocky creek beds and banks in rainforest and cool eucalypt forest, sometimes covering large areas of creek bank.

DISTRIBUTION Qld, NSW, Vic, Tas and SA.

68. *Blechnum nudum*

Blechnum wattsii

Hard Water-fern **Plate 69**

In the Sydney district this species is mostly confined to the Blue Mountains where it is occasionally a dominant fern. The fronds are about a metre long, rising close together on short thick scaly stipes from a thick branched rhizome covered with dark brown or black scales. They are pinnately divided, with a large terminal leaflet. The lateral leaflets (or pinnae) are attached to the rhachis by a very short stalk and, in this respect, resemble 5 other species in the area. In 2 of these (*B. minus* and *B. camfieldii*) the lower pinnae are extremely reduced in size whereas *B. wattsii* has an oblong lamina with the lower pinnae neither significantly reduced nor distant from the others. A third species, *B. indicum* (plate 70), grows on the edges of brackish coastal swamps and lakes from Broken Bay northwards, and is therefore unlikely to cause confusion. The remaining two species, *B. gregsonii* and *B. ambiguum*, have thin, pale to mid-green fronds differing from the rigid, leathery, strongly veined dark green fronds of *B. wattsii*. Unlike these two species, *B. wattsii* has very dissimilar sterile and fertile fronds, the sterile fronds with many pinnae about 25 mm wide, the fertile fronds with few widely separated pinnae no more than 5 mm wide.

HABITAT Rainforest gullies on the lower slopes and banks of creeks, mostly on the tablelands.

DISTRIBUTION Qld, NSW, Vic and Tas (where abundant), SA.

70. *Blechnum indicum*

69. *Blechnum wattsii*

71. *Blechnum ambiguum*

Blechnum ambiguum **Plate 71**

This fern resembles *B. wattsii* and *B. gregsonii* but, unlike those species, is found mainly on the coast. It has erect or hanging fronds up to about 60 cm in length with few to many pairs of shortly stalked pinnae and a large terminal one. The lower sterile pinnae are opposite, the upper ones are alternate. The middle pinnae are 9 to 12 cm long and 22 to 25 mm broad, the upper and lower ones a little shorter. Pinnae are dull mid-green, slightly paler with a silvery sheen beneath, rather thin and pliable, becoming papery when dried and somewhat harsher on older fronds. They taper gradually to an acute or blunt apex and have finely serrated margins, especially towards the apex. Fertile pinnae are similar in length to the sterile ones but only about one third to half as wide. They are well separated on the rhachis, often in opposite pairs and prominently lobed at the base.

HABITAT In open forest and rainforest, usually on sandstone ledges or wet cliff-faces or on the sandy floor of caves near creeks and waterfalls.

DISTRIBUTION Central coast and tablelands of NSW.

72. *Blechnum penna-marina*

Blechnum penna-marina

Alpine Water-fern **Plate 72**

This small fern is common in alpine and elevated non-alpine areas of south-eastern Australia, typically inhabiting the grassy banks of streams or boggy ground amongst rocks. It is uncommon in the Sydney district, being confined to the highest coldest parts of the Main Dividing Range. It has a long-creeping wiry rhizome with the stipes crowded towards the growing end. Fronds are leathery, narrow and 5 to 25 cm high. Sterile fronds have short broad pinnae attached alternately to the rhachis by their wide decurrent bases, the lower pinnae widely spaced and becoming progressively smaller. Fertile fronds are usually taller than the sterile fronds and have longer stipes. Fertile pinnae are widely spaced and narrow, also reducing in size towards the base.

HABITAT In grassy or boggy ground amongst rocks in cold elevated areas (e.g. Boyd Plateau).

DISTRIBUTION NSW, Vic, Tas, NZ, sub-Antarctic islands and South America.

Doodia

A genus of 11 species, 5 in Australia, 3 in the Sydney district. Terrestrial ferns with an erect dark rhizome bearing wiry black roots and dark narrow scales. Stipes tufted, scaly, brown or black. Fronds harsh, leathery, uniform or dimorphic, pinnatifid to pinnate. Pinnae with prickly margins, the veins prominent and forming a network. Sori short, oblong, in 1 or 2 interrupted rows on each side of the midvein. Indusium opening towards the midvein.

73. *Doodia aspera*

Doodia aspera

Prickly Rasp-fern, Rasp Fern **Plate 73**

This is the easiest of the *Doodias* to recognise since it is the only one of the 3 local species in which all the pinnae are attached to the rhachis by their broad bases (except sometimes the lowermost pair is shortly stalked and enlarged). It varies from scattered individuals on the rainforest floor, consisting of a few fronds under 10 cm long, to large dense colonies in moist open forest with fronds often exceeding 40 cm in length. Fronds are a fishbone shape, broadest near the middle, tapering considerably towards the apex and base. Pinnae are stiff and harsh with prickly margins, often an attractive soft to bright pink colour when young, turning pale to dark green at maturity. Fertile fronds are similar to the sterile fronds, and have sori less than 1 mm long in 1 or 2 rows on each side of the midvein.

HABITAT Rainforest, also moist open forest in gullies and valleys on coast and tablelands, amongst rocks or in open ground.

DISTRIBUTION Qld, NSW, Vic and Norfolk Island.

Doodia caudata

Small Rasp-fern **Plate 74**

This is a variable species capable of division into a number of forms, varieties or even separate species. It is uncommon in the Sydney district, usually found growing on the edge of shaded creeks. The fronds are tufted, small and rather weak, varying in size and appearance on the same plant. In local forms the fertile fronds are nearly always much taller, narrower and more erect than the sterile ones. Usually the upper part of the frond is pinnatifid with upturned segments incompletely divided to the midrib or attached by broad expanded bases. The lower third at least, or sometimes most of the frond, is pinnately divided with the pinnae opposite, shortly stalked, toothed on the margins, prominently lobed at the base, and obtuse or rounded at the apex. The sori are elongated, in 2 rows near the midvein, sometimes forming a semicontinuous band covering most of the undersurface.

HABITAT On muddy or sandy banks or amongst rocks on the edge of shaded creeks in open forest.

DISTRIBUTION Qld, NSW, Vic, Tas and New Caledonia.

Doodia media

D. media is a robust, erect species resembling *D. aspera* in habit but, unlike that species, having shortly stalked pinnae in the lower third of the frond. In appearance it is intermediate between *D. aspera* and *D. caudata* and has frequently been confused with both in the past. The stipes are only sparingly scaly and have fewer sharp tubercles than those of *D. aspera*. Young fronds are often an attractive pink colour, turning dark green at maturity. Fertile and sterile fronds are similar, the sterile fronds with short sori in 1 or 2 rows on each side of the midvein. This species is much more common in Victoria and Tasmania than in the Sydney district.

HABITAT Open forest in shady gullies or rocky slopes.

DISTRIBUTION Qld, NSW, Vic, Tas, NZ, Norfolk Island and Polynesia.

75. Marsilea mutica

Marsileaceae

The water ferns in this family and the next grow in freshwater ponds and lagoons or periodically dry clay pans and waterholes. They are distinctive biologically as well as structurally since, unlike the land ferns, they produce micro and macro spores that germinate and grow into separate male and female prothalli.

Marsilea

Marsilea species have an extensive forking rhizome rooted in the mud. Long stalks (stipes) are periodically sent to the surface where 2 pairs of leaflets (pinnae) are produced close together, appearing as a 4-leaf clover floating on the water surface. Spores are borne in sori in sporocarps attached to the rhizome, near the base of the leaf stalks. Each sporocarp consists of a hard conceptacle on a short axis.

Marsilea mutica **Plate 75**

M. mutica is characterised by its relatively large glabrous two-toned leaflets, rounded at the apex and usually bearing a brownish band across the middle. The conceptacles are globular without teeth, and are borne in groups of 2 or 3 on branched stalks.

HABITAT In permanent or occasionally dry freshwater ponds, lagoons, etc.

DISTRIBUTION All states except Tas; also New Caledonia.

Marsilea hirsuta

Short-fruit Nardoo **Plate 76**

This is a much smaller species than *M. mutica* with fronds rarely above 15 cm high rising singly, or a few together, from distant nodes. The leaflets are mostly 10 to 15 mm long, relatively narrow and wedge-shaped with rounded ends. Leaflets and stems are often hairy, especially when young. The conceptacle of the sporocarp is ovoid and 4 to 5 mm long with 2 teeth at the base above the point of attachment to the stalk which is unbranched and shorter than the conceptacle.

HABITAT Shallow muddy pools, lagoons, etc.

DISTRIBUTION All states except Tas. Uncommon in the Sydney district; occurs at Eastern Creek, Doonside.

Azollaceae

Azolla

Azolla is a small but widespread genus of floating ferns, often carpeting the surface of ponds, dams, lakes and the like. Plants have branched stems bearing tiny overlapping sessile leaves in 2 rows, as well as roots which hang below the floating branches. The leaves have 2 unequal lobes, the upper lobe green or reddish and containing the cyanobacteria, *Anabaena azolla* in the central cavity, the lower lobe thin and bearing male and female sporocarps in the axils. Two species occur in Australia, both in the Sydney area.

Azolla pinnata

Ferny Azolla **Plate 77**

The (locally) commoner *A. pinnata* is triangular in outline with regular branches bearing thick, reddish, slightly overlapping leaves. It is most reliably identified by its feathery branched roots. The other species, *A. filiculoides*, is variable in shape due to irregular branching and has thin, broad, much overlapping leaves. It can be identified by its non-branched roots.

HABITAT On the surfaces of slow-moving or stagnant bodies of water, often forming large dense colonies.

DISTRIBUTION All states except Tas, also Japan, tropical Asia and Africa. *A. filiculoides* occurs in all states, NZ and America.

74. Doodia caudata

76. Marsilea hirsuta

77. Azolla pinnata

4. Gymnosperms

78. *Macrozamia communis*

The gymnosperms constitute one of the two divisions of seed-producing plants. They are distinguished from the other group, the angiosperms or flowering plants, by their 'naked' seeds which are borne on the surface of cone scales (megasporophylls) or at the ends of stalks among the leaves. By contrast, the seeds of angiosperms are produced in hard or fleshy fruits of various kinds, and develop from ovules enclosed in a carpel of a flower.

The gymnosperms are a diverse assemblage consisting of both extinct and living orders. The major groups of living gymnosperms are the cycads (*Cycas, Macrozamia*, etc.), ginkgo (one living species, *Gingkgo biloba*), taxads (yews and plum yews), conifers (pines, cedars, spruces, cypresses, etc.) and the gnetophytes (*Ephedra, Gnetum* and *Welwitschia*). Most of these groups were once much more widely spread than at present. Cycads, for example, presently restricted to parts of the tropics and subtropics, once had a pole-to-pole distribution. The ginkgos, restricted in historic times to a small area in eastern China, and now possibly extinct in the wild, were almost worldwide during the Mesozoic Era (65–225 million years ago). At present only the cycads and conifers occur naturally in Australia.

The gymnosperms, like all plants, have alternately an asexual generation called the sporophyte and a sexual generation called the gametophyte. In primitive non-vascular plants (bryophytes), the gametophyte is the dominant conspicuous phase of the cycle. In the ferns and fern allies (pteridophytes) the two phases have a separate existence, although the sporophytic phase is dominant. In the seed-bearing plants (spermatophytes), the gametophytic generation is simplified and reduced, and develops on the sporophyte.

In the gymnosperms, pollen, which carries male gametophyte cells, is usually carried by the wind and may land by chance on naked ovules borne on the same or a different plant. The female megaspore is not shed, as in bryophytes and pteridophytes, but develops within the ovule. Fertilisation follows growth of a pollen tube to the female egg cell and discharge of two male gametes into the tube. In the cycads and ginkgo, the gametes are flagellate spermatozoids which swim down the tube, but in conifers they are non-motile sperm nuclei. After fertilisation, the ovule enlarges and develops into a seed which is eventually released and grows into a new plant bearing male or female sporangia, or both.

Zamiaceae

The Zamiaceae belong to a group of palm-like or fern-like plants referred to as cycads, the living remnants of a large diverse group which replaced the Palaeozoic pteridophytes and seed ferns, flourished for the greater part of the Mesozoic Era, then all but disappeared in the late Cretaceous period some 80 million years ago. Cycads resemble in a number of ways the extinct seed ferns, and may have evolved from them. They are not closely related to the conifers which represent another line of development extending back to the Carboniferous period.

The cycads consist of ten genera and about 100 species which are either regarded as a single family, Cycadaceae, or as three families, Cycadaceae *(Cycas)*, Stangeriaceae *(Stangeria)* and Zamiaceae (*Macrozamia, Lepidozamia, Bowenia* and five extra-Australian genera). They are scattered throughout the warmer regions of the world, growing mostly in dry open country. They are slow growing, and many are reputed to live for several hundred years.

Cycads range in size from a few centimetres to more than 15 m high with a crown that may contain hundreds of leaves up to 5 m long. The leaves are pinnate except for Bowenia which has bipinnate leaves. Plants are either male or female, bearing respectively microsporangia, which produce pollen, or megasporangia (ovules), which develop into seeds. In Zamiaceae the sporangia are borne on sporophylls (modified leaves) arranged spirally into large cones or strobili. The sporophylls are stalked with thickened ends, terminating, in some genera, in a sharp spine. The small rounded microsporangia are clustered on the lower surface of microsporophylls. The large ovules are attached to the inner surface of the megasporophylls on either side of the stalk. Female cones are generally larger than the males and may reach 60 cm in length and 40 kg in weight. At maturity the sporophylls fall away from the central axis of the cone, releasing the large fleshy seeds.

The seeds of cycads contain starchy endosperm

and were frequently eaten by Aborigines after treatment by washing and roasting to destroy poisonous glycosides. Starch is also present in pith in the trunk and contains similar poisonous principles. The consumption of seeds by cattle causes severe gastroenteritis, liver damage, and occasionally death, and the consumption of leaves impairs the central nervous system producing the irreversible condition known as 'zamia staggers'.

Macrozamia

Two species of *Macrozamia* occur in the Sydney area, with two more on the northern and north-western outskirts.

Macrozamia communis

Burrawang **Plate 78**

Due to an early mistake and subsequent confusion of species, this species is treated in older works under the name *M. spiralis* (or *Zamia spiralis*), but is distinct from the plant now bearing that name. It is the commonest *Macrozamia* in New South Wales, and in the Sydney district. Its caudex or trunk is usually partly buried except in shallow soils where it may form a trunk a metre or more tall and 40 to 50 cm across. Plants have from 20 to 100 fronds, each 70 to 150 cm long with numerous dull green, rigid, sharp-pointed pinnae along the rhachis, the pinnae becoming progressively shorter and eventually spine-like towards the base. Older fronds are straight, but new growth, particularly after fire, is sometimes slightly twisted. Plants develop from 1 to about 6 large male or female cones 20 to 50 cm long rising from the centre on short stalks. Male cones are reddish, narrow cylindrical and 6 to 12 cm in diameter, the females bright orange-red and 10 to 20 cm in diameter. Cones are often produced following a bushfire.

HABITAT On well-drained sandy soils of medium fertility in open eucalypt forest, sometimes forming a dense understorey on deep coastal sands.

DISTRIBUTION Coastal districts of NSW from Bega to the Macleay River, extending to the Goulburn River Valley and the foothills of the Blue Mountains.

Macrozamia spiralis **Plate 79**

M. spiralis is relatively restricted in distribution and readily distinguished from the more widespread *M. communis*. It is a smaller plant with 2 to 12 fronds less than a metre long rising from a partly buried caudex about 10 to 20 cm across. The rhachis is spirally twisted about half a turn, more rarely straight or turned through 360°. The pinnae do not become progressively shorter and spine-like towards the base of the frond as in *M. communis*. Each pinnae is abruptly pointed at the apex and reddish, pinkish or yellowish at the base. Male plants produce 1 to 4 cylindrical cones about 15 to 20 cm long and 5 to 6 cm in diameter, the females usually a solitary ovoid cone 12 to 20 cm long and 7 to 9 cm in diameter.

HABITAT Scattered in dry open forest, never in dense stands as in *M. communis*.

DISTRIBUTION The Nepean district from Wallacia and Nortons Basin to Windsor and Maroota, extending through the Putty district to the Goulburn River Valley. Common in the Castlereagh State Forest.

Cupressaceae

The cypress family, Cupressaceae, belongs to the readily identifiable group of plants known as conifers. The group consists of about 500 species divided into 50 genera and six to eight families. Its members are usually, but not always, upright woody evergreen trees or shrubs bearing small male cones, which produce pollen, and large woody female cones, which produce seed on hard or papery scales.

The conifers are an ancient group with primitive ancestors, now extinct, present in the late Carboniferous period, 300 million years ago. They reached their highest point in the Cretaceous period about 100 million years ago when most of the existing families were present, but have since gradually declined in the face of competition from the angiosperms (flowering plants) which arose and expanded rapidly in the Cretaceous period. Roughly two-thirds of conifer species occur in the northern hemisphere where, in cold high latitude or high altitude areas, they form extensive uniform forests, and are still the dominant form of vegetation. Although a few genera such as *Podocarpus* cross the equator, the southern hemisphere conifers are generally distinct from those in the north. In the south they are rarely dominant, growing mostly in mixed hardwood forests and rainforest.

Australia is not renowned for its conifers, having only about 40 species in 11 genera and four or five families. Almost all the species and five of the genera are endemic. The endemic genera consist of one or only a few species each, and have very restricted distributions within the country.

Cupressaceae is the third largest family of conifers, but is the largest in Australia. It is divided into two subfamilies, the Callitroideae with 10 genera and 40 species confined (with one exception) to the southern hemisphere, and Cupressoideae with 9 genera and 90 species confined to the northern hemisphere. The family is distinguished by its scale-like leaves which are borne in whorls of 2 (e.g., *Widdringtonia*), 3 (e.g., *Callitris*) or 4 (e.g. *Cupressus, Diselma*). Although in the first year the leaves of the young plant are needle-like and free, as in the common pines *(Pinus)*, they soon become attached and decurrent on the stem between the internodes so that only the tip remains free. The pollen is borne on microsporophylls (modified leaves) arranged in whorls of 3 or 4 into strobili less than a centimetre long. In local species the seed cones are on short lateral branches and consist of 6 woody megasporophylls or cone scales arranged in a more or less globular or conical shape 1 to 3 cm in diameter. The cones ripen in the second season after fertilisation and eventually open from the top, releasing a dozen or more hard-winged seeds and revealing the columella, a distinctively shaped 3-lobed projection of the axis.

Of the 3 Australian genera *(Callitris, Diselma, Actinostrobus)*, only *Callitris* occurs in the Sydney district.

79. Macrozamia spiralis

80. Callitris rhomboidea

Callitris

Callitris rhomboidea

Port Jackson Pine, Port Jackson Cypress **Plate 80**

Frequently referred to as *C. cupressiformis*, this is by far the commonest native conifer in the Sydney district. It forms a tall narrow shrub or small tree with erect branches drooping at the ends. The leaves, including the decurrent portion, are 2 to 3 mm long and are prominently, but not acutely, keeled along the back. Woody fruiting cones occur in clusters on small branchlets, and remain attached after opening. Mature cones are globular or slightly tapered to the apex. They are less than 2 cm in diameter with 6 cone scales which are alternately large and small. The scales have a short conical point near the middle. The columella is short and 3-lobed. Seeds have narrow wings about 1 mm wide.

HABITAT Widespread on sandstone soils in open forest, usually on hillsides, often in or near rocky creek beds.

DISTRIBUTION Coast and northern and central tablelands of NSW, also Qld, Vic, Tas and SA. Occurs in the Clarence–Wolgan district where it may be confused with *C. endlicheri* and *C. muelleri*.

82. *Callitris endlicheri*

Podocarpaceae

The Podocarpaceae is a predominantly southern hemisphere conifer family, although Podocarpus extends northwards in Asia to China and Japan, and in the Americas to Mexico and the West Indies. Podocarp fossils have been found in middle and lower Jurassic strata 160 to 190 million years old in all the southern continents including Antarctica. At this time these continents existed as one land mass, Gondwanaland, which had extensive podocarp and araucarian forests. Representatives of these families were subsequently dispersed on drifting continental fragments and are today widely separated.

The Podocarpaceae consists of two widespread genera, *Podocarpus* with 110 species and *Dacrydium* with 25, and 4 genera with restricted distributions and 1 to 3 species each. These include small scrambling or shrubby plants as well as many large, often magnificent, forest trees. *Podocarpus* species are important timber-producing trees in Africa, New Zealand and South and Central America.

Members of the family are mostly dioecious (i.e., bear male and female organs on separate plants) and in this respect resemble the cycads rather than other conifers. The (male) pollen-bearing strobili are short and cylindrical to spherical, as with other conifers. The seeds may be borne singly on fleshy or hard sporophylls in tiny cones, or are produced at the ends of short branches above an expanded fleshy or hard axis know as the receptacle. Often the seeds and receptacle are coloured red to purplish black at maturity. The leaves of the various species differ considerably, ranging from tiny scales 1 or 2 mm long to flat linear or oblong structures up to 20 cm in length.

The Podocarpaceae is represented in Australia by 4 genera and 13 species, of which 2 genera and 3 species occur in the Sydney area.

81. *Callitris muelleri*

Callitris muelleri

Mueller's Cypress **Plate 81**

This species is easily confused with *C. rhomboidea* but is slightly larger in most respects and does not have drooping branch tips. The mature leaves are 4 to 10 mm long, including the decurrent part, and are prominently, but not acutely, keeled on the back. Occasionally, spreading non-decurrent juvenile leaves are also present. The woody fruiting cones are solitary or several together on short branchlets, and remain attached after opening. They are rounded or depressed-globular and up to 3 cm in diametr. The 6 cone scales each have a small conical point near the apex. The columella is 1 to 4 mm long and 3-lobed. Seeds have wings 2 to 4 mm wide.

HABITAT Often in exposed rocky places on sandstone soils.

DISTRIBUTION Central and south coast of NSW and the central tablelands where it is fairly common.

Callitris endlicheri

Black Cypress Pine **Plate 82**

This is an inland species, confined in the Sydney district to the Newnes–Wolgan area and the hills north and west of Lithgow. It is a tree with green or glaucous spreading branches, differing from the ascending branches of *C. muelleri* and *C. rhomboidea*. Leaves are 2 to 4 mm long, and prominently, but not acutely, keeled on the back. The woody fruiting cones are solitary or several together on slender clustered branchlets. They are ovoid or globular and less than 2 cm in diameter. They have 6 cone scales, each with a small conical point near the apex, the larger scales usually angled towards the apex, the alternate scales much reduced in size. The columella is short with 3 or 4 parts. The seeds have wings less than 3 mm wide.

HABITAT Mostly on rocky hillsides.

DISTRIBUTION North coast, tablelands, slopes and western plains of NSW, also Qld, Vic.

83. Podocarpus spinulosus

Podocarpus

Podocarpus spinulosus — Plate 83

P. spinulosus is a fairly common scrambling shrub with irregular straggling branches growing to a metre high. The leaves are 3 to 6 cm long, rigid and narrow linear with an acute pungent point. They are dark green above, paler below. Male cones are cylindrical, about 5 mm long, in small sessile axillary clusters. Female cones are produced on separate plants, occurring singly on stalks in the leaf axils. At the mature fruiting stage the stalk forms a glaucous blue-black fleshy receptacle bearing 1, or sometimes 2, globular seeds at the end.

HABITAT In sheltered areas of open forest, usually on sandy soil, occasionally forming a spreading mass several metres across.

DISTRIBUTION Central and south coast of NSW from about Woy Woy to Ulladulla, and in valleys of the Blue Mountains. Also Qld.

84. Podocarpus elatus

Podocarpus elatus

Brown Pine, Plum Pine — **Plate 84**

Although forming a large tree with finely fissured brown bark in northern rainforests, this species rarely exceeds 5 m in height in the Sydney district. It is uncommon in the area, being mostly confined to coastal rainforests. The leaves are oblong-linear, thick, 5 to 15 cm long and 6 to 12 mm wide. New growth is light yellowish green contrasting with the sombre dark green older foliage. Male and female cones are borne on separate trees, the males 2 to 3 cm long and several together in sessile axillary clusters, the females solitary on stalks in the leaf axils. The fruiting receptacle is blue-black, glaucous and fleshy at maturity and bears a globular seed about 1 cm in diameter.

HABITAT More or less confined in the Sydney district to coastal rainforests in soils varying from deep sand to volcanic loams.

DISTRIBUTION From the Illawarra district to north-eastern Qld. Locations include Berkeley Hill and West Dapto near Wollongong, Minnamurra Falls and The Spit at Minnamurra.

85. Microstrobos fitzgeraldii

Microstrobos

Microstrobos consists of 2 species, one *(M. niphophilus)* restricted to alpine areas of Tasmania, the other *(M. fitzgeraldii)* to the spray zone below certain waterfalls in the Blue Mountains.

Microstrobos fitzgeraldii

Dwarf Pine — **Plate 85**

A small shrub with weak drooping branchlets crowded with narrow spreading pointed leaves 2 to 4 mm long. The leaves are spirally arranged and attached by their decurrent bases. Small male and female cones are solitary and terminal on the branchlets, with the sexes borne on separate plants. The male sporophylls are purplish brown, the females woody at maturity and about 6 mm long. This is a protected species, threatened in some locations by erosion and pollution resulting from urban development in the catchments of the waterfalls.

HABITAT On wet rocks and ledges in the spray zone of south-facing waterfalls, occasionally in caves nourished by seepage.

DISTRIBUTION Endemic in the Blue Mountains, where restricted to 6 locations between Wentworth Falls and Bonnie Doon Falls, Katoomba.

5. Angiosperms

The angiosperms, or flowering plants, are the most recently evolved major group of plants. They have already become the largest group both in numbers of species and numbers of individuals, and the dominant vegetation of most of the Earth's surface today. Unlike the gymnosperms, which encompass unrelated evolutionary lines, the angiosperms are a natural group in that they appear to be derived from a single group of related plants. Although their origin is still unclear, and the earliest forms unrecognised, it is likely that they developed from a group of Mesozoic pteridosperms (seed ferns), possibly during the early Cretaceous period about 130 million years ago, split at an early stage into monocots and dicots, then diversified and expanded rapidly during the remainder of the Cretaceous period and the early part of the Tertiary period. It is generally accepted that the angiosperms are not the descendants of any existing gymnosperm; we must look further into the past, possibly to Carboniferous pteridosperms, for a common ancestry.

The angiosperms differ from the gymnosperms in a number of respects of which the most obvious is the possession of flowers. A flower is actually a short shoot bearing specially adapted leaves arranged in whorls. The outer whorl, called the calyx, consists of sepals which are usually green and which protect the developing flower. The second whorl, called the corolla, consists of petals which are often brightly coloured and designed to attract the attention of pollinating insects and animals. The third whorl, called the androecium, consists of stamens which have organs called anthers that produce male microspores (pollen). The inner whorl, called the gynoecium or pistil, consists of one or more units called carpels which are analogous to the megasporophylls of gymnosperms. Each carpel consists of an ovary topped by an upright column called the style with an expanded pollen-receptive tip called the stigma; in different species the parts are fused and modified in various ways. In many cases one or more whorls of the flower are reduced in size or absent entirely.

The possession of a flower provides several advantages which enable angiosperms to compete successfully with other plants and to occupy a much greater diversity of habitats. Pollination of flowers by insects, birds and animals is much more efficient and effective in mixed forests for example, where the chance of successful pollination by wind decreases with increasing variety or complexity of the forest. (Some modern angiosperms such as casuarina are wind-pollinated, but this is a derived, not a primitive, character.) Because the ovules are protected in closed carpels instead of being exposed on the surface of sporophylls, the risk of insect predation is lessened, the ovules can become smaller and simpler, and their development faster. The plant invests far less in the production of seed, or is able to produce many more seeds. In many species the seeds are dispersed by animals and birds which feed on succulent fruit formed from the gynoecium. In others the seeds and fruit are modified for dispersal on the bodies of animals and birds or by wind and water.

The angiosperms ae divided into 2 classes, the dicotyledons and the monocotyledons, which are distinguished by a combination of characters. Evidence indicates that the dicotyledonous state preceded the monocotyledonous state. This is usually interpreted to mean the monocotyledon line separated and developed from some extinct, vessel-less, herbaceous, aquatic or amphibious dicotyledonous angiosperm having flowers with an indeterminate number of parts, one or several free carpels and monocolpate pollen.

The dicots are much more diverse than the monocots in size, habit and lifestyle. About 50% of dicot species are woody trees and shrubs compared to only about 10% of monocots. Woody dicots are usually many-branched but woody monocots such as palms and bamboos are generally unbranched. Herbaceous dicots are mostly small with small leaves but herbaceous monocots may be large with long, narrow or broad strap-shaped leaves arising from a clump on the ground or occasionally from a short caudex. It is generally accepted that the woody habit is primitive in angiosperms, and the herbaceous habit (in which production of seed is maximised compared to production of vegetative mass) is derived. Herbaceous plants colonise bare or disturbed areas faster, produce seeds sooner, expand their geographic range quicker, and exist in more extreme climates than trees and shrubs.

Monocot leaves usually have parallel venation (with veins running lengthwise), leaf bases which sheath the stem, and leaves that are not clearly differentiated into petiole (stalk) and lamina (blade). Dicots, on the other hand, have leaves with reticulate venation (a midvein, diverging secondary veins and a network of minor veins), are rarely sheathed at the base, and usually have a short or long stalk. The floral parts (sepals, petals, stamens) of monocots are typically in sets of three (rarely four) whereas in dicots they are in sets of four or five (rarely three or fewer). As indicated by their common names, the two groups differ in the number of cotyledons (leaf-like food storage organs) on the embryo. There are also differences in root structure, pollen, and the nature of vascular bundles in the stem.

Figure 2

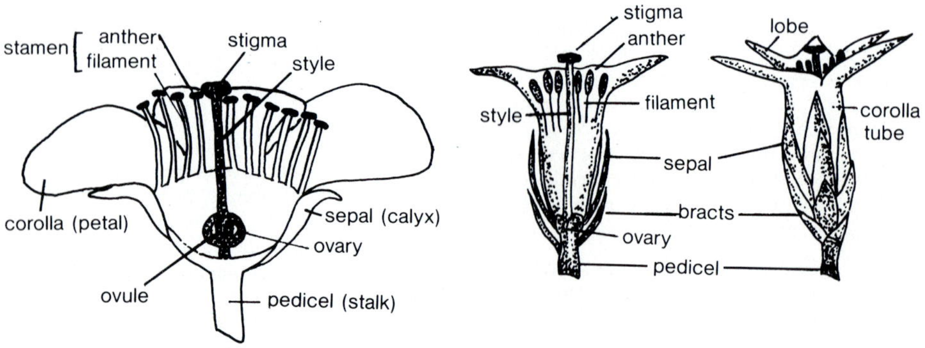

86. Tasmannia insipida

87. Tasmannia lanceolata

Dicotyledons

Winteraceae

The Winteraceae is a small family of six or seven genera found in several types of rainforest in areas fringing the southern Pacific, the Philippines and central America.

The family displays a large number of 'primitive' features such as flowers with an indefinite number of spirally arranged parts which vary in structure, position and degree of diversification. Also, the wood lacks the long water-conducting tubes known as vessels that are characteristic of angiosperm wood structure. The family is therefore of particular interest since, as the least specialised of present-day angiosperm families, it is perhaps the closest to the gymnosperms and the extinct angiosperm forebears.

The Winteraceae is represented in Australia by 2 genera, *Tasmannia* with 5 species on the east coast and in Tasmania, and *Bubbia* with 3 species in Queensland. Two species of *Tasmannia* occur in the Sydney area.

Tasmannia Pepper-bush

Tasmannia insipida

Pepper-bush **Plate 86**

T. insipida can grow to 5 m tall, but shrubs of 1 to 2 m are most common. Its stems are often a distinctive reddish colour. Leaves are oblanceolate, 8 to 16 cm long, blunt at the base and almost stalkless. Oil dots are visible under a lens. There is a smell of oil when crushed and a pepper taste when chewed. White flowers are borne in umbels. Male and female flowers are carried on the one plant, with some flowers which are bisexual. Male flowers are dominated by a mass of stamens beneath which are two strap-like petals. The fruit is an oval berry which changes colour from red to purple as it matures.

HABITAT Shaded gullies and rainforests.

DISTRIBUTION Widespread along the coast and Blue Mountains. Extending from south coast to Qld.

FLOWERING September–December.

Tasmannia lanceolata

Mountain Pepper **Plate 87**

A compact dark shrub growing from 2 to 4 m tall. It is uncommon in the Sydney district as it is essentially a cool mountain species, typical of the highlands of Victoria and Tasmania. The main difference between it and *T. insipida* is that in this species the leaves taper gradually towards the base into a short stalk. Leaves are lance-shaped, 4 to 7 cm long and distinctly paler on the undersurface. Fruit is a purple-black globular berry.

HABITAT Sheltered gullies of the upper mountains.

DISTRIBUTION Blackheath (e.g., Centennial Glen). Also recorded on the south coast, north of the Barrington Tops area, and south to Vic and Tas.

FLOWERING Irregular. Recorded elsewhere as flowering in summer, but no summer flowering observed at Blackheath.

Eupomatiaceae

This family comprises a single genus, *Eupomatia*, with two species; *E. bennettii*, restricted to northern coastal New South Wales and Queensland, and *E. laurina* with a range extending from New Guinea to Victoria.

The creamy or yellow waxy flowers which occur singly in the leaf axils are not differentiated into sepals and petals. Instead the perianth is fused into a lid which falls off at flowering to expose 2 or 3 rows of stamens and a similar number of staminodes all spirally arranged around the rim of the receptacle. Numerous carpels, each containing several ovules, are immersed in the pulpy mass of the broad receptacle. The flowers are pollinated by beetles which feed on the staminodes.

The Eupomatiaceae is one of several primitive Angiosperm families comprising one or a few relict species occurring in the rainforests of Queensland and islands to the east and north.

Eupomatia

Eupomatia laurina

Bolwarra **Plate 88**

Bolwarra is a shrub or small tree which has characteristic weak straggling branches. Shiny, broad lance-shaped leaves, 7 to 12 cm long, are arranged in alternative rows along these branches. The unusual flowers are large (25 mm across) and solitary. Flower buds have a conical lid which falls off to reveal a white flower. The flowers have no petals or sepals and what appear to be petals are modified stamens. The fruit is an oval berry, 2 cm across, with a flat top encircled by a rim.

HABITAT In or near rainforests.

DISTRIBUTION Widespread from Victoria to New Guinea. Fairly common along coastal zone; extending into Blue Mountains where there is suitable habitat. Recordings from Royal National Park, Mt Keira, Macquarie Pass and Gosford district.

FLOWERING December–January, but apparently not flowering every year.

88. Eupomatia laurina

Monimiaceae

Nearly 30 families of plants have flowers with spirally arranged parts and a number of separate carpels which develop into separate fruits, conditions which are recognised as 'primitive' compared to flowers with whorled parts and united carpels. At one time the whole group was placed in a single order, Polycarpicae, but is now divided amongst four or more orders on the basis of similarities and differences in a range of characters. The Monimiaceae is one of from 4 to about 14 families (depending on the system of classification followed) now included in the order of Laurales which, generally speaking, displays fewer 'primitive' characters than the Magnoliales to which the two previous families belong.

The family consists of both tropical and cool-temperate species widely distributed in Australasia, Malesia, Polynesia, South America and Madagascar. It is poorly represented in Africa and absent from India. The diversity of the family is such that it is sometimes divided into as many as 4 families. An unusually large number of the 30 or so genera are monotypic (i.e., consist of a single species). The 6 representatives in the Sydney area are from 6 different genera consisting of rainforest trees or shrubs and a climber or scrambler. The tree species can usually be recognised by their opposite, thick, serrated leaves. *Atherosperma, Doryphora* and *Daphnandra* are sometimes placed in the family Atherospermataceae.

89. Doryphora sassafras

Doryphora

Doryphora sassafras

Sassafras **Plate 89**

Sassafras is a common, medium to tall tree of the rainforest. It is readily identified by its aromatic leaves which when crushed give out a distinctive fragrant smell. The leaves are opposite, 6 to 10 cm long, broad-lanceolate, coarsely serrated and dark glossy green. Flowers are on short stalks in clusters at the leaf axils. The buds are enclosed by 2 bracts which fall off to reveal delicate white flowers, with 6 perianth segments. The flowers are very striking when massed on the branches, the white contrasting with the dark leaves.

HABITAT Coastal rainforest and deep sheltered gullies.

DISTRIBUTION Extending along coast and tablelands from Victorian border to southern Qld. Common along coast and in Blue Mountains (e.g., Upper Causeway in Royal National Park and Jenolan Caves).

FLOWERING Irregular, but mainly August–September.

90. Atherosperma moschatum

Atherosperma

Atherosperma moschatum

Black Sassafras, Southern Sassafras **Plate 90**

Black Sassafras is a tree of fairly high altitudes, where it grows in ferny gullies and along sheltered streams. It is uncommon in the Sydney district. Leaves are 3 to 10 cm long, usually without teeth, have a white or greyish undersurface and are fragrant when crushed. Like *Doryphora sassafras*, buds are enclosed by 2 deciduous bracts, but the flowers are quite different. They are cup-shaped, creamy in colour, turning pink with age and solitary in the leaf axils.

HABITAT Along streams in deep gullies at high altitudes.

DISTRIBUTION Upper Blue Mountains. Recordings include Blackheath, Mt Wilson and Gordon Falls, Leura. Also occurring Tas, Vic, southern highlands and Barrington Tops.

FLOWERING September–October.

Daphnandra

Daphnandra micrantha

Socket Wood **Plate 91**

Although Socket Wood is a common small tree in most rainforests of the north coast, it is uncommon in the Sydney district where it reaches its southern limit. Its leaves are 7 to 10 cm long, opposite, glossy, hairless, dark green above, light green below, ovate to elliptic, with numerous fine teeth. Fine oil dots are present, but often very faint, and the leaves are somewhat aromatic. Flowers are white, on stalks 2 to 4 mm long and borne on a short panicle from the leaf-bases. The inflorescence helps distinguish this species from *Doryphora sassafras* which has only 3 flowers, all clustered together on a short stalk. Fruit is covered with fine brown hairs, enclosed in the persistent floral tube which splits open to release the seeds. The name 'Socket Wood' is derived from the socket-like appearance along the trunk at the joint of each main branch. This genus is undergoing revision and the species in the Sydney district may be separated from *D. micrantha* and given a new name.

HABITAT Rainforest gullies, often on rocky slopes.

DISTRIBUTION Scattered and uncommon. Recorded from Minnamurra Falls and Yarramalong, west of Wyong. Extending along north coast to Qld.

FLOWERING December–February.

91. Daphnandra micrantha

Wilkiea

Wilkiea huegeliana

Wilkiea **Plate 92**

The most striking feature of Wilkiea is its leaves. These are leathery, dark green, 7 to 15 cm long, 3 to 6 cm wide, usually serrated, and with a very strong pattern of veins, especially visible underneath. Flowers are small, unisexual and rather insignificant. After flowering, an enlarged receptacle forms on which the fruits develop. Mature fruits consist of a number of black ovoid berries each about 12 mm long, packed around the receptacle. Wilkiea is usually a shrub but can grow to a small tree.

HABITAT Rainforests or sheltered forested slopes, especially on volcanic soils.

DISTRIBUTION A coastal species extending from near Narooma (NSW) northwards to southern Qld. Recorded from Palm Jungle, Minnamurra, Mt Keira, Gap Creek and Maitland Bay (Bouddi National Park).

FLOWERING December–February.

92. Wilkiea huegeliana

Hedycarya

Hedycarya angustifolia

Native Mulberry **Plate 93**

Native Mulberry is a shrub which bears its male and female flowers on different plants. The leaves are broad, lance-shaped, irregularly toothed, strongly veined, paler underneath, much thinner than those of *Wilkiea* and smaller (5 to 10 cm long). Male flowers are a mass of tiny stamens covering a flat or hemispherical disc; female flowers also sit on a disc. The fruit is a compact collection of small globular yellow-green berries, somewhat resembling a mulberry.

HABITAT In and near rainforests, and bordering on tall sheltered forests.

DISTRIBUTION Extending from Tas to southern Qld. Common along coast and cool gullies of tablelands. Recordings include Royal National Park, Kiama, Mt Irvine and Dog Trap Creek (Gosford).

FLOWERING August–September.

93. Hedycarya angustifolia

Palmeria

Palmeria scandens

Anchor Vine **Plate 94**

Anchor Vine is a rainforest climber which when old may have large woody stems. When young, it resembles a scrambling shrub. The main identifying features are its leaves and fruit. Leaves are opposite (or almost so), broadly elliptic, 5 to 13 cm long, with entire margins, dark green above and rough to the touch. Male and female flowers are separate on the same plant, each in loose clusters from the leaf axils. Male flowers are yellowish, 8 to 9 mm across, like a flattened star, with 5 lobes and numerous anthers. Female flowers are smaller and globular. The floral tube enlarges at fruiting time to about 20 mm across before splitting to reveal bright red globular berries.

HABITAT Coastal rainforests.

DISTRIBUTION Common both south and north of Sydney in suitable habitat. Extending from south coast northwards to Queensland.

FLOWERING Irregular, but mainly June–August.

94. Palmeria scandens

Lauraceae

The Lauraceae is an important component of moist tropical and subtropical forests in South-east Asia, New Guinea and tropical America. Although now largely absent from temperate parts of the northern hemisphere, the Lauraceae was widespread and common in these areas during the warm moist period from the late Cretaceous to the middle Tertiary. With over 2000 species in about 30 genera, it is the largest family in the order Laurales.

The family consists of trees and shrubs of an aromatic nature due to the presence of oils in all parts. Useful products include cinnamon, camphor, essential oils such as bay and cassia, avocado and fine timbers. In Australia species of *Cryptocarya, Endiandra* and *Litsea* from east coast subtropical and temperate rainforests provide timber for plywood and interior work.

Members of Lauraceae can be distinguished from those of Monimiaceae by the leaves which are arranged alternately and are simple, entire, dark glossy green above, paler and sometimes glaucous beneath, often with a noticeably pale yellow mid-vein. The flowers are small with the parts usually in multiples of three. The fruits are black, 1-seeded berries or drupes more or less enclosed in, or seated upon, the floral tube in much the same way as in Monimiaceae.

Cryptocarya

Trees with alternate leaves; flowers small, with 6 perianth segments and 9 perfect stamens in 2 rows, the inner row alternating with staminodes.

Cryptocarya glaucescens

Brown Beech, Native Laurel **Plate 95**

Brown Beech is a common rainforest tree. It is easily identified from beneath the tree as the undersurface of the leaves is blue-grey and there is a conspicuous yellow midrib. Leaves are alternate, elliptic, 6 to 12 cm long, dark above and light below. Flowers are small and in terminal clusters. The fruit also helps in identification. They are black and shiny, about 20 mm across, globular but compressed at the ends and hang in large bunches.

HABITAT Rainforests.

DISTRIBUTION Extends from Narooma to north Qld in coastal areas. Illawarra Escarpment from Cambewarra to Royal National Park and, in the north, Strickland and Wyong State Forests.

FLOWERING October–November.

95. Cryptocarya glaucescens

Cryptocarya microneura

Murrogun, Brown Jack **Plate 96**

This small tree could easily be confused with Brown Beech as it has alternate leaves of similar size and terminal clusters of flowers. However it differs from that species in having leaves which are green on both surfaces and black fruit which are globular and pointed (not compressed) at the ends. Like Brown Beech, the leaves have a clearly visible yellow midvein. The specific name 'microneura' refers to the conspicuous fine network of veins in the leaves.

HABITAT Rainforests.

DISTRIBUTION Uncommon. Extends from near Bateman's Bay northwards to Qld. Isolated occurrences along coastal zone (e.g., Minnamurra Falls, Macquarie Pass, Mt Kembla, Wyong State Forest).

FLOWERING September–November.

96. Cryptocarya microneura

Neolitsea

Neolitsea dealbata

White Bolly Gum **Plate 97**

This tall rainforest shrub or small tree is readily recognised by its large broad pointed leaves, 10 to 20 cm long, which are pale whitish or blue-grey on the undersurface. The specific name 'dealbata' is derived from the Latin word meaning 'whitewashed, covered with a white powder'. The buds, branchlets and veins beneath the leaves are all rusty-hairy. Male flowers are in stalkless clusters along the stems, with conspicuous yellow stamens.

HABITAT Rainforests.

DISTRIBUTION Reaches its southern limit on benches of the Illawarra Escarpment near Mt Keira. More common north of Sydney in Strickland, Olney and Wyong State Forests (e.g., The Basin Camping Area and Gap Creek). Extending to the north coast and northern Qld.

FLOWERING April–June.

Litsea

Litsea reticulata

Bolly Gum

Bolly Gum is a tall rainforest tree with leathery leaves, 4 to 8 cm long. These leaves are rounded at the apex or slightly pointed and have numerous, prominent net veins. As the tree is often large and the leaves out of reach, the grey mottled bark marked with shallow indentations and pink-brown patches may be a clue to its identity. Flowers are small, borne from the leaf axils, with 4 bracts enclosing about 6 flowers. Fruit is an oval black berry.

HABITAT Rainforests.

DISTRIBUTION Uncommon in the area, but recorded from Minnamurra Falls and Wyong State Forest. Extending to the north coast and northern Qld.

FLOWERING May–July.

Endiandra

Trees and tall shrubs with alternate leaves. Flowers with 6 perianth segments and 3 stamens alternating with 3 staminodes, borne in short clusters from leaf axils. Fruit a berry.

97. Neolitsea dealbata

Endiandra sieberi

Corkwood **Plate 98**

Corkwood is a tree with a grey-brown fissured bark, corky in appearance. Its leaves somewhat resemble those of *Cryptocarya*, as they have a distinctive yellow midrib, but they lack the blue-grey undersurface and the conspicuous lateral veins on the upper surface. Leaves are tough, broadly lance-shaped, 5 to 8 cm long, with a drawn-out tip, dark above and slightly paler beneath. A feature of the leaf useful for identification is the presence of a narrow yellow translucent margin. Fruit is a shiny black oval berry, up to 25 mm long, which ripens from March to August.

HABITAT Coastal rainforests, especially in sandy soils.

DISTRIBUTION Extending from Jervis Bay (Long Beach rainforest) northwards to southern Qld. Illawarra region (e.g., Minnamurra Spit, Stanwell Park, Coalcliff), Royal National Park (Palm Jungle), Kurnell and Bouddi National Park (Maitland Bay).

FLOWERING June–October.

98. Endiandra sieberi

Endiandra discolor

Rose Walnut **Plate 99**

A rainforest tree, its leaves glossy, elliptic, 7 to 9 cm long, 4 to 5 cm wide, entire, blunt at the tip, firm, dark green above, lighter green below, with yellow veins. The undersurfaces of young leaves often have a grey-blue waxy bloom. Leaves have characteristic large V-shaped glands (domatia) in the angles of the midvein and main lateral veins on the undersurface. The fruit is black egg-shaped berries, up to 30 mm long, which are ripe in about March.

HABITAT Rainforests.

DISTRIBUTION Gosford district (e.g. Strickland State Forest, Katandra Reserve) extending along the north coast to northern Qld.

FLOWERING Irregular, but chiefly October–November; apparently not flowering in some years.

99. Endiandra discolor

Cassythaceae

This family, consisting of a single genus of leafless twining parasitic shrubs, is closely related to the Lauraceae, and sometimes included in it. The relationship is seen in the structure of the tiny white flowers and the single-seeded fruits which are enveloped in the fleshy receptacle or floral tube. The plants themselves closely resemble another parasite, *Cuscuta*, but are not related to it (for differences, see *Cuscuta*).

Cassytha Dodder, Devil's Twine

These leafless parasitic plants climb over a variety of host plants, attaching themselves by means of suckers along the stems. *Cassytha* seeds germinate in the ground, but roots function only until the stems become parasitic. Older plants are rootless. Stems have the greenish-yellow appearance typical of most parasitic plants. Flowers have 6 perianth segments, but appear to have only 3 petals as the segments are in two whorls, one larger than the other. Fruit is succulent and berry-like.

Cassytha glabella **Plate 100**

This parasitic twiner may be distinguished from other Dodder species by its fine thread-like, hairless stems; these stems are often a reddish colour. Flowers are small, white and hairless, with about 6 clustered together in globular heads on stalks 5 to 7 mm long. Fruit is narrow to egg-shaped and reddish-brown.

HABITAT Widespread in heath and open forest.

DISTRIBUTION Coast to mountains. Occurring in all states, but not NT.

FLOWERING Chiefly November–March, with some flowers at other times.

100. Cassytha glabella

101. Cassytha pubescens

Cassytha melantha

This twiner has the thickest stems of the Dodder found in the Sydney district. These stems are without hairs and do not have a warty appearance. Flowers are small (about 4 mm long), ovoid, greenish-brown outside and cream inside and are in short clusters borne on downy stalks. The individual flowers are covered with blackish hairs. The fruit is green, globular and hairless, turning brown with maturity.

HABITAT Forest and woodland.

DISTRIBUTION Higher Blue Mountains. Not recorded from coastal districts. Extending to western plains, Vic, Tas, SA and WA.

FLOWERING October–December.

Cassytha pubescens **Plate 101**

The stems of this twiner are 1 to 2 mm in diameter (twice as thick as those of *C. glabella*), warty when mature, and usually hairy when young. The specific name 'pubescens' means 'downy' and refers to the stems and to the soft hairy flowers, flower-stalks and fruit. Flowers are in elongated spikes (often in short heads when young). The fruit is a greenish globular berry, 4 to 5 mm long, not egg-shaped (as is *C. glabella*), nor globular and hairless (as is *C. melantha*). This species appears to intergrade with *C. paniculata* and some classifications include both species under *C. pubescens.*

HABITAT Widespread in heath, scrub and open forest.

DISTRIBUTION Coast to mountains. Found throughout most of NSW, except far west, and in all States, except NT.

FLOWERING October–December.

102. Cassytha paniculata

Cassytha paniculata **Plate 102**

This species is now generally regarded as a form of *C. pubescens*, but is here treated separately as the form in the Sydney district has at least two features which make it appear different. Flowers are borne in elongated spikes, which are mostly longer and more branched than those of the usual form of *C. pubescens*. The fruit is globular, with 6 conspicuous raised ridges, hairless or minutely downy. The stems and flowers of this plant are usually hairless, but may be minutely hairy and resembling some of the forms of *C. pubescens*.

HABITAT Widespread in dry forests, woodlands and on shrubby dunes.

DISTRIBUTION Coast (e.g., Kurnell) and lower Blue Mountains (e.g., Erskine Creek). Also occurring on north and south coasts.

FLOWERING October–December.

103. Piper novae-hollandiae

Piperaceae

The Piperaceae is the largest of the two or three families making up the order Piperales. Although there is little information available on the natural history of plants of this order there is general agreement that the Piperales represents only a slight development from the 'primitive' order Magnoliales, from which it is thought to be derived. The Piperaceae is closely related to the Peperomiaceae, and the two are often combined in the family Piperaceae. There are however a number of differences which many botanists consider sufficient to warrant separate family status.

Piper, with about 2000 species, is by far the largest and most important genus in the family and the order. Some of its products are widely used and much appreciated by man. Pepper, one of the earliest spices known, is the dried and ground fruit of *Piper nigrum*, a climbing vine with large, shiny, ovate, prominently veined leaves similar in appearance to the local species *Piper novae-hollandiae*. The fermented drink kava is prepared from the roots of a tropical species of *Piper*, and in the Indomalaysian region the leaves of another species, sprinkled with lime to release the alkaloids, are chewed with betel nut.

Piper

Nine species of *Piper* occur in the rainforests of Queensland and New South Wales, with 1 in the Sydney area.

Piper novae-hollandiae

Pepper Vine **Plate 103**

A rainforest climber with stems 30 to 40 m long. It is usually found covering the trunks of trees or hanging from the branches. It is hairless, with entire, ovate, alternate leaves, 7 to 10 cm long. These are usually dark glossy green and have 5 to 7 conspicuous veins running from the base. Flowers are unisexual; the male flowers are in a cylindrical spike about as long as the leaf, which hangs from a stem at the leaf-axil; female flowers are in a smaller ovoid spike. The fruit is a small drupe.

HABITAT Shaded sites, in rainforest.

DISTRIBUTION Coastal zone (e.g., Minnamurra Falls, Mt Keira, Palm Jungle) extending from the south coast to the north coast and Qld.

FLOWERING June–September.

Peperomiaceae

The Peperomiaceae is a family of small terrestrial or epiphytic succulent herbs. It consists of the widespread genus *Peperomia* with about 1000 species, and 3 genera from the West Indies with 1 or a few species each. The family is closely related to the Piperaceae and is sometimes included in it. Both families have tiny reduced flowers without petals and sepals arranged in narrow cylindrical spikes and more or less embedded in the fleshy axis. The flowers of both develop into 1-seeded fruits. Certain differences in form, stem anatomy and pollen separate the Peperomiaceae from the Piperaceae. Whereas the Peperomiaceae are small herbs, the Piperaceae includes trees, shrubs and climbers. The Peperomiaceae also lacks stipules, which are present in Piperaceae, and has consistently bisexual flowers with 2 stamens instead of unisexual flowers with 4 stamens. The 2 families also differ in their fruits, with Peperomiaceae forming a berry and Piperaceae a drupe.

Peperomia

Five to 7 species of *Peperomia* occur in northern and eastern Australia with 2 species in the Sydney area. Cultivated forms of *Peperomia* are popular indoor pot plants.

104. Peperomia tetraphylla

Peperomia tetraphylla **Plate 104**

The specific name 'tetraphylla' refers to the leaves in whorls of 4, a characteristic feature of this small, semi-succulent rainforest herb. These leaves are hairless, ovate, darky shiny green and almost without stalks; each leaf is 5 to 12 mm long. Stems are 6 to 20 cm long. Flowers are crowded onto terminal spikes which are up to 3 cm long.

HABITAT On shaded rocks and logs in rainforest.

DISTRIBUTION Coast to mountains. Recordings include Minnamurra Falls, Palm Jungle and Strickland State Forest. Widespread throughout coast and tablelands of NSW, but also found on Pacific Islands, New Zealand, Asia and Africa.

FLOWERING November–December.

Peperomia leptostachya **Plate 105**

This is an erect herb with semi-succulent stems to 30 cm tall. Leaves are opposite (not whorled as are those of *P. tetraphylla*), ovate, 1 to 3 cm long, about 1.5 cm wide (sometimes larger), on stalks 5 to 8 mm long, paler underneath and are covered with short soft hairs. Flowers are crowded onto terminal spikes which are 4 to 10 cm long.

HABITAT On shaded rocks in sheltered gullies and rainforest.

DISTRIBUTION Coastal zone, chiefly north of Sydney. Extending along the north coast to Qld; also Pacific Islands.

FLOWERING November–December.

105. Peperomia leptostachya

106. Brasenia schreberi

Cabombaceae

This is one of several families of aquatic herbs comprising the order Nymphaeales which includes water lilies, hornworts and the sacred lotus. The Nymphaeales is 1 of the primitive orders having (usually) an indefinite number of incompletely differentiated floral parts and multiple fused or free carpels forming separate fruits.

The Cabombaceae consists of 2 genera of tropical and subtropical water plants, *Brasenia* with 1 widespread species, and *Cabomba* with 6 species confined to America.

Brasenia

Brasenia schreberi **Plate 106**

Brasenia is a water plant with floating leaves similar to those of the water-lilies. The leaves are round to elliptic, 4 to 12 cm long. The submerged stems may be 1 m long and are rooted in the mud. A characteristic feature is that the stems and lower surfaces of the leaves are covered with a thick layer of mucus. Flowers are small, up to 18 mm long, purplish, on long stems from the leaf axils and float on the surface of the water.

HABITAT Lakes and the backwaters of rivers.

DISTRIBUTION Uncommon in Sydney district. Restricted to Thirlmere Lakes and Colo and Cattai Rivers. Widely distributed northwards to tropical Australia; also occurring in Africa, Asia and America.

FLOWERING January–March.

Ranunculaceae

The fact that the first 2 orders of primitive angiosperms consist almost entirely of rainforest trees and shrubs distributed in the warmer, moister parts of the world tell us something of the climate and the nature of the vegetation of Australia and the world generally for much of the last 100 million years. The Ranunculaceae is represented in Australia by only a few genera, the best known being *Ranunculus*, the buttercup, and *Clematis*, a common and widespread climber. In different ways both are atypical of the family. *Ranunculus* is unusual in having a well-differentiated calyx and corolla with 5 sepals and 5 petals, and *Clematis*, with opposite pinnate leaves, is a departure from the usual radical or alternate lobed and dissected leaves.

The family, which shows signs of development from Magnoliales-like stock, is closely related to the Menispermaceae. These 2 families, with Berberidaceae and a number of small families, make up the order Ranunculales. The aquatic species of *Ranunculus* provide a link with the order Nymphaeales which is composed entirely of aquatic plants.

Clematis

Climbers with opposite, compound leaves and 3 leaflets. Flowers often unisexual; female flowers develop into clusters of fruit with distinctive long silky appendages. Flowers lack true petals; 4 petal-like sepals take their place.

Clematis aristata

Old Man's Beard, Traveller's Joy **Plate 107**

This attractive climber is conspicuous when it adorns forest trees and shrubs with its abundant white flowers and its fluffy masses of bearded fruits. Flowers are up to 4 cm across, with numerous stamens and are borne on long slender stalks from the leaf axils forming compound inflorescences. An important feature of this species is that the stamens have an appendage projecting above the anthers as long again as the anthers. This differs from *C. glycinoides* which only has a short blunt anther appendage. Leaflets are rather firm, dark green and irregularly toothed.

HABITAT Widespread in forest, especially shaded slopes.

DISTRIBUTION Coast to mountains. Extending from the north coast of NSW to Vic, Tas and WA.

FLOWERING October–November.

107. Clematis aristata

108. Clematis glycinoides

Clematis glycinoides

Old Man's Beard **Plate 108**

The general appearance of this species is similar to *C. aristata*. It also is a climber with compound leaflets, masses of white flowers and conspicuous fluffy clusters of fruit. It differs from that species in having leaflets which are thin and mostly entire and flowers which are smaller (about 3 cm across). The stamens have a short, blunt appendage which is not longer than the anther.

HABITAT Forests, especially sheltered areas.

DISTRIBUTION Widespread and common from coast to mountains. Extending to Vic and Qld.

FLOWERING August–October.

Ranunculus Buttercup

Herbs with 5 petals, usually yellow, and numerous stamens. The shape and texture of the fruit is often important in separating species which resemble each other. A number of introduced Buttercups are widely naturalised and common in damp places.

Ranunculus inundatus

River Buttercup **Plate 109**

River Buttercup often grows with its roots in the water and, as it spreads by means of underground stems, it sometimes forms large clumps. Like most Buttercups, its flowers are yellow, with 5 or more petals and numerous stamens; they are about 15 mm across. The leaves are the best identifying feature as they are on long slender stems and are divided into narrow leaf segments. Leaves appear as an almost circular set of ribs on the top of the stem and are up to 6 cm across. The whole plant is without hairs.

HABITAT In permanent water or in swampy soils.

DISTRIBUTION Widespread from coast to mountains. Extending to western slopes, north to Qld, and south to Vic and Tas.

FLOWERING November–January.

Ranunculus rivularis **Plate 110**

R. rivularis is a small weak plant with decumbent stems. Its leaves are hairless, with three wedge-shaped lobes on the end of long stalks. Each lobe is usually broadly toothed at the apex. Flowers are glossy yellow, 8 to 10 mm across, with 5 to 9 petals. Each petal has a shallow nectary pit above the base. The fruitlets are covered with indistinct wrinkles.

HABITAT In and near swamps and along creek margins.

DISTRIBUTION Higher Blue Mountains. Extending through the southern tablelands to Vic and Tas.

FLOWERING September–December.

Ranunculus lappaceus

Common Buttercup **Plate 111**

A softly hairy plant, with erect, branched flower stems up to 50 cm tall. Leaves on long stalks arise from the base of the plant. They are divided into 3 wedge-shaped lobes, which may be stalked. Each lobe is toothed or lobed at the apex. Flower stems have a few narrower stalkless leaves. The golden flowers are 2 to 3 cm across with 5 obovate petals. The fruitlets are smooth and surmounted by a slender coiled beak which is the persistent style.

HABITAT Damp or shaded sites, mainly in open situations on shale or clay.

DISTRIBUTION Coast to mountains, but localised and relatively uncommon. Occurring throughout most of NSW except drier west; also found in Qld, Vic, Tas and SA.

FLOWERING August–December.

109. Ranunculus inundatus

110. Ranunculus rivularis

111. Ranunculus lappaceus

112. Ranunculus plebeius

114. Ranunculus pimpinellifolius

Ranunculus plebeius

Hairy Buttercup **Plate 112**

Hairy Buttercup is a small herb, 10 to 30 cm tall. Its leaves are hairy, with blades up to 7 cm long and about as broad; they are divided into 3 lobes with each lobe having 3 broad teeth. The golden-yellow flowers are about 10 to 12 mm across. A characteristic feature of this species is that its sepals are rolled back, not pressed up against the petals as in closely related species of buttercup.

HABITAT Shady and damp situations in forests or beside streams and swamps.

DISTRIBUTION Widespread from coast to mountains, in locations as diverse as South Creek, Shane Park, and the margins of rainforests at Macquarie Pass and Ourimbah State Forest. Extending from south coast northwards to Qld.

FLOWERING Most of the year, but chiefly December–January.

113. Ranunculus collinus

Ranunculus collinus

Strawberry Buttercup **Plate 113**

This small creeping perennial grows in forests at higher elevations, in damp or shady sites. Its leaves are basal, with 3 leaflets; each leaflet is usually 3 toothed, rather like the leaf of a strawberry. Flowers are up to 20 mm across, solitary on stalks longer than the leaves.

HABITAT Shaded forests in grassy understorey.

DISTRIBUTION Higher Blue Mountains (e.g., Jenolan Caves). Extending through southern tablelands to sub-alpine areas of Vic and Tas.

FLOWERING September–October.

Ranunculus pimpinellifolius **Plate 114**

This is a small downy perennial herb, to 20 cm tall. Leaves, especially those at the base of the plant, are pinnate, with 5 (sometimes 3) leaflets which have broad blunt lobes. Leaflets and stems are softly hairy. Flowers are yellow, about 12 mm across, on spreading stems.

HABITAT Swampy areas, such as bogs beside mountain streams.

DISTRIBUTION Higher Blue Mountains (e.g., Kanangra-Boyd National Park). Extending along tablelands south to Vic and Tas.

FLOWERING December–January.

Menispermaceae

The Menispermaceae is a family of rainforest climbers or lianas with a mainly tropical distribution. A number contain powerful poisons and other physiologically active compounds.

Usually the leaves are large and broadly ovate, sometimes peltate or rounded, with long curved petioles. They are arranged alternately and are without stipules. The greenish white flowers are small and unisexual, with males and females borne in axillary inflorescences on separate plants. Although the number of parts varies, particularly the number of stamens and carpels, the flowers frequently have petals and sepals each in 2 rows of three. The male flowers have 6 stamens which are free or united, and the females have 3 free superior carpels.

Twelve genera occur in Australia, mainly in Queensland. Most of these are widespread in tropical Africa, Asia and New Guinea, but 1 or 2 are endemic. Three genera, represented by 1 species each, occur in the Sydney area. Two extend to the south coast and 1 into eastern Victoria.

115. Stephania japonica var. *discolor*

Stephania

Stephania japonica var. *discolor*

Stephania, Snake Vine **Plate 115**

Stephania is a slender climber with peltate leaves (that is, the stalk joins the leaf underneath, in from the edge and near the centre). Leaves are up to 10 cm wide and 12 cm long, broadly ovate to triangular. The undersurface is pale greyish green and has clear veins radiating out from the stalk. Flowers are in small compound umbels rising from the leaf axils. The clusters of small globular berries are attractive, turning from green to orange as they mature.

HABITAT In and near rainforests. Especially common in sheltered areas near the sea.

DISTRIBUTION Widespread along coast and in mountains, extending to the western slopes. Also occurring in Asia.

FLOWERING December–January.

116. Sarcopetalum harveyanum

Sarcopetalum

Sarcopetalum harveyanum

Pearl Vine **Plate 116**

The leaves of this woody climber are heart shaped, not peltate like the preceeding species. They are broadly ovate, with a long tip and cordate (heart-shaped) at the base. The stalk joins the leaf in the V at the base and 7 clear veins radiate from the stalk. The small pink-yellow flowers are borne in racemes, 3 to 7 cm long, not in umbels as are those of *Stephania*. Fruit is a globular berry, somewhat flattened and pink in colour.

HABITAT In and near rainforests.

DISTRIBUTION Widespread in coastal zone both north and south of Sydney. Extending to Vic and Qld.

FLOWERING October–December.

Legnephora

Legnephora moorei

Roundleaf Vine **Plate 117**

This vine is a much taller climber than *Stephania* or *Sarcopetalum*. Its leaves are seldom borne near the ground and are usually high in the canopy of tall rainforest trees. As the common names suggests, the leaves are rounded, nearly orbicular, but sometimes slightly heart shaped at the base and slightly pointed at the tips. They are a distinctive ashy grey on the undersurface. Fruit is a globular berry, somewhat flattened and red in colour.

HABITAT In and near rainforests, usually in the upper canopy.

DISTRIBUTION Widespread in coastal zone. Common in the Illawarra district, Katandra Reserve near Gosford and Gap Creek (Olney State Forest). Extending from the south coast northwards to Qld.

FLOWERING December–January.

117. Legnephora moorei

Ulmaceae

Four families, Ulmaceae (*Ulmus:* elm), Cannabaceae (*Cannabis:* hemp), Moraceae (*Morus:* mulberry) and Urticaceae (*Urtica:* nettle) constitute the order Urticales, one of several orders with simple, small, wind-pollinated unisexual flowers thought to be derived from the witch-hazel order (Hamamelidales) or from an ancestor of that group.

The Urticales has an extensive fossil record extending back to Upper Cretaceous strata, with many modern genera represented in Eocene deposits in a number of places in the northern hemisphere. This simplicity and long period of stability, however, should not be equated with primitiveness. On the contrary, there is evidence that the simple wind-pollinated flowers are derived from a more complex bisexual condition, and the Urticales is considered an evolutionarily advanced group.

Within the Urticales the family Ulmaceae has the most primitive characters. It is usually divided into 2 fairly distinct tribes or subfamilies, 1 consisting of trees such as the elm with dry winged fruits, the other also of trees but with slightly fleshy fruits with a large stone. The former group has a northern temperate distribution, whereas the latter is chiefly tropical and subtropical.

Two genera from the second tribe are represented in the Sydney area, by 1 species each. One, *Celtis paniculata*, is at its southern limit and is uncommon in the area, but the other, *Trema aspera*, extends to Victoria and is fairly common in sheltered gullies and on the margins of rainforest. The leaves of both species have 3 main veins diverging from the base, a characteristic of this tribe, but not unique to it.

Trema

Trema aspera

Native Peach **Plate 118**

A tall shrub with alternate, toothed, sandpapery, ovate to lanceolate, peach-like leaves, arranged in opposite rows along the branchlets; each leaf is 4 to 9 cm long. Young branchlets are softly hairy and the bark is tough. Flowers are small in short clusters in the leaf axils. An interesting feature of these flowers is that they are a mix of male, female and bisexual. Fruit is a small berry, green at first, becoming shiny black.

HABITAT Margins of rainforest and sheltered eucalypt forest.

DISTRIBUTION Widespread from coast to mountains, but most common near coastal rainforest. Recordings include Watagan Mountains, Girrakool, Bola Creek, Thirlmere Lakes and Minnamurra Falls. Occurring from far south coast northwards along the coast and tablelands to Qld and NT, and westwards to the western slopes; also Mallacoota Inlet (Vic).

FLOWERING December–February.

118. Trema aspera

Celtis

Celtis paniculata

Native Celtis **Plate 119**

Native Celtis is a small to medium-sized tree to 4 m tall. Leaves are alternate, 5 to 8 cm long, ovate, soft, hairless, with numerous fine translucent dots (glands) and 3 prominent veins at the base. Flowers are small, polygamous and either clustered or solitary in the leaf axils. Fruit is an oval purplish-black berry, about 10 mm across.

HABITAT Rainforest and deep sheltered gullies.

DISTRIBUTION Coastal. Uncommon in the Sydney district, where it reaches its southern limit. Recorded from Minnamurra Spit. Extending along the north coast to Qld; also Malesia and Pacific Islands.

FLOWERING January–February.

Moraceae

The Moraceae is another family having inconspicuous unisexual flowers but is distinguished from related families by the presence of milky sap secreted from special latex ducts. The fruits of Moraceae are also distinctive and, although very different in each genus, have a general structure known as an infructescence which is developed, not from a single flower with multiple carpels, but from an inflorescence of tiny flowers, each with 1 carpel. In genera such as *Ficus* (fig) and *Artocarpus* (breadfruit) the fleshy portion in which the seeds are embedded consists of the swollen fused receptacles of the many flowers forming the infructescence. Similarly, in the genus *Morus* (mulberry) after which the family is named, the tiny purple drupelets making up the fruit develop from separate flowers in the inflorescence (a catkin), and in this respect the mulberry differs from the superficially similar raspberry and blackberry (family Rubiaceae) where the drupelets are formed from the free carpels of a single flower.

The family consists of trees and shrubs distributed mainly in tropical and subtropical areas of both hemispheres, often extending to temperate regions. The leaves are simple, sometimes lobed, with entire or finely toothed margins and are subtended by stipules. In *Ficus* the stipules are green and large and protect the new leaves at the ends of the branches in a conical envelope. After a time they fall off leaving a scar around the stem opposite each leaf stalk.

Many of the figs are called strangler figs because of their habit of germinating in the forks or in wounds of other trees then sending down aerial roots which eventually form a network enclosing and strangling the host tree. The Indian banyan in the same genus sends roots down from its branches, and as the tree grows and spreads the roots become stout props supporting the branches away from the main trunk.

The flowers of the fig are borne on the inner wall of the structure we normally regard as the fruit. Although details differ between species, and male and female flowers may be in the same or in different receptacles, a common arrangement has male flowers grouped near the top entrance to the structure. The females are of 2 types, long-styled fertile flowers which form 1 seed each, and short-styled sterile gall flowers in which the pollinating wasp or fly lays its eggs. When the eggs hatch the young insects feed on the sterile flowers, and on reaching maturity, mate. Often the male insects then die, and the females, in forcing their way out the entrance, pick up pollen from the male flowers and carry it to other receptacles, thus ensuring cross-fertilisation. In some cultivated varieties seedless figs are produced without aid of insects and without the normal processes of pollination. Frequently in *Ficus*, and in others in the family, the inflorescences, and later the fruits, grow from the trunk.

The family consists of about 60 genera and 1400 species. *Ficus* with about 800 or 900 species is one of the larger genera in the plant kingdom. Nearly 40 species of fig occur in Australia, 5 in the Sydney coastal region. Figs are relatively common in the area, and many which were left by the early settlers are now large and dominant around farm buildings. The other 3 genera *Maclura, Malaisia* and *Streblus* are uncommon in coastal rainforests and are represented in Australia by 1 species each.

Ficus Fig

Shrubs to large trees with minute flowers enclosed within a hollow ovoid to globular inflorescence popularly known as a 'fig'.

Ficus coronata

Sandpaper Fig **Plate 120**

A bushy small tree with elliptic leaves, 5 to 14 cm long. The upper leaf surface is characteristically rough and like sandpaper to touch. The fruit is oval, about 15 mm long, and densely covered with rough hairs. The specific name 'coronata' refers to the crown of bristles on the top of the fruit.

119. Celtis paniculata

120. Ficus coronata

HABITAT In or near rainforest, mainly along watercourses.

DISTRIBUTION Widespread and common from coast to valleys of Blue Mountains. Extending along the coast from Gippsland (Vic) to North Qld and NT; also southern tablelands and central western slopes.

FRUIT Ripe from January–June.

Ficus obliqua

Small-leaved Fig **Plate 121**

A rainforest giant, a 'strangler' fig which in more luxuriant situations may reach a height of 40 m, have large spreading buttresses and produce aerial roots. Its leaves are evergreen, smooth on both surfaces, elliptic, 4 to 7 cm long, with short leaf stalks (6 to 15 mm). The fruit is in pairs, globular, small (6 to 8 mm across), yellow turning orange, with short thick stalks (or almost stalkless) in the leaf-axils. The small fruit, small and hairless leaves and short leaf-stalk help separate this species from other local fig-trees.

HABITAT Rainforest.

DISTRIBUTION Illawarra region. Recordings include Minnamurra Falls, Jamberoo, Bulli and Bola Creek. Extending along the coast from near Central Tilba (south coast) northwards to northern Qld. Also occurring in New Guinea and South-east Asia.

FRUIT Ripe from April–June.

121. Ficus obliqua

Ficus superba var. *henneana*

Deciduous Fig **Plate 122**

A large strangler fig, 30 m or more tall in better rainforest sites. The trunk is usually buttressed. Unlike other local figs, the leaves are deciduous in late winter. Leaves are ovate-elliptic, green and smooth on both surfaces, 8 to 12 cm long, with leaf stalks 2 to 5 cm long. The fruits are conspicuous and unmistakable when ripe. They are solitary, large and globular (15 to 25 mm across), yellow turning red-purple, spotted, on stalks 5 to 10 mm long.

HABITAT Rainforest.

DISTRIBUTION Illawarra region, which is the southern limit of this species. Recordings include Minnamurra Falls, Jamberoo, Calderwood and Berkeley Hill. Extending along the north coast to Qld and NT.

FRUIT Ripe from December–June.

122. Ficus superba var. henneana

123. Ficus rubiginosa

124. Ficus macrophylla

125. Maclura cochinchinensis

Ficus rubiginosa

Port Jackson Fig, Rusty Fig **Plate 123**

This common spreading shrub or tree is often seen in the early stage of its life as an epiphyte clinging to clefts in rocks or in the forks of trees. Although it may grow to a large tree over 30 m tall, in the Sydney area it is seldom more than 10 m tall. Leaves are ovate-elliptic, rusty coloured and downy below, 6 to 10 cm long, with leaf stalks 1 to 4 cm long. Mature fruits are mostly in pairs, 12 to 20 mm across, yellow turning red, marked with warts, and with a nipple at the tip. The fruits are on short stalks, 2 to 5 mm long. The smaller leaves and shorter fruit stalks separate this species from *F. macrophylla.*

HABITAT Gullies and rocky slopes, or in drier margins of rainforest.

DISTRIBUTION Widespread and common from coast to mountains. Extending northwards along coast from near Bega to Qld. Also occurring on the northern tablelands and western slopes.

FRUIT Ripe from February–July.

Ficus macrophylla

Moreton Bay Fig **Plate 124**

Despite its common name, Moreton Bay Fig is common as far south as the Shoalhaven River. At its best, it is a beautiful large tree, up to 50 m tall, with a spreading crown, aerial roots and massive buttressed trunk. It often towers high above the surrounding rainforest. Leaves are ovate-elliptic, large, 10 to 25 cm long, hairless on both surfaces, but usually red-brown on the undersurface when mature. Leaf stalks are 5 to 10 cm long. Mature fruits are borne singly on thick stalks up to 15 mm long; fruits are globular, about 25 mm across, and orange turning purple, dotted with white.

HABITAT Rainforest.

DISTRIBUTION Illawarra escarpment and coastal plains. Recordings include Kiama, Berkeley Hill and Fig Tree. Widely planted in Sydney parks. Extending from Shoalhaven area northwards along the coast to North Qld.

FRUIT Some fruit ripe all year round.

Maclura

Straggling shrub with rigid thorns and flowers in globular heads.

Maclura cochinchinensis

Cockspur Thorn **Plate 125**

An untidy climber or straggling shrub, sometimes forming thickets, its most conspicuous feature is the presence of stout thorns along the stems, some about 4 cm long. Leaves are alternate, oblong-elliptic, up to 8 cm long. Flowers are in rounded heads, 3 to 6 mm across, which develop into larger yellow globular compound fruits, 12 to 16 mm across.

HABITAT In or near rainforest.

DISTRIBUTION Coastal, especially Illawarra district. Recordings include Minnamurra Spit, Macquarie Pass, Berkeley Hill and Mt Marshall. Extending from the south coast to Qld; also South-east Asia.

FRUIT November–January.

126. Malaisia scandens

Malaisia

A climber with alternate leaves; male and female flowers unisexual and borne on different plants.

Malaisia scandens
Burny Vine, Crow Ash **Plate 126**

An inconspicuous climbing shrub, with rough, leathery, alternate, oblong-elliptic leaves, 4 to 9 cm long. These leaves are irregularly toothed when young, but entire or only with a few teeth near the apex when older. The bark on twigs is very tough. Male flowers are in dense, cylindrical spikes, about 10 mm long, the female flowers in smaller globular heads. The dry fruit is enclosed by the thin red fleshy perianth. A milky sap is sometimes present when the stems or leaves are broken.

HABITAT In or near rainforest.

DISTRIBUTION Coastal, especially Illawarra district. Recordings include Macquarie Pass, Jamberoo, Mt Keira and Bola Creek. Extending from south coast to north coast and Qld; and NT, South-east Asia and Pacific Islands.

FLOWERING Irregular.

Streblus

Small tree with toothed alternate leaves; male and female flowers unisexual and borne on the same plant.

Streblus brunonianus
Whalebone Tree **Plate 127**

A small tree, seldom more than 6 m tall in this area, with very tough bark and branchlets which exude a milky sap when broken. The toughness of the timber made it suitable for use as axe handles and it is sometimes commonly known as Handlewood or Axe-handle Wood. Leaves are alternate, usually lance-shaped and toothed, drawn out to a tip, glossy but often sandpapery to touch. Leaves may vary from 10 to 60 mm long (usually about 25 mm), the juveniles being long and narrow. Male flowers are borne in dense cylindrical spikes, about 30 mm long, from the leaf axils, the female flowers in much shorter spikes. The fruit is a yellow, egg-shaped berry, about 5 mm long, topped by the remains of the style.

HABITAT Rainforest, especially drier margins and along streams.

DISTRIBUTION Coastal areas of Illawarra and Mt Wilson in Blue Mountains. Extending from near Milton to northern Qld.

FLOWERING October–February.

Urticaceae

Although many people may have reason to remember that this is the stinging nettle family, that feature is present in only 1 of the 6 tribes in the family. Urticaceae is closely related to Moraceae, but has only half as many genera and species. It is sparsely represented in Australia by 9 or 10 genera and 16 species, of which a number are herbaceous and uncommon.

Stinging Nettles (*Urtica* spp.) are found in fertile soils, often in disturbed or previously cultivated sites. The one native species, *Urtica incisa*, is also found in rainforests. The Stinging Tree, *Dendrocnide excelsa*, 1 of 4 named species of this genus in eastern Australia, is possibly better known by its former name, *Laportea*. It occurs in coastal rainforests and is readily recognised by its large, soft, rounded leaves on long stalks and its smooth, pale, spongy trunk. Stinging hairs cover the younger branches, leaf stalks and the undersides of the leaves, especially on young trees. When brushed against, the stiff silicified tips of the hollow hairs are easily broken, allowing the injection of an irritant beneath the skin. Folklore has it that juice from the cunjevoi *(Alocasia macrorrhizos)* is an effective antidote against formic and acetic acids in the sting. However these are not the main stinging agents; the effects are too severe and prolonged to be due to these simple, easily denatured compounds. Tests on the Queensland species, *D. moroides*, suggest that the active agents are haemolytic saponins which produce histamine-like reactions.

Dendrocnide Stinging Tree

Dendrocnide excelsa
Giant Stinging Tree **Plate 128**

Giant Stinging Tree is conspicuous as its large, heart-shaped, light-green leaves are in marked contrast to the darker foliage of the surrounding rainforest. It can grow over 30 m tall, but is usually much smaller in this area. Leaves are 10 to 30 cm long (sometimes longer), smooth above, downy underneath. The lower surface also has conspicuous veins. Both branchlets and leaves are covered with rigid hairs which, on contact with the skin, inflict a painful and long-lasting sting. Male and female flowers are on different trees; they are yellowish white in short clusters in the leaf-axils.

127. Streblus brunonianus

128. Dendrocnide excelsa

The Small-leaved Stinging Tree *(D. photinophylla)*, a tree of the north coast and Queensland, reaches its southern limit north of Sydney, but is rare in the area. It has ovate (not heart-shaped), shining, almost hairless leaves, 10 to 15 cm long.

HABITAT Rainforest, on fertile soils.

DISTRIBUTION Escarpment and gullies of Royal National Park, Illawarra and lower Blue Mountains (e.g., Palm Jungle, Mt Keira, Jamberoo). Extending from near Bega to Qld.

FLOWERING December–April.

129. Elatostema reticulatum

Elatostema

Elatostema reticulatum **Plate 129**

This is a straggling plant with very fleshy stems up to 50 cm long. Leaves are dark green, broadly lance-shaped, toothed and up to 15 cm long. They are conspicuously veined. Male and female flowers are borne on different plants. Male flowers are white, in dense heads 10 to 20 mm across, on stalks up to 7 cm long. Female flower heads are smaller and almost stalkless.

HABITAT Rainforest, on wet shaded rocks or along streams.

DISTRIBUTION Coastal gullies (e.g., Minnamurra Falls, Macquarie Pass National Park). Extending from the south coast to the north coast.

FLOWERING November–January.

130. Urtica incisa

Urtica Stinging Nettle

One native species and one introduced species in the area. The introduced species, *U. urens*, is an annual weed in disturbed soils. Its leaves are ovate, 2 to 5 cm long and deeply serrated; its flowers are small, green, with males and females mixed in a short cluster in the leaf axils, not in slender spikes as the flowers of *U. incisa*.

Urtica incisa

Scrub Nettle **Plate 130**

A well-known plant very common outside the Sydney area as a weed of waste places, roadsides and around habitation. It may grow to 60 cm tall. Stems and leaves are covered with rigid stinging hairs. Leaves are opposite, lanceolate, 5 to 10 cm long, coarsely toothed and often with cordate bases. Flowers are small, green and borne in slender spikes from the leaf bases.

HABITAT Disturbed areas in and around rainforest and wetter forest; also waste places. Generally absent from dry sandstone areas.

DISTRIBUTION Widespread from coast to mountains. Occurring throughout NSW, except Far Western Plains; also Qld, Vic, Tas, SA, New Zealand and some Pacific Islands.

FLOWERING Mainly July–January, but irregular.

Casuarinaceae

Casuarinaceae is an isolated family of 4 genera and 66 species of wind-pollinated trees and shrubs occurring in Malesia, New Caledonia, Fiji and the eastern half of Australia. The distinctive appearance of this family is due to the jointed cylindrical branchlets with nodes about a centimetre apart, each node carrying a whorl of reduced leaves appearing as a ring of triangular teeth around the branchlet. Male flowers are carried in simple spikes at the ends of the upper branchlets and sometimes during flowering appear to colour the whole tree brown. Each tiny male flower consists of a single stamen surrounded by 2 perianth lobes which are subtended by 2 scale-like bracteoles. Female flowers grow lower down on the same tree, or on separate trees, in short crowded heads terminating very short lateral branches. Each has 2 long filamentous red stigma or style branches which catch the wind-borne pollen. After fertilisation the ovary develops into a single-seeded nut and the inflorescence becomes a small woody cone due to enlargement of the bracteoles. The 2 bracteoles enclosing each nut form 2 valves which open after maturity to release the fruit. Although the cones of many species are characteristic in shape and size they are frequently malformed owing to abortion or incomplete fertilisation of the head. Generally, the branchlets and the number of teeth in each whorl provide more reliable means of identification than the cones.

131. Casuarina glauca

Reliance on wind-pollination and the peculiar structure of casuarinas suggested to early workers a link with the Gymnosperms, notably the pines or *Ephedra* in the Gnetales, and Engler for example regarded them as the most primitive of the Angiosperms. This is not the current view which regards the casuarinas as having a common ancestry with the Hamamelidaceae (witch-hazels), and the reduction in parts a result of the method of pollination and an extreme adaptation to dry conditions. Certainly a number of species occur in the inland and one, *Casuarina cristata*, Belah, is a feature of the driest parts of the interior, but others are typically littoral *(C. equisetifolia)*, or occur along salty tidal *(C. glauca)* or freshwater *(C. cunninghamiana)* rivers, or even in moist forests *(Allocasuarina torulosa)*.

Although *C. cunninghamiana* is valued for its stabilisation of river banks, and other species make very durable fence posts or are prized for their dense, hard, beautifully figured reddish timbers, casuarinas are frequently poorly regarded by farmers who maintain that they poison the ground. This seems to be the case with some common species around Sydney where little or no plant growth occurs in the mat of fallen branchlets beneath the trees. This may be due to the release of antibiotics in the root zone by actinomycete bacteria associated with nitrogen fixing nodules on the roots, or possibly to inhibitory compounds in the branchlets.

Casuarina

Trees with male and female flowers on separate plants or on the same plant. Differs from *Allocasuarina* in having cones without protruberances and seed-like winged fruits which are pale and dull (not dark and shining).

Casuarina glauca

Swamp Oak **Plate 131**

A tree to 20 m tall, found typically near salt water and along larger streams. It may occasionally be found as a stunted shrub on exposed headlands or in drier areas. Male and female flowers are usually on separate trees. Leaf-teeth are in whorls of 12 to 16 around the stem of each joint. Male spikes are dense, 1 to 4 cm long. Cones are globular to shortly cylindrical, 10 to 15 mm long, 12 mm diameter.

132. Casuarina cunninghamiana

HABITAT Alluvial flats along tidal streams, estuaries, coastal lagoons and along larger streams of Cumberland Plain, occasionally on drier sites.

DISTRIBUTION Coastal zone. Recordings include Mooney Mooney Creek, Towra Point, Wattamolla, Lake Illawarra and Castlereagh State Forest. Extending from the south coast northwards to Qld.

FLOWERING May–August.

Casuarina cunninghamiana

River Oak **Plate 132**

A large stately tree, 20 to 25 m tall, found in narrow belts along rivers and large streams where its roots have access to a regular water supply. Male and female flowers are borne on separate trees. Branchlets are slender and rather erect. Leaf-teeth are in whorls of 8 to 10 around the stem of each joint. Male spikes are short and dense, about 15 mm long. Cones are globular, small, about 10 mm long, 8 mm diameter.

HABITAT Banks of rivers and streams, usually below flood level, but sometimes on silty loams on flats adjacent to rivers.

DISTRIBUTION Coast to mountains. Recordings include Macquarie Rivulet and other Illawarra streams, Nepean River, Grose River and Wombeyan Caves. Found scattered throughout NSW, except far western plains; also Qld and New Guinea.

FLOWERING Some flowers all year round.

Allocasuarina

Shrubs or trees with male and female flowers on separate plants or on same plant. Formerly included with *Casuarina* but reclassified on the basis of small differences in fruits. Genus includes about 46 named species (6 in the Sydney district) and a number of recently discovered unnamed species.

Allocasuarina torulosa

Forest Oak **Plate 133**

Forest Oak is a graceful tree, 12 to 20 m tall. It has hard, deeply furrowed bark, with a corky appearance. The main branches are spreading to erect; smaller branchlets are slender and somewhat drooping. Male and female flowers are borne on separate trees. Leaf-teeth are in whorls of 4 (occasionally 5) around the stem at each joint. Male spikes are 2 to 4 cm long at the ends of the stamens.

133. Allocasuarina torulosa

135. Allocasuarina verticillata

Cones are nearly globular, 18 to 25 mm across, 20 to 30 mm long, with rough protruberances.

HABITAT Forest, usually on sheltered slopes or valley floors in comparatively fertile soils.

DISTRIBUTION Widespread from coast to mountains. Extending along the coast from Nowra northwards to Cairns, Qld; also northern tablelands and central western slopes.

FLOWERING September–December.

Allocasuarina littoralis

Black She-oak **Plate 134**

A common erect tree to about 12 m tall, with hard rugged bark and slender branchlets. It occurs on poorer soils than *A. torulosa*, and may be a stunted shrub on stony heaths. Male and female flowers are frequently on the same tree. Leaf-teeth are in whorls of 6 to 8 around the stem of each joint. Male spikes are slender, about 20 mm long. Cones are cylindrical, the ends truncate (or with a small point), 12 to 30 mm long, with prominent angular valves.

HABITAT Open forest and sandy heath, often on stony ridges.

DISTRIBUTION Widespread from coast to mountains. Found throughout the coastal and tablelands divisions of NSW. Also occurring on the western slopes, north-western plains, Qld, Vic and Tas.

FLOWERING May–June.

134. Allocasuarina littoralis

Allocasuarina verticillata

Drooping She-oak **Plate 135**

A spreading tree with coarse, ribbed drooping branchlets and dark-brown fissured bark. In protected sites it may grow to 10 m tall, but is frequently found on exposed coastal cliffs where it is seldom more than 2 to 3 m tall. Male and female flowers are borne on separate bushes. Leaf-teeth are in whorls of about 11 around the stem of each joint. Male spikes are a characteristic feature, as they are 5 to 10 cm, longer than any other *Allocasuarina* in the area. *A. verticillata* may be confused with *Casuarina glauca*, especially when they grow in similar habitats. *A. verticillata* has larger cones, coarser branchlets and less teeth in a whorl.

HABITAT Along coastline, on exposed cliffs and more sheltered sites, on shale soils; also recorded from stony ridges and rocky outcrops further inland.

DISTRIBUTION Coast from Newport to Mona Vale and Garie to Coledale. Also occurring on the south coast, southern tablelands, western slopes, south-western plains, Vic, Tas and SA.

FLOWERING July–August.

Allocasuarina nana **Plate 136**

A shrubby plant about 1 m tall, with characteristic stiff, erect branchlets. Male and female flowers are usually borne on separate bushes. Leaf-teeth are in whorls of 5 (occasionally 6) around the stem of each joint. Male spikes are deep tan, 5 to 15 mm long, erect and conspicuous on the ends of the short upright stems. Cones are oblong, up to 20 mm long without stalks.

HABITAT Heath, on sandy soils (rarely in open forest).

DISTRIBUTION South of Sydney (e.g., Royal National Park, Barren Grounds), upper Blue Mountains (e.g., Narrow Neck, Newnes State Forest, Clarence) and Southern Highlands (e.g., Wanganderry Plateau). Extending along the south coast and southern tablelands to Vic.

FLOWERING September–November.

Allocasuarina distyla

Scrub She-oak **Plate 137**

A spreading shrub to about 3 m tall. A conspicuous feature of this species is its abundant male flower spikes which give the plant a tan appearance and make it stand out from the surrounding bush. Male and female flowers are usually borne on separate bushes. Leaf-teeth are in whorls of 6 to 8 around the stem of each joint. Small, but noticeable, ridges run along the stems. Male spikes are up to 7 cm long. Cones are cylindrical, with pointed ends, 20 to 30 mm long, with rounded valves which only slightly protrude.

HABITAT Sandy heath and shrubland, especially in deeper sands.

DISTRIBUTION Common in coastal zone (e.g., Brisbane Water, Royal and Macquarie Pass National Parks), but extending to the mountains. Also occurring on the north coast, south coast and southern tablelands.

FLOWERING July–September.

136. Allocasuarina nana

137. Allocasuarina distyla

Allocasuarina paludosa **Plate 138**

A spreading shrub to 2 m tall, uncommon in the Sydney district where it reaches the northern limit of its distribution. Male and female flowers are usually on the same bush. Leaf-teeth are in whorls of 6 to 8 around the stem of each joint. Branchlets lack the longitudinal ridges of *A. distyla* and are noticeably hairy. Male spikes are short (up to 20 mm) and are borne at the ends of upright stems. Cones are smooth, oblong, up to 20 mm long.

HABITAT Wet heath.

DISTRIBUTION Coastal zone south of Sydney (e.g., Curra Moors, Maddens Plains). Extending southwards along the south coast and southern tablelands to Vic; also Tas and SA.

FLOWERING January–February.

Allocasuarina sp. aff. *diminuta* **Plate 139**

At least two forms of *Allocasuarina* related to *A. diminuta* occur in the Sydney district. Both are shrubs to 1.5 m tall. One form occurs along Heathcote Road at Lucas Heights. It is monoecious (unisexual male and female flowers on the same plant), with branches erect, only slightly spreading. Branchlets are 7 to 10 cm long, 0.5 to 1 mm wide and not prominently ridged. There are 6 to 7 teeth in each whorl, with 5 to 10 mm between each whorl of teeth. Male flowers are in spikes up to 4 mm long. The cones are slightly stalked, almost globular, 10 to 12 mm across, or cylindrical, 11 mm across and 17 mm long. Flowers have been recorded between May and July. The nut is 4 to 5 mm long, including the wing.

Another form occurs at West Berrima (Medway). It is dioecious (unisexual male and female flowers on different plants), with branches erect, only slightly spreading. Branchlets are up to 15 cm long, about 1 mm wide and not prominently ridged. There are 7 teeth in each whorl, with 8 to 10 mm between each whorl of teeth. Male flowers are in spikes about 6 cm long. The cones are slightly stalked, almost globular, 8 to 12 mm across. Flowers have been recorded in September and October. The nut is 3 to 4 mm long, including the wing.

HABITAT Woodland and open forest, on sandy soils.

DISTRIBUTION Scattered (as above).

FLOWERING As above.

138. Allocasuarina paludosa

139. Allocasuarina sp. aff. *diminuta* (Berrima)

Nyctaginaceae

This pantropical family is represented in Australia by 2 genera, *Pisonia*, with 2 tree species and 1 climber, and *Boerhavia*, with 2 prostrate or scrambling species. The 2 tree species are soft wooded with thick, large, dark-green leaves and small pale flowers in large leafless panicles. One, *P. umbellifera*, occurs along the coast from Jervis Bay to North Queensland and in New Zealand, Malesia and Pacific Islands. The other, *P. grandis*, is a species of northern Australia, the Malaysian region and coral islands of the Indian and Pacific oceans. Both are known as the Bird-lime Tree because of their long glandular sticky fruits. Their reputed ability to ensnare even small birds or reptiles by this means seems exaggerated; the trees are in fact favourite resting places for a variety of seabirds in northern Australia.

Pisonia

Pisonia umbellifera

Bird-lime Tree **Plate 140**

This is a striking small tree, reaching a height of about 5 m in the Sydney district. The leaves are large, entire, 20 cm or more long, up to 10 cm wide and rounded at the tip. Flowers are funnel-shaped, without petals, pale green and borne in terminal leafless clusters. Fruit is a distinctive shape, narrow cylindrical with 5 ribs, up to 4 cm long and tapering to a narrow neck; it is covered with a sticky gum.

HABITAT Littoral rainforest.

DISTRIBUTION Isolated occurrences at Palm Jungle and Bola Creek in Royal National Park, below the cliff-line of Illawarra Escarpment from Austinmer to Clifton and at Minnamurra Falls. Common in north Qld, including islands of the Barrier Reef, but reaching its southern limit around Jervis Bay. Extending to Malesia and Western Pacific.

FLOWERING September–January.

Aizoaceae

This is a large family containing the common succulent garden plants with showy daisy-like flowers formerly grouped under the name *Mesembryanthemum*, but now split into more than 100 genera. Many of the 2000 species of this complex are native to hot arid areas of South Africa and have developed a variety of ingenious strategies for protection against dessication. The subfamily in which the daisy-like appearance of the flowers is due to numerous petaloid stamens is represented in the Sydney area by one native species each of *Carpobrotus* and *Disphyma* and some introduced species. The common genus *Tetragonia* with small yellow flowers lacking staminodes is sometimes regarded as another subfamily, and sometimes as a separate family, the Tetragoniaceae.

140. Pisonia umbellifera

141. Carpobrotus glaucescens

143. Calandrinia calyptrata

Carpobrotus Pigface

Carpobrotus glaucescens
Pigface **Plate 141**

Pigface is a prostrate spreading plant with fleshy leaves. The flowers are large, up to 6 cm across, with numerous purple strap-like petals (actually petal-like staminodes), white towards the base, and 300 to 400 yellow stamens. Even when not in flower, the plant is easily identified by its thick, fleshy, three-angled leaves, 4 to 10 cm long. The berry-like, longer than wide, red to purple fleshy fruit was eaten by Aborigines; the generic name refers to the fact that the fruits are edible. The leaves were also roasted and eaten and may have been a salt substitute.

HABITAT Coastal sand dunes and rocks near the sea.

DISTRIBUTION Common along coast extending to Qld and Vic. Also Lord Howe Island.

FLOWERING Most of the year.

Disphyma

Disphyma crassifolium
Round-leaved Pigface

This species may at first be mistaken for a small *Carpobrotus*. It is prostrate, rooting at the nodes, with ascending shoots to 20 cm tall and succulent leaves, 2 to 5 cm long. Leaves are not as strongly triangular in cross section as are those of *Carpobrotus*, as the keel is rounded. Flowers are yellow-white, solitary, 2 to 3 cm across, with numerous strap-shaped petal-like staminodes. Unlike *Carpobrotus*, this genus has a fruit which is a capsule on an upright stem; this capsule has 5 valves which burst open when mature to release the seeds.

HABITAT Sand dunes on coastal headlands (also on samphire flats and saline areas in inland Australia).

DISTRIBUTION Rare in Sydney district. Reported from Cape Banks, Botany Bay. A widely distributed species throughout Australia, also South Africa and New Zealand.

FLOWERING May–June.

Tetragonia

Tetragonia tetragonioides
New Zealand Spinach, Warrigal Cabbage **Plate 142**

A semi-erect coastal herb with broad leaves. Stems are thick, rather succulent when young. Leaves are broadly lance to diamond-shaped, up to 8 cm long, paler below and semi-succulent. Both leaves and stems are covered with small glistening blisters. Flowers are small, yellow, stalkless and borne in the leaf axils. Its leaves were cooked as a cabbage or spinach-like vegetable by early Australian explorers, hence its common names.

HABITAT Margins of salt marshes and protected sites near the coast.

DISTRIBUTION Scattered along coast. Also found inland on the western slopes and plains, in Qld, Vic, Tas, Lord Howe Island and New Zealand.

FLOWERING August–December.

142. Tetragonia tetragonioides

Portulacaceae

The portulaca family consists of fleshy or succulent herbs and undershrubs inhabiting the drier parts of the world but lacking the extreme adaptations to aridity present in related families Aizoaceae and Cactaceae. These three families, with 7 or 8 others, constitute the order Caryophyllales which has a number of characteristic structural and biochemical features, though these are not all shared by all members.

The best known of the Australian genera are *Calandrinia* (40 endemic species) with pink or magenta flowers and *Portulaca* (13 endemic species) with usually yellow flowers. Both occur extensively throughout the interior where attractive shrubby species of *Calandrinia*, known generally as parakeelya, provide both food and water for grazing stock. The seeds, leaves and tuberous roots of some *Portulaca* species were Aboriginal food sources.

Three native species of *Calandrinia* and one of *Montia* occur in the Sydney district.

Calandrinia

Calandrinia calyptrata
Pink Purslane **Plate 143**

This small herb has succulent cylindrical basal leaves, 1 to 5 cm long, which are often a pink-red colour. Flowers are pink, with 5 petals; they are borne on stalks about 10 mm long near the ends of stems which rise to about 10 cm from the basal leaves. The other *Calandrinia* species in the Sydney district are similar: *C. eremaea* has finely wrinkled seeds (*C. calyptrata* has smooth shiny seeds); *C. pickeringii* has flowers with usually 4 petals.

HABITAT Dry rocky sites.

DISTRIBUTION Cumberland Plain, Blue Mountains and southern highlands. Also found in Vic, Tas, SA and WA.

FLOWERING September–October.

144. Montia australasica

145. Stellaria pungens

Montia

Montia australasica

Formerly *Neopaxia australasica*
White Purslane **Plate 144**

White Purslane is a low creeping herb with stems bearing tufts of strap-shaped basal leaves, 3 to 12 cm long. Flowers are white, 5-petalled, star-like, on slender stalks. There are 5 stamens, opposite the petals, with orange anthers, and the style is divided into 3 short branches.

HABITAT Wet places, sometimes actually growing in water.

DISTRIBUTION Uncommon in the Sydney district, except at higher altitudes (e.g. Mt Werong). A tablelands species, extending through the alpine zone to Vic and Tas.

FLOWERING September–December.

Caryophyllaceae

This large family of herbs is well represented in temperate regions of the northern hemisphere, but only sparsely represented in the southern. Two-thirds of the genera occurring in Australia, and 26 out of 33 species in the Sydney area, are introduced weeds, many of them with an almost cosmopolitan distribution. Members of the family can often be recognised by a thickening of the stems at the leaf nodes and by the opposite, simple, entire leaves which are sometimes joined at the leaf bases. The flowers are regular (radially symmetrical) and consist usually of a calyx of 5 free sepals, a corolla of 5 free petals, with 10, or sometimes 5, stamens surrounding a superior 1-locular ovary composed of several fused carpels with several styles. Frequently the petals are notched or toothed at the end or (as in *Stellaria)* deeply cut into 2 linear lobes.

Stellaria Starwort

Stellaria pungens

Prickly Starwort **Plate 145**

The generic and common names indicate the main features of this sprawling herb. The white flowers are star-like ('stella'—a star). What appear to be 10 strap-like petals are in fact only 5 petals, each divided almost to the base. Stamens and styles are conspicuous. Leaves are pungent (prickly), narrow and rigid, 6 to 8 mm long. The stems are weak and angular, often draped over other plants or sprawled over rocks.

HABITAT Understorey in tall forest, especially rocky areas.

DISTRIBUTION Higher parts of Blue Mountains. Extending through the tablelands to the western slopes and along the south coast to Vic. Also Qld, Tas and SA.

FLOWERING October–December.

Stellaria flaccida

Forest Starwort **Plate 146**

A common species in sheltered sites, it has opposite, ovate and pointed leaves which are soft, not prickly. The stems and leaves have a weak, drooping appearance, hence the specific name ('flaccidum'—feeble, drooping). Flowers are solitary on long stalks. Petals are white, 6 to 8 mm long, with 5 petals divided to the base so that the flower appears to have 10 petals.

HABITAT Edge of rainforests, and damp shaded situations in gullies.

DISTRIBUTION Widespread from coast to mountains. Recordings include Jamberoo, Bents Basin (Nepean River), Mt Tomah and St Albans-Bucketty. Extending along coast and tablelands to Qld and Vic. Also Tas and SA.

FLOWERING October–January.

Amaranthaceae

The Amaranthaceae consists mainly of herbs and shrubs with a few climbers. It is a fairly large family with about 900 species scattered throughout the world. Although a few attractive species are used horticulturally many others in the family are weeds.

The Amaranthaceae is closely related to the Chenopodiaceae, but differs from it in having flowers with a dry membranous perianth subtended by well developed and sometimes showy bracteoles. Frequently the stamens are joined at the base or fused into a cup around the ovary. Whereas the Chenopodiaceae are mostly wind pollinated the Amaranthaceae are probably pollinated by insects.

The largest Australian genus is *Ptilotus* with many attractive species throughout the dry interior of the continent. One climbing shrub, *Deeringia amaranthoides*, and a dozen or so native and introduced weed species occur infrequently in the Sydney area.

146. Stellaria flaccida

Deeringia

Deeringia amaranthoides

Deeringia **Plate 147**

Deeringia is a hairless climber, often with a shrubby habit; stems are wirey and 1 to 3 m long. Leaves are alternate, broadly lance shaped, entire, 5 to 12 cm long, 6 to 7 cm wide, on stalks 1 to 2 cm long. Flowers are small in loose spikes 10 to 20 cm long, either terminal or in the upper leaf axils. Fruit is a red globular berry, 4 to 5 mm across.

HABITAT In and near rainforest.

DISTRIBUTION Coast to mountains, but uncommon. Recordings include Burning Palms (Royal National Park) and Razorback Mt. Also occurring on the south coast, north coast, northern tablelands, central western slopes, Qld and Asia.

FLOWERING January–March.

147. Deeringia amaranthoides

148. *Alternanthera denticulata*

149. *Nyssanthes erecta*

150. *Sarcocornia quinqueflora*

Alternanthera

Alternanthera denticulata

Lesser Joyweed **Plate 148**

This small hairless herb is mainly found in moist disturbed soils, such as beside roads. It has prostrate or ascending stems to 30 mm long, sometimes rooting at the nodes. Leaves are variable, but mostly opposite, narrow to lance shaped, up to 30 cm long. Flowers are small and are subtended by papery bracts so that they appear as a conspicuous chaffy cluster in the leaf axils.

HABITAT Disturbed clay soils and in moist cleared sites.

DISTRIBUTION Scattered throughout area from coast to mountains. Occurs in every district of NSW and in every other State.

FLOWERING Mostly October–March, but some flowers at other times.

Nyssanthes

Nyssanthes erecta

Plate 149

N. erecta is an erect herb to about 50 cm high, with spreading upper branches. Leaves are opposite, elliptic to lance shaped, 2 to 7 cm long, the upper leaves smaller than the lower. Flowers have 4 stamens and are in stalkless clusters in the upper forks of stems. Perianth segments are rigid and the bracts spiny, spreading and up to 10 mm long. A similar species, *N. diffusa*, also occurs in the area. Its flowers have only 2 stamens and the upper stems are weak and diffuse.

HABITAT Sheltered forested slopes.

DISTRIBUTION Coast to lower mountains, but uncommon. Recordings include Razorback Mt, Kiama and South Creek (Berkshire Park). Also occurring on the south coast, north coast, central western slopes and Qld.

FLOWERING December–February.

Chenopodiaceae

The Chenopodiaceae is the final family in the distinctive order Caryophyllales which includes Nyctaginaceae, Aizoaceae, Portulacaceae, Caryophyllaceae and Amaranthaceae, all represented in the Sydney region by native and introduced species. Features of this order include absence of stipules, an unusual arrangement of cambium tissue in stems and roots, curved ovules and embryos, an uncommon type of placentation in which the ovules are attached to the central axis of the ovary or to the base, and the possession of characteristic pigments known as betalains.

The Chenopodiaceae consists mainly of scurfy herbs or small shrubs inhabiting saline or alkaline soils in both marshy and arid habitats. Members of this family known as saltbushes dominate vast areas of salt pan and salt lake country in south-western New South Wales, the Lake Eyre Basin and the Nullarbor Plain. In the Sydney area the commonest chenopod is *Sarcocornia*, a small leafless succulent herb with upright jointed stems found growing in or near mangrove swamps. *Sarcocornia* and similar leafless succulent plants are sometimes placed in a separate family, Salicorniaceae.

Sarcocornia

Sarcocornia quinqueflora

Samphire, Glasswort **Plate 150**

A sprawling, succulent, leafless plant with the ability to tolerate saline conditions. It is common on estuarine salt-marshes and in some areas, such as Towra Point (Botany Bay), it may be the dominant vegetation over many hectares. Stems are decumbent or erect, up to 25 cm long, green, fleshy and jointed; some often have a reddish appearance. Flowers are minute and are borne in clusters around the joints at the end of the stems. Only the anthers and styles protrude from the joints.

HABITAT Coastal salt-marshes and river estuaries.

DISTRIBUTION Scattered along the coast. Occurring in every State, except NT; also Lord Howe Island, New Caledonia and NZ.

FLOWERING November–February.

Enchylaena

Enchylaena tomentosa var. *tomentosa*

Ruby Saltbush, Plum Puddings **Plate 151**

Ruby Saltbush is a weak shrub seldom more than 50 cm tall, with stems covered with woolly hairs. Leaves are alternate, almost cylindrical, 6 to 16 mm long, and covered with short hairs. Flowers are small and solitary in the leaf axils. Fruit is succulent, 4 to 6 mm across, flattened-globular, at first green, becoming yellow and red; the berries are edible.

HABITAT Sea coasts in Sydney district, but widespread in drier inland areas.

DISTRIBUTION Scattered along coast and tidal inlets (e.g. Minnamurra Spit). Extending to the north and south coasts, and inland from the western slopes. Widely distributed in all mainland States.

FLOWERING Chiefly September–December, but some flowers at other times.

151. *Enchylaena tomentosa*

152. Suaeda australis

Suaeda

Suaeda australis

Seablite **Plate 152**

Seablite is a bushy, hairless plant up to 80 cm tall; it is common in saline conditions. A rather insignificant plant, it is easily overlooked. Its leaves are soft, slightly succulent, semi-terete, incurved and up to 3 cm long. Flowers are minute, stalkless and in elongated clusters almost hidden among the leaves. Mature fruit is enclosed by the globular persistent perianth.

HABITAT Edge of salt-marshes, often on landward side of mangroves.

DISTRIBUTION Scattered along coast. Occurring along the coast of every State, except NT.

FLOWERING September–December.

153. Rhagodia candolleana

Rhagodia

Rhagodia candolleana ssp. *candolleana*

Coastal Saltbush **Plate 153**

This is a weak shrub, seldom more than 1 m tall, with spreading or erect branches. Leaves are elliptic to broadly ovate, 10 to 25 mm long, flat and covered with a grey sheen below. Small whitish flowers are clustered along terminal branchlets. Fruits are red, globular and flattened at the ends.

HABITAT Sea coast, especially on the seaward side of vegetation in protected inlets.

DISTRIBUTION Scattered along coast (e.g., Werrong Beach, Minnamurra Inlet). Extending from north coast to Vic, Tas, SA and WA.

FLOWERING Most of the year.

154. Einadia hastata

Einadia

Einadia hastata

Saloop **Plate 154**

A spreading perennial with stems which may be decumbent or upright (to 40 cm tall). Leaves are opposite, triangular to broadly hastate (spear-shaped with two spreading basal lobes), mostly 10 to 15 mm long and almost as broad at the base. Flowers are in short clusters. Fruit is a small depressed-globular red berry about 3 mm across.

HABITAT Open forest and grassland, usually in clayey soils.

DISTRIBUTION Scattered from coast to mountains. Recordings include Razorback Mt and Yallah near Wollongong. Extending to the western plains, along the north coast and tablelands to Qld and along the south coast to Vic.

FLOWERING October–February.

Polygonaceae

The Polygonaceae is widespread but chiefly distributed in temperate countries of the northern hemisphere. It consists mainly of herbs and shrubs, and contains some ornamentals (e.g., polygonum), a few plants with edible seeds (buckwheat), stalks (rhubarb) or leaves (sorrel) and quite a few invasive weeds (docks, knotweeds, wire weed, spiny emex). About a dozen native species of *Persicaria, Muehlenbeckia* and *Rumex* occur in the Sydney area, most of which are weed-like in appearance.

The Polygonaceae is a fairly isolated group usually placed in a separate order, Polygonales. Its closest relationships are with the Plumbaginaceae. Members of the family have alternate simple leaves with a characteristic membranous sheath at the base of each leaf stalk formed by expansion of the stipules around the stem.

Persicaria Knotweed

Herbs of wet places, including a number of introduced weeds. Flowers small, with 5 almost equal segments.

Persicaria strigosa

Spotted Knotweed **Plate 155**

P. strigosa is a weak herb with ascending flower stems. Long brown stipules form a sheath around the stem above the base of the leaf stalk. The main features which separate this species from other Knotweeds are the presence of stiff, downward-pointing hairs on the branches and the arrangement of the white flowers close together in ovate spikes. Leaves are broadly spear-shaped, 3 to 7 cm long, with a cordate base.

155. Persicaria strigosum

HABITAT Wet places in and near open streams and on river flats.

DISTRIBUTION Widespread from coast to mountains. Extending along the north coast to Qld. Also found in China and India.

FLOWERING December–January.

Persicaria hydropiper

Water Pepper **Plate 156**

A weak annual herb with hairless stems and leaves. The stem-stipules are fringed with long bristles. Leaves are lance-shaped, 3 to 9 cm long, with a short stalk. Flowers are greenish white, often with a little red, and are scattered along slender drooping spikes. A characteristic feature of this species is that the perianth is dotted with prominent glands, easily seen with the aid of a hand lens. The leaves have a burning taste when chewed.

HABITAT Shallow water and wet sites beside streams.

DISTRIBUTION Widespread from coast to mountains. Occurring throughout NSW, except in drier western plains; also Qld, Vic, Europe, Asia and America.

FLOWERING April–May.

156. Persicaria hydropiper

Persicaria decipiens

Slender Knotweed **Plate 157**

Clumps of this slender sprawling plant are common in and beside rivers. Its stems and leaves are hairless, but the sheath around the stem has long fine bristles along its margins. Ascending or spreading stems up to 50 cm high bear pink flowers crowded on slender spikes. The leaves are lance-shaped, almost stalkless, 10 to 15 cm long and often marked on the surface with dark blotches.

HABITAT Shallow water and wet sites beside streams.

DISTRIBUTION Widespread from coast to mountains. Extending along north and south coasts and tablelands, with isolated occurrences further west. Found in every State, NZ and Norfolk Island.

FLOWERING January–May.

Persicaria lapathifolia

Pale Knotweed **Plate 158**

A stout annual herb, with erect hairless stems which may reach 1 m tall. The broad, lance-shaped leaves are 7 to 15 cm long, hairless, shortly stalked and often gland-dotted underneath. The stipules are ribbed, but lacking bristles. Flowers are white turning dull pink, in dense terminal spikes, 2 to 6 cm long.

HABITAT Shallow water and wet sites beside streams.

DISTRIBUTION Widespread from coast to mountains. Found throughout Australia, except WA; also Europe, Asia and America.

FLOWERING December–February.

157. Persicaria decipiens

Persicaria sp. C.

Hairy Knotweed **Plate 159**

An erect perennial, with appressed hairs on the stems and leaves. Leaves are lance-shaped, 6 to 15 cm long and stalkless or almost so, hence the former name, *P. subsessile*. Sheathing stipules are conspicuous, mebranous, pale brown 10 to 15 mm long, ciliate at the end. Flowers are yellowish white (sometimes pink), few and distant along a spike 2 to 5 cm long.

HABITAT Along streams, in shaded and wet sites.

DISTRIBUTION Coastal zone (e.g., Eastern Creek, Shane Park). Extending along north and south coasts; also Qld, Vic, Tas, WA and New Caledonia.

FLOWERING December–January.

158. Pericaria lapathifolia

159. Persicaria sp. C

Muehlenbeckia Lignum

Climbers or shrubs with small unisexual flowers borne in spike-like racemes or terminal panicles. Stems with sheathing stipules.

Muehlenbeckia gracillima

Slender Lignum **Plate 160**

A weak perennial with thin twining stems up to 2 m long. Leaves are heart-shaped, 2 to 5 cm long, thin, with distinctly wavy margins, and stalks 10 to 15 mm long. Flowers are small, creamy and borne in slender racemes.

HABITAT Forest, often in rocky places.

DISTRIBUTION Coast to mountains, but uncommon. Extending from south coast to north coast and Qld.

FLOWERING December–January.

160. Muehlenbeckia gracillima

161. Muehlenbeckia adpressa

Muehlenbeckia rhyticarya

This may be an erect shrub or a robust climber with broadly ovate to round thick leaves, cordate at the base, 3 to 4 cm long, on short stalks 6 to 12 mm long. Sheathing stipules are large, brown and conspicuous. Flowers are small and borne in racemes which are all terminal or with some borne from the leaf axils on the upper part of the stems. The fruits are egg-shaped and wrinkled. A similar species, *M. adpressa* (Plate 161), occurs in isolated areas of the southern Blue Mountains (e.g., Nattai River). It has flowers in racemes which are all axillary and smooth gobular fruit.

HABITAT Forest, usually on rocky slopes near watercourses.

DISTRIBUTION Blue Mountains, but uncommon. Also occurring north and south coasts and southern tablelands, Warrumbungle Mts (north-western slopes), Goulburn Valley (central-western slopes), Qld and Vic.

FLOWERING September–December.

Dilleniaceae

This is a fairly large family of trees, shrubs and climbers showing a considerable number of primitive floral features. The flowers are bisexual and regular with 5 large petals, numerous stamens and a superior ovary with one or several free carpels. The family is represented in Australia by about 120 species of *Hibbertia*, 4 species of *Pachynema*, 2 species of *Tetracera* and 1 of the type genus *Dillenia*. A few species of *Hibbertia* are found outside Australia in New Caledonia, New Guinea and Madagascar. *Hibbertia* is the only genus occurring in New South Wales.

Hibbertia Guinea Flower

Shrubs or climbers with leaves spirally arranged. Flowers with 5 broad, spreading, delicate petals, yellow (in the Sydney district), and often lasting only a day or two before shrivelling. Stamens few or numerous; carpels 2 to 5; fruit a follicle. Individual species are frequently difficult to identify owing to variations in habit, leaf shape and flower form. Species are usually divided into those with stamens which encircle the carpels and those with stamens which are all on one side of the carpels. Features which help identification include the number of stamens (more or less consistent for each species), the hairy or glabrous nature of the carpels, leaf shape and inflorescence.

Figure 3

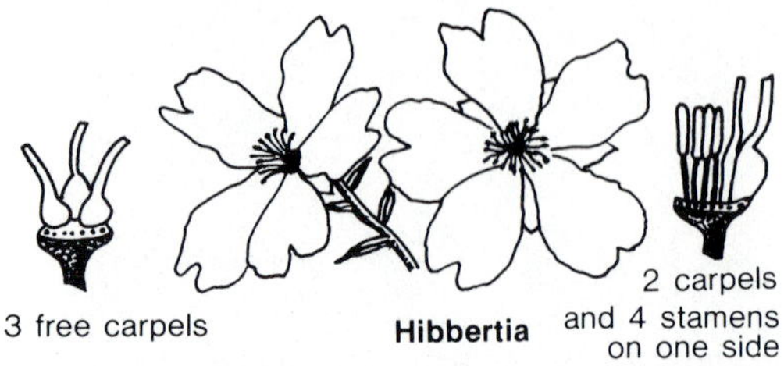

162. Hibbertia scandens

Hibbertia scandens

Climbing Guinea Flower **Plate 162**

Only two species of *Hibbertia* in the Sydney district are climbing or trailing plants. This species is commonly found scrambling over coastal sand dunes or in nearby scrub and forest. It is readily identified by its glossy entire leaves. (The other trailing species, *H. dentata*, has toothed leaves.) Leaves are lance-shaped to obovate, mostly 5 cm long and about 2.5 cm wide. The large open flowers, up to 5 cm across, with their masses of stamens, are conspicuous and very attractive to bees and smaller insects. The fruit is a follicle which splits to reveal orange-red seeds.

HABITAT Near coast on sandy sites or in nearby scrubland and forest; moist forest away from coast.

DISTRIBUTION Mainly coastal, but extending to sheltered mountain gullies. Also south coast, north coast, northern tablelands and Qld.

FLOWERING Flowers recorded in every month of the year.

Hibbertia dentata **Plate 163**

The leaf-margins of this twiner are distinctly toothed, distinguishing it from *H. scandens*. Leaves are also clearly stalked, quite hairy and sometimes purplish when young, becoming hairless at maturity. This is a common plant in moist forests and gullies, and its wiry stems are often seen trailing over lower shrubs or hanging down over shaded rocks. Its flowers are smaller than those of *H. scandens* (about 3 cm across) and have numerous stamens. Sepals, buds and flower-stalks are distinctly hairy. The fruit is a follicle which splits to reveal orange-red seeds.

HABITAT Sheltered forests and rainforest margins.

DISTRIBUTION Widespread from coast to mountains. Extending all along coast from Vic to Qld.

FLOWERING August–November.

163. Hibbertia dentata

Hibbertia saligna **Plate 164**

This upright medium to tall shrub is most readily identified by its long, thin, velvety leaves. These are up to 10 cm long, narrow-lanceolate, pointed and clasping the stem. They tend to be clumped along the stem. The specific name 'saligna' refers to the willow-like leaves. Young branchlets are softly downy. Flowers are large, 3 to 4 cm across, without stalks, and are borne in clusters of leaves. Stamens are numerous.

HABITAT Gullies, creek banks and sheltered forests.

DISTRIBUTION Endemic in the Blue Mountains and parts of the Cumberland Plain. Recorded from the lower Blue Mountains (e.g., Erskine Creek) to the higher mountains (e.g., Mt Wilson, Mt Irvine).

FLOWERING August–November.

164. Hibbertia saligna

Hibbertia diffusa **Plate 165**

H. diffusa is a low, diffuse or prostrate shrub with many weak branches which sometimes form small mats. Flowers are almost stalkless, terminal on the ends of small branchlets, and have petals about 10 mm long. Each flower has 20 to 25 stamens, placed all around the central carpels, not on one side of them. This placement of stamens and the plant's low habit are distinguishing features. (*H. serpyllifolia* is also a prostrate plant, but it has fewer stamens and the carpels are hairy. The carpels of *H. diffusa* are glabrous.) Leaves are very variable. Typically, they are oblanceolate to spoon-shaped, tapering to a short stalk, 5 to 25 mm long, dark green and hairless. The tips may be rounded, slightly notched or have a small mucro.

HABITAT Open forest, especially on shale.

DISTRIBUTION Widespread from coast to central Blue Mountains. Uncommon on sandy soils of coastal zone, but common on shales of Agnes Banks Nature Reserve, Castlereagh State Forest and clays of Menangle. Occurring all along the coast from Vic to Qld; also southern tablelands.

FLOWERING August–February.

165. Hibbertia diffusa

Hibbertia linearis **Plate 166**

The specific name 'linearis' refers to narrow leaves with parallel sides, but this is misleading as the leaves of this species vary from linear to obovate. They may be pointed or rounded at the tips (rarely notched), flat or with margins curved underneath. Most leaves are about 20 mm long. The habit of this species is also very variable; it may be erect or diffuse. Petals are 8 to 10 mm long, and 15 to 20 stamens are all around the carpels. This species is closely related to *H. obtusifolia*, which was formerly named *H. linearis* var. *obtusifolia*, and it is often difficult to separate the two. (For differences, see *H. obtusifolia*.)

HABITAT Heath and scrubland, on sandy soils.

DISTRIBUTION Coastal. Recordings include Brisbane Water National Park, Ku-ring-gai Chase National Park, Heathcote National Park and old dunes at Kurnell. Extending from south coast northwards to Qld, with an isolated record from the central western slopes.

FLOWERING May–October, but some flowers at other times.

Hibbertia obtusifolia **Plate 167**

This is an erect or diffuse shrub, with stems and branchlets covered with short greyish star-shaped hairs or soft erect hairs. This feature helps separate the species from the closely related *H. linearis* which is a hairless plant (except sometimes the young shoots). Leaves are oblanceolate, 10 to 30 mm long, with rounded tips, often notched and with a fine mucro. Flowers are solitary, sessile and terminal on short side branches. Petals are 9 to 15 mm long (somewhat larger than those of *H. linearis*) and there are 30 to 40 stamens around the carpels.

HABITAT Heath and scrubland, on sandy soils.

DISTRIBUTION Widespread from coast to mountains. Found throughout NSW, except in far west; also Qld and Vic.

FLOWERING May–November.

166. Hibbertia linearis

167. Hibbertia obtusifolia

168. Hibbertia monogyna

Hibbertia monogyna Plate 168

This is a shrub up to 50 cm tall, rather similar to *H. diffusa*, but much more erect. The most obvious difference is that the leaves of *H. monogyna* are broad at the tip and narrowed abruptly at the base. Leaves are up to 12 mm long and wedge shaped. The broad tip is often notched, with a tooth on either side of the apex. The flowers have stamens all around the carpels. The carpels are glabrous. A closely related species (as yet unnamed) has been recorded at Cataract River, south-west of Appin.

HABITAT Open forest, on sandy soils.

DISTRIBUTION Mainly Blue Mountains, but extending south to the coastal zone near Nowra. Also recorded along south coast, north coast, on western slopes and in Qld.

FLOWERING August–October.

Hibbertia pedunculata Plate 169

H. pedunculata is easily identified as it has flowers on stalks (peduncles) 10 to 12 mm long, about twice as long as the leaves. It is a diffuse, almost prostrate plant, with narrow shining leaves, 4 to 6 mm long. Flowers are about 12 mm across and have 15 to 20 stamens all around the capels. Ovaries and styles are conspicuously hairy.

HABITAT Open forests and semi-cleared areas, on sandy soils and shales.

DISTRIBUTION Scattered throughout the area, from coast to mountains, but not common. Populations around Berkshire Park (Cumberland Plain) endangered because of housing sub-divisions. Also recorded along north coast, northern tablelands (Mt Kaputar), southern tablelands and western slopes.

FLOWERING November–March.

169. Hibbertia pedunculata

Hibbertia vestita Plate 170

H. vestita has a number of characteristics which clearly distinguish it from other *Hibbertia*. It is a diffuse shrub with spreading branches which are often prostrate. Stems, leaves and sepals are hairy, sometimes with minute hairs, often with longer, softer hairs. It is the only *Hibbertia* in the area with 30 or more stamens placed all around hairy carpels. Leaves are small, 4 to 7 mm long, 1 to 2 mm wide, with margins curved down and under. Flowers are terminal, stalkless and about 20 mm across.

HABITAT Coastal heath, on sandy soils.

DISTRIBUTION North of Sydney, near coast (e.g., Munmorah State Recreation Area). Extending along north coast to Qld; also recorded on north-west slopes and northern tablelands.

FLOWERING September–November.

170. Hibbertia vestita

171. Hibbertia serpyllifolia

Hibbertia serpyllifolia Plate 171

This is a decumbent or prostrate shrub. Its most characteristic features are its small leaves, its hairy carpels and the number of stamens which encircle the carpels. Leaves are 3 to 5 mm long, 1 mm wide; they may be shiny dark green or rough. Leaf margins are curved under and the tips are blunt. Flowers are borne at the ends of short side branchlets; they are stalkless, with 11 to 13 stamens placed around the hairy carpels. Petals are 5 to 6 mm long.

HABITAT Coastal heath on deep sands; heath and scrubland on shallow sandy soils away from coast.

DISTRIBUTION Widespread from coast to mountains. Recordings include old dunes of Royal National Park (Bundeena-Marley) and exposed heaths of Hassans Walls (near Lithgow). Occurring from north coast and northern tablelands southwards to Vic and Tas.

FLOWERING August–October, but as late as February in the mountains.

172. Hibbertia sp. aff. *serpyllifolia*

Hibbertia sp. aff. *serpyllifolia* Plate 172

This small shrub has the general appearance of *H. serpyllifolia*, but close examination reveals differences in the leaf and flower. It is decumbent or prostrate (sometimes upright to 60 cm), with rough, narrow, linear to lanceolate leaves, 8 to 10 mm long and 1 mm wide. The tip is recurved, more or less acute and pungent; the margins are curved under and the undersurface is paler than above. Sepals have fine appressed hairs or are occasionally hairless on the outside. The most obvious difference between this species and *H. serpyllifolia* is that this has many more stamens; 15 to 19 are usual, but some flowers have up to 28 stamens.

HABITAT Open woodland and scrubland, on sandy soils.

DISTRIBUTION Recorded from Menai, Lucas Heights and Woronora Dam road; possibly more widespread in Water Board Catchment and Army Reserve south of Heathcote Road.

FLOWERING November.

Hibbertia hermanniifolia Plate 173

It is not likely that the casual walker or nature lover will see this rare plant in its natural habitat, as it has a very small and localised population. It is an attractive erect shrub, about 1 m tall. Its stems and leaves are covered with matted grey woolly hairs. Leaves are obovate, about 20 mm long, light green above, grey below. Flowers are about 15 mm across, with a mass of stamens around the hairy carpels.

HABITAT Rocky wooded slopes.

DISTRIBUTION Restricted in NSW to Bents Basin State Recreation Area (Nepean River). Also occurs in Vic, where it is very localised.

FLOWERING April–May.

173. Hibbertia hermanniifolia

174. Hibbertia fasciculata

Hibbertia fasciculata **Plate 174**

The specific name of this small shrub comes from the Latin word for 'a bundle' and refers to the bundles of small leaves along the stems. These are mostly 5 to 7 mm long, linear and less than 1 mm wide. Leaf margins never curve downwards or under to the midvein, but sometimes curve upwards so that they are convex on the undersurface. Flowers are without stalks, and are borne on short side branchlets. Most flowers around Sydney have 6 to 8 stamens, all on one side of the carpels, but some plants may have 8 to 12 stamens all around the carpels (e.g., east of Bundeena in Royal National Park). Smaller plants are often covered with soft spreading hairs, but larger plants are usually hairless.

HABITAT Heath, scrubland and dry open forest, on sandy soils.

DISTRIBUTION Widespread and common from coast to mountains. Also recorded from north coast, south coast and southern tablelands.

FLOWERING July–December.

Hibbertia bracteata **Plate 175**

H. bracteata is an erect shrub, with a hairy brown calyx. A number of features help confirm its identity. Its leaves are about 20 mm long and 7 mm wide, shiny green above, lighter below, flat or with margins slightly curved under, rounded at the tip and with a minute point. Flowers are usually borne at the ends of short side branches within a tuft of leaves. About 16 stamens are arranged on one side of the carpel. This species may be confused with *H. nitida*, which also has shining slightly convex leaves. However, *H. nitida* has hairless sepals and only about 11 stamens.

HABITAT Scrubland and open forest, on sandy soils.

DISTRIBUTION Endemic in the Sydney district. Recordings include Brisbane Water National Park, Mt Colah, Muogamarra Sanctuary, Mona Vale, Wisemans Ferry and lower Blue Mountains.

FLOWERING July–December.

Hibbertia nitida

Shining Guinea Flower **Plate 176**

Both the common and specific name of this attractive erect shrub (to 80 cm tall) refer to its conspicuous glossy green leaves. These leaves are oblong or oblanceolate, 12 to 25 mm long, 4 to 8 mm wide, with a pointed tip. Branches are often reddish, making a striking contrast with the shiny leaves. Flowers are stalkless and borne in tufts of leaves along the stem. Sepals are hairless (distinguishing it from *H. bracteata* which has densely hairy sepals). About 11 stamens are arranged on one side of the hairy carpels.

HABITAT Sheltered forested slopes.

DISTRIBUTION Endemic in Sydney's coastal zone, but uncommon. Recordings include Woronora River and Picnic Point.

FLOWERING October–November.

Hibbertia riparia

Erect Guinea Flower **Plate 177**

This small shrub is very variable; its habit, foliage and size depend on where it is growing. Usually, it is an erect shrub to 60 cm tall; branchlets are either covered with short hairs or are hairless. Leaves are narrow-linear, 8 to 16 mm long, blunt, stiff, mostly rough, and have margins which are rolled under to the midvein. Flowers are about 20 mm across and stalkless. Sepals are usually rough. There are 5 to 9 stamens, all arranged on one side of 2 hairy carpels. This species may be confused with *H. cistiflora*, which is also an erect shrub with narrow leaves which are rolled under to the midvein. However *H. cistiflora* has hairless carpels and leaves with a minute yellowish point. Both species are found on sandy soils throughout the area, although *H. riparia* is more common.

HABITAT Heaths, on sandy soils.

DISTRIBUTION Widespread from coast to mountains. Occurring throughout NSW, except far west; also in all other States, except NT.

FLOWERING May–December.

175. Hibbertia bracteata

176. Hibbertia nitida

177. Hibbertia riparia

Hibbertia rufa

This small spreading shrub is closly related to *H. riparia* and was once regarded as a variety of that species. Its most conspicuous features are that its flowers are on slender stalks about 10 mm long, and have only 4 stamens, all on one side of the carpels. Its stems and sepals often have a reddish tinge, hence the specific name ('rufum' is Latin for 'reddish'). Leaves are about 8 mm long, almost sessile, narrow-triangular, with a blunt tip and margins curved under.

HABITAT Heath and damp sedgy sites, on sandy soils.

DISTRIBUTION Chiefly Blue Mountains, but also recorded from coastal zone. Widespread through tablelands; also north coast, western slopes, Qld and Vic.

FLOWERING August–November.

178. Hibbertia cistiflora

Hibbertia cistiflora Plate 178

H. cistiflora is a small shrub, 30 to 80 cm tall. It has narrow leaves, 10 to 15 mm long, about 1 mm wide, with margins curved under; the tip of each leaf has a tiny blunt yellowish point. Flowers are stalkless, with hairless sepals and carpels; less than 12 stamens are arranged on one side of the carpels. The form at Mt Tomah has only 4 stamens. This species was formerly regarded as a variety of *H. riparia* and may be confused with that species. However, *H. riparia* has hairy carpels and obtuse leaves.

HABITAT Heath and damp rocky areas on sandy soils.

DISTRIBUTION Widespread from coast to mountains, but not common. Recordings include French's Forest, Mt Tomah and Gosford. Also occurring in Qld and Vic.

FLOWERING August–September.

179. Hibbertia sericea

Hibbertia sericea
Silky Guinea Flower Plate 179

Although widespread away from the coast and in other States, Silky Guinea Flower is a rare plant in the Sydney district. It is a small spreading shrub to 1 m tall, with branchlets covered with silky hairs. Leaves are oblong, 5 to 10 mm long, 3 to 4 mm wide, round at the tip, with margins curved under. Both surfaces are densely covered with hairs, some short and stellate, others long and soft; older leaves may be almost hairless on the upper surface. Flowers are yellow, about 20 mm across, stalkless and borne at the ends of short side branchlets. The sepals are acute and silky hairy; there are 10 to 16 stamens on one side of the hairy carpels.

HABITAT Open forest, in shallow sandy soils, often in rocky situations.

DISTRIBUTION Rare. Occurring in the higher areas south-west of Sydney (e.g., Wingello State Forest above Tallowa Creek). Recorded from central and southern tablelands, central-western and south-western slopes, western plains, Qld, Vic, Tas and SA.

FLOWERING October–November.

Hibbertia acicularis
Prickly Guinea Flower Plate 180

The pungent-pointed leaves and the flowers on slender stalks are two features which help identify this species. It is a low hairless shrub with a network of branching stems. Its leaves are narrow-linear, up to 10 mm long, sharp-pointed at the tip, with margins that curve down. Leaves tend to stick out at right angles from the stem. Flowers are borne on slender stalks, about 20 mm long, from the leaf axils. There are 4 to 8 stamens, all arranged on one side of two hairy carpels.

180. Hibbertia acicularis

HABITAT Rocky hillsides in open forest and moist coastal heath, on sandy soils.

DISTRIBUTION Widespread from coast to mountains. Recordings include Woronora River Valley, Medway (near Berrima) and Mt Cookem (Blue Mts). Found throughout NSW, except drier western plains; also Qld, Vic and Tas.

FLOWERING May–June in coastal areas, August–November in mountains.

Hibbertia aspera
Rough Guinea Flower Plate 181

Both the common and specific names of this small erect shrub refer to the roughness of the leaves which have a sandpapery feel. This feature combined with the fact that flowers are borne on short stalks, 12 to 15 mm long, makes it easy to identify. It may grow to 60 cm tall, but is more often only a small undershrub, 20 to 25 cm tall. The plant is usually hairy. Leaves are oblanceolate, dark green and rough above, up to 18 mm long and 7 mm wide, with margins curved under. The undersurface of the leaf is covered with grey downy hairs. Petals are 6 to 7 mm long; sepals are almost as long, pointed and hairy. Stamens are arranged all on one side of the hairy carpels.

181. Hibbertia aspera

HABITAT Heath, woodland and forest, on both sandy and shaly soils.

DISTRIBUTION Coastal zone. Extending along coast from Qld to Vic, Tas and SA; also northern tablelands.

FLOWERING August–December.

Hibbertia empetrifolia
Trailing Guinea Flower Plate 182

H. empetrifolia has sprawling stems and a decumbent habit, flowers on slender stems and stamens placed on one side of the carpels. Its leaves are a useful identifying feature. They are smaller (5 to 7 mm long) than those of *H. aspera*, shiny green, sometimes minutely warted or hairy, but mostly hairless. Leaf-margins are curved under. Flowers are small, with petals 5 mm long, but are attractive as they hang on stalks from the leaf axils. There are usually 4 to 6 stamens, all on one side of the hairy carpels.

HABITAT Heath, on sandy soils.

DISTRIBUTION Coastal zone. Recordings include Curra Moors, Ingleside, Maddens Plains and Carrington Falls. Extending along south coast to Vic and Tas; also occurring on western slopes.

FLOWERING September–March, with some flowers at other times.

182. Hibbertia empetrifolia

183. Hypericum gramineum

Clusiaceae

This family of tropical and temperate trees and shrubs is related to the Theaceae, the tea or camellia family. It consists of about 40 genera divided into 5 or 6 subfamilies, 1 of which is sometimes treated as a separate family, Hypericaceae. *Hypericum*, with 2 widespread native species, is the only Australian representative of this segregate group. Both species occur in the Sydney district along with 2 larger exotic weed species. Both the generic name (hyper, above; eikon, image) and the common name, St John's Wort, refer to the ceremony of the ancient midsummer festival, Walpurgisnacht, later the Feast of St John, in which the flowers of certain species were placed above images to ward off evil.

Hypericum St John's Wort

Small perennials, with yellow 5-petalled flowers and numerous stamens. The 2 native species have 20 to 50 stamens; 2 introduced species have 50 to 120 stamens.

Hypericum gramineum

Small St John's Wort **Plate 183**

This is a small, erect (to 30 cm tall), perennial with quadrangular stems. Leaves are opposite, 5 to 25 mm long, heart-shaped at the base, without stalks and stem-clasping. Flowers have 5 yellow petals, about 10 mm long, and 20 to 50 stamens. They are borne on stalks, in small clusters or solitary from the leaf axils, or a single flower may be terminal. Fruit is a 3-valved capsule.

HABITAT Grassland and open forest.

DISTRIBUTION Widespread from coast to mountains. Found throughout NSW and in every other State.

FLOWERING September–February.

Hypericum japonicum

Matted St John's Wort

H. japonicum is a procumbent herb, often forming a soft mat, with stems about 15 cm long. It differs from *H. gramineum* in having leaves which are shortly stalked or, if stalkless, are not stem-clasping. Flowers have 5 yellow petals about 8 mm long and 20 to 30 stamens. They are usually terminal, either solitary or 2 to 3 together.

HABITAT Grassland and open forest, usually in damp places.

DISTRIBUTION Widespread from coast to mountains. Common along coast and tablelands divisions and scattered throughout western districts. Also Vic, Tas and Asia.

FLOWERING September–February.

Elaeocarpaceae

Two genera, *Elaeocarpus* with about 200 species and *Sloanea* with more than 100 account for over 90 per cent of the species in this family. The remainder is shared between 10 other genera. *Elaeocarpus* and *Sloanea* are essentially South-east Asian and Australasian. They are separated on the basis of their fruits; those of *Elaeocarpus* are succulent drupes and those of *Sloanea* are hard capsules covered with bristles.

Elaeocarpus, with 20 endemic Australian species, is found in a variety of forest habitats from the Northern Territory down the east cost to Victoria. One species *(E. reticulatus)* extends to Tasmania. Four species occur near Sydney, although only *E. reticulatus*, the Blueberry Ash, is common.

Sloanea, with 4 endemic species, occurs in coastal rainforests from north Queensland to Batemans Bay in New South Wales. The only species reaching the Sydney area *(S. australis)* is characteristically a fairly large spreading tree leaning across rainforest creeks, often with a number of upright suckers growing from the base.

Elaeocarpus

Trees with alternate, strongly veined leaves, white or pink flowers (often fringed) and clusters of blue-black berries.

Elaeocarpus reticulatus

Blueberry Ash **Plate 184**

This large shrub or small tree is well known because of its masses of pendulous white flowers which are followed by ornamental blue berries. The bell-like flowers have a fringed edge and on some trees have a pink tinge. Leaves are lance-shaped, alternate, up to 12 cm long and finely toothed. A clear network of veins is visible on the leaf, hence the specific name 'reticulatus' (reticulate—forming a network of lines).

HABITAT Sheltered forest and rainforest margins, on sandy soils.

DISTRIBUTION Widespread from coast to mountains. Recorded all along coast and tablelands of NSW; extending from Fraser Island (Qld) to Vic and Tas.

FLOWERING October–December.

184. Elaeocarpus reticulatus

185. *Elaeocarpus kirtonii*

Elaeocarpus kirtonii

Pigeonberry Ash **Plate 185**

This is a tree to 15 m tall, found only in the coastal rainforests. Leaves are lance-shaped, finely toothed and strongly veined. Leaf-blades are 9 to 15 cm long, with a stalk 3 to 5 cm (sometimes longer). In spring, vivid pink new growth makes this tree stand out from its darker green surrounds. White, fringed, bell-shaped flowers are borne on a raceme 10 to 12 cm long. Fruits are round to ovoid blue berries. Further north, this species may be a large tree to 30 m with buttress roots, but it is much smaller in the Sydney district.

HABITAT Rainforest.

DISTRIBUTION Scattered along Illawarra coastal escarpment (e.g., below Mt Keira); also recorded for Gosford district. Two large trees are growing in Sydney's Botanic Gardens. Extending along the coast from near Milton (south coast) to north Qld.

FLOWERING Late January-March.

186. *Elaeocarpus holopetalus*

Elaeocarpus holopetalus

Black Oliveberry **Plate 186**

This tree is not often seen as it grows mainly in isolated gullies in the Blue Mountains. Its leaves are 3 to 7 cm long, tough, strongly veined and finely toothed. The young twigs, flower stalks and leaf undersurfaces are covered with short dense pale hairs. The leaves often turn red before falling. Flowers are white, and the petals, unlike those of other members of this genus, are entire, not fringed or lobed. The fruit is small, black and fleshy like an olive, hence its common name. The botanical name is very descriptive: 'Elaia'—olive; 'carpos'—a fruit; 'holos'—whole; 'petalus'—a petal.

HABITAT Cool mountin gullies.

DISTRIBUTION Blue Mountains, including Gordon Falls (Leura) and Mt Tomah. Extending from northern tablelands to southern tablelands and Vic.

FLOWERING November.

Elaeocarpus obovatus

Hard Quandong, Blueberry Ash **Plate 187**

This is a medium to tall tree, uncommon in the Sydney district where it reaches its southern limit at about Wyong. Leaves are alternate, obovate, about 8 cm long, lobed or shallowly toothed on the upper two-thirds, shiny above and dull below. The veins are visible but not as prominent as those of *E. reticulatus*. The white fringed flowers are borne in long narrow sprays from the leaf axils. Fruit is a succulent blue berry, about 6 mm across, surrounding a warty stone.

HABITAT Along streams of coastal rainforests and in protected sites along coastline.

DISTRIBUTION Coastal zone from Bateau Bay and Wyong northwards to Qld. Recordings include Gap Creek Flora Reserve (Olney State Forest).

FLOWERING September-November.

188. *Sloanea australis*

Sloanea

Rainforest tree with large leaves and a fruiting capsule covered with soft bristles.

Sloanea australis

Maiden's Blush **Plate 188**

This medium to large spreading tree is easy to identify from its large leaves. These are alternate and may be 30 cm long and 12 cm wide, with tooth-like wavy margins; they narrow at the base to be obtuse at the leaf stalk. The popular name refers to the deep red colour of the new leaves. Flowers are creamy-white, about 2 cm across and pendulous on slender stalks. The fruit is a conspicuous feature from about May through the winter months. The capsules are covered with dense soft hairs and open into 4 woody sections. At first they are an attractive orange but turn light brown with age.

HABITAT Rainforest gullies, especially along or near watercourses.

DISTRIBUTION Coastal zone. Widespread below the Illawarra Escarpment as far south as Batemans Bay; north of Sydney recorded from Ourimbah, Strickland State Forest and Gap Creek Flora Reserve. Extending along north coast to north Qld.

FLOWERING October-November.

187. *Elaeocarpus obovatus*

Sterculiaceae

Sterculia, the type genus of this family, was named by Linnaeus after Sterculius, the Roman god of dung and privies, because of the fetid smell from the flowers and leaves of some species. A few family members have economically important species. *Cola* from tropical Africa produces cola nuts used in drinks and medicines. The nuts, which contain caffeine and other stimulants, are also chewed by natives to delay fatigue. The cacao tree, *Theobroma cacao*, originally from tropical America but now cultivated in Africa and elsewhere, is the source of cocoa, cocoa butter and chocolate. These are prepared from roasted seeds of cacao beans which sprout directly from the trunks and major branches of the trees. Several African genera have important timber trees.

About 20 genera occur in Australia, the largest being *Lasiopetalum* (about 30 species), *Rulingia* (23 species) and *Commersonia* (12 species). The most outstanding is *Brachychiton* (12 species) which includes kurrajongs, the Queensland bottle tree and the Illawarra flame tree. The close relationship of Sterculiaceae with the family Bombacaceae—which includes the remarkable baobabs or monkey-bread trees (*Adansonia* sp.) of Africa, Madagascar and the Kimberley region of Australia—is apparent in the resemblance of some *Brachychiton* species to *Adansonia* and other obese members of that group, and in the possession of water storage tissue in the trunk.

Brachychiton

Trees. Flowers unisexual, with male flowers having stamens united into a column bearing 10 to 30 anthers at its summit, and female flowers with 5 free carpels raised on a stalk and surrounded at the base by bundles of staminodes. Flowers are bell-shaped and 5-lobed; they have no petals but are formed by petal-like sepals. Fruits are black boat-shaped pods, hairy inside, containing numerous seeds.

189. *Brachychiton acerifolius*

190. *Brachychiton populneus*

Brachychiton acerifolius

Illawarra Flame Tree **Plate 189**

When in full flower, the Flame Tree is very conspicuous and easily identified. Masses of red flowers, each bell-shaped and 1 to 2 cm long, show up against the darker rainforest canopy. The effect is highlighted by the fact that the tree loses most of its leaves around flowering time. The loss of leaves and flowering cycle are very variable and it is not uncommon to find only part of the tree in flower with most of the leaves still present. Leaves are large, 10 to 20 cm long, usually 3-lobed but entire on some trees. Young saplings have leaves which are larger, more deeply dissected and 5 to 7 lobed. Fruits are black boat-shaped pods, up to 15 cm long. They open to reveal seeds surrounded by fine brown hairs which stick into and irritate the skin when touched.

HABITAT Coastal rainforest. Also grown as an ornamental tree in gardens and parks.

DISTRIBUTION Common in the richer soils and gullies beneath the Illawarra escarpment. Also recorded in the Gosford-Wyong and Gap Creek rainforests north of Sydney. Extending from the Shoalhaven River to Cape York (Qld).

FLOWERING December–January.

Brachychiton populneus

Kurrajong **Plate 190**

Kurrajong is a tree to 15 m, with a thick trunk, grey bark and a dense and spreading canopy. It is predominantly a tree of the rocky hills and sandy red loams of the central west of the State, but does occur as scattered trees in the Sydney district. Like the Flame Tree, its shiny thin leaves are variable in shape. When juvenile, they are distinctly 3 to 5 lobed, but mature leaves are lance shaped or 3 lobed. The centre lobe is long (to 15 cm) and pointed. Flowers are creamy-white bells with red centres. Fruits are black, boat-shaped pods about 5 cm long.

HABITAT Rocky hillsides or scree slopes, especially in drier areas. Also grown as an ornamental tree in gardens and parks.

DISTRIBUTION Scattered, but most common in the west of the area (e.g., Jenolan Caves, Wombeyan Caves). Some trees closer to the coast (e.g., Razorback Mt, Jamberoo, Cobbitty, Cumberland Plain, Bouddi National Park). Found throughout most of NSW except far western plains; also Qld and Vic.

FLOWERING November–January.

Rulingia

Shrubs. Flowers bisexual, with 5 petals and 5-lobed calyx; the petals are usually shorter than the calyx. The 5 stamens are opposite the petals and alternate with 5 entire, linear-lanceolate staminodes. Fruit is a capsule covered with bristles.

Rulingia dasyphylla

Kerrawang **Plate 191**

Kerrawang is an erect open shrub, 1 to 2 m high. The stems and leaf undersurfaces are covered with soft velvety hairs; the upper surface of the leaf is scabrous, or rough, to the touch. Leaves are lance shaped, toothed and up to 8 cm long. Flowers are white, with 5 petals and although small (6 to 8 mm across), they are conspicuous as they are clustered together in loose cymes near the ends of the branches. Fruit is a globular capsule, about 8 mm across, covered with rigid bristles, 2 to 3 mm long.

HABITAT Open forest, mainly on sheltered slopes or in gullies.

DISTRIBUTION Widespread, from coast to mountains. Also occurring north coast, south coast, northern tablelands, western slopes, Qld and Vic.

FLOWERING August–October.

191. Rulingia dasyphylla

Rulingia hermanniifolia — Plate 192

This is a sprawling shrub which grows in large clumps on coastal heaths. Flowers are similar to those of *R. dasyphylla*, but are borne in abundance and seem to almost cover the plant. Leaves are small, 5 to 12 mm long, almost stalkless, dark green above and very wrinkled; the undersurface is densely hairy. Fruit is a capsule, 4 mm across, covered with short bristles.

HABITAT Coastal heath, on sandy soils.

DISTRIBUTION Common in suitable habitats, such as Kurnell Peninsula, Bombi Head (Bouddi National Park) and Deer Pool Track (Royal National Park). Extending to the south coast.

FLOWERING August–September.

192. Rulingia hermanniifolia

Commersonia

Shrubs or small trees. Flowers bisexual, with 5 petals and 5-lobed calyx; the petals are as long as the calyx. The 5 stamens are united at the base and alternate with 5 staminodes which are deeply 3-lobed and as long as the petals. Fruit a capsule covered with soft bristles.

193. Commersonia fraseri

Commersonia fraseri

Commersonia, Brush Kurrajong — Plate 193

This is usually a tall shrub, 2 to 3 m tall, common on rainforest margins. It regenerates vigorously in disturbed areas, such as along roadsides or under power-lines, where the native vegetation has not been totally cleared. Leaves are variable. Young leaves are broad, jagged and softly hairy; adult leaves are broadly lance shaped to heart shaped, with irregular teeth. Flowers are white and massed in loose clusters. Fruit is 15 to 25 mm across, covered with bristles 5 to 10 mm long.

HABITAT Rainforest margins and sheltered gullies.

DISTRIBUTION Common along Illawarra Escarpment (e.g., Macquarie Pass, Mt Keira, Mt Kembla); also north of Sydney in Olney State Forest and Gap Creek Flora Reserve. Extending westwards to Nepean and Hawkesbury Rivers and Blue Mountains. Occurring from Genoa River (Vic) all along coast to southern Qld.

FLOWERING September–October, with a few flowers as late as January.

194. Seringia arborescens

Seringia

Shrub with bisexual flowers. Petals absent, but the calyx is petal-like. Flowers with 5 lobes and covered with closely matted short hairs; 5 stamens alternate with 5 staminodes. Fruit a capsule. *Seringia* may at first be confused with other members of the family. Careful examination of the flowers reveals significant differences. *Rulingia* and *Commersonia* have petals; *Seringia* has no petals only 5 sepals. *Seringia* has 5 stamens alternating with 5 long, narrow, tapering staminodes, which clearly separate it from broad-leaved *Lasiopetalum* species which have minute or no staminodes. *Seringia* differs from *Keraudrenia* in that the calyx does not enlarge or colour after flowering.

Seringia arborescens

Seringia — Plate 194

This is a tall shrub, 2 to 3 m in the Sydney district, with rusty-hairy young branches. Leaves are broadly lance shaped, 8 to 12 cm long, hairless above and with rusty-brown hairs below; margins have rounded teeth and the stalk is short, 5 to 10 mm long. Flowers are in short, stalked clusters. The calyx is angular in bud and velvety brown. Fruit is softly hairy with a short wing on the back.

HABITAT Forest.

DISTRIBUTION Coastal zone, but uncommon; also found in Blue Mountains near Yerranderie (e.g., Byrnes Gap), along Nepean River (e.g., Norton's Basin) and Ku-ring-gai Chase National Park (Campbell's Crater). Extending from south coast northwards to Qld.

FLOWERING October–November.

195. Keraudrenia corollata var. *denticulata*

Keraudrenia

A shrub similar to *Seringia* but with narrower leaves and a calyx which enlarges and colours after flowering. Fruit a capsule covered with long hairs.

Keraudrenia corollata var. *denticulata*

Velvet-flower — Plate 195

This is a compact shrub, 1 to 1.5 m tall, with branches covered with rusty hairs. Leaves are lance shaped, alternate, 6 to 12 cm long, dark green above and very wrinkled and rough to the touch; the undersurface is lighter, with hairs which are often rusty, especially along the conspicuous veins. The leaf margins are broken by small teeth. Flowers have 5 white-pink petal-like sepals, each about 5 mm long

196. Lasiopetalum ferrugineum var. *ferrugineum*

and hairy; anthers are yellow. Flowers are borne in clusters of about 5 flowers near the ends of the branches. Fruit is a densely hairy globular capsule which splits into distinct segments.

HABITAT Forest, in sheltered rocky areas.

DISTRIBUTION Localised and rare. Recorded from Colo River near Putty Road. Also found near Grafton (north coast) and in Qld.

FLOWERING October–December.

Lasiopetalum Rusty-petals

Shrubs, more or less covered with rusty hairs. Flowers with 5 petal-like calyx lobes and 5 stamens. Careful examination of the flower is necessary to distinguish one species from another. *Lasiopetalum* differs from *Seringia* and *Keraudrenia* in not having conspicuous staminodes alternating with the stamens.

Lasiopetalum ferrugineum

Rusty-petals, Rusty Velvet-bush **Plate 196**

This species has two varieties based on leaf shape, with a number of intermediate forms, possibly hybrids, so that one variety blends into the other in some places. Both feature small clusters of flowers with rusty hairs on the outside and creamy-white hairs on the inside. Variety *ferrugineum*. Leaves are variable, from linear to narrow-oblong or lance shaped, 3 to 10 cm long, hairless above and with rusty-brown hairs below. This is a very variable shrub, growing to 2 m in forests but may be prostrate on exposed coastal headlands, such as above Maitland Bay in Bouddi National Park. Variety *cordatum* (Plate 197). The leaves of this variety are broader and have a distinctly cordate (heart-shaped) base. Generally an erect shrub to 1 m tall.

HABITAT Open forest and heath, on sandy soils.

DISTRIBUTION Var. *ferrugineum* is widespread from coast to mountains. It occurs all along the coast from the north coast southwards to Vic; also central western slopes. Var. *cordatum* occurs in the Blue Mountains, including the Nattai area in the south-west and Wollemi area in the north-west. It extends to the north coast, southern tablelands and north-western slopes.

FLOWERING September–November.

197. Lasiopetalum ferrugineum var. *cordatum*

Lasiopetalum rufum

Red Rusty-petals **Plate 198**

The specific name 'rufum' refers to the reddish colour of the inside of the flower. The petal-like calyx lobes are 3 to 5 mm long with the outer part of the calyx brown when immature but pinkish when older. This is an erect or diffuse shrub, differing from *L. ferrugineum* in having narrower leaves (2 to 4 mm wide). Those leaves are about 6 cm long, green and hairless above, white-hairy below. Flowers are in small, drooping, fairly open clusters.

HABITAT Heath, on sandy soils.

DISTRIBUTION Endemic in the Sydney area, from coast to lower Blue Mountains, where it is common.

FLOWERING August–November.

198. Lasiopetalum rufum

199. Lasiopetalum parviflorum

Lasiopetalum parviflorum

Small Rusty-petals **Plate 199**

L. parviflorum is a shrub up to 80 cm tall. Like other members of the genus, its stems are clothed in rusty closely matted hairs and it is easily mistaken for other narrow-leaved species of *Lasiopetalum*. Its most characteristic feature (obvious under a hard lens) is that the inside of the petal-like calyx lobes is hairless. In the Sydney district, only this species and the broader-leaved *L. macrophyllum* have this feature. Leaves are narrow-oblong, 4 to 8 cm long, 4 to 6 mm wide, dark green and smooth above and white underneath. Flowers are white-hairy outside, with calyx segments about 4 mm long and narrow brown bracteoles.

HABITAT Open forest, usually on shale.

DISTRIBUTION Widespread away from coastal sandstone zone, but uncommon. Recordings include Cumberland Plain, Hornsby Plateau, Burragorang and lower Shoalhaven River. Also occurring on south coast, north coast and central tablelands.

FLOWERING August–November.

200. Lasiopetalum macrophyllum

Lasiopetalum macrophyllum

Shrubby Rusty-petals **Plate 200**

Shrubby Rusty-petals may grow to 2 m tall in protected forests or be a dwarf shrub on exposed coastal sites. It has broad ovate to lance-shaped leaves, up to 10 cm long and 15 mm wide, green and hairless above and white-hairy underneath. The inside of the flowers is hairless, a characteristic feature (clearly seen with a lens) which separates this species from the other broad-leaved species, *L. ferrugineum* var. *cordatum*. The outside of the petal-like calyx is white-hairy, stamens are often reddish and the bracteoles are ovate-lanceolate. Flowers are borne in short dense clusters which hang downwards.

HABITAT Open forest to scrubby coastal headlands.

DISTRIBUTION Widespread from coast to mountains, but uncommon. Recordings include Pearl Beach, West Pittwater, Olney State Forest, Nattai and Colo Heights. Also occurring south coast, north coast, southern tabelands, central western slopes, Vic and Tas.

FLOWERING September–October.

202. *Hibiscus heterophyllus*

201. *Lasiopetalum joyceae*

Lasiopetalum joyceae

Joyce's Rusty-petals **Plate 201**

L. joyceae is an attractive slender shrub, 50 to 80 cm tall, closely allied to *L. rufum* and easily mistaken for that species when not in flower. It was first collected in 1915 by Joyce Blakely, the adopted daughter of the botanist W.F. Blakely, from the Mount Colah-Bobbin Head area and named in her honour. It has narrow-oblong leaves, about 8 cm long and 5 mm wide, dark green above and white below. The petal-like calyx lobes are 10 mm long, flesh-coloured to reddish-brown. The features which distinguish this species from *L. rufum* are its longer and broader leaves, the larger flowers (twice the size of those of *L. rufum*) and bracteoles which are broader and closer to the calyx.

HABITAT Open woodland and heath, on sandy soils.

DISTRIBUTION Restricted to Hornsby Plateau south of Hawkesbury River. Recorded within Ku-ring-gai Chase National Park at Mt Colah, Mt Ku-ring-gai, Berowra-Cowan and near Terrey Hills.

FLOWERING August–October.

Malvaceae

Although the family is named after the temperate northern hemisphere genus *Malva*, the largest and best-known genus in the warmer regions of the world is *Hibiscus*. Australia has about 50 native species of *Hibiscus* out of a world total of 300. Closely related are *Alyogyne* (4 endemic Australian species) and *Gossypium* (8 species in Australia, 20 in the world) all with capsular fruits which open by splitting down the centre of each compartment. *Hibiscus* has a greater range of floral colours and a much wider distribution in Australia than the other genera which are restricted to the drier regions of central, southern and western Australia. *Alyogyne* and the Australian species of *Gossypium* nearly always have mauve flowers with dark centres. *Hibiscus* also differs in having an epicalyx of from 5 to 12 bracteoles (consistently 6 in *Alyogyne* and 3 in *Gossypium*) and in having a divided style with 5 spreading lobes (undivided in the others).

Although *Hibiscus* may be better known, *Gossypium* is by far the most important genus in the family since it is the basis of the world cotton industry. Cotton lint is obtained from the white fibres which cover and surround the ripe seeds of the cotton plant. Most of the Australian species have seed hairs but none produces a lint suitable for commerce. The seeds of *Gossypium* are crushed to yield the highly regarded cottonseed oil used in cooking, soap making and cosmetics, and a protein-rich seed meal which is used as a livestock feed.

The two commonest Malvaceae species in the area are *Sida rhombifolia* and *Pavonia hastata*. Once regarded as native species, it is now believed that they are both introduced. No doubt exists however about the purple-flowered monotypic endemic, *Howittia trilocularis* which has a sparse distribution in moist valleys and coastal gullies in New South Wales and Victoria. The genus is separated from *Hibiscus* by the 3-celled ovary, the 3-lobed style and by the absence of an epicalyx. One or two species of yellow-flowered *Abutilon* are found occasionally in moist locations in or near the area.

The features which characterise Malvaceae are most constant and more easily recognised than in most families. The flowers are bisexual and regular with 5 colourful, free, spreading petals. The 5-lobed calyx is frequently subtended by an epicalyx of bracts. The stamens are numerous and prominent, and are joined for most of their length into a long hollow tube attached to the petals and surrounding the ovary and style. The style projects slightly beyond the stamens and is often branched at the end so that if cross-pollination does not occur self-pollination becomes possible by the style branches bending down to touch the anthers. Frequently the stems, leaves and fruit are covered with stiff or soft stellate hairs.

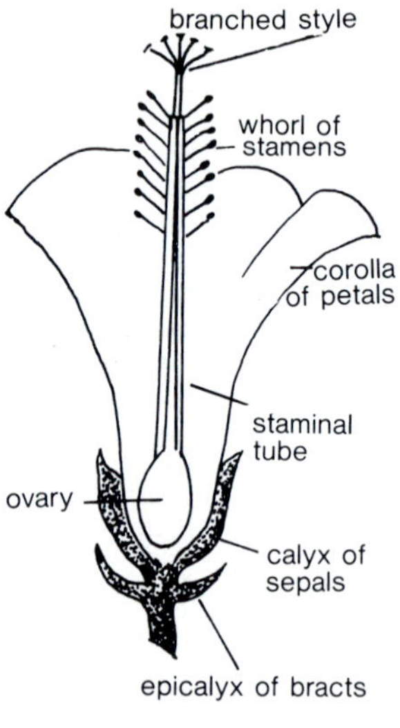

Features of Malvaceae flower

Figure 4

The Malvaceae are possibly the most advanced of the closely interrelated group of 6 or so families (including Elaeocarpaceae and Sterculiaceae) forming the order Malvales. Common features of the group include the regular flowers with free petals which are contorted in bud (i.e., the petals overlap slightly so that each is outside one adjacent petal and inside the other), fusion of the stamens, the presence of an epicalyx, and the stellate indumentum.

Hibiscus

Shrubs or small trees with large 5-petalled flowers and 5 to 12 bracts beneath the calyx forming an epicalyx. Only one species *(H. heterophyllus)* is common, although *H. trionum* (weed-like, with pale yellow flowers and inflated bladder fruits) and *H. diversifolius* (with pale yellow flowers and calyx and fruit covered with rigid hairs) may occasionally be found in disturbed areas.

Hibiscus heterophyllus

Native Rosella **Plate 202**

A tall shrub of the rainforest margins. Its flowers are large, to 15 cm diameter, with white petals often with a pinkish tinge. The base of the petals and the staminal column are a deep crimson colour. Leaves are usually deeply 3-lobed and up to 15 cm long, but some may be almost entire. This variation in leaf-shape is reflected in the specific name ('hetero'—different; 'phyllus'—leaf). The fruit is a densely hairy ovoid capsule with a pointed apex.

HABITAT Rainforest margins and sheltered tall forests.

DISTRIBUTION Coastal zone. Common in Illawarra region (e.g., Kiama, Jamberoo, Berkeley Hills, Macquarie Pass); also Colo River, lower Portland and Wyong State Forest. Extending along north coast and northern tablelands to Qld.

FLOWERING November–January.

Howittia

Shrub. Differs from *Hibiscus* in having no epicalyx.

Howittia trilocularis

Howittia **Plate 203**

Howittia is an erect shrub with a yellowish tomentum covering stems, leaf stalks and buds. The leaves are ovate-cordate to lance shaped, 2 to 10 cm long, with wavy edges. The upper surface of the leaf is green and slightly rough, but the undersurface has whitish or rusty hairs. The flowers are the main identifying feature. They are deep mauve, with purple centres and have clusters of yellow stamens. Each of the 5 petals is about 10 mm long and overlaps in the shape of a cup. The flowers are borne on long slender stalks. The fruit is 3-celled.

HABITAT Forest, mainly on protected slopes.

DISTRIBUTION Widespread from coast to mountains, but uncommon. Recordings include Bundanoon, Pierces Pass (Grose River) and Maitland Bay (Bouddi National Park). Extending from north coast to south coast and Vic; also central western slopes.

FLOWERING August–November.

Abutilon

Shrub with yellow flowers. Differs from *Howittia* in having ovary and fruit which are 5-celled or more.

203. *Howittia trilocularis*

Abutilon oxycarpum

Lantern-bush **Plate 204**

The most obvious feature of this small shrub is its yellow flowers. These are 5-petalled and borne on stalks about 2 cm long arising in the leaf axils. The petals are 5 to 8 mm long and have a cluster of yellow stamens in the centre. The leaves are cordate to lance shaped, with basal lobes. The lower surface of the lead and the leaf stalk are covered with short white hairs. Fruiting capsules are conspicuous. Each capsule is about 8 mm across, and has a flat star-like arrangement on top. Each capsule consists of 5 to 7 fruitlets, each with a 2 mm awn at the upper outer edge. The specific name refers to this feature ('oxy'—sharp; 'carpos'—a fruit).

HABITAT Rainforest margins and drier areas of remnant rainforest.

DISTRIBUTION Coastal zone, but uncommon. Recordings include Upper Kangaroo Valley, Berkeley Hills (Wollongong) and Razorback Mt. Extending from the south coast northwards to Qld and also to drier western slopes around Nyngan.

FLOWERING February–April.

204. *Abutilon oxycarpum*

Flacourtiaceae

This large family with about 1300 species of tropical and subtropical trees and shrubs has few members of economic or horticultural significance. It is represented in Australia by 8 genera (2 endemic) and 17 named species, chiefly in Queensland rainforest and vine forests. One species, *Scolopia braunii*, extends from Cape York to Jervis Bay, south of Sydney. The flowers exhibit a number of primitive features and the family is considered the least advanced in the order Violales.

Scolopia

205. *Scolopia braunii*

Scolopia braunii

Scolopia, Flintwood **Plate 205**

This is a small tree with shiny alternate leaves. Juvenile growth often has spines in the forks of the leaves. Leaves are 4 to 9 cm long, smooth, glossy above and paler underneath. Younger leaves are usually angular rhomboidal (holly-like) irregularly toothed but older leaves are ovate to lance shaped, with margins entire or only obscurely toothed. New growth is a deep tan colour, clearly visible from a distance. Flowers are small, about 4 mm across, white, with 4 petals, 4 sepals and 40 stamens which are about twice as long as the petals. Flowers are borne in short clusters, 1 to 2 cm long. Fruit is a globular berry, about 1 cm across, turning from dark red to black.

HABITAT In and near rainforest.

DISTRIBUTION Scattered along coastal zone. Recordings include Foxground, Berkeley Hill (Illawarra), Palm Jungle (Royal National Park), Maitland Bay (Bouddi National Park), Patonga, Yarramalong and Gap Creek (Olney State Forest). Extending northwards from Jervis Bay to northern Qld.

FLOWERING October–November.

Violaceae

Of the 950 or so species in this family, *Viola* accounts for about 400, *Rinorea* about 340, and *Hybanthus* about 150. The remainder is divided among 19 other genera, most of which are shrubby rather than herbaceous like *Viola* and *Hybanthus*. The family has a worldwide distribution but is mostly restricted in the tropics to cooler mountainous areas.

The flowers of Violaceae may be regular or irregular with 5 sepals, 5 petals and 5 partly fused stamens alternating with the petals. The superior ovary consists of 3 fused carpels forming 1 locule with parietal placentation. Although the flowers of *Viola* are adapted for insect pollination, quite a few species are cleistogamous i.e., undergo self-pollination without opening. The seeds of some species are gathered and dispersed by ants.

The Violaceae is represented in the Sydney district by *Viola* (3 native species), *Hybanthus* (3 species) and *Hymenanthera* (1 species).

206. Hymenanthera dentata

Hymenanthera

Hymenanthera dentata

Tree Violet **Plate 206**

It is difficult to believe that this much-branched, spiny tall shrub with tiny yellow pendulous flowers is in any way related to the delicate herbaceous violets, but a close examination of the flower structure will reveal the similarities which place them in the same family. Tree violet has the characteristic 5 stamens surrounding the ovary with appendages extending beyond the anthers, and nectaries at the base of the anthers. Even without such a close examination, this species is easy to identify. It has rigid branches, with side branchlets ending in thorns, and leaves which are 1 to 3 cm long and irregularly toothed. Flowers are small yellow bells which are crowded along the branchlets. The fruit is a small purplish berry.

HABITAT Near coastal rainforest and along river banks or on rocky slopes in tablelands.

DISTRIBUTION Widespread from coast to mountains. Common around Wombeyan Caves. Occurring throughout NSW, except in drier western area; also Vic, Tas, SA and NZ.

FLOWERING September–December.

207. Viola betonicifolia

Viola Violet

Small herbs with irregular violet flowers on long stalks. Lowest of 5 petals spurred or pouched at the base.

Viola betonicifolia

Showy Violet, Mountain Violet **Plate 207**

This perennial herb is immediately identified by its leaves. They rise from a cluster at the base of the plant on long narrow stalks and have a blade the shape of an arrow-head, broad-lanceolate, up to 6 cm long and cordate at the base. The flowers are violet to purple, 10 to 15 mm across, borne singly on slender stems. The side petals are bearded inside.

HABITAT Open forest, especially in shaded sites.

DISTRIBUTION Widespread from coast to mountains. A common species throughout most of NSW, except dry west. Also occurring in Qld, Vic, Tas, SA and Asia.

FLOWERING September–December.

208. Viola hederacea ssp. *hederacea*

Viola hederacea ssp. *hederacea*

Ivy-leaved Violet **Plate 208**

This small herb spreads by runners. Individual plants are small and are easily recognised by their leaves. These rise from the base of the plant on stalks 2 to 6 cm long. The leaf-blades are kidney-shaped or oval, wider than long and indented where the stalk joins the blade. Flowers are on stems longer than the leaves; they are white to pale violet with darker violet blotches towards the centre. The side petals are bearded inside. Plants in the mountains have darker flowers. *V. hederacea* ssp. *perreniformis* also occurs along the coastal zone from north Qld southwards to Sydney. It is usually a softly hairy plant, with leaf blades which are rounded, and bases deeply cordate; flowers are a uniform pale blue-violet and side petals are distinctly twisted and bearded.

HABITAT Shaded and moist situations in forest and woodland.

DISTRIBUTION Widespread from coast to mountains. Extending to Qld, Vic, Tas and SA; also occuring in Malaysia.

FLOWERING Mainly September–December, with a few flowers at other times.

209. Viola hederacea ssp. *sieberiana*

Viola hederacea ssp. *sieberiana* **Plate 209**

This is a small plant often less than 3 cm tall, with rosettes of leaves from which runners emerge as lateral stems. These root at the nodes to produce new clumps of leaves. Leaves are rounded at the apex, broadly toothed, wedge-shaped at the base, 6 to 12 mm wide, on stalks longer than the leaf blades. Flowers are a uniform pale mauve and, unlike *V. hederacea* ssp. *hederacea*, are hairless inside.

HABITAT Sandy, open, wet situations.

DISTRIBUTION Confined to Sydney district, except for an isolated occurrence in East Gippsland (Vic). Recordings include Mt Tomah, Newnes Plateau, Fitzroy Falls, Kendalls Glen at Gosford and south of Shoalhaven River along Nerriga-Sassafras road.

FLOWERING September–December.

Hybanthus

Hybanthus differs from *Viola* in having one petal much larger and longer than the others, so that it appears at first to have only one petal.

Hybanthus monopetalus

Slender Violet **Plate 210**

This slender plant has weak stems up to 30 cm high. Narrow leaves to 4 cm long are widely spaced along the stems, the upper ones often opposite. Flowers are borne in racemes on long stalks at the end of the stems. The conspicuous lower petal is blue-mauve, 7 to 20 mm long; the side petals are minute, 2 to 3 mm long, lance-shaped and falcate.

HABITAT Open forests and heaths, usually in cooler shady sites.

DISTRIBUTION Widespread from coast to mountains. Occurring throughout NSW, except far west; also Qld, Vic and SA.

FLOWERING September–December.

211. *Hybanthus vernonii*

Hybanthus vernonii **Plate 211**

This small plant closely resembles *H. monopetalus* in its habit and flowers. It differs by having flowers which are solitary on short stalks in the axils of the upper leaves and not on long stalks bearing a number of flowers. The conspicuous lower petal is blue-mauve, 9 to 13 mm long; the side petals are narrow oblong, 1.5 to 3 mm long.

Two subspecies are recognised. *H. vernonii* ssp. *vernonii* has hairless stems and leaves. Upper leaves are narrower than the lower leaves, usually linear, 5 to 45 mm long. *H. vernonii* ssp. *scaber* has rough (scabrous) stems. Its leaves are all lance shaped to obovate, 5 to 15 mm long and rough, especially on veins and margins.

HABITAT Heath and open forest.

DISTRIBUTION Subspecies *vernonii* occurs along coastal zone, south of Muswellbrook to eastern Vic. Subspecies *scaber* occurs along coastal zone south of Clarence River to southern highlands (e.g., Bundanoon).

FLOWERING September–December.

210. *Hybanthus monopetalus*

Passifloraceae

Passiflora with more than 400 species centred in tropical South America is the only genus of importance in this family. Three endemic species occur in Australia with two in the Sydney area where they are much less common than escaped exotic species.

When the first plants of the genus were brought to Europe in the seventeenth century the name 'passion flower' was given in the belief that the parts of the flower contained the elements of Christ's crucifixion and suffering. The 10 perianth segments were said to represent the 10 faithful apostles, the corona of stamen-like filaments was likened to the crown of thorns, the 5 stamens significed the wounds to the hands, feet and side of Christ, and the 3 styles, each with a prominent stigma, were the nails used on the hands and feet. Today, in a less pious age, a more utilitarian function for the various parts can be recognised. The perianth attracts pollinating insects, usually bees, the corona directs them to reserves of nectar at the bottom of the flower, the anthers are positioned to brush pollen onto the backs of insects thus attracted, and the styles, when receptive, bend downwards to collect pollen from the backs of later visitors.

212. *Passiflora herbertiana*

Passiflora Passion-flower

Climbers with tendrils growing from the base of the leaf stalk. Leaves are 3-lobed; fruit is a large berry.

Passiflora herbertiana

Yellow Passion-flower, Native Passion-flower
Plate 212

This is a climber with stems several metres long. Stems, leaf stalks and the undersurface of the leaves are finely hairy. Leaves are up to 12 cm wide, with 3 broad, almost acute lobes. Immediately beneath the leaf blade, at the top of the leaf stalk, are two raised glands. Flowers are pale, white to yellow-orange, about 6 cm across, with a yellow-green centre. The fruit is a typical passion-fruit shape, green, up to 4 cm long, but drier than the common edible passion-fruit *(P. edulis)*.

HABITAT Sheltered forested slopes and rainforest margins, on more fertile soils.

DISTRIBUTION Widespread, but uncommon, from coast to lower Blue Mountains. Also occurring south coast, north coast, northern tablelands and Qld.

FLOWERING October–December.

213. *Passiflora cinnabarina*

Passiflora cinnabarina

Red Passion-flower **Plate 213**

Red Passion-flower is a rare plant in the Sydney district. It is easily distinguished from *P. herbertiana* by its crimson flowers. The plant also is without the fine hairs of *P. herbertiana*. Leaves are about 10 cm wide and deeply lobed; the 3 lobes are usually pointed. There are no glands on the leaf stalk. The specific name is derived from 'cinnabar' (red mercuric sulphide), a reference to the colour of the flower. Fruit is ovoid, greenish-grey and about 2 cm long.

HABITAT Sheltered forest and rocky gullies.

DISTRIBUTION Scattered from coast to mountains. Extending from the northern tablelands to southern tablelands, and along the south coast to Vic; also Tas.

FLOWERING August–October.

Epacridaceae

This largely Australasian family of nearly 400 species in 30 genera is a major component of coastal heathlands in the Sydney region and is also fairly common in the understorey in different types of woodland and forest.

One species, *Trochocarpa laurina*, is a rainforest shrub or tree.

The Epacridaceae are mostly small straggly shrubs, less than a metre, high, with simple narrow rigid and often pungent-pointed leaves arranged alternately or spirally along the stem. Even in the absence of flowers, members of the family are recognisable by the parallel longitudinal venation of the leaves which is most easily seen on the undersurface. The flowers are small and regular with 5 petals and sepals and 5 stamens attached to the petals. The petals are fused for part of their length, forming either a bell-shaped flower or, more commonly, a tubular flower with 5 lobes. In the local species the sepals are surrounded by an indefinite number of bracts which are similar in appearance to the sepals. The flowers of most species possess nectaries at the base of the ovary, and these are particularly well developed in the genera *Melichrus* and *Acrotriche*.

The family consists of 3 tribes: Styphelieae in which the ovary has one ovule in each compartment and develops into a drupe (a succulent fruit with a central stone), Epacrideae with several ovules in each compartment and a dry capsular fruit, and Prionoteae in which the stamens are attached below the ovary. Only the first two tribes are represented on the Australian mainland.

Individual genera may be recognised by the presence and nature of hairs on the corolla, by the length of the corolla tube, and by whether the stamens project beyond the tube or are enclosed within it. *Sprengelia* and *Dracophyllum* are readily recognised by their leaves which sheath the stems. The majority of species have white flowers but species of *Styphelia, Astroloma* and *Melichrus*, and a few *Epacris* have flowers in shades of red, yellow and green. Deep red flowers of *Woollsia* are known, but mostly they are white, sometimes turning pink with age.

Figure 5

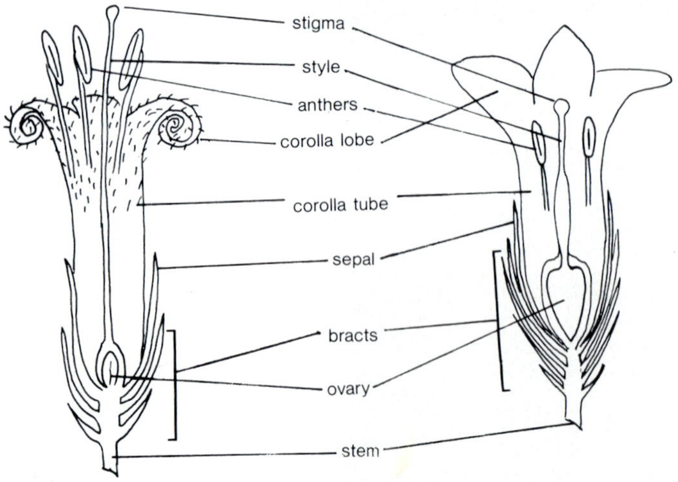

Flower of *Styphelia* (in section) **Flower of *Epacris* (in section)**

215. Styphelia tubiflora

Styphelia Five-corners

Shrubs with rigid, pungent leaves. Flowers with a cylindrical floral tube and 5 stamens projecting beyond the tube. Fruit a drupe.

Styphelia longifolia

Long-leaf Styphelia **Plate 214**

This is an uncommon shrub growing to 2 m tall. Branches are covered with downy to long spreading hairs, especially when young. Leaves are 25 to 50 mm long (mostly 40 mm), 3 to 5 mm wide, lance shaped, concave and taper gradually to a fine point. Flowers are greenish-yellow, with a tube 18 to 25 mm long and stamens extending about 15 mm beyond the tube.

HABITAT Open forest, on sandy soils.

DISTRIBUTION Restricted to small areas north of Sydney (e.g., Katandra Reserve, Mona Vale). Endemic in the Sydney area.

FLOWERING May–June.

214. Styphelia longifolia

Styphelia tubiflora

Red Five-corners **Plate 215**

Red Five-corners is a very common much-branched undershrub in the Sydney district. Its tubular red flowers are very conspicuous as they are in bloom in autumn and winter when few other flowers adorn the undergrowth. Leaves are pointed, spreading, oblong-linear, up to 15 mm long, with margins curved under. Flowers are red, with a slender tube 20 to 25 mm long, with red curled lobes and a greenish-yellow calyx.

HABITAT Heath and open forest, on sandy soils.

DISTRIBUTION Widespread from coast to mountains. Extending to south coast and southern tablelands.

FLOWERING April–August.

Styphelia laeta

Styphelia laeta is a robust shrub, 1 to 2 m tall, with downy-hairy branches. Flowers are creamy yellow (old flowers have a pinkish tinge), tubular, 10 to 20 mm long, with lobes curled back and prominent stamens extending about 10 mm beyond the tube. The leaves are up to 25 mm long, without stalks, finely pointed, erect and crowded along the stems. The shape of the leaf is very variable and three varieties have been recognised in the Sydney district based on leaf shape.

Variety *laeta* (Plate 216) has leaves which are ovate to lance shaped, less than 8 mm wide, and taper rather abruptly to a fine point.

216. Styphelia laeta var. *laeta*

217. Styphelia laeta var. *angustifolia*

218. Styphelia laeta var. *latifolia*

Habitat Heath and open forest, on sandy soils or soils derived from Wianamatta Shales.

Distribution Coast to mountains. Restricted to the Sydney district.

Flowering March–June.

Variety *angustifolia* (Plate 217) has leaves which are narrow lance-shaped, up to 8 mm wide and taper gradually to a fine point.

Habitat Heath and open forest, on sandy soils.

Distribution South of Sydney, Cumberland Plain (e.g., Castlereagh State Forest) and lower Blue Mountains. Also recorded along southern tablelands and north-western slopes.

Flowering March–June.

Variety *latifolia* (Plate 218) has leaves which are 8 to 15 mm wide and taper abruptly to a fine point.

Habitat Open forest, on sandy soils.

Distribution North of Sydney between Hawkesbury River and Gosford (e.g., Warrah Sanctuary, Patonga). Also recorded on central tablelands and north-western slopes.

Flowering March–June.

219. Styphelia viridis

Styphelia viridis

Green Five-corners **Plate 219**

S. viridis is a medium-sized shrub confined to sandy soils close to the coast. It reaches the southern limit of its distribution in the Royal National Park. Its branches are hairless (or nearly so). Leaves are 10 to 25 mm long, broad, lance shaped, tapering to a fine point. Flowers are green, with a tube 16 to 18 mm long, curled hairy lobes, long protruding stamens and conspicuous brown anthers. The habit and leaf shape of this species are similar to *S. laeta*, but it differs from that species in having green flowers, hairless stems and leaves more widely spaced along the branches.

Habitat Old dunes and sandy heaths, near sea.

Distribution Coastal zone from Royal National Park northwards to Qld.

Flowering April–August.

Styphelia triflora **Plate 220**

S. triflora is easily identified by the colour of its flowers. The tube is pink, the curled hairy lobes are yellow and the calyx is a light yellow-green. This colour combination and the fact that the flowers are often clustered together high on the stems and are longer than the leaves immediately attract attention. This is an erect shrub, sometimes 2 m tall. Leaves are up to 25 mm long, erect, broad, lance shaped, tapering to a sharp point.

Habitat Heath and shrubland, on sandy soils.

Distribution Widespread from coast to mountains. Found throughout coastal and tablelands divisions and westward to edge of the western plains; also Qld.

Flowering July–October, but later in Blue Mountains.

220. Styphelia triflora

Astroloma

Small shrubs with tubular flowers 10 to 18 mm long. Lobes at apex of flower tube short; stamens not projecting beyond the tube. Fruit a drupe. The name *Astroloma* is derived from the Greek words meaning 'star-fringe', a reference to the star-like tuffs of hairs found below the middle of the floral tube in most members of the genus.

Astroloma pinifolium

Pine Heath **Plate 221**

This is a low diffuse shrub, with leaves which are crowded, stalkless, narrow-linear, 10 to 18 mm long, and somewhat resembling small pine needles, as the specific name suggests. Flowers are solitary, tubular, about 18 mm long, red, becoming yellow then green towards the tip. The lobes at the tip are small and straight, not spreading. The stamens are enclosed within the flower tube. The flowers lack hairs inside the tube, a feature of other *Astroloma*. The base of the tube is enclosed in bracts. Fruit is a globular drupe about 8 mm across.

Habitat Heath and open forest, on sandy soils.

Distribution Coast to lower Blue Mountains. Also occurring north coast, south coast, southern tablelands, Vic and Tas.

Flowering May–September.

221. Astroloma pinifolium

Astroloma humifusum

Cranberry Heath **Plate 222**

Cranberry Heath is a low shrub with many diffuse branches, often forming small mats over the ground. Leaves are 5 to 12 mm long, spreading, narrow lance shaped to linear, with a sharply pointed tip and minutely fringed margins. The flower tube, unlike that of *A. pinifolium*, is entirely red, 10 to 12 mm long, with tufts of hair inside below the middle and silvery-hairy, spreading lobes at the tip. Flowers are somewhat hidden in the foliage. Fruit is a greenish globular drupe.

Habitat Open forest, on soils derived from sandstone and shale, or on heavier clays.

Distribution Widespread from coast to mountains. Also occurring south coast, southern tablelands, western slopes, south-western plains, Vic, Tas, SA and WA.

Flowering Chiefly February–June, but some flowers at other times.

222. *Astroloma humifusum*

Melichrus

Small shrubs with shallow bell-shaped flowers. Flower tube short, with spreading star-like lobes longer than the tube, and 5 glandular scales alternating with the stamens. Fruit a globular drupe.

Melichrus procumbens

Jam Tarts **Plate 223**

Jam Tarts is a low sprawling shrub with hairy branches. Leaves are narrow lance shaped, 12 to 25 mm long, finely pointed and fringed along the margins with conspicuous hairs. Flowers are creamy, 10 to 12 mm across, rather open, with spreading lobes and a large amount of nectar. A feature of this species is that the flowers face the ground on the underside of the branches. This means that they are often overlooked by humans, but it makes them very attractive to ants which seem to be the main pollinating agent. Fruit is a red drupe, 4 to 5 mm across.

HABITAT Heath and open forest, on sandy soils.

DISTRIBUTION Widespread from coast to mountains. Extending along the north coast and northern tablelands to Qld.

FLOWERING August–September.

Melichrus urceolatus

Urn Heath **Plate 224**

Urn Heath is a small erect stiff shrub, usually about 60 cm tall. Although it occurs extensively throughout the State, it is not common in the Sydney district. Leaves are narrow lance shaped, 10 to 25 mm long, stalkless, concave, spreading and narrowing to a sharp point at the apex. The undersurface of the leaf is clearly marked with parallel veins. Flowers are small, creamy green, with an urn-shaped tube with 5 slightly spreading lobes. They are borne singly and without stalks in the leaf axils. Buds are conical. Fruit is a globular greenish succulent drupe.

HABITAT Forest and woodland, on shallow sandy soils on slopes and ridges.

DISTRIBUTION Western parts of Blue Mountains and the Wedderburn-Appin district. Found throughout NSW, except far western plains; also Vic.

FLOWERING August–November.

223. *Melichrus procumbens*

224. *Melichrus urceolatus*

225. *Leucopogon virgatus*

Leucopogon Beard-heath, White-beard

Shrubs with small white flowers. The 5 star-like lobes are densely covered on the inside with white hairs; the hairs descend into the throat of the flower tube. Stamens enclosed in the tube. Fruit a drupe.

Leucopogon virgatus **Plate 225**

This is a small shrub with wiry branches to about 50 cm tall. The narrow lance-shaped leaves are concave, up to 15 mm long and tapering to a long point. There is a short, but distinct, leaf stalk. Some leaves are spreading, others pressed against the stem. Flowers are in short dense spikes, crowded near the ends of the branches. Sepals are small (2 mm long) and obtuse. The flower tube is also small; the lobes are spreading with conspicuous fluffy white hairs. Buds and sepals are often pink. This species is widely distributed throughout south-eastern Australia, but is not as common in the Sydney district as a number of other *Leucopogon.*

HABITAT Old dunes, heath and open forest, on sandy soils.

DISTRIBUTION Widespread from coast to mountains. Recordings include Bouddi National Park, Castlereagh State Forest and Agnes Banks Nature Reserve. Found throughout coast, tablelands and western slopes divisions; also Qld, Vic, Tas and SA.

FLOWERING September–October.

Leucopogon appressus **Plate 226**

This is a small shrub, with wiry, often straggling, branches. Leaves are stalkless, lance shaped, 5 to 15 mm long, concave, tapering to a long fine point at the apex and with numerous veins spreading from the base. Leaves are pressed against the stem or clustered and spreading on short side branches. Flowers are small, erect and borne singly among the leaves on the side branches. Sepals are obtuse.

HABITAT Heath and open forest, on sandy soils or shales.

DISTRIBUTION Coast and lower Blue Mountains. Recordings include Bangor and Bilpin-Mountain Lagoon. Endemic in central coast and central tablelands.

FLOWERING November–February.

Leucopogon amplexicaulis **Plate 227**

This small undershrub is easily identified because of its stem-clasping, broadly heart-shaped leaves, 12 to 25 mm long, which are fringed with hairs. Although it may be locally common, it is regarded as a plant at risk of extinction because of its small populations and restricted distribution. Stems are reddish, in contrast to the light green leaves. Flowers are in spikes longer than the leaves.

HABITAT Sheltered forested sites, often beneath rocky outcrops.

DISTRIBUTION Restricted to the coastal zone. Recordings include Brisbane Water National Park, Cowan Creek, Muogamarra, St Ives, Oatley Park, Woronora River and Audley. Extending to the south coast.

FLOWERING August–October.

226. *Leucopogon appressus*

227. *Leucopogon amplexicaulis*

228. *Leucopogon lanceolatus*

229. *Leucopogon parviflorus*

Leucopogon lanceolatus var. *lanceolatus*

Lance-leaf Beard-heath **Plate 228**

This tall shrub (to 1 m high) has the longest leaves of any *Leucopogon* in the area. They are up to 40 mm long, 3 to 10 mm wide, narrow lance shaped, with conspicuous longitudinal veins and narrow evenly at both ends. Stems are usually smooth, although young stems may have minute downy hairs. Flowers are white, with a tube only 3 mm long, but are borne in unmistakable loose spikes longer than the leaves.

HABITAT Forest.

DISTRIBUTION Widespread from coast to mountains. Extending all along coast and tablelands divisions from Qld to Vic; also north-western slopes, Tas and SA.

FLOWERING September–December.

Leucopogon parviflorus

Coast Beard-heath **Plate 229**

Coast Beard-heath is rather like a smaller version of *L. lanceolatus*. Its leaves are lance shaped (often wider in the upper half), bluntly pointed, narrow evenly at both ends and are mostly 10 to 20 mm long. Flowers are white and fluffy and are borne in dense spikes from the leaf axils. Unlike the spikes of *L. lanceolatus*, these are generally shorter than the leaves.

HABITAT Coastal sand dunes, sandy heaths and woodlands near the sea.

DISTRIBUTION Common along coast (e.g., Burning Palms, Pearl Beach, Bouddi National Park, Minnamurra Spit). Also occurring south coast, north coast and along coast of all other States (except NT), New Zealand and some Pacific Islands.

FLOWERING July–September.

Leucopogon microphyllus

Small-leaved White-beard **Plate 230**

Both the botanical and common names clearly indicate the main features of this common wiry shrub. It has the smallest leaves of all the local *Leucopogon*, usually 2 to 4 mm long, but occasionally longer on young plants. They are flat, firm, ovate to narrow lance shaped and obtuse. The plant may be erect or straggling, 30 to 80 cm tall, with numerous small side branches. The woolly white flowers are clustered at the ends of these side branches. The corolla tube is only about 3 mm long and is almost covered by the narrow, acute, brownish-red sepals.

HABITAT Heath and open forest, on sandy soils.

DISTRIBUTION Widespread from coast to mountains. Extending all along coast and tablelands divisions; also central-western slopes and Qld.

FLOWERING Chiefly May–November, but some flowers at other times.

Leucopogon exolasius **Plate 231**

A spreading shrub, often over 1 m tall, endemic in the Sydney district. It is identified primarily by its pendulous flowers which are covered with minute hairs on the outside of the corolla as well as the white hairs inside the corolla which are typical of the genus. Flowers are on slender stalks, 2 to 4 mm long, from the leaf axils, with 1 to 3 flowers together. Leaves are narrow, 1 to 2 mm wide, up to 15 mm long, tapering to a long pungent tip; leaf margins are curled underneath. This species may be confused with *L. setiger* which also has pendulous flowers. However, *L. setiger* has flowers on longer stalks, the corolla is hairless outside and its leaves are shorter and broader.

HABITAT Rocky river banks.

DISTRIBUTION Upper Georges River (e.g., The Basin), Woronora River and Grose River.

FLOWERING August–September.

230. *Leucopogon microphyllus*

Leucopogon setiger **Plate 232**

Numerous delicate white bell-shaped flowers hanging on slender stalks below the stems make this spreading shrub easy to identify. Flowers are without hairs on the outside of the tube (which separates this species from *L. exolasius*) and are borne singly or in groups of 2 to 4 on a stalk up to 8 mm long. Leaves are linear to lanceolate, up to 12 mm long, 2 to 4 mm wide, and narrow to a sharp point.

HABITAT Open forest, on sandy soils.

DISTRIBUTION Widespread from coast to mountains, but most common in higher regions away from coast (e.g., Springwood, Blackheath, Bundanoon). Also occurring central-western slopes.

FLOWERING August–October.

231. *Leucopogon exolasius*

232. *Leucopogon setiger*

235. *Leucopogon ericoides*

Leucopogon muticus

Blunt Beard-heath **Plate 233**

An erect shrub, 1 to 2 m tall, with blunt, dull green leaves, up to 15 mm long and about 3 mm wide. These have a broad and rounded apex, with a minute tip, narrow to a short stalk, and are flat or have slightly down-curved margins. This shrub has a characteristic erect habit, with reddish young stems. Flowers are in small clusters in the leaf axils.

HABITAT Heath and open forest, on sandy soils.

DISTRIBUTION Blue Mountains and higher parts of Woronora and Hornsby Plateaux. Also occurring north coast, northern tablelands, southern tablelands, western slopes and Qld.

FLOWERING September–October.

233. *Leucopogon muticus*

Leucopogon juniperinus

Prickly Beard-heath **Plate 234**

This is a low straggling shrub. Leaves are spreading, pungent, narrow lanceolate to linear, 5 to 10 mm long, almost flat or with margins curled under, and tapering to a fine point. The flowers are 5 to 7 mm long, with a floral tube about twice as long as the sepals. Lobes are short so their hairiness is not as conspicuous as in most other *Leucopogon*. Flowers are usually solitary (sometimes a few together) in the leaf axils.

HABITAT Forest, but also found along roadsides and in semi-cleared areas.

DISTRIBUTION Widespread from coast to mountains. Also occurring south coast, north coast, northern tablelands, Qld and Vic.

FLOWERING Chiefly July–October, but some flowers at other times.

234. *Leucopogon juniperinus*

Leucopogon ericoides **Plate 235**

A small shrub, especially common on old coastal dunes and heaths. It may be a slender spreading shrub, 60 to 80 cm tall, or a low, almost prostrate, clumped shrub. Leaves are spreading, up to 8 mm long, narrow-oblong, with margins curved under, and with a short pungent point. Flowers are small, with a corolla 2 mm long, but numerous along the stems in stalkless clusters in the leaf axils. Buds are often pinkish. The length of the flower clearly distinguishes this species from *L. juniperinus.*

HABITAT Old coastal dunes, heath and open forest, on sandy soils.

DISTRIBUTION Widespread from coast to mountains. Recordings include Kurnell, La Perouse, Heathcote National Park, Muogamarra, Brisbane Water National Park and Agnes Banks. Also occurring north coast, south coast, southern tablelands, Qld, Vic, Tas and SA.

FLOWERING July–October.

236. *Leucopogon esquamatus*

Leucopogon esquamatus **Plate 236**

An erect shrub, seldom more than 60 cm tall, with conspicuous erect, lance-shaped leaves which overlap each other. These leaves are flat, often with a slight twist, about 15 mm long, 2 to 4 mm wide, and have a long fine point. Flowers are small, and are either solitary or in pairs in the leaf axils all along the upper part of the stems.

HABITAT Heath, on sandy soils.

DISTRIBUTION Widespread from coast to mountains, but most common on coastal heaths. Also occurring north coast, south coast, southern tablelands, Vic and Tas.

FLOWERING August–September.

Leucopogon fraseri **Plate 237**

A low spreading shrub, 30 to 60 cm tall. It is essentially a tablelands species, and is absent from the coastal zone. Leaves are small, flat, 6 to 10 mm long, mostly elliptic, clumped and pungent-pointed. Flowers are solitary in the leaf axils, with a long flower tube which often extends beyond the leaves. The tube is 6 to 8 mm long, pink-red, topped by short, spreading, woolly white lobes. Sepals are acute, but much shorter than the flower tube.

HABITAT Forest and heath, in rocky areas, usually in soil pockets of flat rock outcrops.

DISTRIBUTION Higher Blue Mountains and southern highlands (e.g., Wingello State Forest). Extending from northern tablelands to southern tablelands; also higher parts of south coast, Vic, Tas and New Zealand.

FLOWERING August–October.

Brachyloma

Erect shrubs with white flowers; tube less than 6 mm long, the throat closed by hairs which descend between the stamens (hence the generic name: 'brachys', short; 'loma', fringe). Fruit is a drupe.

Brachyloma daphnoides

Daphne Heath **Plate 238**

An erect grey-green shrub, up to 1 m tall, with small white tubular flowers which are very sweet scented. These flowers are about 6 mm long, broader at the base than at the tip, with 5 spreading star-shaped lobes. The tips of the lobes are usually orange-brown. There is a ring of hairs inside the throat of the tube. These hairs are not as conspicuous as those of *Leucopogon* as the tube is not as open. Leaves are almost stalkless, 4 to 10 mm long, ovate to lance shaped, flat or concave, with veins diverging from the base. The fruit is a small round drupe, 3 to 4 mm across.

HABITAT Heath and open forest, on sandy soils.

DISTRIBUTION Widespread from coast to mountains. Recordings include Girrakool, Heathcote, Agnes Banks, Castlereagh State Forest and Woodford. Found throughout NSW, except far western plains; also Qld, Vic, Tas and SA.

FLOWERING June–November.

237. Leucopogon fraseri

Lissanthe

Erect shrubs with white flowers; tube less than 6 mm long, hairy inside about the middle. Lobes spreading and without hairs. Ovary with 5 to 7 cells. Fruit a drupe.

Lissanthe strigosa

Peach Heath **Plate 239**

A small rigid shrub about 50 cm tall with narrow spreading sharp-pointed leaves. Numerous small, white, bell-shaped flowers, about 3 mm long, are borne in short clusters among the leaves. These flowers are shortly hairy inside above the middle of the tube, but the 5 spreading lobes are quite hairless. The stamens are conspicuous in the throat of the tube as they are joined to the edge of the tube and slightly protrude. Buds are obtuse and often have a pinkish tinge. Leaves are rigid and sharp pointed, linear, up to 10 mm long and 1 to 2 mm wide. The fruit is round, pale and fleshy.

HABITAT Open forest, often on lateritic and clay soils.

DISTRIBUTION Widespread from coast to mountains. Found throughout NSW, except far western plains; also Qld, Vic, Tas and SA.

FLOWERING July–September.

238. Brachyloma daphnoides

240. Lissanthe sapida

Lissanthe sapida

Native Cranberry **Plate 240**

L. sapida is a much larger plant than *L. strigosa* (up to 1 m tall), with more spreading branches. The white tubular flowers are larger (about 5 mm long) and pendulous; they are borne in loose few-flowered clusters on slender stalks. Leaves are narrow-oblong, 15 to 25 mm long, with a short sharp point. The fruit is round and fleshy, turning bright red when mature.

HABITAT Open forest, often on exposed rocky areas near cliffs.

DISTRIBUTION Blue Mountains (e.g., Mt Tomah, Pierces Pass, Nepean Lookout, Mountain Lagoon); also recorded from Woronora Plateau. Endemic in the Sydney district.

FLOWERING May–September.

Monotoca Broom-heath

Shrubs with small whitish bell-shaped flowers. Floral tube and lobes without hairs. Ovary 1-celled (hence the generic name: 'mono', single; 'tokos', birth). Fruit a drupe.

239. Lissanthe strigosa

241. *Monotoca elliptica*

Monotoca elliptica

Tree Broom-heath **Plate 241**

A very common shrub, especially near the sea. In protected sites, it may grow almost to tree size (3 to 4 m tall), but in exposed cliff-top situations it is a small, densely clumped shrub, 50 to 80 cm tall. Flowers are small white bells with a tube less than 3 mm long and 5 spreading, hairless lobes. Orange-yellow anthers are attached to the tube and are clearly visible between the lobes. There are no hairs in the tube and the style is very short. Unlike *M. scoparia*, the flowers of this species are borne along racemes which may be longer than the leaves. Leaves are elliptic-oblong, 12 to 25 mm long, with a short point. The undersurface of the leaf is pale and marked with clear parallel veins. Fruit is a red, fleshy berry with one compartment.

HABITAT Old coastal dunes, heath and forest, on sandy soils.

DISTRIBUTION Most common along coast near the sea and estuaries, but also recorded in mountains. Extending all along coast from Qld to Vic and Tas; also occurring southern tablelands.

FLOWERING July–September.

Monotoca scoparia

Prickly Broom-heath **Plate 242**

Prickly Broom-heath is a common rigid shrub to 1 m tall. It is sometimes confused with *Lissanthe strigosa*, but the leaves of *M. scoparia* are broader, and the flowers smaller, more open and lacking any hairs in the corolla tube. The ovary and fruit of *Monotoca* have only one compartment, whereas the ovary and fruit of *Lissanthe* have 5 to 7 compartments. Leaves are oblong-linear, 8 to 10 mm long, with abrupt sharp points. The undersurfaces are pale and marked with clear parallel veins. Flowers are white, hairless, with a tube about 2 mm long and 5 spreading lobes. The yellow-orange anthers are conspicuous in the throat of the tube. Flowers are in clusters of 2 to 4 or very small racemes in the leaf axils.

HABITAT Heath and open forest, on sandy soils.

DISTRIBUTION Widespread from coast to mountains. Found all along coast and tablelands divisions and westwards to western slopes; also Qld, Vic, Tas and SA.

FLOWERING April–September.

Acrotriche

Rigid shrubs, with small flowers clustered on the older stems. Flowers small, greenish, with hairs at the tips of the spreading lobes. The generic name means 'tip-hair'. Fruit a small drupe.

243. *Acrotriche divaricata*

Acrotriche divaricata

Ground-berry **Plate 243**

A spreading shrub, 1 to 2 m tall, often wider than its height, with a dense appearance. Leaves are rigid, pungent-pointed, narrow lance shaped, flat and up to 15 mm long. The most conspicuous feature of this shrub is the way the small flowers are b orne in clusters on the older branches, hidden below the leaves. Flowers are small (about 3 mm) and green. A distinctive erect tuft of hair is at the end of each of the 5 corolla lobes. The ovary is green, with 4 to 5 compartments. Another species, *Acrotriche aggregata*, is found in the Blue Mountains (e.g., Yerranderie, Lawson, Blackheath, Colo River). It has 6 to 10 compartments in the ovary and a band of red down the sides of the ovary. Its other features are similar to those of *A. divaricata*.

HABITAT Open forest, on sandy soils.

DISTRIBUTION Widespread from coast to mountains. Extending along coast and tablelands divisions from Qld to Vic.

FLOWERING July–September.

Trochocarpa

Tree with broad leaves up to 5 cm long from rainforest margins. Flowers small in terminal spikes. Fruit a drupe.

242. *Monotoca scoparia*

244. *Trochocarpa laurina*

Trochocarpa laurina

Tree Heath **Plate 244**

A common tree or shrub in protected forests. It is an attractive plant with broad, dark green, glossy leaves and conspicuous light tan new growth. Leaves are entire, elliptic to ovate, pointed, about 5 cm long, glossy on both sides, but paler underneath and with 5 to 7 clear parallel veins. Flowers are rather inconspicuous, small white and tubular, and borne in spikes 2 to 3 cm long. The mature fruit is a dull blue-black berry.

HABITAT In and near rainforest and damp sheltered eucalypt forests.

DISTRIBUTION Coastal, reaching the southern limit of its range in Illawarra district. Recordings include Mt Kembla, Bola Creek, Kelleys Falls, Wondabyne, Gap Creek and Watagan State Forest. Extending along north coast and northern tablelands to Qld.

FLOWERING Chiefly January–April, but some flowers at other times.

Epacris

Shrubs with tubular or bell-shaped flowers. Hairs absent from the inside of the flower tube; stamens inserted at the throat of the tube. Style inserted in a pit in the ovary summit. Fruit a capsule.

245. *Epacris longiflora*

Epacris longiflora
Fuchsia Heath **Plate 245**

A striking small shrub as its tubular flowers are red with white spreading tips (rarely completely white), up to 2 cm long, and are very showy as they hang closely together along the upper stems. Leaves are small, 5 to 12 mm long, crowded, heart-shaped and pointed.

HABITAT Heath and open forest, on sandy soils.

DISTRIBUTION Widespread from coast to mountains, but especially common on coastal heath. Extending along north coast and nothern tablelands to Qld.

FLOWERING Chiefly April–November, but some flowers at other times.

246. Epacris reclinata

Epacris reclinata **Plate 246**

A small shrub with weak, recumbent branches (hence 'reclinata') which trail over sheltered, often damp, rock ledges in the upper Blue Mountains. Flowers are tubular, 8 to 12 mm long, and uniformly red or pink. Leaves are small, 4 to 6 mm long, heart shaped, spreading and pointed.

HABITAT On sheltered sandstone cliff faces in gullies.

DISTRIBUTION Restricted to upper Blue Mountains (e.g., Blackheath, Wentworth Falls, Hassans Walls); extending to central western slopes.

FLOWERING August–November.

Epacris obtusifolia
Blunt-leaf Heath **Plate 247**

A common shrub with erect stems, usually 50 to 80 cm tall, occurring mainly in swampy sites. The white flowers are cylindrical, somewhat bell shaped, and 4 to 8 mm long. The throat of the corolla is open and the 5 anthers are clearly visible around the edge of the tube. The flowers are conspicuous as many are borne together along the top 10 cm of the stems. Leaves are broad linear, shortly stalked, blunt, erect and about 10 mm long.

HABITAT Heath and shrubland, usually in swampy sandy areas.

DISTRIBUTION Widespread from coast to mountains. Very common on wet coastal heaths (e.g., Kurnell, La Perouse, Munmorah State Recreation Area). Extending all along coast and tablelands divisions; also Qld, Vic and Tas.

FLOWERING June–November.

248. Epacris crassifolia

Epacris crassifolia **Plate 248**

This is a very variable shrub with slender stems which trail weakly over sheltered to damp cliff faces. Coarser forms with more rigid stems also occur. Leaf and flower size vary so much that it is often difficult to believe that extreme forms belong to the same species. The extremes often grow side by side (e.g., at Govetts Leap, Blackheath). The smaller form has obovate to oblanceolate leaves, 4 to 5 mm long, 2 mm wide, almost stalkless, scattered with loose hairs and with hairy margins. The flower is tubular, white, about 6 mm long, 2 mm wide, with the tube about the same length as the ring of sepals (3 mm). The leaves of the larger form are quite hairless, up to 7 mm long and 3 mm wide, distinctly stalked. The flower is 20 mm long, 7 mm wide, and much longer than the sepals which are 4 to 5 mm.

HABITAT Damp sheltered sandstone ledges and cliffs in gullies.

DISTRIBUTION Coast to mountains (e.g., Brisbane Water National Park, Wentworth Falls). The weak, small-leaved form occurs only in the Blue Mountains. Extending to south coast, southern tablelands and central western slopes.

FLOWERING November–January.

247. Epacris obtusifolia

249. Epacris coriacea

Epacris coriacea **Plate 249**

This is an uncommon plant, seldom seen because of its narrow distribution and its specialised habitat preference. It grows as an erect, sometimes tall (1 to 2 m) shrub, with flowers which are short white bells borne in the leaf axils and usually slightly longer than the leaves. The 5 spreading star-shaped lobes are about as long as the flower tube. Leaves are crowded, broadly lance shaped, 4 to 8 mm long, shiny, slightly concave and have a short, but distinct, stalk.

HABITAT Sandstone cliffs.

DISTRIBUTION Restricted to a narrow band along Illawarra Escarpment (e.g., Bulli Lookout, Macquarie Pass, above Werrong Beach); also reported from central tablelands.

FLOWERING August–October.

250. Epacris rigida

Epacris rigida **Plate 250**

An erect rigid shrub, 50 to 80 cm tall, found only in the central Blue Mountains. Its leaves are small, 2 to 4 mm long, almost stalkless, elliptic, obtuse and have a prominent thickened mid-rib underneath. Flowers are white, bell-shaped and borne in the axils of the upper leaves. The flower stalks are less than 2 mm long and the spreading lobes are obtuse. This species is very similar to *E. muelleri* which also occurs in the Blue Mountains, and the two species seem to intergrade in some areas. However, in the typical form, *E. rigida* has shorter flower stalks, leaves with thicker keels and tends to flower earlier than *E. muelleri*.

HABITAT Open rocky sites.

DISTRIBUTION Restricted to central Blue Mountains (e.g., Linden, Woodford, Kings Tableland).

FLOWERING September–October.

251. Epacris muelleri

Epacris muelleri
Mueller's Heath **Plate 251**

A spreading or erect shrub, 10 to 30 cm tall, found only in the Blue Mountains. Leaves are thick, blunt, elliptic, 2 to 4 mm long and almost stalkless. Flowers are white, bell-shaped and borne in small clusters in the axils of the upper leaves. The flower stalks are 2 to 5 mm long and the spreading lobes are obtuse. This species seems to intergrade in some areas with *E. rigida*, but differs from that species in having leaves without the prominent thickened keel, long flower stalks, a slightly later flowering time and grows on damp rock ledges.

HABITAT Damp rock faces on sheltered hillsides.

DISTRIBUTION Endemic in the Blue Mountains. Scattered throughout higher areas (e.g., Mt Irvine, Blackheath, Bowens Creek, Newnes).

FLOWERING November–December.

Epacris paludosa
Swamp Heath **Plate 252**

A stiff, erect shrub, usually about 1 m tall, occurring in wet areas. Its leaves are narrow lance shaped, pointed, about 10 mm long and rather upright along the stem. Flowers are white, and have a characteristic narrow floral tube, 1 to 2 mm diameter and up to 6 mm long. Lobes are obtuse and spreading. The style is slightly longer than the floral tube. Flowers are numerous and crowded in leaf axils at the ends of the branches, appearing like dense terminal heads.

HABITAT Swampy sites, on sandy soils.

DISTRIBUTION Woronora Plateau (e.g., Barren Grounds) and Blue Mountains (e.g., Blackheath, Newnes). Also occurring south coast, southern tablelands, northern tablelands, Vic and Tas.

FLOWERING September–December.

252. Epacris paludosa

Epacris calvertiana **Plate 253**

This is an erect shrub, with lance-shaped, pointed, erect, concave leaves, 1 to 2 cm long. Flowers are white or yellowish, with a somewhat swollen floral tube, 2 to 4 mm diameter, and narrow, acute lobes. The style is longer than the floral tube. The swollen floral tube and the acute lobes help distinguish this species from *E. paludosa*. Flowers are borne in the leaf axils, near the ends of the branches. In the southern Blue Mountains, there is a variety with a pink corolla-tube and white lobes, known as variety *versicolor*.

HABITAT Forest, usually on sheltered rocky slopes.

DISTRIBUTION Southern highlands (e.g., Bundanoon, Fitzroy Falls) and higher Blue Mountains (e.g., Blackheath). Extending to south coast and southern tablelands.

FLOWERING September–November.

253. Epacris calvertiana

254. Epacris microphylla

Epacris microphylla
Coral Heath **Plate 254**

An attractive small shrub, 30 to 60 cm tall, often abundant in swampy heaths. The specific name 'microphylla' refers to its characteristic small leaves; these are 3 to 4 mm long, ovate, stalkless, concave and have a short point. Flowers are small, with a broad floral tube about 2 mm long. They are usually white, but buds and flowers often have a pinkish tinge. Lobes are spreading, as long as the tube, and the sepals taper to a point. Stamens are clearly visible in the throat of the floral tube. Although the flowers are small, they are borne abundantly along the upper part of the stems.

HABITAT Chiefly heath, but also open forest, on sandy soils.

DISTRIBUTION Widespread from coast to mountains. Occurring throughout the coast and tablelands divisions, and south-western slopes; also Qld, Vic and SA.

FLOWERING Chiefly June–November, but some flowers at other times.

255. Epacris apiculata

Epacris apiculata **Plate 255**

This straggling shrub is regarded as a rare plant, at risk because of its small and very localised populations. Leaves are pointed, heart shaped, about 8 mm long, and crowded to almost hide the stems. Flowers are white, stalkless and crowded along the upper part of the stem. They have broad, spreading lobes, an open floral tube, orange-yellow anthers and an unusual and characteristic long style.

HABITAT Sheltered, often damp, rock ledges.

DISTRIBUTION Restricted to the Blue Mountains at Wentworth Falls (e.g., Undercliff Walk), Leura and Blackheath (e.g., Centennial Glen).

FLOWERING October–December.

Epacris purpurascens var. *purpurascens*

This is a rigid erect shrub with a few straight branches, sometimes 1 m or more tall. Leaves are 8 to 12 mm long, stiff, ovate, stalkless, with the base curved around the stem. Flowers are pink, with a broad floral tube about 6 mm long and lobes 4 to 5 mm long. The sepals are about as long as the floral tube. Flowers are borne from the leaf axils and are a little longer than the surrounding leaf. A white-flowered form (var. *onosmiflora*) occurs in sheltered damp rocky areas of the Blue Mountains.

HABITAT Damp forest, in sands, shales or rocky sites.

DISTRIBUTION Restricted to coastal zone around Sydney (e.g., Pennant Hills Park, Bankstown, Brisbane Water National Park) where it is uncommon.

FLOWERING September–October.

256. Epacris pulchella

Epacris pulchella — Plate 256

A common medium-sized shrub, usually 50 to 100 cm tall, with wiry branches. It is sometimes mistaken for *E. microphylla*, but it differs from that species in having longer leaves and flowers. On coastal heaths, it flowers in autumn, which distinguishes it from most other *Epacris* species in the Sydney area. Leaves are ovate, stalkless, concave, pointed and 4 to 6 mm long. Flowers are white (sometimes pinkish), borne in the leaf axils and are longer than the leaves. The floral tube is broad, sepals finely pointed and corolla lobes are 2 to 3 mm long.

HABITAT Heath and open forest, on sandy soils.

DISTRIBUTION Widespread from coast to mountains. Occurring throughout the coast and tablelands divisions of NSW; extending to Qld.

FLOWERING Chiefly March–June, but some flowers at other times, especially in mountains.

257. Rupicola sprengelioides

Rupicola

An erect shrub with bell-shaped flowers, without hairs inside. Stamens attached to the base of the floral tube (unlike those of *Epacris*). Fruit a capsule.

Rupicola sprengelioides — Plate 257

This erect shrub is a rare species, with small populations and restricted habitat preference. Its generic name is a reference to its preferred habitat ('rupi', rock; 'colo', to inhabit). Leaves are short (5 to 6 mm long), rigid and crowded along the stems; they have a very short stalk and an obtuse point. Unlike most other members of Epacridaceae, the flowers are open and star-like, with only a very short tube (2 mm long). Flowers are white, 12 mm across, 5 petalled and have orange anthers which encircle the style. Flowers are borne along the stems and are often clustered near the ends of the branches.

HABITAT Rocky cliffs and escarpments.

DISTRIBUTION Restricted to small areas of the Blue Mountains and Burragorang Tableland. Recordings include three locations overlooking Lake Burragorang-McMahons Lookout (Kings Tableland), Nepean Lookout (Erskine Creek) and Burragorang Lookout (Nattai).

FLOWERING November–December.

Woollsia

Monotypic genus consisting of an erect shrub with tubular flowers. Floral tube without hairs inside; lobes twisted in bud but spreading when flower opens.

Woollsia pungens

Woollsia — Plate 258

This medium-sized shrub (usually less than 1 m tall) is a common ingredient of heaths and dry eucalypt forests throughout eastern NSW. Its leaves are crowded, up to 12 mm long, ovate and tapering to a rigid pungent point. Flowers are mostly white, but there are pink forms, especially in shaded sites. South of the area, on Beecroft Peninsula (Jervis Bay), there are plants with deep crimson flowers. The floral tube is long and narrow and topped by 5 broad, spreading, overlapping lobes. When in bud, these lobes are conspicuously twisted. Flowers are borne in attractive clusters in the axils of the upper leaves.

HABITAT Hind dunes, heath and open forest, on sandy soils.

DISTRIBUTION Widespread from coast to mountains. Also occurring south coast, southern tablelands, north coast and Qld.

FLOWERING May–September on coast, later in mountains.

258. Woollsia pungens

Sprengelia

Small shrubs with leaves sheathing at the base around the branches. Floral tube short, with spreading lobes and anthers clustered around the style.

Sprengelia incarnata

Pink Swamp Heath — Plate 259

The terminal clusters of pink flowers of *S. incarnata* often dominate open swampy areas in spring. It is a common upright shrub to 1 m tall, with characteristic stem-sheathing leaves, which are 12 to 20 mm long, and pointed. Flowers are pink (hence 'incarnata'—flesh coloured), or occasionally white, with narrow lobes which are star-like and spreading;

259. *Sprengelia incarnata*

261. *Sprengelia monticola*

the edges of the petals touch, but do not overlap, in bud. Sepals are pale and narrow.

HABITAT Swamp areas, especially on heaths.

DISTRIBUTION Widespread from coast to mountains. Occurring throughout coast and tablelands divisions; also Vic, Tas and SA.

FLOWERING July–October.

Sprengelia sprengelioides — Plate 260

This is a small erect shrub, up to 1 m tall (but usually 30 to 50 cm), with crowded, pointed leaves, 3 to 5 mm long and stem-clasping. Flowers are white, with 5 broad star-shaped lobes. Sepals are broad, green, about 6 mm long and leaf-like, an important feature distinguishing this species from *S. incarnata*. Flowers are borne singly at the end of the stems.

HABITAT Swampy areas, especially in sedgy heaths.

DISTRIBUTION Coastal zone north of Sydney, but uncommon in area. More widespread and common on north coast; extending to Qld.

FLOWERING September–October.

260. *Sprengelia sprengelioides*

Sprengelia monticola

Rock Sprengelia — Plate 261

A small spreading shrub which characteristically clings to sheltered rocks in the Blue Mountains. Its leaves are narrow, 3 to 6 mm long, stem-sheathing and rather spreading. Flowers are white, solitary and star-like. In contrast to the white petals, the orange-brown anthers are clearly visible surrounding the style.

HABITAT Sheltered, often wet, rocks and cliff faces.

DISTRIBUTION Endemic in the Blue Mountains (e.g., Kings Tableland, Wentworth Falls, Blackheath).

FLOWERING August–December.

Dracophyllum

Shrub with long sheathing leaves. Flowers with cylindrical tubes and small lobes.

262. *Dracophyllum secundum*

Dracophyllum secundum — Plate 262

A common small shrub on damp rock ledges where it is easily distinguished by its pineapple-like leaves and long one-sided stems of tubular flowers. Plants are usually 20 to 50 cm tall, but old protected specimens may grow to 2 m or more. Leaves are stem-sheathing, and up to 10 cm long; old leaves fall off leaving scars on the bare branches. Flowers are bell shaped, about 10 mm long, white or pink, and the lobes are small and recurved. Buds are covered by long conspicuous brown bracts.

HABITAT Damp rock ledges on sheltered slopes and in gullies.

DISTRIBUTION Widespread from coast to mountains. Also occurring north coast, south coast and southern tablelands.

FLOWERING July–September, later in the mountains.

Ebenaceae

The Ebenaceae, centred in Indomalesia and tropical Africa, consists of tropical rainforest trees noted for the dark heavy timber called ebony which is obtained from the heartwood of several species of *Diospyros*. Other *Diospyros* species produce edible fruits of which the best known are the American persimmon and the kaki, or Japanese persimmon. The name *Diospyros* means 'fruit of the gods'.

The family consists of 2 genera, the widespread *Diospyros* with over 400 species, and *Euclea* with a few species confined to Africa. The Ebenaceae does not have the milky sap of the related Sapotaceae, and differs also in having unisexual flowers with the sexes borne on different trees. The male flowers usually occur in axillary clusters and the females, which are often larger, singly or a few together. The flowers and fruit of the 2 species occurring in the Sydney area are typical of those of the family.

263. Diospyros australis

Diospyros

Diospyros australis

Black Plum **Plate 263**

Although Black Plum may grow to a height of 20 m in favourable sites further north, in the Sydney district it is a tall shrub or small tree. Its leaves are alternate, not toothed, hairless, 4 to 8 cm long, elliptic and with a blunt or rounded tip. The upper surface is dark green and glossy, the lower surface pale yellowish-green. Flowers are creamy green, with 4 petals about 5 mm long. The fruit is conspicuous; at maturity it is a shiny black berry, up to 20 mm diameter, sitting on a base formed by the persistent calyx. Each berry contains a single seed surrounded by purple flesh, hence the common name.

A north coast species, *D. pentamera*, also occurs in a few locations in the area, such as Minnamurra Falls, Strickland State Forest and Gap Creek Flora Reserve; it has 5 sepals and petals, and its leaves are 3 to 7 cm long and taper to a point. Leaves are alternate along zig-zagging branchlets.

HABITAT Rainforest.

DISTRIBUTION Coastal zone. Recordings include Kangaroo River, Palm Jungle, Bola Creek, Gosford district and Gap Creek Flora Reserve. Occurring from Pebbly Beach (south coast) to north coast and north Qld.

FLOWERING October–December.

Sapotaceae

The Sapotaceae with about 800 species of tropical trees is the largest of the 4 families comprising the order Ebenales. It is a well-defined, homogeneous group in which characters often grade gradually from one state to another leading to difficulties in the delimitation of tribes, genera and species.

Like other members of the Ebenales the family has large simple leathery leaves without stipules, and small regular creamy flowers with usually 4 or 5 partly fused petals. The flowers occur singly or in small clusters, in the leaf axils, or directly on the trunk and larger branches. Sapotaceae is distinguished from the Ebenaceae in having bisexual flowers. The fruits of Sapotaceae are large with a firm flesh containing 1 to 5 smooth shiny seeds with a characteristic dull scar (hilum).

The Sapotaceae is an important component of many tropical forests, and its relative abundance is often regarded as an indication of the age of the forest. Being mostly large trees they are a valuable timber resource in a number of countries. Species of *Achras* (sapodilla), *Calocarpum* (sapote) and *Chrysophyllum* (star apples) are cultivated for their edible fruits. Also the seeds of many species contain edible oils which form an important part of the diet of local people. At one time different species were tapped for latex (guttapercha, balata, chicle), but this use has declined.

Only 2 *(Chrysophyllum, Planchonella)* of the 6 Australian genera extend to New South Wales, and only one species *(Planchonella australis)* occurs as far south as Sydney.

264. Planchonella australis

Planchonella

Planchonella australis

Black Apple **Plate 264**

The Illawarra region is the southern limit of this tall rainforest tree which is found in various types of coastal rainforest north to Queensland. Its leaves are alternate, not toothed, elliptic-obovate, 8 to 10 cm long and somewhat rounded at the tip. They are thick and firm in texture with prominent net veins on both surfaces. A white milky sap is exuded when the branchlets and leaves are broken. Flowers are white, bell-shaped, about 5 mm long, with 5 petals and are borne in clusters along the stems. Fruits are large and conspicuous, plum-like, black and up to 5 cm diameter.

HABITAT Rainforest.

DISTRIBUTION Coastal zone. Recordings include Minnamurra Falls, Berkeley Hill, Palm Jungle (Royal National Park), Ourimbah and Gap Creek Flora Reserve. Extending along north coast to Qld.

FLOWERING November–January.

Myrsinaceae

The Myrsinaceae is morphologically similar to the Primulaceae, and the two, with the smaller segregate family Theophrastaceae, form the order Primulales in most modern systems of classification. However the Primulaceae is herbaceous while the others consist of trees and shrubs, and there is some doubt as to whether the group comes from a common ancestral stock. If the similarities are simply the result of convergence, the Primulaceae would appear to be misplaced and certain similarities to the Caryophyllales might perhaps receive more recognition. This leaves the other 2 families as the order Myrsinales showing clear relationship to the Ebenales and the Theales.

The 2 families Myrsinaceae and Primulaceae differ in both their leaves and fruit. The leaves of Myrsinaceae are thick and leathery with conspicuous glands and resin ducts on the surface. The fruits are fleshy drupes, unlike those of the Primulaceae which are capsules with numerous small seeds.

Two genera occur in the Sydney area, *Rapanea* which is very similar to *Myrsine*, the type genus, and *Aegiceras*, the River Mangrove. Although mangrove communities are ecologically important because of their immense productivity and their effect on siltation and erosion of estuaries and rivers, the Myrsinaceae generally are of little economic significance or horticultural interest.

Rapanea

Shrubs or small trees with flowers in clusters on old wood. Fruit a small globular drupe.

Rapanea variabilis

Muttonwood **Plate 265**

A tall shrub; the leaves are oblanceolate, 4 to 8 cm long, tough, shiny above and may have entire or toothed edges. Flowers are small, 4-lobed, creamy and on short stalks in clusters from the leaf axils along the older branches. The corolla is divided only about half way to the base, a feature which helps separate this species from *R. howittiana*, which has a corolla divided nearly all the way to the base.

HABITAT Sheltered forest and rainforest margins, but occasionally occurring on drier forest slopes.

DISTRIBUTION Coastal zone and lower mountains. Extending from south coast to north coast and Qld; also northern tablelands and central western slopes.

FLOWERING July–September.

265. Rapanea variabilis

266. *Rapanea howittiana*

267. *Aegiceras corniculatum*

268. *Samolus repens*

Rapanea howittiana

Brush Muttonwood **Plate 266**

This is a small tree, less common in the Sydney district than *R. variabilis* as it prefers a rainforest habitat. Leaves are elliptic to oblanceolate, 5 to 10 cm long and glossy. The margins are entire or with fine serrations near the tip. Juvenile leaves are distinctly serrated. New stems and terminal buds are usually covered with dense rusty hairs. Flowers are small, 5-lobed and in clusters on the old wood. The corolla is divided nearly to the base, a feature of which separates it from *R. variabilis.*

HABITAT In and near rainforest.

DISTRIBUTION Scattered along coast (e.g., Strickland State Forest, Ourimbah State Forest, Mt Keira). Also occurring north coast, south coast, central tablelands, southern tablelands, central western slopes and Vic.

FLOWERING August–November.

Aegiceras Mangrove

Only 2 species of mangrove occur in the Sydney district. *Aegiceras* is the smaller and is usually found in association with the taller Grey Mangrove *(Avicennia marina).*

Aegiceras corniculatum

River Mangrove

This is usually a hairless shrub (rarely a tree), easily identified by its obovate leaves with rounded tips (4 to 7 cm long), and by its umbels of white flowers. Fruit is cylindrical, horn shaped and 3 to 4 cm long. It commences sprouting while still attached to the parent plant. River Mangrove grows closer to the land and extends further up rivers and inlets than Grey Mangrove. They are easily separated, as Grey Mangrove has larger lance-shaped leaves, orange flowers and a fruit which is a large compressed capsule which opens into 2 sections.

HABITAT Mud-flats of saltwater estuaries and tidal rivers.

DISTRIBUTION Common along coast, extending well upstream in larger rivers, such as Georges and Hawkesbury Rivers. Extending north from Jervis Bay (south coast) to Qld, NT and WA. Also occurring throughout Indonesia, Malaysia and Lord Howe Island.

FLOWERING June–November; a few flowers at other times.

Primulaceae

This essentially northern temperate family of annual and perennial herbs is poorly represented in Australia. The only members native to the Sydney area are two species of the widespread genus, *Samolus.*

The flowers of *Samolus* are bisexual with a 5-lobed calyx, a corolla with 5 spreading lobes about as long as the corolla tube, and 5 stamens opposite the lobes as with most members of the family. *Samolus* differs from other family members in having filiform staminodes alternating with the stamens (i.e., between the lobes) and in having a half-inferior ovary instead of a superior ovary.

Unlike the Myrsinaceae which is presently included in the same order, the Primulaceae contains many genera of horticultural importance such as *Primula, Cyclamen* and *Soldanella.* The scarlet and blue pimpernels are introduced species of *Anagallis* occurring in the Sydney area.

Samolus Brookweed

Samolus repens

Creeping Brookweed **Plate 268**

Creeping Brookweed is a common small herb in coastal saltmarsh. It has creeping root-like stems from which grow upright stems to 30 cm high. Leaves at the base of the upright stems are mostly 3 to 4 cm long, whilst those along the erect stems are alternate, semi-succulent, narrow lanceolate and 1 to 2 cm long. Flowers are white, about 10 mm across, with 5 petals and are borne singly on long stalks from the upper leaf axils.

Another species, *S. valerandii,* grows on the margins of swamps and rivers. It has thin, ovate leaves, 2 to 3 cm long, and flowers borne in loose racemes.

HABITAT In and near saltmarsh, mangrove swamps and on rock ledges near the sea.

DISTRIBUTION Coast, along sea and tidal inlets. Extending from north coast to south coast; also Vic, Tas, SA and WA.

FLOWERING Chiefly October–January, but a few flowers at other times.

Euphorbiaceae

This is a large cosmopolitan family showing great variation and many peculiarities. Its 5000 species in 300 genera include perennial and annual herbs, shrubs, climbers and trees. Some African Euphorbias are cactus-like plants, others have thick fleshy green branches and stems that function as leaves. Plants from a number of genera have a milky sap or latex which is often poisonous or caustic, blistering the skin and capable of causing temporary blindness. Many have glands, either on the flowers, the inflorescence or the leaf stalks. In some, the ripe fruits explode with a loud noise, scattering the seeds widely.

The flowers of Euphorbiaceae are small and sometimes extremely simple, a characteristic that is more likely a result of reduction than a sign of primitiveness. The flowers are always unisexual with the two sexes either on the same or on different plants. The perianth has 4 to 6 (usually 5) segments, less commonly twice as many when petals are present, or it may be absent entirely. The stamens of male flowers vary from one per flower to many. Female flowers usually have 3 fused carpels with the styles free or joined to each other at the base, and either 1 or 2 ovules in each loculus. The fruit is usually dry and splits at the lines of fusion of the carpels into 2-valved mericarps which later split into 2 to release the seeds. In a few genera such as *Claoxylon* and *Breynia* the fruit is succulent.

The Australian members of this family number about 300 species in 47 genera. Twenty genera occur in the Sydney area; nearly half of them endemic to Australia. The best-known and the most attractive of the local species is *Ricinocarpos pinifolius*, the Wedding Bush, with 4 to 6 narrow white petals. The others tend to have insignificant flowers, although in a few cases where the flowers are clustered *(Pseudanthus)* or in large inflorescences *(Poranthera)*, the overall effect is pleasing enough. Occasionally, as in *Glochidion*, the fruits may attract attention. *Ricinocarpos* species are difficult to cultivate, and none of the Australian species is cultivated to any extent.

Glochidion

Tall shrubs or small trees. Male and female flowers separate but on the same plant. Fruit is most conspicuous, as it is pumpkin-shaped, 12 to 20 mm across and splits to reveal numerous bright red seeds.

269. *Glochidion ferdinandi*

Glochidion ferdinandi

Cheese Tree **Plate 269**

This is a shapely tree, usually 4 to 6 m tall, although it may grow to 15 m. The common name is a reference to the fruits which remain on the tree for most of the year. These are round and flattened on top, to 20 mm across, and divided into segments like segments from a block of cheese. Leaves are alternate and distinctive as they are in 2 rows along the stem like a compound leaf. They are shiny, broadly lanceolate, 5 to 10 cm long, with a clear network of veins and are often somewhat unequal at the base (like the leaflets of *Euroschinus falcata*). Flowers are inconspicuous in the leaf axils.

HABITAT Along tidal inlets, river banks and coastal rainforest margins.

DISTRIBUTION Common in coastal area. Recordings include Sydney Harbour, Port Hacking, Oatley Park, Carrs Park and Minnamurra Spit. Extending from near Ulladulla (south coast) northwards to Qld; also NT and WA.

FLOWERING October.

Breynia

Shrub with male and female flowers separate but on the same plant. Male flowers with 6 fused perianth segments and 3 stamens fused to form a column. Female flowers with 3 styles. Fruit a globular berry.

Breynia oblongifolia

Breynia **Plate 270**

This is a shrub to 2 m tall with spreading branchlets almost parallel to the ground. Its most conspicuous feature is the arrangement of the leaves. These are broadly lanceolate, alternate, up to 4 cm long, paler underneath and arranged in 2 distinct rows along the branchlets. Flowers are small and hang on short, thin stalks from the leaf axils. The fruit is a pendulous round berry about 6 mm across.

270. *Breynia oblongifolia*

HABITAT Shady eucalypt forest and near rainforest.

DISTRIBUTION Widespread and common from coast to mountains. Extending to south coast, western slopes, north coast and Qld.

FLOWERING October–December.

Claoxylon

Tall shrub or small tree with male and female flowers separate but on the same plant. Leaves thin and toothed. Flowers greenish white and borne in racemes. Male flowers with 3 perianth segments and about 20 stamens, female flowers with 3 perianth segments and 3 styles.

271. *Claoxylon australe*

Claoxylon australe

Brittlewood **Plate 271**

Brittlewood is a rainforest shrub or small tree, characterised by its brittle branches and its soft, thin, toothed leaves, which are elliptic to oblong and 10 to 18 cm long. Young leaves are very light green, almost translucent. Two or more raised glands can be found on the leaf stalk just below the leaf blade. Male flowers have prominent white stamens and are borne on racemes 5 to 8 cm long, shorter than the leaves. Female flowers are on shorter racemes. Fruit is a purple-black, 3-lobed berry, about 6 mm across, with 1 red warty seed in each lobe.

HABITAT Rainforest.

DISTRIBUTION Coastal zone. Recordings include Minnamurra Spit, Mt Keira, Palm Jungle (Royal National Park) and Gap Creek Flora Reserve (Olney State Forest). Common from near Bermagui (south coast) along coast to north Qld; also recorded from southern tablelands and central western slopes.

FLOWERING November–December.

Coelebogyne

Shrub with male and female flowers on separate plants. Leaves tough and holly-like. Male flowers with 4 perianth segments and 8 stamens; female flowers with 4 perianth segments and 3 styles.

Coelebogyne ilicifolia

Native Holly **Plate 272**

The similarity of the leaves of this small compact shrub to those of English holly is the main identifying feature. These leaves are leathery, 5 to 9 cm long and have sharp spikes at the tips and along the angled leaf margins. Flowers are very small, only

272. *Coelebogyne ilicifolia*

1 mm long, in racemes from the leaf axils. The fruit is more conspicuous. It is a 3-lobed capsule about 8 mm across, green at first, turning brown.

HABITAT Drier rainforest patches or along eucalypt forest-rainforest margins.

DISTRIBUTION Coastal zone, but uncommon. Recorded from remnant rainforest at Berkeley Hill and Mt Kembla (Illawarra) where it is at its southern limit. Extending along north coast to north Qld; also central western slopes.

FLOWERING November.

273. *Croton verreauxii*

Croton

Tall shrubs or small trees, usually with male and female flowers separate but on the same plant. Flowers with 4 to 5 petals and 4 to 5 sepals; male flowers with many stamens; female flowers with 3 styles, each divided into 2 to 4 branches.

Croton verreauxii

Native Cascarilla **Plate 273**

This is a small tree to 6 m tall. Leaves are similar in shape to those of *Claoxylon*, but smaller, not as soft nor as thin. They are up to 12 cm long, with fine, shallow teeth. Two raised glands can be found on the leaf stalk just below the leaf blade. Young leaves are light green. A characteristic feature is that older leaves turn orange-red, especially in winter, making the tree readiy recognised from a distance. Flowers are small and borne on terminal racemes, 3 to 5 cm long. Male flowers have 10 to 12 yellow stamens. Fruit is a round capsule which splits into segments.

HABITAT In and near rainforest; often a pioneer species on the eucalypt forest-rainforest boundary.

DISTRIBUTION Coastal zone, especially in the Illawarra district where it is common below the escarpment (e.g., Mt Keira, Berkeley Hill, Minnamurra Falls, Jamberoo). Also reported from Illawarra to Qld; also NT.

FLOWERING November–December.

Baloghia

Trees with male and female flowers separate but on the same plant. Leaves opposite and leathery. Petals and calyx with 4 to 5 lobes. Male flowers with an indefinite number of stamens; female flowers with 3 styles, each divided into two.

Baloghia lucida

Brush Bloodwood **Plate 274**

This is a small rainforest tree. Its leaves are its most distinctive feature as they are opposite, thick, tough, up to 12 cm long, elliptic to obovate, very dark above, lighter underneath, tapering at the base to a short stalk and rather blunt at the apex with a small drip-tip. On larger trees, the bark can help in identification; it is mostly light grey to fawn, blotched with darker rough areas. The common name is derived from the dark red sap which exudes when the bark is cut. Flowers are borne in short racemes between the terminal pairs of leaves. Fruit is a large, hard, globular capsule, to 20 mm across, slightly 3-lobed, with a single seed in each lobe.

HABITAT Rainforest.

DISTRIBUTION Coastal zone, especially in the Illawarra district where it is scattered below the escarpment. Recordings include Minnamurra Spit, Jamberoo, Berkeley Hill and Palm Grove (Gosford). Extending from Milton (south coast) to Qld; also found on Lord Howe Island.

FLOWERING September–November.

274. Baloghia lucida

Omalanthus

Shrubs or small trees with male and female flowers separate but on the same plant. Flowers small, with numerous male flowers clustered along a raceme and with a few female flowers on stalks at the base of the raceme. Male flowers with only a few stamens; female flowers with 2 styles.

Omalanthus populifolius

Bleeding Heart, Native Poplar **Plate 275**

This is a tall shrub of sheltered forests. Leaves are broadly ovate, almost triangular, 8 to 13 cm long, soft, tapering to a point and resemble those of the poplar tree. As the leaves get older, they turn red and most plants have a sprinkling of conspicuous red leaves amongst the foliage. There is a prominent network of veins on the paler undersurface of the leaf. Each leaf is on a long reddish stalk. The plant exudes a milky latex when damaged. Flowers are small and arranged in a terminal raceme which is about as long as the leaves. Fruit is a smooth, 2-lobed capsule, 8 to 10 mm across, with a bluish lustre.

A closely related species, *Omalanthus stillingiifolius*, also occurs in the area, but is uncommon. It is restricted to sheltered rocky hillsides in eucalypt forests of the Blue Mountains (e.g., Wheeny Creek), extending along the north coast to Qld. Its leaves are smaller than those of *O. populifolius* (less than 5 cm long), and its fruits have 2 to 6 conical projections.

HABITAT Rainforest margins and sheltered moist gullies, often on semi-cleared slopes on volcanic soils.

DISTRIBUTION Most common along coastal zone. Recordings include Minnamurra Falls, Mt Keira, Bola Creek (Royal National Park) and Olney State Forest. Also recorded from mountains. Extending from south coast northwards to Qld and NT; also northern tablelands, north-western slopes and Lord Howe Island.

FLOWERING August–December.

275. Omalanthus populifolius

276. Phyllanthus gasstroemii

Phyllanthus

Shrubs, with leaves often distichous. Male and female flowers are borne on separate plants. Flowers very small, axillary. Male flowers with 6 perianth segments and 3 stamens; female flowers with 3 styles. Male flowers usually borne in clusters, female flowers solitary.

Phyllanthus gasstroemii

Blunt Spurge **Plate 276**

This is an erect hairless shrub to 1.5 m tall with spreading branches. Leaves are flat, distichous, elliptic to obovate, 1 to 2 cm long, 5 to 10 mm wide, with short stalks, 1 to 2 mm long. Flowers are very small and borne in the axils of the leaves. The filaments of the 3 stamens are fused together. Fruit is a depressed-globular capsule about 4 mm across.

HABITAT Sheltered forest.

DISTRIBUTION Coast to mountains. Recordings include Yarramundi (Nepean River) and Bola Creek (Royal National Park). Extending all along the coast and tablelands divisions and western slopes; also Qld.

FLOWERING August–November.

277. Phyllanthus thymoides

Phyllanthus thymoides

Thyme Spurge **Plate 277**

This small shrub (to 40 cm tall) usually goes unnoticed because it does not bear large showy flowers, but it is a common component of the understorey in dry forests, shrubland and heaths. Its leaves are the main identifying feature. They are arranged alternately, almost in opposite rows, up the stem. Each leaf is narrow-obovate, about 6 mm long, recurved at the tip and sprinkled with a few hairs. The stems are also covered with whitish hairs. Male flowers are minute and are in creamish clusters on slender stalks from the leaf axils; female flowers are larger (2 to 3 mm long) and are borne singly on stouter stalks. The fruit is a hairy globular capsule with 3 lobes, 3 to 4 mm across.

Habitat Open forest and heath, on sandy soils.

Distribution Widespread and common from coast to mountains. Found throughout NSW, except far west; also Qld and Vic.

Flowering July–November.

279. Ricinocarpos pinifolius

Beyeria

Shrubs with well-developed leaves. Male and female flowers borne on separate plants. Male flowers with 5 petals and numerous stamens; female flowers with 3-celled ovary. Fruit an oval greenish capsule.

278. Beyeria viscosa

Beyeria viscosa

Pinkwood **Plate 278**

Pinkwood is a shrub to 3 m tall. It is more typical of the western slopes than of the Sydney district where it is uncommon. Leaves are oblong to lanceolate, about 5 cm long and 10 mm wide. They often exude a sticky resin, hence the specific name 'viscosa' (sticky). Flowers are yellowish, borne in the leaf axils, and the fruit is a conspicuous globular capsule about 8 mm long.

Habitat Rocky slopes and gullies.

Distribution Scattered and uncommon from coast to northern Blue Mountains. Recordings include upper Georges River (The Basin Reserve), Nepean River (Maldon), Cumberland Plain, lower Blue Mountains and Wolgan Valley. Also occurring south coast, southern tablelands, western slopes, western plains, Qld, Vic, Tas and WA.

Flowering September–December.

Ricinocarpos Wedding Bush

Shrubs with male and female flowers separate but on the same plant. Male flowers with 5 to 6 large petals, usually white, and numerous stamens fused into a column. Female flowers with 3 styles, deeply bifid, and a 3-celled ovary. Fruit a globular capsule. The generic name suggests a likeness between these fruits and those of the Castor-oil Plant *(Ricinus)*.

Ricinocarpos pinifolius

Wedding Bush **Plate 279**

An erect shrub sometimes reaching 2 m tall. Its common name comes from the massed display of white flowers which are clustered at the ends of the branches to make a spectacular show. Each flower has strap-shaped spreading petals, 10 to 15 mm long; male flowers have a central column of yellow anthers. Some plants bear pinkish flowers. Leaves are narrow and up to 4 cm long. The specific name refers to the pine-like foliage. Fruits are distinctive, as they are globular, about 12 mm across, and covered with short hard protuberances.

The related *R. bowmanii*, a species from the western slopes and plains, has been recorded along Mill Creek (Dharug National Park). It has sepals with stellate hairs and sand-papery leaves.

Habitat Open forest and heath, on sandy soils. Especially common on old dunes.

Distribution Widespread and common from coast to lower Blue Mountains. Recordings include Kurnell, La Perouse, West Head, Muogamarra Sanctuary, Oatley Park, and Grose Lookout (near North Richmond). Also occurring north coast, south coast, northern tablelands, Qld, Vic and Tas.

Flowering August–November.

Bertya

Shrubs with male and female flowers separate but on the same plant. Leaves well-developed. Male flowers with 5 petal-like perianth segments below a cone-like column covered with stalkless anthers. Female flowers with smaller perianth segments under a 3-celled ovary which is topped by branched styles. Fruit is a small oval capsule.

Bertya pomaderroides **Plate 280**

This shrub grows to about 1 m tall. Leaves are alternate and quite variable. They are 1 to 2 cm long (sometimes longer), flat or slightly curved under, obtuse, sometimes ovate-oblong, sometimes linear-lanceolate; the undersurface is covered with pale hairs. Flowers are small, but very characteristic of the genus. They are borne on a short stalk and have no petals. The male flowers have a column of yellow stamens above petal-like bracts. Stems may be covered with closely matted short hairs or may be quite smooth.

Habitat Slopes of gullies and creek banks, on sandy soils.

Distribution Scattered and uncommon from coast to mountains. Recordings include Woronora River Valley, Douglas Park, Little River (Nattai) and Glenbrook. Also found on southern tablelands.

Flowering June–November.

280. Bertya pomaderroides

Bertya astrotricha **Plate 281**

A shrub 1 to 1.5 m tall, with stems and the undersurface of the leaves covered with loose short stellate hairs, giving the plant a rough appearance. Leaves are oblong to lanceolate, about 35 mm long and 12 mm wide, dark green above and dotted with stellate hairs giving the surface a sandpapery feel. Flowers are similar to those of *B. pomaderroides*, but the perianth segments of female flowers are covered with minute stellate hairs.

HABITAT Sheltered forest.

DISTRIBUTION Restricted to gullies near Mt Colah and Katandra Sanctuary, Mona Vale. Also recorded on north coast.

FLOWERING May–July.

Monotaxis

Small shrubs with male and female flowers separate but on the same plant. Flowers with petals (sometimes minute) and borne in clusters with several male and one female intermixed. Male flowers with 7 to 8 stamens.

Monotaxis linifolia
Monotaxis **Plate 282**

This low shrub sends out numerous slender stems from a perennial root. These are spreading, often trailing over the ground, 15 to 30 cm long. Along these stems, spaced well apart, are a few narrow opposite leaves, about 20 mm long. Flowers are white and borne in small clusters at the ends of the branches between the terminal leaves.

HABITAT Wet ground, on sandy soils.

DISTRIBUTION Widespread from coast to mountains, but not common. Also occurring south coast and southern tablelands.

FLOWERING October–January.

Amperea

Erect shrub with reduced leaves. Male and female flowers separate but on the same plant. Flowers lacking petals.

Amperea xiphoclada
Broom Spurge **Plate 283**

This plant is easily recognised by its stiff, erect, angular and apparently leafless branches which are mostly 30 to 40 cm tall. Leaves, if present, are narrow and scale-like on the flowering stems. Flowers are small and stalkless, with brown and cream male and female flowers mixed in a cluster at the leaf nodes. Fruit is an ovoid capsule.

HABITAT Variety of habitats from coastal dunes and heaths to dry open forests, on sandy soils.

DISTRIBUTION Widespread and common from coast to mountains. Recorded all along coast and tablelands from Qld to Vic; also Tas and SA.

FLOWERING Some flowers all year round.

281. Bertya astrotricha

282. Monotaxis linifolia

283. Amperea xiphoclada

Poranthera

Herbs or small shrubs with conspicuous white flowers in terminal clusters. Male and female flowers separate but on the same plant. Flowers with 5 sepals, 5 petals and 5 stamens. Female flowers with 3 styles, divided into 2 almost to the base.

Poranthera microphylla
Small Poranthera **Plate 284**

This small annual herb is likely to be overlooked unless it is in flower. It is seldom more than 10 cm tall, with spreading, soft, hairless branches, which may be ascending or semi-erect. The leaves are small, mostly 2 to 5 mm long, spoon-shaped to obovate. Individual flowers are white and very small, but are grouped together in showy leafy terminal clusters.

HABITAT Shady areas of forest and woodland.

DISTRIBUTION Widespread from coast to mountains. Found throughout NSW, except in drier western plains and in all States, except SA.

FLOWERING September–December.

Poranthera ericifolia
Heath-leaved Poranthera **Plate 285**

This is a small shrub, usually only 15 to 20 cm high. The stem is upright and usually scaly in appearance from old leaf bases. Spreading branchlets terminate in clusters of many small white flowers. Some flowers have a pinkish tinge. Leaves are crowded, narrow, about 8 mm long and without stalks.

HABITAT Open forest and heath, on sandy soils.

DISTRIBUTION Widespread from coast to mountains. Recordings include Menai, Girrakool and Mount Wilson. Also occurring north coast, south coast and southern tablelands.

FLOWERING September–November.

Poranthera corymbosa **Plate 286**

This is a larger plant than the other species of *Poranthera* in the area. It usually grows to about 60 cm tall, is very upright and has few branches. Leaves are linear, 3 to 5 cm long, with slightly rolled-back edges. Close examination of the base of the leaves reveals stipules which are entire, in contrast to those of *P. ericifolia* which are jagged. White flowers are borne in large terminal clusters. The specific name 'corymbosa' refers to the corymbose arrangement of the flowers. Although the flower stalks originate at different levels on the stems, the flowers are at about the same level.

HABITAT Open forest, on sandy soils.

DISTRIBUTION Widespread from coast to mountains. Also occurring north coast, south coast, northern tablelands, central western slopes, Qld and Vic.

FLOWERING September–November.

Micrantheum

Shrubs with leaves in groups of 3; male and female flowers separate but on the same plant. Petals absent, but with 4 to 6 petal-like sepals. Male flower with 3 to 9 stamens; female flower with 3 styles and 3-celled ovary.

284. *Poranthera microphylla*

285. *Poranthera ericifolia*

286. *Poranthera corymbosa*

Micrantheum ericoides **Plate 287**

This is a small, inconspicuous shrub with tiny solitary flowers. Each flower is only 2 mm across. Male flowers have 3 stamens (sometimes more). Leaves are linear, 8 mm long and 1 mm wide, and are in groups of 3 alternately along the stems. Fruit is an elliptic-oblong capsule, sometimes larger than the leaves and reddish when young. Flowers and fruit may be on a plant at the same time.

HABITAT Open forest and heath, on sandy soils.

DISTRIBUTION Widespread and common from coast to mountains. Also occurring on the north and south coasts.

FLOWERING August–November.

Micrantheum hexandrum **Plate 288**

Unlike *M. ericoides*, this species is a medium-sized bushy shrub with erect branches. Leaves are in groups of 3, clustered along the branches; they are narrow, 10 to 15 mm long, with a minute pointed tip. Flowers are small, white, about 5 mm across and borne on small stalks among the leaves. Male flowers have 6 stamens (sometimes more), a feature referred to in the name 'hexandrum' ('hex'—6; 'andro'—male).

HABITAT Along river banks, amongst rocks.

DISTRIBUTION Upper Georges River (e.g., The Basin Reserve, near Campbelltown) and south to southern highlands. Uncommon in Sydney area, but recorded throughout the tablelands from Qld to Vic and Tas.

FLOWERING July–October.

Pseudanthus

Small shrubs with leaves mostly spirally arranged. Male and female flowers separate but on the same plant. Petals absent, but with 4 to 6 petal-like sepals. Male flowers with 6 stamens and 6 perianth segments; flowers usually clustered. Female flowers solitary with a 3-celled ovary.

Pseudanthus pimeleoides

Pseudanthus **Plate 289**

Pseudanthus is a stiff, upright shrub, growing to about 80 cm tall. Leaves are narrow, crowded, 8 to 12 mm long, and have pointed tips. White flowers are borne abundantly near the ends of the branches. The petal-like perianth segments of the male flowers are 8 to 12 mm long. Female flowers are few and inconspicuous.

Two other *Pseudanthus* occur in the area but are less common. Each is a low shrub with small flowers (perianth segments only 1 to 2 mm long). *P. orientalis* is found on old coastal dunes, and *P. divaricatissimus* (Plate 290) occurs on sandy and rocky heaths in the higher Blue Mountains.

290. *Pseudanthus divaricatissimus*

287. *Micrantheum ericoides*

288. *Micrantheum hexandrum*

289. *Pseudanthus pimeleoides*

HABITAT Common south of Sydney amongst boulders of creek beds and on nearby creek banks. In mountains, it grows on the sandy soils of protected slopes or on exposed rocky cliff tops.

DISTRIBUTION Scattered from coast to mountains. Recordings include Heathcote Creek, Kangaroo Creek (Royal National Park), Upper Georges River, Mt Hay road and above Colo River. Extending along north coast to Qld; also central western slopes.

FLOWERING July–October.

Chamaesyce Spurge

Herbs with milky sap exuding from broken leaves and stems. Leaves opposite. Flowers grouped in terminal clusters with 1 female surrounded by 8 to 15 male flowers, the whole cluster surrounded by a collar of petal-like bracts. A number of introduced species occur in the area, but only 3 native species. This genus was formly known as *Euphorbia.*

Chamaesyce drummondii

Caustic Weed, Flat Spurge **Plate 291**

This is a small hairless prostrate plant, spreading to about 30 cm across. Stems are often reddish. Leaves are opposite, ovate, blue-green, 4 to 8 mm long, with a very short stalk. Flowers are minute and borne in small heads at the ends of short branches. Fruit is a round, hairless, 3-celled capsule. Although the milky sap is reported to have poisoned stock, the plant was used by Aborigines and bushmen for its healing powers, especially for treating snake bite and dysentery.

The closely related *Chamaesyce sparrmanii*, with leaves 1 to 3 cm long, is found on sandy dunes near the sea, where it forms mats of up to 1 m across. *C. dallachyana* has also been recorded in the cracks of a pavement at Granville. It is similar to *C. drummondii* except that its fruit has a fringed appendage pressed against its base and the stipules of opposite leaves touch.

HABITAT Open country, roadsides and disturbed areas. Sometimes occurring as a garden weed.

DISTRIBUTION Coastal zone (e.g., Razorback Mt). Extending to north coast, northern tablelands, southern tablelands; common on western slopes and plains. Found in all mainland States.

FLOWERING September–January.

291. Chamaesyce drummondii

Thymelaeaceae

The only Australian genus of any importance in this family is *Pimelea* with about 65 species shared fairly evenly between east and west. A further 20 or so species occur in neighbouring countries, with New Zealand having fifteen.

The inflorescence of *Pimelea* is not unlike that of the garden plant *Daphne*, another member of the Thymelaeaceae. In both cases the flowers are arranged in racemes which are frequently condensed and terminal. Individual flowers consist of a tube formed from 4 petal-like sepals, with the stamens inserted in the tube around a superior ovary having a single simple style. Petals are absent in both genera.

In *Pimelea* the pedicels are so short that the inflorescence is often referred to as a spike or, when condensed, as a head. Frequently the leaves immediately below the head are broadened or slightly hairy, and protect the head in the bud stage. When different in appearance they are usually referred to as bracts. One Western Australian species, *Pimelea physodes*, called the Qualup Bell, has a pendulous bell-like inflorescence consisting of smallish greenish-yellow flowers surrounded by large red and green bracts which become darker with age. Most Pimeleas however have inflorescences similar to the common eastern species, *Pimelea linifolia*, with white, pink or yellow flowers. Some of the smaller species have quite small, obscure flowers. A common feature of *Pimelea*, and others in the family, is the long, fibrous almost unbreakable bark, a feature which has given rise to names such as lacebark or bootlace bush. More commonly, however, Pimeleas are known as rice flowers.

Pimelea Rice-flower

Pimelea linifolia ssp. *linifolia*

Slender Rice-flower **Plate 292**

The white flower-heads of Slender Rice-flower are a common and conspicuous part of the bushland around Sydney. It is a very variable plant. On coastal sand dunes it is often a low clumped shrub, 30 to 40 cm tall, with abundant flowers, unlike the woodland and forest form which is a slender erect plant with hairless branched stems to 1 m tall and more scattered flowers. Leaves are opposite, narrow-elliptic, about 3 cm long and 2 to 5 mm wide, without prominent veins. Flowers are white and borne in large heads at the ends of the stems. Flowers are occasionally tinged with pink and there is a form with all pink flowers on the cliffs near Bundanoon. Flower-heads are surrounded by 4 smooth ovate bracts.

HABITAT Common in various habitats from coastal dunes to forests.

DISTRIBUTION Widespread from coast to mountains, including most urban bushland areas. Found throughout the State, except far west, and in Qld, Vic, Tas and SA.

FLOWERING Mainly July–October, but some flowers at other times.

292. Pimelea linifolia ssp. *linifolia*

Pimelea linifolia ssp. *linoides* **Plate 293**

This tall leafy shrub was previously known as *P. linoides*, but is now regarded as a sub-species of *P. linifolia*. It grows to 2 m tall and has narrow-elliptic leaves up to 35 mm long.

Also regarded as a subspecies of *P. linifolia* is the former *P. collina*. Its leaves have 2 prominent longitudinal veins near the margins. The bracts also have obvious intramarginal veins. Subspecies *collina* occurs in heaths and forests of the higher Blue Mountains (e.g., Clarence), often in damper areas than ssp. *linifolia* or ssp. *linoides*.

HABITAT Scrubby hillsides and forests on sandstone soils, often in exposed positions.

DISTRIBUTION Blue Mountains. Recordings include Wentworth Falls, Echo Point and Yerranderie. Also recorded on western slopes, north coast and in Vic.

FLOWERING September–October.

293. Pimelea linifolia ssp. *linoides*

294. Pimelea glauca

Pimelea glauca

Smooth Rice-flower **Plate 294**

An erect branched shrub to 1 m tall, with hairless stems which are brownish when young and reddish when older. Leaves are opposite, bluish green, narrow-elliptic, about 20 mm long and up to 5 mm wide. The flowers are white to cream in terminal heads of up to 35 flowers. Four leaf-like bracts are at the base of the flower; the inner bracts are fringed with hairs.

HABITAT Dry open forests and grasslands, mainly on clayey soils.

DISTRIBUTION Widespread from coast to drier western plains but not common. Found throughout the State, except in far west; also Qld, Vic, Tas and SA.

FLOWERING August–December.

Pimelea ligustrina

Tall Rice-flower

A tall leafy shrub, 1 to 2 m tall. It is a variable plant with 3 subspecies recognised in New South Wales; 2 of these occur in the Sydney district. Stems may be downy or hairless. Leaves are opposite, ovate-lanceolate, 2 to 8 cm long, 10 to 25 mm wide, with veins prominent on the undersurface. Flowers are white, in large round heads, about 4 cm across.

One subspecies, *ligustrina* (Plate 295), has 4 to 5 green bracts beneath the flower-head which are hairless on the outside but silky inside; it also has a hairless flower stalk. Another subspecies, *hypericina* (Plate 296), has 6 to 8 hairy bracts and a hairy flower stalk. The latter is the more common of the 2 subspecies around Sydney.

295. Pimelea ligustrina ssp. *ligustrina*

HABITAT Margins of rainforests and in sheltered moist forests.

DISTRIBUTION Subspecies *hypericina* is common along the coast and in the Blue Mountains, extending to northern tablelands. Recordings include Burning Palms, Mt Keira, Sugarloaf Mt, Blackheath and Cambewarra Mt.

Subspecies *ligustrina* is absent from the coastal area around Sydney, but occurs in the Blue Mountains (e.g., Mt Tomah). Also occurring through tablelands division, north coast, south coast, Qld, Vic, Tas and SA.

FLOWERING September–November.

296. Pimelea ligustrina ssp. *hypericina*

Pimelea curviflora

Curved Rice-flower **Plate 297**

P. curviflora is an erect, open, many-branched shrub, up to 1.5 m tall, with stems covered with appressed silky hairs. Leaves are mostly alternate, narrow-elliptic, up to 20 mm long and 2 to 5 mm wide. The upper surface is hairless or with only a few hairs, the lower surface hairy (unlike *P. latifolia*, which has a tangle of short fine hairs on both surfaces). Flowers are yellow to greenish yellow (sometimes red) in clusters of a few flowers, either terminal or appearing lateral due to the ongrowth of the axillary bud below the young flowers. This is a variable species, with two subspecies and a number of varieties recognised in the Sydney district.

Subspecies *curviflora* is a shrub to about 50 cm tall, restricted to the coastal zone around Sydney. Its flower tube is dark red and covered with long appressed hairs. Its fruit is curved.

Subspecies *gracilis* has 3 varieties in the Sydney district. Its flower tube is greenish-yellow, often with dark red lobes, lightly covered on the outside with short hairs. Its fruit is straight (occasionally var. *gracilis* has curved fruit). Subspecies *gracilis* var. *gracilis* is an erect shrub (to 1.5 m tall). Its flower stalk is covered with long upward-pointing hairs. This is chiefly a tablelands shrub, recorded from only one locality around Sydney, south of Dapto. It also occurs in the western plains, southern tablelands, Qld, Vic and SA.

Subspecies *gracilis* var. *sericea* is a smaller shrub (30 to 40 cm tall) with flower stalks covered with short appressed hairs. It occurs in open sandy forests of the southern highlands (e.g., Berrima, Tallong), and also southern tablelands, western slopes, Qld, Vic and SA.

Subspecies *gracilis* var. *subglabrata* has leaves which are hairless above and with only a few short appressed hairs underneath. It occurs in scrub and pasture land, mainly on the central tablelands and western slopes, with one recording from Mt Victoria.

297. Pimelea curviflora ssp. *gracilis*

HABITAT Open forest and shrublands, on sandy soils.

DISTRIBUTION See notes above.

FLOWERING September–January.

Pimelea latifolia

Plate 298

An erect hairy spreading shrub, with yellow or greenish-yellow flowers. Stems, leaves and flowers are densely covered with soft tangled hairs. Both surfaces of the leaf are hairy, separating it from *P. curviflora* which has the upper surface of the leaf hairless (or almost so). Of the 4 subspecies recognised in New South Wales, 2 occur in the Sydney district.

Ssp. *hirsuta* is found in sheltered forests of the Blue Mountains and Hornsby Plateau. It is a small shrub (to 50 cm tall) with elliptical leaves, about 10 to 12 mm long and 6 to 8 mm wide. Both sides of the leaf are covered with short fine tangled hairs which mostly stick out at right angles to the leaf.

Ssp. *elliptifolia* is a small shrub from forested, rocky hillsides. Its leaves are longer and narrower, about 25 to 30 mm long, 6 to 10 mm wide, and are densely covered with longer hairs which usually point upwards.

HABITAT Sheltered forests and hillsides.

DISTRIBUTION Ssp. *hirsuita*. Blue Mountains (e.g., Kurrajong Hill, Colo Heights) and Hornsby Plateau (e.g., Watagan Mts, Wyong, West Pittwater). Also occurring north and south coasts and southern tablelands.
Ssp. *elliptifolia*. Blue Mountains. Also north coast and central western slopes (e.g., Muswellbrook-Sandy Hollow).

FLOWERING September–March.

298. Pimelea latifolia ssp. *hirsuta*

299. Pimelea spicata

300. Wikstroemia indica

Pimelea spicata **Plate 299**

A rare species occurring only in small populations in a few localities where it is endangered. It is a decumbent to erect shrub, about 30 to 40 cm tall, with hairless stems and leaves. Leaves are narrow-elliptic, 5 to 20 mm long. Flowers are white, or sometimes pinkish, 6 to 10 mm long, without an involucre of bracts and borne in terminal spikes up to 14 mm long.

HABITAT Shrubland (e.g., below *Bursaria spinosa*) on clay soils.

DISTRIBUTION Restricted to Campbelltown-Narellan and Bankstown areas, where it is rare. Also recorded at Shellharbour in 1964, where it may now be extinct.

FLOWERING December–February.

Wikstroemia

Similar to *Pimelea*, but greenish-yellow flowers have rudimentary petals and 8 stamens.

Wikstroemia indica

Wikstroemia **Plate 300**

This erect shrub may grow to 1.5 m tall. Leaves are opposite, paler below, oblong to lanceolate, and up to 3 cm long and 1 cm wide. Flowers are borne in few-flowered terminal heads. Individual flowers are cylindrical, 8 mm long, with 8 stamens, and there are small, scale-like petals. Fruit is a red berry.

HABITAT Hind dunes and scrub close to the sea and margins of coastal rainforest.

DISTRIBUTION Coast, chiefly north of Sydney (e.g., Long Reef, Pearl Beach, Bateau Bay). Extending along north coast to Qld; also Pacific Islands and Asia.

FLOWERING November–December.

Droseraceae

The Droseraceae consists of small carnivorous herbs possessing special adaptations for trapping and digesting insects, presumed to be an evolutionary response to growth in waterlogged soils deficient in nitrogen. The sundews *(Drosera)* and venus flytrap *(Dionaea)* are well known and are sometimes cultivated for their special interest. *Drosera* are widespread and common, particularly in Australia which has about 70 of the world total of nearly 90 species. *Dionaea* is confined to North and South Carolina in the USA. Two other monotypic genera which complete this family are less well known, although one, *Aldrovanda*, an aquatic species with similarities to *Dionaea*, occurs in northern Australia.

The carnivorous habit appears to have developed independently in at least 4 groups of unrelated plants. Three of these consist of pitcher plants from the families Cephalotaceae, Nepenthaceae and Sarraceniaceae. The fourth, *Byblis*, with 2 species in northern and south-western Australia, closely resembles the narrow-leaved sundews such as *Drosera binata* and *Drosera indica*. However the 2 genera differ in floral structure as well as in the sticky glands which in *Byblis* do not move to enfold the prey. Also *Byblis* has separate, shorter digestive glands.

Drosera leaves have many fine hairs which produce a sticky fluid and act as insect traps. The fluid glistens in the sun to attract the attention of insects, hence the common name 'Sundew'. The generic name is derived from the Greek word for 'dew' ('drosos'). Any small insect alighting on the leaves is trapped by the fluid and is soon enmeshed by the hairs which bend over it. A digestive enzyme is then secreted to convert the insect protein into simple soluble forms for absorption.

Drosera Sundew

Drosera binata

Forked Sundew **Plate 301**

The leaves of Forked Sundew are on long (6 to 16 cm) hairless stems; they are narrow and strap-like, covered on the upper surface with glistening reddish glandular hairs. Each leaf is forked into 2 segments, which may be divided once or twice again. An erect flower stalk, often longer than the leaves, bears several white flowers each about 25 mm across.

HABITAT Sedge and grassy swamps; often clinging to wet rock ledges and beneath waterfalls.

DISTRIBUTION Widespread from coast to tablelands. Recordings include Curra Moors (Royal National Park), Linden, Blackheath and Brisbane Waters National Park. Extending north to Qld and south to Vic, SA and Tas. Also occurring in Asia.

FLOWERING December–January.

301. Drosera binata

302. *Drosera spathulata*

Drosera spathulata

Common Sundew **Plate 302**

There are 4 species of Sundew in the Sydney district which have a rosette of leaves at the base of the stem, and flowers on a slender leafless stalk rising from near the centre of the rosette. One of these, *D. pygmaea*, is easy to distinguish as it is smaller, has a solitary flower, and has a conspicuous white centre to the reddish rosette. Another, *D. glanduligera*, has numerous orange or red flowers on a glandular-hairy stalk. It is rare in the area, but has been recorded from Castlereagh State Forest. The 2 remaining species, *D. spathulata* and *D. burmanni* (Plate 303) are very similar. *D. spathulata* is very common. It has a rosette of reddish spathulate leaves, with stalks which merge into the blades; the whole leaf is 8 to 20 mm long. Pink or white flowers are arranged along the end of the slender stalk. Each flower has 3 styles, divided almost to the base (this can be seen with a lens). *D. burmanni* differs in having 5 styles which are undivided. The greenish leaves have oval blades which do not merge into the stalk. It is rare in the Sydney district.

Habitat Wet exposed rocks and damp soils.

Distribution Widespread and common from coast to tablelands. Occurring along the coastal zone from north Qld to western Vic. Also in Tas, Papua-New Guinea, NZ and Asia.

Flowering October–February.

Drosera pygmaea

Pygmy Sundew **Plate 304**

Pygmy Sundew is less common than *D.. spathulata*, which it resembles. The rosette of leaves is about the size of a one cent coin, distinctly silver hairy in the centre. Leaves are round, rather than spathulate, and on slender stalks. Flowers are tiny, solitary, white and are borne on the end of a short stalk, 5 to 40 mm tall.

Habitat Heavy moist peaty soils.

Distribution Widespread from coast to tablelands. Extending along the coastal zone from north coast south to Vic. Also WA, SA, Qld, Tas and NZ.

Flowering October–December.

303. *Drosera burmanni*

304. *Drosera pygmaea*

Drosera peltata

Pale Sundew **Plate 305**

Two local species have 2 different types of leaf on the one plant. They each have a basal rosette of leaves (which may die away as the plant matures) and leaves along the stem. These stem-leaves do not look like leaves as they are fringed with long sticky red-brown hairs. The 2 species, *D. peltata* and *D. auriculata*, are separated on the basis of their differing sepals. *D. peltata* has hairy sepals with fringed to toothed margins. White flowers are borne in racemes at the end of erect stems (to 25 cm high).

Habitat Damp sites in grasslands and forests.

Distribution Widespread and common from coast to tablelands. Found throughout NSW (except far west); also Qld, Vic, Tas, WA, SA and Asia.

Flowering Mainly September–December, but sometimes in winter.

305. *Drosera peltata*

Drosera auriculata

Plate 306

This species closely resembles *D. peltata*. The main difference observed in the field is that the sepals of *D. auriculata* are without any hairs. This difference can be easily seen by close examination. Other features are similar, including the glandular stem-leaves and the erect stems. Flowers are pink or white. Some classifications have regarded *D. auriculata* as a subspecies of *D. peltata*, but recent studies show that they are a distinct species. They can be separated by a combination of factors: seeding morphology, peduncle length, sepal margin outline, sepal hairiness, petal length, sepal length and seed length.

Habitat Damp sites in grasslands, forests and gullies.

Distribution Widespread from coast to tablelands. Extending westwards to western slopes, north along the coast to Qld and south along coast and tablelands to Vic. Also Tas and SA.

Flowering Mainly September–December, but sometimes in winter.

306. *Drosera auriculata*

Eucryphiaceae

The distribution of this small family of 1 genus with 3 species in Australia and 3 in Chile is often cited as proof of former close links between the continents of the southern hemisphere. For many years it was customary to invoke the former existence of ocean-spanning land bridges to explain such occurrences but, with our increased understanding of sea-floor spreading and its effect on continental movement, it is now generally accepted that Australia and South America were linked through Antarctica during the early Tertiary period. This allowed the interchange of plant and possibly some animal species before the route was closed by the separation of Australia and the subsequent glaciation of Antarctica.

Two of the 3 Australian species are Tasmanian and the third occurs on the mainland from the Howe Range in eastern Gippsland to just north of Macquarie Pass near Wollongong. The Tasmanian species, *Eucryphia lucida*, has large, white, strongly perfumed flowers and is utilised by honey producers for production of the well-known leatherwood honey. It forms a tree to about 10 m and is sometimes grown as an ornamental. The local species with smaller flowers and pinnate leaves, although not used in horticulture, is nevertheless very attractive when in flower. Its natural habitat is shady gullies in high rainfall coastal ranges.

Eucryphia moorei

Leatherwood, Plumwood **Plate 307**

Leatherwood is a tree to about 15 m tall, uncommon in the Sydney district. The main identifying feature is its leaves. These are pinnately divided, with 3 to 5 sets of leaflets and a longer terminal leaflet. Leaflets are up to 6 cm long, entire, dark green and firm. The undersurface is whitish. Each leaflet ends abruptly with a tiny point. The white flowers are very attractive; they are up to 3 cm across, with 4 broad petals and numerous stamens. Fruit is an oblong or egg-shaped capsule, 8 to 16 mm long.

HABITAT Sheltered gullies and below rocky escarpments.

DISTRIBUTION Illawarra, southern highlands and southern Blue Mountains (e.g., Carrington Falls gorge, old coal mine at Bundanoon). Extending southwards to Vic.

FLOWERING February–March.

307. Eucryphia moorei

Cunoniaceae

Cunoniaceae is a southern hemisphere family with comparatively recent extensions into the northern hemisphere. Its centre of diversification is eastern Australia and Oceania. Australia has 14 genera and 23 species out of a world total of 26 genera and about 300 species. In other words, Australia has more than half the genera but less than one-tenth the species. This is due to the preponderance in Australia of small genera, frequently monotypic or endemic, and the widespread extra-Australian genus *Weinmannia* with 160 species together with 4 other genera in the Pacific area which account for another 90 species.

There is no single distinguishing feature that immediately identifies this family, rather there are a number of features which, taken together, make identification possible. Probably the best means of recognition is the simple or compound leaves with serrated margins and interpetiolar stipules (i.e., small leafy appendages on the stem between the bases of the leaf stalks). In the absence of flowers this will distinguish it from rainforest species of Escalloniaceae, Myrtaceae and Rutaceae which may have opposite leaves (serrated in Escalloniaceae, entire in the others) but are without stipules. Cunoniaceae is also distinguished from closely related Escalloniaceae by the stamens, which in local species number 8 to 10 in the former and 4 to 5 in the latter. Also, in Cunoniaceae the stamens are attached to the edge of a nectar-secreting disc which surrounds the ovary.

The variety of inflorescences in Cunoniaceae tends to increase the difficulty of recognition. Usually the flowers are in panicles or cymes but occasionally they are in heads, as in *Callicoma*, or in racemes or solitary. The small white or pinkish regular flowers usually have 4 or 5 sepals and petals, although sometimes the petals are very small or absent. Occasionally the calyx (i.e., sepals) enlarges and becomes coloured after fertilisation.

The most commercially valuable species in the family is *Ceratopetalum apetalum*, Coachwood, which is a common tree in rainforests and gullies on the coast and tablelands from Batemans Bay to the Queensland border. It, like other tree species in the family, yields fine-grained, pinkish brown timbers formerly prized for coachbuilding, but now used chiefly for plywood, furniture, joinery and pallets. The other New South Wales species in this genus, *C. gummiferum*, known as Christmas Bush, is a popular shrub for cultivation owing to its enlarged red calyces which appear in summer.

The practice in Sydney's early days of wattle and daub construction using pliable branches daubed with clay led to the name Black Wattle for *Callicoma serratifolia*. The name is preserved in Black Wattle Bay, an early source of these shrubs. Other Australian members of the family are insufficiently known to receive much attention, indeed *Acrophyllum australe* is a rare shrub found only in the lower Blue Mountains in a few places which satisfy its specific requirements.

308. Callicoma serratifolia

309. Ceratopetalum gummiferum

Callicoma

Small tree with flowers lacking petals and borne in dense globular heads. The name is derived from 'calli', beautiful, and 'come', hair, a reference to the fine stamens which dominate the flower head.

Callicoma serratifolia

Black Wattle **Plate 308**

Despite the common name, this small tree is not a 'wattle', a name now generally applied to the genus *Acacia*. The flowers do resemble those of wattles as they are a dense, pale yellow head of stamens. The common name is also derived from the fact that stems of this plant were used to construct wattle-and-daub buildings in the early years of Sydney. Leaves are distinctive, being opposite, 5 to 10 cm long, dark green above and pale grey below, with regular and prominent teeth. Fruit is a small capsule, unlike *Acacia* which has pods.

HABITAT Rocky gullies, along sheltered creeks and damp shaded sites, usually on poorer sedimentary soils.

DISTRIBUTION Widespread and common from coast to mountains. Abundant beside some Blue Mountains tracks, such as those of Govetts Leap and Valley of the Waters, Wentworth Falls. Extending from south coast to north coast and Qld.

FLOWERING October–December.

Ceratopetalum

Tall shrubs and trees with opposite leaves. Petals small or absent, but sepals enlarging as they mature. Fruit dry (unlike *Schizomeria* which has a drupe).

Ceratopetalum gummiferum

Christmas Bush **Plate 309**

Christmas Bush is a widely cultivated shrub or small tree, bearing loose clusters of small white flowers in late spring. After flowering the 5 petal-like sepals remain, begin to enlarge and change from white through pink to a rich red colour. By Christmas time, the plant is a mass of red, making it attractive and conspicuous. When not in flower, the leaves help in identification. They are divided into 3, each leaflet lanceolate, 3 to 7 cm long and with toothed margins.

HABITAT Slopes and gullies of open forest, on sandy soils.

DISTRIBUTION Widespread and common from coast to mountains. Also occurring along north coast and south coast.

FLOWERING October–November.

Ceratopetalum apetalum

Coachwood **Plate 310**

The straight trunks and dark foliage of Coachwood are features of many deep shaded gullies in the Sydney district. It is a tall tree, sometimes up to 25 m, with bark which is smooth, whitish-grey and blotched, making it readily recognised. Leaves are opposite, elliptic, 8 to 16 cm long, finely toothed and have a prominent swelling or joint a the base of the leaf-blade. Flowers lack petals, but have 5 petal-like

310. Ceratopetalum apetalum

sepals. Although individually small, the flowers are white and in dense clusters making them very conspicuous. Like the Christmas Bush, the sepals enlarge when fruiting and become red, so that the tree is often able to be picked out from a distance. Coachwood was, and still is, a valuable timber tree. Its wood has been used for coach-building, fruit-cases and flooring. In the war years, it was logged extensively to supply wood for gun stocks.

HABITAT Sheltered gullies, and rainforest on talus slopes or poorer sedimentary soils.

DISTRIBUTION Widespread and common from coast to mountains. Recordings include Minnamurra Falls, Carrington Falls, Mt Keira, Royal National Park, Katoomba and Ourimbah State Forest. Extending from near Batemans Bay (south coast) northwards along coast to Qld.

FLOWERING November–December.

311. Aphanopetalum resinosum

Aphanopetalum

Straggling climber with opposite leaves. Petals minute or absent, but with 4 conspicuous sepals.

Aphanopetalum resinosum

Gum Vine **Plate 311**

This is a hairless climber which at times takes on the appearance of a straggling shrub. Leaves and flowers are similar to those of Coachwood. The shiny, lance-shaped leaves are 4 to 10 cm long and have small regular teeth. Young leaves are light green. Flowers are few in a loose cluster, with 4 greenish petal-like sepals which enlarge with fruiting.

HABITAT Rainforest, sheltered gullies, and drier slopes near rainforest margins.

DISTRIBUTION Common along coastal zone (Palm Jungle, Mt Kembla, Razorback Mt, Olney State Forest); also recorded from Blue Mountains. Occurs all along coast and tablelands divisions; also central western slopes and Qld.

FLOWERING August–October.

Schizomeria

Tree with opposite leaves. Differs from *Ceratopetalum* in having drupaceous fruit.

Schizomeria ovata

Crab-apple **Plate 312**

Crab-apple is a tall tree with a general appearance of Coachwood. The bark is grey, fairly smooth or wrinkled in small trees, but deeply furrowed on older trees. Leaves are opposite, 8 to 18 cm long, irregularly toothed and strongly veined. The flowers seem to resemble those of Coachwood, but close examination reveals 5 petals (2 mm long) which are cut into 3 or 4 teeth at the tip (hence the generic name 'schizo'—I cut, 'meria'—a part). The sepals do not enlarge at fruiting time and, unlike Coachwood, the fruit is a drupe, creamy-white, about 12 mm wide and shaped like an apple.

HABITAT Rainforest and gullies, often associated with Coachwood.

DISTRIBUTION Coastal zone. Recordings include Macquarie Pass, Mt Keira, Royal National Park (Upper Causeway), Hawkesbury River and near Gosford. Extending from near Narooma (south coast) north along coast to north Qld; also northern tablelands.

FLOWERING September–October.

312. Schizomeria ovata

313. Acrophyllum australe

Acrophyllum

A rare shrub, with leaves in whorls and dense clusters of white flowers.

Acrophyllum australe — Plate 313

This small spreading shrub is very localised, growing only on damp sandstone drip-ledges in the central Blue Mountains. Its leaves are in whorls of 3, almost stalkless, rigid and with distinct saw-like teeth. Flowers are small and white, but borne in dense clusters near the ends of the branches. Flowers have narrow petals and conspicuous stamens. The fruiting capsule bears the remains of the 2 styles on its valves.

HABITAT Drip-ledges and wet rock crevices beneath sandstone cliffs.

DISTRIBUTION Endemic in the Springwood-Linden area.

FLOWERING October–November.

Baueraceae

Although the genus *Bauera*, named after the botanical artists Franz and Ferdinand Bauer, is often placed in a separate family, Baueraceae, its distinctive nature is not always recognised at this level and it is sometimes placed in a subfamily of Saxifragaceae, or in Cunoniaceae. An old common name, Dog Rose, is a reminder of the interrelationship of all these groups which are part of the order Rosales. *Bauera* consists of 3 shrub species with opposite sessile leaves and small pink or white flowers borne singly in the leaf axils. The flowers have 4 to 10 sepals and petals and numerous stamens. The fruit is a 2-celled capsule containing a number of small brown seeds.

B. rubioides is a common wiry shrub which grows in moist shady locations in the eastern states and South Australia. *B. capitata*, uncommon in the Sydney area, is a smaller diffuse shrub from the Queensland and northern New South Wales coasts. The other species, *B. sessiliflora* is an erect, somewhat hairy shrub with sessile magenta flowers crowded along the branches and is confined to the Victorian Grampians. *B. rubioides* and *B. sessiliflora* are attractive plants in cultivation, easily propagated from seed or cuttings.

Bauera

Bauera rubioides

River Rose, Dog Rose — Plate 314

Two forms of this species are present in the Sydney area. The typical form (var. *rubioides*) may grow into a large scrambling shrub, 2 m tall, with tangled interwoven branches that sprawl over 3 to 4 m of ground. Its leaves are opposite, each leaf divided into 3 leaflets, appearing to surround the stem. Flowers are pink, occasionally white, 12 to 18 mm across, with 8 to 10 petals and numerous stamens. They are borne on slender stalks which are longer than the leaves.

314. Bauera rubioides var. *rubioides*

315. Bauera rubioides var. *microphylla*

Var. *microphylla* (Plate 315) is a small low shrub, with flowers usually white and about 5 mm across. Leaves are smaller than the typical form. The two varieties intergrade and intermediate forms may be found.

HABITAT Open, wet situations, on sandy soils. Var. *rubioides* grows at its best along creek banks.

DISTRIBUTION Both varieties occur along the coast and in the mountains. Var. *rubioides* is also found along the southern tablelands, in Vic, Tas and SA. Var. *microphylla* occurs on the north coast and northern tablelands. Forms in other States seem to be intermediate between the two varieties.

FLOWERING July–December.

Bauera capitata

This small diffuse shrub to 25 cm tall differs from *B. rubioides* in having flowers without stalks. Flowers are crowded at the ends of short leafy branches. Leaves are divided into 3 leaflets, each leaflet 5 to 6 mm long, usually with a prominent lobe on each side.

HABITAT Wet heath, on sandy soils.

DISTRIBUTION Coastal area, north of Botany Bay but uncommon in the area. Extending along north coast to Qld.

FLOWERING Some flowers most of the year.

Escalloniaceae

This is another group related to the Rosaceae that is recognised as a separate family in some classifications and as a subfamily in others. In some modern systems the group is assigned to Saxifragaceae or to Grossulariaceae, the gooseberry and currant family. Escalloniaceae consists of 18 genera and about 160 species in Australia, Malesia, New Caledonia and South America. About 20 species in 7 or 8 genera occur in Australia, mainly in the rainforests of Queensland and New South Wales. Only 3 species occur in the Sydney area.

Escalloniaceae differs from Cunoniaceae in having alternate or opposite simple leaves without stipules and in having the same number of stamens as sepals and petals (mostly 4 or 5) attached to the floral tube instead of twice as many (mostly 8 to 10) attached to a perigynous disc. In 1 or 2 genera such as *Polyosma* the petals are partly fused into a tube. Unlike the Cunoniaceae, none of the species provides timber of commercial value. Although some have attractive foliage, they are rarely cultivated.

Abrophyllum

Abrophyllum ornans — Plate 316

The botanic name of this tall shrub (to 8 m) refers to its most conspicuous feature, its large attractive leaves. 'Abro' is derived from the Greek word for 'gay' and 'ornans' is Latin for beautiful. These leaves are alternate, thin, up to 20 cm long and 10 cm wide. The undersurface of the leaf is paler than above, and there are distinct raised veins. Flowers are small, greenish yellow and in clusters shorter than the leaves. The fruit is a globular purple-black berry.

HABITAT Cool moist situations, on rainforest slopes or in gullies.

DISTRIBUTION Scattered from the Illawarra northwards to north Qld, extending to lower Blue Mountains. Recordings include Minnamurra Falls, Royal National Park (Bola Creek), Wondabyne, Gap Creek Flora Reserve and Erskine Creek (lower Blue Mts).

FLOWERING November–January.

316. Abrophyllum ornans

Quintinia

Quintinia sieberi

Possumwood **Plate 317**

This is a medium-sized tree with corky bark. Branchlets are often red-brown and terminal shoots and leaf stalks pinkish. Leaves are alternate, tough, elliptic, entire and 5 to 10 cm long. Although individual flowers are small, the inflorescence is very striking, with many 5-petalled white flowers clustered along terminal panicles up to 12 cm long. Fruit is a capsule.

Habitat Rainforest gullies and sheltered forests, especially at higher altitudes.

Distribution Common in mountains; less common near coast. Recordings include Robertson, Katoomba, Mt Victoria, Mt Wilson, Gap Creek Flora Reserve and Bulli. Extending from the Sassafras-Shoalhaven district northwards along tablelands and coast to Qld.

Flowering October–November.

317. Quintinia sieberi

Polyosma

Polyosma cunninghamii

Featherwood **Plate 318**

Both the flowers and the fruits are good identifying features of this small rainforest tree. The flowers are tubular, white to creamy-green, about 10 to 12 mm long with 4 spreading star-like tips. The style projects just beyond the tube. Flowers are strongly scented and the generic name is derived from the Greek words for 'many scents'. The fruit is an ovoid berry, pointed at both ends and strongly ribbed. Leaves are opposite, bluntly pointed, glossy above and paler underneath, coarsely toothed and up to 10 cm long.

Habitat Rainforest.

Distribution Widespread and fairly common along the coast. Recordings include Minnamurra Spit, Mt Keira, upper Hacking River, Strickland State Forest and Gap Creek Flora Reserve. Also recorded in rainforest gullies of Blue Mountains. Extending from south coast along north coast to Qld.

Flowering April–October.

Pittosporaceae

Although this is not a large family its members are quite well known with a number of species, chiefly of *Pittosporum*, grown for their fragrant flowers and attractive foliage. Two other genera, *Hymenosporum* and *Bursaria*, also have sweetly scented flowers.

The largest genus, *Pittosporum*, with about 150 species, extends through Africa, Asia, Australia, New Zealand and Pacific islands to Hawaii. The 7 smaller genera totalling 35 species are endemic to Australia except for *Hymenosporum flavum* which extends to New Guinea and 1 or 2 species of *Citriobatus* in Indonesia and Malaysia.

Members are trees, shrubs or twiners. Some of the shrubs have scrambling or climbing stems. They all have more or less bell-shaped flowers with 5 sepals and 5 petals cohering at the base then spreading from near the middle. The 5 stamens surrounding the superior ovary are attached to the receptacle.

Pittosporum

Shrubs or trees, with alternate leaves. Flowers bell-shaped, with 5 free petals and 5 stamens. Fruit a capsule which splits to reveal a mass of sticky seeds. The generic name is a reference to these seeds ('pitta'—resin; 'sporos'—seed).

Pittosporum undulatum

Sweet Pittosporum, Mock Orange **Plate 319**

This is a small spreading tree with a dense, dark appearance. Its leaves are shiny, up to 12 cm long and hairless. The margins of the leaves are wavy ('undulate'). Flowers are white, bell-shaped and in terminal clusters. They are heavily scented, especially at night, and are pollinated by moths. The clusters of smooth, round, orange-coloured fruits, each about 10 mm across, are very conspicuous. Seeds are reddish and sticky. This hardy and adaptable plant is often grown as a street tree.

Habitat Shaded hillsides, rocky and scrubby gullies, along creek banks and in rainforest.

Distribution Widespread and common from coast to mountains. Often found in remnants of urban bushland where it may dominate gullies. Occurring all along coast and tablelands from south Qld to Vic, Tas and SA; also western slopes.

Flowering August–September.

318. Polyosma cunninghamii

319. Pittosporum undulatum

Pittosporum revolutum

Yellow Pittosporum **Plate 320**

A shrub or small tree, its flowers, leaves and fruit all distinguish it from *P. undulatum*. The flowers are yellow, not white, and its leaves are broader and may be rusty-hairy on the undersurface when young. Fruits are large, about 25 mm across, rough and orange-coloured. They split into 3 parts to reveal bright red seeds. The specific name 'revolutum' refers to the way the petals are rolled back.

Habitat Sheltered forest and rainforest margins.

Distribution Widespread from coast to mountains. Occurring all along coast and tablelands from Qld to Vic; also western slopes.

Flowering September–October.

320. Pittosporum revolutum

Hymenosporum

A small tree distinguished from *Pittosporum* by having flat, winged, non-sticky seeds with an ovate capsule.

321. Hymenosporum flavum

Hymenosporum flavum

Native Frangipani **Plate 321**

This small tree is often grown in suburban gardens because of its attractive bell-shaped flowers, its fragrant perfume and its dark shiny leaves. The flowers are 3 to 4 cm long and borne in loose terminal clusters. As they mature, they change from cream to yellow with a pink throat. Leaves are lanceolate, up to 16 cm long and taper to a blunt point. Fruiting capsules are brown and about 3 cm long. These contain numerous flat-winged seeds, from which the generic name is derived ('hymen'—a membrane; 'sporos'—a seed).

HABITAT Rainforest and along creeks in sheltered gullies.

DISTRIBUTION Uncommon in Sydney district. Recorded from Wheeny Creek, Colo River, St Albans and northwards. Extending along north coast and northern tablelands to Qld and New Guinea.

FLOWERING October–November.

322. Bursaria spinosa

Bursaria

Rigid shrubs, often with spiny branches. Flowers small and white. Fruit a compressed, thin-walled capsule, somewhat heart-shaped. The generic name is derived from 'bursa' (a purse), a reference to the shape of the fruit.

Bursaria spinosa

Blackthorn, Boxthorn **Plate 322**

Blackthorn is a medium to tall thorny shrub with numerous wiry branches. It is extremely variable; the size and shape of the leaves and the degree of spinescence vary with the age and location of the plant. Leaves are usually 10 to 25 mm long, crowded along the branchlets. In sheltered sites, the leaf may be 60 mm long. Masses of small white 5-petalled flowers are borne at the ends of the branches. These are followed by thin brown capsules, 4 to 6 mm long, which, despite their small size, are so abundant that they attract attention. The flowers are fragrant and rich in nectar and attract many insects. The species is a useful honey-producing plant.

HABITAT Forest, river flats, semi-cleared areas and along roadsides, especially on shale or clay soils.

DISTRIBUTION Widespread from coast to mountains; common along roadsides and in farm paddocks of the Liverpool-Penrith-Picton area and on shale cap at Menai. Occurring all along coast and tablelands from Qld to Vic; also western slopes, western plains, Tas and SA.

FLOWERING January–April, but as late as August in sheltered spots.

323. Bursaria longisepala

Bursaria longisepala

Plate 323

This is a much-branched shrub, seldom more than 1 m tall. It differs from *B. spinosa* in having its starry-white flowers scattered along the branches among the spines rather than massed at the ends of the branches. The leaves are clumped and crowded along the branches, seldom more than 8 to 10 mm long. The flattened seed capsule is 5 to 7 mm wide. Spines are numerous, up to 15 mm long. The specific name refers to the sepals which are somewhat longer in this species than in *B. spinosa*.

HABITAT Exposed rocky sites.

DISTRIBUTION Blue Mountains, including Glenbrook, Nattai Valley and the limestone outcrops of Wombeyan Caves. Scattered and uncommon. Also occurring north coast, northern tablelands, southern tablelands and central western slopes.

FLOWERING August–September, as late as December–January in higher mountains.

Citriobatus

Shrubs with rigid thorny branches and round succulent fruits.

Citriobatus pauciflorus

Orange Thorn **Plate 324**

Orange Thorn is a rigid, wiry shrub, seldom more than 1 m tall and often straggling. Its common name indicates its most conspicuous features, its small orange berries, 4 to 8 mm across, and the numerous fine thorns along its branches. Leaves are small, 8 to 12 mm long, 4 to 6 mm wide, stalkless, and usually have a few prickly teeth. In most cases, the white bell-shaped flowers are fairly inconspicuous, being solitary, axillary and with petals less than 4 mm long (hence, the specific name, 'pauci'—few, 'floris'—flower). Each flower has 5 petals, united to above the middle into a tube, and 5 stamens. Some plants bear abundant flowers and fruit. Because of this, and because the leaf shape is very variable, there was some confusion over naming this species. One form was formly known as *C. multiflorus.*

HABITAT In or near rainforest or in tall wet forest, especially on shales and richer soils.

DISTRIBUTION Widespread from coast to mountains. Found all along coast and tablelands divisions. Extending to central western slopes and Qld.

FLOWERING September–December.

324. Citriobatus pauciflorus

325. Billardiera procumbens

Billardiera

This genus originally consisted of a small group of twining plants, with bell-shaped flowers and a plump oblong berry as fruit. Recent revisions have reclassified the former genus *Rhytidosporum* (also known as *Marianthus*) and added it to *Billardiera.* These are small, spreading to prostrate shrubs, with star-shaped flowers and a fruit which is a compressed globular capsule.

Billardiera procumbens **Plate 325**

There has been some confusion about the appropriate classification of this small shrub. Prior to 1972, it was placed in the genus *Marianthus*. Then in 1975, the name was changed to *Rhytidosporum procumbens*, a name which had been give to it by Ferdinand von Mueller in 1860. In a revision of the family in 1978, it was placed in the genus *Billardiera* on the basis that the fruiting capsule was similar to that of the Western Australian *Billardiera villosa*. As the specific name suggests, this species is usually procumbent (spreading over the ground), but it may also be erect to 25 cm tall. Leaves are crowded and narrow, up to 12 mm long and 2 mm wide. Flowers are 5-petalled white stars, sometimes tinged with pink, with 1 or 2 together on stalks up to 5 mm long. The superior ovary in the centre is surrounded by 5 conspicuous stamens. This is an attractive plant when flowering as the flowers are near the ends of the branches and are often interspersed with pink buds. Fruit is a subglobular thinly leathery capsule, about 6 mm across.

HABITAT Dry open forest and heath, on sandy soils.

DISTRIBUTION Widespread from coast to mountains. Found all along the coast and tablelands of NSW; extending to western slopes. Also occurring in Qld, Vic and Tas.

FLOWERING September–November.

Billardiera scandens

Dumplings, Apple-berry **Plate 326**

This is a common little twining plant in the undergrowth of forests. The popular names are derived from the shape of its fleshy cylindrical fruit, which is up to 3 cm long, greenish and hairy (sometimes hairless) and hangs from the end of the stem. Flowers are yellow, bell-shaped and pendulous on slender stalks. Leaves are narrow lanceolate, 3 to 5 cm long, with wavy margins; when young, leaves may have a velvety appearance. Aborigines are believed to have eaten the ripe fruit raw and the green fruits after roasting.

HABITAT Open forest; less common in coastal scrub and heath.

DISTRIBUTION Widespread from coast to mountains. Found all along the coast and tablelands of NSW; extending to western slopes. Also occurring in Qld, Vic, Tas and SA.

FLOWERING September–December.

Crassulaceae

The Crassulaceae is a family of succulent herbs and shrubs related to the non-succulent Saxifragaceae, but unrelated to succulent families in the Caryophyllales. Although widespread in the drier parts of the world, the family has its greatest development in South Africa.

One native genus, *Crassula*, occurs in Australia, with 4 species in the Sydney area. owing to the ability of some Crassulaceae to reproduce from pieces of stem and fallen leaves or plantlets borne on the leaves, 1 or 2 introduced species have invaded areas of bushland around Sydney (e.g., Mother-of-Millions, *Kalanchoe tubiflora*).

326. Billardiera scandens

Crassula

All 4 local species of *Crassula* are insignificant herbs, with opposite succulent leaves. All prefer damp sites, including roadside ditches, muddy soils and mossy rocks.

Crassula sieberiana

Australian Stonecrop **Plate 327**

C. sieberiana is the most common of the 4 local species of *Crassula*. It has numerous small branchlets up to 10 cm tall and a tendency to turn a reddish colour with age. Its flowers are tiny, almost stalkless and arranged in short dense clusters in the leaf axils. This flower arrangement distinguishes it from the similar *C. macrantha* which has flowers in branched clusters longer than the leaves. The other 2 species, *C. helmsii* and *C. purpurata*, have weak stems which float when in water and solitary flowers in the leaf axils. They are difficult to separate in the field, but *C. helmsii* has its carpels subtended by basal scales which are absent in *C. purpurata*.

HABITAT Common on shaded mossy rocks.

DISTRIBUTION Widespread from coast to mountains. Found throughout the State; also in Qld, Vic, Tas, SA, WA, NZ and Lord Howe Island.

FLOWERING September–November.

327. Crassula sieberiana

Rosaceae

The rose family comprises more than 3000 species of trees, shrubs and herbs with a worldwide distribution having maximum development in northern temperate regions. Many varieties have been developed to satisfy man's desire for beauty as well as his need for food. The rose in its various forms has been cultivated since ancient times and has been called the 'queen of flowers'. Other favourites include cotoneaster and the flowering cherries, almonds and plums. The list of genera providing edible fruits includes *Prunus* (almond, apricot, cherry, plum, peach, nectarine), *Rubus* (raspberry, blackberry, loganberry), *Malus* (apple), *Pyrus* (pear) and *Fagaria* (strawberry).

The family is poorly represented in Australia with only 7 genera and about 15 species, and the plants themselves are unimpressive by comparison with the above. The 2 genera with native species in the Sydney area consist of prickly brambles with dry or flavourless compound fruits and some weedy herbs with tiny flowers that develop into burrs. Neither genus is characteristically Australian, and the species themselves may extend beyond Australia to New Zealand or as far as China and Malesia.

In the bramble genus, *Rubus*, reproduction is often apomictic or vegetative which tends to perpetuate variants and make the delineation of species difficult. Similar difficulties exist with the other genus, *Acaena*, due to hybridisation and the variability of forms. *Acaena* shows a number of advanced characters in comparison with the somewhat primitive characters of the better known and more attractive genera. Advanced characters of *Acaena* include the herbaceous habit and small flowers without petals, few stamens and a single carpel, compared to the shrubby habit, relatively large flowers, prominent free petals, numerous stamens and numerous free carpels displayed by genera such as *Rubus* and *Rosa*. *Acaena* does not secrete nectar and appears to have dispensed with the assistance of insects altogether, relying instead on wind pollination. Its prickly barbed burrs are designed for dispersal by animals.

328. Acaena ovina

329. Acaena novae-zelandiae

Acaena Burr

Perennial herbs with pinnate leaves. Fruits covered with barbed spines; these are the burrs which stick to walkers' socks and sheep's wool.

Acaena ovina

Sheep's Burr **Plate 328**

This is an erect shrub, 10 to 50 cm tall, with leaves chiefly clumped at the base of the stem but a few smaller leaves along the stems. Leaves are compound, somewhat fern-like, with many opposite leaflets, 10 to 20 mm long, which are toothed. Leaves are hairy on the undersurface, especially on the veins, and less hairy (sometimes hairless) above. Flowers and burrs are borne in terminal spikes at the end of an erect, almost leafless stalk. The fruitlets have short, irregular spines, with 3 to 6 longer than the others. This species appears to hybridise with *A. novae-zelandiae*.

Two closely related species, *A. agnipila* and *A. echinata*, are also widespread in grassland and open forest. They differ in small details which are difficult to distinguish in the field. *A. agnipila* has spines which are regular. *A. echinata* has less hairy leaves, with spines thickened at the base and flowers more scattered than *A. ovina*.

HABITAT Grassland and open forest.

DISTRIBUTION A tablelands species, absent from the coast. Recordings include Coxs River and Berrima. Extending throughout tablelands division to western slopes; also Qld, Vic, Tas, SA and WA.

FLOWERING October–January.

Acaena novae-zelandiae

Bidgee-widgee **Plate 329**

This creeping, mat-forming herb differs from other *Acaena* species in the area in having its flowers and burrs borne in dense globular heads, not in terminal spikes. Leaves consists of 3 to 4 pairs of deeply toothed leaflets. Each flower-head is about 2 cm across. Seed-heads turn from red to brown as they dry. Prickles are 4 in number, up to 10 mm long and nearly equal in length.

HABITAT Grasslands, roadsides and open forest.

DISTRIBUTION Widespread and common from coast to mountains. Occurring all along coast and tablelands divisions, extending to north-western slopes; also Qld, Vic, Tas, SA and New Zealand.

FLOWERING November–February.

Rubus Bramble

Prickly shrubs and climbers, including the introduced blackberry and bramble. Flowers with 5 broad petals and 5 persistent sepals; stamens numerous. Fruit with succulent segments clustered together into an aggregate fruit like a blackberry or raspberry.

Rubus hillii

Broad-leaf Bramble **Plate 330**

This is a scrambling shrub, often with long stems resembling a climber. It is the only *Rubus* species in the area with simple leaves. These leaves are broadly ovate to heart-shaped, 6 to 10 cm long and wide, with finely toothed margins and rusty-hairy underneath. Flowers are pink (occasionally white). Fruit is a red globular raspberry, about 12 mm across. These fruits were part of the diet of the Aborigines and a medicinal drink to relieve stomach upset was made by soaking the leaves in warm water.

HABITAT Sheltered forests, and rainforest margins or clearings.

DISTRIBUTION Widespread along coastal zone. Extending along coast from Qld to Vic and SA.

FLOWERING November–February.

331. Rubus parvifolius

330. Rubus hillii

Rubus parvifolius

Native Raspberry, Small-leaf Bramble **Plate 331**

A scrambling shrub with stems to 60 cm long. Its leaves are compound, with 5 or more leaflets, 5 to 30 mm long, deeply wrinkled and often deeply toothed. Leaflets have characteristic white, closely matted hairs on the undersurface. Flowers are pink to red. Fruit is a fairly large globular cluster of succulent red fruitlets, like a raspberry.

HABITAT Shaded forested slopes and damp creek banks, often in clay soils.

DISTRIBUTION Widespread from coast to mountains. Recordings include Royal National Park (Werrong), Brisbane Water National Park (Girrakool), Menangle and Cumberland Plain (Eastern Creek, Shane Park). Occurring throughout NSW, except drier western plains; also Qld, Vic, Tas, SA and China.

FLOWERING October–December.

332. *Rubus rosifolius*

Rubus rosifolius

Forest Bramble, Rose-leaf Bramble **Plate 332**

Forest Bramble is a weak shrub regarded by some botanists as a hybrid of *R. hillii* and *R. parvifolius*. Leaves are compound, with 5 to 7 leaflets which are larger than those of *R. parvifolius*, mostly 3 to 8 cm long. The terminal leaflet is larger than the others, being about 8 cm long. Leaves are green on both surfaces, lacking the white hairs of *R. parvifolius*, and have toothed margins. Petals are usually white. The red raspberry fruit is oval in shape, rarely globular. These fruits are edible but less succulent than those of *R. parvifolius*.

Habitat Moist eucalypt forests and rainforest margins.

Distribution Widespread from coast to mountains. Extending all along coast from Qld to Vic; also Asia and Malesia.

Flowering Most of the year.

Rubus sp. aff. *moorei*

Bush Lawyer **Plate 333**

A 'bush lawyer' was someone from whom it was difficult to escape once he got his hands on you. Early bushmen who walked into this climbing, prickly plant, and had difficulty extracting themselves must have been acutely aware of the similarity of the situation. The leaves branch from the stem like fingers on a hand, with 5 leaflets, rather like the leaves of a blackberry. Each leaflet is 7 to 12 cm long, with toothed margins. The branches and leaf stalks are covered with numerous prickles. White flowers are borne in loose clusters from the leaf axils and the fruit is dark red.

Habitat Sheltered forests and margins of rainforests.

Distribution Widespread along coast. Extending from south coast along north coast to Qld.

Flowering October–January.

333. *Rubus* sp. aff. *moorei*

Mimosaceae

In older classifications (and 1 or 2 modern ones) all pod-bearing plants were included in 1 large family, Leguminosae. It is, however, convenient when dealing with the large Australian contingent of pod bearers to divide them into 3 families, Mimosaceae (wattles), Caesalpiniaceae (cassias) and Fabaceae (peas).

In Mimosaceae the flowers are small and regular with 4 or 5 valvate petals, an equal number of fused sepals and prominent stamens. The stamens are either definite in number (as in *Mimosa*) or indefinite (as in *Acacia*). In the other 2 families the flowers are irregular with 5 imbricate petals, free or fused sepals and a definite number of stamens (usually 10). The feature common to all, the legume or pod, is a fruit formed from a single carpel which usually splits at maturity along 2 sutures to release the seeds which are borne in a row along the inside of the ventral suture.

There are 10 genera of Mimosaceae in Australia but only 1, *Acacia*, is widespread and well known. Australian acacias are known popularly as wattles, a name that was originally applied to *Callicoma serratifolia* because of its use in wattle and daub construction, but later associated with the acacias, presumably because of their similar flower heads and the suitability of some species for this type of construction. Although the wattle is Australia's floral emblem it is not confined to this country, and may not have originated here. Acacias are also a feature of the dry African landscape and are widely dispersed in the tropics and subtropics. It is estimated that there are about 1200 species of *Acacia* with more than 700 native to Australia where they occur in habitats ranging from rainforests to the driest parts of the interior.

The genus has been divided variously into sections or subgenera, but since in every case the majority of the Australian species fall into one group little would be gained here by a detailed consideration.

It only needs be said that Australian acacias are distinct from those found in other parts of the world which are characterised by having bipinnate leaves and often vicious spines or thorns. Australian acacias rarely have spines, although in many cases the 'leaves' are sharply pointed.

A common feature of the Australian acacias is a modification of the bipinnate leaves in which the lamina is contracted and the stalk is expanded, taking over the functions of photosynthesis and respiration. The expanded leaf stalks are called 'phyllodes'. All species have bipinnate leaves in the seedling stage and the change to phyllodes may occur after the first few leaves, or may be delayed in some species until the plant is 2 or 3 m tall.

Leaf-like phyllodes differ from true simple leaves in a number of respects. Firstly they are flattened in the vertical plane giving them an upper and lower margin rather than an upper and lower face. In many cases they have 2 or 3 prominent nerves or veins

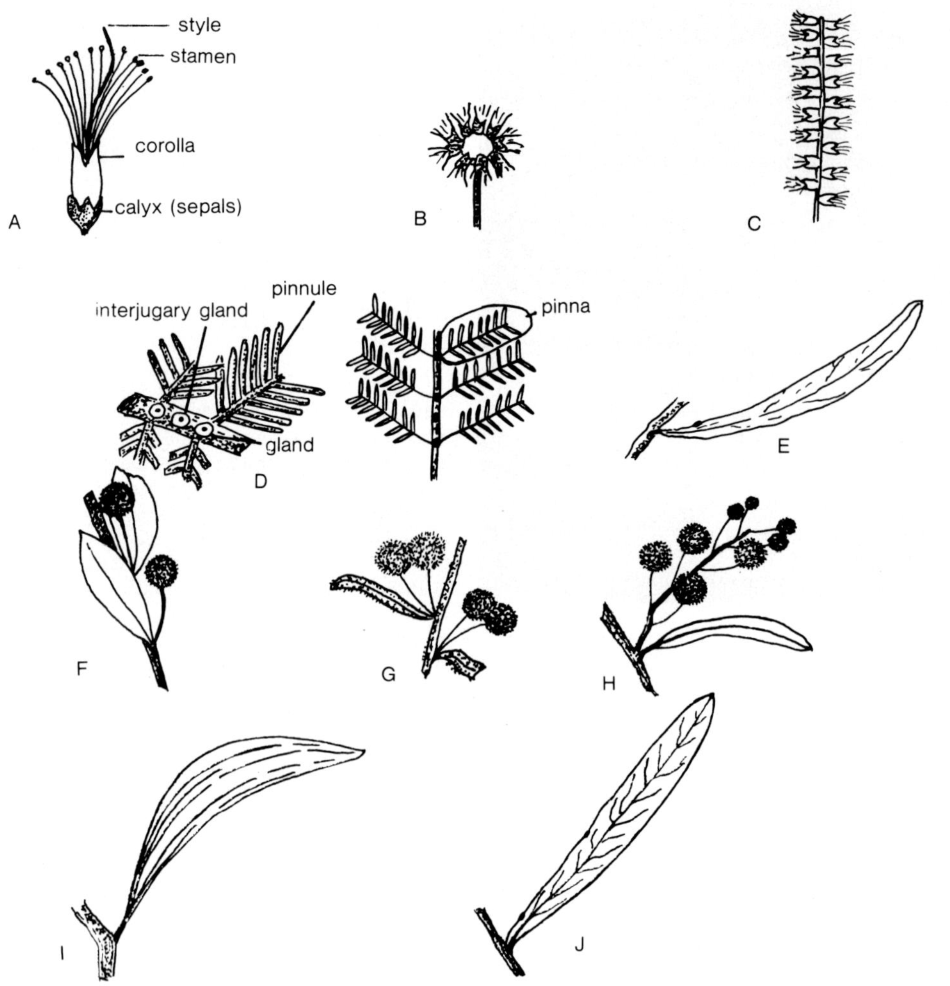

Figure 6

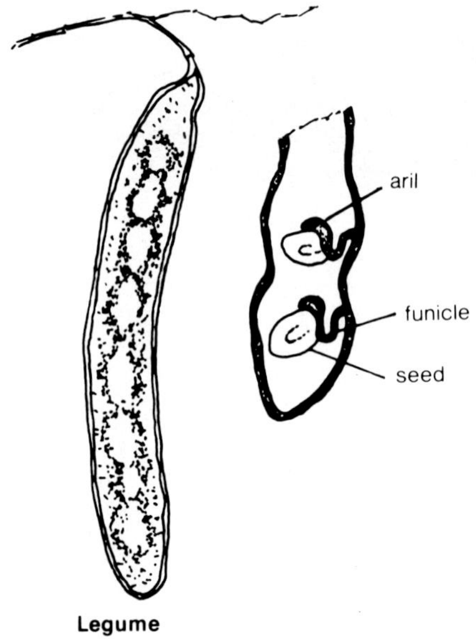

Figure 7

running lengthwise instead of a mid-vein with secondary lateral veins and minor reticulate veins. Usually the upper margin has one or more tiny depressed or raised gland. Similar glands occur on the rhachis (axis or mid-rib) of the bipinnate leaves. Although in the Australian species these glands do not have an obvious function, they are useful in identification since they vary from species to species in structure, number and position.

In many African species the glands secrete nectar which provides food for ants living in association with the plant. In some cases the ants are provided a home in large hollow spines, and in return discourage insects and animals from eating the foliage. Perhaps in the absence of a large browsing fauna in Australia these features were unimportant and were gradually lost in the Australian acacias, or possibly, in the fragmentation of Gondwanaland, Australia retained a different primitive acacia stock which did not develop spines and thorns. The little that is known suggests that diversification and expansion in Australia proceeded from north to south and that acacias were not common or widespread until a general decline in soil fertility and the development of arid habitats became significant in the latter part of the Tertiary period.

The number of flowers in each head or spike may assist in identification. The number may vary from 4 to 40 between different species but is fairly consistent within each species. It is easiest to count the number of flowers when the plant is in bud. The size and shape of the pea-like pods (legumes) can also be an aid in identification. When pods open they expose and release the seeds which are joined to the pod by a stalk called the *funicle.* The thickened part of the funicle which grows partly around the seed is called the *aril.* Careful examination of the funicle is often necessary for positive identification of some species.

Identification of wattles

Identification of wattles is best accomplished by first grouping them on the basis of foliage and inflorescence which are, on the whole, clear-cut and consistent features. The tiny wattle flowers (Figure 6A) are either grouped in globular heads (Figure 6B) or arrange in spikes (Figure 6C). Twelve or thirteen species in the Sydney area have bipinnate leaves (Figure 6D) and flowers in heads. The combination of bipinnate leaves and flowers in spikes is rare and does not occur in the Sydney area, but 9 species have phyllodes (Figure 6E) and flowers in spikes. About 45 species which have phyllodes and flowers in globular heads can be divided into smaller groups according to whether the phyllodes are pungent pointed and whether the flower heads occur singly in the leaf axils (Figure 6F), in small clusters (Figure 6G), or in racemes (Figure 6H). The largest group with flower heads in racemes can be further divided depending upon the size of the phyllodes, whether the phyllodes have 1, 2, or several longitudinal veins (Figure 6I), or are penniveined (Figure 6J). Positive identification within groups often depends on a combination of minor differences between species and may at times be difficult.

Acacia Wattle

Over 60 species of wattle occur naturally in the Sydney district, with up to 10 additional species occurring on the fringes of the area. For the purpose of identification, they can be divided into the following groups based on foliage and inflorescence.

Group 1 Wattles with bipinnate leaves.
Group 2 Wattles with cylindrical flower spikes.
Group 3 Wattles with globular flower heads; phyllodes with pungent points.
Group 4 Wattles with globular flower heads and flowers borne singly, in pairs or clusters (not in racemes); phyllodes not pungent pointed.
Group 5 Wattles with globular flower heads borne in racemes; phyllodes under 7 cm long.
Group 6 Wattles with globular flower heads borne in racemes; phyllodes more than 7 cm long.

Group 1

Watles with bipinnate leaves. Bipinnate leaves consist of a central stem or midrib (the rhachis) from which spread pairs of leaflets (pinnae) which are themselves divided into tiny leaf segments (pinnules). Along the rhachis is often a series of glands which at times secrete a sticky nectar attracting ants and other insects. The arrangement of these glands is an important identifying feature for some species.

Acacia elata

Cedar Wattle **Plate 334**

Cedar Wattle grows to a medium-sized tree up to 20 m tall. It has large leaves, somewhat resembling those of a pepper tree, but much darker. The pinnae are widely spaced and carry 8 to 14 pairs of pinnules, each up to 5 cm long. Young foliage is a distinctive light cream in contrast to the darker foliage. Flower heads are large, creamy-yellow, and in showy terminal clusters.

HABITAT Grows best in shaded gullies and along rainforest margins, but also found in more open areas and along roadsides near tall eucalypt forests.

DISTRIBUTION Widespread from coast to mountains. Also occurs on north coast and northern tablelands.

FLOWERING December–February.

Acacia schinoides **Plate 335**

This tall shrub to small tree usually has a straggly habit. Its drooping leaves are light green with broad pinnules about 12 mm long and 2 to 6 mm wide. New growth is very striking as the tips are a bronze colour. Another conspicuous feature is the silvery-grey colour of the trunk, branches and branchlets. The specific name is a reference to its foliage being like a pepper tree (*Schinus).*

HABITAT Sheltered forests and along gully creeks.

DISTRIBUTION North of Sydney. Locally common north of Mooney Mooney Creek (e.g., Watagan State Forest and Olney State Forest). Endemic in the central coast and north coast.

FLOWERING December–March.

334. Acacia elata

335. Acacia schinoides

Acacia terminalis

Sunshine Wattle **Plate 336**

A very common shrub around Sydney, conspicuous because it flowers through the autumn and winter months and has relatively large flower heads. These flowers vary in colour from pale yellow to deep gold. Flower heads consist of only 6 to 15 flowers, and are widely spaced along racemes which exceed the leaves. Branchlets are often reddish in colour. There are 8 to 16 pairs of pinnules; pinnules are about 10 mm long and 2 mm wide, dark green above and much paler underneath.

HABITAT Dry open forest and scrubland, on sandy soils; often found in remnants of urban bushland.

DISTRIBUTION Widespread and common from coast to mountains. Also occurring north coast, northern tablelands, central western slopes, south coast, Vic and Tas.

FLOWERING March–July, with some flowers as late as September.

336. Acacia terminalis

337. *Acacia irrorata*

Acacia irrorata ssp. *irrorata* — Plate 337

This is a small tree, usually 5 to 12 m tall, with branchlets which are ridged and rough to touch. Young shoots and leaves are golden yellow, but there are other bipinnate wattles with this feature. This is one of a group of feather-leaved wattles which are often difficult to positively identify. There are 2 major differences between *A. irrorata* and other bipinnate wattles in the area. First, the foliage is green, not silvery, with pinnules covered with a light sprinkling of hairs (best seen with a hand lens). Secondly, glands are usually absent from the rhachis, or if present are only at the base of the upper pairs of pinnae. The pod is black, rough to touch, 5 to 10 cm long and covered at first with short pale hairs.

HABITAT Ranging from open forest to rainforest margins, on sandy soils; often near creeks.

DISTRIBUTION Widespread from coast to mountains. Also occurring north coast, northern tablelands, south coast, western slopes. Qld and Tas.

FLOWERING November–January.

Acacia jonesii — Plate 338

This is a straggly, open shrub, 0.5 to 3 m tall. Each leaf has a smaller number of pinnae than most other bipinnate wattles in the area. There are usually 5 to 6 pairs of pinnae, sometimes up to 9 pairs, each with 12 to 18 pairs of pinnules. A conspicuous gland is located on the rhachis between the pairs of pinnae. Stems and rhachis are usually covered with short grey hairs, but may be quite hairless. Flowers are few, only 8 to 12 in a head; they are golden-yellow. The inflorescence is a raceme which may be longer or shorter than the foliage.

HABITAT Forested slopes and ridges, on sandy soils.

DISTRIBUTION Southern highlands (e.g., Bundanoon and Wingello State Forest) and southern Blue Mountains. Extending to south coast and southern tablelands.

FLOWERING September–early October.

338 *Acacia jonesii*

Acacia parramattensis

Sydney Green Wattle — Plate 339

This small tree is common south of Sydney and westwards to the lower Blue Mountains. Its trunk is brown-black and the tips of the young foliage are greenish yellow. Stems and branchlets have minute ridges (0.2 mm high). Branchlets, rhachis and glands are usually covered with short greyish hairs. Leaves are very finely divided, with 20 to 35 pairs of pinnules, 3 to 5 mm long. Glands are present at the base of the pairs of pinnae and usually, but not always, between successive pairs of pinnae. Flowers are pale yellow to almost white. This species tends to flower later than *A. decurrens* and *A. mearnsii*, with which it could be confused.

HABITAT Open forest, on both shales and sandstone soils.

DISTRIBUTION Widespread from coast to mountains. Recordings include Brisbane Water National Park, Loftus, Richmond, lower Kedumba Creek and St Albans. Extending to south coast, southern tablelands and central western slopes.

FLOWERING October–February.

339. *Acacia parramattensis*

340. *Acacia decurrens*

Acacia decurrens

Black Wattle, Green Wattle — Plate 340

This is a medium-sized tree, with dark grey to black bark which contains a high percentage of tannin, once widely used in the preparation of leather. It commences flowering in mid-winter, before *A. mearnsii, A. parramattensis* and *A. irrorata,* and for this reason it is sometimes referred to as 'Early Green Wattle'. The most obvious differences between this species and other bipinnate wattles is that it has pinnules which are more widely spaced and branchlets with conspicuous wing-like ridges up to 2 mm high. Glands occur at the base of each pair of pinnae and 1 or 2 on the rhachis below the lowest pair of pinnae, but never between successive pairs of pinnae. Flower heads are brilliant yellow and abundant, making the tree very showy.

HABITAT Open hilly country and open forest, mainly on shales.

DISTRIBUTION Scattered from coast to mountains. Abundant beside south-western expressway near Bargo. Extending to south coast and southern tablelands. Naturalised in other areas of NSW and other States.

FLOWERING Late July-September.

341. *Acacia pubescens*

Acacia pubescens

Downy Wattle — Plate 341

This is a bushy shrub, to 3 m tall, regarded as endangered because of its narrow range of distribution. The specific name 'pubescens' refers to the dense white hairs which cover the branchlets, flower stalks and rhachis. Leaves are light green, without hairs, and pinnae and pinnules appear crowded. Pinnae are short with only 7 to 10 pairs and pinnules are small, 2 to 4 mm, with 6 to 20 pairs. Flowers are numerous, small and brilliant yellow, borne in slender racemes of 10–20 flower heads; the inflorescence exceeds the leaves.

HABITAT Open forest and scrub, on shale and gravelly clay.

DISTRIBUTION Restricted to a few areas around Sydney; uncommon. Recordings include Menai, Liverpool, Salt Pan Creek and Kemps Creek.

FLOWERING August-September.

Acacia filicifolia

Fern-leaf Wattle — Plate 342

Fern-leaf Wattle is a small tree with a dark trunk and characteristic blue-grey branchlets. The rhachis has numerous hairy glands over its whole length. The tip of the young foliage is a light cream colour. The

342. *Acacia filicifolia*

pinnules are narrow, very crowded and there may be 50 or more pairs. It closely resembles *A. irrorata*, but *A. filicifolia* has more glands on its rhachis and the branchlets are smooth, not rough. It differs from *A. parvipinnula* in having a brown or grey trunk (sometimes green when young). Flowers are bright yellow, with 22 to 30 in each head. The pod is straight-sided, to 12 cm long and blue-grey in colour.

HABITAT Open forest and woodland, mainly on alluvial flats but also on stony hillsides.

DISTRIBUTION North-western Hornsby Plateau and Blue Mountains (e.g., upper Wolgan Valley and lower Kedumba Valley); also recorded on Woronora Plateau. Extending from south coast northwards to north coast and Qld; also northern tablelands and western slopes.

FLOWERING August-September.

343. *Acacia fulva*

Acacia fulva

Velvet Wattle **Plate 343**

A. fulva is restricted in the Sydney district to a small patch on the slopes of Mount Wareng east of Howes Valley on the north-west Hornsby Plateau. There it is a small tree with branchlets and leaves covered with velvety silver-grey hairs. New growth is a distinctive chestnut colour. Flowers are bright yellow and buds are covered with dense chestnut hairs. Pods are up to 11 cm long, dark brown and densely woolly.

HABITAT Tall eucalypt forest

DISTRIBUTION Known only from two localities: Mt Wareng and Gloucester Buckets (west of Gloucester).

FLOWERING April-May.

Acacia parvipinnula

Silver-stemmed Wattle **Plate 344**

This is a tall shrub or tree to 10 m tall. Three features, when combined, help separate it from other bipinnate wattles. First, the bark of the trunk, the branches and branchlets are very silvery. Secondly, there are only 14 to 18 flowers in each head. Thirdly, there are numerous hairy glands occurring irregularly along the rhachis. The pinnules are small and crowded, with 14 to 30 pairs. Young foliage has a distinct creamy colour. Flower heads are light yellow.

HABITAT Open forest, mainly on shale but sometimes on lateritic sandstones.

DISTRIBUTION Cumberland Plain, Blue Mountains, Hornsby Plateau (e.g., Colo Heights) and a few scattered trees on shale outcrops south of Sydney (e.g., Menai). Also occurring north coast and central western slopes.

FLOWERING September-December.

344. *Acacia parvipinnula*

Acacia mearnsii

Black Wattle **Plate 345**

A medium-sized tree with a black trunk. Branchlets, leaf rhachis, leaf stems and the undersurface of the pinnules are densely hairy. Numerous hairy glands are scattered irregularly along the rhachis. Young growth is downy yellow and conspicuous. Flowers are pale yellow, with 20 to 30 in each head. Pods are hairless, leathery and 5 to 12 cm long.

HABITAT Open forest and undulating cleared land, on various soils; preferring moderately rich shale soils.

DISTRIBUTION South of Sydney and Blue Mountains. Common on coastal plain south of Wollongong where it has recolonised cleared land; also recorded on Wianamatta Shales near Aylmerton and in sheltered gullies along Nattai River. Extending to south coast, southern tablelands, Vic and Tas. Naturalised in other States.

FLOWERING October-November.

345. *Acacia mearnsii*

Group 2

Wattles with cylindrical flower spikes

Acacia oxycedrus **Plate 346**

This is a tall rigid shrub and the only wattle in the area having both flower spikes and pungent-pointed phyllodes. Phyllodes are flat, rigid and very sharp, 2 to 3 cm long, with 4 or more conspicuous parallel veins. Flowers are bright yellow and borne in dense spikes about 2 to 3 cm long. The pod is woody, cylindrical and 5 to 10 cm long.

HABITAT Open forest, on sandy soils.

DISTRIBUTION Restricted to the Hornsby Plateau and lower Blue Mountains. Common north of Sydney (e.g., West Head, Girrakool), and on ridge tops around Faulconbridge in the Blue Mountains. Also occurring in Vic and SA.

FLOWERING July-October.

346. *Acacia oxycedrus*

347. *Acacia longissima*

Acacia longissima **Plate 347**

This species is readily recognised as it is a much-branched shrub, with drooping branches and long narrow phyllodes. The phyllodes are up to 30 cm long and only 2 to 3 mm wide with one clearly defined central vein. Flowers are a light cream and loosely scattered along the flower spike. Flower spikes are rather sparse on the branches. At first glance, the long leaves and pendulous appearance may suggest *A. floribunda*, which is much more common, but that species bears abundant flower spikes and has phyllodes with several prominent parallel veins.

HABITAT Forested slopes, on sandy soils.

DISTRIBUTION Coast and Cumberland Plain. Common in and around Katandra Reserve, Mona Vale. Extending all along the coast from the south coast northwards to Qld.

FLOWERING Mainly December-January, but some flowers at other times.

348. Acacia maidenii

349. Acacia obtusifolia

350. Acacia longifolia var. *longifolia*

Acacia maidenii

Maiden's Wattle **Plate 348**

Maiden's Wattle is a small tree with distinctive phyllodes. These are long, to 20 cm, 8 to 12 mm wide, somewhat falcate (sickle-shaped), with many parallel veins and no obvious main vein. The phyllodes are very tapered at each end. Flower spikes are pale yellow, clustered together but not abundant or showy. Pods are narrow, cylindrical and very twisted, a feature which helps identification.

HABITAT Forest on deeper soils and on margins of rainforest.

DISTRIBUTION Widespread from coast to mountains. Common along foothills of Illawarra escarpment (e.g., Macquarie Pass and below Mt Keira). Also occurring north coast, northern tablelands, central western slopes, south coast, Qld and Vic.

FLOWERING January-May.

Acacia obtusifolia **Plate 349**

This erect shrub is easily confused with *A. longifolia.* The most striking feature separating the two is their flowering times. *A. obtusifolia* does not begin flowering until December by which time *A. longifolia* is usually finished flowering. The phyllodes are 8 to 20 cm long and up to 2 cm wide, thick and with several parallel veins. The specific name refers to the obtuse or blunt apex of the phyllode. (*A. longifolia* has a pointed tip, at least when young). New growth is often reddish. The pale yellow flower spikes are about 4 cm long and showy when in full bloom. Flowers are loosely and often irregularly spaced along the spike, in contrast to the more densely packed flowers of *A. longifolia.*

HABITAT Forest and woodland, often along creeks.

DISTRIBUTION Widespread from coast to mountains. Extending all along coast and tablelands from Qld to Vic; also central western slopes.

FLOWERING December-February.

Acacia longifolia var. *longifolia*

Sydney Golden Wattle **Plate 350**

Sydney Golden Wattle is one of the best-known native plants of the Sydney district. Its masses of brilliant yellow flower spikes make it very conspicuous. It is very common in some areas and is a familiar shrub or small tree (2 to 4 m tall) along roadsides. Phyllodes are 8 to 16 cm long and up to 2 cm wide, with several parallel veins. Most phyllodes are pointed at the apex, not rounded like *A. obtusifolia.* Flowers are golden yellow and borne in dense spikes in the leaf axils.

HABITAT Forest and woodland, on sandy soils.

DISTRIBUTION Widespread and common from coast to mountains. Also occurring north coast, northern tablelands, south coast, southern tablelands and Vic.

FLOWERING June-November.

Acacia longifolia var. *sophorae* **Plate 351**

This variety of *A. longifolia* is found chiefly on coastal sand dunes where it may form large dense clumps. Its deep roots and many branches help prevent the sand from drifting and it is an important agent in the stabilisation of many dune systems. The plant has adapted to these harsh conditions by having smaller phyllodes than var. *longifolia.* They are obovate to oblong, 5 to 9 cm long, up to 36 mm wide, broadly obtuse at the apex but with a short mucronate tip, at least when young.

HABITAT Coastal sand dunes.

DISTRIBUTION Common along coastline. Recordings include Bouddi National Park., Kurnell, Royal National Park and Minnamurra Spit. Extending all along the coast from Qld to Vic, Tas and SA.

FLOWERING July-October.

351. Acacia longifolia var. *sophorae*

Acacia floribunda

Sally Wattle **Plate 352**

A graceful small tree with pendulous branches. Phyllodes are variable, between 5 to 12 cm long, sometimes quite narrow (2 mm), but often much broader (12 mm). This means that they are broader than the phyllodes of *A. longissima* but usually narrower than *A. longifolia.* Flowers are pale yellow and borne in loose pendulous spikes up to 6 cm long. The specific name 'floribunda' refers to the abundant flowers which make this species so attractive and a common garden plant.

HABITAT Forested slopes, sheltered gullies, and along creek banks.

DISTRIBUTION Widespread from coast to mountains. Recordings include Brisbane Water National Park, Royal National Park and Bents Basin, Nepean River. Extending all along the coast from Qld to Vic; also northern tablelands and central western slopes.

FLOWERING August-September.

352. *Acacia floribunda*

Acacia dorothea

Dorothy's Wattle **Plate 353**

Dorothy's Wattle is an erect shrub, with phyllodes which are lanceolate to slightly falcate; they have a conspicuous mid-vein and a small recurved point at the apex. A prominent gland occurs midway along the margin. Phyllodes and branchlets are covered with small white hairs. Flower spikes are very short. Flowers are bright yellow and compact. Each spike is borne on a sturdy raceme which is shorter than the phyllodes. The pod is densely hairy, 3 to 4 cm long, and contains only 2 to 3 seeds.

HABITAT Open forest and woodland, on sandy soils.

DISTRIBUTION Endemic in higher Blue Mountains (e.g., Lithgow, plateau of Newnes State Forest and Boyd River-Kanangra Walls) and Budderoo National Park.

FLOWERING September-October.

353. *Acacia dorothea*

354. *Acacia binervia*

Acacia binervia

Coast Myall **Plate 354**

An attractive medium-sized tree, easily recognised by its grey-green (glaucous) foliage and its large (to 6 cm) deep-yellow flower spikes which are produced in great abundance. The specific name 'binervia' is rather misleading as the phyllodes have several parallel veins. This species was previously known as *A. glaucescens,* a more suitable name as it refers to the colour of the phyllodes. Branchlets and phyllodes are covered with short dense soft hairs and new foliage is a light green colour. Phyllodes are sickle-shaped, 10 to 15 cm long and 5 to 20 mm wide.

HABITAT Forested slopes and along watercourses.

DISTRIBUTION Mainly a coastal species, extending to Blue Mountains. Common along Georges River, Nepean River and Wheeny Creek. Planted and naturalised in some locations, such as Kurnell. Extending along coast south from Port Macquarie to south coast. Also occurring along tablelands and central western slopes.

FLOWERING September–October.

Group 3

Wattles with globular flower heads; phyllodes with pungent points.

Acacia ulicifolia

Prickly Moses **Plate 355**

A very common shrub around Sydney on sandy soils of low fertility. Its phyllodes are short and needle-like, somewhat flattened, 8 to 14 mm long, rather crowded along the stem and slightly swollen at the base. Flower heads are on long slender stalks, 8 to 15 mm long. Flowers are pale cream, unlike those of *A. echinula* and *A. brownei* which have deep yellow-orange flowers. The pod is 3 to 5 cm long, 3 mm wide, curved and evenly constricted between the seeds. The name 'Prickly Moses' is a corruption of 'prickly mimosa' and is applied to a number of wattles with sharp phyllodes.

HABITAT Open forest, woodland and heath, on sandy soils.

DISTRIBUTION Widespread and common from coast to mountains. Found throughout NSW, except drier western plains; also Qld, Vic and Tas.

FLOWERING June–September, with some flowers in April and May.

355. *Acacia ulicifolia*

Acacia echinula **Plate 356**

This rigid bushy spiny shrub is easily confused with *A. brownei.* Phyllodes are terete, crowded along the branches, 5 to 12 mm long and do not widen at the base. The terete nature of the phyllodes help distinguish this species from *A. ulicifolia* and *A. brownei,* which have somewhat angular phyllodes. The phyllodes narrow abruptly at the apex to a sharp point. Flower heads are fairly large and deep orange-yellow. Flower stalks are thicker (0.3 to 0.5 mm) than those of *A. brownei.* The pod is 2 to 4 cm long and about 5 mm wide, shorter and wider than that of *A. ulicifolia.*

HABITAT Open forest and heath, on shallow soils of low fertility.

DISTRIBUTION Widespread from coast to mountains. Also occurring north coast and south coast.

FLOWERING June–August.

356. Acacia echinula

357. Acacia brownei

Acacia brownei **Plate 357**

A combination of 2 features helps distinguish this small diffuse prickly shrub from *A. echinula* and *A. ulicifolia.* Unlike *A. ulicifolia,* it has deep yellow-orange flower heads; unlike *A. echinula,* its phyllodes are compressed and somewhat angular (not terete) and widely spaced along the branches. Phyllodes are 10 to 25 mm long, about 1 mm wide, with a central vein and a sharp point. Flower stalks are more slender (0.1 to 0.2 mm) than those of *A. echinula.* The pod is 3 to 4 cm long, about 3 mm wide and curved.

HABITAT Open forest and heath, on low fertility soils, including sandstones, Wianamatta shales and alluvial sands.

DISTRIBUTION Widespread from coast to mountains, including shales at Menai, sandstones at Oatley Park and alluvium at Agnes Banks Nature Reserve. Occurring throughout NSW, except for drier western plains; also Qld and Vic.

FLOWERING July–September.

358. Acacia genistifolia

Acacia genistifolia

Spreading Wattle **Plate 358**

This is a hairless shrub to about 2 m tall, with angular branches and long, rigid, needle-like phyllodes. The common name and former botanical name, *A. diffusa,* refer to its spreading habit, but it is very variable and may be quite erect. 'Genistifolia' suggests a similarity of its phyllodes to the leaves of the introduced Broom (*Genista*). Each phyllode is 20 to 25 mm long, terete and tapers to a sharp point. Flower heads are pale to bright lemon-yellow. A characteristic feature is that flower heads are in groups of 2 or 3 borne on long slender stalks (5 to 20 mm) from the leaf axils.

HABITAT Open forest, on sandy soils.

DISTRIBUTION On edge of Sydney district in southern Blue Mountains and southern highlands. Extending through southern tablelands and south coast to Vic and Tas; also western slopes and south western plains.

FLOWERING June–October, but sometimes flowering earlier.

Acacia asparagoides **Plate 359**

This rigid, hairless shrub could at first be mistaken for *A. ulicifolia,* as its phyllodes are short, 10 to 20 mm long, rigid, with a sharp point, a central vein and are slightly broader at the base because of the presence of a gland. However, the flower heads of this species are on very short stalks, appearing stalkless, an important identifying feature. Flower heads are borne singly or in small clusters from the leaf bases.

HABITAT Open forest, on sandy soils.

DISTRIBUTION Restricted to upper Blue Mountains from Glen Davis to Lawson; uncommon. Recordings include Newnes, Blackheath and Hassans Walls (Lithgow). Also recorded in northern tablelands.

FLOWERING August–October.

359. Acacia asparagoides

Acacia paradoxa

Kangaroo Thorn, Hedge Wattle **Plate 360**

Although this wattle is widespread throughout Australia, it is uncommon in the Sydney district. It is a dense to open spreading shrub to 3 m tall. Branches are angular and often hairy. Phyllodes are 10 to 20 mm long, variable in shape, but often ovate to lanceolate, with one wavy margin and a pungent point. Beneath the leaves are rigid spine-like stipules to 12 m long. The former name of this species, *A. armata,* was a reference to its being armed with spines. Flower heads are deep golden-yellow, on solitary stalks, 10 to 13 mm long.

HABITAT Woodland, open forest and scrub.

DISTRIBUTION Uncommon in Sydney district, occuring in widely scattered locations, such as Colo Heights, Pierces Pass (Grose River) and Yerranderie to Lake Burragorang. Found throughout NSW, except for drier western plains; also Qld, Vic, SA, WA and introduced into Tas.

FLOWERING August–October.

360. Acacia paradoxa

Acacia gunnii

Ploughshare Wattle **Plate 361**

The common name of this small, rigid, prickly undershrub is a reference to the unusual shape of its phyllodes. These are small, to 10 mm long, finely pointed, with a broad base which gives them a triangular or ploughshare appearance. A prominent vein runs along the lower margin of the phyllode. Flower heads are whitish yellow, usually solitary and on slender hairy axillary stalks up to 10 mm long. The pod is 2 to 3 cm long, 4 to 5 mm wide, thin in texture and constricted between the seeds.

HABITAT Open forest, in sandy soils or on rocky hillsides.

DISTRIBUTION Uncommon in Sydney district. Found on higher parts of Blue Mountains and southern highlands. Also occurring south coast, northern tablelands, southern tablelands, western slopes, south-western plains, Qld, Vic, Tas and SA.

FLOWERING August–September.

361. *Acacia gunnii*

Acacia bynoeana — Plate 362

This low shrub has phyllodes which are flattened and have several parallel veins. They are shiny-resinous, slightly curved, sharp pointed and 1 to 4 cm long. Apart from *A. trinervata,* which is a tall hairless shrub, no other wattle around Sydney has this type of phyllode. Branchlets are covered with rough hairs. Flower heads are creamy-yellow and borne on very short stalks, usually 2 to 3 mm long. The pod is small, 2 to 3 cm long, 2 to 3 mm wide and contains only a few seeds.

HABITAT Heath and woodland.

DISTRIBUTION Scattered from coast to mountains; uncommon. Recordings include Agnes Banks Nature Reserve, Castlereagh State Forest, Hazelbrook-Bell and Wheelbarrow Ridge, Lower Portland. Endemic in the Sydney district.

FLOWERING December–March.

362. *Acacia bynoeana*

Acacia trinervata

Three-veined Wattle — Plate 363

The chief identifying feature of this diffuse, hairless shrub (1 to 2 m tall) is its phyllodes, which are flattened, narrow, rigid, spreading, 1 to 4 cm long and pungent pointed. Two or three parallel veins run the length of the phyllode (hence its specific name). Flower heads are rather large, bright yellow and borne on slender stalks about 2 cm long near the ends of the branchlets.

HABITAT Open forest, on sandy soils.

DISTRIBUTION Endemic in the lower Blue Mountains and western Hornsby Plateau. Common south of Glenbrook.

FLOWERING March–June.

363. *Acacia trinervata*

Acacia amblygona — Plate 364

This is a low spreading shrub, often prostrate and seldom more than 60 cm tall. Branches are covered with stiff hairs. Phyllodes are ovate, lance-shaped or almost triangular, asymmetrical, 6 to 12 mm long, stiff and sharp-pointed. Most phyllodes have 3 prominent veins, with 1 main vein which is off-centre although narrow leaves may only have a main vein. Flower heads are small and borne singly on stalks about 8 mm long. Buds are enclosed by a conspicuous bracteole. Pods are up to 6 cm long, 3 to 4 mm wide and usually much curved.

HABITAT Open forest.

DISTRIBUTION Fringes of the Sydney area, such as north and north-western Hornsby Plateau and Wollemi National Park. Extending to Hunter and Goulburn valleys, north coast, central western slopes, western plains and Qld.

FLOWERING July–October.

364. *Acacia amblygona*

Acacia quadrilateralis — Plate 365

This rare and endangered wattle is a small to medium shrub with needle-like phyllodes to 8 cm long which grow very erect. The phyllodes are terete or somewhat square in cross-section. Flower heads are borne on slender stalks less than 10 mm long so that the

365. *Acacia quadrilateralis*

flowers are held close to the stems. The pod is straight, up to 5 cm long and 4 to 5 mm wide.

HABITAT Sandy coastal heath.

DISTRIBUTION Rare in Sydney area. Recorded from Bombi Moor in Bouddi National Park and La Perouse. Recorded also on north coast, northern tablelands and Qld.

FLOWERING August–October.

Group 4

Wattles with globular flower heads and flowers borne singly, in pairs or clusters (not in racemes); phyllodes not pungent pointed.

366. *Acacia baueri* ssp. *aspera*

Acacia baueri — Plate 366

A small undershrub of heathlands, seldom reaching 30 cm tall and often decumbent and hidden amongst the vegetation. Its phyllodes are in whorls of 6 to 9, up to 15 mm long and only 1 mm wide. Flower heads are small, bright yellow and borne on slender stalks. The pod is linear, slightly curved, 2 to 4 cm long and only 2 mm wide. Two subspecies occur in the area, but both are uncommon. Subspecies *baueri* has stems and phyllodes which are hairless and smooth. Subspecies *aspera* has minute rough hairs on stems and phyllodes.

HABITAT Ssp. *baueri* occurs on damp coastal heath. Ssp. *aspera* occurs on drier heaths further inland.

DISTRIBUTION Ssp. *baueri*. Coastal zone. Extending along the north coast to Qld.
Ssp. *aspera*. Blue Mountains heaths (e.g., Kings Tableland and Wentworth Falls); also near Wilton, east of Picton.

FLOWERING December–January.

367. *Acacia gordonii*

Acacia gordonii — Plate 367

This rare wattle is a shrub to about 1.5 m tall with stems and phyllodes covered with rough hairs. Its phyllodes are crowded, scattered or irregularly whorled, narrow linear, 10 to 15 mm long and curved at the tip. Flower heads are golden yellow, 10 to 12 mm across, on a sturdy hairy flower stalk, up to 15 mm long; they are borne singly from the axils of the upper phyllodes. The seed pod is flattened, straight, about 4 cm long and 10 mm wide.

HABITAT Open forest, on sandy soils.

DISTRIBUTION Restricted to central Blue Mountains (e.g., North Linden, Kings Tableland).

FLOWERING August–September.

Acacia juncifolia var. *juncifolia*

Rush-leaf Wattle — Plate 368

The most outstanding feature of this tall shrub is its long erect phyllodes. They may be up to 15 cm long. Although they appear terete from a distance, closer examination shows them to be slightly flattened; one faint vein runs along each side of the phyllode, ending in a short point. Flower heads are bright yellow, 5 to 8 mm across, and borne singly or in pairs on short fine stalks, 4 to 10 mm long.

HABITAT Open forest, on sandy soils or amongst rocks.

DISTRIBUTION Uncommon in Sydney district. Recorded above Nepean River at Mulgoa, Glenbrook and on Hornsby Plateau north of Hawkesbury River. More common on drier western slopes and north-western plains; also Qld.

FLOWERING August–September.

368. *Acacia juncifolia*

Acacia ptychoclada — Plate 369

This wattle is common in the upper Blue Mountains in wet swampy conditions. Its phyllodes are very narrow, 1 to 1.5 mm wide, 5 to 6 cm long and have a small pointed tip which is not pungent. Phyllodes are upright or slightly spreading. Flower heads are pale yellow and on hairy stalks much shorter than the phyllodes. The inflorescence is rather sparse. The pod is straight sided, 4 to 5 cm long and about 5 mm wide.

HABITAT Swamps and drip ledges.

DISTRIBUTION Restricted to the Blue Mountains between Woodford and Mt Victoria. Recordings include Lawson, Leura and Blackheath.

FLOWERING September–March.

369. *Acacia ptychoclada*

370. *Acacia elongata*

Acacia elongata

Swamp Wattle — Plate 370

This erect shrub (to 2 m tall) is very striking in late winter and spring when its showy large yellow flowers are massed together along the stems. Each flower head is borne on a slender stalk, up to 12 mm long, which is characteristically covered in silky-white hairs. Flower heads are usually in clusters at the leaf axils. Phyllodes are flat with several veins, 7 to 13 cm long, 2 to 4 mm wide, and end in a short point.

HABITAT Open forest, woodland and heath, on sandy soils; most commonly in wet areas and along creeks.

DISTRIBUTION Widespread from coast to mountains. Extending from north coast and northern tablelands to south coast and southern tablelands; also western slopes and Vic.

FLOWERING July–September.

Acacia verniciflua

Varnish Wattle — Plate 371

Varnish Wattle is uncommon in the Sydney district. It is found mainly in isolated situations on the fringes of the area. It is an erect, often bushy shrub with phyllodes which are very variable. They are usually narrow lanceolate, sometimes slightly curved, and have 2 main veins. They vary in length from 4 to 10 cm and in width from 5 to 12 mm. The most conspicuous feature of this species is the glossy varnish-like sheen on the phyllodes, which makes them quite sticky, especially when young. Flower heads are deep golden-yellow, fairly large and borne on stalks 5 to 8 mm long in groups of two or more. The pod is straight, narrow, 6 to 8 cm long and usually sticky.

HABITAT Dry forest and woodland, on sandy soils.

DISTRIBUTION Scattered through western and southern Blue Mountains (e.g., near Yerranderie, Newnes Glow-worm Tunnel and Glen Davis). More common on western slopes, southern tablelands and south-western plains; also Vic, Tas and SA.

FLOWERING September–November.

371. *Acacia verniciflua*

372. Acacia ixiophylla

Acacia ixiophylla **Plate 372**

This is a medium-sized shrub, somewhat resembling *A. verniciflua*, as it has a glossy varnish-like (viscid) appearance. Branchlets are usually densely hairy. Phyllodes are lance shaped, 1 to 3 cm long, viscid, with several parallel veins (unlike *A. verniciflua* which has 2 main veins). The tip is acute to obtuse, with a small mucro. Flower heads are deep yellow, borne on stalks up to 5 mm long; they may be borne singly, in pairs or even in short racemes of 2 to 3 flowers. The pod is curled and twisted, 2 to 5 cm long and sticky.

HABITAT Open forest and woodland, on sandy soils.

DISTRIBUTION Glen Davis area, where it is common. Extending to western slopes and north-western plains (e.g., Pilliga scrub); also Qld and WA.

FLOWERING August–October.

Acacia hispidula **Plate 373**

The specific name 'hispidula' is a reference to the short stiff hairs which cover the branchlets and phyllodes of this small shrub, giving it a sandpapery feel. Phyllodes are small, 1 to 2 cm long, broad, very rough, with small teeth and a pointed apex. Flower-heads are a very light yellow, borne on single stalks, widely spaced along the stems. The pods are ovate, brown, small and contain only 1 or 2 seeds.

HABITAT Open forest and woodland, often on dry rocky slopes.

DISTRIBUTION Widespread along coastal zone. Recordings include Loftus and Woronora Dam. Also occurring south coast, north coast, northern tablelands, north-western slopes and Qld.

FLOWERING Most of the year.

373. Acacia hispidula

374. Acacia uncinata

Acacia uncinata

Wavy-leaf Wattle, Round-leaf Wattle **Plate 374**

This is an attractive tall shrub to 3 m tall. The common names indicate its main identifying feature. Phyllodes are variable, round or oval, hairy or hairless, often (but not always) with wavy margins, up to 25 mm long but sometimes only 7 to 8 mm long as on forms at Glen Davis. Margins appear thickened, the main vein is off centre and there is a distinct point. Abundant flower heads are borne singly or in pairs, and as they are on long stalks, they exceed the phyllodes and are very showy. The pod is flat, oblong, broad, 4 to 6 cm long and about 2 cm wide.

HABITAT Open forest, on sandy soils.

DISTRIBUTION Widespread away from the coastal zone. Recordings include western Blue Mountains (Megalong Valley, Glen Davis), Nattai River and St Albans-Bucketty Road. Also occurring south coast, southern tablelands, north coast, western slopes and Qld.

FLOWERING November–April.

375. Acacia stricta

Acacia stricta

Hop Wattle **Plate 375**

This shrub, to 2 m tall, is readily identified by its erect branches and its upright phyllodes. The specific name 'stricta' is derived from the Latin word for straight or stiff. Phyllodes are 5 to 12 cm long and up to 15 mm wide, rounded at the apex; sometimes there is a small point on the rounded apex. Flower heads are borne on short stalks from the leaf axils. They are usually in opposite pairs up the stem, but may be single or in clusters. The pod is straight and 3 to 5 cm long.

HABITAT Open forest, often on dry hillsides.

DISTRIBUTION Widespread from coast to mountains. Extending all along coast and tablelands from Qld to Vic; also south-western slopes, Tas and SA.

FLOWERING August–September.

376. Acacia myrtifolia

Group 5

Wattles with globular flower heads borne in racemes; phyllodes under 7 cm long.

Acacia myrtifolia

Myrtle Wattle **Plate 376**

This small erect shrub is widely distributed throughout Australia and is very common around Sydney. Several features make it easy to identify. The angular branchlets often have a distinctive reddish colour, and the young growth and immature pods are often red. The pale cream flower heads are quite large but only have 3 to 6 flowers in each head (this is obvious in bud). Phyllodes are thick, somewhat lanceolate, 2 to 5 cm long, with a blunt point. The pod is up to 7 cm long, 3 to 4 mm wide, thick and woody, curved with wavy margins.

HABITAT Open forest and heath, on sandy soils.

DISTRIBUTION Widespread from coast to mountains. Also occurring north coast, northern tablelands, south coast and in all States, except NT.

FLOWERING May–August, but later in higher areas.

377. *Acacia prominens*

Acacia prominens

Gosford Wattle **Plate 377**

This is a tall shrub or tree to about 15 m tall. Phyllodes are 2 to 6 cm long, narrow lanceolate, distinctly blue-grey in colour, and have a prominent gland on the margin just below the centre (hence the specific name). Flower heads are small but abundant, bright lemon-yellow and borne on racemes slightly longer than the phyllodes. Because of its attractive habit, foliage and abundant flowers, Gosford Wattle is often grown in parks and gardens.

HABITAT Sheltered forests and along creeks in gullies.

DISTRIBUTION Common north of Hawkesbury River (e.g., Mooney Mooney Creek and St Albans-Bucketty Road); also recorded near Mulgoa (Nepean River). Endemic in central coast and north coast.

FLOWERING July–October.

378. *Acacia fimbriata*

Acacia fimbriata

Fringed Wattle **Plate 378**

Fringed Wattle is a bushy shrub or small tree, often with drooping branches. Phyllodes are narrow lanceolate, 2 to 5 cm long and 4 to 6 mm wide, tapering to a fine non-pungent point. Both the botanical and common names refer to the margins of the phyllodes which are fringed with fine hairs. These are difficult to see with the naked eye and need to be examined with a good lens. Flower heads are small but a brilliant yellow and borne on racemes which exceed the phyllodes. When in flower this is a most attractive plant and it is often grown in native gardens.

HABITAT Forest, often on protected slopes.

DISTRIBUTION Widespread from coast to mountains, but uncommon. Recordings include Erskine Creek (lower Blue Mountains), Cumberland Plain, Wondabyne (Brisbane Water National Park), near Goodmans Ford (Wollondilly River) and St Albans-Bucketty road. Also occurring north coast, northern tablelands, southern tablelands, north-western slopes and Qld.

FLOWERING August–September.

Acacia decora

Western Golden Wattle **Plate 379**

An attractive, densely flowered shrub, 1 to 2 m tall, with angular branches. Phyllodes are narrow oblong to lanceolate, 2 to 4 cm long, rounded or pointed at the apex, and tapering at the base to a small hairy stalk. A prominent gland is usually present near or above the base and a second gland is sometimes present higher up. Flower heads are bright yellow on stout golden-hairy stalks, 3 to 6 mm long; they are massed together near the ends of the branches in showy racemes usually longer than the phyllodes. The pod is flat, curved or straight, up to 10 cm long and 4 to 9 mm wide.

HABITAT Open forest and woodland, in sandy or rocky situations.

DISTRIBUTION North-western fringe of Sydney district (e.g. Colo Plateau, Wollemi National Park). Extending to Hunter and Goulburn valleys, northern tablelands, north coast, western slopes and western plains; also Qld and Vic (Warby Range).

FLOWERING June–September.

Acacia buxifolia

Box-leaf Wattle **Plate 380**

An erect shrub to 2 m tall, its phyllodes are broadly lanceolate, 15 to 25 mm long and up to 12 mm wide, pale green or greyish, with a small point at the tip. The midvein is slightly to one side. Flowers are golden-yellow, only 8 to 12 in each head, but flower heads are abundant and showy. Flower heads are borne in racemes 3 to 5 cm long.

HABITAT Rocky hills and dry open woodland.

DISTRIBUTION Widespread away from coastal zone. Common near Little Hartley and near Newnes. Occurring throughout NSW, except on south coast and far western plains; also Qld and Vic.

FLOWERING September–October.

379. *Acacia decora*

380. *Acacia buxifolia*

Acacia linifolia

Flax Wattle **Plate 381**

A small shrub with slender branches, common in sandy soils around Sydney; it frequently occurs in remnant bushland areas and by the roadside. It is easily identified by its short and narrow phyllodes, 2 to 4 cm long and 1 to 3 mm wide, which are often closely packed along the slender branches. Flowers are a light cream, with 6 to 12 flowers in each head. They are borne in racemes which are densely packed and usually near the ends of the branches.

HABITAT Open forest, on sandy soils.

DISTRIBUTION Common in most bushland areas of the coastal zone; also recorded from lower Blue Mountains, but uncommon.

FLOWERING Chiefly January–April, but some flowers at other times.

381. *Acacia linifolia*

382. *Acacia hamiltoniana*

383. *Acacia amoena*

Acacia hamiltoniana

Hamilton's Wattle **Plate 382**

This species may at first be confused with *A. linifolia* as it is also a shrub with small narrow phyllodes. However the phyllodes of *A. hamiltoniana* are slightly longer (3 to 6 cm) and wider (3 to 4 mm), have thicker margins and are covered with minute dots. Another difference is that the smaller branches of this species are reddish brown and often ribbed. There are about 20 flowers in each head. Flower heads are deep golden-yellow and are in racemes which do not exceed the phyllodes.

HABITAT Open forest, on sandy soils.

DISTRIBUTION Blue Mountains (e.g. Bell to Mt Wilson), but uncommon. Extending south to southern tablelands (e.g. Sassafras-Nerriga).

FLOWERING August–September.

Acacia amoena

Boomerang Wattle **Plate 383**

This is an attractive small shrub to 2 m tall, with a specific name which means 'pleasant' or 'delightful'. Phyllodes are a distinctive shape; broader at the apex than at the base. Each is 4 to 7 cm long and up to 10 mm wide usually with a minute point at the apex and 1 to 3 prominent glands on the thickened upper margin. The flower heads are light yellow and in racemes which do not extend much beyond the leaves.

HABITAT Open forest, often on rocky hillsides.

DISTRIBUTION Blue Mountains, especially in the south-west of the area. Common along Wombeyan Caves road beyond Wollondilly River. A tablelands species extending from northern tablelands to southern tablelands and Vic; also central western slopes.

FLOWERING August–September.

Acacia clunies-rossiae

Kowmung Wattle **Plate 384**

This wattle is seldom seen because of its very restricted distribution in rugged country in the southern Blue Mountains. Because of its narrow range, it is regarded as 'at risk' or 'vulnerable'. It is a bushy shrub or small tree with narrow grey-green phyllodes which are densely hairy. Phyllodes are 4 to 6 cm long, up to 8 mm wide, and have a short point at the apex. Smaller branches are very hairy and young foliage is greyish-white. Flower heads are golden yellow and borne in racemes which usually do not exceed the phyllodes.

HABITAT Forest and rocky forested hillsides.

DISTRIBUTION Restricted to valleys of Kowmung River and adjacent Coxs River, and Mt Cookem-Yerranderie area of the Blue Mountains.

FLOWERING August–September.

Acacia chalkeri

Chalker's Wattle **Plate 385**

Chalker's Wattle is restricted to the limestone outcrops of the Wombeyan Caves, where it is locally common. It is a small bushy shrub with angular branches. Phyllodes are a similar shape to those of *A. amoena,* oblanceolate, mostly about 4 cm long and 4 to 8 mm wide, with a short point at the apex. Unlike *A. amoena,* the phyllodes of this species have only one gland which is rather inconspicuous. Flowers may be sparse or abundant. The pod is flat and straight, 6 to 9 cm long and about 7 mm wide.

HABITAT Rocky woodland, on limestone.

DISTRIBUTION Restricted to Wombeyan Caves.

FLOWERING November–December; a few flowers at other times.

384. *Acacia clunies-rossiae*

385. *Acacia chalkeri*

Acacia obtusata

Blunt Wattle **Plate 386**

This is an uncommon erect shrub with angled or flattened branchlets. Phyllodes are variable, but are mostly oblanceolate, 5 to 7 cm long (sometimes shorter or longer), 2 to 12 mm wide and much narrowed towards the base. The specific name refers to the rounded ends of the phyllodes. A prominent vein runs closer to the upper than to the lower margin and a prominent gland is present below the centre of the upper margin. Flower heads are bright yellow and borne in short racemes about half as long as the phyllodes.

HABITAT Woodland and open forest.

DISTRIBUTION South-western fringes of the Sydney district (e.g., Wombeyan Caves). Extending from central tablelands to southern tablelands and Vic.

FLOWERING August–October.

386. *Acacia obtusata*

Group 6

Wattles with globular flower heads borne in racemes; phyllodes more than 7 cm long.

Acacia rubida

Red-stemmed Wattle **Plate 387**

This is a bushy shrub, usually 1 to 3 m tall, sometimes larger, with angular branchlets which are often a conspicuous reddish colour. The specific name 'rubida' (reddish) is a reference to this characteristic. Foliage is unusual as the plant retains its bipinnate leaves on the lower branches while also bearing broad lance-shaped or sickle-shaped phyllodes. These phyllodes are up to 15 cm long, have a prominent gland on the margin below the centre and are often reddish, especially when young.

387. *Acacia rubida*

Flower heads are deep yellow and dense along a raceme which is shorter than the phyllodes.

HABITAT Forest and woodland.

DISTRIBUTION Scattered throughout Blue Mountains, southern highlands (e.g., Wingello), Burragorang Tableland (e.g., Sheehy's Creek road) and Woronora Plateau (e.g., Lucas Heights). Also occurring south coast, southern tablelands, western slopes, Vic and Qld (near Stanthorpe).

FLOWERING July–September.

Acacia falcata

Sickle Wattle **Plate 388**

Sickle Wattle is a tall open shrub, usually occurring on shale soils. Its phyllodes are falcate, 7 to 12 cm long and up to 20 mm wide, blue-grey to light green. The main vein is off-centre and the marginal gland is obscure or absent. Flower heads are small, pale yellow and borne on short racemes. The pod is 5 to 10 cm long, about 6 mm wide, blue-grey and slightly constricted between the seeds.

HABITAT Forest or cleared land, usually on shales.

DISTRIBUTION Widespread along coastal zone and lower Blue Mountains. Recordings include Milperra, Menai, Agnes Banks and Nepean Lookout (Glenbrook). Also occurring north coast, south coast, northern tablelands, central western slopes and Qld.

FLOWERING April–July.

388. *Acacia falcata*

389. *Acacia obliquinervia*

Acacia obliquinervia **Plate 389**

This tall shrub or small tree has angular branchlets which may be purplish-red or blue-grey. Phyllodes have the main vein off-centre like those of *A. falcata,* but they differ in other ways. They are extremely variable in shape, sometimes sickle-shaped, sometimes broadly oblanceolate-elliptic, from 5 to 17 cm long and 10 to 40 mm wide. The tip is rounded and a network of fine veins is visible running from the midrib. A prominent gland occurs near the base. This gland may have a distinct nerve connecting it to the midvein as in *A. falciformis* and *A. penninervis* but usually it does not. Flower heads are golden yellow and the stems of the racemes are often distinctly zig-zagged.

This species is distinguished from *A. falciformis* in having hairless flower stalks, whereas those of *A. falciformis* are covered with velvety golden hairs. It is distinguished from *A. penninervis* by its gland which is less than 10 mm from the base of the phyllode, and usually not connected to the midvein, whereas that of *A. penninervis* is usually more than 10 mm from the base and is connected to the midvein.

HABITAT Forest.

DISTRIBUTION Western and south-western Blue Mountains (e.g., Boyd River to Kanangra Walls). Also occurring southern tablelands, central western slopes and Vic.

FLOWERING September–October.

Acacia falciformis

Broad-leaf Hickory **Plate 390**

This is a small tree to 12 m tall with young branches covered in golden or silvery soft hairs. Phyllodes are falcate, usually broadest about the centre and tapering at both ends (unlike those of *A. obliquinervia* and *A. penninervis* which are broader above the centre). They vary in size from 8 to 17 cm long and 17 to 35 mm wide. The apex is usually rounded. A prominent marginal gland is indented above the base of the phyllode and connected to the midvein by a distinct nerve. A clear network of fine veins is visible running from the midrib. Flower heads are pale yellow and borne on stalks which are covered in velvety golden hairs (*A. obliquinervia* and *A. penninervis* have hairless flower stalks). Flower heads are in racemes shorter than the phyllodes. The bark on older trees is dark and corrugated; it is rich in tannic acid.

HABITAT Forest.

DISTRIBUTION Blue Mountains, including Mt Wilson-Mt Irvine, Megalong Valley and Yerranderie. Found throughout NSW, except western plains; also Qld and Vic.

FLOWERING October–January.

390. Acacia falciformis

Acacia penninervis

Mountain Hickory, Hickory Wattle

Mountain Hickory is a tall shrub or tree to 12 m tall with branches which are often angular and reddish. Phyllodes are oblong to broadly lanceolate, usually widest above the centre, 5 to 15 cm long and 8 to 25 mm wide (occasionally wider). The apex may be blunt or have a short tip. A prominent marginal gland is indented some distance from the base (usually more than 10 mm) and is connected to the midvein by a distinct nerve. There is a strong central vein and a clear network of smaller veins running from the midvein (penniveins). New growth is often a reddish colour. Flower heads are pale yellow, on smooth stalks; they are borne in loose terminal racemes. This species could be mistaken for *A. obliquinervia* and *A. falciformis* (see *A. obliquinervia* for differences).

HABITAT Forest

DISTRIBUTION Chiefly in western Blue Mountains on fringes of Sydney district, although recorded for central coast. Occurring throughout coast and tablelands divisions, extending to western slopes and north-western plains; also Qld and Vic (near Seymour).

FLOWERING October–February.

Acacia suaveolens

Sweet-scented Wattle **Plate 391**

This is a familiar wattle to many people as it is widespread and common around Sydney and it bears its pale cream flowers during autumn and winter when few other species are in bloom. It is an erect shrub with a few angled and flattened branches. New growth is distinctly pinkish. Phyllodes are narrow, slightly curved, rather erect, 6 to 12 cm long and 4 to 10 mm wide. There is a distinct gland near the base of the phyllodes. The main identifying feature is that the unopened racemes are enclosed in large brownish deciduous bracts. Each raceme is shorter than the phyllodes and the flowers are sweetly scented.

HABITAT Open forest, woodland and heath, on sandy soils.

DISTRIBUTION Widespread from coast to mountains. Also occurring south coast, southern tablelands, north coast, Qld, Vic, Tas and SA.

FLOWERING March–August on coast, later at higher altitudes.

Acacia saliciformis **Plate 392**

This is an attractive shrub or small tree with a smooth blue-grey trunk and pendulous branches. The specific name is a reference to its likeness to the willow (*Salix*) because of its drooping foliage. Phyllodes are elliptic-oblong, 6 to 12 cm long, up to 15 mm wide and often slightly curved. Young shoots are often reddish. A prominent gland is located well above the base (15 to 40 mm) and is connected to the midvein by a distinct nerve. The apex of the phyllode is pointed. Flower heads are pale yellow and borne on racemes shorter than the phyllodes.

HABITAT Sheltered and open forest, on sandy soils.

DISTRIBUTION North-western Hornsby Plateau. Common along Putty Road and around Colo Heights. Also occurring southern tablelands.

FLOWERING March–August.

393. Acacia hakeoides

Acacia hakeoides

Hakea-leaf Wattle **Plate 393**

Hakea-leaf Wattle is a spreading shrub to 2 m tall. Phyllodes are oblong to oblanceolate, 5 to 12 cm long, 5 to 10 mm wide, slightly curved, rather thick, with a blunt rounded apex and gradually narrowing at the base to a short stalk 1 to 2 mm long. A conspicuous gland is located on the upper margin; it may be from near the base to near the middle of the phyllode. Flower heads are bright yellow and are borne in short racemes (2 to 5 cm long) of 6 to 12 flower heads. The pods are up to 10 cm long, curved and much constricted between the seeds so they appear like strings of beads.

HABITAT Open forest, on sandy soils.

DISTRIBUTION Glen Davis and north-western part of Sydney district. Extending to western slopes, southern tablelands and western plains; also Qld, Vic, SA and WA.

FLOWERING July–September.

391. Acacia suaveolens

392. Acacia saliciformis

394. Acacia binervata

Acacia binervata

Two-veined Hickory **Plate 394**

This is a common wattle, especially on the edges of rainforest where it grows to a medium-sized tree. The dense foliage and two-veined phyllodes make it easy to recognise. Phyllodes are broadly lanceolate, 7 to 13 cm long and 10 to 30 mm wide, dark green, clearly 2 veined and have a conspicuous marginal gland just below the middle. Flower heads are pale yellow and borne in short racemes. The pod is thin, flat, straight, oblong, up to 14 cm long and 12 mm wide.

HABITAT Sheltered gullies and rainforest margins.

DISTRIBUTION Widespread from coast to mountains. Common in the Illawarra district. Also occurring north coast, northern tablelands and south coast.

FLOWERING September–October.

396. Acacia implexa

395. Acacia melanoxylon

Acacia melanoxylon

Blackwood **Plate 395**

Blackwood grows to a tall tree, usually 8 to 15 m tall, but up to 30 m in favourable situations. Its timber is highly prized for ornamental woodwork. Blackwood may be confused with *A. implexa* as they are both trees with broad many-veined phyllodes and twisted seed pods. However, they differ in a number of ways. The phyllodes of *A. melanoxylon* are broadly lanceolate, 7 to 12 cm long, 9 to 20 mm wide, with longitudinal veins and a small blunt point. The bark is dark grey and furrowed; branchlets are slightly angular. Juvenile growth is hairy and young phyllodes are often large. Racemes consists of 3 to 5 flower heads and are shorter than the phyllodes. Flowers appear in spring, whereas *A. implexa* usually flowers in late summer and early autumn. If the pods are ripe and the seeds exposed, another clear difference can be seen. The funicle is red and folded around the seed; the funicle of *A. implexa* is paler and folded under the seed.

HABITAT Sheltered forest, on moist and fairly rich soils.

DISTRIBUTION Widespread from coast to mountains, but more common at higher altitudes of Blue Mountains and southern highlands. Recordings include Blackheath, Bundanoon and Mt Keira. Extending all along coast and tablelands from Qld to Vic; also western slopes, Tas and SA.

FLOWERING August–November.

Acacia implexa

Hickory **Plate 396**

Hickory is a small tree found scattered throughout the coast and tablelands regions. Its phyllodes have several parallel veins and the pods become very twisted ('implexa' means 'tangled', probably a reference to the pods). A combination of features separates it from *A. melanoxylon* which it superficially resembles. The phyllodes are thinner, more pointed and more curved than those of *A. melanoxylon*. They are also usually longer (to 18 cm) and narrower (6 to 16 mm), although this is not consistent. Branchlets are round rather than angular. New growth is light green. Flower heads are borne on rather long and slender stalks in clusters or short racemes of 4 to 8 flower heads. The funicle is pale and folded under the seed, not encircling it as in *A. melanoxylon*.

HABITAT Open forest and woodland, on rocky hillsides and alluvial flats.

DISTRIBUTION Widespread from coast to mountains. Occurring throughout NSW, except far western plains; also Qld and Vic.

FLOWERING January–March; sometimes at other times of the year.

397. Pararchidendron pruinosum

Pararchidendron

Pararchidendron pruinosum

Snow-wood, Stinkwood **Plate 397**

An uncommon rainforest tree with large pom pon flowers which change from white to greenish yellow with age. The flowers differ from those of *Acacia* as the stamens are united into a tube, not free. The foliage is distinctive, with large bipinnate leaves about 40 cm long, usually with 3 pairs of pinnae bearing leaflets 2 to 8 cm long. The pod becomes twisted into a circle when ripe and opens to reveal an orange to deep red inner lining. The presence of numerous red seed pods on the ground is often the first indication that this tree is nearby. The common name 'Stinkwood' is derived from the strong but not unpleasant smell given off when the tree is cut.

HABITAT Rainforest.

DISTRIBUTION Scattered throughout coastal rainforest patches (e.g. Jamberoo), but uncommon. Reaches its southern limit in the Illawarra region. Extending northwards along the coast to north Qld. Also central western slopes, New Guinea and Asia.

FLOWERING October–December.

Caesalpiniaceae

The Caesalpiniaceae may be regarded as being intermediate in floral structure between the regularity of Mimosaceae and the marked irregularity of Fabaceae. Within Caesalpiniaceae itself there is a range extending from *Cassia*, in which the petals are nearly equal, to *Cercis* (the Judas tree), in which the structure approaches that of the pea flowers. In both Caesalpiniaceae and Fabaceae the petals are imbricate (overlapping), but in the former the upper petal is innermost and the basal petal is outermost and slightly enlarged, whereas in Fabaceae the upper petal is largest and outermost and the lower 2 petals are partly fused and enclosed by 2 lateral petals. The number of stamens in Caesalpiniaceae varies from 2 to 10 but is constantly 10 in Fabaceae.

Caesalpiniaceae is a fairly large group of about 150 genera and nearly 3000 species and includes many well-known ornamental trees and shrubs from tropical countries as well as a number which provide useful products such as tanning agents, resins, medicinals, dyes, stains and timber. The family is represented in Australia by 11 genera and about 70 native species plus a number of introduced species, mainly of *Cassia*, which have become naturalised. Ten small, relatively unknown genera occur almost exclusively in northern Australia but the 40 or so

399. Cassia aciphylla

native species of *Cassia* are widespread except in Tasmania. Four native species and 5 introduced species of *Cassia* occur in the Sydney area.

Cassia

Shrubs with pinnate leaves, yellow flowers with overlapping petals, 10 stamens, a curved stigma and a pea-like pod. The 4 species occurring in the Sydney area are rare or uncommon. Introduced Cassias from South America and Africa are often found on the margins of rainforest or in gullies where they appear to be natives. All these introduced species have a mixture of perfect stamens and sterile stamens (staminodes) and all have flat leaflets, not curved down or rolled under to the undersurface. The only native species with staminodes is *C. barclayana*, a rare plant in the area; it has 4 to 6 pairs of flat leaflets, up to 1 cm wide, 6 stamens (4 short and 2 long) and 4 staminodes.

Cassia odorata

Southern Cassia **Plate 398**

Southern Cassia is a shrub to 2 m tall, with leaves 5 to 10 cm long, divided into 6 to 11 pairs of leaflets. These leaflets are very variable; they may be oblong to linear, 10 to 25 mm long and 2 to 6 mm wide. They are paler on the undersurface and may be flat or with margins which curl downwards, but not rolled under to touch the undersurface. Tips are rounded, notched or have a small non-pungent point. A hair-like gland is present on the leaf stem between pairs of leaflets. The yellow flowers have 10 stamens, all perfect, without staminodes.

HABITAT Sheltered forested slopes.

DISTRIBUTION Widespread from coast to mountains (e.g. Douglas Park); uncommon. Also occurring south coast, north coast, north-western slopes and Qld.

FLOWERING September–December.

398. Cassia odorata

Cassia aciphylla

Australian Senna **Plate 399**

Australian Senna is a shrub to 1.5 m tall. It is similar to *C. odorata* but differs in having leaflets which have acute and almost pungent pointed tips and margins which are rolled inwards to touch or almost touch the undersurface. There are 5 to 12 pairs of leaflets, each 15 to 30 mm long and 2 to 3 mm wide. A hair-like gland is present on the leaf stem between pairs of leaflets. The yellow flowers have 10 stamens, all perfect, without staminodes.

HABITAT Forest and woodland.

DISTRIBUTION Widespread, but uncommon near the coast. Chiefly western Blue Mountains (e.g. Glen Davis) and south-west of area (e.g. limestone outcrops at Wombeyan Caves). Also occurring north coast, northern tablelands, southern tablelands, western slopes, Qld and Vic.

FLOWERING October–March.

400. Cassia eremophila var. *zygophylla*

Cassia eremophila var. *zygophylla*

Desert Cassia **Plate 400**

This is a spreading bushy to erect shrub to 2 m tall, with leaves 4 to 8 cm long, divided into 1 to 2 pairs of narrow leaflets 3 to 4 cm long and 3 to 5 mm wide. There is a small gland between the lowest pair of leaflets. The leaf stalk is cylindrical and 10 to 25 mm long. Flowers are yellow and borne in short racemes in the leaf axils

HABITAT Woodland and open sandy areas.

DISTRIBUTION Chiefly a species of the drier western slopes and plains, reaching the Glen Davis area; also occurring northern tablelands, Qld, SA and NT.

FLOWERING August–October.

Fabaceae

Over the years the peas have been known by a number of names. Bentham, for example, in *Flora Australiensis* included them with other pod-bearing plants in the family Leguminosae. In many later classifications this very large group was divided into 3 separate families, Mimosaceae (wattles, etc.), Caesalpiniaceae (cassias) and Papilionaceae (peas). Recently, the name of the latter group was changed to Fabaceae, a name derived from the broad bean, *Vicia faba,* a human food source since Neolithic times. The pod-bearing plants are regarded here as comprising the order Fabales which is divided into 3 families, Mimosaceae, Caesalpiniaceae and Fabaceae.

Although the family is third largest in the plant kingdom after the orchid and daisy families, it contains many more plants useful to man. Many leguminous plants such as peas, beans, lentils, soybeans and peanuts are rich in protein and minerals, and have been cultivated from the earliest times. Some species provide timber, others produce gums such as balsam, and dyes such as indigo, or resins, oils and fibres. The roots of derris yield an insecticide, and those of *Glycyrrhiza* yield liquorice. A number have been developed into highly prized ornamentals. Most leguminous plants develop root nodules in association with *Rhizobium* bacteria and are capable of converting atmospheric nitrogen into more complex nitrogenous compounds. Hence plants such as clover and lucerne can be used for feeding livestock or ploughed back into the soil to provide nitrogenous fertiliser.

Figure 8

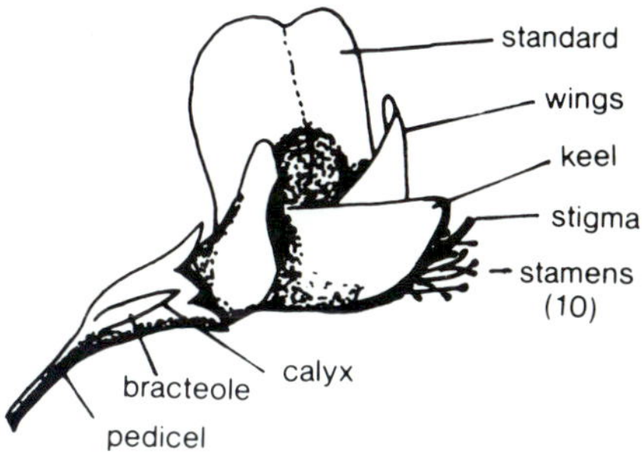

The Fabaceae is closely related to Rosaceae, the rose family. Pea flowers have 5 petals and 10 stamens and, although often highly modified, have a basic structure consisting of an erect petal, called the standard, surmounting 2 lateral petals known as wings, which partly cover 2 lower petals called the keel. The keel encloses the gynoecium (ovary, style and stigma) and the stamens.

The family has been divided into from 11 to 20 or more tribes, mainly on the basis of leaf and pod type and the extent of fusion of the stamens. Although most tribes contain native Australian genera the majority of Australian genera and species are in the most primitive tribe, which is characterised by having 10 free stamens and simple leaves. In this tribe pollination is achieved after the weight of an insect presses down the wings and keel exposing the anthers and stigma. In more advanced tribes 9, or all 10, stamens are fused for part of their length and the pollination process is often more complex. Leaves of the Fabaceae are either simple, as in *Dillwynia,* digitate as in *Gompholobium,* or pinnate as in *Indigofera,* but never bipinnate as in the Caesalpiniaceae and Mimosaceae.

The Fabaceae is about the third or fourth largest Australian plant family. All told the Australian representatives number about 90 genera and 800 to 900 species. This contrasts markedly with New Zealand where the family is sparsely represented. A large number of Australian species are poisonous, the poisonous species coming from various genera and containing a variety of poisons.

There are 40 genera in the Sydney region divided into about 170 species of which many are introduced weeds. A high proportion of the native species are from the Australian endemic genera *Daviesia, Pultenaea, Mirbelia, Dillwynia, Oxylobium, Phyllota, Jacksonia* and *Kennedia.* Mostly the peas are found on sandstone soils in heath and eucalypt forest, although a few such as *Pultenaea villosa* and *Indigofera australis* nearly always grow on heavier clay soils. Because of their ability to fix nitrogen, the pea plants are pioneers. They help to bind dunes on the coast and they germinate quickly and grow thickly after fire. Their role is to increase soil fertility and provide protection for plants in the next stage of succession and, in keeping with this role, they are relatively short-lived. In the absence of fire they tend to disappear from the understorey in about 15 years.

The following simple key can be used to find the genus of most of the pea plants in the area. Six characteristics, each with 2 alternatives, are listed in the key. Start with the first characteristic and select the alternative that best describes the plant under consideration. Work through the key until the correct genus is found. Species can then be identified from the illustrations and descriptions. All the common native species are illustrated.

1. Flower colour

a) Flowers not orange or yellow.

i) climbing or prostrate plants	*Kennedia*	flowers large, red to burgundy.
	Hardenbergia	flowers violet.
	Glycine	flowers pale blue to mauve.
ii) shrubs	*Mirbelia*	leaves simple with strong reticulate venation.
	Hovea	leaves linear to oblong, without strongly marked veins.
	Indigofera	leaves pinnate with numerous leaflets.
	Gompholobium	leaves digitate with 3 or 5 leaflets.

b) Flowers orange to yellow, often tinged red or red centred .... go to 2.

2. Stamens

a) All 10 stamens joined for most of their length.

Bossiaea	flowers small, or leaves small or undeveloped, or stems flattened.
Platylobium	wiry shrub with large flowers and large (>2 cm), prominently veined leaves, stems round.

b) All 10 stamens free .... go to 3.

3. Ovules

(Select an old faded flower, remove petals and stamens, hold the immature pod to the light. The ovules—developing seeds—can be seen through the pod.)

a) Ovules 4 or more.

Oxylobium	leaves simple, mostly opposite. Flowers orange-yellow.
Gompholobium	leaves digitate with 3 or 5 leaflets, or pinnate. Flowers pale to mid yellow.
Mirbelia	leaves simple; pod divided by a partition into 2 cells. Flowers yellow or yellow and red.

b) Ovules 2 .... go to 4.

4. Leaves

a) Leaves undeveloped.

Jacksonia	large shrub. Branchlets grey, angular. Flowers orange.
Viminaria	large shrub. Branchlets green, round. Flowers yellow.
Sphaerolobium	small shrub less than 30 cm high. Pods round.

b) Leaves developed normally .... go to 5.

5. Seed Pods

a) Pods inflated, triangular.

Daviesia	flowers many, small, red centred. Leaves variable, usually with raised veins, sometimes pungent-pointed.

b) Pods not triangular .... go to 6.

6. Bracteoles

a) bracteoles present on or below calyx.

Pultenaea	bracteoles 1 to 3 mm long, brown, scale-like, at base of calyx or attached to it (Fig. 9a). Standard 1 lobed, narrow or rounded and notched at the top (Fig. 9b).
Phyllota	bracteoles large, green, leafy, enclosing or exceeding the calyx (Fig. 9c). Standard more pointed than above, 1 lobed (Fig. 9d).
Dillwynia	bracteoles small, deciduous, on stem below calyx (Fig. 9e). Standard broad, 2 lobed (Fig. 9f).

b) Bracteoles absent.

Aotus	calyx hairy (Fig. 9g). Standard 1 lobed.
Dillwynia	calyx not hairy. Standard 2 lobed (Fig. 9h).

Figure 9

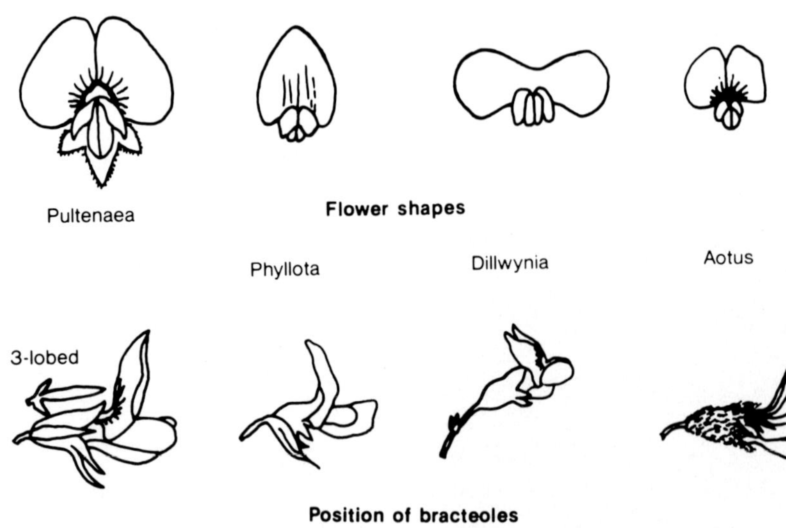

Pultenaea Bush-pea

A large genus of over 100 species, with about 34 species in the Sydney district. They are shrubs with simple entire leaves. Flowers are yellow or red, with 10 free stamens. The placement of the bracteoles is an important identifying feature; they are attached to the calyx or close under it (unlike those of *Dillwynia* which are on the stem well below the calyx). Bracteoles are brown, rarely green and stipules are prominent on stems. Pod is egg-shaped, flat or swollen and often beaked.

Pultenaea daphnoides

Large-leaf Bush-pea **Plate 401**

This is an attractive shrub up to 3 m tall, although it is often more stunted in exposed situations. The plant is easily identified by the shape of its leaves. These are narrow at the base and broad (to 10 mm) and rounded at the apex, 12 to 30 mm long, with a short soft point. Short branchlets terminate in a dense head of yellow flowers, 2 to 3 cm across. The standard has a red centre and the keel is dark. The calyx is silky-hairy as is the ovary and lower part of the style.

HABITAT Sheltered forests and gullies, occasionally shrubland and protected sites on heath.

DISTRIBUTION Widespread and common from coast to mountains. Extending along the coast southwards from Port Stephens to Vic; also southern tablelands, Tas and SA.

FLOWERING August–November.

401. Pultenaea daphnoides

402. Pultenaea retusa

Pultenaea retusa

Plate 402

A small shrub with numerous short spreading branches. Its leaves are 5 to 15 mm long, oblong to wedge-shaped, with a characteristic notch in the apex (occasionally the leaves have an obtuse apex, but the specific name 'retusa' refers to the typical notched apex). Flowers are in small terminal heads, 8 to 12 mm across, on short side branchlets. Flowers are yellow with a red centre and dark keel. The calyx, ovary and base of the style is silky-hairy.

HABITAT Scrub and forest, on both clayey and sandy soils.

DISTRIBUTION Widespread from coast to mountains, but not common. Recordings include Belmont, Kurnell, Menai, Robertson and Yerranderie. Extending along the coast from Qld to Vic; also northern tablelands.

FLOWERING August–November.

Pultenaea paleacea

Plate 403

This is a low-growing shrub to 50 cm tall, with close-silky hairs on the smaller branches. A combination of features helps separate it from similar members of the genus with terminal flower heads. Leaves are 15 to 20 mm long, about 2 mm wide, narrow-linear to oblanceolate, concave above, with margins curled under and the apex with a short, almost pungent point. Flowers are surrounded by conspicuous persistent papery-brown bracts which completely conceal the calyx. The calyx is covered with downy white hairs. Ovary and style are also silky-hairy.

HABITAT Heath, often on damp sandy soils.

DISTRIBUTION Coastal zone north of Wyong. Recorded from Morisset, Doyalson and Munmorah State Recreation Area. Restricted to central and lower north coast.

FLOWERING August–November.

403. Pultenaea paleacea

404. Pultenaea rosmarinifolia

Pultenaea rosmarinifolia

Plate 404

This is a shrub to 1 m tall with upper stems covered with downy white hairs. Leaves are linear, 2 to 3 cm long, dark green, wtih a conspicuous depressed midvein and a small down-curved point; they are widely spaced along the stem, not crowded. Flowers are in dense heads on the ends of weak branchlets. This species differs from *P. paleacea* in being a more upright plant and in having bracts around the flower heads which are silvery-hairy not papery-brown.

HABITAT Open forest and heath, on sandy soils.

DISTRIBUTION Coastal zone north of Sydney. Common in Dharug National Park and Brisbane Water National Park. Also occurring south coast (Jervis Bay-Milton).

FLOWERING August–October.

405. Pultenaea polifolia

Pultenaea polifolia **Plate 405**

A weak shrub with spreading branches seldom more than 60 cm tall. Younger stems, the undersurface of leaves and the calyx are covered with spreading soft, grey hairs. Leaves are linear, paler underneath, about 2 cm long, and have a distinctive stiff, reflexed point. Flowers are yellow with a red keel and in dense terminal heads, surrounded by overlapping hairy bracts. Bracteoles are dry and membranous, tending to be 3 or more toothed, and have long spreading hairs.

HABITAT Heath and open forest, on sandy soils.

DISTRIBUTION Widespread, but uncommon. Recorded from Ku-ring-gai Chase National Park, St Ives, Hornsby, Leura, Hampton and Mt Werong. Also occurring northern, southern and central tablelands, south-western slopes, Qld and Vic.

FLOWERING August–October.

Pultenaea linophylla **Plate 407**

This is a small erect shrub to about 80 cm tall, with small terminal flowerheads. The main identifying feature is the shape of the leaf. This is linear to wedge-shaped, up to 25 mm long, notched at the apex (emarginate) with a tiny hardened point curved downwards. The undersurface of the leaf is covered with short grey hairs. Flowers are yellow-red with a silky-hairy calyx. The bracts of the flower heads are shorter than the calyx and silky-hairy.

HABITAT Open forest and heath, on sandy soils.

DISTRIBUTION Chiefly along coastal zone but also southern highlands. Recordings include Berowra, Ku-ring-gai Chase and Royal National Parks, Kurnell, Fitzroy Falls, Hilltop and Yerranderie. Extending from north coast southwards to Vic.

FLOWERING August–November.

Pultenaea stipularis

Fine-leaf Bush-pea **Plate 408**

An erect shrub to 2 m tall. Leaves are crowded, dark green, narrow-linear and up to 4 cm long. Stems are covered with closely pressed brown stipules up to 1 cm long, a conspicuous feature, hence the specific name 'stipularis'. Flowers are in dense terminal heads about 3 cm across, with individual flowers separated by brown chaffy bracts.

HABITAT Sheltered forest, on sandy soils.

DISTRIBUTION Restricted to central coast and northern part of south coast. Recorded from Hornsby, Cheltenham, St Ives, Oatley Park, Royal National Park and Heathcote National Park.

FLOWERING August–October.

406. *Pultenaea capitellata*

Pultenaea capitellata **Plate 406**

A prostrate or trailing shrub from the higher parts of the Blue Mountains. Leaves are nearly flat, ovate to elliptic, up to 6 mm long, dark green and glossy above, paler below, and with a sharp reflexed point, about 1 mm long. Younger stems and the calyx are covered with soft grey hairs. Flowers are yellow with a red keel and are in terminal heads, surrounded by overlapping hairy bracts. Clusters of seed pods are conspicuous as they are covered with long greyish hairs.

HABITAT Heath and woodland, on damp soils.

DISTRIBUTION Higher Blue Mountains (e.g. Ruby Creek on Yerranderie road and Mt Werong). Extending along southern tablelands to Vic.

FLOWERING September–October.

407. *Pultenaea linophylla*

408. *Pultenaea stipularis*

409. *Pultenaea scabra*

411. *Pultenaea microphylla*

413. *Pultenaea brunioides*

Pultenaea scabra

Rough Bush-pea **Plate 409**

This is a small shrub, seldom more than 1 m tall. The specific name 'scabra' is a reference to the rough appearance and feel of the whole plant. Stems are covered with short whitish or dense browish spreading hairs, and the stipules (along the stem beneath the leaves) are also brown and spreading. Leaves are also hairy on the undersurface; the upper surface may be slightly hairy or hairless and covered with small rough tubercles. Leaves are wedge-shaped, 3 to 15 mm long, with a V-shaped notch at the apex (sometimes apex straight or truncated) and a small point, curved downwards. Yellow and red flowers are borne in small heads at the ends of side branchlets; bracts are silky-hairy. Plants from the Colo Plateau and north-western Hornsby Plateau have very short leaves, 3 to 8 mm long, and deeply notched at the apex. They are sometimes referred to as var. *biloba* (Plate 410).

HABITAT Open forest, on sandy or clayey soils.

DISTRIBUTION Widespread from coast to mountains, but most common in cooler forests of Blue Mountains and north-western Hornsby Plateau. Also occurring north coast, south coast, southern tablelands, central western slopes, Vic and SA.

FLOWERING September–November, but some flowers until late autumn.

Pultenaea microphylla

Spreading Bush-pea **Plate 411**

A low, much-branched shrub, often semi-prostrate in exposed positions. Young stems have downy-white hairs but older stems are hairless. Leaves are small, up to 8 mm long, 1 to 2 mm wide, linear to oblanceolate, and pointed (occasionally truncated). The tip is usually curved downwards. Flowers are yellow and red, almost stalkless, and are borne 2 or 3 together at the ends of short branches or in the leaf axils.

HABITAT Open forest, on thin gravelly and sandy soils.

DISTRIBUTION Higher parts of the Blue Mountains (e.g. Wolgan Gap and Glen Davis); also recorded from Cumberland Plain. Essentially a tablelands species, extending westwards to western slopes and northwards to Qld.

FLOWERING September–November.

Pultenaea incurvata **Plate 412**

The leaves of this small shrub are very distinctive and enable it to be easily identified. They are about 1 cm long, narrow with a blunt tip; margins are incurved giving a concave, canoe-like appearance. These leaves are upright along the stems with their tips curved in towards the stem. The upper surfaces of both the leaves and the stems are covered with long spreading white hairs. Flowers are crowded into dense terminal heads. The style of each flower is hooked. This species is regarded as 'at risk' because of its narrow distribution.

HABITAT Wet heaths and swampy ground.

DISTRIBUTION Endemic in the Blue Mountains, between Kings Tableland and Clarence.

FLOWERING September–October.

Pultenaea brunioides **Plate 413**

An upright shrub which only occurs in the upper part of the Blue Mountains and southwards to the southern tablelands. This distribution, when combined with the nature of its leaves and flower-heads, helps in its identification. Leaves are very small and narrow, 5 to 10 mm long, thick, rather triangular in cross-section, rough and with a warty appearance when viewed with a lens. Flowers are in terminal heads. The buds and calyces are densely covered in conspicuous white hairs. The ovary and fruit are also hairy. The style of each flower is hooked.

HABITAT Heath and scrubland, on sandy soil.

DISTRIBUTION Higher Blue Mountains (e.g. Kings Tableland, Narrow Neck, Blackheath). Extending to southern tablelands and Jervis Bay.

FLOWERING October–November.

Pultenaea divaricata **Plate 414**

This is a small shrub to 1 m tall, found in wet places. Its most conspicuous feature is its leaves which are slender, spreading, terete, up to 15 mm long, grooved on the upper surface and rough to the touch. Flowers are yellow with red markings, and borne in terminal heads of about 8 flowers. The calyx is covered with greyish hairs and the ovary is hairy to the base.

HABITAT Wet ground, on sandy soils.

DISTRIBUTION Blue Mountains (e.g. Mini-ha-ha-Falls and Frederica Falls), southern highlands (e.g. Carrington Falls) and higher parts of Woronora Plateau (e.g. above Macquarie Pass).

FLOWERING September–November.

410. *Pultenaea scabra* var. *biloba*

412. *Pultenaea incurvata*

414. *Pultenaea divaricata*

415. Pultenaea villifera

Pultenaea villifera **Plate 415**

A bushy shrub, 1 m or more tall. Its most conspicuous feature is that its stems and leaf-margins are covered with long weak greyish hairs. Leaves are alternate, crowded, lanceolate, up to 2 cm long, flat or slightly concave, and taper to a fine sharp point. Three veins are usually visible on the undersurface. Flowers are yellow, almost stalkless, in the axils of the upper leaves. The ovary is hairy to the base.

HABITAT Sheltered and open forests, on sandy soils.

DISTRIBUTION Endemic in central coast and south coast. Rare in the Sydney district. Recorded from lower Blue Mountains and Minnamurra Falls. Common behind Hyams Beach, Jervis Bay, south of the Sydney district.

FLOWERING April–September.

Pultenaea hispidula **Plate 416**

The term 'hispid' means a dense covering of short stiff hairs, a conspicuous feature of this small shrub. Both branches and leaves are covered with rusty-grey hairs. Leaves are narrow, about 6 to 7 mm long, and curved inwards. The short hairs make the leaves rough to touch. Flowers are in loose heads on the ends of short leafy branchlets.

HABITAT Open forest and sheltered gullies.

DISTRIBUTION Scattered and uncommon. Recorded from coast to mountains. Recordings include Bangor (Still Creek), Manly Dam and near Beacon Hill. Extending along south coast to Vic.

FLOWERING September–October.

416. Pultenaea hispidula

417. Pultenaea viscosa

Pultenaea viscosa **Plate 417**

Although 'viscosa' means 'sticky', the most noticeable feature of this medium shrub is its hairiness. Stems, leaves and calyx are all covered with weak grey hairs, and the ovary is hairy to the base. The leaves are narrow, up to 2 cm long, often curved slightly upwards; they are channelled above because of the incurved margins. The flowers have short stalks and are borne in loose heads on the ends of short leafy branchlets.

HABITAT Open forest, on sandy soils.

DISTRIBUTION Scattered and uncommon. Recorded from coast to mountains mainly to the south and south-west of Sydney. Recordings include Belmore Falls and Devlins Creek, Pennant Hills Park. Extending along south coast to Vic.

FLOWERING September–October.

Pultenaea paludosa

The specific name 'paludosa' refers to the swampy habitat preferred by this small (to 40 cm tall) shrub. Its leaves are flat, hairless, narrow-ovate, up to 8 mm long and 3 mm wide; the leaf-tip may be either acute or obtuse. Flowers are small (6 mm long), yellow and red, with a hooked style; they are crowded into terminal heads. The bracts beneath the flower-heads are ovate to linear, leaf-life, 3 mm long and hairy. The calyx is also hairy. Stipules are small (1 mm or less) or absent.

HABITAT Swampy ground in heath and scrub.

DISTRIBUTION Coastal zone, but uncommon. Recordings include Brisbane Water, Heathcote and Royal National Parks and plateau above Macquarie Pass. Extending from Myall Lakes southwards to Vic.

FLOWERING September–October.

418. Pultenaea blakelyi

Pultenaea blakelyi **Plate 418**

A tall shrub to 2 m tall, with narrow-oblong, flat leaves, smooth above, 2 to 3 cm long, which taper to a blunt point. Leaves are opposite, an important identifying feature which distinguishes this species from *P. flexilis* with which it is easily confused. The yellow flowers have a small red centre and are borne near the ends of side branches. The calyx is covered with silky hairs and the ovary is softly hairy to its base (unlike *P. flexilis,* in which the ovary is hairless, or only the upper half downy).

HABITAT Forest, especially protected slopes and gullies.

DISTRIBUTION Widespread from coast to mountains, but most common south of Sydney (e.g., Royal National Park, Illawarra Escarpment, Mt Keira). Also occurring north coast, south coast and southern tablelands.

FLOWERING September–November.

419. Pultenaea cunninghamii

Pultenaea cunninghamii — Plate 419

This is the most easily identified of all the local *Pultenaea*. Its leaves are in whorls of 3 encircling the stem, broadly ovate, tapering to a long pungent point, 12 mm long and equally as wide at the base. Leaves are hairless but sometimes blue-grey or reddish in colour. Flowers are borne in small clusters in the leaf axils. It is a spreading shrub, usually about 1 m tall.

HABITAT Open forest, on sandy soils or rocky slopes.

DISTRIBUTION North-west of Sydney (e.g., Colo Heights and Wollemi National Park), extending to Hunter Valley. Also found southern tablelands, western slopes, Qld and Vic.

FLOWERING October–January, with a few flowers as late as April.

420. Pultenaea flexilis

Pultenaea flexilis — Plate 420

This erect shrub grows to 3 m tall. Its leaves are alternate, flat, hairless, narrow-oblong, 1 to 3 cm long, about 2 mm wide and have a small but distinct point. Flowers have a short stalk and are borne along the stems or are clustered in the upper leaf axils. The calyx and the ovary are without hairs (sometimes the upper half of the ovary is downy). This species closely resembles *P. blakelyi* in habit, foliage and inflorescence. It differs from that species in having alternate leaves and hairless calyx and ovary.

HABITAT Open and sheltered forests, often in gullies.

DISTRIBUTION Widespread and common from coast to mountains. Also occurring south coast, north coast, northern tablelands, central western slopes and Qld.

FLOWERING August–October.

421. Pultenaea elliptica

Pultenaea elliptica — Plate 421

An attractive erect shrub with upright elliptical (narrow-oval) leaves crowded along the stems. These leaves are slightly concave, 7 to 12 mm long, obtuse or with a pointed apex and have a clear midrib. Juvenile leaves are covered with characteristic long weak hairs. Older leaves are mostly without hairs. Flowers are reddish orange to yellow; the keel and back of the standard are red. Flowers are almost stalkless and are borne in clusters at the end of leafy branchlets.

HABITAT Open forest, on sandy soils.

DISTRIBUTION Widespread from coast to mountains. Also occurring north coast and south coast.

FLOWERING September–December.

422. Pultenaea subspicata

Pultenaea subspicata — Plate 422

This is a low shrub with sprawling stems (sometimes prostrate), from the higher parts of the Blue Mountains. Brown stipules, 2 to 4 mm long, are noticeable on the stem. Leaves are up to 10 mm long, linear to narrow-elliptic, curved inwards, with a blunt tip. The upper surfaces of juvenile leaves have long loose hairs. Flowers are in terminal heads. The bracts and the calyx are hairy giving a greyish appearance. Flowers may be yellow with a little red, or reddish orange.

HABITAT Open forest and heath.

DISTRIBUTION Higher Blue Mountains (e.g., Hartley, Mt Irvine—Mt Wilson, Clarence). Extending along the tablelands from New England (Apsley Falls) southwards to Vic; also south-western slopes.

FLOWERING October–December.

423. Pultenaea aristata

Pultenaea aristata — Plate 423

The specific name 'aristata' means tapering to a long bristle, like the awn of wheat. In this species, it is a reference to the long sharp point which makes the leaves unmistakable. This is a small shrub with long brown stipules (5 to 6 mm) pressed against the stems. Leaves are linear, to 12 mm long, curved inwards with a 2 to 3 mm tip. The upper surfaces of juvenile leaves have long loose hairs. Flowers are in terminal heads, surrounded by long bristly bracts. The calyx has long loose whitish hairs.

HABITAT Wet heath and drier scrub.

DISTRIBUTION Endemic in Woronora Plateau where it has a narrow range between Helensburgh and the Illawarra Escarpment above Macquarie Pass.

FLOWERING September–October.

424. *Pultenaea echinula*

426. *Pultenaea canescens*

Pultenaea echinula Plate 424

This medium-sized shrub is found only in a few isolated areas in the upper Blue Mountains. Stems have a shaggy appearance. The specific name 'echinula' refers to the long, almost prickly, protuberances which cover the older branches. Stipules are brown and 3 mm long. Leaves are terete, grooved above, about 12 mm long, with a distinct, but short, pointed tip. Young leaves have long weak hairs; older leaves are rough and warty. Flowers are in compact terminal heads; the calyx has greyish spreading hairs.

HABITAT Forested slopes, on sandy soils.

DISTRIBUTION Endemic in the Blue Mountains, where it has been recorded from Kings Tableland and Cyrils Rocks, Mt Victoria.

FLOWERING August–September.

Pultenaea glabra Plate 425

As the specific name implies, this is a glabrous (hairless) erect shrub. Brown stipules (4 to 5 mm long) are pressed flat against the stems. Leaves are narrow, spreading up to 20 mm long, curved inwards, with a short point. Flowers are yellow, with only a little red, in compact terminal heads. This species might at first be confused with *P. stipularis,* but it does not have the conspicuous brown chaffy bracts around the flowers that are a feature of *P. stipularis.*

HABITAT Open forest and scrub, on protected slopes.

DISTRIBUTION A rare species, regarded as vulnerable because of its small scattered populations. Occurring only in the Blue Mountains between Wentworth Falls and Katoomba (e.g., slopes above Wentworth Falls).

FLOWERING September–October.

Pultenaea canescens Plate 426

It is often difficult to clearly separate this species from *P. incurvata* or *P. subspicata,* as all 3 are erect Blue Mountains shrubs with dense flower-heads and narrow incurved hairy leaves. However, *P. canescens* is the only one of the 3 which has 3-lobed bracteoles. These bracteoles are about 7 mm long and located on the calyx. Leaves are crowded, erect, narrow-oblong, to 12 mm long, blunt and curved inwards; conspicuous grey hairs are usually present, especially on the undersurface and near the tips. Stems are covered with brown stipules, 4 to 5 mm long. Flowers are in dense terminal heads. This is a variable species with 3 subspecies in the area.

HABITAT Open forest, usually on wet or damp sandy soils.

DISTRIBUTION Restricted to higher Blue Mountains (e.g., Clarence, Mt Irvine).

FLOWERING September–November.

425. *Pultenaea glabra*

427. *Pultenaea parviflora*

Pultenaea parviflora Plate 427

A small erect shrub, regarded as rare and endangered because of its restricted distribution. Suburban subdivision places it at risk. Leaves are crowded along the stems; they are spoon-shaped, 2 to 5 mm long, with a small down-curved tip. Flowers are small, 5 to 7 mm long, sparse, and borne near the ends of the branches in the leaf axils. The calyx is without hairs. Bracteoles are deeply 3-lobed and united with the calyx-tube.

HABITAT Open forest, on heavier clay soils.

DISTRIBUTION Restricted to the western part of Cumberland Plain (e.g., Castlereagh, St Marys).

FLOWERING September–December.

428. *Pultenaea villosa*

Pultenaea villosa Plate 428

A softly hairy, leafy shrub, erect or spreading, but usually with graceful drooping branches. Stems are covered with long soft hairs and the smaller side branchlets are crowded with small (3 to 6 mm long) narrow-oblong leaves, which are conspicuously hairy on the undersurface. Flowers are abundant in the leaf axils. Bracteoles are deeply 3-lobed on the softly hairy calyx-tube.

HABITAT Forest and woodland, mainly on heavier soils.

DISTRIBUTION Coast and Cumberland Plain (e.g., Menai, Castlereagh). Also occurring south coast, north coast, northern tablelands and Qld.

FLOWERING August–October.

429. *Pultenaea ferruginea* var. *deanei*

431. *Phyllota squarrosa*

Pultenaea ferruginea var. *deanei* Plate 429

This is a robust erect shrub, 1 to 2 m high, softly hairy in general appearance. Leaves are 8 to 12 mm long, 4 to 5 mm wide, obovate and covered with soft white hairs. Flowers are relatively large, with a standard up to 12 mm long, and are crowded near the ends of the branchlets; the calyx is distinctly hairy. Variety *deanei* is the more common variety in the area; it is larger in all its parts than var. *ferruginea*. This species differs from *P. villosa* (which is also softly hairy) in having spoon-shaped leaves and larger flowers concentrated near the ends of the branchlets.

HABITAT Sheltered forest, on sandy soils.

DISTRIBUTION Coastal zone north of Sydney from Ku-ring-gai Chase National Park through Brisbane Water National Park to the Gosford area. Also recorded for north coast and south coast.

FLOWERING September–October.

Pultenaea ferruginea var. *ferruginea* Plate 430

This variety is smaller in all its parts than var. *deanei*. It is seldom more than 1 m tall. Leaves are 3 to 7 mm long, about 3 mm wide, obovate and covered with soft white hairs. Flowers have a standard up to 10 mm long.

HABITAT Sheltered forest, on sandy soils.

DISTRIBUTION Colo Plateau and Hornsby Plateau (e.g. Colo Heights, Wisemans Ferry) and lower Blue Mountains (e.g. Grose Vale, Glenbrook, Blaxland). Extending to Goulburn River National Park (Coxs Gap).

FLOWERING October–November.

430. *Pultenaea ferruginea* var. *ferruginea*

Phyllota

Shrubs, with alternate, simple, entire leaves. Flowers yellow or yellow and red, with 10 stamens fused to the petals at the base but free above. Bracteoles large, green and attached to the stalk just below the calyx. Stipules minute. Differs from *Pultenaea* in the nature of the bracteoles and stipules. Pod egg-shaped, small, often enclosed in the calyx.

Phyllota squarrosa Plate 431

A low shrub to 30 cm tall, its stems are hairy on the upper sections. Leaves are linear, 9 mm long, with distinct yellow recurved tips. Flowers are usually yellow (sometimes reddish) and borne in groups near the ends of the branches. Bracteoles at the base of the calyx are green, lanceolate and about 8 mm long. The stamens are mostly fused to the petals. The ovary is covered with long, weak, silky hairs.

HABITAT Heath and dry open forest, on sandy soils.

DISTRIBUTION Higher Blue Mountains, between Katoomba and Newnes State Forest (e.g. Blackheath, Mt Victoria, Clarence) and Kanangra Walls. Also occurring in higher parts of south coast and southern tablelands.

FLOWERING November–December.

Phyllota grandiflora Plate 432

This erect shrub (to 1 m tall) is easily recognised by its compact, almost terminal, leafy flower heads containing large bright yellow flowers. Each flower has a long green bracteole at the base of the calyx. As these bracteoles are larger than the calyx, it gives the appearance of flowers mixed with leaves. Flowers are 12 to 15 mm long. The calyx and ovary are covered with long loose hairs. Young stems are also hairy. Stipules are minute, unlike those of *Pultenaea* which are prominent. Leaves are up to 20 mm long, narrow, dark green above, paler below; the edges of the leaves are rolled back. This species is similar to the more common *Phyllota phylicoides* but is larger in all its parts.

HABITAT Open forest, on sandy and rocky slopes.

DISTRIBUTION Restricted to coastal zone around Sydney. Recorded from Port Jackson to Hawkesbury River, Appin-Bargo and near Bucketty. Common in the Berowra-Cowan Creek section of Ku-ring-gai Chase National Park.

FLOWERING September–December.

432. *Phyllota grandiflora*

433. Phyllota humifusa

Phyllota humifusa **Plate 433**

This prostrate shrub is regarded as an endangered species because of its restricted distribution which is mostly outside national parks and reserves. Stems are white-hairy when young. Leaves are linear, 3 to 8 mm long, obtuse to acute. Flowers are deep orange-yellow to red and borne in leafy spikes towards the ends of the branches. Bracteoles at the base of the calyx are identical in appearance to the foliage leaves. The calyx may be hairless or have scattered short silky hairs. The 10 stamens are mostly fused to the petals. The ovary is covered with long, weak, silky hairs.

Habitat Dry open forest, on deep sandy soils.

Distribution Endemic in the southern highlands. Recorded between Mittagong and Bullio and near Paddys River, west of Penrose.

Flowering October–December.

Phyllota phylicoides

Common Phyllota **Plate 434**

This erect shrub is one of the common yellow pea-flowers in the Sydney district. Its stems are often softly hairy, at least in the upper part, and are crowded with dark narrow leaves, 5 to 20 mm long. Flowers are in dense yellow terminal clusters either from the leaf axils or near the ends of the branches. Each flower is about 10 mm long. The calyx is covered with long white hairs which give the flower heads a fluffy white appearance when in bud. Bracteoles are at the base of the calyx; they are green, lanceolate and longer than the calyx tube.

Habitat Heath, shrubland and woodland, on sandy soils.

Distribution Widespread from coast to mountains. Common along coast in sandstone regions of national parks. Also occurring south coast, north coast, northern tablelands, north-western slopes and Qld.

Flowering August–October.

Aotus

Shrubs. The generic name is derived from two Greek words, 'a' (without) and 'oetos' (ear), a reference to the absence of bracteoles on or below the calyx. This helps distinguish the genus from *Pultenaea*, *Phyllota* and *Dillwynia*, which all have bracteoles. Flowers have 10 free stamens and a calyx with the 2 upper lobes broader and more or less united.

434. Phyllota phylicoides

Aotus ericoides

Common Aotus **Plate 435**

A common small shrub, occurring in a variety of plant communities, but especially common on deeper fixed sands along the coast. Stems are covered with greyish or rusty hairs. Dark green leaves are scattered along the stems. These are narrow, 6 to 12 mm long and have edges which curl back to the undersurface. Flowers are yellow, with red centre markings, and are borne abundantly all along the stems from the leaf axils. The flower has an orbicular standard, a style which is not hooked, a hairy calyx, and is without bracteoles. The ovary is covered with long white hairs and is conspicuous when the petals fall. A low compact form found on deep coastal sands at La Perouse and elsewhere is now regarded as a new species.

Habitat Heath, old dunes, scrubland and open forest, on sandy soils.

Distribution Widespread from coast to mountains. Extending all along coast and tablelands from Qld to Vic; also Tas and SA.

Flowering August–November.

435. Aotus ericoides

436. Dillwynia retorta ssp. A

Dillwynia Eggs and Bacon, Parrot-pea

Shrubs, with linear or terete leaves with a groove along the upper surface. Flowers yellow and red with 10 free stamens. Bracteoles small and on the flower stalk well below the base of the calyx. Style hooked near the tip. Pod egg-shaped or rounded, inflated. Some species are very variable and some confusion has arisen about the correct classification of varieties and subspecies.

Dillwynia retorta ssp. A

Eggs and Bacon **Plate 436**

A number of pea-flowers with yellow standards and red centres share the common name 'Eggs and Bacon', but *D. retorta* is so widespread and common that it is often the species which is being referred to when that name is used. It is a variable shrub, with a confusing history of classification. A number of different forms and species named by early botanists were lumped together under *D. retorta*. Two of these subsequently were raised to species status (*D. phylicoides* and *D. parvifolia* var. *trichopoda*), but recent revisions have classified them as subspecies. The common form is a spreading, twiggy to erect shrub, 1 to 2 m tall, with crowded, narrow, hairless leaves, 5 to 10 mm long, most of which are conspicuously twisted. Stems are lightly sprinkled with hairs. Flowers are mostly bright yellow with red centres, but may be orange-yellow and red. They have broad ear-like standards, short stalks and are borne singly or in clusters near the ends of the branches. Each cluster of flowers may be on a short common stalk or, as in some Blue Mountains plants, on stalks 10 to 15 mm long.

Habitat Heath, shrubland and open forest, on sandy soils.

Distribution Widespread from coast to mountains. Extending along north coast to Qld.

Flowering July–November.

437. Dillwynia retorta ssp. B

Dillwynia retorta ssp. B — Plate 437

This is a prostrate to low shrub from the higher mountains and associated valleys. Branches, leaves and calyx are covered with short rigid hairs, giving the plant a rough appearance and touch (occasionally only the tips of the leaves have these stiff hairs). Leaves are 3 to 8 mm long and conspicuously twisted. Flowers are mostly stalkless and clustered together towards the ends of the branches. This species was formerly known as *D. phylicoides*.

HABITAT Open forest and woodland, on sandy soils.

DISTRIBUTION Higher Blue Mountains, mainly in open valleys (e.g. Coxs River, Hartley). Extending throughout the tablelands and western slopes from Qld to Vic; also north coast.

FLOWERING September–November.

438. Dillwynia retorta ssp. C

Dillwynia retorta ssp. C — Plate 438

There is some dispute about the correct classification of this diffuse shrub, with some botanists regarding it as a subspecies of *D. retorta* and others as a variety of *D. parvifolia*. Its leaves are twisted and only 3 to 4 mm long, much smaller than those of *D. retorta*. Flowers are arranged in terminal umbels or corymbs, on stalks more than 6 mm long, unlike the inflorescence of *D. parvifolia*. The plant is seldom more than 60 to 80 cm tall and often ground hugging, unlike *D. retorta* which can be 2 m tall.

HABITAT Open forest, on thin sandstone soils with ironstones.

DISTRIBUTION Lower Blue Mountains, Mittagong-Berrima, Menai and coastal area north of Sydney. Also occurring north coast, south coast, central tablelands and north-western slopes.

FLOWERING September–December.

439. Dillwynia parvifolia

Dillwynia parvifolia — Plate 439

This species is similar to *D. retorta* and was formerly included in that species, but is a much smaller diffuse shrub, seldom more than 80 cm high. The specific name 'parvifolia' refers to the characteristic small twisted leaves which are only 2 to 3 mm long. Flowers are arranged singly or in pairs at the ends of short (3 to 6 mm) side branchlets or terminal on main stems; they may be stalkless or on short (2 to 3 mm) thin stalks.

HABITAT Open forest or cleared areas, on heavier clay soils.

DISTRIBUTION Cumberland Plain (e.g. Liverpool-Fairfield) where its survival is threatened by urban sub-division, Bangor-Menai, Bankstown (Salt Pan Creek) and The Basin near Campbelltown. Also occurring southern tablelands and central western slopes.

FLOWERING September–October.

Dillwynia ramosissima — Plate 440

The specific name 'ramosissima' means 'much branched', an appropriate name for this medium shrub which has a complex of small, stiff, crowded side branches. Many of these small branches end in spines, a characteristic feature of this species. Leaves are cylindrical, straight (not twisted), 3 to 5 mm long and have a small tip which is curved downwards. Typical 'eggs and bacon' flowers are borne singly or sometimes in pairs near the ends of the side branches.

HABITAT Open forest, on sandy soils.

DISTRIBUTION Widespread, but absent from coastal national parks. Recordings include Lapstone Hill (lower Blue Mountains), Mt Tomah, Mittagong and Bundanoon. Extending to south coast and southern tablelands.

FLOWERING August–October.

440. Dillwynia ramosissima

441. Dillwynia glaberrima

Dillwynia glaberrima

Smooth Parrot-pea — Plate 441

Although this species occurs throughout south-eastern Australia, it is not a well-known plant in the Sydney area. This is partly because of its preference for deep sands and alluvium which limits its distribution and partly because it is often confused with the more common *D. retorta*. It differs from that species in having leaves which are straight (not twisted) and slightly longer (to 15 mm), and hairless branches. It is a low spreading shrub, usually 50 to 80 cm tall, with spreading leaves and flowers borne in clusters at the ends of the branches. A prostrate form with pubescent stems occurs at La Perouse and Kurnell.

HABITAT Open forest and heath, on deep sands and alluvium.

DISTRIBUTION Coast and Cumberland Plain (e.g. Agnes Banks National Park). Extending along the coast the Qld to Vic; also Tas and SA.

FLOWERING September–November.

Dillwynia tenuifolia — Plate 442

A low spreading shrub which resembles *D. glaberrima* and is found in similar habitat on the alluvium of the Cumberland Plain. It differs from *D. glaberrima* in having stems which are covered with minute soft hairs and flowers which are mostly solitary with a short stalk and borne on the ends of the short side branches (not in clusters) Leaves are clustered, 4 to 10 mm long, spreading to erect, with a small tip which is curved downwards.

HABITAT Open forest, on alluvium.

DISTRIBUTION Cumberland Plain (e.g. Agnes Banks Nature Reserve, Castlereagh State Forest) and lower Blue Mountains. Extending to north coast.

FLOWERING August–October.

442. Dillwynia tenuifolia

443. Dillwynia sericea

Dillwynia sericea

Showy Parrot-pea **Plate 443**

An erect shrub to 1 m tall with stems covered with conspicuous silky-white hairs, especially on young stems. Leaves are linear to cylindrical, grooved on the upper surface, up to 12 mm long, straight, not twisted and covered with minute lumps. Flowers are yellow with red markings, stalkless and borne along the branches singly or in pairs. The flowers often appear to be in a cylindrical spike clustered below the new growth of leaves. The outer surface of the calyx is often covered with silky hairs and the pod is sparsely hairy.

HABITAT Open forest and woodland.

DISTRIBUTION Higher Blue Mountains and southern highlands (e.g. Joadja Creek, Mittagong). Also occurring northern tablelands, southern tablelands, western slopes, north-western plains, Qld and Vic.

FLOWERING September–December.

Dillwynia sericea ssp. B **Plate 444**

There has been some confusion surrounding the correct classification of the plant formerly known as *D. rudis*, but it is now regarded as a subspecies of *D. sericea*. In the field, it may also be confused with *D. floribunda* var. *teretifolia*. Basically, it is a darker-flowered form of *D. sericea*. The flowers have a deep red-brown keel and orange-red standard and wings. Stems are covered with loose silky hairs when young and have a shaggy appearance when old. Leaves are hairy when young but rough when older, up to 15 mm long, terete, straight or curved.

HABITAT Heath and shrubland, on dry sandy soils and laterites.

DISTRIBUTION Widespread from coast to mountains. Also occurring south coast, southern tablelands, western slopes and Vic.

FLOWERING August–October.

Dillwynia floribunda var. *floribunda*

Plate 445

This shrub (to 1 m tall) is common in coastal bushland around Sydney. Its erect stems, crowded near the tops with 'eggs and bacon' flowers are a colourful feature of open swampy ground in the national parks both north and south of Sydney. The flowers are usually in pairs in the leaf axils, but packed closely together in what appears to be a terminal inflorescence. Flowers may cover 25 cm of the end of the stem, but the inflorescence averages 10 cm. The young branches are usually without hairs; the calyx is often slightly hairy, but may be hairless. Leaves are crowded, narrow, 5 to 15 mm long, covered with tiny lumps and somewhat rough. Most plants have leaves which are conspicuously twisted, but other plants have straight leaves.

HABITAT Common in swampy heaths; less frequently found on drier sandy soils of headlands, forested gullies and shrubland.

DISTRIBUTION Coastal zone. Recordings include Brisbane Water National Park, Muogamarra Nature Reserve, Mt White, Royal National Park, Agnes Banks and Carrington Falls. Extending from south coast northwards to Qld.

FLOWERING July–October.

Dillwynia floribunda var. *teretifolia*

Plate 446

This variety tends to be a more robust shrub than variety *floribunda*. Its leaves are more cylindrical, less twisted, almost smooth and the groove or channel is more obvious on the upper surface. Leaves are 10 to 15 mm long. These features seem to be a response to the drier soils on which this variety occurs.

HABITAT Dry shrubland, open forest and heath, on thin sandy soils.

DISTRIBUTION Scattered from coast to mountains. Recordings include Menai-Lucas Heights, where it is endangered because of urban sub-division.

FLOWERING August–October.

444. Dillwynia sericea ssp. B

445. Dillwynia floribunda var. *floribunda*

Dillwynia juniperina

Prickly Parrot-pea **Plate 447**

This is a rigid shrub, usually about 1 m tall, with many spreading branchlets. Its spreading, rigid, needle-like leaves with fine pungent points help distinguish it from other *Dillwynia* species in the area. These leaves are up to 15 mm long, triangular or somewhat concave in cross-section and grooved on the upper surface. Flowers are borne in small clusters at the ends of the branches. The calyx is U-shaped at the base (not V-shaped as most other *Dillwynia* species in the area) and covered with soft

446. Dillwynia floribunda var. *teretifolia*

447. Dillwynia juniperina

white hairs. The petals are persistent and their remains adhere to the pod.

HABITAT Open forest, on dry sandy soils.

DISTRIBUTION Widespread from coast to mountains, but not common. Recordings include Wallacia, Casula, Bundanoon and Wingello State Forest. Extending all along the tablelands from Qld to Vic, also south coast and western slopes.

FLOWERING August–October.

Dillwynia acicularis Plate 448

A tall shrub, 1 to 2 m high, its most characteristic feature is that its leaves are erect or lie closely along the branches. These leaves are terete, up to 3 cm long, sharply pointed, but (unlike *D. juniperina*) not pungent. Flowers are borne in clusters at the ends of branches. The calyx is U-shaped at the base. The pod is usually surrounded by the persistent remains of the petals.

HABITAT Open forest, on sandy soils.

DISTRIBUTION Scattered throughout Hornsby Plateau and Woronora Plateau, but not common. Extending from central tablelands to southern tablelands and central western slopes.

FLOWERING September–October.

448. Dillwynia acicularis

449. Daviesia virgata

Daviesia Bitter-pea

Shrubs, with alternate leaves, usually tough and rigid, often pungent-pointed. Flowers yellow with red centres, with 10 free stamens. Pods triangular. As no other pea-flowers in the area have such distinctive pods, they are the best feature with which to identify the genus.

Daviesia virgata

Slender Bitter-pea Plate 449

This is an upright to twiggy shrub to 1.5 m tall. Its wiry branches bear tough rigid leaves, spaced well apart on the stem. They are mostly linear, obtuse to acute, 3 to 9 cm long, somewhat wrinkled and with conspicuous veins. Upper leaves are often smaller, less numerous or sometimes completely absent. Flowers are yellow, with red-brown centres and are borne abundantly on slender stalks in small clusters in the axils of the upper leaves or scattered along almost leafless branches. Fruit is a hairless, triangular pod, 6 to 10 mm long, containing only one seed. Some botanists regard this species as a poor form of *D. mimosoides*.

HABITAT Open forest, especially on rocky slopes.

DISTRIBUTION Higher Blue Mountains (e.g. Wolgan Gap, Jenolan Caves) and southern highlands (e.g. Tallong, Wingello). Also occurring northern and southern tablelands, central western slopes, Vic and SA.

FLOWERING October–November.

Daviesia latifolia

Broad-leaf Bitter-pea Plate 450

The broad, strongly veined leaves immediately identify this bushy shrub even when it is not in flower. These leaves are ovate to almost lanceolate,

450. Daviesia latifolia

up to 8 cm long and 2 to 5 cm wide. A network of fine veins is prominent. Numerous small yellow flowers with red-brown centres are borne in racemes smaller than the leaves (3 to 6 cm long) from the leaf axils. The contrast of dense yellow flower clusters and broad green leaves makes this plant very attractive when in full flower.

HABITAT Open forest, on sandy soils.

DISTRIBUTION Higher parts of Blue Mountains and southern highlands. Recordings include Hartley, Coxs River, Clarence, Narrow Neck and Bundanoon. Occurring throughout the tablelands and western slopes; also south coast, Vic and Tas.

FLOWERING September–November.

Daviesia mimosoides Plate 451

This is a shrub to 2 m tall with attractive drooping branchlets. The specific name 'mimosoides' indicates that its leaves resemble those of a wattle ('Mimosa'). They are flat, oblanceolate to narrow lanceolate, usually about 8 cm long (but sometimes longer), slightly curved, pointed but not spine-tipped and without a strong network of veins. Young leaves are soft, tan-coloured and hang drooping at the ends of the branches. Numerous small yellow flowers with red-brown centres are borne in racemes much smaller than the leaves (about 2 cm long) all along the upper branches.

HABITAT Forest, on deep sands.

DISTRIBUTION Restricted distribution in Sydney area. Locally common on old dunes at Kurnell. Also recorded in Blue Mountains and Nattai area (e.g. Blue Gum Creek). Extending to Hunter River and Goulburn River Valley; also south coast, southern tablelands, western slopes, Qld and Vic.

FLOWERING July–November.

451. Daviesia mimosoides

452. *Daviesia corymbosa*

453. *Daviesia ulicifolia*

454. *Daviesia squarrosa*

Daviesia squarrosa var. *squarrosa*

Plate 454

D. squarrosa is a shrub with softly hairy stems. It may be a small undershrub or a taller bush to 1.5 m tall with spreading branches. Leaves are small, 6 to 10 mm long, heart shaped, hairless (or with short hairs on the margins), stalkless and tapering to a fine pungent point. The midvein of each leaf is prominent. Flowers are small, about 6 mm long, yellow with red centres and attractive as they are borne all along the stem from the leaf axils. Each flower is on a slender stalk which is slightly longer than the leaves.

HABITAT Open forest, on heavier soils.

DISTRIBUTION Scattered along coast and tablelands. Recordings include Howes Valley, Maroota, Macdonald Valley, Castlereagh State Forest, Wedderburn, Megalong, Glenbrook and Thirlmere. Also occurring north coast, south coast, southern tablelands and north-western plains.

FLOWERING July–November.

Daviesia corymbosa

Plate 452

This is a medium-sized shrub, usually about 1 m tall. Leaves are variable, but usually lanceolate to narrow ovate, 6 to 8 cm long, with a clear network of fine veins. Yellow and red flowers are borne in corymbose racemes and this distinguishes it from other broad-leaved bitter-peas such as *D. latifolia*. The term 'corymbose' refers to a flower arrangement where all the flower stalks begin at different levels on a raceme but the flowers are borne at about the same level.

HABITAT Open forest, on sandy soils.

DISTRIBUTION Widespread from coast to mountains. Recordings include Colo Heights, Wentworth Falls, Lawson, Heathcote, Menai, Buxton and Thirlmere Lakes. Extending to Nowra–Sassafras district (south coast–southern tablelands).

FLOWERING August–November.

Daviesia ulicifolia

Gorse Bitter-pea **Plate 453**

An erect shrub to about 1 m tall, with angular, short branchlets which end in a spine. The small rigid leaves are also sharp-pointed. These leaves are ovate to narrow lanceolate, 5 to 15 mm long and stalkless. Flowers are yellow with deep red centres and are borne on short stalks from the leaf axils, either singly or in groups of 2 to 4.

HABITAT Open forest, on sandy soils or on shale cappings.

DISTRIBUTION Widespread from coast to mountains. Found throughout NSW and in all other States (but not NT).

FLOWERING August–November.

Daviesia acicularis

Plate 455

A shrub to 1 m tall, its leaves rigid, pungent-pointed, narrow lance shaped to linear, 2 to 3 cm long, upright, and with margins which are conspicuously rolled or curved towards the undersurface. Branchlets are not spine-tipped as are those of *D. ulicifolia*. Flowers are yellow with red-brown centres, about 7 mm long, and are usually borne singly on short stalks in the leaf axils.

HABITAT Open forest, on sandy soils.

DISTRIBUTION Scattered from coast to mountains. Recordings include Agnes Banks Nature Reserve, Mangrove Creek, The Basin at upper Georges River and Mt Wilson road. Also occurring south coast, north coast, northern tablelands, western slopes and Qld.

FLOWERING August–November.

455. *Daviesia acicularis*

457. *Chorizema parviflorum*

458. *Oxylobium cordifolium*

Daviesia alata — Plate 456

D. alata is easily identified as it is a low straggling shrub without any leaves (they are reduced to minute scales). The branches are flat with a raised wing down the centre. Flowers are clustered together in a small raceme whose stalk is very short. Flower colour is variable. South of Sydney in the Royal National Park and nearby areas, flowers are yellow with red centres. In the upper Blue Mountains, flowers are a deep orange with red and yellow markings. This species may be confused with *Bossiaea ensata*, another leafless, flat-stemmed pea-flower. The most striking difference is that *D. alata* has flowers in clusters, while *B. ensata* has single flowers. The filaments are all free in *Daviesia*, but joined into a tube in *Bossiaea*. Only the genus *Daviesia* has triangular pods.

HABITAT Open forest, woodland and heath, on sandy soils.

DISTRIBUTION Widespread from coast to mountains. Recordings include Warrah (Brisbane Water National Park), Lucas Heights, Heathcote and Royal National Parks, Carrington Falls, Leura and near Mt Wilson. Extending to lower north coast and south coast (near Nowra).

FLOWERING August–November.

456. *Daviesia alata*

Chorizema

A genus of 16 species of shrubs, predominantly occurring in Western Australia, with only one species on the east coast. Leaves alternate. Flowers with stamens all free, usually orange or red and borne in terminal racemes. Ovary pubescent, with 8 or more ovules. Fruit is an inflated pod.

Chorizema parviflorum — Plate 457

This is a shrub, usually 30 to 40 cm tall, with a stout rootstock and slender, ascending, angular stems. Leaves are alternate, linear to obovate, spreading, 10 to 30 mm long, with a conspicuous midrib and a downy undersurface. Flowers are yellow with a red centre in loose clusters at the ends of the branches. Small bracteoles are on the flower stalk well below the calyx and are usually deciduous. The pod is 5 to 8 mm long, almost stalkless.

HABITAT Heath, open forest and grassland, on heavy clay soils.

DISTRIBUTION Uncommon in Sydney district. Recorded from Cumberland Plain (e.g. Castlereagh State Forest) and Pleasure Point, Georges River. Old records from Ingleburn, Rooty Hill and Newport-Barrenjoey. Extending along coast northwards to Qld.

FLOWERING September–December.

Oxylobium

Shrubs. Leaves simple, mostly opposite or in whorls of 3. Flowers yellow-orange with 10 free stamens. Pod egg-shaped, inflated and not divided by a longitudinal partition as those of *Mirbelia*.

Oxylobium cordifolium — Plate 458

This is a small prostrate shrub with downy stems. Leaves are small, 5 to 8 mm long, in whorls of 3, ovate to heart-shaped and curved downwards at the tip and along the margins. Flowers are orange with red central markings, with a standard 6 to 8 mm long and are borne in short terminal clusters. The calyx is hairy. The pod is egg-shaped, up to 10 mm long, almost stalkless and softly hairy.

HABITAT Heath, especially in damp sandy soils, and on rocky outcrops.

DISTRIBUTION Chiefly coastal. Uncommon around Sydney where its habitat has been reduced. Recorded from Kurnell and La Perouse. Extending to south coast (e.g. Jervis Bay, Milton). Also recorded from Wingello State Forest and from central and southern tablelands.

FLOWERING October–November.

Oxylobium pulteneae — Plate 459

A low spreading plant from the north-western part of the Sydney district. Its flowers are orange-red in short, attractive, terminal clusters. Leaves are mostly in whorls of 3, up to 10 mm long and 4 mm wide,

459. *Oxylobium pulteneae*

distinctly obtuse at the base, and almost stalkless. Margins are curved downwards. The upper surface of the leaf is rough, the lower surface downy. The pod is egg-shaped and downy.

HABITAT Open forest, on dry or wet sandy soils.

DISTRIBUTION North-western Hornsby Plateau (e.g. Howes Valley) to Hunter Valley. Extending to central western slopes (Goulburn River National Park).

FLOWERING October–November.

Oxylobium arborescens

Tall Shaggy-pea **Plate 460**

Although this tall shrub (to 2 m) is widespread throughout south-eastern Australia, it is fairly uncommon in the local area. Leaves are in whorls of 3, or irregularly opposite, up to 5 cm long, narrow elliptic, the upper surface hairless and dark green, the undersurface silky-hairy. Flowers are bright yellow with a red centre, and are clustered in short racemes at the ends of the branches. The fruit is an inflated, hairy pod, 7 to 8 mm long, which has a distinct curved point on the end.

HABITAT Sheltered forest and gullies.

DISTRIBUTION Scattered along highlands escarpment around Fitzroy Falls, Wingello and Jamberoo Mountain; also reported from Glenbrook (in 1909). Extending along southern tablelands to Vic and Tas.

FLOWERING October–November.

Oxylobium ellipticum **Plate 461**

This is a shrub to 2 m tall with downy stems. It is common in subalpine areas, but seldom seen in the Sydney district. Leaves are elliptic, pungent-pointed, in whorls of 3 to 4, tough, 1 to 3 cm long, 2 to 8 mm wide, green and hairless above, with pale silvery hairs below and margins rolled under. Flowers are orange-yellow and clustered near the ends of the branches. The calyx is silky-hairy. Pods are about 8 mm long and silky-hairy.

HABITAT Heath and woodland, in rocky areas.

DISTRIBUTION Higher Blue Mountains (e.g. Clarence, Boyd River, Mt Werong). Extending along tablelands from Barrington Tops southwards to Kosciusko district, Vic and Tas. Also recorded in Qld.

FLOWERING October–December.

Oxylobium ilicifolium

Native Holly, Prickly Shaggy-pea **Plate 462**

Native Holly is a shrub, occasionally reaching 3 m in height, but usually much smaller. Both the botanical and common names are references to its prickly leaves ('ilici', holly-like). These leaves are opposite, 2 to 4 cm long, strongly veined, with 3 or more lobes each with a pungent point. Leaves are very variable and the number of lobes and size depend on the habitat. Flowers are yellow with a red keel; the standard is about 8 mm long. They are borne either in the leaf axils or in short terminal racemes.

HABITAT Forest, often on sheltered hillsides, on sandy soils.

DISTRIBUTION Widespread from coast to mountains. Also occurring south coast, southern tablelands, north coast, northern tablelands, central western slopes and Qld.

FLOWERING September–November.

460. Oxylobium arborescens

461. Oxylobium ellipticum

462. Oxylobium ilicifolium

Oxylobium scandens var. *scandens*

Plate 463

A prostrate plant with long trailing stems and opposite entire leaves. These leaves are elliptic to ovate, 2 to 5 cm long, with a small point at the tip, strongly veined, dark green above and lighter underneath. Flowers are deep yellow with a red centre and are clustered together in terminal racemes so that the plant is very conspicuous at flowering time.

HABITAT Dry open forest on heavier soils.

DISTRIBUTION Coastal zone. Recordings include Thirlmere, Tahmoor, Cumberland Plain (Shane Park) and lower Blue Mountains. Occurring from south coast northwards to Qld.

FLOWERING August–December.

463. Oxylobium scandens

464. Oxylobium aciculiferum

Oxylobium aciculiferum **Plate 464**

The specific name 'aciculiferum' is derived from 'akis' or 'ake' meaning 'a sharp point' or 'thorn', and is very appropriate for this tall shrub (to 2 m). Not only do the leaves have a long, needle-like tip, but beneath the leaves there are spines about 6 mm long (actually spine-like stipules). Leaves are broad-ovate, 10 to 12 mm long, narrowing into a sharp point, and crowded along the branches. Flowers are yellow, with a red ring near the centre of the standard, and are borne solitary or a few together in the leaf axils.

HABITAT Sheltered forest.

DISTRIBUTION North of area, mainly in Watagan Mountains. Extending northwards along coastal highlands and tablelands to Qld. Also recorded on south coast west of Milton and west of Bodalla.

FLOWERING December–April.

Mirbelia

Small shrubs. Leaves simple, with strong reticulate venation, often in whorls of 3 but also alternate or opposite. Flowers with 10 free stamens, pink, reddish-purple or yellow and red. Pod egg-shaped, inflated and divided into 2 by a longitudinal partition.

465. *Mirbelia rubiifolia*

466. *Mirbelia speciosa*

Mirbelia rubiifolia Plate 465

This low diffuse shrub is common in open situations where it trails through the undergrowth. Several pink-lilac flowers are borne together in the leaf axils, and even though each flower is small, the many clusters along the slender branches make the plant conspicuous. Leaves are readily identified. They are in whorls of 3, stiff, narrow, pungent pointed, up to 25 mm long and have margins curved down. There is a conspicuous network of veins on the dark green upper surface. Fruit is an inflated pod about 5 mm long.

HABITAT Heath and open woodland.

DISTRIBUTION Widespread and common from coast to mountains. Recordings include sandy heaths of coastal national parks, heavy clay soils near Tahmoor and open woodland near Ingar and Mt Tomah in the Blue Mountains. Extending all along coast and tablelands; also Qld.

FLOWERING September–November.

Mirbelia speciosa Plate 466

M. speciosa is a much more upright shrub than *M. rubiifolia*. Its large purple flowers immediately attract attention in the undergrowth during winter months. Flowers are borne along erect terminal branches. The centre of the flower is darker purple with 2 yellow spots. Upper branchlets and the calyx of each flower are sprinkled with white hairs. Leaves are linear, up to 25 mm long, blunt or pointed, without a conspicuous network of veins, and the margins are curved beneath to the undersurface. Lower leaves are in whorls of 3, but upper leaves are usually alternate.

HABITAT Heath and woodland, on sandy soils.

DISTRIBUTION Scattered along coastal area. Common south of Sydney from Royal National Park through to Waterfall, Darkes Forest and Wedderburn. Also occurs at West Head (Ku-ring-gai Chase National Park) and north of Hawkesbury River (e.g. Mt White-Kariong). Extending along north coast and northern tablelands to Qld.

FLOWERING July–August.

Mirbelia platylobioides Plate 467

The similarity of the leaves of this species to those of *Platylobium formosum* is reflected in the specific name of this prostrate shrub. However *Platylobium* has stamens all united into a tube, whereas *Mirbelia* has its stamens all free. Leaves are ovate to broadly lanceolate, about 4 cm long, stiff, strongly veined, shiny above and silky-hairy below. They may be opposite or alternate. Flowers are a bright yellow, with a red centre, relatively large, and are borne in small clusters. The calyx, branches and pod are all covered with silky hairs.

HABITAT Heath, on sandy soils.

DISTRIBUTION Higher Blue Mountains (e.g. Blackheath, Wentworth Falls, Mt Wilson, Newnes) and southern highlands (e.g. Bundanoon and Wombeyan Caves road). Also occurring southern tablelands and Kandos-Rylstone area.

FLOWERING August–October.

467. *Mirbelia platylobioides*

Mirbelia baueri **Plate 468**

M. baueri is a very small shrub, seldom reaching 30 cm, and often forming prostrate mats amongst the undergrowth. Its leaves are needle-like, pungent-pointed, up to 15 mm long with margins curled under. They are scattered along the downy branches, sometimes in whorls of 3, but often not in whorls. Flowers are an orange-red colour with some yellow, short-stalked, 10 to 12 mm long, and are borne in the leaf axils near the ends of the branches.

HABITAT Heath, on sandy soils or rocky areas.

DISTRIBUTION Higher areas away from coast (e.g. Wentworth Falls-Katoomba, Penrose-Wingello). Extending to Nowra, Jervis Bay and Sassafras-Mongarlowe.

FLOWERING October–November.

Gompholobium Wedge-pea

Shrubs, with leaves digitate (usually 3 leaflets) or sometimes pinnate (to 30 leaflets). Flowers yellow (except for *G. uncinatum*), with 10 free stamens. Pod round, inflated, but not divided by a longitudinal partition.

Gompholobium latifolium

Broad-leaf Wedge-pea **Plate 469**

This is an erect shrub, sometimes 1 to 2 m tall, but usually smaller. Leaflets are in groups of 3, flat or with margins slightly curved under, 3 to 5 cm long and 3 to 5 mm wide. The flower is completely yellow and large; the standard is 25 to 30 mm long, smaller in the mountains. Flowers are borne on short stalks in the axils of the upper leaves. This species may be confused with *G. grandiflorum*, also an erect shrub with large yellow flowers, but *G. latifolium* can be distinguished by the keel of the flower which is conspicuously fringed with white hairs and by the broader leaflets.

HABITAT Open and sheltered forests, on sandy soils.

DISTRIBUTION Widespread from coast to mountains. Especially common north of Sydney (e.g. Brisbane Water National Park). Also occurring north coast, northern tablelands, south coast, Qld and Vic.

FLOWERING August–November.

Gompholobium huegelii **Plate 470**

This is a low diffuse shrub, seldom more than 50 to 60 cm tall, with 3 narrow leaflets, 8 to 20 mm long, with margins curved to the undersurface and an acute apex, often curved down, with a minute point. Flowers are yellow, with a standard about 15 mm long, and a keel which is fringed with conspicuous white hairs. The calyx is very dark and conspicuous in bud. Flowers are borne on stalks up to 2 cm long near the ends of the branches.

HABITAT Woodland and open forest, on sandy soils.

DISTRIBUTION Higher Blue Mountains. Recordings include Mt Victoria, Mt Wilson, Hartley, Clarence and Newnes State Forest. Extending throughout the tablelands and eastern edge of western slopes; also south coast, Vic and Tas.

FLOWERING September–November.

468. Mirbelia baueri

469. Gompholobium latifolium

470. Gompholobium huegelii

Gompholobium uncinatum **Plate 471**

A bushy to diffuse shrub, up to 30 cm tall, from the western parts of the Blue Mountains. It is the only *Gompholobium* in the area with orange-red flowers. It may be confused with one of the red-flowered *Oxylobium* species, but is distinguished from them by its distinctly 3-foliolate leaves (that is, 3 leaflets spread out from a common stalk). Leaflets are about 7 mm long, dark green and linear. Stems are covered with warty lumps when old. Flowers are about 8 mm long, orange-red with a yellow patch at the base, on stalks about 5 mm long.

HABITAT Heath and open forest, on sandy soils or among rocks.

DISTRIBUTION Higher Blue Mountains. Recordings include Clarence, Newnes State Forest and Mt Victoria. Also recorded from Marulan to Tallong. Extending to north coast, northern tablelands and central western slopes.

FLOWERING November–December.

471. Gompholobium uncinatum

Gompholobium minus

Dwarf Wedge-pea **Plate 472**

A diffuse shrub with many branches, sometimes up to 30 cm high, but usually more mat-like. Its stems are downy. Leaflets are in threes, linear, 6 to 9 mm long and less than 1 mm wide. The main identifying feature is the flower bud which is distinctly rigid, unlike other local members of the genus whose flower buds are smooth. Flowers are plain yellow and borne on short stalks at the ends of the branches. The flower does not have a conspicuously hairy keel as does *G. huegelii*. *G. minus* may at first be confused with *G. glabratum*, but *G. glabratum* does not have ridged buds and it has 5 leaflets.

HABITAT Heath and open forest, on sandy soils.

DISTRIBUTION Extending along coast and coastal highlands from Toronto-Tuggerah south to Sassafras-Braidwood area; also southern highlands (e.g.

472. *Gompholobium minus*

Mittagong, Wingello), Cumberland Plain (Castlereagh State Forest) and higher Blue Mountains (Colong-Mt Werong).

FLOWERING September–November, often later in higher areas.

Gompholobium grandiflorum Plate 473

Despite the specific name which implies that the flowers of this species are large, they are in fact a little smaller than those of *G. latifolium*, with which it is often confused. This is an erect shrub up to 1 m tall, with leaflets in threes. A pair of opposite leaflets often looks like a ring of narrow leaves around the stem. Leaflets are narrow-linear, 20 to 25 mm long, about 1 mm wide, with a sharp but non-pungent point and with margins curled under. Flowers are yellow, with a standard 15 to 25 mm long. The keel is not conspicuously hairy like that of *G. latifolium*.

HABITAT Open forest and heath, on sandy soils.

DISTRIBUTION Widespread from coast to mountains. Recordings include Brisbane Water, Heathcote and Royal National Parks, Bargo, Mt Banks, Mt Tomah and Ingar. Extending to south coast, east and west of Nowra.

FLOWERING August–November.

473. *Gompholobium grandiflorum*

474. *Gompholobium glabratum*

Gompholobium glabratum Plate 474

This is the only member of the genus in the area which has 4 to 7 leaflets (usually 5) arranged pinnately on a short central axis. It is a low shrub, seldom more than 30 cm tall with weak stems and is common in sandy heaths and shrubland. Leaflets are narrow-linear, 4 to 10 mm long, and about 1 mm wide. Stems are covered with small warts, giving them a rough appearance. Flowers are yellow, with a standard 8 to 10 mm long and are borne near the ends of the branches.

HABITAT Open forest and heath, on sandy soils.

DISTRIBUTION Widespread along coast and coastal highlands. Also occurring in Blue Mountains (e.g. Lawson, Leura, Woodford, Jenolan Caves). Extending along coast from Port Stephens to Conjola.

FLOWERING August–November.

Gompholobium pinnatum Plate 475

This is a small shrub with slender stems which are often weak and tangled through the undergrowth. It is easy to identify as it is the only *Gompholobium* in the area with 20 to 30 leaflets arranged pinnately on a long central axis. The leaf is up to 5 cm long and each leaflet is narrow-linear, 10 to 12 mm long and about 1 mm wide. Flowers are yellow with a light red centre.

HABITAT Open forest and heath, on sandy soils.

475. *Gompholobium pinnatum*

DISTRIBUTION A coastal species, but uncommon and generally absent from sandstone communities of national parks around Sydney. Recorded from deep sands at Agnes Banks and Castlereagh State Forest. Extending from Jervis Bay northwards to Qld.

FLOWERING August–December.

Viminaria

This genus is represented by only 1 species which occurs along the coast of both south-eastern and south-western Australia. The name 'viminaria' means 'twiggy', a reference to its leafless stems. Unlike *Jacksonia*, this genus has ovate and hairless pods and a calyx with short and equal lobes.

476. *Viminaria juncea*

Viminaria juncea

Native Broom, Golden Spray **Plate 476**

Native Broom is an erect shrub sometimes 2 to 3 m tall. Its branches are round, green and apparently leafless (leaves are reduced). Masses of yellow flowers are borne along the ends of pendulous branches, justifying its common name, Golden Spray. Flowers are yellow, often with a red centre.

HABITAT Damp places, such as drainage lines of wet heaths and swampy margins.

DISTRIBUTION Coast to lower Blue Mountains. Occurring all along coast from Qld to Vic; also SA and WA.

FLOWERING September–November.

Jacksonia

This genus is endemic in Australia, with most of the 46 species occurring in south-west Western Australia. Only 1 species is found in the local area. It is characterised by its leafless angular branches, oblong-elliptic hairy seed pod and a calyx which is deeply divided into 5 lobes.

477. Jacksonia scoparia

Jacksonia scoparia

Dogwood **Plate 477**

A tall leafless shrub up to 3 m high, with angular branches which are minutely downy, giving them a grey appearance. Branches are often drooping and bear a profusion of flowers, making it very showy and noticeable. Flowers are orange-yellow with red centres. The standard is orbicular; the calyx is 5-lobed and lightly covered with silky hairs. The fruit is an oblong pod, about 10 mm long, flattened and covered with long whitish hairs. Dogwood may be confused with the other leafless pea-flower, *Viminaria juncea*, but *Viminaria* has round green stems and grows exclusively in wet places.

HABITAT Woodland and open areas, mainly on old sediments; absent from shallow sandstone soils.

DISTRIBUTION Widespread from coast to mountains, but generally absent from coastal national parks. Recordings include Castlereagh, Moorebank and Lapstone Hill. Occurring throughout most of NSW, except southern tablelands and far western plains; also found in Qld.

FLOWERING October–November.

Sphaerolobium

This small genus is endemic in Australia. One species occurs in the local area. It is characterised by its leafless branches (sometimes with scale-like rudimentary leaves), globular pods and the shape of the calyx which has 5 lobes, the 2 upper lobes united and longer than the 3 lower lobes.

Sphaerolobium vimineum **Plate 478**

This is a small shrub, seldom more than 50 cm tall, with round, leafless, erect stems. These are usually unbranched and bear numerous flowers like a spike near the ends of the stems. Flowers in the Sydney area are usually all yellow with a dark green calyx, although some flowers have a hint of red-orange. Clusters of flowers may be 6 to 7 cm long and are conspicuous amongst the grass and sedges of their swampy habitat. The pod is small, 3 to 4 mm across and globular.

HABITAT Damp sandy heath and open swampy places.

DISTRIBUTION Widespread from coast to mountains. Recordings include Munmorah State Recreation Area, Brisbane Water National Park, Menai, Bowens Creek (Mt Irvine) and Ingar. Occurring all along coast and tablelands of NSW, and in all other States (except NT).

FLOWERING September–November.

478. Sphaerolobium vimineum

479. Hovea linearis

Hovea

Shrubs, usually with hairy stems. Flowers purple with 10 stamens joined into a sheath open on the upper side. Anthers of different lengths, with 5 long alternating with 5 short. Calyx densely hairy. Pod globular and inflated. It is often difficult to separate one species of *Hovea* from another and naming has been confused for many years. For positive identification, it is necessary to closely examine the nature of the stem hairs, and the position and shape of the floral bracts and bracteoles.

Hovea linearis

Narrow-leaf Hovea **Plate 479**

A small procumbent to erect shrub seldom reaching 1 m tall. Stems are sparse and more or less covered in dense, long, upward-pointing hairs. Leaves are narrow-linear, 5 to 7 cm long, about 3 mm wide, dark green above, lighter green below and with a conspicuous midvein. Unlike the other *Hovea* species, this plant does not have the brown or grey closely matted hairs on the leaf undersurface, but is hairless or only sparsely hairy. Mauve flowers are clustered in the leaf axils. The calyx is brown and densely hairy and the pod is globular, swollen and usually without hairs.

HABITAT Heath and open forest, on sandy soils.

DISTRIBUTION Widespread from coast to mountains. Found throughout NSW, except western plains; also Qld, Vic, Tas and SA.

FLOWERING July–August on coast, as late as October in mountains.

480. *Hovea longifolia*

481. *Hovea purpurea*

Hovea longifolia

Long-leaf Hovea, Rusty Pod **Plate 480**

A tall shrub, often 1.5 to 2 m in height. Stems are covered with short, dark to brownish, closely matted hairs, with longer grey hairs projecting above the short hairs or with the longer hairs loosely appressed and covering the short hairs. Leaves are narrow-linear to oblong, 2 to 8 cm long and 2 to 4 mm wide, glossy above and with close-matted curled hairs below; the margins are curled under. Flowers are mauve in clusters of 2 to 3; the flowers have short stalks covered with long brown and grey hairs. A small black hairy floral bract is located about 4 mm below the bracteoles at or near the base of the flower stalk. The placement of this bract helps separate this species from *H. purpurea* which has its floral bract less than 2 mm under the bracteole. Other differences between these 2 species are that the stem hairs of *H. purpurea* are longer (villous) and its flowers are almost stalkless.

HABITAT Shaded moist slopes and creek banks.

DISTRIBUTION Central coast (e.g. Douglas Park, Woronora River, Thirlmere Lakes), Lower Blue Mountains (e.g. Erskine Creek) and southern highlands (e.g. Bundanoon-Wingello). Extending northwards to Port Stephens.

FLOWERING August–September.

Hovea purpurea

Velvet Hovea **Plate 481**

This species has been the subject of misapplied names and some confusion as to its status. It now includes plants formerly described as *H. pannosa*, *H. longifolia* var. *pannosa* and *H. rosmarinifolia* var. *villosa*. It is an erect shrub, 1 to 3 m tall, characterised by stems which are covered predominantly by long, spreading weak hairs. Leaves are narrow-elliptic, 3 to 4 cm long, the upper surface with fine veins visible, the lower surface brownish, with long loose hairs and some curled ones; the margins are curled under. Flowers are mauve in small clusters. Flower stalks are very short and usually obscured by the spreading hairs. The calyx, like other parts, is covered with soft spreading hairs. A floral bract is located 1 to 2 mm below the bracteoles. This species is easily confused with *H. longifolia* (see that species for differences).

HABITAT Open and sheltered situations, on sandy, often rocky, slopes.

DISTRIBUTION Widespread from coast to mountains. Extending along coast from north Qld to Vic; also southern tablelands and central western slopes.

FLOWERING August–September.

Hovea lanceolata

Lance-leaf Hovea **Plate 482**

This is a shrub to 2 m tall characterised by stems which are covered in tightly curled grey hairs, giving a felt-like appearance, and leaves which are almost flat and lanceolate. These leaves are 3 to 8 cm long, 8 to 15 mm wide, dull green above (greyish when young), grey-brown and felted on the undersurface. Flowers are mauve and borne in a few-flowered cluster from the leaf axils or on a raceme up to 5 cm long. A floral bract is located close below the bracteoles. The broader leaves of this species clearly distinguish it from other *Hovea* species in the area.

HABITAT Open forest, especially on rocky outcrops.

DISTRIBUTION Uncommon in area, being confined to a few areas in the higher Blue Mountains and north-west Hornsby Plateau. Chiefly a plant of western slopes, where it is common; also occurring north coast, northern tablelands, north-western plains, Qld and SA.

FLOWERING July–August.

482. *Hovea lanceolata*

Platylobium

Of the 4 members of this genus occurring along Australia's east coast, only one is found in New South Wales. Platylobium means 'broad pod', a reference to the characteristic flat, oblong pods which open along the lower join. Flowers are large, with stamens all united into a tube. Leaves are opposite and strongly veined.

483. *Platylobium formosum* ssp. *formosum*

Platylobium formosum ssp. *formosum*

Handsome Flat-pea **Plate 483**

This is a common, slender, somewhat straggling shrub, usually about 1 m tall, sometimes larger. Leaves are broad, opposite, ovate to heart shaped (cordate), 3 to 4 cm long, about 2 cm wide and strongly veined. Flowers are yellow with a red centre and fairly large. The calyx is 8 to 10 mm long, standard about 18 mm long. Flowers are borne singly or 2 together, in the leaf axils. The ovary and fruit are softly hairy on the edges. A smaller-flowered variety known as subspecies *parviflorum* also occurs in the area, but is less common. It has a calyx 4 to 5 mm long and standard 5 to 8 mm long with narrow ovate leaves which are not cordate. They are longer but narrower than those of ssp. *formosum*. The ovary and fruit are hairless or almost so. It grows in sheltered situations north of Sydney (e.g. Hornsby, Cumberland State Forest, Pennant Hills Park) and in the Blue Mountains (Mt Solitary, Leura Falls).

HABITAT Sheltered forest.

DISTRIBUTION Widespread from coast to mountains. Also occurring north coast, south coast, southern tablelands, south-western slopes, Qld, Vic and Tas.

FLOWERING August–November.

Bossiaea

A genus of 42 species of shrubs, 12 in the Sydney district. Stems often flattened. Leaves small and

484. *Bossiaea lenticularis*

sometimes absent. Flowers yellow with red centres; stamens all united into a tube. Fruit a flattened pod.

Bossiaea lenticularis **Plate 484**

This species is readily recognised as it is the only local *Bossiaea* with opposite and round leaves. These are 3 to 7 mm across, stalkless and well-spaced on the branchlets. The whole plant is hairless, diffuse, less than 1 m tall and with numerous tough slender stems. Flowers are larger than the leaves and are borne on small stalks in the leaf axils.

HABITAT Open forest, on sandy soils or wet places.

DISTRIBUTION Restricted to the Sydney district where it is scattered from coast to mountains. Recordings include Mt Tomah, Kings Tableland, Clarence, Maroota, Thirlmere Lakes and Colo Heights.

FLOWERING August–October.

485. Bossiaea prostrata

Bossiaea prostrata **Plate 485**

This is a small procumbent undershrub with a woody rootstock and weak stems. Leaves are alternate, often rather distant, narrow ovate to circular, 3 to 15 mm long (usually about 10 mm), with stalks up to 5 mm long. The leaf base is symmetrical, rounded or slightly heart-shaped; the apex is acute or obtuse, with or without a small point. Stipules are 1 to 2 mm long, distinctly shorter than the leaf stalk. The yellow and red flowers are borne singly or 2 together on stalks up to 15 mm long from the leaf axils near the end of the stems. A pair of bracteoles is located on the flower stalk below (or rarely at) the middle, often at the base of the stalk; they are deciduous and sometimes apparently absent. This species may be confused with *B. neo-anglica* and *B. buxifolia*, but is more widespread and common in the area than either of those species (see description of those plants for differences).

HABITAT Heath and open forest.

DISTRIBUTION Widespread from coast to mountains. Recorded from the exposed coastal headlands of Bouddi National Park, open forests on shale at Menai, Munmorah State Recreation Area and Mittagong district. Specimens from western outskirts of Sydney (Marayong-Scheyville) approach the closely related *B. buxifolia* in general appearance. Also occurring north coast, south coast, southern tablelands, western slopes, Vic, Tas and SA.

FLOWERING August–September.

486. Bossiaea neo-anglica

Bossiaea neo-anglica **Plate 486**

This is a small, sprawling, often prostrate, shrub with softly hairy young stems. Leaves are alternate, closely spaced, ovate, 2 to 7 mm long, with very short stalks and distinctly darker above than below. The leaf base is often asymmetrical and heart shaped; the apex is obtuse (occasionally acute) with a recurved non-pungent point. Stipules are about 2 mm long, and exceed the leaf stalk. The yellow and red flowers are borne singly in the leaf axils near the ends of the branches on stalks 5 mm or less in length (shorter than those of *B. prostrata* or *B. buxifolia*). A pair of bracteoles is located on the flower stalk near the middle; they are persistent, not deciduous. This species closely resembles *B. buxifolia* but differs in the length of the flower stalk and in the position of the bracteoles on the flower stalk.

HABITAT Open forest, on sandy or shaley soils, often in sheltered sites.

DISTRIBUTION Essentially a species of northern and central tablelands. Restricted in the Sydney area to southern highlands (e.g. Thirlmere Lakes, Hilltop-Oakdale, Fitzroy Falls, West Berrima, Belanglo State Forest); also occurring north coast, north-western slopes and Qld.

FLOWERING September–October.

487. Bossiaea buxifolia

Bossiaea buxifolia

Matted Bossiaea **Plate 487**

This is a diffuse shrub, usually decumbent, with loosely matted stems reaching 50 to 60 cm in height. Stems are sprinkled with soft, short white hairs. Leaves are alternate, closely spaced, elliptic to ovate or almost rounded, 2 to 5 mm long (variable even on the same plant) with very short stalks and distinctly darker above than below. The undersurface of young leaves has scattered white hairs. The leaf base is often slightly asymmetrical and heart shaped or rounded; the apex is obtuse or acute with a minute recurved point. Stipules are 1 to 2 mm long, brown and usually exceed the leaf stalk. The yellow and red flowers are borne singly at the ends of short lateral branches on stalks often 10 mm long (much longer than the leaves). Bracteoles, often unevenly paired, are located on the upper half of the flower stalk; they are persistent, not deciduous. This species differs from *B. prostrata* in its larger stipules and the position of the bracteoles. It differs from *B. neo-anglica* in its longer flower stalk and the position of the bracteoles.

HABITAT Open forest, woodland and heath.

DISTRIBUTION Widespread, but rare along the coast (e.g. Sandy Point, Georges River). More common on tablelands (e.g. Lithgow, Kanimbla Valley, Jenolan Caves, Wingello). Also occurring north coast, south coast, northern tablelands, southern tablelands, western slopes and Vic.

FLOWERING September–October.

488. Bossiaea obcordata

Bossiaea obcordata

Spiny Bossiaea **Plate 488**

Spiny Bossiaea is a small, rigid, twiggy shrub, about 50 cm high. Its distinguishing characteristics are its alternate, heart-shaped leaves and spine-tipped branchlets. The leaves are small, 2 to 6 mm long, notched at the top or with a small down-curved point. Flowers are about 8 mm long, yellow with a deep red keel, borne in the leaf axils.

HABITAT Open forest and shrubland, on sandy soils.

DISTRIBUTION Widespread from coast to mountains. Recordings include Brisbane Water National Park, Mt Colah, Pennant Hills Park, Glenbrook, Thirlmere Lakes, Bundanoon and Belmore Falls. Occurring all along coast and tablelands of NSW; also north-western slopes and Vic.

FLOWERING August–October.

489. Bossiaea rhombifolia

Bossiaea rhombifolia **Plate 489**

This is a hairless shrub, about 1 m tall, with round or slightly flattened stems which are not spine-tipped. Branchlets are erect and arranged on opposite sides of the main stem alternately and on the same plane. Leaves are rhomboid-shaped, about as broad as long (5 to 8 mm) and point upwards to appear almost pressed against the stem. Flowers are yellow with red centres and are borne singly in the leaf axils on short stalks.

HABITAT Open forest, on sandy soils.

DISTRIBUTION Widespread from coast to mountains. Common in the Nepean River area and lower Blue Mountains (e.g. Agnes Banks, Bents Basin, Erskine Creek, Grose Vale). Also recorded at Colo Heights, Wilberforce, Glen Davis and Thirlmere Lakes. Extending along north coast, northern tablelands and north-western slopes to Qld.

FLOWERING August–October.

490. Bossiaea kiamensis

Bossiaea kiamensis **Plate 490**

This is a bushy medium-sized shrub, 1 to 2 m tall. Younger stems are somewhat flattened and covered with a soft white down. Leaves are opposite (an important identifying feature), ovate to elliptic, 10 to 20 mm long, shortly stalked, hairless above and silky-hairy underneath. The apex is acute or obtuse, with a short point. Flowers are solitary on stalks less than 5 mm long from the leaf axils.

HABITAT Heath and shrubland, especially on damp sheltered sites.

DISTRIBUTION Southern parts of Illawarra Escarpment and southern highlands. Recordings include Belmore Falls, Carrington Falls, Minnamurra Falls, Fitzroy Falls, Barren Grounds and gullies near Penrose. Extending to south coast and southern tablelands.

FLOWERING September–October.

491. Bossiaea heterophylla

Bossiaea heterophylla **Plate 491**

A variable shrub, usually less than 1 m tall. Stems are flattened and bear small alternate leaves arranged in opposite rows along the stem. Leaves are sometimes variable on the one plant. They may be obovate, lanceolate or narrow-linear, and 1 to 3 cm long. Flowers are relatively large, 7 to 15 mm long, yellow with a deep red keel, and borne singly on a short stalk from the leaf axils. They are a conspicuous part of the sandstone flora in autumn months.

HABITAT Open forest, woodland and old dunes, often on dry sandstone ridges.

DISTRIBUTION Widespread and common from coast to mountains. Found all along NSW coast, extending into south-eastern Qld.

FLOWERING April–June.

Bossiaea stephensonii **Plate 492**

The most striking feature of this small erect shrub is the presence of large (5 to 8 mm) leaf-like, triangular stipules at the base of the leaves. These stipules are green, turning grey-brown with age, erect and pressed against the stem. Stems are flattened and often covered with loose whitish hairs. The leaves are also sprinkled with long white hairs, especially when young. Leaves are alternate, oblong to lanceolate, 5 to 20 mm long, with a small soft point and are arranged in opposite rows along the stems. Flowers are yellow with red centres and are borne singly along the stems.

HABITAT Open forest, on sandy soils.

DISTRIBUTION Restricted to a narrow zone north and south of Sydney close to the coast. Common in Royal National Park between Loftus and Heathcote and north of Hawkesbury River around Patonga, Wondabyne, Doyalson and Tuggerah. Also recorded on lower north coast.

FLOWERING August–September.

492. Bossiaea stephensonii

493. Bossiaea ensata

Bossiaea ensata **Plate 493**

There are 2 local species of *Bossiaea* which have flattened, leafless stems. Leaves are actually reduced to minute scales. Sometimes a few fully developed leaves can be seen on juvenile growth. The main difference between the 2 species is their size and habit. *B. ensata* is a straggling shrub with weak stems, 30 to 50 cm long and 7 to 8 mm wide. Its flowers are solitary along the stems, 7 to 10 mm wide, on stalks 2 to 5 mm long. The pod is also on a small but distinct stalk. This species is sometimes confused with *Daviesia alata* which is also a leafless straggling plant. The most obvious difference is that the flowers of *D. alata* are in clusters, not singly as in *B. ensata*.

HABITAT Sandy heath and shrubland, often in deep sands.

DISTRIBUTION Widespread and common on coastal sandstone areas; less common in Blue Mountains, but recorded from Leura and Kings Tableland. Also occurring south coast, north coast, northern tablelands and Qld.

FLOWERING August–November.

Bossiaea scolopendria **Plate 494**

The leafless stems of this species are strong and erect, unlike those of *B. ensata*. They may grow to 1 m tall and are up to 10 mm wide. Flowers are 10 to 15 mm long, solitary along the stems, with stalks 1 to 3 mm long. The pod is almost without a stalk, in contrast to the distinctly stalked pod of *B. ensata*.

HABITAT Open forest, woodland and heath, on sandy soils.

DISTRIBUTION Restricted to the coastal area from Jervis Bay in the south to the Gosford district. Most common north of Sydney. Recordings include Somersby, Muogamarra Sanctuary, West Head, Terrey Hills and La Perouse.

FLOWERING August–October.

494. Bossiaea scolopendria

Goodia

Two species of *Goodia* occur in New South Wales. They are both tall shrubs with leaves of three rounded leaflets, and bear numerous bright yellow flowers. The stamens are joined to form a tube which is open on the upper side. One species occurs in the Sydney district.

Goodia lotifolia

Golden-tip **Plate 495**

A striking shrub, 2 to 3 m tall. It is not often seen as it has a scattered distribution and grows on sheltered slopes and in forested gullies. Leaves are on a long stalk and are divided into 3 leaflets, which are entire, obovate and 15 to 18 mm long. Leaves and stems are hairless but often have a blue-grey tinge when young. Flowers are yellow with a red-brown centre, about 10 mm long, and are borne in loose racemes which may be 10 cm long.

HABITAT Moist forested gullies and sheltered slopes.

DISTRIBUTION Widespread along central coast, mainly south of Sydney. Also southern Blue Mountains and southern highlands. Recordings include Douglas Park, Thirlmere Lakes, Fitzroy Falls, Tallong, Bundanoon and Kanangra-Jenolan Caves. Found throughout coast and tablelands divisions and eastern parts of western slopes; also Qld, Vic and Tas.

FLOWERING September–October.

495. Goodia lotifolia

496. Indigofera australis

Indigofera

This is a large genus, with about 700 species worldwide. About 30 occur in Australia, mainly in the north-east; only 1 species is found in the local area. The genus is characterised by its pink-purple flowers, with 9 stamens united into a tube and the upper 1 free, and anthers tipped with a small gland. The pod is cylindrical, with seeds separated by cellular partitions.

Indigofera australis

Native Indigo **Plate 496**

A slender erect shrub, usually 1 m or more tall, with leaves about 7 cm long, consisting of numerous leaflets. These leaflets are oblong to obovate, about 15 mm long, hairless on the upper surface, but sometimes downy underneath. Flowers are pink to lilac, numerous and borne in a raceme from the leaf axil. This raceme is about as long as the leaves. Fruit is a narrow cylindrical pod, 25 to 40 mm long.

HABITAT Sheltered forest, usually on shales or rocky slopes.

DISTRIBUTION Widespread from coast to mountains. Recordings include Royal National Park, Brisbane Water National Park, Wollemi, Agnes Banks, Erskine Creek, Thirlmere Lakes and Wombeyan Caves road. Occurring throughout NSW and in every other State (except NT).

FLOWERING August–September.

497. Swainsona galegifolia

Swainsona Darling Pea

About 28 species of *Swainsona* occur in New South Wales. Most are plants of the drier western slopes and plains. Only 1 could be regarded as a plant of the Sydney district, although *S. monticola* and *S. oroboides* have been recorded on the fringes of the area. The genus is characterised by its mauve to purple flowers, with 9 stamens united and 1 free, anthers without a small gland, and its pinnate leaves.

Swainsona galegifolia

Smooth Darling Pea **Plate 497**

This is a hairless shrub, with trailing or erect stems to about 1 m tall. Leaves are 5 to 8 cm long, divided into 8 to 11 pairs of leaflets. Each leaflet is obovate, 6 to 25 mm long and 2 to 8 mm wide. Flowers are pink to mauve and borne on short stalks along a raceme which is about 20 cm long. Fruit is a hairless, lime-green, leathery, swollen pod, 25 to 40 mm long, with a beak up to 15 mm long.

HABITAT Sheltered forested slopes and river banks.

DISTRIBUTION Higher Blue Mountains (e.g. Jenolan Caves). Widespread throughout tablelands, western slopes and western plains; also north coast, south coast and Qld.

FLOWERING October–December.

Canavalia

Climber occurring along coastlines. Leaves divided into leaflets. Flowers pink-mauve and pod large.

Canavalia maritima

Canavalia **Plate 498**

This trailing plant is more typical of tropical and subtropical coastlines of northern Australia and Asia than of the Sydney coast. It is rare in this area as development has overtaken its habitat and it is near the southern limit of its distribution. Its stems are 2 to 3 m long and trail over the beach sand or weathered coastal cliffs. Leaves are divided into 3 leaflets, each orbicular to obovate and 6 to 9 cm long; the leaf stalks are distinctly thickened at the base of the blade. Flowers are pale pink to mauve and

498. Canavalia maritima

are borne in racemes on long stalks from the leaf axils. The pod is a large bean, about 5 cm long, with a conspicuous rib on the upper margin.

HABITAT Sea coast, including sandy beaches and nearby shales of cliff-lines.

DISTRIBUTION Rare; restricted in the Sydney area to Long Reef where it occurs near the base of chocolate shale cliffs with a northern aspect. Extending northwards along coast to Qld, NT and WA. Also Pacific Islands (e.g. Lord Howe Island), Africa and Asia.

FLOWERING February–May.

Kennedia

Twining and prostrate plants with leaves divided in 3. Flowers red (elsewhere blue and black-yellow), relatively large, with 9 united stamens and 1 free.

Kennedia rubicunda

Dusky Coral-pea **Plate 499**

A robust climber with stems which twine over and around small shrubs and trees. Its large attractive red flowers, about 4 cm long, hang freely in a showy display. The calyx is silky brown. Leaves consist of 3 leaflets which are very variable in size and shape; they may be ovate to orbicular and 1 to 10 cm long. Pods are flat, 5 to 10 cm long and covered with brown hairs.

HABITAT Found in most plant communities; especially common in protected sandy shrubland and woodland near the coast.

DISTRIBUTION Widespread along coastal zone. Extending to Blue Mountains. Found all along the coast from Qld to Vic.

FLOWERING July–November, but some flowers at other times.

499. Kennedia rubicunda

500. Kennedia prostrata

Kennedia prostrata

Running Postman **Plate 500**

A prostrate plant, with downy stems which trail over sandy soils, or sometimes form small mats. Leaves consist of 3 leaflets, obovate to orbicular, 15 to 20 mm long; the leaflets usually have conspicuously wavy margins. Flowers are scarlet with a yellow centre, about 18 mm long. They are borne singly or in pairs from the leaf axils. The pod is cylindrical and 4 to 5 cm long.

HABITAT Heath, woodland and old dunes.

DISTRIBUTION Widespread from coast to mountains, but most common near coast. Also occurring north coast, south coast, southern tablelands, south-western slopes, Vic, Tas and SA.

FLOWERING July–October.

Hardenbergia

Twining plant with violet flowers, large simple alternate leaves and flat seed pod, about 4 cm long.

Hardenbergia violacea

Purple Twining-pea **Plate 501**

This is a common twiner found widely throughout the area. It may be a wiry trailing plant with long stems which twist through the undergrowth, over rocks and roadside banks, or it may scramble over shrubs and ascend small trees. A bushy erect form occurs on the clay soils west of Liverpool and around Campbelltown and Parramatta. Leaves are alternate, simple (not divided into 3 leaflets like the other twining pea-flowers), firm, broadly ovate, strongly veined and up to 9 cm long. Flowers are violet, with a small yellow centre, and are borne in a raceme. These racemes are sometimes very numerous and conspicuous.

HABITAT Forest and woodland, in sandy or clayey soils, or rocky situations.

DISTRIBUTION Widespread from coast to mountains. Found throughout NSW; also Qld, Vic and SA.

FLOWERING July–October.

501. Hardenbergia violacea

Glycine

Weak twining plants with pale blue-mauve flowers. Leaves divided into 3 leaflets. In addition to the 2 species mentioned below, 3 other species have been recorded for the area in recent years; as yet, they are unnamed.

Glycine clandestina

Twining Glycine **Plate 502**

This is a slender trailing plant with stems covered with small white backward-pointing hairs (sometimes almost hairless). It is usually found climbing over fallen branches, around grass stems or on low bushes. Three varieties have been recognised; all occur in the Sydney district. Leaves consist of 3 shortly-stalked leaflets which vary greatly in their size and shape. They may be narrow linear to broad lanceolate, 1 to 4 cm long and 4 to 10 mm wide. All 3 leaflets are on short and equal stalks or the terminal leaflet is stalkless. This characteristic separates it from *G. tabacina* which has the terminal leaflet on a longer stalk than the other 2 leaflets. Flowers are pale pink to mauve and borne on slender stalks in loose racemes in the upper leaf axils.

HABITAT Grassland, heath, woodland and forest; especially common on shaded hillsides and in coastal forests.

DISTRIBUTION Widespread from coast to mountains. Occurring throughout NSW, except for drier western plains, and in all states, except WA.

FLOWERING Some flowers all year round.

502. Glycine clandestina

503. Glycine tabacina

Glycine tabacina **Plate 503**

This slender trailing plant has similar features to *G. clandestina*, except that the terminal leaflet is on a longer stalk than the other 2 leaflets. Flowers are often darker in colour and the racemes are longer, much exceeding the leaves. This species is less widespread than *G. clandestina*.

HABITAT Open forest and partly cleared areas, preferring clay soils.

DISTRIBUTION Widespread from coast to mountains. Common on heavier soils (e.g. Campbelltown, Menangle, Blacktown, Casula), but generally absent from Hawkesbury sandstones. Found throughout NSW, except far western plains; also Qld, Vic and the Pacific Islands.

FLOWERING September–November.

Lotus

Herbs, with leaves divided into 5 leaflets. Flowers in umbels with 9 stamens united and 1 free and the filaments alternately long and short. At least 4 species of European yellow-flowered *Lotus* are naturalised in the area, but there is only 1 native species.

504. Lotus australis

Lotus australis

Australian Trefoil **Plate 504**

An erect herb to 60 cm tall. Leaves consist of 5 narrow leaflets borne on a flattened stalk; 3 of the leaflets are near the end of the stalk and are 1 to 3 cm long; the other 2 near the base are shorter and resemble stipules. Flowers are white to pink, about 15 mm long, and borne in umbels of 3 to 8 flowers; the umbel stalk is longer than the leaves. Small leafy bracts are at the base of the umbel. Fruit is a cylindrical pod, 2 to 5 cm long and about 3 mm wide, straight at first but spirally twisted at maturity.

HABITAT Grassland and open woodland, especially on clay soils.

DISTRIBUTION Widespread from coast to mountains, but uncommon. Recordings in the Sydney district are chiefly from the Cumberland Plain (e.g. Blacktown, Camden, Campbelltown). Found throughout tablelands and western slopes divisions, north coast, south-western plains and in all other States.

FLOWERING October–November.

Desmodium Tick-trefoil

Trailing shrubs with 3-foliolate leaves. Flowers with 9 stamens united and 1 free. Fruit a lomentum (a bead-like pod with distinct gaps between the seeds, breaking into 1-seeded segments when mature).

505. Desmodium varians

Desmodium varians

Slender Tick-trefoil **Plate 505**

The variable leaf shape of this small trailing plant is reflected in its specific name. Leaves are divided into 3 leaflets. These leaflets are from 6 to 25 mm long and may be rounded or narrow-oblong. Stems are slender, up to 50 cm long, hairless or slightly downy. Flowers are small (4 to 5 mm long), pink, pea-like and borne in small clusters on slender stalks at the ends of the branches. The fruit is characteristic of the genus and distinctive. It is a pod, 12 to 25 mm long, with the lower margin scalloped to form 5 or 6 separate segments.

HABITAT Grassland, woodland and open forest.

DISTRIBUTION Widespread from coast to mountains. Common, but often overlooked because of its trailing habit. Occurring throughout NSW, except in drier western plains; also Qld, Vic and Tas.

FLOWERING December–February, but a few flowers at other times.

Desmodium rhytidophyllum

Rusty Tick-trefoil **Plate 506**

This prostrate trailing plant is much more robust and coarser than *D. varians*. Its leaflets are larger (2 to

506. Desmodium rhytidophyllum

6 cm long), ovate to elliptic, with the terminal leaflet longer than the other two. The upper surface is prominently veined. Stems are rusty-hairy all over. Flowers are mauve and borne in ascending racemes at the ends of the branches. The lower edge of the pod is scalloped to form separate segments and the pod is on a stalk at least as long as the calyx (about 5 mm).

Large Tick-trefoil (*D. brachypodum*) also occurs along the coastal zone but it is less common. It differs from *D. rhytidophyllum* in having a more erect habit, hairless stems (or almost so) and leaflets with blunt tips.

HABITAT Open forest, in sandy soils.

DISTRIBUTION Coastal zone. Extending from the south coast northwards to Qld.

FLOWERING December–February.

Zornia

Small trailing herb with 2-foliolate leaves. Flowers with 9 stamens united and 1 free. Seed pod a lomentum.

Zornia dyctiocarpa

Zornia **Plate 507**

A decumbent (sometimes ascending) plant immediately distinguished by its pair of ovate to linear leaflets which are borne on long stalks. Each leaflet is up to 25 mm long. Flowers are yellow and red, small, 6 to 8 mm long, and almost enclosed in a pair of leafy bracts. Several flowers are borne near the end of thin stems. The fruit is unusual. It is about 15 mm long and divided into 3 to 6 hemispherical lobes which are covered with hooked bristle.

HABITAT Open forest, woodland and grassland, especially on heavier soils.

DISTRIBUTION Widespread from coast to mountains (e.g. Doonside, Shane Park). Also occurring south coast, southern tablelands, north coast, northern tablelands, western slopes, Qld and NT.

FLOWERING October–December.

507. Zornia dyctiocarpa

Proteaceae

The history of the Proteaceae is intimately associated with the breakup of the Mesozoic southern continent Gondwanaland and subsequent separate development of the flora of each of the southern continents. Its distribution today is still mainly southern although it extends into the northern hemisphere in equatorial America, Africa, Malesia and from India through China to Japan. Major development of the family has occurred in South Africa and eastern and south-western Australia. Relationships, however, are much closer between the Australian and temperate South American members (4 out of 8 South American genera shared with Australia) than of either with the African (no genus in common). The Asian extension is due to the large genus *Helicia* which also occurs in New Guinea and eastern Australia.

Fossil pollen records show that major groups were widespread and common in Australia, New Zealand and Antarctica, possibly as early as 75 million years ago, but certainly by the early Tertiary period. These groups appear to have greater affinity with present-day tropical and subtropical rainforest tribes than with common sclerophyllous types like *Banksia, Hakea* or *Grevillea.* To explain this it has been suggested that the sclerophyllous Proteaceae were also present at this time, occupying drier lower-nutrient sites, but remained a minor element through most of Tertiary times until the sudden decline in conditions at the end of the Miocene era 5 to 10 million years ago. It was this decline that eliminated all but 2 species of the New Zealand Proteaceae. On the other hand, New Caledonia with a different geological history remains remarkably rich in Proteaceae. Of about 40 present-day Australian genera (75 in the world) a dozen are typically found in rainforests. Only 2, *Hakea* and *Grevillea,* occur throughout the drier parts of the interior.

The Proteaceae has traditionally been divided into the sub-families Proteoideae and Grevilleoideae, but in a recent revision the number was increased to 5 to accommodate a small number of genera that do not fit quite so well with the others. This classification divides the genera present in the Sydney area in the following way. (Genera listed together are in the same tribe.)

Subfamily	Genus
Persoonioideae	*Persoonia*
Proteoideae	*Symphionema, Isopogon,* *Petrophile, Conospermum*
Grevilleoideae	*Telopea, Lomatia, Stenocarpus* *Grevillea, Hakea, Helicia,* *Xylomelum, Lambertia,* *Banksia*

In the first 2 sub-families the flowers occur singly in the inflorescence and the fruit is indehiscent with 1 seed. In Persoonioideae the fruit is a fleshy drupe with a large woody seed; in Proteoideae it is a nut with a small wingless seed. In the Grevilleoideae, which shows more primitive features, the flowers are in pairs in the inflorescence and the fruit is a follicle which splits in 2 at maturity to release 2 to many large, flat, winged seeds.

Proteaceae flowers are normally bisexual with 4 perianth lobes united to various degrees and reflexed at the tips. The anthers are frequently attached directly to the top of the lobes, and the long bent style which protrudes through a slit in the perianth is held captive in the bud stage. At the time of flowering the style is released (Figure 10) and carries pollen from the anthers for transfer to pollinators such as birds, insects and small marsupials attracted by copious nectar secreted by glands present in some genera. When it is first released the stigma is unreceptive to pollen, but later after the flower has given up its pollen the stigma matures for fertilisation to occur.

Relatively few Proteaceae flowers set seed. A *Banksia* flower spike, for example, may have more than a thousand individual flowers but not more than 20 or 30, sometimes none, develop into woody follicles set into the rhachis of the spent spike. Genera having thin-walled follicles release their seed annually but in those with thick woody follicles the valves usually remain unopened until the branch dies or the heat of a passing fire causes them to dry out and shrink. Seeds released by fire benefit not only from nutrients in the ash bed but also by a reduction in competition from established plants and the destruction of pathogenic micro-organisms and their byproducts in the top layers of the soil.

Seed dispersal by wind is aided in subfamily Grevilleoideae by the wing on the seed and in some Proteoideae by hairs surmounting the seed. Seeds of *Persoonia* are slow and difficult to germinate and may require the destruction of inhibitors with time or by passage through the gut of birds which act as disseminators.

Figure 10

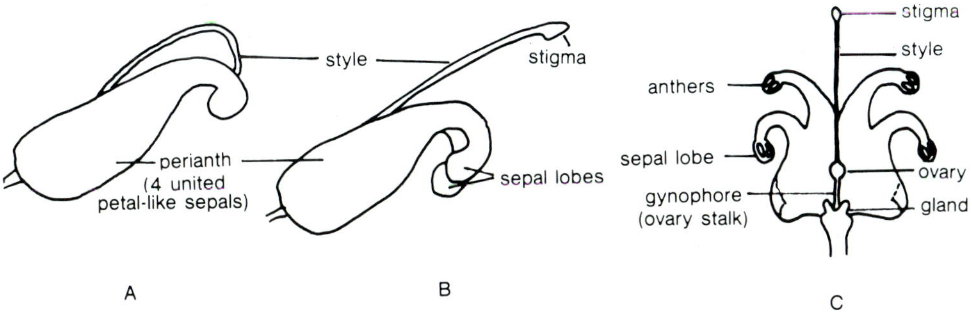

A typical Proteaceae flower

Opened *Grevillea* flower

508. Persoonia laurina

Persoonia Geebung

Geebungs are shrubs with entire and mostly alternate leaves. The genus is easily identified by its distinctive flowers and fruit. The flowers are yellow, 10 to 15 mm long, cylindrical in bud and when open, have 4 spreading segments. The fruit is an oval, fleshy berry-like drupe, usually green, but sometimes turning purple or red. The old style remains attached to the fruit. Fruits of some species of Geebung were once eaten by the Aborigines. They have a rather insipid and somewhat astringent taste.

Species are separated on a combination of factors. It is often necessary to examine the ovary to see if it is hairy or hairless. Strip away the yellow petals and perianth segments and look at the immature fruit at the base of the style. Although the ovary is small, it is quite obvious if it is villous. The habit of the plant, the nature of the leaf, the hairiness of the perianth and whether the flower is erect, or nodding, all help in identification.

Persoonia laurina

Laurel Geebung **Plate 508**

A small erect shrub, easily recognised when in bud or flower as its flower tube is covered with dense rusty hairs. The flowers are borne in stalked clusters in the leaf axils. When not in flower, this species is still conspicuous because of its broad, opposite leaves which are ovate to lanceolate, light green and up to 10 cm long. New growth is covered with short dense rusty hairs. The ovary is densely villous. Fruit is relatively large and turns purple with age. This is a variable species and is divided into 3 subspecies, 2 of these occurring along the coast; all 3 occur in the tablelands.

HABITAT Forest and heath.

DISTRIBUTION Widespread from coast to mountains. Also recorded from north coast and southern tablelands.

FLOWERING October–December, sometimes later in mountains.

509. Persoonia hirsuta

Persoonia hirsuta

Hairy Geebung **Plate 509**

The stems and flowers of this spreading shrub are covered with long coarse hairs and the leaves have short stiff hairs. This hairiness is the most characteristic feature of the species. Leaves are small, 6 to 12 mm long, oblong to narrow and crowded along the stems. Leaf margins are curled under. Flowers are borne along the branchlets and exceed the leaves. The ovary and the fruit are both distinctly hairy. Two subspecies have been described from the coastal area; one of these extends to the Blue Mountains.

HABITAT Open forest and heath, in sandy soils.

DISTRIBUTION Restricted to the Sydney district. Widespread but not common. Recordings include Royal National Park, Jannali Reserve and Glenbrook Native Gardens.

FLOWERING November–January.

510. Persoonia chamaepitys

Persoonia chamaepitys

Prostrate Geebung, Mountain Geebung **Plate 510**

Clumps of this prostrate trailing shrub are a common sight in the upper Blue Mountains where it forms mats of light green foliage which may be 1 to 2 m across. Leaves are hairless, soft, needle-like, pointed but not pungent, 10 to 15 mm long and crowded along the stems. Flowers are in small clusters, terminal or from the leaf axils near the ends of the branches. The ovary is villous.

HABITAT Open forest and heath, in sandy soils.

DISTRIBUTION Blue Mountains and higher parts of Woronora Plateau. Common from Katoomba to Mt Victoria and Clarence. Extending to central western slopes.

FLOWERING December–January.

Persoonia myrtilloides

Myrtle Geebung **Plate 511**

Myrtle Geebung is a spreading shrub up to 2 m tall, regarded as vulnerable because of its small scattered populations. Leaves are flat, lanceolate, 2 to 4 cm long, with a small point. A midvein is prominent. Young leaves and stems may be covered with downy grey hairs, but adult leaves are hairless. The flowers are nodding, on short, softly-hairy stalks; the ovary is hairless. Each of the 4 petals (perianth segments) has a small 2 mm terminal appendage. Two subspecies are recognised.

HABITAT Dry forest and scrub.

DISTRIBUTION Upper Blue Mountains (e.g. Blackheath, Newnes State Forest). Extending to central western slopes.

FLOWERING October–January.

511. Persoonia myrtilloides

Persoonia nutans

Nodding Geebung **Plate 512**

P. nutans is an uncommon spreading shrub, usually about 1 m tall. Its leaves are narrow, 15 to 30 mm long, 1 to 2 mm wide. Its most conspicuous feature is its pendulous flowers and fruit, which hang on fine stalks about 10 mm long. Each segment of the flower is tipped with a small appendage, about 2 mm long. The flower and the ovary are hairless. A number of subspecies are recognised.

HABITAT Dry forest and scrub, in sandy or shale soils.

DISTRIBUTION Cumberland Plain (e.g. Castlereagh State Forest and Agnes Banks Nature Reserve), and lower Blue Mountains (e.g. Glenbrook). Other subspecies occur along tablelands and in Qld.

FLOWERING December–March.

512. Persoonia nutans

513. Persoonia oxycoccoides

Persoonia oxycoccoides **Plate 513**

This geebung is very variable in its habit and leaf shape. Despite obvious differences in general appearance, it is regarded by some botanists as a subspecies of *P. nutans*. In its typical form it is a low leafy shrub with a few ascending branches. New stems are often a reddish colour and may be downy or hairless. Leaves are small, 8 to 20 mm long and rigid; they are variable in shape and may be round or oblong to lanceolate. Flowers are nodding on short stalks.

HABITAT Open forest and sandy heath, often along roadsides and on margins of wet places.

DISTRIBUTION Upper Blue Mountains (e.g. Newnes State Forest) and southern highlands (e.g. Carrington Falls). Extending to northern tablelands.

FLOWERING December–January.

Persoonia oblongata **Plate 514**

This is an attractive, light-green, spreading shrub with a limited distribution. It is easily identified by its ovate to lanceolate leaves, which are 2 to 5 cm long, pointed but not pungent. Flowers and fruit are solitary and pendulous on slender stalks, 1 to 2 cm long. The flower segments lack the small terminal appendages of *P. nutans*. Flowers, ovary and fruit are all quite hairless.

HABITAT Dry open forest.

DISTRIBUTION Restricted to lower Blue Mountains (e.g. Glenbrook Lagoon and Kurrajong-Wheeny Creek road).

FLOWERING March–June.

514. Persoonia oblongata

515. *Persoonia lanceolata*

517. *Persoonia linearis*

516. *Persoonia levis*

Persoonia lanceolata

Lance-leaf Geebung **Plate 515**

An erect, tall shrub (often compact and bushy on coastal dunes). Its hard persistent bark, erect flowers and fruits, and lanceolate leaves all help distinguish it from other *Persoonia.* Leaves are 3 to 7 cm long, 1 to 2 cm wide, with a distinct tip; they are without veins, except for the midvein. Young branches and flower stalks are often covered with soft white hairs. Flowers are borne in the leaf axils on very short stalks. The fruit is almost stalkless and erect. The ovary and fruit are hairless. A form with blue-green leaves occurs in the southern Blue Mountains.

HABITAT Open forest and heath, on sandy soils; common on fixed coastal dunes.

DISTRIBUTION Widespread from coast to mountains. Common on dunes at Kurnell and Bundeena-Marley. Extending to north coast.

FLOWERING Most of the year.

Persoonia levis

Broad-leaf Geebung **Plate 516**

This medium to tall shrub is very common in the area. Its large, bright-green leaves usually make a striking contrast to the duller foliage of the surrounds. Also conspicuous is its loose, flaky bark which is a deep red under its duller outer surface. Leaves are very variable. They range from 8 to 20 cm long and 1 to 4 cm wide; they are usually curved (falcate), but may be broadly lanceolate or obovate. Lateral veins are clear and almost parallel to the midrib. Branchlets are often bright red, contrasting with the greenness of the leaves. Flowers are erect, solitary or a few together on short stalks.

HABITAT Dry open forest and heath, on sandy soils.

DISTRIBUTION Widespread from coast to mountains. Also occurs along north and south coasts, on southern tablelands and in Victoria.

FLOWERING Mainly September–December, but some flowers at other times.

Persoonia linearis

Narrow-leaf Geebung **Plate 517**

This attractive shrub can grow quite tall (to 5 m), but is usually only 2 to 3 m tall. Its leaves are long, (2 to 8 cm), flat and narrow (less than 3 mm wide). The bark is rough and somewhat flaky. Branchlets and the outside of the flower are covered with downy hairs. Flowers are borne near the ends of the branchlets.

HABITAT Forests.

DISTRIBUTION Widespread and common from coast to mountains. Extending throughout coast and tablelands divisions and central western slopes; also Vic and Qld.

FLOWERING December–July.

518. *Persoonia chamaepeuce*

Persoonia chamaepeuce

Dwarf Geebung **Plate 518**

This is a prostrate or decumbent shrub with stems up to 25 cm long. Stems are hairless (sometimes downy when young) and reddish. Leaves are flat, about 15 mm long, 1 to 2 mm wide, fairly thick, pointed and spreading all around the stem. Flowers and fruit are erect and borne on stout stalks from the leaf axils. The ovary is hairless.

HABITAT Rocky areas.

DISTRIBUTION Higher Blue Mountains. A tablelands species, extending from northern tablelands to southern tablelands and Vic.

FLOWERING August–October.

519. Persoonia mollis (forest form)

Persoonia mollis

Soft Geebung **Plate 519**

Soft Geebung is a very variable species with 3 subspecies in the area. It may be a tall shrub to 2 m with lanceolate hairy leaves up to 7 cm long in sheltered situations, a shrub with linear hairless leaves, strongly recurved (formerly called *P. caleyi*), or a low undershrub, with leaves less than 2 cm long, in exposed coastal scrubland. There is even a prostrate form along the Wombeyan Caves road. Despite this diversity, most plants have some or all of the typical features. Young shoots and the flowers are softly hairy and the edges of the leaves are curled under. Leaves are 2 to 7 cm long, paler underneath, either silky or rough above, and either blunt or acute at the ends. The hairs of the young shoots and leaves are often a rusty-brown colour.

HABITAT Varied, including sheltered slopes near rainforests, forests and exposed coastal scrub.

DISTRIBUTION Coast south of Sydney and Blue Mountains. Extending to south coast and southern tablelands.

FLOWERING June–August, with some flowers at other times.

Persoonia pinifolia

Pine-leaf Geebung **Plate 520**

A large attractive shrub, with spreading branches. Its most conspicuous features are its crowded pine-like leaves and the clusters of red-green fruits which hang from the tips of the branches like bunches of grapes. Leaves are 3 to 5 cm long and grooved on the upper surface. Flowers are in dense racemes up to 15 cm long at the ends of the branches; shorter leaves are between the flowers. Clusters of fruit are so heavy that they weigh down the branchlets.

HABITAT Forest, on sandy soils.

DISTRIBUTION Restricted to coastal zone around Sydney.

FLOWERING March–May.

Persoonia acerosa **Plate 521**

A small spreading shrub with crowded, narrow, upward-pointing leaves. These are 10 to 15 mm long, grooved on the upper surface and curved inwards towards the tip. This species is easy to identify when in flower as its flowers are crowded along the upper branchlets, not at the tips like those of *P. pinifolia*. The leaves between the flowers are the same as the rest of the foliage (unlike *P. pinifolia*).

A similar species, yet unnamed, (Plate 522), is found on the Hornsby Plateau (e.g. Ourimbah State Forest, Mt White). It differs from *P. acerosa* in having leaves 15 to 25 mm long, grooved on the undersurface, with a tip curved downwards. A small shrub found in dry eucalypt forests in the upper Blue Mountains (e.g. Newnes State Forest, near Lithgow), formerly known as *P. angulata*, is now believed to be a hybrid of *P. acerosa* and *P. levis*. Its main identifying feature is its narrow concave leaves; they are almost 4 cm long, 3 to 4 mm wide, incurved at the tip, pointed and rather erect. *P. acerosa* has similar leaves, but that species has dense clumps of flowers quite unlike the sparse flowers of this hybrid.

HABITAT Dry open forest, woodland and heath, on sandy soils.

DISTRIBUTION (*P. acerosa*) Upper Blue Mountains (e.g. Kings Tableland).

FLOWERING March–May.

520. Persoonia pinifolia

Symphionema

Small shrubs often found in damp sandy soils. Leaves divided into narrow segments. Flowers small, yellow and in narrow spikes. Fruit an oblong nut about 2 mm long.

522. Persoonia sp. (Hornsby Plateau)

521. Persoonia acerosa

Symphionema montanum

Mountain Symphionema **Plate 523**

A small shrub, seldom more than 60 cm tall, with a spreading bushy appearance. Its leaves are 2 to 4 cm long, divided into narrow flat segments. The flat leaf-segments separate this species from *S. paludosum* which has terete leaf-segments. Flowers are small, with 4 pale yellow to white spreading lobes and bright yellow anthers; they are borne in spikes above the leaves.

HABITAT In and near wet heath.

DISTRIBUTION Upper Blue Mountains, Woronora Plateau and southern tablelands (e.g. Woodford, Kings Tableland, Narrow Neck).

FLOWERING September–November.

523. Symphionema montanum

524. Symphionema paludosum

Symphionema paludosum

Swamp Symphionema **Plate 524**

Mainly a coastal species, it is similar to Mountain Symphionema, but is usually more sprawling and less bushy. The most obvious difference is that the leaf-segments are almost terete (cylindrical) rather than flat. Although the tips are acute, they are not pungent-pointed. Flowers are 1 to 2 cm long, pale yellow to white, and borne in narrow spikes.

HABITAT In and near wet heath.

DISTRIBUTION Mainly coastal, but extending to Blue Mountains. Also occurring on north and south coasts.

FLOWERING August–October.

525. Isopogon anethifolius

Isopogon Drumsticks

Shrubs, with flowers crowded into dense globular heads. The common name is derived from the distinctive rounded woody cones, which, in NSW, distinguish this genus from *Petrophile* which has oblong-ovoid cones. The scale-like bracts at the base of the flowers are deciduous, whereas those of *Petrophile* are persistent on the cone and open to release the fruits. The botanical name is derived from 'isos', 'pogon' (equal, beard) a reference to the hairs on the fruit.

Isopogon anethifolius

Narrow-leaf Drumsticks **Plate 525**

An erect shrub, 1 to 2 m tall. It is easily identified as it is the only *Isopogon* whose leaves are divided into narrow terete (needle-like) segments. Leaf-segments point upwards. In winter, new growth is a striking tan-red colour. The stems of the plant are also often reddish. Flowers are yellow, in dense, terminal, stalkless, globular cones.

HABITAT Shallow sandy soils, in dry eucalypt forest and heath.

DISTRIBUTION Common from coast to mountains. Extending to south coast and southern tablelands.

FLOWERING September–December.

526. Isopogon anemonifolius

Isopogon anemonifolius

Broad-leaf Drumsticks **Plate 526**

An erect shrub, 1 to 2 m tall. Leaves are deeply dissected into 3 narrow segments which are again divided and lobed at the tip. Leaf-segments are flat, not needle-like, and 3 to 5 mm wide. Leaves are spreading, and although the segments appear to be pointed, they are not pungent. Flowers are yellow, with cones up to 20 mm across.

HABITAT Shallow sandy soils, in dry eucalypt forest and heath.

DISTRIBUTION Widespread and common from coast to mountains. Extending from south coast and southern tablelands northwards to Qld.

FLOWERING Chiefly September–November, but some flowers at other times.

527. Isopogon prostratus

Isopogon prostratus

Prostrate Drumsticks **Plate 527**

I. prostratus was formerly regarded as a variety of *I. anemonifolius*. It closely resembles that species except that its leaf-segments are narrower (about 2 mm wide) and are ridged near the ends on the undersurface. The most conspicuous feature of this species, recognised by its specific name, is its prostate, ground-hugging habit. Although the branches are prostrate, the leaves grow upright. Flowers are yellow and the cones are usually smaller (12 to 16 mm across) than those of *I. anemonifolius*

HABITAT Sandy soils in eucalypt forests.

DISTRIBUTION Higher Blue Mountains, where it is restricted to small areas north of Clarence in Newnes State Forest. Also south coast and southern tablelands, extending into Vic.

FLOWERING November–January.

528. Isopogon dawsonii

Isopogon dawsonii

Dawson's Drumsticks **Plate 528**

This attractive tall shrub (1 to 3 m) has a limited distribution in the Sydney district where it is seldom seen in its natural habitat, although it is frequently grown in native gardens. Its divided leaf-segments are flat and point upwards. Flower tubes are white, tipped with yellow and pink; they are silky-hairy, an important identifying feature. The drumstick fruiting cone is also covered with distinctive silver downy hairs.

HABITAT Rocky slopes and cliffs (in Sydney area).

DISTRIBUTION Confined to the escarpment near the junction of the Nepean River and Glenbrook Creek. Also found in sandy soils in Glen Davis-Wolgan-Rylstone district and Goulburn River valley.

FLOWERING August–October.

Isopogon fletcheri

Fletcher's Drumsticks **Plate 529**

A rare plant occurring only on sheltered cliff-faces in the upper Blue Mountains. Its population is considered at risk, mainly from bushfires which could destroy its habitat. The fires of 1982/83 burnt many of the plants and considerably damaged their habitat, but some plants have regenerated. No other *Isopogon* in the area has entire leaves, so it is easily identified. The leaves are about 8 cm long and taper towards the stem. Flowers are terminal on stout erect stems. The whole plant is rarely more than 1.5 m tall.

HABITAT Sheltered cliff-faces, amongst ferns and other small plants on damp ledges.

DISTRIBUTION Restricted to upper Blue Mountains (e.g., Govett's Leap, Blackheath).

FLOWERING October–November.

529. *Isopogon fletcheri*

Petrophile Conesticks

Conesticks are erect shrubs with stiff divided leaves and flowers in dense cone-like spikes. The bracts of these flower spikes harden after flowering to form woody egg-shaped to oblong cones. Conesticks are a common ingredient of the sandstone flora around Sydney. Although 'petro-phile' means 'rock-lover', all 4 species also grow in shallow sandy soils in dry eucalypt forests and heaths.

Petrophile pulchella

Conesticks **Plate 530**

This species is an erect shrub to 3 m tall, with yellow-flowered terminal cones. These may be stalkless or with short stalks, solitary or in small clusters. Although the cones are terminal, the stems continue to grow, sending new branchlets around the cone and new flowers are borne above the old cones. The fruiting cone is up to 5 cm long and about 2 cm wide. Leaves are very divided, erect but not pungent-pointed. Young shoots are usually without hairs, but sometimes are hoary-downy. The main features which separate this species from others in the genus are its terminal cones and non-pungent leaves. In some areas, there are intermediate forms between this species and *P. sessilis* and *P. canescens*

HABITAT Dry eucalypt forest and heath, on rocky slopes and in shallow sandy soils.

DISTRIBUTION Coast and Blue Mountains. Extending to south and north coasts.

FLOWERING Mainly December–March, but some flowers at other times.

530. *Petrophile pulchella*

531. *Petrophile sessilis*

Petrophile sessilis

Prickly Conesticks **Plate 531**

The most characteristic feature of this stiff erect shrub is its spreading, rigid and pungent leaf-segments. This feature distinguishes it from *P. pulchella* and *P. canescens*. The specific name 'sessilis' refers to the stalkless flower-cones, which separate it from *P. pedunculata*. Young shoots are softly hairy, but the leaves are hairless.

HABITAT Dry eucalypt forest and heath, on rocky slopes and in shallow sandy soils.

DISTRIBUTION Coastal zone south of Sydney and Blue Mountains. Common in Royal National Park, Heathcote National Park and Agnes Banks Nature Reserve. Extending to southern tablelands.

FLOWERING December–January.

532. *Petrophile canescens*

Petrophile canescens **Plate 532**

As this species is a tall shrub (to 3 m), with stalkless cones and spreading-ascending, non-pungent leaf-segments, it is often confused with *P. pulchella*. It differs from that species in having downy young shoots and greyish-downy leaf-segments (especially when young). Leaves are 2 to 5 cm long, with rounded or flattened segments, 3 to 9 mm long. Unlike *P. sessilis*, its leaf-segments are not pungent pointed. Fruiting cones are up to 25 mm long and 15 mm across. This species is usually found in deeper sands rather than shallow sands or rocky areas preferred by other conesticks.

HABITAT Deep sands in open forest.

DISTRIBUTION Mainly Blue Mountains, but also recorded in sands of Castlereagh State Forest. Extending from southern tablelands, to northern tablelands, north coast, north-western slopes and Qld.

FLOWERING December–February.

533. *Petrophile pedunculata*

Petrophile pedunculata

Stalked Conesticks **Plate 533**

This is the only species of Conesticks in the area with stalked cones. It is a tall, hairless shrub with leaves which are spreading, up to 16 cm long and not pungent-pointed. The flower-cones are 15 to 25 mm long and borne in the axils of the leaves. Unlike the other 3 conesticks which have downy grey bracts on the young cones, this species has smooth and hairless bracts.

HABITAT Shallow sandy soils in dry eucalypt forest and heath.

DISTRIBUTION Blue Mountains and south to southern highlands (e.g. Wingello, Carrington Falls). Extending to south coast.

FLOWERING November–January.

Conospermum Coneseeds, Smokebush

Shrubs with flowers which are small and tubular with irregular lobes appearing as 2 lips. The lower lip has 3 narrow spreading lobes and the upper lip is broader and hooded. Most local species (except *C. tenuifolium*) have white flowers borne in dense terminal clusters, making them very conspicuous. Fruit is a nut, about 3 mm long, covered with hairs.

Conospermum longifolium

Long-leaf Coneseeds **Plate 534**

A variable species divided into 3 subspecies according to the shape of the leaves. These subspecies are also separated in terms of distribution, although there is some overlapping and intergrading, especially in the north-west and south-west of the Sydney district. All subspecies have erect stems, long flat leaves and dense terminal white flower-heads.

534. A. Conospermum longifolium ssp. *longifolium*

Subspecies *longifolium* (Plate 534A) has oblanceolate leaves (broad above the centre, tapering to a long stalk). These are 5 to 25 cm long, more than 8 mm wide and are mostly at the base of the flower stem.

DISTRIBUTION Coast and Hornsby Plateau north of Sydney. Common in Ku-ring-gai Chase and Brisbane Water National Parks. Extends to the Putty area, where it intergrades with ssp. *mediale*.

Subspecies *augustifolium* (Plate 534B) has thin narrow leaves less than 4 mm wide. These may be spaced along the stems.

DISTRIBUTION South of Sydney in the Georges River-Waterfall-Appin area.

Subspecies *mediale* has leaves whose width is between 4 and 8 mm, that is, mid-way between the width of the other subspecies.

DISTRIBUTION West of Sydney, from Putty in the north to the Moss Vale area, including the central Blue Mountains (e.g. Mt Solitary to Glenbrook, Kurrajong to Mt Tomah). Also Nowra and Jervis Bay district.

HABITAT (All subspecies) Sandy soils, especially in dry open forest and heath, often on stony ridges and slopes.

FLOWERING (All subspecies) July–November.

534. B. Conospermum longifolium ssp. *angustifolium*

535. Conospermum tenuifolium

Conospermum tenuifolium

Sprawling Coneseeds **Plate 535**

This is the only Conospermum in the area which is not an upright shrub with white flowers. It is a slender sprawling plant often found in large clumps in damp sandy soils. Narrow leaves, 8 to 16 cm long, almost cylindrical but channelled above, curl upwards from almost prostrate stems. Blue or lilac flowers are borne in small heads on erect stalks, 18 to 50 cm long; these stalks are slender and reddish in colour.

HABITAT Wet sandy soils.

DISTRIBUTION Widespread from coast to mountains (e.g. Heathcote, Mt Wilson, Mt Hay, Girrakool). Extending to south coast.

FLOWERING September–January.

Conospermum taxifolium

Coneseeds **Plate 536**

This is an erect shrub with large terminal clusters of white flowers. The leaves are crowded, less than 3 cm long and very variable. Because of this variability, some confusion has arisen as to whether this is 1 species with a number of forms, or whether the group can be split into more than 1 species. Some botanists divide the complex into (a) plants with elliptic to broad-lanceolate leaves more than 4 mm wide (*C. ellipticum*), (b) plants with linear to terete leaves, 5 to 15 mm long and about 1 mm wide (*C. ericifolium*), and (c) plants with intermediate leaves, flat, lance-shaped to linear, 1 to 3 cm long, about 2 mm wide and often somewhat twisted (*C. taxifolium*). These 3 forms tend to intergrade and it is often difficult to distinguish one from the other, hence the preference for regarding them as a single species with different leaf forms. *C. taxifolium* is also known to hybridise with *C. longifolium* resulting in an intermediate form.

HABITAT Sandy soils in dry open forest and heath.

DISTRIBUTION The type form (*C. taxifolium*) occurs on coastal heaths from Port Jackson south to Kurnell. The narrow-leaved form (*C. ericifolium*) occurs north of Port Jackson in woodland and heath, through West Head to near Toukley; a similar form also occurs in the Nowra-Jervis Bay area. The broad-leaved form (*C. ellipticum*) occurs in coastal sites from the North Head of Port Jackson south to Jervis Bay, with an isolated occurrence in Bouddi National Park, north of Broken Bay. The species extends along coast, tablelands and western slopes divisions and to Qld, Vic and Tas.

FLOWERING July–October.

536. Conospermum taxifolium (narrow leaf form)

537. *Telopea speciosissima*

Telopea Waratah

Erect shrub with beautiful dense red flower-heads. Fruit a large woody follicle to 10 cm long, containing 10 to 20 winged seeds. The Gippsland Waratah (*T. oreades*), a tree from Victoria and the southern tablelands, reaches its northern limit just beyond the Sydney district. The Monga Waratah (*T. mongaensis*), from sheltered forests east of Braidwood, is now thought to be a small variety of *T. oreades*.

Telopea speciosissima

Waratah **Plate 537**

The Waratah is one of our best-known plants and is the floral emblem of the State of New South Wales. Its botanical name is most appropriate as it means 'most beautiful plant seen from afar' ('tele'—far; 'opsis'—sight). Flower-heads are terminal, up to 15 cm wide and are made up of a dense cluster of flowers surrounded by long, bright-red bracts. Leaves are up to 15 cm long and 4 cm wide, stiff , strongly veined and with toothed edges. The boat-like fruit hangs on the plant for a long time after release of the seeds and is quite conspicuous.

HABITAT Forest, on sandy soils.

DISTRIBUTION Common and widespread from coast to higher Blue Mountains. Also occurs on south coast and northern tablelands (Gibraltar Range).

FLOWERING September–October near coast; October–December in mountains.

Lomatia

Shrubs with yellowish-white flowers closely resembling those of *Grevillea*. Perianth tube slit on lower side when in bud; the style protrudes from the perianth slit until released. Three glands present at the base of the ovary-stalk (*Grevillea* has only a single gland). Fruit a thin-walled follicle, but differing from the fruit of *Grevillea* in having many winged seeds in 2 rows, separated by a yellow powder and papery plates.

538. *Lomatia silaifolia*

Lomatia silaifolia

Crinkle Bush **Plate 538**

A common erect stiff shrub on sandy soils in the area. It is easily identified by its much-divided, light green, spreading leaves. These leaves are occasionally divided in a simple pinnate manner, but more often they are twice or thrice pinnately divided. The leaf segments are narrow and usually sharply toothed. The creamy-white flowers are in pairs along erect stems well above the leaves.

HABITAT Heath and open forest, on sandy soils.

DISTRIBUTION Common and widespread from coast to mountains. Extending along north coast and northern tablelands to Qld; also western slopes.

FLOWERING November–February.

539. *Lomatia myricoides*

Lomatia myricoides

Long-leaved Lomatia **Plate 539**

This is the only Lomatia in the area which grows into a tall shrub or a small tree to 4 m tall. Its leaves are up to 20 cm long and 2 cm wide; the margins are usually coarsely toothed. Flowers are greenish yellow and borne from the leaf axils in loose clusters shorter than the leaves.

HABITAT Gullies, creek margins and sheltered forest.

DISTRIBUTION Widespread from coast to mountains. Recordings include Brisbane Water National Park, Blackheath, Royal National Park, Minnamurra Falls and Paddys River. Extending along south coast and southern tablelands to Vic.

FLOWERING December–January.

540. Lomatia ilicifolia

Lomatia ilicifolia

Holly-leaved Lomatia **Plate 540**

This is a shrubby plant with tough upright stems bearing large coarsely serrated leaves. These are 10 to 18 cm long, 3 to 5 cm wide, with a clear network of veins visible on the upper surface. Young leaves are a distinct light tan colour. The flowers are yellowish white and resemble those of *Grevillea*; they are borne near the ends of erect stems well above the leaves. The fruit is a black, thin-walled follicle.

HABITAT Forest, on sandy soils.

DISTRIBUTION Blue Mountains and southern highlands. Extending along south coast and southern tablelands to Vic.

FLOWERING November–January.

541. Stenocarpus salignus

Stenocarpus

Rainforest trees with alternate leaves. Flowers borne in umbels. Fruit a follicle; seeds flat, with a membranous wing at the lower end. Two species in NSW, but the larger Fire-wheel Tree (*S. sinuatus*) does not occur south of the Bellinger River, although it is widely planted in parks and gardens.

Stenocarpus salignus

Scrub Beefwood **Plate 541**

This rainforest tree may grow to a height of 30 m along the north coast, but around Sydney it is mostly a large shrub or small tree. Leaves are entire, broadly lanceolate, up to 10 cm long, dark green and glossy on the upper surface and paler underneath. Most leaves have 3 to 5 prominent longitudinal veins and a network of minor veins. The specific name 'salignus' is from the Latin for 'willow-like', referring to the appearance of the leaves. Flowers resemble those of *Grevillea*; they are white, with 15 to 20 individual flowers in an umbel borne either at the ends of the branches or in the leaf axils. Fruit is a narrow leathery follicle 5 to 10 cm long.

HABITAT Sheltered rainforest gullies.

DISTRIBUTION Coastal rainforest, such as Mt Kembla, Minnamurra Falls, Royal National Park and Gap Creek (Olney State Forest). Also suitable sites further inland, such as Kurrajong, Colo River and Glen Davis. Extending from Milton (south coast) northwards to Qld.

FLOWERING October–January.

Grevillea

Shrubs or trees usually with alternate leaves. Individual flowers with 4 perianth segments forming a tube; the segments roll back to one side. The style protrudes in a curve from a slit on one side of the perianth before being fully released. The end of the style, known as the 'stigmatic disc', carries pollen which is rubbed off by the pollinating agent. The disc then becomes sticky and receptive to pollen from another flower. Fruit is a thin follicle, boat-shaped, with the remains of the style forming a tail, containing 2 wingless seeds. For identification purposes, the Grevillea of the Sydney district can be divided into those species with long comb-like racemes of flowers and those with short umbel-like racemes, often called Spider-flowers because of the conspicuous clusters of styles. Other features useful for identification are flower colour, leaf shape and the nature of the ovary (whether it is hairy or hairless, stalked or sessile).

542. Grevillea longifolia

Grevillea longifolia **Plate 542**

This attractive spreading shrub, 2 to 5 m tall, is easily identified by its red flower combs and its long thin saw-toothed leaves. These leaves are about 20 cm long and 15 mm wide, serrated almost half-way to the mid rib. Occasionally some leaves are without serrations. The undersurface of the leaves is a distinct silky-grey. This species may be confused with *G. aspleniifolia* but each occurs in different habitats. The branchlets of *G. longifolia* are angular and silky, its pollen presenter is broadly rhomboidal and its leaf serrations are not as deep as are those of *G. aspleniifolia*

HABITAT Banks of perennial streams.

DISTRIBUTION South of Sydney in the Heathcote Creek and Georges, Woronora and Nepean River systems, as far south as Appin and the lower Cataract River. Also in the Springwood-Lawson area of the Blue Mountains. Endemic in the area.

FLOWERING July–November, but also recorded as flowering in April.

Grevillea aspleniifolia **Plate 543**

This large bushy shrub closely resembles *G. longifolia* as it has red flower combs and long narrow leaves. Some confusion has developed over the naming of both species and it is common to find *G. longifolia* cultivated under the name *G. aspleniifolia*. *G. aspleniifolia* may be distinguished by its rounded, densely downy branches and its obovate to rhomboidal pollen presenter (broadly rhomboidal in *G. longifolia*). Its leaves are up to 25 cm long, deeply serrated or often entire; the undersurface is covered with closely matted brownish-white curled hairs. This species prefers a rocky habitat on slopes, whereas *G. longifolia* most commonly is found beside streams. Despite its striking foliage and flowers, it is not often encountered outside cultivation as it occurs well away from roads and habitation.

HABITAT Rocky ridges and slopes in woodlands.

DISTRIBUTION Blue Mountains and southern highlands, including Nattai, Wollondilly, lower Coxs, Kowmung and Shoalhaven Rivers' catchments. Extending to southern tablelands.

FLOWERING Chiefly July–November, but some flowers in late summer and autumn.

543. Grevillea aspleniifolia

544. *Grevillea rivularis*

Grevillea rivularis

Carrington Falls Grevillea **Plate 544**

G. rivularis has a very restricted natural distribution and is mostly known to the public through specimens in native gardens. Outside cultivation it is only found at Carrington Falls in the southern highlands, where it is a straggling to spreading shrub up to 2 m high. It forms thick clumps along the banks of the upper Kangaroo River above the causeway at the top of the falls. Flower combs are very attractive as they are terminal and a delicate pink colour with long erect styles tipped with lime green. Leaves are stiff and deeply divided into narrow sharply pointed lobes. It is readily distinguished from other *Grevillea* by its divided leaves, hairless perianth and conspicuously stalked ovary.

HABITAT In sandy soil and alluvium on banks of small streams.

DISTRIBUTION Restricted to Carrington Falls, near Robertson

FLOWERING Late August- September.

Grevillea acanthifolia ssp. *acanthifolia*

Plate 545

This is a common plant in swampy ground of the higher Blue Mountains. It is a straggling shrub, dwarfed or to 2 m tall, with red branchlets and stiff divided leaves. Leaves are about 12 cm long, with broad leaflets which are themselves divided into 3 sharply pointed lobes. The undersurface of the leaves is hairless. Flower combs are red or deep pink, with a silvery-downy perianth (best seen when in bud).

HABITAT Wet sandy and peaty soils.

DISTRIBUTION Endemic in upper Blue Mountains (e.g. cliff walks at Blackheath).

FLOWERING September–November; some flowers at other times.

545. *Grevillea acanthifolia*

546. *Grevillia laurifolia*

Grevillea laurifolia **Plate 546**

The broad entire ovate leaves and the trailing habit clearly distinguish this *Grevillea* from others in the Blue Mountains. This is a common plant, forming mat-like clumps up to 3 m in diameter. Long trailing stems bear dense red flower combs. Leaves are always undivided, 3 to 12 cm long and ovate; they are shiny green above, with clearly defined veins, and silky below.

This species is known to hybridise with *G. acanthifolia* to form the small shrub (Plate 547) formerly known as *G. gaudichaudii*. This hybrid combines the features of both plants. It has red to pink flower combs. Leaves are usually deeply lobed, but each lobe is ovate and entire, not divided again as are the leaves of *G. acanthifolia*. Some plants may have undivided leaves, but the tendency is for them to divide into lobes. This hybrid is also restricted to the upper Blue Mountains.

HABITAT Open sites, on sandy soils.

DISTRIBUTION Restricted to the Blue Mountains (e.g. Mt Wilson, Ingar, Wentworth Falls, Blackheath, Hartley and Clarence).

FLOWERING October–December.

547. *Grevillea* X *gaudichaudii*

548. *Grevillea triternata*

Grevillea triternata **Plate 548**

This plant is easily recognised by its divided pungent leaves and its small creamy flower clusters. It is a small prickly shrub, fairly common throughout the drier north-western slopes but not often encountered in the Sydney district. The specific name refers to the division of the leaf into 3 narrow (2 mm) linear, sharply pointed segments, which are, in turn, also usually divided into 3. Flowers are borne in clusters about 2 cm long, which are in the leaf axils and are much shorter than the leaves.

HABITAT Sandy soils of woodlands and rocky sites.

DISTRIBUTION Valleys of upper Blue Mountains and southern highlands (e.g. Mittagong, escarpment of Wingecarribee Gorge, West Berrima). Also recorded from northern tablelands, north-western slopes and central western slopes.

FLOWERING September- November.

549. *Grevillea caleyi*

Grevillea caleyi **Plate 549**

This large bushy shrub, to 3 m tall, is unmistakable as its branches, leaf undersurfaces, flower stalks and inflorescences are densely covered with soft long rusty hairs. Leaves are fern-like, to 15 cm long, divided to the mid-rib, with between 12 and 15 pairs of lobes sticking out like a series of fish-bones from the mid-vein. New leaves are a striking tan-red colour. Flowers are deep red in terminal combs and the fruit is conspicuously striped.

HABITAT Open forest, on sandy soils.

DISTRIBUTION Restricted to only 2 localities in the Sydney area where it is endangered—Terrey Hills to Mona Vale and Bulli to Appin.

FLOWERING October–December.

550. *Grevillea buxifolia*

Grevillea buxifolia ssp. *buxifolia*

Grey Spider-flower **Plate 550**

This is a common shrub, growing to 2 m tall. Flower heads are conspicuous as they are grey, terminal and often abundant. The greyness is the result of a dense covering of hairs on the perianth tube and style. The perianth tube is 4 to 5 mm long and the style has a small (0.5 to 4 mm) but distinct horn-like appendage extending beyond the round stigmatic disc. (A form growing around Gospers Mt, Wollemi, has no horn-like appendage.) Young branchlets are covered with spreading brown hairs. Leaves are lanceolate to ovate, mostly under 2 cm long and crowded along the branches. This type form intergrades with subspecies *phylicoides* in some areas, especially near Wisemans Ferry.

HABITAT Open forest and heath, on sandy soils.

DISTRIBUTION Widespread from coast to mountains. Extending to south coast.

FLOWERING September–November.

Grevillea buxifolia ssp. *phylicoides*

Plate 551

This subspecies is a smaller version of the type form of *G. buxifolia*. It is seldom more than 1 m tall. The greyness of the flowers, the hairy young branches and the small appendage on the stigmatic disc are all similar to those of the larger *G. buxifolia*. However, the perianth tube is smaller (only 2 to 4 mm long) and the stigmatic disc is oblong to elliptic, not round. Leaves are narrower, almost linear, and 16 to 30 mm long. The difference in leaf shape and the smaller flower heads are the most obvious difference between this subspecies and the type species.

HABITAT Heath and open forest, on sandy soils.

DISTRIBUTION Common in lower Blue Mountains west to about Wentworth Falls, with a recording near Oakdale, west of Camden. Endemic in the Sydney district.

FLOWERING August–November.

551. *Grevillea buxifolia* ssp. *phylicoides*

552. *Grevillea buxifolia* ssp. *sphacelata*

Grevillea buxifolia ssp. *sphacelata*

Plate 552

The distribution of this subspecies overlaps with that of the typical form of *G. buxifolia* in coastal areas and because they both have terminal clusters of grey flowers, they are frequently confused. The subspecies is a smaller shrub, seldom reaching 1 m high, with branchlets which have appressed, rather than spreading, hairs. Its flowers are smaller; the perianth tube is only 2 to 3 mm long. The appendage at the tip of the style is inconspicuous and the stigmatic disc is oval. Leaves are very variable, often linear and about 2 cm long, but shorter and obovate near the coastline.

HABITAT Heath and open forest, on sandy soils.

DISTRIBUTION Coast to lower Blue Mountains. Common in Royal and Heathcote National Parks. Extending to south coast and northern tablelands.

FLOWERING August–November.

Grevillea baueri **Plate 553**

This is primarily a species from the higher country of southern and central New South Wales. In the local area, it is found in the southern highlands where it occurs as a small shrub, less than 1 m tall. It is easily identified by its flowers and leaves. Flowers are red, with a hint of green. The outside of the perianth tube is without hairs, but the inside and the style are sprinkled with short hairs. The ovary is covered with long dense hairs. Leaves are elliptic, up to 25 mm long, without stalks and with a small pointed tip; they are usually tightly packed along the stems.

A form with rough sand-papery leaves, well known in cultivation, is now known as ssp. *asperula*.

HABITAT Open forest, often in damp sandy soils.

DISTRIBUTION Higher areas to the south and south-west, such as around Bundanoon and Joadja Creek,

553. *Grevillea baueri*

West Mittagong. Extending to south coast and southern tablelands.

FLOWERING June–December.

Grevillea mucronulata

Green Spider-flower **Plate 554**

Green Spider-flower is a common shrub to 1 m tall found on both shaded hillsides and more exposed positions. Its flowers are green, sometimes with dark markings, and sprinkled with hairs on the inside and outside of the perianth tube. The style is often pink or almost black and is topped by a green disc. The ovary is densely hairy. Leaves are very variable. They may be round to elliptic or narrow lanceolate, 5 to 20 mm long, flat or somewhat cupped, lighter underneath and with a small pointed tip. New growth is often a distinct tan colour.

HABITAT Open forest and woodland, on sandy soils.

DISTRIBUTION Widespread from coast to mountains. Common in small reserves along Georges and Woronora Rivers, in Castlereagh State Forest and Blue Mountains. Extending to south coast and central western slopes.

FLOWERING April–October.

554. *Grevillea mucronulata*

555. Grevillea obtusiflora

Grevillea obtusiflora — Plate 555

This small shrub (rarely 1 m tall) is regarded as a rare plant, at risk because it only occurs in small scattered populations. Leaves are narrow lanceolate, up to 3 cm long, 4 to 5 mm wide, with margins curled under and the undersurface a lighter shade than the top surface. Flowers are creamy-green, downy outside, with pink styles covered with longer weak hairs. The perianth segments at the opening of the tube are distinctly obtuse.

HABITAT Open forest, on sandy soils.

DISTRIBUTION Restricted to a few isolated localities in the Blue Mountains, such as ridges in Kedumba Valley and open hillsides near Mt Cookem, Coxs River. A subspecies (ssp. *granulifera*) occurs at Mt George on the north coast.

FLOWERING September–November.

556. Grevillea arenaria

Grevillea arenaria — Plate 556

This is a very variable shrub with several recognised forms. Typically it is a tall spreading shrub, with oblong to elliptic leaves, 2 to 4 cm long, rounded at the ends but with a small pointed tip. On some plants, the perianth and style are a delicate yellow-green with pink lobes, on others the perianth is red. Sometimes the flowers are densely hairy outside, at other times only slightly so. The lobes of the perianth taper to a point, in contrast to most other local Grevilleas which have blunt, rounded lobes. Flowers are borne at the ends of the branchlets, with 1 to 6 flowers grouped together.

HABITAT Gullies, on sandy soils, often near rocky water-courses.

DISTRIBUTION Nepean and Nattai Rivers district (e.g. Hilltop, Little River) and southern Blue Mountains (e.g. Wombeyan Caves). Extending along the north and south coasts and throughout the tablelands.

FLOWERING April–September.

557. Grevillea shiressii

Grevillea shiressii — Plate 557

This is a tall spreading shrub or small tree, regarded as endangered because of its very restricted distribution. Fortunately, it is also grown in a few native gardens so its future seems assured. Its graceful habit, broad leaves and unusual flowers make it an ideal garden plant and unmistakable when seen in its natural setting. Leaves are up to 16 cm long, 2 to 4 cm wide, narrow lanceolate, dark green and distinctly veined. Flowers are borne in loose clusters in the leaf axils. They are pale mauve with long greenish-yellow lobes. The style is greenish, about 2 cm long, and appears as an extension of the long (15 mm) reddish stalk of the ovary.

HABITAT Forest in gullies, usually along banks of streams.

DISTRIBUTION Restricted to Mullet Creek and Mooney Creek on the edge of Brisbane Water National Park.

FLOWERING August–December.

558. Grevillea juniperina

Grevillea juniperina

Prickly Spider-flower — Plate 558

A number of Grevilleas are known popularly as 'Prickly Spider-flower', and this low shrub, with its crowded spreading needle-like leaves, clearly deserves the name. Its specific name also refers to the pungent leaves, as 'juniper' is a prickly-leaved shrub. Leaves are 1 to 2 cm long, linear and with margins curled under to the undersurface. Flowers are borne in clusters among the sharp leaves along short branchlets. On the tablelands, where this species is common, the flowers are mostly pink or red, and this colour form is commonly planted in gardens and used for landscaping. However, the flowers of those plants occurring on the Cumberland Plain are light yellow, with greenish tips. The ovary is always without hairs and slightly stalked.

HABITAT Open forest and cleared areas, on shale.

DISTRIBUTION Cumberland Plain, Southern Highlands (e.g. Tallong Dam) and Blue Mountains. Common in and around Castlereagh State Forest and in the Rooty Hill-Plumpton area, where it is endangered because of housing developments. Forms of this species are found throughout the tablelands and all along the coast of NSW.

FLOWERING September–November; some flowers as late as January.

Grevillea speciosa ssp. *speciosa*

Red Spider-flower — Plate 559

G. speciosa is a bushy shrub to 2 m tall, with clusters of red spider-like flowers which hang on stalks from the ends of side branches. Flower stalks may be only 5 mm long, or up to 5 cm long in the Gosford form. The perianth tube is 8 to 10 mm long. The red styles project well beyond the perianth and are 20 to 25 mm long. Leaves are ovate to lanceolate sometimes 5 cm long, but more often about 2 cm, paler below and with a small point. Young branches are silky-hairy. This subspecies may be confused with ssp. *oleoides* with which it intergrades. Both have red flowers. However, ssp. *oleoides* has longer leaves, stalkless flower clusters and is not found north of Sydney.

HABITAT Heath and open forest, on sandy soils.

DISTRIBUTION North of Sydney to Gosford and Bucketty districts. Common on North Head and Dobroyd Head (Sydney Harbour) and in Ku-ring-gai Chase and Brisbane Water National Parks. Endemic in the Sydney district.

FLOWERING June–September.

559. *Grevillea speciosa*

560. *Grevillea speciosa* ssp. *oleoides*

Grevillea speciosa ssp. *oleoides*

Red Spider-flower **Plate 560**

This is an erect shrub to 2 m tall, with stalkless clusters of 12 to 16 red spider-like flowers borne in the axils of the leaves or from the ends of side branches. The perianth tube is 8 to 14 mm long and the long styles project well beyond the perianth. Leaves are narrow lanceolate 5 to 10 cm long, and up to 15 mm wide. The undersurface of the leaf is silky-hairy. Narrow leaf forms with more open and paler flowers occur in the Blue Mountains.

HABITAT Heath, scrubland and open forest, on sandy soils; often in the rocky beds and sandy banks of creeks.

DISTRIBUTION Coast to Blue Mountains. Common south of Sydney in Royal and Heathcote National Parks and along Erskine Creek, lower Blue Mountains. Endemic in the Sydney district.

FLOWERING July–October.

Grevillea molyneuxii **Plate 561**

This is a small shrub (to 60 cm tall) with a very localised distribution; it is apparently confined to the vicinity of rock platforms on the edge of Wingello State Forest. Originally discovered in 1933,

561. *Grevillea molyneuxii*

the location was lost until rediscovered by Bill Molyneux in 1976. Its leaves are linear, 2 to 4 cm long, sharp pointed with close silvery hairs underneath. Flowers are a brilliant red in short racemes of 10 to 12 flowers. The perianth is 7 to 9 mm long, style 20 to 24 mm long, ovary stalk about 2 mm long and the ovary is hairless. A ring of white hairs is located just above the middle of the tube. Flower clusters are borne on very short stalks from the leaf axils.

HABITAT Margins of rock platform, in shallow, often damp soils.

DISTRIBUTION Restricted to small areas above Tallowa Creek, Wingello State Forest.

FLOWERING September–November.

Grevillea sericea

Pink Spider-flower **Plate 562**

A common and colourful component of the sandstone flora, it is well known because it is widely distributed and bears flowers in most months of the year. It is a compact shrub to 2 m tall, with short clusters of pink flowers which are sprinkled with silky hairs. The style is 10 to 15 mm long. Sometimes the flowers are massed at the ends of the leafy branches; at other times, just a few flowers are borne. Some forms have very dark pink flowers (such as those on Wheelbarrow Ridge between Colo Heights and Upper Portland). Leaves are narrow, usually about 3 cm long, but ranging between 1 and 6 cm; they are densely silky-hairy below. This species hybridises with *G. diffusa* to produce a shrub similar in habit to *G. sericea* but with smaller, darker and more compact flowers. The hybrid is common in the vicinity of Winifred and Anise Falls in Royal National Park.

HABITAT Widespread in a variety of habitats, on sandy soils.

DISTRIBUTION Coast to mountains. Extending to central western slopes.

FLOWERING Mostly July–November, but some flowers at other times.

562. *Grevillea sericea*

Grevillea parviflora

Small Spider-flower **Plate 563**

It is often difficult to distinguish this species from *G. linearifolia* and *G. sericea*. In general, this is a small shrub, usually less than 1 m tall. Flowers are pale pink or almost white, with a shorter style than that of *G. sericea* (less than 10 mm). Leaves are up to 4 cm long, narrow (1 to 2 mm), pointed, with edges curled under to give the impression of grooves along the undersurface. The perianth tube is only minutely hairy inside below the middle, unlike *G. linearifolia* whose perianth is densely bearded inside.

HABITAT Dry open forest, on sandy soils.

DISTRIBUTION Scattered throughout Blue Mountains and southern highlands (e.g. Carrington Falls); less common nearer coast (recorded from Heathcote National Park) Also occurring south coast, northern tablelands, southern tablelands, central western slopes and Vic.

FLOWERING August–January.

563. *Grevillea parviflora*

564. *Grevillea diffusa*

566. *Grevillea oldei*

568. *Grevillea linearifolia*

Grevillea diffusa **Plate 564**

A variable plant, mostly occurring as a low shrub to 1 m tall, but sometimes growing very tall along protected creeks. It has dense clusters of deep red to almost black flowers with short styles (6 to 10 mm); flower clusters hang from a stalk which may be short or quite long. A number of distinct forms have been recognised. The typical form is from the Royal and Heathcote National Parks. Its leaves are green, oblong to narrow lanceolate, about 3 cm long and silky-hairy underneath.

In the Menai-Holdsworthy area, south to Appin, is a form with hairy leaves which give it a silky-grey appearance. It occurs on both lateritic and sandstone soils, growing only to about 50 cm tall.

Further south, along the ridges and hilltops around Wollongong, is the 'Bulli' form, a plant to 50 cm tall, with longer leaves and longer hairs and a distinctly shaggy appearance.

North of Sydney, at Mt White, another distinct form is found. It is a small shrub to 50 cm tall, with long (to 6 cm), narrow, hairless leaves and flower clusters on very long stalks. It is known as subspecies *filipendula* (Plate 565).

East of Rylstone, beyond the north-western boundary of the Sydney district and possibly elsewhere in the northern Blue Mountains, subspecies *evansiana* occurs. It is a shrub to 1 m tall with very dark red flowers sprinkled with grey hairs; leaves are oblong-elliptic, 2 to 4 cm long, and silver-grey underneath.

HABITAT Open forest, on ridge tops, rocky or sandy slopes and in gullies.

DISTRIBUTION All forms except ssp. *evansiana* restricted to the coastal zone north and south of Sydney.

FLOWERING July–November.

565. *Grevillea diffusa* ssp. *filipendula*

Grevillea oldei **Plate 566**

This small decumbent shrub is closely related to *G. diffusa* and, prior to its achievement of species status in 1986, was regarded as a very localised form of *G. diffusa*. Stems are spreading and sprinkled with long loose hairs. Leaves are narrow elliptic to almost triangular, up to 10 mm long, paler underneath, with a short mucro and are often twisted. Leaves have a tendency to cluster in groups of 2 or 3. Flowers hang from the ends of branchlets on stalks about 15 mm long. They are deep red, with a perianth tube about 5 mm long and style 10 mm long. There is a tuft of white hairs inside the tube near the middle. The ovary is hairless. Although restricted in its natural distribution, this species is commonly sold in nurseries and is easy to grow.

HABITAT Open forest.

DISTRIBUTION Restricted to the Mt Penang area near Gosford.

FLOWERING September–October.

Grevillea australis

Alpine Spider-flower **Plate 567**

This is a subalpine or high tablelands species which reaches its northern limit in the plateau areas of the southern Blue Mountains. There it is a low erect shrub, seldom more than 60 cm tall with stiff elliptic to narrow oblong leaves, up to 3 cm long, with a short mucro, margins curved under and with silky-grey or brownish hairs on the undersurface. Flowers are white to pale pink, in small heads. The perianth tube is silky-hairy on the outside and 3 to 5 mm long.

HABITAT Heath and open woodland, on sandy soils.

DISTRIBUTION Southern Blue Mountains (e.g. Kanangra Walls). Extending south along southern tablelands to Vic and Tas.

FLOWERING November–December.

567. *Grevillea australis*

Grevillea linearifolia **Plate 568**

This is a medium-sized shrub, to 3 m tall, rather open in appearance, often with slender drooping branchlets. The specific name 'linearifolia' indicates one of its chief identifying features, its long linear leaves. These are up to 10 cm long and only 2 to 3 mm wide, and are silky-hairy underneath. Flowers are usually white (occasionally pink, as in specimens from Middle Head, Sydney Harbour), terminal, with a perianth tube 5 to 6 mm long and style 8 to 10 mm long. This species appears to intergrade with *G. parviflora*. In its typical form, it differs in having longer leaves and larger flowers than *G. parviflora* and its perianth tube is densely bearded inside near the middle.

HABITAT Open forest, on sandy soils, often in gullies.

DISTRIBUTION A coastal species, extending from south coast northwards to Qld. Most common locally north of Sydney, especially between Terrey Hills and Hornsby.

FLOWERING July–October.

Hakea

Shrubs, mostly with needle-like leaves, but 2 species in the area with flat leaves. Flowers on short stalks in clusters from the leaf axils. Flowers usually white (occasionally yellow or pink) with segments of perianth tube rolling back to one side, becoming free. Fruit a distinctive woody follicle containing 2 flat winged seeds and splitting into 2 solid parts.

Hakea teretifolia

Dagger Hakea **Plate 569**

Dagger Hakea is a common dense rigid shrub of the coastal district. It often forms impenetrable thickets in wet soils of heaths or swampy gullies and responds vigorously to fire. The leaves are needle-like, sharp-pointed, 2 to 5 cm long. Flowers are white, with a softly hairy perianth, and are borne in great profusion along the branchlets. This species is distinguished from other Hakeas by the shape of the fruits, which are small (30 mm long), narrow, with a long dagger-like point, rather like a bird's beak.

HABITAT Scrubland and heath, in both dry and wet soils.

DISTRIBUTION Widespread from coast to mountains; especially common on coastal heaths. Extending from the north coast southwards to Vic and Tas; also southern tablelands.

FLOWERING January–April.

570. Hakea sericea

569. Hakea teretifolia

Hakea sericea

Needle-bush, Silky Hakea **Plate 570**

Needle-bush is a common tall shrub to 3 m high. It is a well known component of the Sydney sandstone flora, and its abundant white flowers borne in late winter and spring make it very conspicuous. When not in flower, the large old fruits and the short needle-like leaves (2 to 6 cm long) are easily recognised. It may be confused at first with 2 other needle-leaved Hakea which grow in similar habitats, *H. propinqua* and *H. teretifolia*. However, the woody fruits are different for each species. Those of *H. sericea* are ovate-globular, about 2 cm across and have a short smooth beak. The fruit of *H. propinqua* is also ovate-globular, but larger (2 to 3 cm across), more warty and has no beak. *H. teretifolia* has fruit which is long and pointed like a bird's beak.

Habitat Heath, scrubs and dry open forest.

Distribution Widespread from coast to mountains. Extending from Coffs Harbour south along the coast to Vic and Tas; also on southern tablelands.

Flowering June–September.

571. Hakea gibbosa

Hakea gibbosa

Plate 571

A compact upright shrub, 1 to 2 m tall, with sharp needle-like leaves, 4 to 8 cm long. What distinguishes it from other needle-leaved *Hakea* is the presence of long hairs covering its stems and young leaves, giving the whole plant a downy appearance. Flowers are white or often cream and are clustered along the stems beneath the leaves. Fruit is up to 30 mm across, distinctly beaked, flattened, not globular, mostly smooth with only a few warts.

Habitat Heaths and scrubs, on sandy soils.

Distribution Restricted to coastal area around Sydney, where it is common in the national parks.

Flowering June–September.

572. Hakea constablei

Hakea constablei

Plate 572

This *Hakea* has a very restricted distribution so it is unlikely that it would be mistaken for any other similar species. Its young stems are covered with short whitish hairs but the plant does not have the very compact hairy appearance of *H. gibbosa*. It is a medium-sized shrub with needle-like leaves, 5 to 10 cm long, and small white flowers, typical of the genus. The woody fruit is almost 40 mm in diameter; fruits are warty and do not have any beak at the tip.

Habitat Rocky outcrops and dry forest on sandy soils.

Distribution Endemic in the Mt Wilson-Bell area of the Blue Mountains.

Flowering October–December.

573. *Hakea bakeriana*

Hakea bakeriana — Plate 573

When in flower, this medium-sized, hairless shrub is unmistakable as it has pink flowers with long protruding styles borne along the older branches below the foliage. On some plants, the flowers are orange with the tips of the perianth segments yellow. Leaves are thin, needle-like and 5 to 7 cm long. Fruit is relatively large, 4 to 5 cm across, rough and usually has a short beak.

HABITAT Scrub and open forest, in sandy soils.

DISTRIBUTION Hornsby Plateau (e.g. Dharug National Park and near Morisset). Extending to the north coast.

FLOWERING May–July.

574. *Hakea propinqua*

Hakea propinqua — Plate 574

The most conspicuous feature of this shrub (1 to 2 m tall) is its large woody fruits. These are 2 to 3 cm across, rough and deeply wrinkled and warty. They lack the beak which is present on the fruit of *H. sericea*. Leaves are needle-like and a little shorter than most other needle-leaved *Hakea* (2 to 4 cm long). Flowers are borne along the ends of the branchlets. On the coast these flowers are white but there are yellow forms in the higher Blue Mountains (e.g. around Blackheath).

HABITAT Scrubland and dry open forest, on sand or gravel.

DISTRIBUTION Widespread from coast to Blue Mountains. Also found on south coast.

FLOWERING June–October, flowering later in the mountains.

575. *Hakea microcarpa*

Hakea microcarpa

Small-fruited Hakea — Plate 575

The specific and common names of this small stiff shrub indicate its most conspicuous feature. The woody fruit is small, about 15 mm long and 6 to 7 mm wide, smooth, elongated rather than globular, and often has tiny horns near the tip of each valve. Leaves are slender needles up to 8 cm long. Flowers are small, white and clustered in the leaf axils. This is chiefly a tablelands species and is uncommon in the Sydney district.

HABITAT Swampy heaths and wet areas in open forests

DISTRIBUTION Higher Blue Mountains and southern highlands (e.g. Wombeyan Caves road). Also occurs along north coast, south coast, and all along the tablelands from Qld to Vic and Tas, extending to western slopes.

FLOWERING September–December.

Hakea dactyloides

Broad-leaved Hakea — Plate 576

This is a shrub to 3 m tall with flat, rather stiff leaves, 5 to 10 cm long, with 3 or more prominent longitudinal veins. The leaves vary greatly in width, from 5 to 25 mm; broad-leaved plants tend to grow in forests and narrow-leaved plants grow in more exposed sites. Numerous flowers are borne in clusters along the branchlets. They are white, sometimes tinged with pink, and have silky-hairy stalks. New foliage is often a striking tan colour. Fruits are ovate, slightly warty, 12 to 25 mm across and have a small straight beak.

HABITAT Heath and forest, often in gullies, on sandy soils.

DISTRIBUTION Widespread from coast to mountains. Extending from Qld to Vic and westwards to the western slopes.

FLOWERING September–November.

576. *Hakea dactyloides*

Hakea salicifolia

Willow-leaved Hakea — Plate 577

H. salicifolia may vary in size from a small shrub to a small tree. It is generally larger in protected gullies. A small form with shorter and narrow leaves and smaller fruit occurs along Heathcote Creek. The typical form has flat, light green, lanceolate leaves, 5 to 12 cm long, without prominent longitudinal veins or only the main vein distinct. Flowers are white and the flower stalks and perianth are without hairs. Fruits are ovate, about 2 cm across, covered with prominent warts and have a distinct beak. This species is commonly planted as a street tree in Sydney suburbs.

HABITAT In gullies and protected places, especially along creeks; occasionally in drier woodlands.

DISTRIBUTION Widespread from coast to mountains. Extending from south coast northwards to Qld; also northern tablelands.

FLOWERING September–November.

577. *Hakea salicifolia* (forest form)

578. *Xylomelum pyriforme*

Xylomelum

Small tree with flowers in dense spikes and a distinctive woody, pear-shaped fruit.

Xylomelum pyriforme

Woody Pear **Plate 578**

Both the common and botanical names of this small tree highlight its conspicuous large woody pear-shaped fruits. ('Xylon'—wood; 'melum'—fruit; 'pyriforme'—pear-shaped.) These are 6 to 9 cm long, and split to release 2 large winged seeds. Leaves are lanceolate, up to 20 cm long, opposite, leathery and strongly veined. Juvenile leaves are coarsely toothed. Flowers are small but tightly crowded on tan spikes, 5 to 8 cm long. Young shoots, inflorescence and new fruits are covered with long rusty velvety hairs, and young leaves are thin and reddish.

HABITAT Open forest, on sandy soils, often in slightly sheltered sites, or tall coastal heaths.

DISTRIBUTION Widespread from coast to mountains. Extending along north coast to Qld.

FLOWERING October–November.

579. *Lambertia formosa*

Lambertia

Shrubs, with leaves in whorls of 3 and terminal clusters of red flowers. Fruit woody, with a short beak and a horn on each valve.

Lambertia formosa

Honey Flower, Mountain Devil **Plate 579**

A medium-sized shrub, with conspicuous clusters of 7 erect tubular red flowers up to 5 cm long, enclosed by overlapping bracts. Each flower is rich in nectar. A style projects above the tube. Leaves are rigid, pungent pointed, 3 to 5 cm long, in whorls of 3 and paler underneath. The name 'Mountain Devil' comes from the shape of the woody fruits which are beaked and horned.

HABITAT Heath, scrub and open forest, on sandy soils.

DISTRIBUTION Widespread and common from coast to mountains. Occurring throughout coast and tablelands divisions of NSW.

FLOWERING Chiefly September–May, but some flowers at other times.

Banksia

Banksias are woody shrubs or medium-sized trees with conspicuous cylindrical flower spikes. Individual flowers are tubular in bud, splitting into 4 segments, releasing the rather wiry style which may be straight or hooked. Flowers are borne in stalkless pairs on the thick stem of the spike. Fruit is a woody follicle set into the stem of the spike. Often only a few fruits are produced from hundreds of flowers; withered unfertilised flowers often cling to the spike giving it a shaggy appearance. The follicle opens (often after fire) to release 2 seeds and leaving 2 woody 'lips' on the spike.

580. *Banksia ericifolia*

Banksia ericifolia

Heath Banksia **Plate 580**

The long orange-red flower spikes of Heath Banksia are a conspicuous feature of the coastal bushland in autumn and winter. This is a large shrub to small tree, usually bushy, with dense branches and foliage. It often forms dense impenetrable thickets with *Hakea* and *Allocasuarina* species. Leaves are linear and crowded, up to 15 mm long, with margins curled to the undersurface. Flower spikes are up to 15 cm long. They are rich in nectar and visited by many different honeyeaters. The style which extends beyond the flower has an orange hook at the tip, contrasting with the red of the spike.

HABITAT Scrub and heaths, on sands.

DISTRIBUTION Coastal zone, from Hastings River district to south coast. Less common on central and southern tablelands.

FLOWERING April–August.

Banksia spinulosa var. *spinulosa*

Hairpin Banksia **Plate 581**

The typical form of *B. spinulosa* is a small, multi-stemmed branching shrub, 1 to 2 m tall, with a lignotuber. It has narrow, upright leaves, 3 to 8 cm long. These leaves vary in their width and number of teeth, but the typical leaf is 1 to 2 mm wide, with margins rolled back often to the midrib, a notched tip and a few teeth along the margins near the end. Flower spikes are up to 15 cm long, golden-yellow in colour, with dark red or blue-black styles with yellow tips. These styles are hooked at the tip, hence the common name. Flower spikes sit on the junction of leafless branches, usually emerging above the foliage to be very conspicuous.

HABITAT Forest and scrub, on sandy soils.

DISTRIBUTION South of Brisbane Water National Park and westwards to the Blue Mountains. Between Sydney and Newcastle there is a gradual change to var. *collina*. Extending along the coast and tablelands to Vic.

FLOWERING March–September.

581. *Banksia spinulosa* var. *spinulosa*

Banksia spinulosa var. *collina*

Hill Banksia **Plate 582**

This variety is similar in appearance to var. *spinulosa*, as it is a multi-stemmed shrub with a lignotuber. The only difference is in its leaves which are broader, with margins curled under but not touching the undersurface (recurved not revolute) and are usually serrated throughout. Occasionally the margins are entire. This variety is distinguished from var. *cunninghamii* by the presence of the lignotuber and the white undersurface of its leaf. Near the Hawkesbury River, which is the southern limit of its range, intermediate forms between var. *collina* and var. *spinulosa* are found.

HABITAT Low coastal scrub, open forest and heath, on sandy soils.

DISTRIBUTION North of Hawkesbury River. Extending to north coast and northern tablelands.

FLOWERING April–August.

582. Banksia spinulosa var. *collina*

Banksia spinulosa var. *cunninghamii*

Plate 583

This variety is regarded by some botanists as a separate species (*B. cunninghamii*) and by others as part of the *B. spinulosa-collina* complex. It has a number of characteristics which separate it from those 2 varieties. The most obvious is that it is a robust shrub or small tree to 5 m tall (not a small branching shrub), is single-stemmed at the base and has no lignotuber. Branchlets are usually very hairy. Its narrow leaves are usually entire or almost so, but some populations, such as at Mt Wilson, have toothed leaves. The leaf has a close pale-brown downy undersurface, without obvious venation. The flower spike has attractive styles which are deep purple, almost black, with a pale hooked tip.

HABITAT Sheltered hillsides amongst tall understorey.

DISTRIBUTION Blue Mountains (e.g. Blackheath, Lawson, Hassans Walls) and southern highlands (e.g. Fitzroy Falls, West Berrima).

FLOWERING April–July.

583. Banksia spinulosa var. *cunninghamii*

Banksia marginata

Silver Banksia, Honeysuckle **Plate 584**

Silver Banksia is a tall shrub to small tree (often dwarfed on open heaths). It is widespread throughout eastern Australia in the higher rainfall areas, occurring from sandy coastal heaths to rocky subalpine situations. It is common in the Sydney district and its characteristic small yellow flower spikes, rich in nectar, have given it the popular name of 'Honeysuckle'. These flower spikes are 4 to 10 cm long and 4 to 6 cm wide. The styles are not hooked but straight or slightly curved upwards. The stigma on the end of the style is small and thin. Leaves are narrow, 2 to 6 cm long and 3 to 10 mm wide, cut off squarely at the tip, either toothed or entire and distinctly white-velvety on the undersurface. The withered flower parts adhere for a long time to the woody cone.

HABITAT Heath and dry open forest.

DISTRIBUTION Widespread from coast to mountains. Found throughout NSW, except in drier western plains. Also in Vic, Tas and SA.

FLOWERING February–July.

584. Banksia marginata

Banksia oblongifolia **Plate 585**

This is a common low shrub on shallow sandstone soils and exposed rocky ridges around Sydney. It can be readily identified when new growth is present as the young shoots are covered with dense rusty hairs. The main veins on the undersurface of the leaves are also covered with rusty hairs, especially when young. Leaves are elliptic-oblong, 5 to 10 cm long, silver-backed, usually toothed but sometimes entire. Leaves spread out, rather than stick upwards as do those of *B. paludosa*. Flower spikes are greenish yellow, up to 10 cm long and 6 to 7 cm wide. Styles are 3 cm long, without hooks and stick out at right angles to the flower spike or point slightly downwards.

HABITAT Dry heaths and rocky situations.

DISTRIBUTION Widespread and common along coastal zone. Extending along north coast to Qld.

FLOWERING March–July.

585. Banksia oblongifolia

Banksia conferta var. *penicillata* **Plate 586**

This tall shrub is restricted to a few localities in the Blue Mountains around Newnes-Wolgan Valley. This population is separated by about 600 km from the closest occurrence of *B. conferta* var. *conferta* on Lamington Plateau and Glasshouse Mountains in Queensland. The Newnes population was recorded in 1906 but neglected since then and confused with *B. paludosa*; only since 1970 have further specimens been examined. Variety *penicillata* is an open shrub to about 4 m tall without a lignotuber. Leaves are obovate, about 8 to 10 cm long, with numerous teeth along the margins. The new growth is covered with tan downy hairs. Flower spikes are up to 19 cm long, crowded with flowers which are grey and brown in bud, opening to yellow. The racemes of flowers cling to the old spikes and the numerous (often 100) small follicles mostly remain closed until burnt.

HABITAT Dry open forest on shaly loam, steep rocky slopes and around sandstone outcrops.

DISTRIBUTION Upper Blue Mountains (e.g. along Newnes tunnel road).

FLOWERING April–July.

586. Banksia conferta var. *penicillata*

587. *Banksia integrifolia*

Banksia integrifolia

Coast Banksia **Plate 587**

As the common name suggest, *B. integrifolia* occurs all along the coast wherever the sandy soils are consolidated and relatively deep. It is a common tree around beaches and in woodlands near tidal inlets, where it may grow to 15 m tall. There is also a form, known as var. *compar,* which occurs in scattered localities in the forests of the upper Blue Mountains (e.g., near Mt Wilson). Adult leaves are mostly entire (hence the specific name 'integer'—whole), oblanceolate, 5 to 10 cm long and 15 mm wide, tapering to a short stalk. They are dark green and smooth above, silvery underneath. Sometimes the leaves are irregularly toothed. Flower spikes are relatively small, about 10 cm long, yellow, with straight styles. The woody valves of the fruit protrude from the spike and are usually open as the seeds are released as soon as they are mature.

HABITAT Hind dunes near sea and coastal forests on sandy soils; also forests of tablelands.

DISTRIBUTION Coast, extending from Port Phillip (Vic.) north to Fraser Island (Qld). Var. *compar* occurs on tablelands from Mt Wilson to Qld.

FLOWERING Chiefly January–June, but some flowers at other times.

588. *Banksia paludosa*

589. *Banksia robur*

Banksia paludosa

Swamp Banksia **Plate 588**

This small shrub grows to about 1 m tall and has coarsely toothed elliptic to obovate leaves. It may be initially confused with *B. oblongifolia,* but there are significant differences. *B. paludosa* is restricted to wet sandy places whereas *B. oblongifolia* prefers dry rocky situations. The young branches of *B. paludosa* are hairless; those of *B. oblongifolia* are covered with dense rusty hairs. The main veins on the undersurface of the leaves of Swamp Banksia lack any rusty hairs. Leaves are almost spoon-shaped (obovate-spathulate), 4 to 8 cm long, silver underneath, toothed mainly at the tip, tapering to the base and tend to point upwards rather than spreading out. Flower spikes are yellowish, up to 10 cm long but relatively narrow (3 to 4 cm wide). Styles are small, about 16 mm long, without hooks, and stick out at right angles to the flower spike.

HABITAT Sandy wet heaths.

DISTRIBUTION Woronora Plateau and Illawarra (e.g., Maddens Plains, Carrington Falls) and near Glen Davis. Extending to south coast and southern tablelands.

FLOWERING April–August.

590. *Banksia serrata*

Banksia robur

Swamp Banksia, Large-leaf Banksia **Plate 589**

Swamp Banksia is a shrub, 1 to 2 m tall, common in drainage lines and hanging swamps of coastal heaths. Its large spreading leathery leaves make identification easy. These are up to 25 cm long and 8 cm wide, irregularly toothed with a distinctive yellow midvein. The undersurface of the leaf is strongly veined and usually covered with rusty-brown hairs. Branches are usually covered with grey downy hairs. Flower spikes are green in bud, becoming cream-green and up to 15 cm long; the style is spreading and not hooked.

HABITAT Swampy areas and drainage lines, especially in upper catchments of creeks in heath and open forest.

DISTRIBUTION Coastal zone. Extending along north coast to Qld.

FLOWERING January–July.

Banksia serrata

Old Man Banksia, Saw Banksia **Plate 590**

The tough gnarled appearance of this tree, with its rough-barked trunk, its leathery leaves and bristly old flower spikes has led to its common name, 'Old Man Banksia'. It is a common species on poor sandy soils and although usually a tree or a tall shrub, it may be dwarfed on exposed coastal heaths to less than 1 m tall. Old Man Banksia closely resembles *B. aemula* and they are difficult to separate without close examination. However, *B. serrata* is much more widespread and common in the Sydney area (see *B. aemula* for the main differences). *B. serrata* usually has one main stem. Its leaves are stiff, bluntly toothed, up to 16 cm long, 2 to 4 cm wide and green on both surfaces. Flower spikes are large, up to 16 cm long and 10 cm wide, grey in bud but greyish-cream in full flower. The straight styles have cylindrical stigmas, 2 mm long, the main botanical feature which separates this species from *B. aemula.*

HABITAT Dry open forest, heath and coastal hind dunes, on sandy soils.

DISTRIBUTION Coast to mountains. Common along coast, extending from Qld to Vic and Tas.

FLOWERING December–June; often flowering at other times on heath.

591. *Banksia aemula*

Banksia aemula

Wallum Banksia **Plate 591**

B. aemula is a tall shrub to 4 m, typically found on old coastal dunes. It is common on the north coast of NSW, but is not found south of Botany Bay. Further inland, on the low stable sand deposits south of Richmond, some specimens are 7 m tall. This species closely resembles *B. serrata* and careful examination of the flower of each is necessary for positive identification. The most important botanical difference is that in *B. aemula* the stigma on the end of the style is ovoid in shape and about 1 mm long. The habit of this plant is more spreading and branched from the base (not one main stem) and the serrated leaves are rather narrower (1 to 2 cm wide) than *B. serrata*. Follicles are larger and the flowers pale lemon-yellow.

HABITAT Coastal sand dunes and deep stable sand deposits.

DISTRIBUTION Coast. Extending from south-eastern Qld southwards to La Perouse (northern headland of Botany Bay). Also Agnes Banks Nature Reserve, south of Richmond.

FLOWERING March–June.

Lythraceae

Although this is a mainly tropical group, the type genus, *Lythrum*, is common in northern temperate regions. Two widespread species of *Lythrum* occur in the Sydney area but are by no means common, and the family is perhaps better known in this country by the crepe myrtle (*Lagerstroemia indica*) from China.

Lythrum

The flowers of *Lythrum* are unusual in being heterostylous, which in this case means that the styles, which may be long, medium or short, occur with stamens that are respectively medium and short, long and short, and long and medium. Flowers of the one type are borne on separate plants.

Lythrum salicaria

Purple Loosestrife **Plate 592**

The erect leafy, hairy stems of this perennial grow to 1 m tall. Leaves are mostly opposite, stalkless, lanceolate pointed, 2 to 5 cm long. Flowers are light purple or deep pink, clustered in leafy bracts at the ends of the stems. Individual flowers have petals about 8 mm long and a tubular fused calyx about 5 mm long. Purple Loosestrife is easily distinguished from Lesser Loosestrife (*L. hyssopifolia*), as that species is a weak hairless annual, little more than 30 cm tall, with small pinkish flowers borne singly (not in clusters) in the leaf axils.

HABITAT Edges of swamps and watercourses.

DISTRIBUTION Widespread from coast to mountains, but not common. Extending to western plains, and along coast and tablelands north to Qld and south to Vic and Tas. Also occurring in temperate regions beyond Australia

FLOWERING January–March.

Myrtaceae

Myrtaceae is a large family of about 100 genera and 3000 species. It is widespread in tropical and subtropical regions, and is an important element in Australian and tropical American forests. It is represented in Europe and Mediterranean countries by the common myrtle, *Myrtus*, a shrubby plant with fragrant leaves, which had particular significance in the feasts and rituals in classical times, and which gives the family its name.

The family consists of evergreen shrubs and trees with (usually) opposite entire leaves dotted with translucent oil glands. These glands, which characterise the family, are also present on young stems, flowers, fruit, and sometimes also the bark. Some members yield spices such as clove and pimento, others yield essential oils such as bay and eucalyptus, and some produce edible fruit such as guava. Many are used as ornamentals and some, such as the eucalypts, are a valuable timber resource.

Myrtaceae is divided into 2 subfamilies: Myrtoideae with fleshy berries; and Leptospermoideae, a predominantly Australian group with capsular or nut-like fruits. In the Sydney area, members of the first group are commonly found in or near rainforests or moist eucalypt forests, whereas those of the second group inhabit drier and less fertile areas.

Myrtaceous flowers have 4 or 5 petals and sepals (the parts are fused in *Eucalyptus*) and numerous, often conspicuous, stamens. The flower arrangement is frequently cymose, although reduction of some of the parts in the inflorescence of *Eucalyptus* gives a structure resembling an umbel. In some cases, such as in *Leptospermum*, *Baeckea* and *Micromyrtus* the flowers are solitary in the upper leaf axils and the petals are prominent. In a few cases, such as in *Callistemon* and *Melaleuca*, the flowers are arranged in showy terminal spikes and the long filaments of the stamens provide a bottlebrush appearance.

The 2 genera *Angophora* and *Eucalyptus* have many features in common, and it is now recognised that the differences between them are no more distinctive or fundamental than differences within *Eucalyptus*.

592. *Lythrum salicaria*

Acmena

Tree with opposite leaves. Flowers with 4 petals and 4 deciduous sepals and numerous free conspicuous stamens. Fruit a purple-white globular berry.

593. *Acmena smithii*

Acmena smithii

Lillypilly **Plate 593**

Lillypilly is a common small tree with attractive glossy leaves and abundant white to mauve fruits. It is widely cultivated. Near the sea cliffs it may be a compact shrub, 1 to 2 m tall, but it will grow to 20 m or more in sheltered rainforest. Leaves are opposite, broadly lanceolate, 4 to 8 cm long, glossy above, with a long blunt point. These leaves are similar to those of *Syzygium australe* and flowers or fruit are required for positive identification. Flowers are creamy-white or pinkish, small but numerous in branched clusters. There are 4 petals, 4 sepals and numerous stamens about 3 mm long. Fruits are often very showy in early winter. They are round, 10 to 20 mm across, and crowned by the circular rim of the calyx.

A smaller-leaved variety (var. *minor*) has also been recorded on the central coast. Its leaves are elliptic, 3 to 6 cm long; its fruits are also smaller (6 to 15 mm across).

HABITAT Coastal rainforest and sheltered places.

DISTRIBUTION Widespread from coast to mountains. Extending from Gippsland (Vic) to north Qld in most coastal and escarpment rainforests.

FLOWERING November–February.

594. Syzygium australe

Syzygium

Large shrubs or trees with opposite leaves. Flowers with 4 petals and 4 sepals, persistent or deciduous, and numerous free conspicuous stamens. Fruit a red or blue egg-shaped berry.

Syzygium australe

Brush Cherry **Plate 594**

Brush Cherry is usually a small tree, 5 to 8 m tall, with a dense crown of opposite, lanceolate glossy leaves. These leaves are about 6 cm long, with either a short fine point or bluntly pointed, the upper surface dark green, the underside paler. Young leaves and stems are often a reddish colour. Flowers are a mass of white stamens. Two features distinguish this species from *S. oleosum*. Its sepals are of unequal length and its fruit is a red oval berry.

HABITAT In or near rainforest from littoral sands to sheltered gullies, especially near watercourses.

DISTRIBUTION Widespread in Illawarra and Gosford districts. Recordings include Cambewarra, Minnamurra, Berry, Mt Keira, Berkeley Hill, Kurnell and Gosford. Extending from near Batemans Bay (south coast) along the coast to north Qld.

FLOWERING January–April.

595. Syzygium oleosum

Syzygium oleosum

Blue Cherry **Plate 595**

An attractive small rainforest tree, rather similar to *S. australe*, but less common in the Sydney area. Leaves are opposite, lanceolate, up to 10 cm long and with a long narrow point at the tip. Oil dots are more numerous and conspicuous than those of *S. australe*. Flowers are a mass of creamy white stamens. Sepals are almost equal. Fruit is a distinctive purplish-blue berry when ripe, urn-shaped, with succulent pulp surrounding a single mauve-pink seed.

HABITAT In or near rainforest, often in sheltered moist gullies.

DISTRIBUTION Rare in the Illawarra district (e.g. Mt Kembla) and uncommon in Gosford-Wyong district (e.g. Strickland State Forest). Extending from Jervis Bay along the coast to north Qld.

FLOWERING December–March.

596. Rhodamnia rubescens

Rhodamnia

Tall shrub or small tree of sheltered situations. Leaves broad and distinctly 3-veined. Fruit a berry.

Rhodamnia rubescens

Brush Turpentine, Scrub Stringybark **Plate 596**

Brush Turpentine is a rainforest tree. Although it may reach a height of 25 m in northern NSW, around Sydney it is seldom taller than 6 to 7 m. Its trunk has reddish-brown brittle bark of stringy appearance. Branchlets are downy towards the ends, especially when young. Leaves are opposite, broadly lanceolate and up to 12 cm long. Three conspicuous parallel veins run the length of the leaf from which its former name, *R. trinervia*, was derived. Flowers are white and arranged in clusters from the leaf axils. They have numerous free stamens arranged in several whorls. Fruit is a berry, about 6 mm across, crowned by the remains of the calyx lobes; its colour changes from green to red to glossy black.

HABITAT Rainforest and rainforest margins; a common pioneer species around the edge of eucalypt forest.

DISTRIBUTION Chiefly coastal zone, but also recorded in mountains. Recordings include Minnamurra Spit, Macquarie Pass, upper Hacking River, Palm Jungle, Olney State Forest and Lithgow. Extending from about Milton on the south coast northwards to Qld; also northern tablelands.

FLOWERING September–October.

Austromyrtus

Shrubs or small trees with opposite leaves. Flowers with 5 petals and 5 sepals; stamens numerous, free and in several rows. Fruit a berry.

597. Austromyrtus tenuifolia

Austromyrtus tenuifolia

Narrow-leaf Myrtle **Plate 597**

This is a graceful shrub restricted to the Sydney district. Leaves are opposite, narrow, up to 4 cm long with edges curled under. The lower surface of the leaf is softly downy, especially when young. Flowers are white (often pinkish in bud), 5-petalled with numerous free stamens and grouped together on short stalks in the leaf axils. Fruit is an indigo berry.

Less widespread is *Austromyrtus acmenioides* (Scrub Ironwood), a small tree from Qld and the north coast which reaches its southern limit in the Illawarra rainforests. It has flat, elliptic to lanceolate leaves up to 8 cm long.

HABITAT Damp sheltered places in forests, often beside streams.

DISTRIBUTION Endemic in Sydney district. Scattered from coast to mountains. Recordings include Girrakool (Brisbane Water National Park), Pennant Hills Park, Glenbrook Creek and Grose River.

FLOWERING November–December.

598. *Micromyrtus ciliata*

599. *Micromyrtus blakelyi*

600. *Micromyrtus minutiflora*

Micromyrtus

Low shrubs with small opposite leaves. Flowers small, cup-shaped, with 5 petals; stamens 5, each one opposite a petal (10 stamens in 2 rows in some western species). Ovary 1-celled; fruit a nut.

Micromyrtus ciliata

Fringed Heath-myrtle **Plate 598**

M. ciliata is a low spreading shrub. Leaves are very small (2 to 3 mm long), opposite, each pair at right angles to those adjacent, crowded, overlapping, dotted with oil glands and triangular in cross-section. The leaf edges are minutely fringed. Flowers are white (sometimes pinkish), with 5 petals and 5 stamens opposite the petals. Stamens and style are only 1 mm long. Each flower has 4 ovules per ovary. Although the flowers are small (3 to 4 mm across) and borne singly in the upper leaf axils, they are abundant and make the plant very attractive. Small differences separate this species from *M. blakelyi* and *M. minutiflora*. *M. blakelyi* has a fringed leaf keel, fringed margins of sepals, and stamens and style above 2 mm long. *M. minutiflora* has a narrower floral tube and 2 ovules per ovary. Both *M. blakelyi* and *M. minutiflora* are rare species with restricted distributions.

HABITAT Variety of habitats, including sandy coastal heaths and rocky slopes.

DISTRIBUTION Coast and lower Blue Mountains (to Springwood). Also occurring in western slopes (e.g. Warrumbungle Mountains), Qld, Vic and SA.

FLOWERING August–September.

Micromyrtus blakelyi **Plate 599**

A study of the genus *Micromyrtus* in 1983 referred to this species as 'rare, probably endangered, possibly extinct'. The authors found a number of plants growing on an area of flat sandstone within Muogamarra Sanctuary between Cowan and Hawkesbury River, a location recorded by W.F. Blakely in 1918. It is a low, cushion-like shrub, 30 to 60 cm high, occurring as dense masses in cracks and depressions on these flat rocks. Leaves and flowers are similar to those of *M. ciliata*. The main differences are that the leaf keel and sepals are clearly fringed with tiny hairs, and the stamens and style are above 2 mm long.

HABITAT Flat sandstone outcrops.

DISTRIBUTION Endemic in the Hawkesbury River area. Small populations recorded in Muogamarra Sanctuary and Canoe Grounds.

FLOWERING October–November.

Micromyrtus minutiflora **Plate 600**

A rare and endangered species, vulnerable because of its small populations which only occur on old river sediments around Castlereagh (Cumberland Plains). These areas are gradually being cleared as settlement becomes more intense. It is a small shrub, seldom 1 m tall, with opposite, linear leaves, about 3 mm long, which are crowded and overlapping on the short side branches but appressed to the stem on older branches. Leaf edges are fringed with minute hairs. Flowers are similar to those of *M. ciliata*, but the floral tube is narrow and funnel-shaped. Close examination under a lens reveals only 2 ovules per ovary, an important distinguishing feature.

HABITAT Open forest, on consolidated river sediments.

DISTRIBUTION Restricted to Cumberland Plain, mainly around Castlereagh.

FLOWERING September–December.

Calytrix

Shrub with leaves spirally arranged. Flowers with 5 spreading petals. The 5 sepals develop conspicuous long bristle-like appendages (calyx-awns).

Calytrix tetragona

Fringe-myrtle **Plate 601**

This bushy shrub is a common component of the Sydney sandstone heath community. Its attractive, often abundant, star-like flowers and distinctive red calyces with long awns make it noticeable and easy to identify. Leaves are variable, linear, 4 to 8 mm long, finely toothed, triangular in section and point upwards. Flowers are in terminal heads, each flower with 5 white to pink spreading petals and numerous stamens. The petals are deciduous, leaving the 5 red sepals which have long calyx-awns.

HABITAT Heath and shrubland, on sandy soils.

DISTRIBUTION Widespread from coast to mountains. Found throughout NSW, except far western plains, and in all States, but not NT.

FLOWERING July–December.

Darwinia

Shrubs with narrow leaves, usually triangular in section. The flower is a narrow tube, almost enclosed at the top by incurved calyx lobes consisting of 5 small petals and 5 petal-like sepals. Style is erect and protrudes beyond the floral tube. Two bracteoles, sometimes deciduous, sometimes persistent, are at the base of the floral tube. The distinction between the species often depends on a close examination of the flower, including the size of the floral tube, the length of the style, and the nature of the bracteoles.

601. *Calytrix tetragona*

602. *Darwinia fascicularis* ssp. *fascicularis*

Darwinia fascicularis ssp. *fascicularis*

Plate 602

The specific name of this upright shrub (to 2 m tall) is a reference to the Latin word 'fascis', a bundle. A Roman symbol of authority was a bundle of rods with an axe in the middle (the political term 'fascist' is derived from the same source). A characteristic feature of this species is its crowded bundles of flowers, which contain more flowers than other *Darwinia* species in the area. The number of flowers per bundle varies, even on the one plant. North of the Hawkesbury River, the average is 6 flowers per bundle; in Ku-ring-gai Chase the average is 11 per bundle; south of Sydney the average is 14 per bundle. Flowers are white, turning red with age and styles are straight. Leaves are almost terete, not compressed as are the leaves of the other species in the area. The floral tube is 5 to 7 mm long; style 12 to 18 mm long; the bracteoles are deciduous and 3 to 5 mm long.

Four natural hybrids have been recorded: *D. fascicularis* x *D. procera* and *D. fascicularis* x *D. glaucophylla* at Piles Creek near Kariong; *D. fascicularis* x *D. grandiflora* and *D. fascicularis* x *D. diminuta* near Helensburgh.

HABITAT Shrubland and sandy heath.

DISTRIBUTION Common from Gosford to Bulli within 30 km of coast. Recordings include Brisbane Water, Royal and Heathcote National Parks and Hilltop-Nattai.

FLOWERING June–September, with some flowers at other times.

603. *Darwinia fascicularis* ssp. *oligantha*

Darwinia fascicularis ssp. *oligantha*

Plate 603

This subspecies is found only in the central Blue Mountains, where it is a decumbent shrub, sometimes rooting along the branches. Leaves are terete, like those of subspecies *fascicularis*, but the flowers are mostly in clusters of 4. Flowers are white, turning to red with age, and styles are straight. The size of the floral tube, style and bracteoles are similar to those of ssp. *fascicularis*.

HABITAT Heath, on sandy soils.

DISTRIBUTION Restricted to the Blue Mountains from Mt Banks to Kings Tableland and Woodford.

FLOWERING October–February.

604. *Darwinia taxifolia* ssp. *taxifolia*

Darwinia taxifolia ssp. *taxifolia* **Plate 604**

This is a decumbent shrub (occasionally erect) from the central Blue Mountains. Its leaves are compressed, opposite, 5 to 12 mm long. Flowers are deep red, 2 to 4 (mostly 3) in a cluster. A characteristic feature is that the bracteoles around the flowers are persistent and extend beyond the floral tube. The floral tube is 5 to 6 mm long; bracteoles 5 to 8 mm long; styles 6 to 12 mm long.

HABITAT Heath, on sandy soils.

DISTRIBUTION Endemic in central Blue Mountains, above 915 m including Blackheath, Narrow Neck, Echo Point (Katoomba), Leura and Kings Tableland (Wentworth Falls).

FLOWERING September–December.

Figure 11

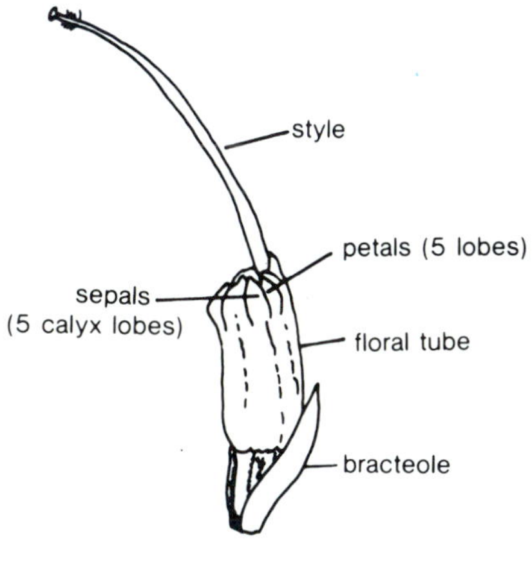

A *Darwinia* flower

605. *Darwinia taxifolia* ssp. *macrolaena*

Darwinia taxifolia ssp. *macrolaena*

Plate 605

This subspecies has larger floral parts than ssp *taxifolia*. It is also a decumbent to erect shrub with compressed, opposite leaves and occurs in rocky heaths. Flowers are deep red, in a cluster of 3 or 4. The floral tube is 6 to 9 mm long. Bracteoles are persistent, 7 to 14 mm long, extending beyond the floral tube. The style is conspicuous as it may be 15 to 24 mm long.

HABITAT Heath, on sandy soils.

DISTRIBUTION Restricted to the southern highlands (e.g. Moss Vale, West Berrima, Sutton Forest), southern tablelands and coastal plateaux around Nowra (e.g. Sassafras-Nerriga road).

FLOWERING October–November.

Darwinia grandiflora **Plate 606**

Despite the specific name, this *Darwinia* does not have larger flowers than other members of the genus. It is a prostrate shrub, rooting along the branches; smaller leafy branchlets grow up from the prostrate stems. Its leaves are compressed, opposite, 8 to 18 mm long, and triangular in cross-section. Flowers are in clusters of 4 or 5. The floral tube is white-pink when young with a contrasting deep red style, but becomes dark red with age. The bracteoles are mostly deciduous, 4 to 8 mm long; if present, they are shorter than the floral-tube. A characteristic feature of this species and *D. diminuta* is that the calyx-lobes are almost as long as the petals (both enclosing the top of the floral tube). The floral tube is 7 to 12 mm long; style 12 to 20 mm long.

HABITAT Wet heath, on sandy soils.

DISTRIBUTION Endemic in the area between Waterfall and Dapto. Recordings from Helensburgh, Darkes Forest, Heathcote National Park, Maddens Plains, Mt Ousley and West Dapto.

FLOWERING July–August.

606. *Darwinia grandiflora*

607. *Darwinia diminuta*

Darwinia diminuta Plate 607

This is a rare and vulnerable small shrub, restricted to 2 isolated locations north and south of Sydney. Unlike *D. grandiflora* it is usually an upright plant, to about 1 m tall. Its leaves are compressed, 6 to 11 mm long, triangular in cross-section, in pairs alternately at right angles on short side branches. Flowers are in clusters of 2 to 5 at the end of short side branches. The floral tube is white when young, becoming pink-red with age. The bracteoles are deciduous, 3 to 4 mm long; if present, they are shorter than the floral tube. Calyx lobes are almost as long as the petals (both calyx and petals enclose the top of the floral tube). The floral tube is 3 to 5 mm, style 5 to 9 mm long.

HABITAT Damp or dry sandy heath.

DISTRIBUTION Restricted to 2 areas—Sutherland to Bulli (including Lucas Heights, Heathcote, Helensburgh) and Terrey Hills to Manly.

FLOWERING September–January.

Darwinia camptostylis Plate 608

A small upright shrub (about 30 cm tall) from the south of Sydney. This species is readily identified by its very short style which is bent at right angles. Leaves are compressed, 6 to 17 mm long, triangular in cross-section, in pairs alternately at right angles on short side branches. Flowers are small in clusters of 2 to 4. The floral tube is white-cream when young and the style is white tipped with red. Bracteoles are shorter than the floral tube, yellow-green or brownish, and often persist after the flower opens. The floral tube is 3 to 6 mm long; style 2.5 to 5 mm long; bracteoles 3.5 to 6 mm long.

HABITAT Heath, on sandy soils.

DISTRIBUTION Recorded from Little River (Nattai catchment), but mainly a species from south of Shoalhaven River (e.g. Jervis Bay) to far south coast (e.g. Green Cape) and Gippsland (Vic).

FLOWERING August–September.

608. *Darwinia camptostylis*

609. *Darwinia leptantha*

Darwinia leptantha Plate 609

A small upright shrub (about 80 cm tall) which may be confused with *D. camptostylis* as their distributions overlap near Nowra. *D. leptantha* has narrower, slightly curved flowers, smaller deciduous bracteoles and longer styles which are not bent at right angles. Leaves are compressed, 7 to 11 mm long, triangular in cross-section, in pairs alternately at right angles on short side branches. Flowers are small, in clusters of 2 to 5, almost hidden by the leaves. The floral tube is white, becoming redder with age. Bracteoles are shorter than the floral tube and usually fall off before the flowers open. The floral tube is 4 to 6 mm long; style 4 to 6.5 mm long; bracteoles 2 to 3.5 mm long.

HABITAT Heath, on sandy soils.

DISTRIBUTION Coast from Laurieton in the north to Clyde River in the south. Uncommon near Sydney, but recorded from North Head (Sydney Harbour) and Long Bay. More common in Jervis Bay area (Currarong—Beecroft Peninsula) and Myall Lakes National Park.

FLOWERING April–September.

Darwinia glaucophylla Plate 610

A rare species, vulnerable because its distribution is limited to a small area between Gosford and the Hawkesbury River. It is a prostrate plant, often forming extensive mats over rocky areas or in wet sandy soils. Short branchlets grow erect from the prostrate stems, seldom more than 10 to 15 cm tall. Leaves are compressed, 8 to 17 mm long, triangular in cross-section and in pairs alternately at right angles. Leaves have a characteristic bluish (glaucous) colour and many have a reddish tinge. Flowers are in clusters of 2 to 4, white when young, turning pink-red with age, with a long straight style. Bracteoles are red-brown and fall off as the flower ages. The floral tube is 7 to 8 mm long; style up to 16 mm long; bracteoles 4 to 6 mm long.

HABITAT Heath.

DISTRIBUTION Restricted to small areas around Calga, Kariong and Mr Kariong (Brisbane Water National Park).

FLOWERING July–November.

610. *Darwinia glaucophylla*

611. *Darwinia biflora*

Darwinia biflora Plate 611

A rare species, vulnerable because its distribution is limited to Wianamatta shale and laterite deposits in a small area centred around Hornsby. It is an upright shrub to 80 cm tall. Leaves are compressed, 6 to 10 mm long, triangular in cross-section and in pairs alternately at right angles. The flowers are the main feature for identification as they are unusual and unique amongst local *Darwinia* species. They are green, usually in groups of 2 and surrounded by 2 broad red bracteoles which are almost as long as the floral tube. The style is green and straight. The bracteoles persist until the flower opens, then fall off. The floral tube is 5 to 8 mm long; style is 10 to 14 mm long; bracteoles 4 to 8 mm long.

HABITAT Open forest and shrubland, on shale and lateritic sandstone.

DISTRIBUTION Restricted to small areas around Cowan, Mt Colah, Hornsby and Pennant Hills Park.

FLOWERING July–November.

Darwinia peduncularis Plate 612

A rare species, vulnerable because its distribution is limited to small areas between Hornsby and the Hawkesbury River and an isolated occurrence in the Blue Mountains. It is a spreading shrub up to 1.5 m tall. Leaves are compressed, 7 to 12 mm long,

612. *Darwinia peduncularis*

triangular in cross-section, in pairs alternately at right angles and less crowded than most other species of *Darwinia*. Flowers are creamy-white, sometimes with pink petals, in groups of 2 (rarely 4) and surrounded when in bud by 2 broad red bracteoles, which fall off as the flower opens. The main feature for identification is that the flowers are on stalks 4 to 7 mm long. The style is bent. The floral tube is 9 to 12 mm long; style is 6 to 10 mm long; bracteoles 4 to 8 mm long.

HABITAT Open forest, on or near rocky outcrops.

DISTRIBUTION Restricted to Hawkesbury River-Hornsby area (including Elouera Reserve, Brooklyn, Berowra and Gundar Trig, Mt Ku-ring-gai) and Kings Tableland (Waratah Trig) in Blue Mountains.

FLOWERING July-August; as late as October in Blue Mountains.

613. Darwinia procera

Darwinia procera **Plate 613**

Although *D. procera* is regarded as a rare species because of its limited distribution, it is fairly common in certain localities within its narrow range. It is an erect shrub, sometimes 2 m tall. Leaves are compressed, 10 to 25 mm long, triangular in cross-section and in pairs alternately at right angles. Old leaf bases form distinct ridges on the stems. Flowers are yellow with dark red lobes on top, in groups of 4 to 8, with straight white or red styles and short (1 to 2 mm) flower stalks. Bracteoles are red-brown, persisting until flowering when they fall off. The floral tube is 5 to 8 mm long; style 14 to 20 mm long; bracteoles 8 to 11 mm long.

HABITAT Forested slopes, on sandy soils.

DISTRIBUTION Restricted to Gosford-Manly area, including Girrakool (Brisbane Water National Park) and Berowra (Ku-ring-gai Chase National Park).

FLOWERING July-October.

Backhousia

Trees with opposite leaves. Flowers with numerous free stamens arranged in several whorls. Sepals persistent and petal-like.

614. Backhousia myrtifolia

Backhousia myrtifolia

Grey Myrtle **Plate 614**

A common tree, often dominating rainforest streams, it may attain a height of 10 m or more. Large trees are often slightly buttressed and have brown, fissured and rather flaky bark. Young stems and leaves are softly hairy. Leaves are opposite, lanceolate, 4 to 7 cm long. Flowers are greenish yellow, with small (3 mm) petals, numerous stamens (10 mm long) and conspicuous sepals (6 mm long). They are arranged in small terminal clusters. After flowering, the sepals are persistent, resembling 5 green petals. Unlike *Acmena* and *Syzygium*, the fruit is a dry capsule enclosed within the enlarged calyx.

HABITAT Rainforest, usually in sheltered gullies along the banks of streams, but also growing on drier hillsides.

DISTRIBUTION Widespread from coast to mountains. Extending from near Bermagui (south coast) along coast to Qld; also southern tablelands and central western slopes.

FLOWERING November-December.

615. Choricarpia leptopetala

Choricarpia

Shrub to small tree with opposite leaves. Flowers arranged in compact globular heads consisting of numerous free stamens and small (2 to 3 mm) hairy floral tubes. Fruit a capsule with 2 segments.

Choricarpia leptopetala

Brown Myrtle **Plate 615**

Brown Myrtle is an attractive shrub or tree to 10 m tall, with conspicuous 4-angled branchlets and rusty-brown new growth which help identification when it is not in flower. Leaves are opposite, ovate to lanceolate, up to 12 cm long, smooth and green above, greyish on the undersurface. Flowers are a mass of stamens in compact globular heads on stalks about 3 cm long, superficially similar to the flowers of *Callicoma*. The name of *Choricarpia* literally means 'dancing fruit', probably a reference to the way the fruit capsule sways on its long stalks.

HABITAT Rainforest margins, especially along watercourses.

DISTRIBUTION Watagan Mountains (e.g. Yarramalong, Gap Creek Flora Reserve) and Colo River northwards; an isolated occurrence at Stanwell Park (Illawarra). Extending along north coast to Qld.

FLOWERING August–September.

Syncarpia

Tree with fibrous bark. Leaves opposite and flowers in compact heads. Fruit consists of characteristic hard woody capsules united into a head. The name 'Syncarpia' literally means 'fruits united together'.

616. *Syncarpia glomulifera*

Syncarpia glomulifera

Turpentine **Plate 616**

Turpentine is a common tall tree with greyish fibrous bark. Because the timber is fairly resistant to termites and is hard and durable, it has been extensively logged in many areas. Leaves are opposite, thick, elliptic to lanceolate, up to 12 cm long, and greyish on the undersurface. Flowers are a mass of stamens in globular clusters on long stalks.

HABITAT Chiefly on good moist soils on forested slopes or rainforest margins, but also on drier shallow soils.

DISTRIBUTION Widespread from coast to mountains. Extending from near Ulladulla (south coast) northwards along coast and lower tablelands to Qld.

FLOWERING October–December.

Tristania

Following a review of *Tristania* in 1982, this genus was reduced to a single species. Other plants of the Sydney district formerly within the genus were reclassified as *Tristaniopsis*. Brush Box (formerly *Tristania conferta*), a common Sydney street tree, was renamed *Lophostemon confertus*. *Tristania* differs from *Tristaniopsis* in having opposite leaves.

617. *Tristania neriifolia*

Tristania neriifolia

Water Gum **Plate 617**

This small shrub (to 3 m tall) is endemic in the Sydney district, where it is found along creeks. Leaves are opposite, narrow lanceolate, and up to 9 cm long. Flowers are in terminal clusters, with 5 yellow petals and erect stamens; 3 to 4 stamens are fused at the base to form a bundle. The fruit is a thin-walled capsule.

HABITAT Creek beds and banks, on sandy soils or among rocks.

DISTRIBUTION Widespread and common from coast to mountains. Recordings include Woronora River, Royal and Brisbane Water National Parks and Lawson.

FLOWERING December–January.

Tristaniopsis

A genus of about 30 species, with 2 in the Sydney district. Leaves are alternate, flowers yellow and fruit is a capsule.

618. *Tristaniopsis laurina*

Tristaniopsis laurina

Kanooka, Water Gum **Plate 618**

A common tree along streams in shaded gullies where it may grow to 15 m tall. Its bark is scaly in older trees, smooth in younger trees, often blotched with grey, white and brown. Leaves are alternate (not opposite like those of *Tristania neriifolia*), lanceolate and up to 14 cm long. Flowers are terminal clusters, with 5 yellow petals and stamens which are in conspicuous short bundles. The fruit is a capsule with a rounded summit extending beyond the calyx.

HABITAT Creek banks of sheltered gullies and rainforest.

DISTRIBUTION Coastal zone. Extending along coast from Vic to Qld.

FLOWERING December–February.

Tristaniopsis collina

Mountain Water Gum

This is a recently described species (1982). It had previously been confused with *T. laurina*, but differs in a number of ways. It is usually a shrub to 3 m tall or sometimes has a mallee-like form on drier exposed ridges. It is rarely found beside streams, preferring drier slopes and ridges. Larger specimens have fibrous bark. Leaves are elliptic, smaller than those of *T. laurina*, and are often drawn out into a narrow tip. Oil dots in the leaf are more numerous. The undersurface of the leaf is usually hairless.

HABITAT Forested slopes and plateaux.

DISTRIBUTION Coastal zone. Common along Illawarra escarpment. Extending from south coast to north coast and Qld.

FLOWERING December–February.

Baeckea Heath-myrtle

Small shrubs with opposite leaves. Flowers resembling those of *Leptospermum*, with 5 to 20 stamens in a single ring. Fruit a woody capsule with 2 or 3 chambers; sepals usually persistent.

Baeckea densifolia **Plate 619**

A small shrub with characteristic leaves which are bluntly needle shaped, 3 to 6 mm long with 2 rows of oil glands. Leaves are arranged in alternate pairs, often closely appressed to the stem or on short (to 10 mm) side branchlets. Flowers are white, about 6 mm across, with 8 stamens (sometimes 9 to 12).

HABITAT Rocky heath, woodland and open forest.

DISTRIBUTION Widespread from coast to mountains.

619. *Baeckea densifolia*

Also occurring north coast, northern tablelands, western slopes, north-western plains and Qld.

FLOWERING December–May.

Baeckea virgata

Twiggy Heath-myrtle, Tall Baeckea **Plate 620**

B. virgata prefers a sheltered environment where it grows as a spreading shrub up to 2 m tall. Its small white stalked flowers are borne in characteristic umbel-like clusters of 3 or 4 flowers. Leaves are opposite, narrow-oblong, up to 25 mm long, with a distinct midvein.

HABITAT Sheltered forest and creek banks, occasionally in drier open forest.

DISTRIBUTION Scattered throughout coastal zone and Blue Mountains, but uncommon. Recordings include Wingello State Forest, Lower Portland and lower Colo River. Extending along the coast from Qld to Vic; also NT.

FLOWERING Chiefly December–January, but also observed flowering in August.

621. *Baeckea ramosissima*

620. *Baeckea virgata*

Baeckea ramosissima ssp. *ramosissima*

Rosy Heath-myrtle **Plate 621**

A low-growing, mainly decumbent shrub, easily identified by its rosy-pink flowers. These are 6 to 10 mm across, with 10 stamens, 5 between the petals and 5 opposite the centre of the petals. Leaves are opposite, narrow, 4 to 7 mm long and 1 to 2 mm wide.

HABITAT Heath and open forest in both damp and exposed dry sites.

DISTRIBUTION Common along coastal zone. Extending from Coffs Harbour (north coast) to Royal National Park; also Vic, Tas and SA. Subspecies *prostrata* occurs on the far south coast and in Vic and Tas.

FLOWERING June–January.

Baeckea linifolia

Swamp Baeckea **Plate 622**

B. linifolia is typically found along watercourses and near waterfalls where its characteristic drooping branches are readily identified. Flowers are white, small (about 5 mm across), but often abundant in the leaf axils of the pendulous branches. Leaves are opposite, narrow or needle-like and 5 to 15 mm long.

HABITAT Damp gullies and along watercourses.

DISTRIBUTION Widespread from coast to mountains. Extending from Qld along north coast, south coast and southern tablelands to Vic.

FLOWERING January–February, with a few flowers at other times.

622. *Baeckea linifolia*

623. *Baeckea brevifolia*

Baeckea brevifolia

Short-leaved Heath-myrtle **Plate 623**

This is a small undershrub, seldom reaching 1 m tall, typically found growing in clumps in shallow depressions of bare sandstone. Protected from bushfires in such sites, it may take on a gnarled appearance. As the specific name suggests, the leaves are tiny, 2 mm or less, opposite, and somewhat triangular in cross-section. Flowers are white and about 4 mm across.

HABITAT Rocky heath, often on flat sandstone outcrops.

DISTRIBUTION Widespread from coast to mountains. Recordings include Hilltop, Mt Banks, West Head and Brisbane Water National Park. Extending to south coast and southern tablelands.

FLOWERING Irregular; flowers observed July–August and February–April.

624. Baeckea imbricata

Baeckea imbricata **Plate 624**

A small undershrub, most commonly found along the very edge of coastal cliffs in damp sandy soils. Its leaves are tiny (4 mm long and 3 mm wide), rounded, crowded along the upright stems and almost overlapping. The small white flowers are stalkless and solitary from the base of the leaves.

HABITAT Damp sandy sites, heath and swamp margins.

DISTRIBUTION Widespread from coast to mountains; most common along sea-cliffs (e.g. Kurnell, Curra Moors). Extending to north coast, south coast and southern tablelands.

FLOWERING February–May.

Baeckea diosmifolia **Plate 625**

A small shrub to 1 m tall. Its leaves are small (2 to 4 mm long), narrow-ovate, opposite and concave. Young leaves are minutely hairy and fringed (clearly seen with a hand lens), a characteristic which separates this species from other small-leaved *Baeckea*. Flowers are white, about 5 mm across, and solitary in the leaf axils. Sepals are also minutely fringed.

HABITAT Heath, especially in rocky areas.

DISTRIBUTION Widespread from coast to mountains. Recordings include Brisbane Water National Park, Canoe Grounds, Agnes Banks, Macquarie Pass and McCarrs Creek (Ku-ring-gai Chase National Park). Extending to south coast, southern tablelands, north coast and central western slopes.

FLOWERING September–December.

625. Baeckea diosmifolia

Baeckea utilis

Mountain Heath-myrtle **Plate 626**

This small shrub (to 1 m tall) is a plant of the high country. It is quite common in the sub-alpine zone of the Snowy Mountains, but only reaches the south-western fringe of the Sydney district. Its leaves are 2 to 4 mm long, about 1 mm wide, narrow-oblong to elliptic, concave above and strongly keeled. Flowers are white, 4 to 6 mm across and borne singly in the leaf axils. The sepals and young leaves are entire, lacking the fringe of minute hairs of *B. diosmifolia*.

HABITAT Heath or open forest, often near swamps or in damp soils.

DISTRIBUTION Higher Blue Mountains. A tablelands species extending from northern tablelands to south tablelands and Vic; also south-western slopes.

FLOWERING December–February.

626. Baeckea utilis

Leptospermum Tea-tree

Tea-trees are shrubs or small trees with alternate leaves. Flowers have 5 round or wedge-shaped petals spreading around a central cup, with an erect style in the centre; 5 sepals lie between the petals. Numerous stamens are in a single ring around the edge of the cup. Fruit is a distinctive woody capsule, with 3 to 12 chambers. The botanical name is a reference to the many small seeds produced by the genus. This genus is currently undergoing revision. It is expected that some species will have name changes, and a number of new species will be named.

627. Leptospermum laevigatum

Leptospermum laevigatum

Coast Tea-tree **Plate 627**

This tall shrub is common on coastal headlands and hind dunes along the coast of south-eastern Australia, where it often forms dense thickets. Its bark is hard. Leaves are up to 2 cm long, grey-green, obovate and with a short blunt tip. Flowers are white, up to 15 mm across; they usually have a distinct reddish ring at the base of the stamens. The woody capsule is distinctive because it splits open into 8 to 12 compartments, the only tea-tree in the area with so many openings.

HABITAT Hind dunes near sea, and coastal headlands on sands.

DISTRIBUTION Widespread along coast. Extending from north coast southwards to Vic, Tas and SA.

FLOWERING July–October.

Leptospermum parvifolium

Small-leaf Tea-tree **Plate 628**

As the name suggests, the most obvious feature of this shrub is its small leaves. These are only 5 to 6 mm long and 1.5 mm wide, narrow-obovate in shape. This is usually a small shrub, seldom more than 1 m tall, with hard bark. Its flowers are white or pinkish, up to 18 mm across; the floral tube is covered with long weak hairs. The woody capsule is small (3 mm across), domed and splits into 3 to 5 compartments.

HABITAT Open forest, on sandy soils.

DISTRIBUTION Widespread from coast to mountains. Recordings include Moorebank, Maroota and Narrow Neck (Katoomba). Also occurring south coast, northern tablelands, western slopes and north-western plains.

FLOWERING September–November.

628. Leptospermum parvifolium

629. Leptospermum attenuatum

Leptospermum attenuatum

Flaky-barked Tea-tree **Plate 629**

This is a common but variable shrub, often growing 3 to 4 m tall, and easily identified by its rough flaky bark. Leaves are 6 to 12 mm long, oblong-elliptic and blunt, or narrow and pointed in the more robust Blue Mountains form. Flowers are white and up to 12 mm across, often with a reddish ring at the base of the stamens. The floral tube and fruit are covered with long weak grey hairs. The thin-walled capsule is 5 to 6 mm across and slightly domed.

HABITAT Dry open forest and scrubland, on sandy soils.

DISTRIBUTION Widespread and common from coast to mountains. Extending all along coast and tablelands from Qld to Vic; also central western slopes.

FLOWERING August–December.

630. Leptospermum sphaerocarpum

631. Leptospermum sp. aff. *sphaerocarpum*

Leptospermum sphaerocarpum

Round-fruited Tea-tree **Plate 630**

This is a spreading shrub to 2 m tall with hard bark. Leaves are narrow lanceolate, about 12 mm long, with edges often rolled upwards and inwards at the tip. Flowers are pink to white, and relatively large (12 to 15 mm across). The floral-tube is covered with long silky hairs. The specific name 'sphaerocarpum' refers to the spherical fruits, a characteristic feature of this species. They are round or egg-shaped, about 10 mm across. An unnamed species with close affinities to *L. sphaerocarpum* (Plate 631) is found along Colo Gorge. It has deep red petals.

HABITAT Sandy heath, woodland and open forest.

DISTRIBUTION Blue Mountains (e.g. Newnes Plateau and Wollangambe) and north-west Hornsby Plateau. Extending to north-western slopes.

FLOWERING October–December.

632. Leptospermum emarginatum

Leptospermum emarginatum **Plate 632**

This is 1 of 2 species of tea-tree in the area having leaves which are emarginate, or notched at the tip. Both grow along creek banks, and therefore may be confused. *L. emarginatum* is a tall shrub with hard bark. Its leaves are 15 to 35 mm long and up to 6 mm wide. The woody capsule is small (4 to 5 mm). Flowers are about 6 mm across, and usually paired or clustered at the base of the leaves. The size of the leaves and capsules and the clustered flowers separate it from the other species, *L. obovatum*.

HABITAT Margins of creeks and in rocky creek beds.

DISTRIBUTION Locally common in the Nattai and Nepean catchments and lower Blue Mountains (e.g. Erskine Creek). Extending along south coast to Vic.

FLOWERING November–January.

Leptospermum obovatum **Plate 633**

A shrub up to 3 m tall, with hard bark and abundant flowers borne along the ends of its somewhat drooping branches. It grows in thickets on creek banks and nearby flats. Its leaves are 6 to 12 mm long and less than 6 mm wide with rounded or notched tips. The woody capsule is about 7 mm

633. Leptospermum obovatum

across. Flowers are up to 10 mm across and usually solitary. The smaller leaves, larger capsules and solitary flowers separate this species from *L. emarginatum*.

HABITAT Creek banks and nearby flats.

DISTRIBUTION Locally common on southern highlands (e.g. Paddys River) and higher Blue Mountains (e.g. Boyd River). Extending along southern tablelands and south-western slopes to Vic.

FLOWERING November–December.

Leptospermum epacridoideum **Plate 634**

This is a small, much-branched shrub of coastal heathlands. Its small (3 to 4 mm), rounded, obtuse leaves are crowded along the branches. Flowers are white (sometimes pinkish), about 10 mm across, sparse, and terminal on short side branches. The capsule is domed and about 8 mm across.

HABITAT Heath, on sandy soils.

DISTRIBUTION Far south of coastal region. More common in the Jervis Bay-Beecroft Peninsula district. Endemic in the central and south-coast of NSW.

FLOWERING August–September.

634. Leptospermum epacridoideum

635. Leptospermum polygalifolium

Leptospermum polygalifolium

Yellow Tea-tree **Plate 635**

Yellow Tea-tree is common throughout the area, especially in damp sandy soils. A medium-sized shrub with hard bark, its leaves are narrow-elliptic, 8 to 20 mm long, green, hairless and with conspicuous oil glands. Flowers are up to 14 mm across, usually creamy-white with green centres, but some varieties have distinctly yellow buds and flowers, hence the common and former specific name, *L. flavescens*. The fruit is domed, 8 mm across, hairless, with 5 compartments.

A larger leaf form is now known as *L. morrisonii* (Plate 636). Leaves are narrow-elliptic, 20 to 30 mm long, with 3 conspicuous longitudinal veins. New leaves are pinkish and covered with soft hairs; older leaves are green, hairless and have many small oil dots (clearly visible through a lens). Flowers are white and about 20 mm across.

HABITAT Open forest, woodland and heath, especially on damp sandy soils.

DISTRIBUTION Widespread and common from coast to mountains. Extending from south coast and southern tablelands northwards along coast and tablelands to Qld; also north-western slopes, western plains, Lord Howe Island and Indonesia. It is even found high on Mount Kinabalu (4100 m) in Sabah.

FLOWERING September–December, sometimes as late as February in the mountains.

636. Leptospermum morrisonii

637. Leptospermum juniperinum

Leptospermum juniperinum

Prickly Tea-tree **Plate 637**

The leaves of Prickly Tea-tree are pungent-pointed, 6 to 12 mm long, 1 to 1.5 mm wide, in clusters, narrow-lanceolate and channelled above. This shrub grows to about 2 m tall and has hard bark. Flowers are white, 10 mm across, usually solitary and without stalks. Capsules are domed, with 5 compartments, 6 to 8 mm across.

HABITAT Heath, dry woodland and scrubland, on sandy soils.

DISTRIBUTION Widespread from coast to mountains. Recordings include Maddens Plains, Brisbane Water National Park and Katoomba. Also occurring south coast, north coast, southern tablelands and Qld.

FLOWERING October–January.

Leptospermum rotundifolium

Round-leaf Tea-tree **Plate 638**

This spreading shrub, once regarded as a round-leaved variety of *L. scoparium*, grows to about 2 m and has a hard bark. Its conspicuous features are its small rounded leaves and its large pink flowers. The leaves are about 6 mm wide, grooved along the centre, and curved down at the tip. Flowers are attractive, about 12 mm across, ranging from white to deep pink, with a dark green centre and creamy-yellow stamens. Capsules are domed and about 10 mm across.

HABITAT Open forest and scrubland, on sandy soils.

DISTRIBUTION Blue Mountains, higher parts of Woronora Plateau and southern highlands. Also occurring south coast and southern tablelands.

FLOWERING November–December.

Leptospermum squarrosum

Pink Tea-tree **Plate 639**

Pink Tea-tree is a tall spreading shrub with crowded narrow rigid leaves, 8 to 15 mm long. It bears conspicuous attractive flowers up to 16 mm across. Buds are pink, although petals vary from nearly white to rich pink. Sepals are often a deeper pink and are clearly visible between the petals. Flowers are mostly borne along the older branches. Capsules are about 12 mm across and hairless.

HABITAT Woodland, scrubland and heath, on sandy soils.

DISTRIBUTION Common on coastal sandstone areas, such as Brisbane Water and Royal National Parks; extending to Blue Mountains. Also occurring south coast and southern tablelands.

FLOWERING January–April, with some flowers as late as September.

638. Leptospermum rotundifolium

639. Leptospermum squarrosum

640. *Leptospermum arachnoides*

Leptospermum arachnoides Plate 640

This is a low stiff spreading shrub often with short leafy branches growing from the upper side of horizontal stems. Leaves are needle-like or concave-linear, about 10 mm long and 1 to 2 mm wide. Flowers are solitary, white, about 8 mm across with a woolly floral tube. The capsule is domed and covered at first with a mat of grey-brown hairs, but later becoming hairless. The decumbent habit, needle-like leaves and hairy floral tube are the chief identification features.

HABITAT Heath and exposed areas, often in damp sandy soils.

DISTRIBUTION Widespread from coast to mountains. Found throughout NSW, except for drier western plains; also Qld.

FLOWERING November–December.

Leptospermum lanigerum

Woolly Tea-tree Plate 641

The common and specific names of this erect shrub (to 3 m tall) refer to the conspicuous dense hairs on the floral tube which persist on the young fruit capsules. Its leaves are oblong, up to 20 mm long, tipped with a point, softly hairy when young. Flowers are white, solitary and about 15 mm across. Capsules are domed, hairy becoming hairless, up to 12 mm across.

HABITAT In wet or waterlogged areas, including creek banks, swamp margins and damp rocky sites.

DISTRIBUTION Widespread from coast to mountains. Most common in Blue Mountains. Extending south along south coast and southern tablelands to Vic; also Tas and SA.

FLOWERING October–November.

641. *Leptospermum lanigerum*

642. *Leptospermum macrocarpum* (red form)

Leptospermum macrocarpum Plate 642

Although once regarded as a variety of *L. lanigerum*, this shrub has enough differences to be regarded as a species in its own right. Its flowers and capsules are much larger than those of *L. lanigerum*, and it grows on drier sites. Flowers are up to 25 mm across, usually lemon-yellow but occasionally a beautiful deep pink or red. Buds, floral tube and young fruit capsules are softly hairy; the capsules become glabrous with age. The abundant capsules are 15 mm or more across. The new growth is a brilliant red-tan, an attractive and conspicuous feature even when the plant is not flowering.

HABITAT Open forest, on sandy or rocky sites.

DISTRIBUTION Localised and uncommon. Although recorded from central and south coast, most recordings have been made along or north of the Bells Line of Road in the Blue Mountains. The lemon-yellow form recorded between Mt Wilson and Zig-zag Railway; pink form recorded along Mountain Lagoon road; red form recorded on cliff-tops above Colo River.

FLOWERING October–December.

Leptospermum myrtifolium

Grey Tea-tree Plate 643

Grey Tea-tree is a small shrub with greyish, elliptical, obtuse leaves up to 12 mm long. The rather inconspicuous flowers are white and about 10 mm across. Buds, floral tubes and young capsules are covered with short grey hairs. Fruit capsules are domed, about 6 mm across, becoming hairless with age.

HABITAT Heaths and waterlogged sandy soils.

DISTRIBUTION Western Blue Mountains (e.g. Hartley Valley, Black Range and Newnes Tunnel). A tablelands species, extending from northern tablelands south to Vic; also south-western slopes.

FLOWERING September–January.

643. *Leptospermum myrtifolium*

Kunzea

Kunzeas are shrubs with alternate leaves. Flowers are arranged in fluffy terminal heads or short leafy spikes, with numerous free stamens. Each flower has 5 petals and 5 sepals shorter than the stamens. The fruit is a capsule surmounted by the persistent sepals. The genus is named after a German botanist, Gustav Kunze.

Kunzea capitata

Pink Kunzea Plate 644

There are 2 pink-flowered Kunzeas in the area. *K. capitata* is a slender erect shrub to 1.5 m tall with oblanceolate-obovate leaves, up to 9 mm long and 3 mm wide. The leaves are distinctly 3-veined, a feature which distinguishes this species from *K. parvifolia* which has smaller 1-veined leaves. Buds are conspicuous as they are gathered in terminal heads and have hairy calyces. Stamens are up to 6 mm long, pink with yellow anthers. A white-flowered form occurs in the southern Blue Mountains.

HABITAT Heath and open forest, on sandy soils.

DISTRIBUTION Widespread from coast to mountains. Common in national parks around Sydney. Also occurring south coast, southern tablelands, north coast and Qld.

FLOWERING September–October.

644. *Kunzea capitata*

645. *Kunzea parvifolia*

Kunzea parvifolia

Small-leaved Kunzea **Plate 645**

A spreading shrub to 1.5 m tall. It bears abundant rosy-pink flowers in terminals heads. Leaves are small, up to 4 mm long and 1.5 mm wide, linear to oblanceolate. Only the midvein is visible (unlike the leaves of *K. capitata* which have 3 visible veins). The calyx is usually hairless. Stamens are 2 to 3 mm long with yellow anthers.

HABITAT Heath and woodland, in shallow soils.

DISTRIBUTION Blue Mountains and southern highlands. Also occurring south coast, southern tablelands, northern tablelands, western slopes and Vic.

FLOWERING September–November.

Kunzea rupestris

Rock Kunzea

Rock Kunzea is a rare and endangered shrub occurring in small populations in rocky areas of Hornsby Plateau. It was collected from Canoe-lands (near Maroota) in 1927, but has not been recorded in recent years. It has creamy-white flowers in terminal heads and crowded narrow oblong leaves, 10 to 14 mm long. The calyx is covered with long white hairs. Flowering recorded in October, probably continuing through summer.

646. *Kunzea cambagei*

Kunzea cambagei **Plate 646**

A rare and vulnerable species because of its restricted range. It occurs only in damp sandstone country in the southern Blue Mountains where it is usually a decumbent or prostrate ground cover. Leaves are elliptic to oblong, about 5 mm long, slightly concave or almost flat. The calyx is covered with white hairs. Flowers are creamy-white in flower-heads at the ends of short side branches.

HABITAT Heath and scrubland, in damp sandstone areas.

DISTRIBUTION Endemic in southern Blue Mountains (near Mt Werong) and adjoining southern highlands (e.g. West Berrima).

FLOWERING September–November.

647. *Kunzea phylicoides*

Kunzea phylicoides **Plate 647**

This unusual plant was once classified as a *Baeckea*, then as a *Kunzea*, then a *Leptospermum* and most recently it was reclassified as a *Kunzea*. It exhibits characteristics which link it with all 3 genera. It is a large shrub, with hard bark, occurring chiefly beside mountain streams. Leaves are variable, narrow lanceolate, about 10 mm long and clustered. Small white flowers are borne abundantly in long sprays at the ends of the branches. Unlike most tea-trees, the stamens are longer than the petals and spreading. Flowers and fruit are on stalks 4 to 6 mm long. Another identifying feature is the flat or concave top of the small woody capsule.

HABITAT Open forest, among rocks, especially beside streams.

DISTRIBUTION Blue Mountains, but uncommon. Extending along the tablelands to Qld and Vic; also south coast, western slopes and SA

FLOWERING December–January.

648. *Kunzea* sp. (Mt Cookem)

Kunzea sp. **Plate 648**

This unnamed Kunzea was discovered by the authors growing on the rocky top of a cliff-line above the Coxs River at Mt Cookem in the central Blue Mountains. It is an upright shrub to 1 m tall with tiny concave ovate leaves, 3 mm long and 1.5 mm wide; each leaf has a minute stalk and the leaf base envelops the stem. Flowers are cream, in heads at the ends of short side branches. The calyx and new shoots are covered with soft short hairs.

HABITAT Rocky cliff-tops.

DISTRIBUTION Apparently endemic in the Mt Cookem area of the Blue Mountains and Axehead Mountain near Yerranderie.

FLOWERING October–November, possibly into summer.

Kunzea ambigua

Tick Bush **Plate 649**

Tick Bush is one of the commonest large spreading shrubs of the Sydney district, typically occurring in dry woodlands and heaths. It is found in most patches of urban bushland and in areas which have regenerated after clearing. Old specimens can be 5 m tall. Leaves are narrow, to 10 mm long, and crowded on short side branches. Flowers are white and abundant along the axils of the upper leaves. They are heavily scented and attract many small insects. This is the only Kunzea in the area that does not have flowers in terminal heads.

HABITAT Heath, woodland and dry open forest.

DISTRIBUTION Widespread from coast to mountains. Also occurring northern tablelands, north-western slopes, southern tablelands, south coast, Vic and Tas.

FLOWERING Mid October–December.

649. *Kunzea ambigua*

Callistemon Bottlebrush

Bottlebrushes are well known as a number of species and hybrids are popular as garden and street plants. Leaves are alternate and flowers are arranged in leafless spikes. The stamens are conspicuous as they are numerous, long, colourful and free (not joined into bundles as are the stamens of *Melaleuca*). The fruit is a woody capsule, crowded into cylindrical clusters which remain on the old wood for some years. The botanical name means 'beautiful stamen'.

Callistemon salignus

Willow Bottlebrush **Plate 650**

Willow Bottlebrush is typically found along creek banks and in damp places. It is a small tree, sometimes as tall as 6 m, with papery bark. Leaves are 6 to 10 cm long, up to 10 mm wide, flat acute, with a prominent midvein underneath. New growth is a characteristic reddish-purple colour. Flowers are lemon-yellow in spikes up to 8 cm long. This species could be confused with *C. paludosus*, but the latter has leaves and flower spikes which are only half the size of *C. salignus*.

HABITAT Swamps, creek banks and other damp places.

DISTRIBUTION Coast and lower mountains (e.g. Bents Basin, Nepean River) but uncommon. Also occurring south coast, north coast, northern tablelands and Qld.

FLOWERING September–October.

650. *Callistemon salignus*

651. *Callistemon paludosus*

Callistemon paludosus

Swamp Bottlebrush **Plate 651**

Swamp Bottlebrush is a tall shrub, 2 to 4 m tall. As its name suggests, it is found along the banks of streams and in swampy places. Leaves are narrow lanceolate, up to 6 cm long and 4 mm wide, flat and tipped with a distinct point. New growth is bronze coloured. Flowers are creamy, sometimes tinged with pink, in spikes up to 4 cm long (half the length of the flower spikes of *C. salignus*).

HABITAT Swamps, creek banks and other damp places.

DISTRIBUTION Western part of the Cumberland Plain, southern highlands and Blue Mountains. Found throughout NSW, except in far western plains; also Vic, Tas and SA.

FLOWERING January–May.

Callistemon pallidus

Lemon Bottlebrush Plate 652

This is an erect shrub, 2 to 3 m tall, occurring in rocky places. Its leaves are narrow-elliptical, up to 5 cm long and 10 mm wide, flat, stiff, and tipped with a distinct point. New growth is a soft, yellow silver. Flowers are lemon-yellow in spikes which are 3 to 8 cm long. The shape of the leaves distinguishes this species from other yellow-flowered bottlebrushes.

HABITAT Among rocks, in damp sites.

DISTRIBUTION Higher Blue Mountains (e.g. Boyd Plateau). A tablelands species, extending from Qld to Vic and Tas, but also found on north coast, south coast and western slopes.

FLOWERING November–December.

652. *Callistemon pallidus*

653. *Callistemon sieberi*

Callistemon sieberi **Plate 653**

A compact stiff small shrub, occurring in damp depressions or along water courses in higher mountain and sub-alpine zones. Its leaves are narrow, sharp, short, up to 20 mm long and 2 to 3 mm wide. Flowers are yellow in short spikes which are terminal or almost so. The short sharp leaves distinguish this species from other yellowish-flowered bottlebrushes.

HABITAT Damp places, usually along creek banks or in wet soils in forests.

DISTRIBUTION Higher Blue Mountains (e.g. Clarence). Extending along the tablelands from Qld southwards to Vic; common in sub-alpine areas of Snowy Mountains.

FLOWERING December–January.

Callistemon pinifolius

Pine-leaved Bottlebrush **Plate 654**

The only bottlebrush in the area with greenish yellow flower-spikes and stiff, needle-like leaves, *C. pinifolius* grows about 2 m tall. Leaves are up to 8 cm long, channelled above or almost terete, less than 2 mm wide and pungent pointed.

HABITAT Damp sandy sites in open forest and woodland.

DISTRIBUTION Coastal zone. Recordings include Casula and Agnes Banks. Also occurring on western slopes.

FLOWERING September–November.

654. *Callistemon pinifolius*

656. *Callistemon linearifolius*

657. *Callistemon rigidus*

658. *Callistemon citrinus*

655. *Callistemon shiressii*

Callistemon shiressii **Plate 655**

C. shiressii is another bottlebrush with white to creamy-yellow flowers. However, unlike the others, the rhachis (central stem of the flower spike) is covered with dense woolly hairs. New growth and calyces are also softly hairy. A shrub to 3 m tall, its leaves are lanceolate, dark, acute, up to 4 cm long and 8 mm wide, with distinct midvein. Flowers are in a spike about 4 cm long.

HABITAT Open forest on shale and protected river slopes.

DISTRIBUTION Uncommon in area. Recorded from rocky river banks near Colo. Absent from Blue Mountains. Also occurring north coast, central western slopes and Qld.

FLOWERING September–October.

Callistemon linearifolius **Plate 656**

There are 5 red-flowered bottlebrushes in the area with leaves which are linear to lanceolate. In this species, the leaves are narrowly lanceolate, pointed, 9 to 14 cm long and up to 12 mm wide, firm but not stiff. The small net-like veins of the leaves are clearly visible between the main and lateral veins, a characteristic feature. The flower spike may be 12 cm long. The opening of the mature fruiting capsule is more restricted than that of *C. citrinus* or *C. subulatus*, with which this species could be confused.

HABITAT Open forest and scrubland, in damp places.

DISTRIBUTION Endemic in the coastal zone around Sydney.

FLOWERING October–November.

Callistemon rigidus

Stiff Bottlebrush **Plate 657**

Stiff Bottlebrush grows on damp sandy sites in scrublands and heaths. Its leaves are narrow, stiff, pointed, up to 10 cm long and 3 to 6 mm wide. The leaf margins are thickened by ridges formed by the lateral veins. The network of reticulations between the veins is visible under a lens, but only the midrib is conspicuous. New leaves are densely covered with silky appressed hairs. Flowers are red in a spike up to 10 cm long. The opening of the mature fruiting capsule is restricted. The narrow aperture of the fruit and the wider leaves distinguish this species from *C. linearis*.

HABITAT Heath and scrubland, on damp sandy sites.

DISTRIBUTION Endemic in the coastal area around Sydney.

FLOWERING October–November.

Callistemon citrinus

Red Bottlebrush **Plate 658**

This is the most widespread and common of the red-flowered bottlebrushes in the Sydney area. Leaves are lanceolate, usually with a distinct rigid tip, up to 7 cm long and about 8 mm wide. Only the midrib is prominent. Young growth is flesh-coloured and softly hairy. Flowers are deep red in spikes up to 12 cm long. Fruiting capsules are 6 to 7 mm across (larger than those of *C. subulatus* and more flattened at the top and less restricted than those of *C. linearifolius* and *C. rigidus*).

HABITAT Prefers wet sites near streams and swamps, but tolerates drier sites.

DISTRIBUTION Widespread and common from coast to mountains. Recordings include Brisbane Water and Royal National Parks, Macquarie Pass, Lawson, Mt Irvine and Agnes Banks. Also occurring north coast, south coast, southern tablelands, western slopes, Qld and Vic.

FLOWERING March–May, also October–December.

659. Callistemon subulatus

Callistemon subulatus Plate 659

C. subulatus is often confused with *C. citrinus*, but has narrower leaves, smaller flower-spikes and capsules and is restricted in its distribution. Leaves are up to 5 cm long, 2 to 4 mm wide, with a pungent point. Flowers are red in spikes up to 6 cm long. Fruiting capsules are 4 to 5 mm across and without a restricted opening.

HABITAT Along streams.

DISTRIBUTION Woronora Plateau (e.g. The Needles, Woronora River), Upper Kangaroo Valley, southern highlands and central tablelands. Extending southwards along south coast and southern tablelands to Vic and SA

FLOWERING October–February.

Callistemon linearis

Narrow-leaved Bottlebrush Plate 660

A spreading shrub about 2 m tall, common in wet sandy soils or rocky creeks around Sydney. Its leaves are stiff, narrow (2 mm or less), linear, channelled above, up to 12 cm long and pungent pointed. Flowers are red in spikes about 10 cm long. Young growth is silky-hairy. Fruit capsules are almost globular, with a small (2 mm) opening.

HABITAT Wet sandy soils and along rocky streams.

DISTRIBUTION Widespread from coast to mountains. Recordings include Brisbane Water and Ku-ring-gai Chase National Parks, Kedumba Valley, Agnes Banks and Heathcote National Park. Also occurring north coast, south coast, northern tablelands, western slopes and north-western plains.

FLOWERING October–December.

660. Callistemon linearis

Melaleuca Paper-bark

Melaleuca species range from small shrubs to tall trees. Many are swamp-loving plants. Leaves may be either opposite or alternate. Flowers are in bottlebrush-like spikes or terminal heads. Stamens are long and conspicuous, joined at the base into 5 bundles, each bundle opposite one of the 5 small petals. Fruit is a woody capsule; these are clustered or scattered along the old wood. Despite the common name, many species do not have papery bark. The botanical name means 'black-white', an apparent reference to the variation in the colour of the bark.

661. Melaleuca erubescens

Melaleuca erubescens

Rosy Paper-bark Plate 661

A medium-sized shrub, typically found in damp soils on the western slopes. It is uncommon in the Sydney district. Flower colour is a poor guide to identification as it may range from pink to rose or lilac. Flowers are borne in spikes, 2 to 4 cm long, appearing below the tip of a shoot. Leaves are mostly alternate, about 10 mm long, narrow (1 mm wide), almost needle-like, with the tips curved down. Capsules are arranged in oblong clusters.

HABITAT Watercourses or open forest which periodically becomes water-logged, usually on shale or clay.

DISTRIBUTION Shale patches on Cumberland Plain (e.g. Castlereagh State Forest), Bankstown district (e.g. Picnic Point) and north-west of Sydney district around Newnes. Also occurring western slopes, western plains and Qld.

FLOWERING November–January.

662. Melaleuca squamea

Melaleuca squamea Plate 662

The terminal bright pink oval flower heads immediately identify this small to medium shrub. It has corky bark and stiff, erect, powdery-white branches. The crowded leaves are narrow lanceolate, pointed, 12 mm long and with margins curved upwards. The leaves, or sometimes just the leaf tips, point upwards along the stems. Capsules have wavy openings and are arranged in short clusters.

HABITAT Typically found in wet coastal heath, but also recorded from a wet shaded site below the Blue Mountains' cliff-line.

DISTRIBUTION Chiefly coastal (e.g. Royal National Park); also recorded at Centennial Glen, Blackheath. Uncommon in Sydney district. Also occurring north coast, Vic, Tas and SA.

FLOWERING August–November.

Melaleuca hypericifolia Plate 663

This is a small spreading shrub, with characteristic opposite leaves spaced evenly along the arching stems. Leaves are broad-lanceolate, up to 4 cm long, with a prominent midrib. Flowers are red, in spikes 3 to 5 cm long, borne on the older wood. Capsules have erect calyx-teeth around the rim of the opening.

HABITAT Damp sites in sandy heath and woodland; also coastal headlands.

DISTRIBUTION Common along coastline. Extending to Blue Mountains and south coast.

FLOWERING October–January.

663. Melaleuca hypericifolia

664. *Melaleuca thymifolia*

Melaleuca thymifolia

Thyme Honey-myrtle **Plate 664**

The flowers of *M. thymifolia* make it unmistakable. They are pink or mauve (occasionally white) with curled stamens and are in irregular clusters on the older wood. It is a small spreading shrub with opposite lanceolate leaves 5 to 12 mm long. Capsules have erect calyx teeth around the rim of the opening.

HABITAT Damp places in forest and heath

DISTRIBUTION Widespread and common from coast to mountains. Extending along the coast from south coast to Qld; also central western slopes.

FLOWERING Mainly October–January, but some flowers at other times.

Melaleuca capitata **Plate 665**

A small shrub (to 2 m) occurring most commonly south of the Shoalhaven River in places such as Beecroft Peninsula and Budawang Mountains. There is a limited occurrence in rocky heaths around Bundanoon. Flowers are yellow and arranged in conspicuous rounded heads. Leaves are narrow-oblong, 10 to 25 mm long, somewhat pungent. Capsules are 4 to 6 mm across, arranged in loose clusters. The larger flower heads and capsules help separate this species from the more widespread and common *M. nodosa*.

HABITAT Heath, often in damp sandy or rocky sites.

DISTRIBUTION Southern highlands (e.g. Bundanoon) and southern Blue Mountains. More common south of the Shoalhaven River along southern tablelands and south coast.

FLOWERING November–December.

665. *Melaleuca capitata*

666. *Melaleuca nodosa*

Melaleuca nodosa **Plate 666**

A small stiff shrub, easily identified by its terminal yellow rounded flower heads and tight globular clusters of capsules. Leaves are narrow, flattened or needle-like, with a pungent point, up to 20 mm long. Flower heads are 10 to 15 mm across. The rounded clusters of capsules are very conspicuous.

HABITAT Common in various coastal communities, including hind dunes, scrubland, heath and headlands.

DISTRIBUTION Widespread along coastal zone. Recordings include Munmorah State Recreation Area, Royal National Park and Castlereagh State Forest. Extending along north coast and northern tablelands to Qld; also western slopes and SA

FLOWERING September–November.

Melaleuca squarrosa

Scented Paper-bark **Plate 667**

Scented Paper-bark is a dense shrub to 2 m tall with a corky bark. Leaves are small, opposite, shiny, ovate to almost orbicular, about 7 mm long and 5 mm wide. Flowers are yellow and pleasantly scented, in attractive terminal spikes up to 4 cm long. Capsules are small with a wavy opening.

HABITAT Damp places in heath and woodland.

DISTRIBUTION Mainly coastal, south of Sydney (e.g. Illawarra Escarpment and Macquarie Pass), but also recorded from Blue Mountains. Extending along south coast to Vic, Tas and SA.

FLOWERING September–October.

667. *Melaleuca squarrosa*

668. *Melaleuca deanei*

Melaleuca deanei **Plate 668**

M. deanei is regarded as a rare and endangered species because of its vulnerable small scattered populations. It is a medium shrub with papery bark found on wet sandy heaths. Its leaves are alternate, elliptic, up to 2 cm long. Flowers are creamy yellow in terminal spikes up to 6 cm long. Although the mature plant is hairless, new stems are covered in short white hairs. Also the receptacle and the petals are silky-hairy on the outside. Capsules are about 7 mm across, with a smooth opening.

HABITAT Wet heath, on sandy soils.

DISTRIBUTION Endemic in the coastal zone around Sydney. Uncommon. Recorded from Brisbane Water National Park, Lucas Heights and Heathcote National Park.

FLOWERING September–November.

669. *Melaleuca sieberi*

Melaleuca sieberi — Plate 669

Chiefly a species of north coast swamps, with a limited distribution in the northern part of the Sydney district. It is a tall tree (to 5 m) with papery bark. Leaves are narrow lanceolate, 4 to 12 mm long. Flowers are white, in fluffy loose terminal spikes about 20 mm long. The calyx and new growth are hairy. Capsules are small (2 to 4 mm across), with a wavy opening.

HABITAT Swampy areas, in deep wet sands.

DISTRIBUTION Coast north of the sandstone zone (e.g. Lake Macquarie). Extending northwards along the coast to Qld.

FLOWERING September–October.

670. Melaleuca armillaris

Melaleuca armillaris — Plate 670

A tall shrub with rough hard bark. It is especially common on coastal headlands and in damp places near the coast, where it may form dense thickets. The dark green leaves are narrow (less than 1 mm wide), 12 to 25 mm long, with tip slightly curved downwards. Flowers are white, in spikes 3 to 7 cm long. Capsules are globular, about 5 mm across.

HABITAT Damp heaths and coastal headlands.

DISTRIBUTION Most common near the sea (e.g. Kurnell), but also recorded in mountains. Extending all along the coast from Qld to Vic and Tas. Also occurring northern tablelands and central western slopes.

FLOWERING September–November.

671. Melaleuca ericifolia

Melaleuca ericifolia

Swamp Paper-bark — Plate 671

An erect shrub to 3 m tall with a flaky or corky bark. It could be confused with *M. armillaris*, but has shorter leaves and flower spikes. Leaves are narrow-linear, crowded, mostly 10 to 15 mm long and 1 mm or less wide. Flowers are creamy white in spikes about 20 mm long and 15 mm across. Capsules are about 3 mm across.

HABITAT Wet places, especially around coastal lagoons and swamps.

DISTRIBUTION Scattered along coastal zone. Extending from Qld to Vic and Tas.

FLOWERING September–November.

672. Melaleuca ternifolia

Melaleuca ternifolia — Plate 672

This uncommon bushy medium-size shrub could be confused with *M. ericifolia*, but its distribution is very restricted and there are differences in the foliage and inflorescence. The specific name refers to the leaves being in groups of 3; leaves are narrow-linear, crowded, upright, 8 to 9 mm long and less than 1 mm wide. The pale lemon-coloured flowers are in tight terminal spikes, narrower than those of *M. ericifolia*. The styles protrude conspicuously from the spike, the staminal claw is very short, and the top of the ovary densely covered with white hairs. Capsules are small (3 mm across) and tightly packed into oblong clusters about 22 mm long.

HABITAT Wet areas

DISTRIBUTION On the fringe of the Sydney district around Tallong. Extending in isolated populations through southern tablelands and far south coast to Vic.

FLOWERING October–November.

673. Melaleuca decora

Melaleuca decora — Plate 673

M. decora is a tree with papery bark. It is common on the clay soils of Cumberland Plain and is one of the dominant trees in damper areas of Agnes Banks Nature Reserve and Castlereagh State Forest and along Richmond Road. Its leaves are flat, lanceolate, 10 to 15 mm long, 1 to 1.5 mm wide, with only one conspicuous vein. The sweet-smelling flowers are creamy white, in loose irregular spikes 3 to 5 cm long, usually below the end shoot. Capsules are small (about 2.5 mm across).

HABITAT Swampy ground or heavy clay soils subjected to occasional flooding.

DISTRIBUTION Mainly shale or clay areas of Cumberland Plain. Extending all along coastal zone from south coast to Qld.

FLOWERING November–January.

674. Melaleuca styphelioides

Melaleuca styphelioides

Prickly-leaved Paper-bark — Plate 674

M. styphelioides is a handsome small tree with papery bark. Its leaves are small (10 to 15 mm long) and ovate to lanceolate, tapering with a slight twist to a sharp point. New growth is softly hairy. Flowers are creamy, in spikes up to 2 cm long below the growth of new shoots. Capsules are 3 to 4 mm across, in oblong cluster; the remains of the calyx-lobes are often found around the rim of the aperture.

HABITAT Sheltered moist situations.

DISTRIBUTION Widespread along coast, extending into gullies of Blue Mountains. Also occurring south coast, north coast, north-western slopes and Qld.

FLOWERING November–December.

Melaleuca linariifolia — Plate 675

A small tree to 8 m tall with papery bark. Its leaves are opposite, narrow lanceolate, up to 35 mm long and 3 mm wide. Flowers are white, in characteristic feathery spikes. Capsules are small, about 3 mm across.

HABITAT In or near swamps, especially on clay or shale soils.

DISTRIBUTION Widespread from coast to mountains. Extending all along NSW coast and to southern tablelands.

FLOWERING October–December.

675. Melaleuca linariifolia

676. *Melaleuca quinquenervia*

Melaleuca quinquenervia

Broad-leaved Paper-bark **Plate 676**

This tall paper-bark is a common tree around the margins of lagoons and swamps along the coast northwards to Queensland. Leaves are up to 7 cm long and 5 to 18 mm wide, with 5 longitudinal veins, 3 of which are usually quite prominent. Flowers are white, in terminal spikes up to 5 cm long.

HABITAT Coastal swamps and brackish lagoons.

DISTRIBUTION Widespread along coast. Recordings include Towra Point (Botany Bay), Centennial Park and Dee Why Lagoon. Extending northwards along the coast to Qld. Also occurring in New Guinea, New Caledonia and America.

FLOWERING February–May, with some flowers through winter.

Eucalyptus and Angophora

Although the earliest flowering plants appeared in the mid Cretaceous period about 130 million years ago there is no reliable fossil evidence of eucalypts until the mid-to-late Tertiary period, 25 to 30 million years ago. Their early development and differentiation is clouded and may always remain so. Probably they were an insignificant part of the Australian vegetation, occupying low-fertility sites during most of the Tertiary period until severe climatic deterioration at the end of the Miocene period, 5 million years ago, created conditions under which they became the dominant vegetation over much of the continent. In recent times they have spread northwards into Timor, New Guinea, Indonesia and the Philippines.

The feature that distinguishes *Eucalyptus* is the presence on the flower bud of a small cap, or operculum, which falls off to release the stamens when the flower opens. (It was this feature that the botanist L'Heritier had in mind when he named the genus from the Greek 'eu'-well; 'kalyptos'-covered.) In the closely related genus *Angophora,* the bud is protected by free (i.e. separate) sepals, and the flower has 4 or 5 small white petals outside the prominent ring of stamens.

Eucalyptus and *Angophora* also differ in leaf arrangement. The leaves of *Angophora* are in opposite pairs but the adult leaves of *Eucalyptus* appear to be arranged alternately. Close examination, however, shows that this is caused by elongation of the nodal region around each leaf pair resulting in an arrangement in which each leaf pair, separated by a short intranode, is more or less at right angles to the next leaf pair and separated from it by a longer internode. Hence the adult leaf arrangement in *Eucalyptus* is actually a modification of the juvenile opposite arrangement and is not so different from that of *Angophora* as first appears.

In the majority of *Eucalyptus* species the operculum consists of fused petals and sepals (corolla and calyx). Either the 2 fall off together at anthesis (flowering) or the outer operculum is shed sometime before anthesis leaving a scar around the bud at the line of detachment. However, in about 90 species petals are absent, and the operculum consists only of fused sepals. In a dozen western and northern species the operculum is formed from the corolla, and the calyx consists of tiny free sepals which persist on the floral tube as in *Angophora.* In other cases the free sepals are shed at an early stage or are carried at the top of the operculum. Thus the definitive character of *Eucalyptus* shows variation in structure and development.

There is also considerable variation in other characters. Blakely recognised 25 variations in the anthers and developed a system of classification based on these differences. There are variations in the arrangement of stamens in the bud and ovules in the ovary. At least 5 or 6 cotyledon types can be recognised. There are major differences in seed structure and in leaf characteristics. The transition from the juvenile to the adult leaf form and arrangement also varies considerably.

Eucalyptus also has several types of inflorescence. Bloodwoods have flowers in large, showy, terminal, compound corymbose clusters similar to *Angophora.* Ironbarks and boxes have a distinctive type of inflorescence described as compound terminal panicles, but the majority of eucalypts have flowers in simple umbel-like clusters in the leaf axils. Differences in bark, buds and fruit are often readily apparent and provide the easiest means of recognition of many eucalypt species.

The complex pattern of similarities and differences brings into question the traditional separation of the group into just 2 genera, *Angophora* and *Eucalyptus,* since this attaches undue significance to the absence of an operculum by failing to similarly recognise other major differences within the group.

In 1971 Pryor and Johnson* published a classification of the eucalypts based on phenetic (structural), genetic and biotic characteristics dividing *Eucalyptus* into 7 subgenera each of which is subdivided into sections and series. Many of the subdivisions in the Pryor and Johnson classification correspond closely to Series in Blakely's 'Key to the Eucalypts'. The primary divisions correspond to fertility barriers which prevent members of one group interbreeding with those of another. In due course the subgenera could be raised to generic rank equal to *Angophora.* Some details of the classification are shown in the following table.

Subgroups in the Sydney Area

Inflorescence terminal, compound

1. Flowers in broad flat corymbose clusters. Fruit usually more than 10 mm across.
   a) Operculum absent (bud), petals present (flower). Fruit thin, ribbed ANGOPHORA
   b) Operculum present (bud), petals absent (flower). Fruit large, woody, with insert valves. CORYMBIA
2. Flowers in narrow panicles. Fruit usually less than 10 mm across. SYMPHOMYRTUS

Inflorescence axillary, simple, very rarely compound

1. Scar present on mature buds due to early detachment of outer operculum. SYMPHOMYRTUS
2. Bud scar absent. MONOCALYPTUS

Angophora

Angophora is a well-defined group of 7 species centred in Queensland and New South Wales. It has traditionally been separated from *Eucalyptus* by the absence of an operculum although it shares many other features. In *Angophora* the developing flower is protected by the free (i.e., separate) sepals which are retained after the flower opens on the rim of the floral tube. *Angophora* flowers have 5 or 4 broad-based white petals outside the ring of stamens in contrast to *Eucalyptus* in which the petals are either fused and form part of the operculum or are absent. *Angophora* also has a characteristic fruit in the form of a ribbed capsule of thin texture with 4 or 5 teeth on the rim, differing from those of most eastern eucalypts which are neither ribbed or toothed.

The identification of individual *Angophora* species is relatively easy owing to obvious differences between species in size, habit, habitat, leaves, flowers, fruit and bark. Furthermore species with intermediate characteristics are rarely seen. In the Sydney area *A. bakeri* x *A. costata* and *A. hispida* x *A. costata* hybrids occur very infrequently.

Subgenus/Genus	*Total Species*	*Species in Sydney Area*	*Major Groups in Sydney Area*
Angophora	7	5	angophoras
Blakella	9	—	
Corymbia	33	3	bloodwoods, spotted gum
Eudesmia	15	—	
Gaubaea	2	—	
Idiogenes	1	—	
Monocalyptus	91	42	stringybarks, ashes, peppermints, scribbly gums
Symphomyrtus	285	46	mahoganies, gums, boxes, ironbarks

*Pryor, L.D. and Johnson, L.A.S., *A Classification of the Eucalypts*, Canberra, Australian University Press.

677. *Angophora hispida*

Angophora hispida

Dwarf Apple **Plate 677**

Angophoras have been called Apples since colonial times because of the supposed resemblance of this species to the apple tree. It is the smallest of the angophoras, usually 1.5 to 2.5 m high, rarely exceeding 5 m, and is a rather straggly tree with loose, flaky rough bark and stiff, broad, stalkless leaves in opposite pairs. The leaves are dull yellowish green above and greyish beneath, obtuse at the tip and usually heart shaped at the base. Young branches, leaves and inflorescence are covered with reddish hairs. Flowers and fruit are typical of the angophoras, although larger than in other species. The flowers have 5 small white petals outside several whorls of creamy stamens. The fruit, about 15 to 19 mm across, is ribbed on the outside with 4 or 5 tiny raised teeth on the rim.

HABITAT Common on dry sandstone sites.

DISTRIBUTION Confined to the central coast of New South Wales.

FLOWERING October–December.

Angophora subvelutina

Broad-leaved Apple **Plate 678**

Although the leaves of the Broad-leaved Apple resemble those of the Dwarf Apple, it is not difficult to distinguish between the 2 species since the former is a much larger, densely crowned tree reaching 25 m in height, and is usually found on heavy alluvial soils. The leaves frequently have a velvety appearance and are pointed at the apex. The flowers are small and the thin fragile fruits are only 7 to 9 mm across. The tree is rough barked and sometimes resembles the Rough-barked Apple, *A. floribunda*, but differs from it in the stalkless adult leaves with heart-shaped bases.

HABITAT In woodland and open forest on heavy loamy soils, often near watercourses.

DISTRIBUTION Coast and tablelands of NSW and south-east Qld. In the Sydney district it occurs mainly in the Nepean area from Camden to Richmond.

FLOWERING October–December.

678. *Angophora subvelutina*

Angophora floribunda

Rough-barked Apple **Plate 679**

A medium to large spreading tree with a short furrowed fibrous-barked trunk, frequently with lower crooked branches hanging close to the ground. Its habit and light greyish green colour distinguish it from *Eucalyptus* species. The adult leaves are ovate to lanceolate, acutely pointed, 5 to 12 cm long and about 25 mm broad. The leaf stalks, about 10 mm long, and the tapering leaf bases distinguish this species from *A. subvelutina*. The flowers are rather small. The thin papery fruits are prominently ribbed and about 7 to 10 mm across.

HABITAT Mostly found on deep alluvial flats, along watercourses and on undulating country, in open forest and woodland.

DISTRIBUTION Extends from eastern Victoria to Queensland. In NSW it occurs on the coast and tablelands; also north-western slopes and plains.

FLOWERING October–December.

679. *Angophora floribunda*

680. *Angophora bakeri*

Angophora bakeri

Narrow-leaved Apple **Plate 680**

This is an attractive compact rough-barked tree attaining 15 m, but often much smaller when growing on shallow sandy soils. The species is readily identified by its drooping narrow lanceolate leaves which are mostly less than 8 cm long and 10 mm wide, on short stalks. Trees shoot readily after fire and the trunks are often fire-blackened. Although infrequently cultivated, the species is a good hardy ornamental suitable for light sandy soils.

HABITAT Patchy occurrence on dry sandy soils.

DISTRIBUTION Occurs on the central and northern coasts, but centred in the Sydney district. Large examples are common at Agnes Banks Nature Reserve.

FLOWERING December–January.

Angophora costata

Smooth-barked Apple **Plate 681**

The Smooth-barked Apple is a characteristic tree of the Sydney area where it is also known as Red or Rusty Gum. On poor sandstone sites it is short-trunked and irregularly branched with a large open crown. The branches are often twisted and gnarled. On better sites it may grow to 25 m with a straight

681. *Angophora costata*

bole, a metre or more in diameter, extending for half its height. The tree is easily recognised by its smooth deciduous bark, which is shed in oval or elongated patches during November leaving a smooth dimpled pinkish brown surface. The trunk and branches turn greyish with age and are often stained with dark red kino exudate. As with other angophoras, the flowers are borne in corymbose clusters at the ends of the branches. Buds and bud stalks are covered with short bristles. The slightly woody ribbed fruits are 12 to 14 mm across with deeply enclosed valves and large flat seeds.

Habitat Grows in a variety of conditions and sites, usually in open forest in association with *Eucalyptus* species and a shrub understorey.

Distribution Three forms or subspecies are recognised. The common Sydney form with strongly ribbed fruit occurs in coastal areas of NSW and the lower Blue Mountains. A form with broader leaves and larger fruit occurs in the Putty-Howes Valley district. A third form with non-ribbed fruit occurs in northern NSW and Qld.

Flowering Mostly November–December. Much more prolific some years than others.

Corymbia

Corymbia is a diverse, widespread group of species known as bloodwoods because of numerous gum veins in the wood. The group occurs in western, northern and eastern Australia, mostly on poor soils, but is absent from Tasmania and from most of Victoria and South Australia.

An interesting aspect of the Pryor and Johnson classification is the recognition of a closer relationship of *Corymbia* and *Blakella* to *Angophora* than to other eucalypts. The 3 groups are readily recognisable by the inflorescence of large, terminal, corymbose flower clusters. In *Corymbia* the flowers develop into large woody gumnut-shaped fruits with deeply enclosed valves. The others have similar but thinner, somewhat papery fruits.

Other similar charactistics of *Corymbia* and *Angophora* include the large oblong anthers, the large round cotyledons, the broad flat seeds, the pollen, the presence of bristly glands and trichomes (hairs) on the juvenile growth and inflorescence, the leaves with fine, regular, close-set veins and low oil content, the pinene-type of oil in the leaves, some features of wood anatomy, abundant kino, and freedom from attack by the lerp-forming psillid, *Glycaspis*.

Corymbia hybrids are uncommon since the species are comparatively stable and well defined.

Eucalyptus gummifera

Red Bloodwood **Plate 682**

This is a common tree in sandstone areas of the coast and eastern slopes of the Blue Mountains. Although typically a slender single-trunked or multi-trunked tree about 15 m high, it varies from a mallee form in coastal heaths to a medium or large tree on deep sandy gravels and loams. Its short-fibred, friable rough bark and large woody urn-shaped fruits with descending disc and deeply enclosed valves are characteristic of the bloodwoods. Leaves are paler on the undersurface and have close parallel veins at a large angle (about 65 degrees) to the midrib. The inflorescences are large and showy with mostly

682. A. *Eucalyptus gummifera* (fruit)
B. *Eucalyptus gummifera* (flowers)

7-flowered clusters in a terminal corymbose arrangement. Buds lack a scar since the 2 parts of the operculum fall together at flowering.

Habitat Usually on poor soils with other eucalypts in low to tall open forest.

Distribution Coastal districts from north-eastern Vic to south-eastern Qld and central tablelands of NSW below 900 m.

Flowering Mainly February–March but as late as May at higher elevations.

Eucalyptus eximia

Yellow Bloodwood **Plate 683**

Yellow Bloodwood is usually a spreading gnarled tree less than 20 m high, instantly recognisable by its rough, tessellated, soft, flaky, yellow-brown bark extending to the small branches and its conspicuous terminal inflorescences of creamy white flowers in mostly 7-flowered clusters. Adult leaves are stalked, lanceolate or falcate, up to 20 cm long and 25 mm wide, and of a similar grey-green colour on both sides. The flowers and fruit, unlike those of the Red Bloodwood, are sessile or nearly so, and buds have a scar around the rim of the floral tube due to early detachment of the outer operculum. Red and Yellow Bloodwoods do not grow intermixed, and do not appear to hybridise, possibly because of genetic differences and their different flowering times.

Habitat Poor sandstone soils, especially on the rim of escarpments.

Distribution Locally common on the upper slopes, escarpments and plateaux of the Nepean and Hawkesbury River districts including the lower Blue Mountains. Extending from south of Nowra to the Hunter Valley.

Flowering Mainly October.

683. A. *Eucalyptus eximia* (bark)
B. *Eucalyptus eximia* (flowers)

Eucalyptus maculata

Spotted Gum **Plate 684**

Inclusion of this tall, straight-boled, smooth-barked forest tree in the bloodwoods may at first seem surprising. The close relationship, however, can be readily seen in the leaves, buds and fruit which are typical of this group. The leaves have close parallel lateral veins at a large angle to the midrib, and the fruits are relatively large (14 mm x 12 mm diameter), woody, ovoid to urn-shaped with a descending disc and deeply enclosed valves. As with other species in the same section (*E. citriodora* and *E. henryi*), the inflorescence consists of 3-flowered clusters arranged in terminal corymbose panicles. The characteristic mottling of the trunk is cause by shedding of the dark grey weathered bark in irregular flakes exposing new cream-coloured bark beneath.

Habitat In open to tall open forest on well-drained, moderately heavy soils.

Distribution Extends from Bega to Bundaberg, with an isolated occurrence in eastern Victoria. In the Sydney area it grows on shale soils at Appin, in the Camden-Bringelly-Liverpool district, and on the coast near Nowra, West Pittwater and north of Gosford.

Flowering May–September.

684. A. Eucalyptus maculata (trunk)
B. Eucalyptus maculata (flowers/fruit)

Monocalyptus

Monocalyptus is an isolated group of about 90 species characterised, as the name suggests, by a single calycine operculum, i.e., the operculum is formed from the calyx only since petals are absent. *Monocalyptus* is usually found on low-nutrient acidic soils, frequently occupying distinctive sites, sometimes in nearly pure stands. Most, if not all, are dependent on mycorrhizal associations in the soil, and some are very sensitive to *Phytophthera* and other root pathogens. The group is well represented in Tasmania and the colder parts of the eastern highlands, but is not found in areas having less than 600 mm rainfall per annum, and has not spread to neighbouring countries to the north.

The monocalyptus group includes the stringybarks, bastard mahoganies, ashes, mallee ashes, snowgums, peppermints and scribbly gums. Stringybark and scribbly gum species are often difficult to identify owing to the small differences between some species, frequent hybridisation, and the occurrence in some areas of intergrades, i.e. populations of trees with character states between those of recognised species.

685. A. Eucalyptus umbra ssp. *umbra* (tree)
B. Eucalyptus umbra ssp. *umbra* (fruit)

Eucalyptus umbra

Bastard or Broad-leaved White Mahogany
Plate 685

E. umbra is a small to medium rough-barked tree reaching its southern limit in the northern suburbs of Sydney. It and the closely related *E. acmenoides* are called White Mahoganies because their leaves and bark resembles that group. Their buds and fruit, however, show an affinity to the stringybarks. Adult leaves are firm, glossy and concolorous, 10 to 14 cm long and 20 to 25 mm wide, sometimes larger in ssp. *carnea*, and taper gradually to a fine point. They are dark green in ssp. *umbra* and bluish green in ssp. *carnea*., The bark, which persists to the small branches, is fibrous and fissured, but not as coarse as a stringybark. Flowers and fruit are shortly stalked in clusters of 7 to 11, though not tightly packed like a stringybark. Buds have a rounded, beaked operculum and lack a bud scar. The fruit, about 6 to 10 mm across, is hemispherical with a narrow to broad flat disc and valves at about rim level.

HABITAT The ssp. *umbra* grows on sandy saline flats or rocky ridges near salt water. The ssp *carnea* is mostly found on stony soils on hillsides.

DISTRIBUTION The ssp. *umbra* occurs from Middle Harbour northwards to north Qld. The ssp. *carnea* occurs from Tuggerah—Wyong northwards to Bundaberg (Qld).

FLOWERING August–September (ssp. *umbra*). November–January (ssp. *carnea*).

686. A. Eucalyptus acmenoides (tree)
B. Eucalyptus acmenoides (fruit)

Eucalyptus acmenoides

White Mahogany **Plate 686**

Although closely related to *E. umbra*, this is a much larger tree preferring more fertile sites, often near rainforest. It is uncommon in the Sydney area, occurring only as scattered specimens. The grey-brown bark is rough and fibrous on trunk and branches. The leaves, unlike those of *E. umbra*, are thin, distinctly paler on the undersurface, and frequently have irregular or wavy margins. They are about 10 to 14 cm long and lanceolate, tapering to a fine point. Fine reticulate veins are very clear between the lateral veins which are regular and parallel. Inflorescences are axillary and with 7 or more flowers. The fruit, about 7 mm across and hemispherical or barrel-shaped, is pedicellate (stalked) with a narrow obscure disc, a thin rim and 4 valves at rim level.

HABITAT Fertile soils on ridge or hillsides, in open and tall open forest.

DISTRIBUTION Parramatta River and Gosford—Wyong northwards to north Qld.

FLOWERING November–January.

687. A. Eucalyptus muelleriana (tree)
B. Eucalyptus muelleriana (buds/fruit)

Eucalyptus muelleriana

Yellow Stringybark **Plate 687**

Yellow Stringybark is a very large tree with a number of distinctive features making identification easier than for most stringybarks. The long-fibred bark on trunk and branches is grey-brown on the outside but, when freshly cut, is distinctly yellow near the cambium. The leaves are paler on the undersurface (unusual for a stringybark), lanceolate-falcate, 10 to 15 x 2 to 3 cm, tapering to a fine tip and an oblique base. Flowers are on pedicels 1 to 3 mm long in clusters of 7 to 11. The buds and fruit, consequently, are not as tightly packed as in other stringybarks. The fruits, about 10 mm across, are truncate-globose with a flat disc and enclosed or slightly exsert valves.

HABITAT On slopes of sheltered moist valleys of the coast and ranges south of Sydney.

DISTRIBUTION Wollongong to Wilsons Promontory. Recorded from below Macquarie Pass and west of Dapto. Common in forests south of Batemans Bay.

FLOWERING November–March.

688. Eucalyptus macrorhyncha (fruit)

Eucalyptus macrorhyncha ssp. *macrorhyncha*

Red Stringybark **Plate 688**

Red Stringybark is a straight-trunked tree to 20 m high occurring in the western parts of the area. Bark and leaves are typical of the stringybarks; the bark deeply fissured, coarsely stringy and reddish brown weathering to grey; the leaves up to 14 cm long, leathery, equally green on both sides with distinct irregular lateral veins and an intra-marginal vein about 2 mm from the edge. Buds and fruit, however, are characteristic. The buds, on short stalks in axillary clusters of 7 to 11, have a beaked, or occasionally conical, operculum about as long as the floral tube. The fruits are large, about 10 to 12 mm across, distinctly stalked in fairly open clusters and more or less globular in shape due to the broad convex ascending disc with 3 broad exsert valves.

HABITAT In open forest and woodland on well-drained, moderately fertile clayey soils in foothills and undulating country.

DISTRIBUTION Tablelands and western slopes of NSW and Vic, with an isolated occurrence in SA.

FLOWERING January–April.

Eucalyptus blaxlandii

Blaxland's Stringybark **Plate 689**

This stringybark is fairly common on the higher parts of the Blue Mountains and the adjacent Hartley and Kanimbla Valleys. On dry shallow soils it is medium-sized, short-boled and spreading, but on deep fertile

689. A. Eucalyptus blaxlandii (tree)
B. Eucalyptus blaxlandii (buds/fruit)

soils at Mt Wilson it is a tall, straight, densely crowned tree exceeding 30 m in height. Unlike most stringybarks, its bark is furrowed and fibrous on the trunk and larger branches but smooth on the upper branches. The leaves, buds and fruit, however, are typical. The thick glossy leaves are 6 to 12 × 2.5 cm, oblique at the base with irregular lateral veins and a distant intramarginal vein. Buds are sessile in clusters of 7 or more and have a rounded or obtusely conical operculum. The fruits are 9 to 12 mm across, tightly clustered and domed with a convex ascending disc and usually 4 exsert valves.

HABITAT In open to tall open forest on sandy, granitic or basaltic soils.

DISTRIBUTION Central and southern tablelands of NSW (Mt Coricudgy to Clyde Mountain).

FLOWERING October–November.

Eucalyptus camfieldii

Heart-leaved Stringybark, Camfield's Stringybark **Plate 690**

This is an endangered species largely represented by a few stands in the Ku-ring-gai Chase and Royal National Parks. It is a mallee or small tree 1 to 4 m high resembling stunted forms of, and closely related to, *E. capitellata*. The two are most easily distinguished by the juvenile leaves—round or heart-shaped and about 2 cm long in *E. camfieldii*, ovate

690. A. Eucalyptus camfieldii (tree)

B. Eucalyptus camfieldii (buds/fruit)

to broad lanceolate and much larger in *E. capitellata*. The adult leaves of *E. camfieldii* are dark green, thick, leathery, and broadly lanceolate or with a rounded mucronate tip. The bark is brown and scaly-fibrous on stems and small branches. Buds and fruit are sessile and crowded on short peduncles. The buds are angular and 6 to 7 mm long; the fruits compressed-spherical, 4 to 6 mm long × 6 to 9 mm across, with a broad convex disc and 3 or 4 slightly exsert valves.

HABITAT On shallow sandstone soils bordering coastal heath in association with other stunted or mallee eucalypts.

DISTRIBUTION Restricted to plateaux between Bulli Pass and Gosford.

FLOWERING Irregular.

Eucalyptus capitellata

Brown Stringybark **Plate 691**

Depending upon the conditions, Brown Stringybark varies from a shrubby stunted tree 2 m high to a thin, often misshapen, tree about 15 m high. In dwarfed coastal forms, the dark green glossy leaves are short thick and broad, but in taller specimens they are broadly lanceolate up to 14 x 3 to 3.5 cm. Buds are sessile, crowded and angular with a short conical operculum. They are in clusters of 7 or more on broad flattened peduncles up to 10 mm long. The fruits are compressed-spherical, broader than tall, and often misshapen at the base due to crowding. The disc is broad and reddish, flat or convex, and the valves are enclosed or the tips exsert.

691. Eucalyptus capitellata (flowers/fruit)

HABITAT Usually on shallow infertile sandstone soils, rarely on shales.

DISTRIBUTION NSW central coast to Port Stephens, mostly north of Sydney; e.g. St. Ives district (tall form), Munmorah State Recreation Area (stunted form).

FLOWERING December–February.

692. A. Eucalyptus agglomerata (tree)

B. Eucalyptus agglomerata (buds/fruit)

Eucalyptus agglomerata

Blue-leaved Stringybark **Plate 692**

This species usually has a tall straight trunk, even when growing on sandy soils, and is a useful timber tree. It grows on valley slopes and is recognisable from a distance by the bluish canopy emerging from the yellow-bronze canopies of other trees. The thick, furrowed, long-fibred bark extends to the smaller branches. The leaves are relatively large for a stringybark, about 14 x 3 cm, broad lanceolate, with the intramarginal vein 3 to 5 mm from the margin. The sessile buds are crowded into axillary umbels of 11 or more, and the rounded fruits are tightly packed into globular clusters. Fruits are 8 to 9 mm across, broader than long, and flattened on the sides due to crowding. A broad convex disc encloses 4 small valves in the narrow opening.

HABITAT On a variety of soil types on moderate to steep slopes of the coast and ranges.

DISTRIBUTION From the mid north coast to beyond the Vic border, inland to Mudgee and Rylstone, and on the central and southern tablelands.

FLOWERING April–July.

693. Eucalyptus eugenioides (flowers)

Eucalyptus eugenioides

Thin-leaved Stringybark **Plate 693**

In the Sydney district this tree grows on clay soils in lower rainfall areas of the coastal plain, on the lower east-slopes of the Blue Mountains and in the Kanimbla, Megalong and Hartley Valleys. It is a medium-sized forest tree often resembling, and at times intergrading with, *E. globoidea*. The two can be distinguished by their buds and fruits which are sessile in *E. globoidea*, shortly stalked in *E. eugenioides*. Adult leaves are oblique at the base, lanceolate, about 12 cm x 25 mm, sometimes narrower, and rather thin. The operculum is conical and pointed, but not angular as in *E. capitellata*. Fruits are about 8 mm across and round at the base, not compressed and misshapen like those of *E. capitellata*. The disc is reddish, broad and more or less flat.

HABITAT In open forest on moderately fertile soils.

DISTRIBUTION Coast and tablelands of NSW, extending to south-eastern Qld.

FLOWERING September–November on the coast, later on the highlands.

Eucalyptus globoidea

White Stringybark **Plate 694**

E. globoidea occurs on the coast near Sydney, where it is reasonably common, also on the southern highlands, the lower Blue Mountains and in Megalong Valley. It is a medium-sized tree, similar in appearance to several other stringy barks in the

694. A. *Eucalyptus globoidea* (tree)

B. *Eucalyptus globoidea* (buds/fruit)

area, but recognisable by its juvenile leaves (seen on seedlings and coppice growth), and its sessile crowded buds and fruit. The seedling and juvenile leaves have rounded bases and crinkled irregular margins with tufts of fine stellate hairs on the edge, veins and undersurface. They are broader than those of *E. oblonga*, usually exceeding 2 cm in width, and change at an early stage through intermediate to adult leaves which are lanceolate to broad lanceolate, glossy green above, duller and slightly paler beneath and rather thin for a stringybark. Buds are broadly spindle-shaped in sessile clusters of 11 to 15. The fruits, either sessile or on very short pedicels obscured by the fruit, are 6 × 8 mm, hemispherical to almost globular with a broad flat disc and valves slightly enclosed or at rim level.

HABITAT On sandy and gravelly loams and on shale capping over sandstone soil; e.g., Menai oval and Kurrajong to Bilpin.

DISTRIBUTION Coast and tablelands of NSW and eastern Vic.

FLOWERING January–June.

Eucalyptus oblonga ssp. *oblonga*

Narrow-leaved Stringybark, Sandstone Stringybark **Plate 695**

E. oblonga and *E. capitellata* are the 2 common stringybarks inhabiting poor, shallow, sandy soils on the coast near Sydney. However, in some places such as the Menai–Lucas Heights area, the stringybarks show considerable variability indicating a complex and obscure parentage which may include either or both of the above, *E. globoidea*, *E. camfieldii*, *E. eugenioides* and possibly others. Often 'pure' forms of likely antecedents are absent from the immediate vicinity. *E. oblonga* is also common on sandstone soils throughout the Blue Mountains and hybridises there with other species.

E. oblonga is a slender irregularly branched tree with 1 to several trunks, 20 to 30 cm in diameter growing to about 5 to 8 m tall from the one rootstock. The seedlings and juvenile leaves are pale on the undersurface, scabrous, crinkled, irregularly toothed on the margins, and less than 2 cm wide. The dark green thick glossy adult leaves are also narrow (1.5 to 2.5 cm across). Buds are sessile or shortly stalked in clusters similar to those of *E. globoidea*, but the fruit is generally smaller (5 to 7 × 4 to 8 mm) and has an ascending (rather than flat) disc and a narrower mouth.

695. A. *Eucalyptus oblonga* (tree)

B. *Eucalyptus oblonga* (buds/fruit)

HABITAT In dry open forest on poor sandstone soils.

DISTRIBUTION From the Shoalhaven to the upper Hunter, extending to the central and north-western slopes.

FLOWERING January–April. Hybrids often flower winter to spring.

696. *Eucalyptus ligustrina* (buds/fruit)

Eucalyptus ligustrina

Privet-leaved Stringybark **Plate 696**

This stringybark is recognisable by its small narrow adults leaves which are thick, obscurely veined, dark green on both sides, and about 4 to 7 × 0.1 to 0.2 cm. It is a mallee or small densely-crowned tree less than 6 m high, often only 2 to 3 m, with criss-crossed matted-fibrous bark on the trunk and main branches and flaky or smooth bark on the upper branches. Buds and fruit are small, sessile and crowded in clusters of 7 or more. Buds are often yellowish, and the filaments of the flowers are yellow. Fruits are hemispherical to truncate-globular and about 5 × 5 mm.

HABITAT Often forming thickets in exposed, shallow, periodically damp sites on sandstone.

DISTRIBUTION Occurs on the coastal escarpment near Wollongong (e.g., Sublime Point to Brokers Nose, and West Dapto), on the Blue Mountains above Wentworth Falls (e.g., near Sunset Rock on Kings Tableland, Narrow Neck, Newnes Plateau) with a small occurrence near Glen Innes.

FLOWERING May–June, occasionally in spring.

697. *A. Eucalyptus pilularis* (tree)

B. Eucalyptus pilularis (buds/fruit)

Eucalyptus pilularis

Blackbutt **Plate 697**

Blackbutt is one of the more common and conspicuous trees in the coastal ranges of NSW and probably the most important timber tree in coastal forests. It is a medium to tall tree with a stocking of dark finely fibrous rough bark at the base, or occasionally covering most of the trunk. The rough bark is shed in strips leaving the upper trunk and limbs a smooth greenish white. On good sites the trunk is straight for nearly two-thirds the tree height. The juvenile leaves on seedlings and saplings are opposite for many pairs, sessile, rounded at the base and distinctly paler on the undersurface but the adult leaves are alternate, stalked, lanceolate and concolorous. Flowers are in axillary clusters of 7 to 15 on flattened peduncles 10 to 20 mm long. Buds have a pointed or beaked operculum and are without a scar. The fruits are hemispherical or truncate-globular, about 10 × 10 mm, on pedicels up to 5 mm long, with a level or descending disc and slightly enclosed valves.

HABITAT In tall open forest in valleys, often on deeper sandy soils on slopes above creeks or rainforest.

DISTRIBUTION Coast and foothills of ranges in NSW and south-eastern Qld.

FLOWERING September–March.

Eucalyptus obliqua

Messmate **Plate 698**

This species occurs on the Southern Highlands near Robertson and from Mittagong to Wingello, but is never common in the Sydney area. It is a large, sometimes massive, tree with a thick, fibrous, rough, brown-grey bark extending to the smaller branches. Although the bark is somewhat stringy, the species is more closely related to the ashes than the stringybarks. The juvenile leaves are discolorous, very broad, oblique at the base and abruptly pointed at the tip, becoming concolorous, narrower and tapering gradually in the intermediate and adult stages. Adult leaves are 10 to 15 × 2 to 3 cm, broad lanceolate and glossy dark green. Inflorescences are axillary and with 7 to 15 flowers. The buds have a short rounded operculum with a tiny raised point at the top; the fruits are ovoid to globular-truncate, about 8 to 9 mm across with a descending disc and 3 or 4 enclosed valves.

HABITAT Chiefly on deep heavy soils.

DISTRIBUTION In high rainfall areas of the tablelands. Extending from Tas to south-eastern Qld; also SA.

FLOWERING December–March.

698. *Eucalyptus obliqua* (buds/fruit)

Eucalyptus fastigata

Brown Barrel **Plate 699**

This species is one of the mountain ash group, a major component of tall open forests in cool moist areas of south-eastern Australia. It occurs on the tablelands in the Sydney district and is a large tree with a rough fibrous brown bark peeling from the upper trunk and branches in fine ribbons which remain attached and hanging in the canopy. Adult leaves are up to 18 × 2.5 cm, lanceolate, with fine lateral veins at a very small angle to the midrib. The species is most readily identified by the inflorescence in which the clusters of 7 to 15 flowers are paired

699. *A. Eucalyptus fastigata* (tree)

B. Eucalyptus fastigata (buds/fruit)

in the leaf axils. Buds are about 4 mm across with a conical pointed operculum. Fruits are 6 to 8 × 5 to 7 mm, cone-shaped or pear-shaped, tapering to the stalk, with a broad slightly rounded disc and 3 slightly exserted valves.

HABITAT In deep fertile soils in cool moist areas of the tablelands.

DISTRIBUTION Eastern Vic to northern tablelands of NSW. In the Sydney area it occurs in the Bowral–Mittagong district, at Macquarie Pass, on basalt caps at Mt Wilson, Mt Tomah and Mt Irvine, at Browns Gap and Hassans Walls near Lithgow, and on the main range near Jenolan Caves.

FLOWERING December–February.

Eucalyptus oreades

Blue Mountains Ash **Plate 700**

E. oreades is a common, tall, straight-shafted tree in valleys and sheltered hillsides on the Blue Mountains above Springwood, but is shorter and more spreading on the plateau in the Clarence–Wolgan area. The old bark is deciduous in long strips leaving a smooth white upper trunk and branches and a small stocking of dark rough bark at the base with strips hanging from it. Adult leaves, up to 17 × 2.5 cm, are lanceolate with the acute lateral venation characteristic of the ash group. Buds are in inflorescences of 7, often curve inwards and lack

700. *A. Eucalyptus oreades* (trunk)
B. Eucalyptus oreades (buds/fruit)

a scar. These features distinguish the species from other tall eucalypts in the area such as *E. cypellocarpa*. The fruits are sessile or shortly stalked, about 10 × 10 mm, barrel-shaped (ovoid-truncate) with a descending disc and 4 or 5 enclosed valves.

HABITAT Mainly in open forest on deeper soils derived from Hawkesbury sandstone, on rocky slopes at the edge of basalt caps and in tall open forest bordering rainforest.

DISTRIBUTION Centred in the Blue Mountains with isolated occurrences on the ranges to the Qld border.

FLOWERING January–February.

Eucalyptus luehmanniana

Yellow-top Ash **Plate 701**

In contrast to the larger members of the ash/snow gum group which frequently form extensive even-aged single-species forests in mountain areas of south-eastern Australia, the mallee ashes usually have very restricted distributions.

E. luehmanniana is one of about a dozen mallee ashes occurring in the Sydney area. Found only on the coast, and rather small, it is closely related to *E. oreades*. It occurs in small thickets with closely spaced stems 3 to 6 m high and less than 10 cm in diameter. Lower stems are smooth and whitish but the young branchlets are yellow, 4-sided and narrowly winged. The leaves are large (15 to 20 × 3 to 5 cm), falcate, thick, grey-green both sides with obscure venation, and are attached by flattened, twisted, yellow petioles 3 cm long. Buds and new growth are often glaucous. Inflorescences are axillary and 7 to 15 flowered on broad flattened striate peduncles. Buds have a pointed operculum. The fruits are rather large (about 12 × 12 mm), ribbed and more or less cylindrical with a wide flattish rim and enclosed valves.

HABITAT On low fertility soils derived from Hawkesbury sandstone in association with other small eucalypts and hard-leaved shrubs.

DISTRIBUTION In scattered stands from near Nowra to Gosford. It occurs in the Ku-ring-gai Chase and Royal National Parks and south of Waterfall.

FLOWERING July–November.

701. *A. Eucalyptus luehmanniana* (tree)
B. Eucalyptus luehmanniana (flowers)

702. *Eucalyptus consideniana* (buds/fruit)

Eucalyptus consideniana

Yertchuk **Plate 702**

Although a member of the ash group, this tree looks more like a peppermint. It occurs uncommonly in the lower to middle parts of the Blue Mountains, becoming more common on the southern highlands and beyond. It is a small to medium crooked tree with rough, finely furrowed grey over yellow bark persistent to the small branches. Adult leaves are leathery and fairly large (up to 18 × 3 cm), with irregular lateral veins at a small angle to the midrib. Inflorescences are 7 to 15 flowered on slightly flattened peduncles about 12 mm long. The operculum is short and rounded with a small point. Fruits are about 8 mm across, cone or pear shaped, with a broad slightly rounded disc and usually 4 slightly exserted valves. *E. consideniana* may resemble smaller specimen of *E. sieberi*, but this species has fruits with a broad flat disc and usually 3 slightly enclosed valves.

HABITAT Occurs in open forest on poor skeletal sandy or clayey soils.

DISTRIBUTION On the coast and tablelands from Gippsland to the Blue Mountains. Recordings include Warrimoo, Hazelbrook and Springwood.

FLOWERING November–December.

Eucalyptus sieberi

Black Ash, Silvertop Ash **Plate 703**

E. sieberi is related to *E. consideniana*, and has a similar distribution, but is much more common in the Sydney area. Depending upon location, it is a medium to tall tree of variable form. The bark of young trees is orange brown, rough, thin, and flaky, but on mature trees it becomes dark brown to black, thick, hard and furrowed like an ironbark. The branches are mostly white and smooth, the young branchlets shiny red. Adults leaves are lanceolate, up to 15 × 2.8 cm, concolorous, fairly thick, glossy and smooth with obscure irregular veins at a small angle to the midrib. The buds and fruit taper gradually to the petiole (stalk) in a manner typical of the ashes. The operculum is short and rounded with a small point, and the fruit is pyriform (pear-shaped), about 8 mm across with a broad flat red disc and usually 3 slightly enclosed valves. The flowers have noticeably white (rather than creamy) filaments.

HABITAT On a wide range of soils and sites in open forest, tall open forest and woodland.

DISTRIBUTION Coast to mountains. Stunted trees are common from Waterfall to Bulli Pass (flowering late October). Large examples occur at Mt Keira lookout. Also Darling Causeway, Hartley and granite slopes of Jenolan Gorge in the Blue Mountains. Extending

703. A. *Eucalyptus sieberi* (tree)
B. *Eucalyptus sieberi* (buds/fruit)

from near Melbourne to the Watagan Mountains near Wyong; also Tas.

FLOWERING September–January.

Eucalyptus multicaulis

Whipstick Mallee Ash **Plate 704**

This species is closely related to *E. sieberi*, differing mainly in its mallee habit, its smooth stems with slightly rough short-fibred bark at the base, and its slightly smaller buds and fruit. It is an uncommon multi-stemmed tree up to 6 m high, occurring in scattered locations in the Sydney area.

HABITAT On dry sandstone ridges and hillsides.

DISTRIBUTION Blue Mountains (Faulconbridge to Woodford), central coast near Sydney (Bulli to Gosford) and near Jervis Bay.

FLOWERING Variable.

704. A. *Eucalyptus multicaulis* (tree)
B. *Eucalyptus multicaulis* (buds/fruit)

Eucalyptus pauciflora ssp. *pauciflora*

Snow Gum, White Sally **Plate 705**

Despite the specific name, which means few flowered, this species flowers profusely in late spring and summer. It is a small to medium tree with usually a short trunk and spreading branches. It has a smooth white bark mottled with cream and grey, and often marked with insect scribbles. Sometimes there is a little persistent rough bark at the base of the tree. The leaves of this species are characteristic. Juvenile leaves are very broad, thick and greyish; the adult leaves are narrower and glossy green with oblique lateral veins running almost parallel to the midrib. The cup-shaped fruits are about 10 mm across with a broad level or descending disc and usually 3 slightly enclosed valves.

HABITAT In cold areas of the eastern dividing range in a variety of habitats.

DISTRIBUTION NSW, Vic, Tas and Mt Gambier, SA. Common in the higher Blue Mountains, mainly in the Clarence–Wolgan area, Wallerawang to Hampton and on the Boyd Plateau, with a few occurrences near Bell and in the Hartley and Kanimbla Valleys. Fairly common in the Berrima–Marulan and Moss Vale–Bundanoon districts.

FLOWERING October–April.

705. *Eucalyptus pauciflora* (flowers/fruit)

Eucalyptus gregsoniana

Mallee Snow Gum, Wolgan Snow Gum **Plate 706**

This small snow gum is closely related to *E. pauciflora* and was formerly regarded as a subspecies of it. It differs in having a mallee form; that is, it has several thin stems arising from a lignotuber or rootstock. It is usually 2 to 3 m in height but may occasionally reach 5 m. This species also has narrow leaves, less than 25 mm wide, and very short peduncles on the flowers and fruit. In its areas of occurrence it is usually scattered rather than in pure stands, often in association with clumps of *E. stricta* from which it is readily distinguished by its taller appearance. The mallee form is adopted by several normally tree-like species under conditions of ecological stress.

HABITAT In low scrub in skeletal sandstone soils on hillsides and plateaux at high altitudes.

DISTRIBUTION From Mt Wilson to the Clarence district northwards to the Wolgan Valley; also in the Budawangs south of Nowra.

FLOWERING November–December.

706. *Eucalyptus gregsoniana* (flowers/fruit)

707. *Eucalyptus dendromorpha* (budsflowers)

Eucalyptus dendromorpha

Giant Mallee Ash, Budawang Ash **Plate 707**

This eucalypt has been classified as a variety of several different species, but is now considered to be a distinct, possibly relict, species. It occurs on specific sites in scattered locations in central and south-eastern NSW as a small to tall tree with a rough flaky or scaly bark which is shed in ribbons leaving a smooth greenish or greyish white surface. Its tree-like form distinguishes it from closely related *E. stricta* and *E. obtusiflora*. Apart from this, it differs from *E. stricta* in its much larger leaves which are about 12 × 2.5 cm, lanceolate-falcate and covered with oil dots, and from *E. obtusiflora* in its acutely pointed operculum. Fruits resemble those of the latter species, and are globular-truncate or urn-shaped, about 10 mm across with a thin rim, a descending oblique disc, and 4 enclosed valves.

HABITAT Moist slopes above cliffs on sandstone.

DISTRIBUTION Blue Mountains from Wentworth Falls to Blackheath, also Wynne's Rocks at Mt Wilson, Illawarra escarpment from Mt Keira to Barren Grounds Reserve, Fitzroy and Carrington Falls, Sugarloaf Mountain and Budawang Ranges.

FLOWERING July–September, sometimes later.

Eucalyptus obtusiflora

Port Jackson Mallee **Plate 708**

Of the dozen or so mallee ashes in the Sydney area, this is the commonest on barren sandstone ridges near the coast. However, it is by no means the only mallee eucalypt in these places since a number of tree species such as *E. haemastoma, E. gummifera, E. capitellata* and *E. punctata* adopt this habit on shallow, infertile sites. *E. obtusiflora* is distinguished from these and other mallees on the coast by its thin bark which is shed in long narrow strips leaving smooth greenish white stems, its short, broad, thick, glossy leaves which are often held erect, its 7-flowered axillary clusters on broad flattened peduncles, its club-shaped buds with a short rounded or obtusely conical operculum, and its rather large barrel-shaped or urn-shaped fruits, 10 to 11 mm across, with a narrow rim, a descending oblique disc and 4 deeply enclosed valves.

HABITAT On shallow sandstone soils in heath and shrubland.

DISTRIBUTION A patchy distribution on coast and tablelands from Jervis Bay to Broken Bay. Fairly common in the Royal National Park.

FLOWERING August–December.

708. A. *Eucalyptus obtusiflora* (tree)
B. *Eucalyptus obtusiflora* (buds/fruit)

Eucalyptus burgessiana

Faulconbridge Mallee Ash **Plate 709**

This is a recently described species restricted to a few dry sandstone ridges in the lower Blue Mountains and at Mt Tomah. It is closely related to *E. obtusiflora* and *E. stricta* but is larger in all respects. It is a thin-stemmed mallee up to 5 m high with a smooth grey bark which peels in strips leaving a tan-coloured new surface. The plentiful juvenile or intermediate leaves are a useful identifying feature. They are alternate, grey-green, thick, large (up to 20 × 7 cm), on long twisted stalks (up to 25 mm) and have numerous oil glands. The adult leaves are similar but smaller (up to 14 × 2.5 cm) with a prominent midvein and very obscure lateral veins. Inflorescences are 7 to 11 flowered on flattened peduncles 15 to 20 mm long. The fruits are similar to those of the above related species, but up to 13 × 13 mm in size.

HABITAT Confined to sandstone ridges and plateaux.

DISTRIBUTION In scattered populations north of the highway in the Springwood–Faulconbridge–Linden area of the Blue Mountains and on the eastern slopes of Mt Tomah.

FLOWERING September–December.

709. A. *Eucalyptus burgessiana* (tree)
B. *Eucalyptus burgessiana* (fruit)

Eucalyptus stricta

Blue Mountains Mallee Ash **Plate 710**

E. stricta is common and widespread on the Blue Mountains, but is also found in a few exposed places on the coast. It is a small many-stemmed mallee, 1 to 3 m high, occurring in small colonies with *Allocasuarina nana* and other small hard-leaved shrubs on windswept sandstone plateaux. It is closely related to *E. obtusiflora* but differs in having narrower leaves and smaller buds and fruit. The adult leaves are thick, dark green, glossy, about 9 × 1 cm and taper gradually to a point. The club-shaped buds are in mostly 7-flowered clusters and have a slightly rough surface due to numerous oil glands. Fruits are less than 10 mm across and slightly constricted below the rim (i.e., urn-shaped). As with other members of this group, the rim is thin, the disc descending and the valves deeply enclosed.

HABITAT Exposed rocky sandy sites.

DISTRIBUTION Blue Mountains, sandstone plateau between Sydney and Wollongong, near Moss Vale, Nerriga–Braidwood district, near Bega and Mt Wellington (Vic).

FLOWERING December–April.

710. *A. Eucalyptus stricta* (tree)
B. Eucalyptus stricta (buds/fruit)

Eucalyptus apiculata
Narrow-leaved Mallee Ash **Plate 711**

This species does not occur on the coast or Blue Mountains although very narrow-leaved forms of *E. stricta* in these areas could easily be confused with it. It is taller than *E. stricta*, often reaching 6 m in height. The leaves are long and very narrow (8 to 12 × 0.5 to 0.7 cm) with a narrow recurved point at the apex, lanceolate, rather shiny and leathery, and without visible venation. Buds and fruit are smaller than those of *E. stricta* but are otherwise similar.

HABITAT In open forest on sandstone ridges and tops.

DISTRIBUTION Berrima district and Lithgow to Rylstone.

FLOWERING Probably August–October.

Eucalyptus rupicola
Cliff Mallee Ash **Plate 712**

This is another recently described mallee ash confined to a few scattered populations on specific sites in the Blue Mountains. It is closely related to *E. stricta, E. apiculata* and *E. obtusiflora* but is smaller and has smaller buds and fruit. The foliage is rather dense with the leaves held erect at the top of thin grey-brown stems. Young leaves in the crown are often pinkish, but mature adult leaves are greyish green, about 6 to 7 × 0.4 to 0.7 cm, narrow lanceolate with the apex sometimes drawn out into a hooked point. The small flowers are in axillary clusters of 7 and partly hidden among the leaves, whereas the fruits are mostly on the stems immediately below the leaves. The fruits are globular-urceolate (rounded and slightly constricted below the rim), about 7 × 7 mm, with a thin rim, a descending disc and enclosed valves.

HABITAT Confined to rocky slopes immediately above cliff edges of the Jamison, Kedumba, Megalong and Grose Valleys in the Blue Mountains.

DISTRIBUTION Sites include Kings Tableland (e.g., Sunset Rock), Sublime Point and Gordon Falls, Leura, and Pulpit Rock near Blackheath.

FLOWERING May–July.

711. *A. Eucalyptus apiculata* (tree)
B. Eucalyptus apiculata (buds/fruit)

712. *A. Eucalyptus rupicola* (tree)
B. Eucalyptus rupicola (flowers)

713. *A. Eucalyptus stellulata* (tree)
B. Eucalyptus stellulata (buds/fruit)

Eucalyptus stellulata
Black Sally **Plate 713**

The name 'stellulata' means little star, and refers to the bud clusters. This species does not occur on the Blue Mountains plateau but is found in a few places in the Hartley and Kanimbla Valleys. It also occurs further west on the higher parts of the Dividing Range on the Boyd Plateau, and to the south between Berrima and Marulan. The species is easily recognised by a number of features. It is a small to medium-sized woodland tree, often short-trunked and spreading. The bark is dark, rough and hard at the base; the upper trunk and branches smooth, shiny and olive green to greyish. The juvenile leaves are opposite, sessile, rounded and very broad, contrasting with the adult leaves which are alternate, stalked, lanceolate or elliptic and about 6 to 8 × 2 cm with several longitudinal veins. The buds are in tight star-shaped axillary clusters of up to 20 or more. They are acutely conical, tapering at each end, sessile and narrow, sometimes curiously expanded at the top of the floral tube. The fruits are small, about 5 × 5 mm, with a descending disc and enclosed valves.

HABITAT In cold wet places at high altitudes, often near streams and swampy flats.

DISTRIBUTION Along the Great Dividing Range from the Victorian Alps to the Qld border.

FLOWERING April–October and summer.

714. *A. Eucalyptus moorei* (tree)

B. Eucalyptus moorei (flowers/fruit)

Eucalyptus moorei

Little Sally, Narrow-leaved Sally **Plate 714**

This species was once regarded as a narrow-leaved form of *E. stellulata* but differs in its frequent mallee habit, its narrow lanceolate juvenile leaves less than 1 cm wide, and its smaller adult leaves (about 7 × 1 cm). It is fairly common on the higher Blue Mountains, often forming thickets on exposed damp sandstone slopes above streams. It can grow as a single-trunked tree to 7 or 8 m but is more often only 2 or 3 m in height. Stems are smooth and pale with strips of old bark at the base. The adult leaves are thick, dark green and glossy, and usually held erect. The leaf surface is dotted with oil glands, and the veins are faint except for the midrib. Buds and fruit are similar to those of *E. stellulata* but the clusters usually have fewer flowers.

HABITAT On moist sandstone slopes in cold elevated areas.

DISTRIBUTION Blue Mountains from Wentworth Falls to Clarence (e.g., Narrow Neck and Minne-Ha-Ha Falls, Katoomba), southern and northern tablelands and Snowy Mountains.

FLOWERING January–April.

Eucalyptus radiata ssp. *radiata*

Narrow-leaved Peppermint **Plate 715**

The peppermints are a small distinctive group related to the ashes and with a similar preference for cool moist sites on the tablelands of south-eastern Australia. They are also closely related to the scribbly gums, but are easily distinguished from this group by their rough short-fibred bark and the strong peppermint smell of the leaves when crushed.

Narrow-leaved Peppermint is common throughout the higher parts of the Blue Mountains and on the southern highlands in the Mittagong, Berrima, Moss Vale district. In the former area it is a medium-sized forest tree with pendulous branches, but in the latter it is more often a short-trunked spreading woodland tree with a large dense canopy. It can be confused with *E. elata* but differs from it in the rough grey bark which covers the trunk and large branches. The leaves are up to 15 × 1.5 cm, narrow lanceolate and rather thin with a typical peppermint venation, i.e., with fine, sparse lateral veins at a very acute angle to the midrib, and intramarginal veins distant from the edges. Inflorescences are also typical of the group, with 11 to 20 or more flowers radiating from the centre of the flower cluster.

HABITAT In open forest on a variety of soils, but mostly sandy with a heavier base.

DISTRIBUTION Eastern Vic and on the tablelands and south coast of NSW.

FLOWERING October–January.

715. *A. Eucalyptus radiata* (tree)

B. Eucalyptus radiata (flowers)

716. *Eucalyptus elata* (budsflowers)

Eucalyptus elata

River Peppermint **Plate 716**

E. elata and *E. radiata* are similar in appearance, leaves, buds and fruit. Close to Sydney, the former species occurs in the Nepean–Hawkesbury area, but does not extend far into the Blue Mountains (it is reported from Mt Tomah). Since the latter does not occur below Woodford, and is commonest on the higher parts of the mountains, the two are unlikely to be confused in these areas. Both species, however, are common on the southern highlands beyond Mittagong, but are distinguishable by their bark which is darker, harder and fissured in *E. elata* and is shed from the upper trunk and branches in long hanging ribbons leaving a smooth white surface. Although the inflorescences are similar, those of *E. elata* are often larger (up to 40 flowers) and have slightly angular peduncles and very slender pedicels. Buds are small, club-shaped and sometimes indented (but not scarred) at the base of the operculum. The rounded fruits are 4 to 5 mm across with a level or descending disc.

E. elata may also resemble *E. smithii*, but is readily distinguished from it by differences in buds and fruit.

HABITAT In open and tall open forest on fertile alluvial soils, on well-drained slopes near rivers and creeks and on undulating country.

DISTRIBUTION Coastal ranges from Gippsland (Vic) to Putty near Singleton. Recorded in the Sydney district from Bents Basin, Bowral, Thirlmere Lakes, Mt Kembla, Upper Colo and Macdonald Valley.

FLOWERING August–November.

717. A. *Eucalyptus dives* (tree)

B. *Eucalyptus dives* (flowers/fruit)

Eucalyptus dives

Broad-leaved Peppermint **Plate 717**

E. dives is a small to medium-sized tree with short-fibrous peppermint-type bark extending to the larger branches. The smaller branches are smooth and greyish. It has glossy, broad lanceolate leaves like those of the common Sydney Peppermint *(E. piperita)*, and the trees are not dissimilar. They can be distinguished, however, by the thick blue-grey (glaucous) broad heart-shaped juvenile leaves of the former and the numerous oil glands and strong peppermint odour of the leaves at all stages. The leaves of *E. piperita*, by comparison, have little oil and odour. The two also differ in their buds and fruit. *E. dives* has club-shaped buds with a rounded operculum and hemispherical fruits with a fairly broad flat or slightly convex disc, whereas *E. piperita* has spindle-shaped buds with a pointed conical operculum and barrel-shaped fruits with a narrow opening and a small descending disc.

HABITAT In open forest and woodland on dry hillsides and poor stony ridges.

DISTRIBUTION On the Great Dividing Range from central NSW to central Vic. It occurs at higher elevations beyond Mt Victoria in the Blue Mountains, with significant stands on the north-western section of Boyd Plateau.

FLOWERING September–November.

Eucalyptus piperita ssp. *piperita*

Sydney Peppermint **Plate 718**

This tree is of historical interest, being the species from which eucalyptus oil was first distilled by Considen, Surgeon of the First Fleet, and also the tree known as the Explorers Tree, marked by Blaxland, Wentworth and Lawson in 1813 during the first crossing of the Blue Mountains.

Sydney Peppermint is one of the commonest trees on sandy soils near Sydney and throughout the Blue Mountains. Its form is variable, depending upon location and depth of soil. At its best it is a straight-trunked, open-crowned tree about 25 m in height, but it is commonly about 15 m high, short-boled and spreading, becoming irregularly branched and misshapen on exposed rocky sites. The rough fibrous bark covers the trunk and sometimes the larger branches; the upper part of the tree is usually smooth or with ribbony bark. The thin, broad, stem-clasping juvenile leaves are green above and distinctly paler on the undersurface but become alternate, thicker, stalked and concolorous in the transition to the intermediate stage. Adult leaves are often oblique at the base and taper to a fine point at the apex.

HABITAT In open forest on sandstone soils in a variety of situations, but preferring cool, moist valley slopes.

DISTRIBUTION NSW central tablelands and coast from Bullahdelah to the Illawarra district. Intergrades with ssp. *urceolaris* in the south.

FLOWERING November–January.

718. A. *Eucalyptus piperita* (trunk)

B. *Eucalyptus piperita* (buds/fruit)

719. *Eucalyptus piperita* ssp. *urceolaris* (buds/fruit)

Eucalyptus piperita ssp. *urceolaris*

Urn-fruited Peppermint **Plate 719**

In the area immediately south of Macquarie Pass this tree intergrades with the Sydney Peppermint, and, although formerly regarded as separate species, the two are now classified as subspecies of *E. piperita*. Ssp. *urceolaris* is a larger tree than ssp. *piperita*, although this may be partly attributable to better growing conditions. The two are similar in form, bark and leaf characteristics, but differ in the buds and fruit. In ssp. *urceolaris* the operculum is longer and more pointed, and the fruits are slightly larger and urn-shaped (concave and contracted below the rim) rather than barrel-shaped (convex below the rim) as in ssp. *piperita*. In both subspecies the rim is thin, the opening narrow, the disc descending, and the valves deeply enclosed.

HABITAT In open to tall open forest on light sandy soils of low to medium fertility and depth.

DISTRIBUTION Batemans Bay to southern highlands of NSW.

FLOWERING January–February.

Eucalyptus haemastoma

Scribbly Gum **Plate 720**

The scribbly gums form a complex in which differences between the species are not always clear-cut. This is one of 3 closely related species occurring on the central coast of NSW. It is common in areas of Hawkesbury sandstone near Sydney and varies from small and mallee-like on shallow sites to medium-sized and lightly crowned under better conditions. The branches and the short, frequently deformed trunk are covered with a smooth patchy white to yellowish or greyish bark marked by insect scribbles. The species has thick broad leathery leaves up to about 15 × 3 cm. The buds are club-shaped with a short rounded or broadly conical operculum; the fruits about 9 × 9 mm, hemispherical or pear shaped with a thick rim and a broad flat or slightly

720. A. Eucalyptus haemastoma (tree)
B. Eucalyptus haemastoma (flowers)

convex reddish disc. The broader juvenile and adult leaves, the large buds and fruit, and the broad disc distinguish this species from others in the complex. Although the specific name refers to the red colour of the disc, that feature is not unique to this species.

HABITAT Mostly on dry skeletal sandstone soils in open forest and scrub.

DISTRIBUTION Eastern part of NSW central coast. The species intergrades with *E. racemosa* north and south of Sydney and precise boundaries are difficult to ascertain.

FLOWERING Usually November–March, occasionally at other times.

Eucalyptus rossii

Scribbly Gum, White Gum **Plate 721**

This scribbly gum occurs mainly on the western side of the tablelands on the margins of the area under consideration. It is a small to medium-sized tree, usually 10 to 20 m tall with a short, often crooked, trunk and a large irregularly branched crown. The trunk is smooth to the base and usually mottled cream to grey due to irregular shedding of the old bark. It is usually prominently marked with insect scribbles and lacks the powdery bloom characteristic of *E. mannifera* ssp. *mannifera*. Seedling leaves are opposite for a few pairs; the juveniles are alternate, stalked and lanceolate to broad-lanceolate. Adult leaves are narrow-lanceolate sometimes falcate, 7 to 15 × 1.0 to 1.5 cm, greyish green and concolorous. *E. rossii* has smaller buds and fruit than *E. haemastoma*, rather like those of *E. sclerophylla*. They are about 5 × 6 mm on slender peduncles and pedicels. The buds have a rounded (i.e., hemispherical) operculum differing from the conical operculum of *E. racemosa*.

HABITAT In woodland on sandy plateaux, ridges and slopes in lower rainfall areas, often in exposed situations.

DISTRIBUTION Capertee and Wolgan Valleys. Distributed along the western slopes and parts of the tablelands from the Qld border almost to Vic.

FLOWERING December–February.

Eucalyptus sclerophylla

Hard-leaved Scribbly Gum **Plate 722**

Formerly regarded as a subspecies of *E. haemastoma*, this is the common scribbly gum found throughout the Blue Mountains and in parts of the southern highlands, occurring mostly as a fairly small short-trunked twisted tree similar in appearance to *E. haemastoma.* On alluvial sands at Agnes Banks near Richmond and on Permian sandstone south of Nowra it is a larger open-crowned tree. *E. sclerophylla* has thick, broad, leathery leaves similar to those of *E. haemastoma*, and small buds and fruit resembling in size and appearance those of *E. racemosa*. The trunk is conspicuously marked by insect scribbles caused by tunnelling of insect larvae between the layers of bark, and exposed when the older outer bark flakes off. This is usually a good identifying feature, but not infallible since species other than scribbly gums may be similarly marked. *E. sclerophylla* can be distinguished from *E. mannifera* ssp. *gullickii*, with which it grows on poor stony soils in the Blue Mountains, by the absence of a bud scar in the former species, and the more tapering fruits with an ascending disc and exsert valves in the latter.

HABITAT In open forest on shallow to deep sandy soils.

DISTRIBUTION On the central and south coast and tablelands from Howes Valley near Putty to the Jervis Bay and Sassafras districts.

FLOWERING December–March.

721. A. Eucalyptus rossii (tree)
B. Eucalyptus rossii (flowers)

722. A. Eucalyptus sclerophylla (tree)
B. Eucalyptus sclerophylla (buds/fruit)

Eucalyptus racemosa

Scribbly Gum, Snappy Gum **Plate 723**

This species prefers deeper moister soils than *E. haemastoma*, and is consequently usually taller and less gnarled. It differs from the latter species in its thinner, narrower juvenile and adult leaves (up to 15 × 1.5 cm), smaller buds and smaller fruits (about 5 × 6 mm) with a narrower disc. Since the two species occupy different distinctive habitats, and normally flower at different times, opportunities for hybridisation are limited. Nevertheless, trees regarded as intergrades, with intermediate character states, are common in some areas from which one or both of the species may be absent.

HABITAT On deeper sandy and silty soils derived from Hawkesbury sandstone.

DISTRIBUTION Confined to the central coast of NSW. Recordings include Mona Vale and Picnic Point (Georges River). 'Pure' forms tend to occur north and north-west of Sydney. Intergrades are common south of Sydney Harbour.

FLOWERING August–November.

723. *Eucalyptus racemosa* (buds/fruit)

Symphomyrtus

Symphomyrtus occurs in all parts of the continent in a wide range of environments, but more often on deeper, fertile soils than the other groups. It is divided into 11 sections of which 6 occur in the Sydney area. Hybridisation occurs frequently within sections, and occasionally between sections.

In all sections except one, the outer operculum is shed at an early stage of growth leaving a bud scar which is a convenient means for distinguishing *Symphomyrtus* species from others of similar appearance. In the Sydney area only 3 box species *(E. moluccana, E. bosistoana, E. melliodora)* and one ironbark *(E. sideroxylon)* shed the double operculum in one piece at flowering, and therefore lack a bud scar.

The first of the 6 sections in the Sydney area consists of gums and mahoganies and includes some of the most magnificent and valuable trees in east coast forests. The section grades from the blue gum group through the grey gums, which are intermediate in habit and habitat, to the red gums which grow on characteristic sites in drier open woodland.

Symphomyrtus also includes a large section of medium-to-large rough and smooth-barked trees occurring throughout the highlands of south-eastern Australia and contributing much to the character of the countryside in these areas. A key feature of this group is the juvenile leaves which in some cases are carried in opposite pairs on sizable trees and often provide a useful first step in the identification of species. Another section includes the ironbarks and boxes which are typically woodland trees found on medium fertility soils in moderate rainfall areas. This section is easily recognised by both the bark and the distinctive type of inflorescence in which flower clusters occur at the ends of branchlets where leaf development is suppressed.

Two isolated *Symphomyrtus* species, *E. squamosa* and *E. microcorys*, also occur in the Sydney region.

Eucalyptus deanei

Mountain Blue Gum, Round-leaved Gum **Plate 724**

This is 1 of 5 species comprising the Eastern Blue Gum group of smooth-bark and rough-barked trees forming part of a larger closely related group of gums and mahoganies having characteristic petiolate, broad juvenile leaves and discolorous leaves at all stages with fine, but clearly marked, regular wide-angled lateral veins. The group includes some of the most magnificent and valuable timber trees in the country.

724. A. *Eucalyptus deanei* (tree)
B. *Eucalyptus deanei* (buds/fruit)

E. deanei is a tall shaft-like tree of great beauty found in sheltered gullies and valleys of the Blue Mountains. It has a smooth white bark throughout with patches of cream and pale grey-blue, occasionally with a little rough, flaky bark at the base. The juvenile leaves are rather large (6 to 15 × 4 to 8 cm) and ovate to orbicular. The large ovate to broad lanceolate intermediate leaves are quite common on mature trees among the adult leaves which are narrower and smaller (8 to 14 × 2 to 3.5 cm). Flowers occur in simple axillary inflorescences of 7 to 11 on angular or flattened peduncles. Although similar in appearance to *E. saligna* and *E. grandis*, this species can be recognised by its cleaner base, the rounded juvenile leaves, the presence of broad intermediate leaves in the canopy, the stalked buds with a hemispherical-apiculate operculum, and the small bell-shaped fruits with 3 or 4 slightly exserted short broad valves which point outwards after opening.

HABITAT Tall open forest in sheltered valleys (e.g., near Gosford, Blue Gum Forest in the Grose Valley), on deeper loams of the eastern slopes of the ranges (e.g., Kurrajong to Bilpin), and in soils derived from volcanic intrusions (e.g., Murphy's Glen).

DISTRIBUTION From Picton Lakes district to the Hunter River, and in the McPherson Ranges on the NSW/Qld Border.

FLOWERING April–August.

Eucalyptus saligna

Sydney Blue Gum **Plate 725**

E. saligna is an attractive tall shaft-like tree resembling, and closely related to, *E. deanei* and *E. grandis*, but with a more coastal distribution than the former and a more southerly distribution than the latter. In coastal gullies near Gosford, where their ranges overlap, *E. saligna* can be distinguished from *E. deanei* by its stocking of rough persistent bark at the base of the trunk, its smooth greyish-green or bluish-green upper trunk and branches, its sessile buds with a long conical operculum and its pear-shaped fruits with an obscure descending disc and 3 or 4 long, narrow pointed erect or out-curved valves. Further north, *E. grandis* can be recognised by its often glaucous fruits with broad, blunt in-curved valves. In the area from Wollongong to Batemans Bay, *E. saligna* intergrades or hybridises with the rough-barked *E. botryoides* producing trees in a variety of habitats with bark, buds, fruit and leaf characteristics ranging between the extremes of the 2 species.

HABITAT Tall open forest on deep, moist but well-drained alluvial loams.

DISTRIBUTION South coast of NSW to south-eastern Qld with a few isolated occurrences to the Qld central coast, extending to the tablelands in the north.

FLOWERING January–April.

725. A. *Eucalyptus saligna* (tree)
B. *Eucalyptus saligna* (buds/fruit)

726. A. *Eucalyptus botryoides* (tree)
B. *Eucalyptus botryoides* (buds/fruit)

Eucalyptus botryoides

Bangalay, Southern Mahogany **Plate 726**

E. botryoides is an adaptable tree occurring in a variety of habitats, though never very far inland or more than a few hundred metres above sea level. It is most common on, or behind, coastal sand dunes as a dominant species in association with banksia, tea-tree and other salt-tolerant plants, or sometimes with a temperate rainforest understorey. It also occurs on sandy soils along river valleys in near-pure stands or with other tall eucalypts. In most situations it is a medium-sized spreading tree with a dense crown of thick glossy leaves. As with other blue gums and mahoganies, the leaves are dark green above and paler below with close regular fine veins at a large angle to the midrib. The bark on the trunk and large branches is brownish, flaky-fibrous, fairly soft and thick with longitudinal fissures. The buds and fruit are sessile, or almost so, in clusters of 7 on broad flattened peduncles, and intermediate in size between those of the smooth-barked blue gums and the true mahoganies.

HABITAT On poor sandy soils, rarely far from salt water.

DISTRIBUTION From Bairnsdale (Vic) to slightly north of Newcastle (NSW).

FLOWERING Mainly December–March, also spring.

Eucalyptus robusta

Swamp Mahogany **Plate 727**

E. robusta and *E. botryoides* are both medium-sized, rough-barked trees belonging to the Eastern Blue Gum group, and may be regarded as links between that group and the mahoganies. Both species are fairly common on the NSW coastal margin, the former mainly on swampy, sandy or clayey soils, the latter on sand dunes or well-drained sandy soils. Both have thick, soft spongy bark, although that of *E. robusta* is more reddish and extends to the smaller branches. They also have similar leaves, but are distinct in floral characteristics. The buds of *E. robusta* are larger (up to 20 × 8 mm) with a long-beaked operculum (conical in *E. botryoides*) and occur in clusters of 9 to 15 (7 in *E. botryoides*) on broad flat peduncles. The fruits are also larger (up to 18 × 12 mm), cylindrical, elongated and noticeably stalked (sessile in *E. botryoides*). Usually the 4 valves are joined at their tips forming a very distinctive cross in the mouth of the fruit.

HABITAT On saline swampy ground near estuaries and lagoons.

DISTRIBUTION From about Nowra (NSW) to Rockhampton (Qld).

FLOWERING April–September.

727. *Eucalyptus robusta* (budsflowers)

Eucalyptus pellita

Large-fruited Red Mahogany **Plate 728**

E. pellita is never common in the Sydney district, and appears to be confined to sandstone ridges and hillsides a few kilometres either side of Broken Bay. In this area it is a small to medium-sized irregular tree with thick, rough fibrous bark on the trunk and branches, ordinary in appearance, but instantly recognisable by its large buds and fruit. These occur in usually 7-flowered (sometimes 3) axillary clusters with flattened peduncles. The buds are often more than 10 mm across, and usually have a short stout stalk (pedicel). The operculum varies from rounded with a short point to conical with a long beak, and is often broader than the calyx tube. The bud scar is very noticeable. The fruits are about 10 to 15 mm across with a prominent concave bud scar on the rim surrounding a flat disc and 4 strongly exserted valves.

HABITAT In open forest with a shrub understorey on sandstone ridges and hillsides near the sea.

DISTRIBUTION Occurs sporadically from about Bega to Wyong (NSW), and in north Qld. Locations in the Sydney district include Foleys Hill and McCarr's Creek near Mona Vale, Patonga, above Maitland Bay in Bouddi National Park, Kincumber, Narara and Wyong.

FLOWERING August–November.

728. *Eucalyptus pellita* (buds)

Eucalyptus notabilis

Blue Mountain Mahogany **Plate 729**

E. notabilis could easily be confused with 2 other eucalypts, *E. pellita* and *E. resinifera*. In the Sydney district it is distinguishable from the first by its different, non-overlapping, distribution; *E. notabilis* on the lower slopes of the ranges and neighbouring areas of the coastal plain, *E. pellita* on the extreme coastal margin. Although *E. resinifera* is mainly coastal, its range is not always distinct from that of *E. notabilis*, and it is safer to rely on differences in floral features to distinguish the two. Generally the

729. *Eucalyptus notabilis* (buds/fruit)

buds and fruit of *E. notabilis* are larger and on much shorter stalks (pedicels) than those of the other. Buds are 7 to 8 mm across (5 to 6 mm in *E. resinifera*), and fruits about 9 to 10 mm (5 to 8 mm in *E. resinifera*). The operculum in *E. notabilis* is obtusely conical and often broader, but rarely longer than the calyx tube, whereas in *E. resinifera* it is drawn out to a fine, often curved, tip and considerably longer than the calyx tube. Both are medium to fairly large trees with a rough, soft, fibrous longitudinally fissured red-brown bark extending to the small branches.

HABITAT On sandy or clayey soils of medium fertility and depth.

DISTRIBUTION Lower slopes of Blue Mountains and nearby plain (e.g., Glenbrook Park, Springwood, Kurrajong to Bilpin, Glenorie), Colo and Putty districts, and isolated occurrences on the northern tablelands and in south-eastern Qld.

FLOWERING January.

730. *Eucalyptus resinifera* (buds/fruit)

731. A. *Eucalyptus punctata* (trunk)
B. *Eucalyptus punctata* (buds/fruit)

732. A. *Eucalyptus longifolia* (trunk)
B. *Eucalyptus longifolia* (buds/fruit)

Eucalyptus resinifera

Red Mahogany **Plate 730**

It appears from Smith's original description of this species in 1790 that the epithet, *resinifera*, which means resin-bearing, is a mistaken reference to kino, which is exuded profusely from *Angophora costata* and the bloodwoods, but not from the mahoganies. The species was later redefined by Bentham, but for many years included other mahoganies.

Red Mahogany is a medium to tall densely crowned tree of the east coast and nearby foothills. It is a typical mahogany with persistent soft fibrous reddish-brown fissured bark on the trunk, extending to the small branches. The heavy, dark-red timber is not unlike that of Jarrah, and is used for general building purposes. The leaves are alternate, stalked and discolorous with wide-angled (i.e., transverse) venation at all stages. Juvenile and intermediate leaves are long and broad with a more or less rounded base. The dark glossy green adult leaves are smaller and narrower (up to 16 × 3 cm), and taper at each end. Points of distinction between this species and *E. notabilis* are given in the previous description. Red Mahogany, and the mahoganies generally, can be distinguished from the stringybarks, which they superficially resemble, by their dark green discolorous leaves, the transverse leaf venation, and the prominent bud scar.

HABITAT In open to tall open forest on sandy or clayey sheltered slopes and valley flats.

DISTRIBUTION Coastal zone from Jervis Bay (NSW) to north Qld.

FLOWERING November–January.

Eucalyptus punctata

Grey Gum **Plate 731**

This is the only member of the grey gum group occurring in the Sydney district; the others have a more northerly distribution. It is most common on the coast, but also occurs in the Blue Mountains. It varies from a small, occasionally mallee-like tree on very poor sites to a medium-sized woodland or forest tree. It is usually recognisable by its thin mottled matt or granular bark which is shed in large irregular flakes exposing bright orange new bark which fades to a light grey then weathers to a dark grey. The adult leaves are glossy dark green above and pale below with numerous fine transverse veins, indicative of the close relationship between the grey gums and the mahoganies and eastern blue gums. The inflorescences are mainly 7-flowered on flattened or angular peduncles. In ssp. *punctata* the buds usually have conical opercula about as long as, or slightly longer than, the calyx tube. The fruits are about 10 mm across and rounded at the top with short exsert valves.

HABITAT In open forest with other eucalypts on sandy, occasionally rocky, soils or sandy soils with a clay base.

DISTRIBUTION Jervis Bay to the upper Hunter Valley on the coast and tablelands.

FLOWERING December–March.

Eucalyptus longifolia

Woollybutt **Plate 732**

E. longifolia is at times a rather unattractive tree with rough dirty-grey flaky-fibrous bark on the trunk and larger branches, and long narrow dull greyish green pendulous leaves adding to the general sombre appearance. Under good conditions on moist heavy soils it can grow to a tall straight-trunked tree about 30 m high, but is often much shorter and spreading with low crooked branches. It is instantly recognisable by its large, long-stalked pendulous buds and fruit in clusters of 3. The clusters hang on slender peduncles 15 to 30 mm long, with individual buds, flowers and fruits on pedicels 5 to 20 mm long. The buds are double, cone shaped and about 20 × 10 mm with a faintly ribbed calyx tube. The creamy flowers are attractive and large. Fruits are cup-shaped or cylindrical, 10 to 15 mm across, with the rim formed by the broad bud scar sloping upwards to the narrow staminal ring. The disc is descending and the valves are usually below rim level.

HABITAT In open to tall open forest on moist alluvial soils derived from shales, on flats more often than on hills.

DISTRIBUTION Coastal zone (e.g., Dapto, Milperra, Castlereagh State Forest). Occurring from Vic and far south coast of NSW to near Newcastle.

FLOWERING February–March.

733. A. *Eucalyptus squamosa* (trunk)
B. *Eucalyptus squamosa* (buds)

Eucalyptus squamosa

Scaly Bark **Plate 733**

Scaly Bark is a small to medium-sized straggly tree with a short trunk and crooked pendulous branches, endemic to the sandstone plateaux of the Sydney district. It belongs to a large group of eucalypts with narrow Y-shaped (i.e., bisected) cotyledons, common in south-western and southern Australia, but with only 3 widely separated species in eastern Australia. These 3 relict species point to a former wider distribution of this group, or its forebears, before increasing aridity of the continent caused isolation resulting in separate development in the east and west.

The grey bark is rough and scaly or flaky on the trunk and branches, and leaves are greyish green, dull and thin, about 11 × 1.5 cm, tapering to a fine point. Flower clusters have 7 to 13 flowers, and are often paired in the leaf axils. The buds have a conical operculum, and the small bell-shaped fruits have 3 exsert valves. Both are frequently disfigured by warty outgrowths.

HABITAT Sandstone ridges and plateaux.

DISTRIBUTION Central coast of NSW. Occurs in the Royal, Heathcote, Blue Mountains and Ku-ring-gai Chase National Parks.

FLOWERING Irregular, but mainly August–October.

734. *Eucalyptus parramattensis* (buds/fruit)

Eucalyptus parramattensis

Drooping Red Gum **Plate 734**

This is the first of the red gums, a group of mainly eastern woodland trees with red timbers characterised by their concolorous long narrow lanceolate leaves with conspicuous oblique veins at 30° to 50° to the midrib, axillary inflorescences with from 3 to 20 flowers per cluster, buds with a conical operculum longer than the floral tube, and small rounded fruits with a wide convex disc and exsert valves. The juvenile leaves are opposite for a few pairs only, then become alternate. They are shortly petiolate, large, broad and frequently subglaucous. They are usually retained until the trees reach 2 to 3 m in height, and in some species are carried on square stems.

E. parramattensis is smaller than other red gums in the area, usually 8 m or less in height with a slender trunk of poor form, though occasionally reaching 20 m. It can be identified by its drooping branches, 7-flowered inflorescences, and buds with a conical operculum only slightly longer than the calyx tube. The bark is smooth, mottled white to grey and dark grey.

HABITAT On low-lying flat clayey soils, sandy soils with a clay subsoil and poor alluvial sandy soils.

DISTRIBUTION Scattered throughout the Sydney district. Recordings include Agnes Banks, Hammondville, East Hills, Thirlmere, Glenbrook–Lapstone and Wallaby Swamp along Putty Road. Also recorded in the Mudgee district.

FLOWERING November–January.

Eucalyptus amplifolia

Cabbage Gum **Plate 735**

Differences in habitat provide a good starting point for distinguishing this species from the closely related Forest Red Gum, *E. tereticornis*. Although both occur on heavier clayey alluvial soils, the former is usually seen on creek flats and poorly drained low-lying areas whereas, the latter shows a preference for well-drained but moist hillsides. Possibly because of its more open habitat, *E. amplifolia* frequently forks close to the ground, developing several upright slender trunks. Large broad ovate to orbicular dull green juvenile leaves are borne on quadrangular stems on saplings up to 2 or 3 metres high beneath the parent trees. Intermediate and adult leaves are narrower (15 to 20 × 2.5 to 3.5 cm) than the juvenile leaves, although broader than those of *E. tereticornis*. The species also differ in floral characteristics. Inflorescences of *E. amplifolia* are mostly 11 to 15 flowered, compared to mostly 7 to 11 flowered for *E. tereticornis*, and buds are narrower (about 4 mm wide) and the fruits smaller (about 5 × 6 mm). The timber of *E. amplifolia* is greatly inferior to that of

735. A. *Eucalyptus amplifolia* (tree)
B. *Eucalyptus amplifolia* (buds)

E. tereticornis, conforming to the general rule that eucalypt timbers grown in waterlogged ground have poor durability and strength.

HABITAT In small stands on low-lying poorly drained clay soils.

DISTRIBUTION Coast and tablelands of NSW. Common in the Sydney district west of Liverpool (e.g., Badgerys Creek, South Creek, Eastern Creek).

FLOWERING November–January.

Eucalyptus tereticornis

Forest Red Gum **Plate 736**

Extending from eastern Victoria to Port Moresby in New Guinea, *E. tereticornis* has the greatest latitudinal range of any Australian tree. It is a medium-sized to tall forest tree with a good straight trunk, ascending major branches and a large open crown. Although taller and more erect than most, it is a typical red gum with a smooth or matt mottled white, grey and dark bluish grey trunk caused by the shedding of old bark in large irregular sheets or flakes. Frequently there is a little rough dark bark near the base. Juvenile leaves are bluish green, ovate to broad lanceolate and somewhat narrower than those of *E. amplifolia*. They are similarly carried on quadrangular stems on saplings 2 to 3 m high. The adult leaves are also narrower, being glossy green on both sides, lanceolate and up to 20 × 2.5 cm.

736. Eucalyptus tereticornis (buds)

Inflorescences have 7 to 11 flowers, frequently 7. The operculum is large, horn shaped and sometimes expanded near the base. The fruit is larger than in *E. amplifolia*, about 7 × 9 mm, and ovoid in shape due to the broad domed disc and strongly exserted valves.

HABITAT In open situations on well-drained but moist alluviums often with a clay subsoil. Mostly on slopes and hillsides in the Sydney district.

DISTRIBUTION On coast and tablelands from Victoria to New Guinea. Common on clay soils west of Liverpool and on the coast and foothills in the Wollongong area.

FLOWERING August–November.

Eucalyptus michaeliana

Brittle Gum, Hillgrove Gum **Plate 737**

This is a rare species of interest because of its disjunct distribution in 3 widely separated locations in eastern New South Wales and also because of its unusual inflorescence consisting of a short axillary raceme of 3-flowered to 7-flowered clusters. It is a small to medium-sized forest tree with a smooth mottled creamy white and grey trunk resembling that of *E. maculata*. The leaves are about 15 × 3 cm, lanceolate and pale green on both surfaces with faintly marked veins. The buds are ovoid and stalked; the fruits cup-shaped or barrel-shaped, about 5 × 5 mm with a thick rim, a level disc and 3 or 4 small valves at rim level.

HABITAT In open forest on moist sandy soils.

DISTRIBUTION Uncommon between St Albans and Bucketty. Also in the Hillgrove–Enmore district near Armidale and near Mt Ballon in the McPherson Ranges.

FLOWERING About June.

Eucalyptus camphora **Plate 738**

Mountain Swamp Gum, Broad-leaved Sally

This species belongs to a small group of gums inhabiting cold elevated regions of south-eastern Australia. The group is part of a large section of about 50 species characterised by their kidney-shaped or bilobed cotyledons, but differing from the majority in having petiolate (stalked) seedling or juvenile leaves, at first opposite, but becoming alternate after a few flowers. *E. camphora* is a mallee or small tree occurring uncommonly in the Sydney district in the higher, colder parts of the tablelands. It can be recognised by its dull greyish green elliptical leaves with a small mucronate tip and distinct camphor odour when crushed. The leaves are up to 14 × 3 cm in size, but are often much shorter. The small white flowers are in axillary clusters of 7 on terete peduncles. Buds have a conical or beaked operculum slightly longer than the calyx tube. The small, funnel-shaped fruits are about 6 × 6 mm, with a flat or slightly raised disc and 3 or 4 exsert valves.

HABITAT On the banks of creeks and in cold elevated areas with poor drainage.

DISTRIBUTION Discontinuous distribution on the tablelands from Qld to Vic. Occurs near Rylstone, in Megalong Valley, Paddys River at Penrose, and near Marulan.

FLOWERING January–April.

737. A. Eucalyptus michaeliana (tree)

B. Eucalyptus michaeliana (buds)

738. A. Eucalyptus camphora (tree)

B. Eucalyptus camphora (buds/fruit)

Eucalyptus ovata

Swamp Gum **Plate 739**

E. ovata is closely related to *E. camphora* and grows under similar conditions in periodically waterlogged flats and cold swampy ground. It is widely distributed in south-eastern Australia, reaching its northern limit on the central tablelands of NSW. In parts of its range it is a medium to tall tree with a straight trunk and a dense crown, but in exposed conditions on the southern highlands of NSW it is usually less than 15 m in height with spreading branches. Bark is shed in strips from the branches and most of the trunk leaving a smooth whitish surface and a short stocking of rough grey bark at the base. The glossy green ovate or lanceolate leaves, up to 15 × 3 cm, often with undulate margins and acuminate tips, distinguish this species from *E. camphora*. The buds and fruit are in 7-flowered clusters and similar to those of *E. camphora* except that they are slightly larger and the valves of the fruit are at rim level or slightly below, rarely exserted like those of *E. camphora*.

HABITAT On poorly drained ground at high elevations.

DISTRIBUTION From the central tablelands southwards to Vic, Tas and SA. Occurs locally in the Wingello, Tallong, Marulan district, and in the higher Blue Mountains west of Coxs River.

FLOWERING March–June.

739. A. Eucalyptus ovata (tree)
B. Eucalyptus ovata (buds/fruit)

740. A. Eucalyptus aggregata (tree)
B. Eucalyptus aggregata (buds/fruit)

Eucalyptus aggregata

Black Gum **Plate 740**

Black Gum is so named because of the curly-flaky dark bark which covers the trunk and branches giving the tree a dark dirty appearance. It is a small to medium-sized densely foliaged tree with twisted branches. It is related to the previous two species, and is found under similar conditions in cold damp alluvial flats and frosty hollows. Like others in the series, it has fairly small, slightly discolorous opposite then alternate stalked juvenile leaves. The adult leaves are lanceolate, about 10 × 1.8 cm, medium green, dullish and concolorous with faint, widely spaced very oblique lateral veins and an intramarginal vein distant from the edge. The tiny buds, flowers and fruit are shortly stalked and aggregated or crowded in clusters of 7. The buds taper to both ends, and the fruits are hemispherical, 4 to 5 × 4 to 5 mm, with a well-defined rim, a flat disc and 3 or 4 exsert valves. The dark flaky bark throughout, the leaf venation and the tiny crowded fruit clusters provide a positive identification for this species.

HABITAT Cold damp flats and hollows on the tablelands.

DISTRIBUTION Tablelands of central and southern NSW with an isolated occurrence in Vic. Common near Wallerawang. Also found in Hartley Valley and near Berrima.

FLOWERING December–February.

Eucalyptus mannifera ssp. *mannifera*

Brittle Gum **Plate 741**

E. mannifera is a complex of different forms, variously named in the past, but now regarded as subspecies which are distinguished by differences in juvenile and adult leaves and features of the buds and fruit. Because the complex is undergoing evolutionary change, intergrades between the recognised subspecies are fairly common. Some added difficulties in identification may arise as a result of hybridisation with members of the viminalis series. Subspecies *mannifera* is a medium-sized

741. A. Eucalyptus mannifera ssp. *mannifera* (tree)
B. Eucalyptus mannifera ssp. *mannifera* (fruit)

slender tree with a smooth white powdery bark mottled with irregular dark pinkish-grey patches of older bark. It has broad elliptical to almost orbicular greyish juvenile leaves with cordate bases, and rather broad dull green lanceolate adult leaves 10 to 25 × 2 to 3.5 cm. The fruit is 6 to 7 mm across with a domed ascending disc and 3 short, broad, slightly exsert valves. This subspecies has broader leaves and larger fruit than ssp. *maculosa* which has a similar distribution. It is readily distinguished from *E. viminalis* and *E. rubida* by its 7-flowered inflorescences.

HABITAT In woodland and open forest in a variety of situations but usually on shallow soil.

DISTRIBUTION Western and southern parts of tablelands and south-western slopes, extending to Vic.

FLOWERING October–December.

Eucalyptus mannifera ssp. *maculosa*

White Brittle Gum **Plate 742**

This is 1 of 5 recognised subspecies of *E. mannifera*, a polymorphic complex of trees occurring fairly commonly on dry mountain slopes of NSW and north-eastern Victoria. It is an attractive tree, suitable for cultivation in cool areas, with a smooth white powdery trunk and branches, narrow leaves and a pendulous habit. The trunk usually has some purplish patches of older bark but is smooth to the

742. *Eucalyptus mannifera* ssp. *maculosa* (fruit)

base. Juvenile foliage is linear to narrow lanceolate, 2 to 3 cm wide, shortly stalked, opposite at first but soon becoming alternate, greyish and dull but not glaucous. The adults leaves are dull green on both sides, narrow lanceolate, 6 to 15 × 0.8 to 1.5 cm. The fruit is similar to that of ssp. *mannifera*, but only 4 to 5 mm across. Where their ranges overlap this tree could be confused with the scribbly gum, *E. rossii*, which also has narrow leaves and small fruits, but can be distinguished by the relative lack of insect scribbles on the bark, the presence of a bud scar, and the 3 short, broad, slightly exsert valves of the fruit.

HABITAT In dry eucalypt forest on the tablelands on sandstone and granite soils.

DISTRIBUTION On the central and southern tablelands of NSW and in north-eastern Vic.

FLOWERING October–December.

743. *Eucalyptus mannifera* ssp. *gullickii* (fruit)

Eucalyptus mannifera ssp. *gullickii*

Mountain Spotted Gum **Plate 743**

This subspecies is common on sandstone plateaux and ridges of the Blue Mountains from about Lawson upwards. It is smaller and much less graceful than ssp. *maculosa* but has a similar light open crown. The bark is white or pale grey with a few insect scribbles, mostly smooth, but sometimes with a fair amount of rough older bark near the base. Juvenile leaves are pale green, dull, oblong and almost sessile, and the adult leaves are grey-green, slightly glossy, lanceolate and 10 to 20 × 1.5 to 2.5 cm. The fruit is 6 to 7 mm across, similar in size to that of ssp. *mannifera*, but tapering more to the base (i.e., obconical rather than hemispherical) and on shorter thicker pedicels. This species can be distinguished from the scribbly gum, *E. sclerophylla*, with which it is often associated, by the bud scar and the obconical fruit.

HABITAT On shallow sandstone soils and damp slopes above streams.

DISTRIBUTION Tablelands of NSW. Common in the Blue Mountains (e.g., Clarence). Also occurs on the plateau in the Kiama and Nowra districts.

FLOWERING December–March.

744. A. *Eucalyptus bridgesiana* (tree)

B. *Eucalyptus bridgesiana* (buds/fruit)

Eucalyptus bridgesiana

Apple Box **Plate 744**

This species belongs to the large series, Viminales, consisting of trees from cool moist regions of south-eastern Australia characterised by kidney-shaped cotyledons and more or less opposite and sessile juvenile leaves which are retained for a considerable time on young trees. It is one of the apple boxes which show some resemblance, particularly in their bark, to the Rough-barked Apple, *Angophora floribunda*. In these trees the pale grey bark is rough, soft, friable, tessellated and persistent over the whole trunk to the small branches. The tree is a typical woodland tree with a short bole and a large spreading crown with crooked branches. The juvenile leaves are broadly heart shaped and glaucous with finely toothed margins, at first sessile and opposite but soon becoming shortly stalked and sub-opposite. Adult leaves are alternate on long stalks, narrow lanceolate, up to 25 × 3.5 cm, dark green and slightly glossy. Buds and fruit are in 7-flowered inflorescences on slightly flattened peduncles. The fruits are hemispherical, about 7 mm across, with a fine concave ring near the top surrounding the thick rim.

HABITAT In open forest and woodland in deep moist loamy soils on the lower slopes.

DISTRIBUTION Widespread on the tablelands and western slopes, extending to the coast in places, from Qld to Vic. Occurs beyond the Blue Mountains plateau (e.g., in the Hartley–Lithgow–Capertee region, also near Wombeyan Caves).

FLOWERING January–May.

Eucalyptus cypellocarpa

Mountain Grey Gum **Plate 745**

E. cypellocarpa, a tall robust tree with a smooth or slightly rough greyish-white bark, resembles other large eucalypts occurring with it in the Blue Mountains and adjacent western valleys. The name *'cypellocarpa'* means cup-fruit, and refers to the goblet shape of the fruit which, although useful as an identifying feature, is not unlike that of *E. oreades* and others of the mountain ash group. The two species are readily distinguished, however, by the flower buds, which in *E. cypellocarpa* are elongated, rather large (up to 12 × 5 mm) and angular with ribs or angles continuing from the pedicel to the floral tube. Unlike those of *E. oreades* (which belongs to *Monocalyptus*), they have a broad shallow scar. The 2 species can also be distinguished by the leaves. At the seedling and sapling stage those of *E. cypellocarpa* are sessile, opposite or sub-opposite and distinctly paler on the undersurface for many pairs in contrast to those of *E. oreades* which are shortly stalked, opposite and discolorous for a few pairs only. The intermediate and adult leaves of *E. cypellocarpa* are alternate, stalked, dark glossy green on both sides and frequently very large (15 to 35 × 2 to 5 cm). The large intermediate leaves and 7-flowered clusters distinguish this species from *E. viminalis* and *E. dalrympleana*.

HABITAT Usually in tall open forest on sheltered slopes and deeper valley soils.

DISTRIBUTION Coastal forests of southern NSW and eastern Vic and tablelands of both states, becoming scattered towards the western and northern extremes of its range. Locally at Mountain Lagoon, Mt Tomah, Mt Wilson, higher parts of Blue Mountains, and Hartley, Megalong and Kanimbla Valleys.

FLOWERING December–February.

745. A. *Eucalyptus cypellocarpa* (tree)

B. *Eucalyptus cypellocarpa* (buds/fruit)

746. *Eucalyptus globulus* ssp. *maidenii* (buds/fruit)

Eucalyptus globulus ssp. *maidenii*

Maiden's Gum **Plate 746**

In addition to differences most clearly seen in the buds and fruit, the southern blue gums have many characteristics in common and, although once regarded as separate species, are now classified as subspecies of *E. globulus*. They are medium to very tall forest trees with a smooth greyish-white trunk and branches, and often a short stocking of rough bark at the base. Young trees have opposite, sessile, thick, broad, glaucous juvenile leaves borne for many pairs on winged square stems. The transition to alternate, stalked, long, narrow, dark green, lanceolate intermediate and adult leaves is distinct. In ssp. *maidenii* the inflorescences are 7-flowered (3-flowered or 1-flowered in the other 3 subspecies) on broad flattened peduncles. Branchlets and inflorescences are sometimes glaucous. The almost sessile buds are angular or ribbed and fairly large (about 10 × 7 mm), although smaller than in the other subspecies, and much less warty. The fruits are cone-shaped, about 10 × 10 mm, with a thick rim, a broad disc and 3 or 4 slightly exsert valves.

Ssp. *bicostata*, with sessile, glaucous 3-flowered inflorescences, occurs at Jenolan Caves, Nullo Mountain near Rylstone, and in other isolated stands on the ranges of NSW and Vic.

HABITAT On sandy or clayey loams in mountain valleys and on slopes.

DISTRIBUTION Tablelands and coast of southern NSW extending to the central tablelands in the Wingello–Tallong district.

FLOWERING March–September.

Eucalyptus quadrangulata

Coast White Box, White-topped Box **Plate 747**

Although not a true box, this tall forest tree has a fine, firm, grey, box-like bark. It has opposite, sessile, stem-clasping, discolorous juvenile leaves borne for many pairs on square (quadrangular) winged stems, and adult leaves which are narrow-lanceolate with scalloped edges due to leaf glands at points along the edges. The inflorescences are axillary and 7-flowered on slightly flattened peduncles. Fruits are cone shaped with a level disc and 3 or 4 exsert valves.

E. quadrangulata is readily distinguished from boxes such as *E. rudderi* by the juvenile and adult leaves, the axillary inflorescences and the exsert valves. It is distinguished from many in the viminalis group by the rough bark and the 7-flowered inflorescences. It differs from other closely related species in its scalloped adult leaves and its rough bark on the upper trunk and branches. Its bark is neither ribbony like that of *E. smithii* and *E. viminalis*, nor coarsely fibrous, loose and fissured like that of *E. macarthurii*. It is most easily distinguished from the apple boxes by its tall erect habit and its sessile leaves and square stems at the juvenile stage.

HABITAT Open and tall open forest on fertile heavy soils.

DISTRIBUTION On the eastern slopes and parts of the adjacent plateau from the Qld border to about Nowra. Occurs locally from Stanwell Park southward (e.g., at Bulli, Mt Kembla, West Dapto, Macquarie Pass, Robertson, Hilltop, Colo Vale and Bundanoon).

FLOWERING January–March.

Eucalyptus macarthurii

Paddy's River Box, Camden Woollybutt **Plate 748**

E. macarthurii is restricted to the southern central tablelands of NSW but is fairly common on flat country near streams and swamps in this area. It is usually a large woodland tree with a stout straight trunk and low side branches. The trunk and larger branches have a coarsely fibrous, thick, loose bark which is shed imperfectly from the upper branches. Juvenile leaves, seen on saplings and sucker growth from stumps, are opposite, sessile, pale green and heart shaped for many pairs before the transition to alternate, stalked broad-lanceolate intermediate leaves, and finally to long narrow-lanceolate adult leaves. The presence of geranyl acetate and geraniol in the oil gives a strong geranium scent to the foliage and freshly cut bark and is a useful identifying feature. Buds and fruit are nearly sessile in 7-flowered axillary inflorescences. The fruits are hemispherical to cone shaped, about 5 × 6 mm, with a broad ascending disc and usually 3 slightly exserted valves.

HABITAT Heavy alluvial moist soils in fairly cold areas of the tablelands.

DISTRIBUTION From about Mittagong to Goulburn. Originally described from trees near Camden. Also reported from Jenolan Caves. Common in the Berrima, Burradoo, Paddy's River area.

FLOWERING November–January.

747. *Eucalyptus quadrangulata* (buds/fruit)

748. A. *Eucalyptus macarthurii* (tree)
B. *Eucalyptus macarthurii* (buds/fruit)

Eucalyptus smithii

Gully Gum **Plate 749**

In coastal gullies and slopes near Wollongong *E. smithii* is a tall shaft-like tree with a light crown, but in open areas of the southern highlands it is shorter and more branched with a denser crown. With hard dark furrowed bark on the lower trunk, and smooth white upper trunk and branches, it could at times be confused with *E. elata*. The two, however, are not closely related, and differ considerably in their buds

749. A. *Eucalyptus smithii* (tree)
B. *Eucalyptus smithii* (buds/fruit)

and fruit. *E. smithii* also resembles *E. viminalis* in the shedding of old bark in ribbons which are left hanging in the canopy, and in its narrow lanceolate juvenile and adult leaves. However, the extent of persistent rough bark differs, with most of the trunk covered in *E. smithii*, but usually only the base in *E. viminalis*. On closer examination the 2 are readily distinguished by the buds and fruit which are in 7-flowered inflorescences in *E. smithii*, and 3-flowered in *E. viminalis*. Also the fruits of *E. smithii* are about 5 × 6 mm and distinctly stalked whereas those of *E. viminalis* are about 6 × 8 mm and sessile or shortly stalked.

HABITAT In tall open forest on coastal slopes, and in open forest on the tablelands, usually on deep moist clayey soils.

DISTRIBUTION Central and southern tablelands and coast of NSW (e.g., Mt Kembla, West Dapto, Robertson, Hilltop, Mittagong), extending to Vic.

FLOWERING January–March.

750. A. *Eucalyptus viminalis* (tree)
B. *Eucalyptus viminalis* (buds/fruit)

Eucalyptus viminalis

Ribbon Gum, Manna Gum **Plate 750**

The name Ribbon Gum refers to the way the bark is shed in narrow strips from the branches and most of the trunk, often remaining hanging in the canopy and branch forks. The name Manna Gum refers to the sweet white exudate stimulated by insect attack on the leaves. Several other eucalypts also produce manna. This species is widely distributed on the coast and tablelands of south-eastern Australia, and shows considerable variation over its range. It is usually an attractive tall forest tree or, in open country, a shorter spreading shady tree, with a mainly smooth white bark, pendulous branches with long narrow-lanceolate leaves and the flowers in threes. It is often difficult to distinguish from *E. dalrympleana* and *E. rubida*, particularly the former, and the only reliable means of identification is the juvenile leaves which can usually be seen on seedlings and saplings growing near the parent tree. The juvenile leaves of *E. viminalis* are opposite, sessile, pale green, lanceolate, usually rounded at the base and about 5 to 12 × 1.5 to 3 cm, much narrower than the orbicular to ovate juvenile leaves of the other 2 species. Differences between *E. viminalis* and *E. smithii* are discussed under that species.

HABITAT In tall open forest, open forest and woodland in moist but well-drained soils in cool mountains and valleys.

DISTRIBUTION From the NSW/Qld border to the Mt Lofty Range in SA, and in Tas. Occurs as very tall trees on basalt at Mt Wilson, and in the Megalong, Hartley and Kanimbla Valleys. Reported occurrences in the Nepean district probably refer to *E. benthamii.*

FLOWERING January–May, and at other times.

751. *Eucalyptus baeuerlenii* (buds)

Eucalyptus baeuerlenii

Baeuerlen's Gum **Plate 751**

This is a rare species with an unusual restricted distribution. It is found only in a limited area in the Valley of the Waters at Wentworth Falls in the Blue Mountains, and in a small area north of Sugarloaf Mountain near Braidwood on the southern tablelands. On favourable sites it grows to a small thin-trunked tree with a smooth mottled cream or greenish brown bark, but on shallower soils of the upper slopes it is a small to tall mallee. The tree is readily identified by the 3-flowered inflorescences and the buds which have a long, tapering, angular floral tube and a short operculum abruptly narrowing to an acutely pointed beak. The fruit is about 10 mm across, sessile and bell-shaped with a slightly flared rim and 3 or 4 large exsert valves.

HABITAT Moist, cool rocky sandstone slopes.

DISTRIBUTION Endemic in Wentworth Falls and Sugarloaf Mountain.

FLOWERING September–October.

Eucalyptus benthamii

Benthams Gum, Camden White Gum **Plate 752**

E. benthamii is a rare, possibly endangered, species with mature trees restricted to sandy flats of the Nepean River and Kedumba Creek, a tributary of Coxs River, and a few small plants in Reedy Creek and Cedar Creek in the central Blue Mountains. It was formerly more widespread, with stands in the Burragorang Valley, at The Oaks and discontinuously along the Nepean from Camden to the Grose River junction, but land clearing and flooding of the Burragorang Valley by the Warragamba Dam have reduced its natural occurrence. The stand of about 100 trees at Bent's Basin is about 30 to 40 m tall with ages of individuals ranging from about 20 to well over 100 years. There are few, if any, young trees. The age distribution appears to be related to periods of major flooding which deposit silt and stimulate germination. Fire causes death of young trees and extensive epicormic growth of juvenile shoots on the trunk and branches of older trees, but does not produce significant germination of seed.

752. A. *Eucalyptus benthamii* (tree)

B. *Eucalyptus benthamii* (juvenile leaves)

E. benthamii is closely related to *E. dorrigoensis*, endemic in the Dorrigo district, and sometimes considered a variety of *E. benthamii*. It is also related to, and resembles, *E. viminalis*, but differs in having broad greyish to subglaucous heart-shaped juvenile leaves, 7-flowered inflorescences, and smaller buds and fruit. The fruits are about 5 mm across, bell-shaped with a thick rim, a flat or slightly ascending disc and 3 or 4 slightly exserted valves.

HABITAT On sandy flats or ridges near streams.

DISTRIBUTION Lower Kedumba Creek and Bent's Basin, with isolated trees between Wallacia and Camden and in Reedy and Cedar Creeks.

FLOWERING April–May.

Eucalyptus dalrympleana ssp. *dalrympleana*

Mountain Gum **Plate 753**

With a centre of distribution in the high mountains further south, *E. dalrympleana* is uncommon on the central tablelands. Except on the poorest sites, it is a tall, straight-trunked tree resembling both *E. viminalis* and *E. rubida*, though usually much larger than the latter species. The branches and most of the trunk have a smooth white bark blotched with various colours due to patchy shedding of the old bark. The three species have buds and fruit of similar appearance in axillary 3-flowered inflorescences, hence these are useful for identifying the trio, but not for distinguishing between them. This is best achieved by examining the juvenile leaves which are distinctive: green, slightly discolorous, narrow, lanceolate and rounded at the base in *E. viminalis*; pale green, concolorous, broad, orbicular to ovate, stem-clasping in *E. dalrympleana*; and greyish blue, glaucous, concolorous, very broad, orbicular to ovate and stem-clasping in *E. rubida*. The juvenile leaves of all 3 species are sessile and horizontal, becoming shortly stalked and vertical in the transition to the intermediate stage.

A 7-flowered subspecies of *E. dalrympleana* occurs on the northern tablelands of NSW, with some individuals having both 3-flowered and 7-flowered inflorescences.

HABITAT On granite, clay and sandstone soils at high elevations, sometimes dominant in tall open forest on the southern alps.

DISTRIBUTION Main dividing range of NSW and Vic (e.g., Boyd Plateau and near Wombeyan Caves and Jenolan Caves). Also Tas. Uncommon on the Blue Mountains from Blackheath to Clarence.

FLOWERING March–June.

753. A. *Eucalyptus dalrympleana* (tree)

B. *Eucalyptus dalrympleana* (buds/fruit)

754. A. *Eucalyptus rubida* (tree)

B. *Eucalyptus rubida* (juvenile leaves)

Eucalyptus rubida

Candlebark **Plate 754**

Although *E. rubida* grows to a tall straight-trunked tree it is more usually a low-branched spreading tree of medium height, sometimes forked near the base. It is mostly smooth-barked, with the old bark shedding imperfectly in flakes and strips leaving patches of dead bark and a mottled white surface which in summer develops a pink or reddish tint on the sunny side of the tree. Frequently the bark is marked by dark horizontal scars caused by burrowing insects.

E. rubida has a similar range to *E. dalrympleana*, but occurs on drier, shallower soils at lower elevations in less mountainous terrain, often in the frost hollows of open mountain valleys. Mature specimens can often be identified by the conspicuous large orbicular to ovate glaucous juvenile foliage of young shrubby offspring growing nearby. The adult foliage varies from dull green to slate grey and is sometimes glaucous. Inflorescences are 3-flowered and similar to those of *E. viminalis* and *E. dalrympleana*, although the fruits are slightly smaller. Buds are more or less sessile and glaucous with a short broad conical operculum. Fruits are hemispherical or cup-shaped, about 5 × 5 mm, slightly domed with a thick rim and 3 or 4 slightly exsert valves.

HABITAT In woodland and open forest in cold parts of the tablelands on a variety of soil types.

DISTRIBUTION Tablelands and western slopes of south-eastern Australia. Occurs locally on the main dividing range, in the Megalong, Hartley and Kanimbla Valleys, on cold dry sites on Boyd Plateau and in many places on the southern highlands from Mittagong to the Marulan–Tallong area.

FLOWERING November–April.

Eucalyptus pulverulenta

Silver-leaved Mountain Gum **Plate 755**

The name 'pulverulenta' refers to the powdery white bloom on the leaves, young stems and inflorescences

755. *A. Eucalyptus pulverulenta* (tree)

B. Eucalyptus pulverulenta (budsflowers)

756. *A. Eucalyptus cinerea* (tree)

B. Eucalyptus cinerea (buds)

of this unusual species which retains its juvenile foliage into the mature flowering stage. Looking more like a wattle than a eucalypt, it is usually a straggly shrub with long weak stems branching at odd large angles and bearing rows of opposite silver-grey round leaves. Like *E. moorei* and *E. baeuerlenii*, it occurs in a restricted area in the Blue Mountains and some hundreds of kilometres away in the southern tablelands, suggesting that it was once much more widespread than now. Indeed, the species is no longer present in some localities between Lithgow and Bathurst for which there are early records of its occurrence.

E. pulverulenta is easily recognised by its thick, glaucous, opposite, sessile leaves, about 5 cm long and wide, borne at right angles along the stems. It is quite striking when in flower; often the stems carry large buds in 3-flowered clusters in the upper leaf axils, then flowers for several leaf pairs below them and, lower down, fruit capsules at various stages of development. The cup-shaped fruits are sessile on short peduncles, about 10 × 8 mm, and often retain their glaucous bloom at maturity. Older, larger trees shed their bark to the base in long strips leaving smooth pale bronze-grey stems.

HABITAT Rocky outcrops on hillsides.

DISTRIBUTION Above the Coxs River near the road from South Bowenfels to the main Jenolan Caves road. Also at several places near Cowra Creek east of Bredbo (southern tablelands).

FLOWERING May–November.

Eucalyptus cinerea ssp. *cinerea*

Argyle Apple **Plate 756**

The botanical name of this species means 'ash-coloured' in reference to the glaucous bloom of the juvenile foliage, which under natural conditions is mostly retained on mature trees. The common name was earned by the resemblance of its brown ribbony fibrous bark to that of the Rough-barked Apple, *Angophora floribunda*, and its common occurrence in Argyle County, an early name for the Goulburn district. It is closely related to *E. pulverulenta*, but is much larger and, because of the rough trunk, is not likely to be confused with it. Juvenile leaves are thick, glaucous, opposite and sessile, at first orbicular and about 5 × 5 cm, later becoming broad heart shaped to ovate in gradual transition to intermediate foliage. Under cultivation, *E. cinerea* sometimes produces adult leaves which are alternate, shortly stalked and broad lanceolate, although still rather thick and glaucous. Inflorescences are 3-flowered in the upper leaf axils; the buds sessile and glaucous with a short conical operculum, and the fruits broadly pear-shaped or cone-shaped, about 9 × 9 mm, with a broad convex disc and 3 or 4 exsert valves.

HABITAT In open forest and woodland on the lower slopes or hilly or undulating country, usually on rather infertile siliceous soils.

DISTRIBUTION Discontinuous on the dividing range from near Bathurst to near Yass. Common from Berrima to Marulan and eastward to the Tallong-Wingello district.

FLOWERING October–December.

Eucalyptus moluccana

Grey Box **Plate 757**

E. moluccana, the commonest of the boxes in the Sydney district, is associated with Forest Red Gum *(E. tereticornis)*, Spotted Gum *(E. maculata)* and ironbarks on clay soils west of Liverpool. The compact, grey, fibrous, tessellated bark is persistent on the lower half of the trunk and shed in ribbons from the upper trunk and branches leaving a smooth, pale-grey surface. The leathery, dull-green juvenile leaves, found on saplings a metre or so high, are alternate, stalked, ovate and often quite large, reaching 15 × 10 cm. The broad lanceolate adult leaves are glossy green, concolorous and about 10 × 3 cm. Buds and fruit are in 7-flowered clusters in terminal panicles. Because the inner and outer parts of the operculum fall together at flowering, the buds lack the scar normally seen in the *Symphomyrtus* group. This distinguishes *E. moluccana* and *E. bosistoana* from other trees in the area with box-like bark. These 2 can be distinguished from each other by their fruits, which are 3 to 4 valved in *E. moluccana* and 5 to 6 valved in *E. bosistoana.*

HABITAT In open forest and woodland in moist, but well-drained, moderately fertile undulating country with a clay soil or subsoil.

DISTRIBUTION Coast and tablelands of NSW and Qld. Common on Wianamatta shale west of Parramatta, Cabramatta and Campbelltown to the Nepean but absent from Tertiary and Quaternary formations in the driest areas from Penrith to Windsor.

FLOWERING January–April.

757. A. Eucalyptus moluccana (tree)

B. Eucalyptus moluccana (fruit)

Eucalyptus bosistoana

Coast Grey Box, Bosisto's Box **Plate 758**

Typically a forest species exceeding 30 m with a tall slender trunk and a small erect crown. However, in the Sydney district, at the northern limit of its range, it is usually smaller and lower branched. It resembles *E. moluccana* in its half-barked trunk; the lower part with light grey thin finely fibrous bark, the upper part smooth and white or light grey. It has alternate, stalked, ovate to orbicular juvenile leaves, similar to those of *E. moluccana* though not so broad and leathery. The adult leaves are dull green, dotted with oil glands, lanceolate, 10 to 20 × 1 to 2 cm, somewhat longer and narrower than those of *E. moluccana*. As in that species, the outer operculum is retained until flowering, hence the buds lack a scar. Fruits are barrel shaped with a descending disc and enclosed or level valves like most other boxes, but are distinctive in having 4 to 7 (usually 5 to 6) valves.

HABITAT Uncommon in open forest with *E. longifolia* and other eucalypts on heavier clay soils.

DISTRIBUTION From Bairnsdale (Vic) to Sydney, mainly coastal but extending to the eastern slopes and tablelands. Locations near Sydney include Bringelly, Picton, Tahmoor, Bargo, West Dapto, Wingello, Tallong, Marulan and the Wombeyan Caves road near Wollondilly River.

FLOWERING November–March.

758. Eucalyptus bosistoana (fruit)

Eucalyptus fibrosa ssp. *fibrosa*

Broad-leaved Ironbark **Plate 759**

E. fibrosa is easily recognised, and common in the drier parts of the Nepean–Hawkesbury area. It occurs both as a medium to tall dominant species with a good trunk to half the tree height, and as a shorter tree of poor form growing with melaleuca and other eucalypts such as *E. parramattensis*. The bark is dark grey and furrowed into broad flat strips, usually softer and more flaky than that of other ironbarks. Saplings 1 to 2 m tall carry alternate, stalked, dull green, large, leathery, ovate juvenile leaves 5 to 10 cm broad. Adult leaves are thinner, grey-green, lanceolate and about 2 to 2.5 cm broad.

759. A. Eucalyptus fibrosa (tree)

B. Eucalyptus fibrosa (buds)

Flowers are in clusters of 7 to 11 in mostly terminal panicles and, like those of *E. crebra*, have stamens which are all fertile (e.g., have anthers). The long buds taper to the top and bottom and have a scar where the narrow operculum meets the slightly broader floral tube. The cone-shaped or pear-shaped fruits are about 10 mm across, and unusual for an ironbark in having exsert valves.

HABITAT In open forest on ridge tops and on flat, poorly drained clay, silt or sandy soils, with a grassy or thin shrubby understorey.

DISTRIBUTION Scattered distribution from Bodalla (NSW) to Rockhampton (Qld). Recordings include the Kemps Creek–Mulgoa district, the dry area between Penrith and Richmond, Castlereagh State Forest, East Hills and Georges River near Campbelltown.

FLOWERING November–February.

760. A. Eucalyptus crebra (tree)

B. Eucalyptus crebra (buds/fruit)

Eucalyptus crebra

Narrow-leaved Ironbark **Plate 760**

On clay soils west of Sydney, *E. crebra* is a medium-sized tree with a slender trunk up to two-thirds of the tree height. It is instantly recognisable by the combination of hard blackish bark and narrow pendulous leaves; the bark coarsely furrowed, impregnated with kino, and persistent to the small branches, the leaves dull green or grey-green, up to 2.5 cm wide at the juvenile stage and less than 1.5 cm wide in the adult stage. Buds and fruit are on fairly long stalks in clusters of 7 to 11, mostly in terminal panicles. The small cup-shaped fruits are 4 to 5 mm across with a thin rim and 3 or 4 enclosed valves. *E. crebra* can be distinguished from *E. beyeri*, an uncommon species, by the stamens; all fertile in *E. crebra*, the outer ones reduced to staminodes (i.e., without anthers) in *E. beyeri.*

HABITAT Usually in woodland on the higher slopes of undulating country. In the Sydney district more or less confined to clay soils derived from Wianamatta shale.

DISTRIBUTION Extending from south coast to northern Qld; also western slopes. On the coastal plain near Sydney to the lower slopes of the Blue Mountains near Kurrajong and Glenbrook.

FLOWERING Mostly August–November.

761. *Eucalyptus rudderi* (tree)

Eucalyptus rudderi

Rudders Box **Plate 761**

E. rudderi has a scattered distribution on the coastal ranges of central and northern New South Wales where it is typically a tall straight-trunked lightly-crowned tree inhabiting steep rocky slopes of protected gullies. It has light grey fibrous box-type bark on the trunk and branches, and thin, densely reticulate lanceolate leaves 6 to 12 × 2 to 3 cm. The flowers are in clusters of 7 arranged in terminal panicles. Like those of a number of related boxes and ironbarks, the stamens are inflexed in bud (i.e., downturned with the anthers at the sides and base of the bud cavity) and the outer ones are without anthers. Since the outer operculum is shed before flowering, the mature buds have a bud scar. This combination of bark, leaf and floral characteristics enables positive identification of the species. In the absence of buds and flowers, *E. rudderi* can be distinguished from other box-barked trees with a similar habitat and range by the small fruits which resemble those of *E. crebra.*

HABITAT Tall open forest, usually on rocky, sandy soils on the lower parts of steep gullies.

DISTRIBUTION Shoalhaven River to near Taree. Recorded from Kangaroo Valley, Picton, Thirlmere, the Warragamba catchment area and the Watagan Mountains.

FLOWERING February–March.

762. *Eucalyptus bauerana* (tree)

Eucalyptus bauerana

Blue Box **Plate 762**

This is a medium-sized woodland tree, similar to *E. polyanthemos*, but mainly coastal in distribution. Although uncommon in the Sydney district, it is readily recognised by its light to dark grey fibrous-flaky bark on trunk and branches and its dense round crown of bluish-green juvenile leaves. These are thin, concolorous and broad ovate with wavy margins, sometimes becoming lanceolate in the upper crown. Flowers are in clusters of 7 in terminal panicles. As with *E. rudderi*, the stamens are inflexed in bud, and the outer ones are without anthers. The buds are club shaped with a scar; the fruits cone shaped on short stalks, about 6 mm across, with a thin flared rim, a descending disc and 3 or 4 enclosed valves.

HABITAT In small stands on river flats and heavy or loamy soils.

DISTRIBUTION Bairnsdale (Vic) to the Putty district with a disjunct occurrence near Armidale. Occurs at Cobbitty, 2.5 to 4.5 km west of Picton on the road to Mowbray Park, on river flats at Moorebank (New Brighton Golf Links) and in Liverpool Cemetery.

FLOWERING August–September and possibly later.

Eucalyptus paniculata

Grey Ironbark **Plate 763**

E. paniculata is common and widespread in the Sydney district on heavy soils or loamy soils with a clayey base. It is a medium to tall forest tree with a hard, deeply furrowed, light to dark grey, but not black, bark on the trunk and branches. It is perhaps most easily recognised by the leaves which are paler on the undersurface at all stages. The juvenile leaves are alternate, stalked, ovate to broad lanceolate and up to 6 cm wide. Adult leaves are thin, green, lanceolate, up to 18 × 3 cm and finely but densely veined. Floral features are also fairly distinctive. Buds are usually angular near the stalk, and the short conical operculum is usually narrower than the floral tube. Stamens are inflexed in bud (downturned so that the anthers touch the sides and base of the bud cavity), and the outer ones are reduced to staminodes. The fruits are about 8 × 6 mm, with a thick rim, a descending disc and 5 slightly enclosed valves.

HABITAT Open forest on heavy soils as a dominant species or in association with other eucalypts.

DISTRIBUTION Coastal areas from Bega to Coffs Harbour.

FLOWERING May–August.

763. A. *Eucalyptus paniculata* (tree)

B. *Eucalyptus paniculata* (buds/fruits)

764. A. *Eucalyptus beyeri* (tree)
B. *Eucalyptus beyeri* (buds)

Eucalyptus beyeri ssp. *beyeri*

Beyer's Ironbark **Plate 764**

This is an uncommon species endemic in central-eastern New South Wales and resembling *E. crebra* in a number of characters, but more closely related to *E. paniculata*. It is a small to medium-sized woodland tree with a large spreading crown and hard, dark furrowed bark on the trunk and branches. Juvenile and adult leaves are grey-green and only slightly broader than those of *E. crebra*. Buds and fruit are similar in size and appearance to those of *E. crebra*. However, the operculum is narrower than the floral tube, the stamens are inflexed in bud and the outer stamens are reduced to staminodes, indicating membership of the *paniculata* group rather than the *crebra* group. The anther structure is also characteristic of the *paniculata* group.

HABITAT Woodland in undulating country on sandy or clayey soils.

DISTRIBUTION Scattered locations from near Nowra to the Hunter River Valley (e.g., Bargo, Kingswood, Windsor, Wheeny Creek).

FLOWERING August–October.

Eucalyptus melliodora

Yellow Box **Plate 765**

Yellow Box is a medium-sized to tall graceful woodland tree of the tablelands and western slopes with a trunk less than half the tree height and a spreading crown of fine pendulous grey-green leaves. The trunk bark is very variable, often yellow-brown or red-brown and fibrous or flaky towards the base, but at times thick, coarse and dark over most of the trunk. Usually the inner bark is yellow beneath the rough outer bark. Juvenile and adult leaves are small, narrow and light green or grey-green with a pleasant cineol odour when crushed. They are densely but faintly veined except the intramarginal vein which is distinct and distant from the leaf edge. Inflorescences are 7-flowered, axillary and crowded towards the ends of the branchlets. Stamens are inflexed in bud, the outer ones barren. Buds and fruit are fairly small, the buds lacking a scar since the 2 parts of the operculum fall together. The fruits are hemispherical on slender stalks, about 6 × 6 mm, and often have the remains of the staminal ring inside the mouth, obscuring the disc and the 5 enclosed valves.

HABITAT In woodland on undulating country.

DISTRIBUTION Western Vic to south-eastern Qld, mainly on the tablelands and slopes. Absent from the Blue Mountains, but found west of Lithgow, south of Hartley and along the Jenolan Gorge.

FLOWERING September–February.

765. *Eucalyptus melliodora* (buds/fruit)

Eucalyptus sideroxylon

Mugga Ironbark **Plate 766**

E. sideroxylon is closely related to *E. melliodora* and is similarly distributed in the eastern States except for its scattered occurrence on the clay soils west of Sydney. It is a medium-sized woodland or forest tree with usually a short trunk of poor form and a large irregular crown with pendent branchlets bearing dull greyish-green leaves. The bark on the trunk and larger branches is thick, hard, deeply furrowed and dark brown to black due to kino impregnation. Inflorescences are characteristic with buds, flowers and fruit hanging on long slender stalks (pedicels) in clusters of 7, the clusters axillary and also on long slender stalks. Stamens are inflexed in bud and the outer ones are without anthers. Flower colour is white or creamy in local trees but pink or red forms are known and are frequently cultivated. The fruits are cup-shaped or barrel-shaped, about 10 × 9 mm with a descending disc and 5 enclosed valves often partly hidden by the dark remains of the staminal ring (a ring of tissue at the base of the stamens).

HABITAT In open forest with other ironbarks on clayey or gravelly low ridges.

DISTRIBUTION From Vic to south-eastern Qld. Locally at Holsworthy, Moorebank, Liverpool, Parramatta, Kemps Creek, St Marys and Richmond.

FLOWERING April–October, sometimes later.

766. *A. Eucalyptus sideroxylon* (tree)
B. Eucalyptus sideroxylon (flowers/fruit)

767. *A. Eucalyptus microcorys* (bark)
B. Eucalyptus microcorys (flowers/fruit)

Eucalyptus microcorys

Tallowwood **Plate 767**

E. microcorys is an isolated species which does not hyrbidise with any other eucalypt. It is common in tall coastal forests from near Newcastle to Queensland and recognisable by its soft, fibrous, light-brown spongy bark on trunk and branches. The leaves at all stages are thin, pale on the underside and finely toothed on the edges. Adult leaves are about 10 × 2.5 cm, lanceolate, glossy dark green above and dull below with fine but distinct diverging lateral veins. Inflorescences are 7 to 11 flowered and crowded towards the ends of the branchlets. The buds taper gradually to the stalk and have a very short rounded operculum marked by cross sutures formed by fusion of the sepals. The stamens tend to be in 4 clusters rather than evenly spaced around the rim, and the outer ones are without anthers. Fruits are cone shaped on slender stalks, about 8 × 6 mm, with a descending disc and 3 or 4 broad valves at about rim level.

HABITAT Tall open forest bordering rainforest on deep, moist, fertile soils.

DISTRIBUTION Coastal ranges from Cooranbong (NSW) to Maryborough (Qld). Locally at Cooranbong, Dora Creek and the Watagan Mountains.

FLOWERING August–October.

Onagraceae

The Onagraceae is a widespread family of herbs, or rarely shrubs, with its greatest representation in Mexico and south-western United States. Many of the genera have flowers with 4 sepals, 4 petals and 8 stamens above an inferior ovary with 4 chambers. In some genera, such as *Epilobium*, the floral tube and ovary are greatly exaggerated in length.

The largest Australian genus, *Epilobium* (willow-herb), is widespread throughout the country and the world. It is uncommon in the Sydney region though relatively common in some of the colder, higher parts of the country. Several species of *Oenothera* (evening primrose) with tubular yellow or pinkish flowers are common introduced roadside weeds. The best-known ornamental genus in the family is *Fuchsia*, a native of central and south America, New Zealand and Tahiti.

Four species of *Epilobium* and 1 native species of *Ludwigia* occur in the Sydney district.

Epilobium Willow-herb

Perennial herbs, with 4 petals, 4 deciduous sepals and 8 stamens. Fruit a long, terete capsule, opening from the summit downwards into 4 valves, revealing seeds with tufts of long hairs.

768. *Epilobium hirtigerum*

Epilobium hirtigerum

Hoary Willow-herb **Plate 768**

This is an ascending perennial herb, 30 to 50 cm tall, with stems covered with long spreading simple hairs mixed with shorter hairs. Leaves are mostly alternate, with a few opposite basal leaves, oblanceolate to narrow lanceolate, 2 to 6 cm long and minutely toothed along the margins. Young leaves are covered with short white hairs. Flowers are erect, white or pink, and borne singly in the upper leaf axils. The fruit is a narrow, downy, cylindrical capsule up to 6 cm long.

HABITAT Damp, shady places, such as river flats and margins of lakes and lagoons.

DISTRIBUTION Coast to mountains, but not common. Scattered and localised throughout most of NSW, except north coast and far western plains. Occurring in all other States, except NT; also NZ, Indonesia and America.

FLOWERING Chiefly December–January, but some flowers at other times.

769. Epilobium gunnianum

Epilobium gunnianum

Mountain Willow-herb **Plate 769**

A perennial herb with decumbent or erect stems to 30 cm tall. Unlike *E. hirtigerum*, its stems lack the long spreading simple hairs and have short glandular hairs or are almost hairless. Leaves are opposite, without stalks, broadly lanceolate, 15 to 30 mm long, toothed and hairless. Flowers are erect and pink; the capsule is up to 7 cm long. A key feature for positive identification is that the seeds have a pale rim or wing, a feature absent from other Willow-herbs in the area.

HABITAT Damp areas, along streams and margins of swamps.

DISTRIBUTION Blue Mountains (e.g., Blackheath). Extending along tablelands; also Vic, Tas and NZ.

FLOWERING December–January.

770. Epilobium billardierianum

Epilobium billardierianum

Smooth Willow-herb **Plate 770**

Smooth Willow-herb is an erect herb, usually about 30 cm tall but sometimes taller. It is a variable species, with stems sometimes covered with short glandular hairs or hoary-downy to almost hairless. Leaves are toothed, blunt and often stem-clasping. Flowers are liliac, pink or sometimes white, erect and borne in the upper stems. The fruiting capsule is up to 10 cm long. Three subspecies occur in the Sydney district.

Ssp. *billardierianum* has hairless, elliptic to lanceolate leaves, up to 6 cm long, and toothed. Mainly occurring in wet areas of coastal sand dunes.

Ssp. *hydrophilum* has hairless, narrow-ovate to lanceolate leaves, up to 2 cm long and coarsely toothed. Flowers have petals 12 to 18 mm long. Mainly occurring in damp inland places.

Ssp. *cinereum* has greyish leaves with minute soft erect hairs, narrow oblong to linear, up to 3 cm long and coarsely toothed. Flowers are small, with petals 4 to 5 mm long. Occurring in drier sites than the other 2 subspecies.

HABITAT Damp areas along streams, behind sand dunes and swamp margins.

DISTRIBUTION Ssp. *Billardierianum.* Central and south coasts. Extending to Vic, Tas and SA.

Ssp. *hydrophilum.* Coast to mountains. Occurring throughout tablelands and north coast; also Qld, Vic and NZ.

Ssp. *cinereum.* Coast to mountains. Occurring throughout coast, tablelands and western slopes and in all other States, except NT; also NZ and Lord Howe Island.

FLOWERING December–February.

Ludwigia

Decumbent or erect herbs with yellow flowers occurring in and near ponds and streams.

Ludwigia peploides ssp. *montevidensis*

Water Primrose, Clove-strip **Plate 771**

Water Primrose is typically found in the mud of ponds or slow-moving streams. Its prostrate spongy stems creep across the mud, rooting at the nodes, or float on the water. Leaves are alternate, glossy-green, variable in shape, obovate to ovate, 3 to 6 cm long, tapering into a stalk 10 to 20 mm long. Two small bladder-like stipules are at the base of the leaf. Flowers are solitary, cup-shaped, with 5 bright yellow petals, up to 25 mm across. Fruit is a 5-celled cylindrical capsule, up to 25 mm long, and somewhat curved.

HABITAT In streams, ponds and lagoons, in shallow water.

DISTRIBUTION Coastal zone (e.g., Wollongong, Minnamurra, Bong Bong Dam). Also occurring western plains, north-western slopes, Qld, Vic, SA and America.

FLOWERING January–April.

771. Ludwigia peploides

Haloragaceae

This is a small family of aquatic and terrestrial herbs or shrubs which, although widespread in temperate and subtropical regions of the world, has its major development in Australia. The family as presently defined (with *Gunnera* in a separate family) consists of 8 genera, 6 of which occur in Australia, with 3, possibly 4, in the Sydney area.

Myriophyllum is an easily recognised genus of aquatic herbs with finely divided submerged leaves, usually in whorls. *Gonocarpus* and *Haloragis* are herbs or small scabrous shrubs with small toothed leaves and tiny green or brown flowers with the parts mostly in 4 or multiples thereof. The 2 genera are distinguished mainly by their ovaries and fruits which are 1-locular and 1-seeded in *Gonocarpus* and mostly 4-locular and 4-seeded in *Haloragis.*

772. Gonocarpus teucrioides

Gonocarpus Raspwort

Small plants, often scabrous. Each plant carries separate male and female flowers. Flowers small, solitary in the axils of bracts and borne in racemes.

Gonocarpus teucrioides

Raspwort **Plate 772**

G. teucrioides is a rather inconspicuous plant with erect stems, usually 30 to 40 cm tall. What makes it noticeable is the rough sandpaper nature of its stems and leaves. These leaves are opposite, 5 to 20 mm long, broadly lanceolate, coarsely toothed and stalkless. The upper leaves are reduced to bracts. Flowers are small and borne in racemes at the ends of the stems.

HABITAT Sandstone soils in open forest, scrub and heath.

DISTRIBUTION Widespread and common from coast to mountains. Recordings include Mangrove Creek, Brisbane Water National Park, Woronora River and Heathcote National Park. Extending along coast and tablelands divisions north to Qld and south to Vic and Tas.

FLOWERING December–January.

Gonocarpus micranthus ssp. *micranthus*

Creeping Raspwort **Plate 773**

This species is common but may be overlooked because of its prostrate habit and lack of striking inflorescence. Its stems are about 10 cm long, creeping and rooting at the nodes. Leaves and stems are usually green, but often turn a reddish colour. Leaves are mostly opposite, round to heart shaped, and 4 to 9 mm long. Flowers are minute (1 mm diameter) and borne in racemes at the ends of the stems.

HABITAT Wet ground and margins of swamps.

DISTRIBUTION Coast to mountains. Also recorded on south coast, southern tablelands, northern tablelands, western slopes, Qld, Vic, Tas, SA, New Guinea, New Zealand and Asia.

FLOWERING Most of the year.

774. Myriophyllum propinquum

Myriophyllum Water-milfoil

Aquatic plants, most of which take root in the mud of shallow water. Each plant carries separate male and female flowers. Flowers small, in the axils of the upper leaves. Confusion has arisen as to the correct naming of many species; a revision of the genus may result in the splitting of species and some name changes.

Myriophyllum propinquum

Common Water-milfoil **Plate 774**

This perennial aquatic herb is common and widespread throughout Australia. In some districts, its stout spongy stems can form dense masses, blocking channels and covering large areas of swamps. Around Sydney there are few habitats suitable for such luxuriant growth. Bright green leafy stems stand 15 to 20 cm above the surface of the water. Leaves are narrow, up to 2 cm long, entire or toothed, and in whorls of 4 or more. Submerged leaves are feathery.

HABITAT In shallow water of swamps, lagoons and slow-moving streams.

DISTRIBUTION Widespread from coast to mountains. Recorded throughout NSW and in all other States, except NT; also NZ.

FLOWERING October–January.

773. Gonocarpus micranthus

775. Myriophyllum pendunculatum

Myriophyllum pedunculatum

Mat Water-milfoil **Plate 775**

This dwarf hairless herb forms extensive mats on the sandy bottom of pools and swamps; it is often submerged by a few centimetres of water. Stems are weak, slightly fleshy, creeping, rooting at the nodes, and mostly less than 2 cm tall. Leaves are opposite (a characteristic feature), linear-terete, 5 to 10 mm long, with a darker blunt tip. Submerged leaves are sometimes divided into a few thread-like segments. Flowers are small and solitary in the upper leaf axils. Male and female flowers are borne separately, sometimes on the same plant with the male above and the female below, or on different plants. Female flowers are stalkless, without petals, but with 4 feathery styles and are subtended by 2 small bracteoles. Male flowers are on stalks 2 to 4 mm long, with 4 reddish petals about 2 mm long, 8 stamens and 2 small bracteoles near the base. Stems and leaves often turn reddish as pools dry out.

HABITAT In shallow water, and margins of pools and swamps.

DISTRIBUTION Blue Mountains (e.g., Clarence). A tablelands species extending from northern tablelands to southern tablelands, Vic and Tas. Also New Zealand and New Guinea.

FLOWERING December–January.

776. Haloragodendron lucasii

Haloragodendron

Haloragodendron lucasii **Plate 776**

This is a rare shrub, once regarded as possibly extinct, and last seen in 1926 until rediscovered in 1986 by a St Ives resident. Unlike other local members of the family, which are herbs or inconspicuous shrubs, this is a shrub to 1 m tall. Its stems are 4-angular and hairless. Leaves are opposite, narrow-elliptic, 25 to 40 mm long, toothed, stalkless or nearly so and leathery. Flowers are white, with 4 petals (twisted in bud), 4 styles and 8 stamens. They are borne in terminal spike-like clusters.

HABITAT Sheltered gullies, often as the base of rock faces, with fern understorey.

DISTRIBUTION Rare and restricted to the Sydney district. Recorded from Wollangambe gorges and Davidson Park (St Ives).

FLOWERING November.

779. *Dodonaea viscosa* ssp. *angustifolia*

Sapindaceae

Although the relationship may not appear obvious, the Sapindaceae is closely related to the Rutaceae, and in some classifications both are included in the order Sapindales.

The family consists of trees, shrubs and some climbers with a predominantly tropical and subtropical distribution. The Australian contingent of 27 genera and 150 species is a small fraction of the world total of 145 genera and 2000 species. *Dodonaea*, by far the largest of the Australian genera, accounts for nearly half the Australian species. Unlike most of the others, which occur in east coast rainforests but have their major development in the islands to the north, *Dodonaea* is widespread and almost entirely Australian. Its popular name, Hop Bush, is a reference to the inflated winged fruits which, both in appearance and taste, resemble the cultivated hop used in brewing, and in the early days of settlement were frequently utilised for that purpose. The fruits of a number of *Dodonaea* species are large, colourful and decorative, although the flowers are insignificant.

Many of the Sapindaceae contain saponins and have been used by natives as arrow and fish poisons. Because saponins lather in water, some species have common names like soapberry or soapnut and are used for washing. Some tropical and Asian species have edible fruits such as rambutan, longan and lychee. In Australia the barks of *Alectryon tomentosus* and the Foam-bark Tree, *Jagera pseudorhus*, are used by the Aborigines as fish poisons. The fruits of *Atalaya* and 2 or 3 other Australian trees are edible.

Dodonaea Hop Bush

Shrubs with alternate, simple or pinnate leaves. Flowers small and insignificant, but fruit a conspicuous capsule, with 2 to 6 papery wings, often red, purple or brownish.

777. *Dodonaea truncatiales*

Dodonaea truncatiales **Plate 777**

This is an erect shrub, to 3 m tall, with angular branches. Leaves are narrow lanceolate, 5 to 10 cm long, up to 13 mm wide, entire or with wavy-toothed margins, hairless and somewhat viscid. Flowers and fruit are in axillary clusters, on stalks 4 to 14 mm long. The fruit capsule is distinctive. It is 4-lobed with wings wider (horizontally) than long; wings are 5 to 10 mm wide.

HABITAT Dry eucalypt forest on sandstone, often near rivers.

DISTRIBUTION Nattai River, Nepean River and northern Blue Mountains. Recordings include Little River (Nattai), Douglas Park (Nepean), Green Gully (Glen Davis) and Mt Yengo (Howes Valley). Also occurring south coast and eastern Vic.

FLOWERING July–August.

Dodonaea triquetra

Common Hop Bush **Plate 778**

D. triquetra is one of the most common shrubs in the local area, especially abundant after bushfires, often predominating until other shrubs become established. It is an erect, hairless shrub, 1 to 3 m tall, with elliptic to lanceolate leaves, 5 to 10 cm long, 1 to 3 cm wide, hairless, with a pointed apex and narrowing to a short stalk. Flowers are small and insignificant; male and female flowers are on different plants. Fruit is a capsule with 3 wings, usually about 1 cm long, green when young, turning purple-brown.

HABITAT Open forest, on slopes and ridges, on sandy soils.

DISTRIBUTION Widespread from coast to mountains. Extending along coastal zone from south-east Qld to eastern Vic; also north-western slopes.

FLOWERING Chiefly July–October, but often at other times.

778. *Dodonaea triquetra*

Dodonaea viscosa ssp. *angustifolia*

Sticky Hop Bush **Plate 779**

D. viscosa is a very variable species especially in its leaf shape. A number of populations once regarded as separate have now been classified as subspecies of *D. viscosa*. These subspecies often intergrade so that the distinction is not always clear. Subspecies *angustifolia* is an erect shrub, often 2 to 3 m tall, with narrow lanceolate leaves, 6 to 10 cm long, 5 to 10 mm wide, acute at the apex and narrowing to a stalk 6 to 18 mm long. The margins are entire or slightly wavy. Leaves are often glossy, with a varnished appearance. Fruit is green, turning to red and light brown, with 3 (rarely 4) vertical wings.

HABITAT Open forest, often in rocky areas.

DISTRIBUTION Scattered from coast to mountains. Most common in valleys and on slopes west of the Blue Mountains, but also found along Illawarra Escarpment (e.g., west of Dapto) and in Nattai catchment. Occurring from south-east Qld to eastern Vic along coast, tablelands and western slopes. Also in Africa, Asia, Pacific region and Americas.

FLOWERING August–October.

780. Dodonaea viscosa ssp. *angustissima*

Dodonaea viscosa ssp. *angustissima*

Narrow-leaf Hop Bush **Plate 780**

An erect shrub, often 1 to 2 m tall, with linear leaves, slightly sticky, 3 to 9 cm long and 1 to 4 mm wide, acute to blunt at the apex, usually with a short point and stalkless. The margins have small irregular teeth and are slightly curved downwards. Fruit is green, turning red to purplish, with 3 (rarely 4) vertical wings. This subspecies intergrades with ssp. *spatulata*.

HABITAT Open woodland, often near rocky areas; occurs on sandplains in inland Australia.

DISTRIBUTION A very common shrub west of the Great Dividing Range, but uncommon in the Sydney area, where it is restricted to valleys of the Blue Mountains (e.g., near Goodman's Ford, Wollondilly Valley). Occurring throughout NSW, except along coastal zone; also Qld, Vic, Tas, SA and WA.

FLOWERING September–November.

781. Dodonaea viscosa ssp. *cuneata*

Dodonaea viscosa ssp. *cuneata*

Wedge-leaf Hop Bush **Plate 781**

A compact spreading shrub, to about 2 m tall, with leaves which are distinctly wedge shaped, slightly sticky, 1.5 to 3 cm long and 5 to 9 mm wide, truncate (cut off squarely) or obtuse at the apex, usually ending in a short point and stalkless or almost so. The margins may be entire or have 2 or 3 irregular teeth. Fruit is green, turning red or purple, with 3 vertical wings. This subspecies intergrades with ssp. *spatulata* in the Sydney district.

HABITAT Open forest, on sandy loam or clay soils.

DISTRIBUTION Common on slopes and plains west of the Great Dividing Range and in upper Hunter Valley (e.g., Lees' Pinch), but uncommon in the Sydney area. Recorded from Cumberland Plain (e.g., Castlereagh, Llandillo) and the Narellan–Campbelltown district. Also occurring in south-east Qld, Vic and SA.

FLOWERING August–October.

Dodonaea viscosa ssp. *spatulata*

This erect to spreading shrub grows to about 3 m tall and has the greatest amount of leaf variation of all the subspecies of *D. viscosa*. Leaves are usually obovate to spoon-shaped, occasionally elliptic, slightly sticky, 2.5 to 7.5 cm long and 6 to 16 mm wide, obtuse or rounded at the apex, ending in a short point and stalkless or with a stalk 1 to 10 mm long. The margins may be entire, irregularly toothed or wavy. This subspecies is similar to ssp. *cuneata*, with which it intergrades, but differs in the size of the leaf and the rounded tip.

HABITAT Open forest, mostly in sandy loam.

DISTRIBUTION Coast to mountains, but uncommon. Extending to semi-arid western slopes and plains. Also south-eastern Qld, Vic, Tas, SA and WA.

FLOWERING September–November.

Dodonaea falcata

Thread-leaf Hop Bush **Plate 782**

This is an erect sticky shrub to 2 m tall; its branches are covered with minute, soft, erect hairs. Leaves are narrow-linear, 2.5 to 5 cm long, about 1 mm wide, concave or channelled above, falcate, entire, obtuse or curved downwards at apex and stalkless. Flowers are in 3 to 4 flowered terminal clusters. Fruit is a 4-winged capsule, up to 12 mm long and 14 mm wide, sparsely covered with minute hairs; fruit turn red with age. *D. falcata* is often confused with *D. filifolia*, a Queensland species, but that plant has hairless branches and straight and longer leaves.

HABITAT Dry open forest.

DISTRIBUTION Restricted in Sydney district to Cumberland Plain (e.g., Castlereagh State Forest, Kenthurst). Extending northwards to Hunter Valley (e.g., Putty Road), north-western slopes (Pilliga Scrub) and Darling Downs in Qld.

FLOWERING September–December.

782. Dodonaea falcata

Dodonaea camfieldii

Camfield's Hop Bush **Plate 783**

D. camfieldii is easily identified as it is the only prostrate Hop Bush in the local area. Occasionally it may grow to 30 cm tall, but its spreading stems are mostly ground-hugging. Leaves are small, 1 to 3 cm long, oblong, stalkless, somewhat stem-clasping, with wavy margins, sometimes entire or irregularly lobed. Flowers are solitary or paired on stalks 2 to 4 mm long. Fruits are 4-winged, 10 to 15 mm long, 9 to 12 mm wide, changing to red or indigo with age.

HABITAT Open forest and woodland, on sandy soils.

DISTRIBUTION Restricted to the central and south coasts from Brisbane Water National Park to Jervis Bay–Nowra district. Recordings include Wondabyne (Mullet Creek), Kangaroo Creek (Royal National Park) and Heathcote Road (Lucas Heights).

FLOWERING September–November.

783. *Dodonaea camfieldii*

Dodonaea megazyga — Plate 784

This is an erect shrub or small tree, 2 to 4 m tall, with distinctive large compound leaves up to 15 cm long. There are 19 to 31 leaflets (sometimes as few as 13) along a winged rhachis. Leaflets are lanceolate, acute, entire, up to 25 mm long; one leaflet is terminal. Flowers are in axillary clusters on short stalks. The fruit is a 3-winged capsule, with wings wider than deep, turning red with age.

HABITAT Rainforest margins or tall forests.

DISTRIBUTION Mainly a species from the north coast and northern tablelands, extending to Lamington National Park (Qld); reaching its southern limit in Olney State Forest, west of Cooranbong.

FLOWERING September–November.

784. *Dodonaea megazyga*

Dodonaea pinnata

Pinnate Hop Bush — Plate 785

A spreading shrub, to 1 m tall, readily recognised by its pinnate (compound) leaves and the covering of dense long hairs on its branches, stalks and both surfaces of the leaflets. Each leaf is up to 35 mm long and has between 10 and 16 leaflets (rarely more), each 4 to 9 mm long, entire or slightly toothed at the apex. Flowers are solitary on softly hairy stalks, 4 to 7 mm long. Fruit is a 4-winged capsule, with wings 3.5 to 6 mm wide. *D. pinnata* closely resembles *D. multijuga* and *D. boroniifolia*, but is the only one to have short hairy flower stalks, solitary flowers and 4-winged capsules.

HABITAT Open forest, on sandy soils.

DISTRIBUTION Restricted to coastal zone around Sydney (e.g., Bobbin Head, Brisbane Water National Park, Maroota, Nepean River near Mulgoa), extending to north-west of Hornsby Plateau (e.g., Colo River, Culoul Range, Wiseman's Ferry).

FLOWERING September–December.

785. *Dodonaea pinnata*

Dodonaea multijuga — Plate 786

An erect shrub to 1.5 m tall, with pinnate (compound) leaves. Branches and leaves may be covered with short rough or downy hairs, but the upper surfaces of the leaves are often hairless. Each leaf is up to 48 mm long and has between 16 and 28 leaflets (rarely more), each 4 to 7 mm long, broadly obovate, sometimes minutely toothed at the apex. Flowers are in axillary clusters, on stalks 7 to 14 mm long. Fruit is a 3-winged capsule, with wings 3 to 4 mm wide. *D. multijuga* is distinguished from *D. pinnata* by its longer flower stalks, its clusters of flowers and its 3-winged capsules, and from *D. boroniifolia* by its 3-winged capsules and the larger number of leaflets.

HABITAT Open forest, on sandy soils, often near watercourses.

DISTRIBUTION Widespread from coast to mountains, but not common. Recordings include Erskine Creek (lower Blue Mountains) and Deepwater Park (Georges River). Occurring from Sassafras–Nerriga and Bateman's Bay on the south coast north to south-eastern Qld.

FLOWERING September–March.

786. *Dodonaea multijuga*

Dodonaea boroniifolia

Fern-leaf Hop Bush — Plate 787

A spreading or erect shrub, usually about 1 m tall, with slightly downy branches and a general varnished appearance because of its viscid leaves. Leaves are compound, with each leaf up to 30 mm long, bearing 5 to 11 leaflets, all opposite except for the germinal leaflet. Leaflets are 3 to 7 mm long, angular-obovate, entire, with edges curved under and with 3 to 6 lobes at the apex. Flowers are paired or 3 together, on stalks 4 to 8 mm long. Fruit is a 4-winged capsule, with wings 2.5 to 5 mm wide. *D. boroniifolia* differs from *D. pinnata* in having toothed leaflets and 2 to 3 flowers (not solitary) and from *D. multijuga* in having fewer leaflets and a 4-winged capsule.

787. *Dodonaea boroniifolia*

HABITAT Open forest, on sandy soils.

DISTRIBUTION Widespread but uncommon in coastal districts. More common in valleys of Blue Mountains and Colo Plateau. Found throughout NSW, except far western plains; also Qld and Vic.

FLOWERING September–November.

Diploglottis

Diploglottis cunninghamii

Native Tamarind — Plate 788

A tall rainforest tree, to about 20 m high, its large pinnate leaves are unmistakable. They may be 35 cm or more long, with 3 to 8 pairs of large leaflets. These leaflets are elliptic, unequal at the base, 10 to 30 cm long, hairless above but covered with soft hairs on the undersurface. Young branches, stalks and inflorescences are also covered with dense rusty-hairy panicles in the forks of the leaves. Fruit is a brown hairy capsule, 2 to 3 lobed, containing a large seed which is enclosed by an orange-yellow juicy pulp. These fruits are somewhat acidic and were used for making jam and drinks by early settlers.

HABITAT Rainforest.

DISTRIBUTION Illawarra escarpment (e.g., Minnamurra Falls, Macquarie Pass, Mt Keira) and Gosford–Wyong district (e.g., Strickland State Forest). Extending along north coast to southern Qld.

FLOWERING September–November.

788. *Diploglottis cunninghamii*

789. Guioa semiglauca

790. Cupaniopsis anacardioides

791. Alectryon subcinereus

Guioa

Guioa semiglauca

Guioa **Plate 789**

Guioa is a small tree, usually 8 to 12 m tall. It is fairly common in protected coastal areas, but may also be found on headlands near the sea where it is a smaller plant. Leaves are pinnate-compound, with 2 to 6 elliptic to obovate leaflets. The main identifying feature is that the leaflets are distinctly bluish-grey on the undersurface (hence the specific name, 'semiglauca'). They are 4 to 8 cm long, entire, usually obtuse or with a short blunt tip and have a lump at the base of the leaflet where it joins the main stalk. Flowers are small, yellowish green and borne in dense clusters in the leaf axils or sometimes terminal. The fruit is a blue-green capsule, 8 to 12 mm wide with 2 to 3 wing-like lobes; each lobe contains a single glossy black seed.

Habitat Rainforest and in protected coastal gullies; sometimes found on more exposed areas near rainforest.

Distribution Coastal zone. Recordings include Upper Kangaroo Valley, Minnamurra Spit, Burning Palms, Calga and Razorback Mountain. Extending from south coast to north coast and Qld.

Flowering October–November.

Cupaniopsis

Cupaniopsis anacardioides

Tuckeroo **Plate 790**

Tuckeroo is common north of Sydney and reaches its southern limit on the shores of Port Hacking. It is a tree with a dark dense crown, tall in protected sites (e.g., Warumbul) but with a short spreading habit in more exposed situations (e.g., Jibbon Beach). Leaves are pinnate-compound with 6 to 10 entire leaflets which are dark green above and paler beneath, but not bluish-grey as is the undersurface of the leaves of the Guioa. Leaflets are broad-ovate, 6 to 12 cm long, leathery, blunt or notched at the apex and are on a short swollen stalk. Flowers are small, greenish white and borne in branched clusters in the leaf axils. Fruit is an orange-yellow capsule, 12 to 16 mm long, with 3 lobes; each lobe contains 1 black seed enclosed in an orange pulp.

Habitat Sheltered places on coastal dunes near the sea or estuaries.

Distribution Scattered along coast from Port Hacking northwards (e.g., Towra Point, Nielsen Park, Tuggerah Lake). Extending along north coast to Qld; also NT.

Flowering May–July.

Alectryon

Alectryon subcinereus

Native Quince **Plate 791**

A tall shrub or small tree to about 8 m tall. Like other rainforest trees in this family, it has compound leaves. Each leaf consists of 4 to 6 leaflets. These are entire or toothed in the upper half, elliptic to lanceolate, 5 to 12 cm long, dull green and paler beneath. The veins are distinct on both surfaces but more conspicuous on the undersurface. The stalks of the leaflets are short (3 mm) and there is a prominent bump where they join the main stalk. Flowers are small, pink and in loose clusters shorter than the leaves. The fruit is unusual and a good identifying feature. It consists of 2 rounded lobes (sometimes 1 or 3) each 8 to 10 mm across, which split open to reveal a dark seed surrounded by bright red pulp.

Habitat In and near rainforest.

Distribution Most common along coastal zone (e.g., Cambewarra, Macquarie Pass, Hacking River, Razorback Mountain, Gosford, Wyong), but also recorded at Jenolan Caves. Extending along coast from south coast to north coast and Qld; also central western slopes.

Flowering November–December.

Anacardiaceae

Anacardiaceae, like the related family Sapindaceae, has alternate or rarely opposite pinnate or simple leaves without stipules. The flowers are bisexual or sometimes unisexual with 5 sepals, 5 free petals and usually 5 or 10 stamens. The stamens are inserted outside a fleshy disc which surrounds a superior ovary. The ovary has from 1 to 5 chambers, each of which produces a single seed. Many members of Anacardiaceae have resinous tissues on the leaves or in other organs which secret poisonous exudates. Exudates from a number of *Rhus* species are severe skin irritants.

The family, with about 600 species in 80 genera, is mainly tropical and subtropical in distribution. Useful products include mango, cashew and pistachio. None of the Australian species, which are few in number, is particularly useful. Only 1 of the 2 New South Wales species occurs as far south as Sydney where it is an uncommon tree on the edge of littoral rainforests.

Euroschinus

The genus *Schinus* includes the pepper tree, a naturalised introduction from South America, well known as a shade and ornamental tree in country towns. *Euroschinus* means the 'south-east schinus', a reference to its distribution and close relationship to the pepper tree.

Euroschinus falcata

Ribbonwood **Plate 792**

Under ideal conditions, Ribbonwood may reach a height of 40 m, but in the local area it is seldom half that height. Leaves are compound, with a central stalk bearing 4 to 10 leaflets, 5 to 7 cm long, ovate to lanceolate, with a long pointed tip. A characteristic feature of the leaflets is that they are asymmetrical

792. *Euroschinus falcata*

at the base. Flowers are pink with cream margins, with 5 petals, only 2 mm long, but numerous in clusters shorter than the leaves. Fruit is a black, egg-shaped berry about 8 mm long. Ribbonwood belongs to the same family as mango, and the sap which exudes from the cut bark and from broken leaf-stalks smells like mangos.

HABITAT In and near rainforest.

DISTRIBUTION Scattered along base of Illawarra Escarpment (e.g., Calderwood, Minnamurra Falls), Palm Jungle in Royal National Park and Gosford–Wyong district (e.g., Bouddi National Park). Extending along north coast to Qld.

FLOWERING November–December.

Meliaceae

This is a family of tropical and subtropical trees and shrubs renowned for fine timbers such as cedar, rosewood and mahogany. From a worldwide total of more than 600 species in 50 genera the family is represented in Australia by only 30 species in 11 genera. Most of the Australian species are endemic and occur in the rainforests of Queensland and northern New South Wales. Only 4 species, each from different genera, occur in the Sydney area. The 3 illustrated species may be seen in or near rainforests throughout the area, but the fourth, *Dysoxylum fraseranum*, occurs only from Wyong northwards.

Members of this family generally have large pinnate leaves (bipinnate in *Melia*) and flowers in panicles. Usually the stamens number 8 to 10 and are united into a tube. The exception is *Toona* which has 4 to 6 separate stamens. The flowers, bark and wood of some species are pleasantly perfumed but in others they have a disagreeable oniony smell. Both red cedar *(Toona australis)* and white cedar *(Melia azedarach)* are deciduous in winter. Although trees of this family are attractive, only *Melia* has been cultivated extensively as an ornamental. Attempts to grow *Toona* in plantations for timber have failed owing to attack of the growing shoots by the small hairy caterpillars of the cedar tip moth.

Red Cedar, with its soft and light yet durable and insect-resistant timber, was formerly prized for furniture and quality interior work. However, owing to ruthless exploitation during the last century, very little new cedar reaches the market today.

Toona

Toona australis

Red Cedar **Plate 793**

Red Cedar is the 'red gold' of the early timber-cutters, eagerly sought after because of its highly valued timber. Although once fairly common, few large specimens remain except in remote areas. Around Sydney some old trees may reach 35 to 40 m but usually it is only half this tall. It is a deciduous tree with an open spreading crown. In winter the leaves brown and fall, leaving bare branches to make it conspicuous amongst the dark green rainforest canopy. New shoots are pink. Leaves are compound, with 4 to 8 pairs of leaflets along a common stem. Leaflets are ovate to lanceolate, pointed, 6 to 12 cm long, shortly stalked and are unequally sided at the base. Flowers are small (5 mm long), white or pinkish, and borne abundantly in long open terminal clusters. Fruit is an oval capsule, and clusters of dried capsules, 20 to 25 mm long, are clearly visible on the tree in later summer and autumn. Each capsule splits into 5 valves, releasing winged seeds. It is easy to confuse a small Red Cedar with *Polyscias murrayi*. The most obvious differences are that *P. murrayi* has longer leaves, up to 15 pairs of leaflets, a terminal leaflet (missing in Red Cedar) and fruit which is succulent and 2 to 3 lobed.

HABITAT Rainforest, especially along gullies and streams.

DISTRIBUTION Illawarra and Wyong districts. Extending along the coast from Milton to north Qld; also New Guinea and Norfolk Island.

FLOWERING October–November.

793. *Toona australis*

794. *Synoum glandulosum*

Synoum

Synoum glandulosum

Scentless Rosewood, Bastard Rosewood **Plate 794**

A common shrub or small tree in rainforests and sheltered situations. Its most conspicuous feature is its fruit which ripen between September and December. These are red, 2 to 3 cm across, 3 lobed, and contain a number of cells each with 2 seeds on either side of a fleshy red aril. The generic name *Synoum* is a reference to these 2 seeds ('syn', together; 'oum', an egg). Leaves are compound, with 2 to 4 pairs of leaflets along a common stalk and a terminal leaflet. Leaflets are elliptic to oblanceolate, tapering to a blunt point, 3 to 10 cm long, dark green above and paler underneath. The base of the leaf is distinctly swollen where it joins the stem. Flowers are white to deep pink, 4-petalled (sometimes 5) and borne in small clusters about 2 cm long in the leaf axils.

HABITAT In and near rainforest and sheltered gullies.

DISTRIBUTION Coastal zone. Extending along coast from Bega (south coast) to northern Qld.

FLOWERING March–July.

795. *Melia azedarach*

Melia

Melia azedarach var. *australasica*
White Cedar **Plate 795**

A spreading deciduous tree found naturally in good soils, especially near rainforests. It is often planted as an ornamental and shade tree and it is sometimes difficult to know whether a specimen is the result of natural distribution or of cultivation. The bark is grey-brown and somewhat furrowed and it was used by Aborigines as a fish poison. The timber of larger trees is prized for its attractive texture and is used for cabinet work. Leaves are large and compound (bipinnate). Leaflets are ovate, pointed, entire or toothed and 2 to 5 cm long. The foliage is prone to attack by caterpillars of the white cedar moth. Flowers are lilac, with stamens united into a purple tube; they are strongly scented. Flowers are borne in long loose clusters which are slightly shorter than the leaves. Fruit is an oval berry which turns yellowish with age and is reputed to be poisonous.

HABITAT Rainforest margins.

DISTRIBUTION Scattered along coast, especially Illawarra district; also Cumberland Plain to Lapstone Hill (where it probably has been planted). Extending along coast from Shoalhaven River to northern Qld; also central western slopes and WA.

FLOWERING October–December.

Rutaceae

Named after the poisonous medicinal plant, *Ruta graveolens*, this large family of shrubs and trees includes the citrus group (oranges, lemons, limes, grapefruit, mandarins, tangerines, pomelos) as well as many ornamental species.

Most members of the family contain glands, seen as small translucent dots on the leaves, which release strongly aromatic oils when the leaves are crushed. Apart from this feature, however, the members exhibit such great variation that family characteristics are difficult to define. The flowers are commonly bisexual, regular and arranged in cymes. They usually have 4 or 5 sepals and petals with either an equal number of stamens alternating with the petals, or twice as many stamens as petals. The flowers often have a conspicuous fleshy disc surrounding the carpels which, in most Australian members, number 4 or 5 and are united only at the base. These usually form a capsular fruit, although those members with united carpels may form a drupe or berry. Frequently only one seed develops in each carpel.

Rutaceae consists of about 150 genera and 1600 species divided amonst 12 tribes. Forty-one genera and 320 species in 6 tribes occur in Australia. The family is distributed mainly in the tropics and temperate regions of the southern hemisphere, with only 1 tribe extending to temperate regions of the northern hemisphere. Of the 6 Australian tribes, the largest, Boronieae, with nearly three-quarters of the Australian species, is almost exclusively Australian. This tribe, which includes the popular and well-known genera *Boronia, Eriostemon, Correa, Phebalium* and *Zieria*, is found in a variety of habitats in all states. Two other tribes present in the Sydney region occur in east coast rainforests while the remaining 3 have a more northerly or inland distribution.

796. Zieria laevigata

Zieria

Shrubs with opposite 3-foliolate leaves. Flowers usually white (sometimes pink) with 4 petals and 4 stamens.

Zieria laevigata **Plate 796**

This is a common small shrub, 50 to 60 cm tall, with hairless stems which are slightly angular. The 3 leaflets are linear, 2 to 4 cm long and less than 3 mm wide, acute, with margins curved to the undersurface. Flowers have pale pink or white petals, about 5 mm long; sepals are short and broad and anthers have a tiny white reflexed point (best seen with a lens). Flowers are in the leaf axils on stalks shorter than the leaves.

HABITAT Open forest and heath, on sandy soils.

DISTRIBUTION Widespread from coast to mountains. Also found south coast, north coast and northern tablelands.

FLOWERING July–October.

Zieria pilosa
Hairy Zieria **Plate 797**

The specific name 'pilosa' refers to the sprinkling of long simple hairs on the stems of this small shrub, a feature which easily separates it from *Z. laevigata*. Its stems are round (not angular) and leaflets linear to oblong, 7 to 25 mm long, acute or obtuse, with margins slightly curved under. Flowers are white, small, and not very conspicuous as they are borne on a very short stalk in the leaf axils. Petals are only 3 mm long; an anther appendage is present but is minute. The sepals are narrow, acute (the only *Zieria* to have such sepals) and hairy.

HABITAT Open forest and heath, on sandy soils.

DISTRIBUTION Widespread from coast to mountains. Extending to north coast and south coast.

FLOWERING May–October, but a few flowers at other times.

797. Zieria pilosa

Zieria cytisoides
Downy Zieria **Plate 798**

Downy Zieria is found in rocky situations throughout south-eastern Australia. It is an erect compact shrub, often over 1 m tall, with leaves and stems covered with soft downy hairs giving the whole plant a velvety appearance. In some areas, the shrub has distinctly greyish foliage. The 3 leaflets are obovate, up to 3 cm long and paler and more densely hairy on the undersurface. Flowers may be white or pink in small but dense clusters at the end of a stalk (10 to 25 mm) from the leaf axils. The 4 carpels are hairy, a feature which distinguishes this species from *Z. compacta*.

HABITAT Open forest, mainly on rocky slopes.

DISTRIBUTION Blue Mountains, from Nepean River gorges (e.g., Bents Basin) to upper mountains (e.g., Narrow Neck, Newnes). Widespread throughout NSW, except drier western plains, also Qld and Vic.

FLOWERING August–October.

798. *Zieria cytisoides*

800. *Zieria robusta*

Zieria smithii

Sandfly Zieria **Plate 801**

This tall hairless shrub is an erect plant to 2 m tall with glandular stems. It emits a strong odour when its leaves are crushed. The leaflets are ovate, 3 to 5 cm long and 4 to 10 mm wide, and have margins which curl under. The length of the leaflets is the main difference between this species and the larger *Z. arborescens* and the leaves intergrade in some areas. Flowers are white (sometimes pale pink), about 8 mm across and are borne in loose clusters in the leaf axils.

HABITAT Forest, in sheltered situations.

DISTRIBUTION Widespread from coast to mountains, but generally absent from the sandstone soils of coastal national parks. Also occurring south coast, north coast, northern tablelands, Qld and Vic.

FLOWERING August–October.

Zieria murphyi

Murphy's Zieria **Plate 799**

Z. murphyi is a softly hairy shrub to 1 m tall. Leaflets are narrow-elliptic, about 4 cm long, blunt at the apex and covered with soft short hairs. Flowers are pale pink to white, with petals about 3 mm long and borne in clusters about as long as the leaves from the leaf axils. The sepals are distinctly hairy.

HABITAT Open forest, in sheltered sites, often just below clifflines.

DISTRIBUTION Endemic in Sydney district. Recorded from near Mt Tomah and in southern highlands at Penrose.

FLOWERING August–September.

Zieria robusta

Plate 800

This is an erect bushy shrub with many prominent tubercles on its stems. Leaflets are obovate, 5 to 12 mm long, rounded at the apex but with a short tip, lighter and sometimes slightly hairy underneath. Flowers are small, white to pinkish, and borne in stalked clusters of a few flowers or solitary from the leaf axils.

HABITAT Protected rocky hillsides.

DISTRIBUTION Higher Blue Mountains (e.g., Kanangra Walls). Also on north-western slopes and in Vic.

FLOWERING September–October.

802. *Zieria arborescens*

799. *Zieria murphyi*

801. *Zieria smithii*

Zieria arborescens

Stinkwood, Tree Zieria **Plate 802**

This is the largest of the local Zieria and may grow to a small tree 3 to 5 m tall. The name 'Stinkwood' comes from the strong and unpleasant odour of the leaves and timber. This species is like a large variety of *Z. smithii*. The leaflets are narrow-elliptic, up to 10 cm long and 25 mm wide, dark green above. Flowers are white, with petals 4 to 5 mm long and borne in clusters on stalks shorter than the leaves.

HABITAT Sheltered forested gullies and near rainforest.

DISTRIBUTION Widespread from coast to mountains. Recordings include Jamberoo Pass, Olney State Forest, Mt Tomah, Mt Wilson and Newnes. Also occurring south coast, north coast, northern tablelands, Qld, Vic and Tas.

FLOWERING September–November.

803. *Zieria granulata*

Zieria granulata — Plate 803

An erect shrub 1 to 3 m tall, it is conspicuous because of its very narrow leaflets which are covered with prominent glands. Branches are hairless, rough and glandular. Leaves consist of 3 narrow linear leaflets, 15 to 30 mm long and 1 mm wide, with margins curved under. The globular glands are so prominent that they make the margins look irregular. Flowers are white, relatively small (7 mm across) and borne in clusters on stalks shorter than the leaves.

HABITAT Basaltic hillsides near rainforest.

DISTRIBUTION A south coast species with a very restricted distribution in the Sydney district. Recorded west of Kiama (Saddleback Mt).

FLOWERING August–October.

Zieria compacta — Plate 804

This small shrub (to 60 cm tall) is regarded by some botanists as a subspecies of *Z. fraseri*, a Queensland and north coast species. Its young branches, leaf stalks, flower stalks and the undersurface of the leaves are covered with short dense soft hairs. Older stems and the upper surface of the leaves are hairless. The leaflets are narrow-elliptic, about 25 mm long and 8 mm wide. Flowers are either white or pink, with hairless carpels. They are borne in dense clusters from the leaf axils, usually shorter than the leaves.

HABITAT Open forest, especially on hillsides and rocky slopes.

DISTRIBUTION Chiefly Blue Mountains, but also parts of Woronora Plateau, such as gullies near Douglas Park and Wingecarribee River. Also recorded from south coast, north coast, northern tablelands, western slopes and Qld.

FLOWERING August–October.

804. *Zieria compacta*

Boronia

Shrubs, with opposite leaves; flowers pink to white with 4 petals and 8 stamens.

Boronia ledifolia

Sydney Boronia, Ledum Boronia — **Plate 805**

Sydney Boronia, with its numerous showy pink flowers borne in profusion in late winter, is a striking feature of the Sydney sandstone country. Despite its common name, it is found beyond the Sydney district. Leaves may be simple or divided into 3 leaflets. They are variable in size and shape from oblong-linear to elliptic, 1 to 3 cm long and 2 to 4 mm wide, shiny green above, paler below and have margins which curl under. Flowers are pink, with petals 5 to 10 mm long.

HABITAT Open forest and heath, on sandy soils.

DISTRIBUTION Widespread from coast to mountains, but most common in coastal sandstone communities. Extending along south coast to Vic; also southern tablelands and central western slopes.

FLOWERING July–September.

805. *Boronia ledifolia*

806. *Boronia ruppii*

Boronia ruppii

Rupp's Boronia — **Plate 806**

This is a shrub to about 60 cm tall with a strong odour when crushed. Leaves are divided into 3 or 5 leaflets (sometimes undivided). Each leaflet is elliptic, flat, 7 to 12 mm long, 4 to 8 mm wide, usually with the terminal leaflet longer than the side leaflets. Flowers are pink, with petals 6 to 10 mm long. There are 4 long and 4 short stamens and filaments are sprinkled with spreading white hairs. Flowers are solitary or in groups of 2 or 3 on stalks to 10 mm long from the leaf axils.

HABITAT Open forest, in sandy soils, often in rocky sites.

DISTRIBUTION North-western margin of Sydney district (e.g., Glen Davis–Newnes). Extending to central-western slopes (Goulburn River National Park) and north-western slopes (near Barraba).

FLOWERING September–October.

807. Boronia algida

Boronia algida **Plate 807**

This is a small, often scrambling shrub, usually less than 50 cm tall, but sometimes up to 1 m. Stems have scattered soft hairs when young. Leaves are divided into 5 thick leaflets, each obovate or spoon shaped, 3 to 6 mm long and about 2 mm wide. Flowers are light pink, with pointed petals 5 to 6 mm long, borne singly in the leaf axils or occasionally terminal.

HABITAT Open forest.

DISTRIBUTION Chiefly a high country plant, occurring throughout the tablelands, including the higher Blue Mountains. Also recorded south-western slopes, north coast, south coast and Vic.

FLOWERING September–December.

Boronia mollis
Soft Boronia **Plate 808**

Soft Boronia derives its name from the covering of long brown spreading hairs on the branches and leaf-stalks. It is a tall robust shrub to 2 m high. The leaves are divided into broad oblong-elliptic leaflets; the terminal leaflet is larger than the others (up to 4 cm long). Flowers are dark pink and sepals narrow; flowers are borne in clusters in the leaf axils. The foliage has an unpleasant odour when crushed.

HABITAT Sheltered forested slopes and gullies.

DISTRIBUTION Coastal zone around Sydney and Cumberland Plain. Extending to the lower north coast.

FLOWERING August–October.

809. Boronia fraseri

Boronia fraseri
Fraser's Boronia **Plate 809**

Fraser's Boronia is similar in appearance to *B. mollis*, except that it lacks the long hairs on the stems and leaf-stalks. It is also a tall shrub of coastal gullies. Intermediate forms exist between the 2 species. The stems of Fraser's Boronia are smooth, sometimes reddish and warty. Its sepals are broadly ovate, not narrow as those of *B. mollis*.

HABITAT Sheltered forested slopes and gullies.

DISTRIBUTION Restricted to the coastal zone around Sydney and lower Blue Mountains.

FLOWERING September–October.

Boronia nana var. *hyssopifolia* **Plate 810**

This is a small trailing shrub, often only 10 to 15 cm tall. Leaves are narrow-ovate, 6 to 7 mm long and covered with short stiff hairs (best observed with a lens). Flowers are white with petals 4 to 6 mm long; anthers have minute appendages. Flowers are borne in pairs from the upper leaf axils or are terminal.

HABITAT Woodland and open forest, amongst grassy understorey.

DISTRIBUTION Upper Blue Mountains (e.g., Newnes State Forest, Kanimbla Valley); uncommon. Extending along southern tablelands and south-western slopes to Vic and Tas.

FLOWERING November–December.

808 Boronia mollis

810. Boronia nana

Boronia polygalifolia
Milkwort Boronia **Plate 811**

Uncommon in the area, it is a weak hairless shrub, often with decumbent or trailing stems, 15 to 20 cm long. Leaves are opposite, narrow, to 12 mm long, with acute tips. Flowers are pink, often pale, and borne singly from the leaf axils on short stalks. This species is similar to *Boronia nana* var. *hyssopifolia* except that *B. nana* has stems covered with short stiff hairs and occurs in the upper Blue Mountains.

HABITAT Open forest among grassy understorey on heavier soils and shales.

DISTRIBUTION Cumberland Plain (e.g., Castlereagh State Forest, Shane Park) and Hornsby Plateau (e.g., Warrah Sanctuary). Recorded throughout the coastal and tablelands division of NSW.

FLOWERING September–February.

811. Boronia polygalifolia

Boronia floribunda
Pale-pink Boronia **Plate 812**

This is a beautiful shrub (to 1 m tall) with abundant pale pink flowers (as the specific name 'floribunda' indicates). The pink petals contrast with conspicuous hairy yellow-white anthers. The top of the stigma is broad and rounded, a feature which separates this species from *B. pinnata*. Leaves are divided into

812. Boronia floribunda

numerous, narrow-oblong leaflets, 5 to 25 mm long; they emit a pleasant aromatic scent when crushed.

HABITAT Open forest and heath, on sandy soils.

DISTRIBUTION Restricted to the coastal zone around Sydney and Blue Mountains (e.g., Heathcote National Park, Newnes State Forest, Kings Tableland); scattered throughout this area, but not common.

FLOWERING September–November.

Boronia microphylla

Small-leaved Boronia **Plate 813**

This is an attractive small shrub. It is usually about 60 cm tall, but may reach 1 m in sheltered sites. Leaves consist of 9 to 11 leaflets, obovate to oblong, mostly 5 to 7 mm long, which have blunt or indented tips. Flowers are pink, with petals 5 to 6 mm long and are borne in groups of 3 in the upper axils of the leaves.

HABITAT Open forest, on sandy soils.

DISTRIBUTION Upper Blue Mountains (e.g., Katoomba, Bell, Mt Wilson), Nattai Tablelands (e.g., Hilltop, West Berrima). Also occurring south coast, north coast, northern tablelands, central western slopes and Qld.

FLOWERING October–December.

813. Boronia microphylla

814. Boronia pinnata

Boronia pinnata

Pinnate Boronia **Plate 814**

An erect, hairless shrub usually 1 m tall, with smooth leaves divided into 5 to 9 leaflets. These leaflets are entire, acute, oblong-linear, about 15 mm long and 2 mm wide, and rather thick. They are slightly aromatic when crushed. Flowers are pink and numerous; they are borne in clusters in the upper axils of the leaves. This species is easily confused with *B. thujona*, but differs in that *B. thujona* has a very strong odour and has leaflets with small serrations, not entire like those of *B. pinnata.*

HABITAT Open forest and heath, on sandy soils.

DISTRIBUTION Widespread from coast to mountains. Extending to north coast and south coast.

FLOWERING August–October.

Boronia thujona **Plate 815**

There are a number of similarities between this species and *B. pinnata* which may lead to confusion in identification. Both are tall hairless shrubs with pinnately divided leaves and masses of pink flowers. However, this species is readily recognised by the strong (some say, unpleasant) odour of the crushed leaves. A close examination of the leaves also reveals features which help identification. They are divided into 9 to 13 thin narrow leaflets which are minutely serrated and marked by conspicuous marginal oil glands. Flowers are usually deep pink, often cupped rather than fully open, and are borne in loose clusters from the leaf axils.

HABITAT Forest, usually but not always, in sheltered sites, such as shady slopes.

DISTRIBUTION Common south of Sydney from Royal National Park to southern highlands (e.g., Barren Grounds, Belmore Falls, Bundanoon). Also recorded from Hornsby Plateau and Kings Tableland in Blue Mountains. Extending to south coast and north coast.

FLOWERING September–November.

815. Boronia thujona

816. Boronia rigens

Boronia rigens

Stiff Boronia **Plate 816**

A small shrub, rarely more than 40 cm high, which is easily overlooked as its flowers are small, white and almost hidden in the leaf axils. In some areas, such as West Head and Mt Colah, there are forms whose flowers do not open, but which self-pollinate (a characteristic known as 'cleistogamous'). The upright stems are softly downy and leaves and stems often have a brownish appearance. Leaves are divided into 3 narrow leaflets, 3 to 4 mm long, which are distinctly grooved on the upper surface and have recurved tips.

HABITAT Exposed lateritic soils, sandy heath, open forest and old dunes.

DISTRIBUTION Widespread from coast to mountains. Recordings include La Perouse, West Head, Bangor, Mt Colah and Evan's Lookout (Blackheath). Extending along southern tablelands; also Qld.

FLOWERING July–October.

Boronia anemonifolia

Sticky Boronia **Plate 817**

A strong-smelling, stocky shrub to 1 m tall, its stems are lightly clothed with hairs and its small leaves

817. Boronia anemonifolia

sometimes have a sticky feel. Leaves are divided into 3 leaflets (sometimes 5), 3 to 8 mm long; each leaflet usually has 3 broad lobes or teeth at the apex. Flowers are small, delicate pink to almost white and are borne singly or in pairs in the axils of the upper leaves.

HABITAT Open forest, on sandy soils.

DISTRIBUTION Upper Blue Mountains (e.g., Narrow Neck) and southern highlands (e.g., Belmore Falls, Bundanoon, West Berrima). Also recorded at Girrakool in Brisbane Water National Park. Extending all along the coastal and tableland divisions; also Vic and Tas.

FLOWERING September–November.

818. Boronia anethifolia

Boronia anethifolia

Narrow-leaved Boronia **Plate 818**

A small open shrub to about 50 cm tall, with 3 to 9 narrow acute leaflets, mostly 5 to 8 mm long and 1.5 mm wide. The softly pointed leaflets distinguish it from *B. microphylla* and *B. anemonifolia*. Some of these leaflets may be further divided into 3 entire secondary leaflets. Flowers are small, pale pink to white, and are clustered together in tight bunches in the leaf axils.

HABITAT Open forest, on sandy soils.

DISTRIBUTION Blue Mountains (e.g., Erskine Creek) and some scattered localities in central coast (e.g., Nortons Basin and Leets Vale, Hawkesbury River). Uncommon. Also occurring north coast, northern tablelands and western slopes.

FLOWERING August–September.

Boronia serrulata

Native Rose **Plate 819**

An unmistakable small shrub, one of the gems of the Sydney sandstone flora. Its erect stems have distinctive broad-ovate leaves, 10 to 12 mm long, which are minutely serrated, acute, crowded and point upwards along the stem. Flowers are bunched together at the ends of the stems and are dark pink cups with yellow anthers. Petals are about 10 mm long and conspicuously overlap in bud.

HABITAT Heath, in damp or dry sandy soils.

DISTRIBUTION Restricted to a narrow coastal band from Brisbane Water National Park (Kariong) to Royal National Park (Waterfall). Reported also from military reserve near Kentlyn, upper Georges River.

FLOWERING August–November.

819. Boronia serrulata

Boronia parviflora

Swamp Boronia **Plate 820**

B. parviflora is 1 of 3 local Boronias which grow in swampy conditions. *B. deanei* is rare and restricted in its distribution. *B. barkeriana*, could be confused with *B. parviflora* but tends to be a larger, more showy plant. As the specific name *'parviflora'* suggests, this species has small flowers (petals 3 to 4 mm long). These are pale pink and borne on short stalks (4 mm), singly or a few together, at or near the ends of the branches. The pointed sepals are about the same length as the petals and are clearly visible between the petals. Leaves are entire, or finely serrated at the top, narrow, up to 15 mm long and 2 to 5 mm wide, well-spaced and often curved and reddish. Stems are erect and seldom more than 30 cm tall.

HABITAT Swampy ground in heath and shrubland.

DISTRIBUTION Widespread from coast to mountains. Also occurring north coast, south coast, southern tablelands, Qld, Vic, Tas and SA.

FLOWERING Some flowers all year round.

820. Boronia parviflora

Boronia barkeriana

Barker's Boronia **Plate 821**

Barker's Boronia has a more scattered distribution than the closely related *B. parviflora*. It resembles that species but generally has larger flowers and longer leaves. Flowers are deep to light pink, with petals 6 to 8 mm long. Flowers are terminal or near the ends of the branches, often numerous and in clusters on the end of stalks 15 to 18 mm long. Leaves are lanceolate to broad-elliptic, 15 to 30 mm long and always minutely serrated.

821. Boronia barkeriana

HABITAT Swampy ground in heath and shrubland.

DISTRIBUTION Scattered from coast to mountains. Recorded from Blue Mountains (e.g., Blackheath) and Illawarra Escarpment (e.g., Barren Grounds, Jamberoo Pass). Also occurring south coast, southern tablelands and Qld.

FLOWERING September–December.

Boronia deanei

Dean's Boronia **Plate 822**

This is a rare plant with a restricted distribution. It is an erect shrub, to 1 m tall, with a strong odour when crushed. Leaves are narrow, soft, semi-terete, 3 to 7 mm long, blunt at the tip, and warty below because of the presence of oil glands. Flowers are deep to pale pink (sometimes white) and are borne singly or 2 to 3 together at or near the ends of short side branches. A form from Fitzroy Falls has flowers borne from the leaf axils in the upper branches as well as terminal.

HABITAT Swampy ground in heath and shrubland.

DISTRIBUTION Restricted to small areas in higher Blue Mountains north of Clarence and to Kanangra–Boyd National Park. An isolated occurrence at Fitzroy Falls.

FLOWERING September–November.

822. Boronia deanei

823. Correa alba var. *alba*

Correa

Shrubs with opposite leaves. Petals 4, united into a tube (becoming free in *C. alba*) and 8 stamens. Calyx cup-shaped (without lobes).

Correa alba var. *alba*

White Correa, Coastal Correa **Plate 823**

C. alba is a low coastal shrub, often found growing at the very edge of the cliff-line, exposed to wind and salt spray. It is a compact shrub, with thick rounded leaves, 2 to 4 cm long, white or rusty underneath with closely matted short hairs. The white flowers are 10 to 15 mm long; they are not developed into a long tube as in most Correas, but the petals roll back as the flower expands.

HABITAT Rocky coastal cliff tops and sandy sites near the sea.

DISTRIBUTION Coast-line. Extending all along the coast from north coast to Vic and Tas.

FLOWERING May–September.

Correa reflexa var. *reflexa*

Common Correa **Plate 824**

A common small shrub, it is very variable and has a number of distinct forms. Around Sydney, it is usually 50 to 80 cm tall, with opposite heart-shaped to elliptic leaves up to 3 cm long. These leaves are stem-clasping, dark green and rough from a covering of short hairs. A pair of leaves may sometimes enclose the cup-shaped calyx and base of the flower. The flower is a cylindrical tube, 2 to 3 cm long, drooping from the ends of branchlets. It may be coloured yellow-green or red with yellow-green tips. There does not appear to be any geographical pattern in the flower colour, but it is uncommon to find both colour forms growing near each other.

HABITAT Forest and heath, on sandy soils.

DISTRIBUTION Widespread from coast to mountains. Found throughout coastal, tablelands and western slopes divisions; also Qld, Vic, Tas and SA.

FLOWERING May–September.

824. A. Correa reflexa (red)

B. Correa reflexa (yellow-green)

Correa lawrenciana var. *macrocalyx*

Mountain Correa **Plate 825**

Mountain Correa is a common plant in the high country of the southern tablelands and Victoria, but this variety is much less common and less widely distributed. It is a shrub about 1 m tall in the Sydney area, but taller elsewhere. Leaves are ovate, 4 to 6 cm long, 2 to 3 cm wide, dark above, lighter below with yellow-brown stellate hairs. The flower is a narrow yellowish-green tube, 20 to 25 mm long, covered with dense velvety hairs. Stamens project conspicuously beyond the tube and the calyx is covered with stellate hairs.

HABITAT Sheltered forest.

DISTRIBUTION Restricted in the Sydney district to a small area near Patonga. Also recorded on north coast.

FLOWERING May–June.

825. Correa lawrenciana var. *macrocalyx*

Philotheca

This genus of 6 species is endemic in Australia and consists of heath-like shrubs with 5 mauve petals and 10 stamens. The genus differs from *Eriostemon* in having the stamens fused together at the base (not wholly free).

Philotheca salsolifolia

Philotheca **Plate 826**

Very variable in its habit, Philotheca is usually an open shrub up to 1 m tall, with scattered flowers, but it may be a low ground-hugging plant on exposed heaths or a round bushy shrub covered with flowers, as it is on old sand dunes at Kurnell. Its leaves are alternate, crowded, narrow-linear to terete, 4 to 12 mm long and dotted with numerous oil glands. Flowers are mauve with stamens united at the base, free above and conspicuously hairy. Flowers are mostly solitary at the ends of the branches.

HABITAT Sandy and rocky heaths, and on old coastal dunes.

DISTRIBUTION Widespread and common from coast to mountains. Recordings include Brisbane Water National Park, West Head, Royal National Park, Agnes Banks and Narrow Neck. Very common on deep coastal sands (e.g., Kurnell, La Perouse, Jibbon Head). Also occurring north coast, south coast, southern tablelands, western slopes and north-western plains.

FLOWERING July–December.

826. Philotheca salsolifolia

828. Asterolasia asteriscophora

829. Crowea saligna

Asterolasia

Shrubs covered with dense stellate and closely matted hairs. Leaves alternate. Flowers with 5 petals and 10 free stamens. Calyx lobes very small, often hidden beneath the hairs.

Asterolasia correifolia

Star-bush **Plate 827**

Star-bush derives its name from the dense covering of star-shaped closely matted hairs which cover the whole plant except for the upper surface of mature leaves. It is also an appropriate reference to the white, 5-petalled, open star-like flowers, with 10 conspicuous yellow stamens. It is a tall shrub, sometimes up to 2 m. Leaves are elliptic to lanceolate, up to 8 cm long and 2 cm wide, the upper surface green and somewhat resembling the leaf of *Correa*, the undersurface densely hairy. Flowers are 5 to 6 mm long, and usually several in a cluster.

HABITAT Sheltered forested hillsides and moist gullies.

DISTRIBUTION Hornsby Plateau (e.g., Cowan Creek, Girrakool, Olney State Forest). Extending north to Hunter River; also recorded on southern tablelands.

FLOWERING September–October, with a few flowers at other times.

Asterolasia asteriscophora **Plate 828**

This is a slender branching shrub to 1.5 m tall. Its branches, buds and centre-vein on the underside of the leaf are covered with closely matted brown stellate hairs. Leaves are rounded, obovate to elliptic, 10 to 17 mm long, 7 to 10 mm wide, green and rough above, lighter underneath with silvery-brown hairs. Flowers are bright yellow stars, with petals 4 to 9 mm long and 10 prominent stamens with orange anthers. They are borne singly from the leaf axils on short stalks.

HABITAT Sheltered forested gullies.

DISTRIBUTION Restricted to small areas near Penrose, Wingello and West Berrima (Wingecarribee River). Extending to central tablelands and northern tablelands (where it is rare); also Vic.

FLOWERING October.

827. Asterolasia correifolia

Crowea

Shrubs with alternate leaves. Flowers bright pink with 5 petals and 10 free stamens. Similar to *Eriostemon* except that the anthers of *Crowea* have long bearded appendages.

Crowea saligna **Plate 829**

This slender low shrub is restricted to the Sydney area (although a single collection was made in 1892 from Pine Mountains near the upper Murray River, Victoria). It is well known as its pink 5-petalled flowers bloom in autumn and winter when Boronias and Eriostemons are normally not in flower. Leaves are broad, narrow-elliptic, about 5 cm long, acute and with a distinct midvein. Petals are bright pink, 12 to 18 mm long. The flower resembles that of *Eriostemon* and this species was formerly called *Eriostemon crowei*. This species has often been confused with *C. exalata*, but can be distinguished from that species by the angular wing on the stem, its hairless branches and sepals and its broader leaves.

HABITAT Sheltered forest, on sandy soils.

DISTRIBUTION Restricted to the coastal zone around Sydney. Recordings include Bouddi National Park, Terrey Hills, Oatley Park, Royal National Park and Bargo.

FLOWERING February–June.

830. *Crowea exalata*

Crowea exalata Plate 830

C. exalata is more widely distributed than *C. saligna*, occurring throughout most of eastern NSW and southwards to central Vic. The species varies considerably in leaf and flower size and in hairiness. It is a small shrub, usually about 60 cm tall, with round to slightly angular stems which are somewhat glandular hairy. Leaves are narrow-obovate to linear, 2 to 4 cm long, the apex rounded or shortly pointed. Flowers are solitary and terminal on the ends of short stems. Petals are pink and about 10 mm long; sepals are softly downy to almost hairless.

HABITAT Forest, on rocky slopes or in sheltered gullies.

DISTRIBUTION Widespread along coast south of Hawkesbury River and along Woronora Plateau. Less common in the Blue Mountains. Also occurring south coast, southern highlands, northern tablelands (Mt Kaputar area) and Vic.

FLOWERING December–June.

831. *Eriostemon australasius*

Eriostemon

Shrubs with alternate leaves. Flowers pink to white with 5 petals and 10 free stamens forming a chamber around the ovary. Filament hairy (hence the name: 'erion'—wool; 'stemon'—stamen) but the anthers lacking the bearded appendages of *Crowea*.

Eriostemon australasius

Pink Wax-flower **Plate 831**

An erect shrub with attractive showy pink flowers which herald the coming of spring. Leaves are lanceolate to linear, 3 to 8 cm long, light-green, with an inconspicuous midvein. Petals are pink, with a waxy texture and up to 15 mm long. Flowers are solitary in leaf axils and often abundant.

HABITAT Forest, heath and coastal dunes, on sandy soils.

DISTRIBUTION Widespread and common from coast to mountains. Most common along coastal zone. Recordings include Bouddi, Brisbane Water and Royal National Parks and Wedderburn. Extending along coast from south coast to Qld.

FLOWERING August–October.

832. *Eriostemon myoporoides*

Eriostemon myoporoides

Native Daphne, Long-leaf Wax-flower **Plate 832**

Native Daphne is an attractive bushy shrub up to 2 m tall, with hairless green stems densely covered with raised oil glands. It is so varied that 6 subspecies have been identified. Of these, only the type subspecies *myoporoides* is found in the local area. This has leaves which are flat, narrow oblong-elliptic, greyish to dark green, 4 to 10 cm long, 5 to 8 mm wide, dotted with oil glands, and with a prominent midrib and pointed apex. In the Blue Mountains there is a form with leathery, broadly elliptic leaves. The white flowers and pink buds are in clusters of 2 to 5 flowers, on a common stout stalk from the leaf axils. Native Daphne may be distinguished from closely related species by being completely hairless and by the presence of the thickened stalks bearing the flowers. In parts of the Blue Mountains (e.g., Lapstone, Faulconbridge, Linden) and at Tahmoor, Native Daphne appears to hybridise with *E. hispidulus*, and in the lower Hawkesbury hybrids with *E. buxifolius* occur.

HABITAT Forest, chiefly on sheltered slopes and valleys.

DISTRIBUTION Scattered throughout the Blue Mountains (e.g., Coxs River) and Nepean River–Tahmoor area (e.g., Bargo River, Bents Basin). Also occurring south coast, southern tablelands, central western slopes, Qld and Vic.

FLOWERING August–October.

Eriostemon trachyphyllus Plate 833

This tall shrub is closely related to *E. myoporoides*. It is hairless and densely covered with raised oil glands. Leaves are stalkless, oblong-elliptic, 3 to 4 cm long, 5 to 8 mm wide, pointed at the apex and with prominent oil glands which give the surface a bumpy appearance. The specific name is a reference to these leaves ('trachys'—rough, shaggy; 'phyllus'—a leaf). The main difference between this species and *E. myoporoides* is that the flowers are solitary or in clusters of 3, each on its own stalk 6 to 12 mm long from the leaf axil.

HABITAT Forest, on rocky hillsides.

DISTRIBUTION Scattered and uncommon. Recorded from Illawarra Escarpment west of Dapto, from Howes Valley, north-western Hornsby Plateau and west of mountains at Evans Crown (Tarana). Also occurring south coast, north coast, southern tablelands, central western slopes and Vic.

FLOWERING September–October.

833. *Eriostemon trachyphyllus*

Eriostemon hispidulus Plate 834

A small undershrub, less than 1 m high, with stems covered with short stiff hairs. Leaves are stalkless, narrow obovate to linear, 10 to 20 mm long, 3 to 4 mm wide, with a straight or down-pointing point. The leaf margins are curved downwards (a characteristic feature) and appear wavy because of the prominent glands. Flowers are white to pale pink and are borne singly on a short stalk from the leaf axils. In the Blue Mountains (Glenbrook, Lapstone, Faulconbridge) and at Tahmoor, a long-leaved form of *E. hispidulus* is found which is believed to be a hybrid between this species and *E. myoporoides*.

HABITAT Open forest, on sandy soils.

DISTRIBUTION Endemic in the Sydney district. Occurring from Nattai Tableland (e.g, Colo Vale, Thirlmere Lakes) northwards through Blue Mountains (e.g., Woodford, Linden, Glenbrook) to Wisemans Ferry and Colo Heights.

FLOWERING September–October.

834. Eriostemon hispidulus

Eriostemon scaber ssp. *scaber* Plate 835

This is a small undershrub, usually less than 60 cm tall, with round stems which are not warty, but softly hairy. It differs from *E. hispidulus* in having leaves which are narrow, almost terete, deeply channelled or flat above and convex below. These leaves are 10 to 15 mm long, 1 mm wide, acute and very warty. Flowers are pale pink, solitary and borne from the leaf axils.

A broad-leaf form known as ssp. *latifolius* (Plate 836) occurs to the south of the Sydney area, from Nowra to Nerriga, and has also been recorded at Bundanoon and Wingello State Forest. It has stems which are conspicuously warty, and oblong-elliptic leaves up to 3 cm long and 2 to 5 mm wide, which are concave above, smooth or only slightly warty. Subspecies *scaber* hybridises with *E. buxifolius* where the 2 grow together south of Botany Bay.

HABITAT Open forest and heath, on sandy soils.

DISTRIBUTION Restricted to the vicinity of Georges and Woronora Rivers. Recordings include Oatley Park, Jannali Reserve, Kurnell, Heathcote National Park and The Basin (Upper Georges River). Extending to south coast.

FLOWERING August–November.

835. Eriostemon scaber ssp. *scaber*

836. Eriostemon scaber ssp. *latifolius*

837. Eriostemon buxifolius ssp. *buxifolius*

Eriostemon buxifolius

Box-leaf Wax-flower **Plate 837**

This is a small undershrub usually less than 1 m tall with branchlets which are round and not warty, and attractive pale pink flowers and deep pink buds borne singly from the leaf axils. Leaves are stalkless, less than twice as long as broad, 6 to 12 mm long, flat or concave above and smooth, keeled and somewhat glandular below. The apex has a small abrupt point. Two subspecies have been identified, each of which has a narrow distribution pattern.

Ssp. *buxifolius* has orbicular to broadly elliptic leaves, cordate (heart-shaped) at the base with entire margins. Petals are about 12 mm long.

HABITAT Heath, often on exposed headlands, on sandy soil.

DISTRIBUTION Restricted to coastal sandstone areas between Dee Why and Port Hacking (e.g., Dobroyd Point, Watson's Bay, North Head, Kurnell) and south of the local area at Point Perpendicular, Jervis Bay.

FLOWERING August–November.

Ssp. *obovatus* (Plate 838) has obovate leaves, 12 mm long, 4 mm wide, flat at the base, with margins indented because of warts.

HABITAT Heath, on sandy soils.

DISTRIBUTION The northern side of Broken Bay and Brisbane Water National Park (e.g., Wondabyne, Patonga, Calga); also recorded on south coast.

FLOWERING August–November.

E. buxifolius hybridises with other *Eriostemon*. In the lower Hawkesbury area, hybrids with *E. myoporoides* occur. The conspicuous *Eriostemon* which brightens the sandy heaths of Royal National Park in late winter and spring is regarded as a hybrid of *E. buxifolius* and *E. scaber*. It differs from *E. buxifolius* ssp. *buxifolius* in having narrower leaves and smaller (10 mm) broader petals.

838. Eriostemon buxifolius ssp. *obovatus*

839. Eriostemon obovalis

Eriostemon obovalis Plate 839

A small shrub from the higher Blue Mountains. Its stems are softly downy and not warty. Leaves are the main identifying feature. These are obovate to spoon-shaped, thick, rounded at the apex, but narrowing to the base, the midvein faint, smooth above and somewhat warty below. Flowers are white, tinged with pink, solitary, borne in the leaf axils among the leaves.

HABITAT Exposed sandstone heaths.

DISTRIBUTION Endemic in the higher Blue Mountains (e.g., Wolgan Gap, Mt Banks, Mt Wilson, Blackheath).

FLOWERING October–December.

Phebalium

Shrubs with alternate leaves. Flowers with 5 petals, usually yellow (sometimes white) and 10 free erect stamens.

Phebalium dentatum Plate 840

A tall slender shrub of sandstone gullies. It lacks the thin scurfy scales of most of the other *Phebalium*, but its young stems have minute star-shaped hairs. Leaves are long and narrow, 6 to 8 cm long, 3 to 4 mm wide, with conspicuous oil dots, dark green above, and greenish-white below. The margins are curved under and despite the name 'dentatum' (toothed), they are only sometimes minutely toothed. Flowers are yellow and abundant. They are borne in short stalked umbels in the leaf axils and often hide the stems with their masses of stamens projecting beyond the petals.

HABITAT Forest, in sheltered gullies.

DISTRIBUTION Widespread on Hornsby Plateau, Woronora Plateau, southern highlands and lower Blue Mountains. Recordings include Girrakool, Pennant Hills Park, Bundanoon Creek, Heathcote Creek and Erskine Creek. Extending to north coast and northern tablelands.

FLOWERING August–September.

Phebalium squamulosum

Scaly Phebalium **Plate 841**

A shrub to 2 m tall, often smaller and compact, with branches, undersurface of leaves and flower stalks more or less covered with small brownish scales. Flowers are yellow to white in terminal clusters. The species shows considerable variation and 10 subspecies have been identified. Four of these occur in the Sydney area.

Ssp. *squamulosum* is the most widely distributed and common of the subspecies. Its leaves are narrow oblong, up to 5 cm long, shiny green above and with a mixture of silvery grey hairs and brownish scales (rusty spots) on the undersurface. The apex of the leaf may be rounded, truncated with or without a small depression. Petals are yellow inside and rusty outside.

HABITAT Forest, woodland and heath, on sandy soils.

DISTRIBUTION Widespread from coast to mountains.

FLOWERING August–November (earlier on coast, later in mountains).

Ssp. *argenteum* (Plate 842) has petals which are creamy-white (not yellow like those of ssp. *squamulosum*) and covered on the outside with silvery scales. Leaves are elliptic to obovate, 15 to 70 mm long, with a rounded apex. The upper surface of young leaves has star-shaped hairs and the lower surface is pale and silvery. In the Georges River–Hacking River area, this subspecies grades into a thin-leaved form of ssp. *squamulosum*.

HABITAT Forest, woodland and heath, usually on sandy soils, but also on Narrabeen shales at Mooney Mooney Creek and clayey banks of sea-cliffs at Garie Beach.

DISTRIBUTION Coastal zone, north and south of Sydney. Extending along the coast from Port Stephens (north coast) to south-eastern Vic; also recorded from central tablelands.

FLOWERING August–November.

Ssp. *lineare* (Plate 843) is a narrow-leaved form mainly occurring north of the Sydney district from the upper Hunter River northwards to Murrurundi, although it has been recorded from near Badgerys Crossing on Shoalhaven River on the southern edge of the local area. Its leaves are 10 to 20 mm long, and about 1 mm wide. A midvein is visible and the apex is obtuse. Flowers are yellow, numerous, on slender stalks and the calyx is covered with reddish-brown scales.

HABITAT Forest, on rocky slopes.

DISTRIBUTION Shoalhaven River (Badgerys Crossing). Also upper Hunter Valley and central western slopes.

FLOWERING August–October.

Ssp. *ozothamnoides* (Plate 844) is a small compact shrub often regarded as a separate species *(P. ozothamnoides)* because of its characteristic small rounded obovate leaves. These leaves are under 1 cm long, shiny green above and whitish below. The tips are rounded and the midrib is inconspicuous. When young, the upper surface is sprinkled with a few star-shaped hairs. Flowers are yellow and borne in terminal clusters on short side branches.

HABITAT Heath and exposed rocky sites.

DISTRIBUTION Upper Blue Mountains (e.g., Hassans Walls, Mt York). Extending from northern tablelands to southern tablelands and Vic.

FLOWERING September–November.

840. Phebalium dentatum

841. Phebalium squamulosum ssp. *squamulosum*

842. Phebalium squamulosum ssp. *argenteum*

843. Phebalium squamulosum ssp. *lineare*

844. Phebalium squamulosum ssp. *ozothamnoides*

Phebalium lamprophyllum **Plate 845**

P. lamprophyllum is a shrub to 1 m tall from the tablelands of New South Wales and Victoria. Its young stems are softly downy, without scurfy scales. Leaves are short, obovate to almost rounded, up to 6 mm long and 5 mm wide. A characteristic feature is that the leaves are very shiny (hence the specific name, 'lampro', like a lamp). Flowers are in terminal clusters with creamy-white petals and stamens projecting well beyond the petals.

HABITAT Heath and woodland, on sandy soils or rocky sites.

DISTRIBUTION Higher Blue Mountains between Lithgow and Glen Davis. Extending to the east of Rylstone and along the southern tablelands to Vic.

FLOWERING October–January.

845. Phebalium lamprophyllum

Phebalium diosmeum **Plate 846**

This species could not be confused with any other *Phebalium* as it bears abundant deep yellow flowers at the top of its erect branches and has small crowded linear leaves. It is a bushy shrub, often more than 1 m tall with long loose hairs on its stems. Leaves are up to 15 mm long, rounded at the apex and with margins curved to the undersurface. Flowers are in clusters and present a mass of attractive yellow stamens to pollinating insects.

HABITAT Heath, on sandy soils.

DISTRIBUTION Coast to mountains. Most common south of Sydney (e.g., Carrington Falls, Barren Grounds, Wingello). Extending along south coast and southern tablelands to Vic.

FLOWERING September–November.

846. Phebalium diosmeum

847. *Phebalium squameum*

Phebalium squameum

Satinwood **Plate 847**

Satinwood is a tall shrub or small tree of sandstone gullies. Its branches, leaf stalks and flower stalks are covered with small scurfy scales. Leaves are flat, lanceolate, 8 to 10 cm long, light green and hairless above and conspicuously silvery-white below. Flowers are white and are borne in clusters on short scaly stalks from the leaf axils.

HABITAT Sheltered gullies, on sandy soils.

DISTRIBUTION Widespread from coast to mountains. Recordings include Pennant Hills Park, Thirlmere Lakes and Bundanoon Gorge. Extending along the coastal zone from Qld to Vic and Tas.

FLOWERING September–October.

Geijera

Trees or shrubs with alternate leaves. Flowers with 4 or 5 sepals and petals and 4 or 5 stamens.

848. *Geijera latifolia*

Geijera latifolia

Brush Wilga **Plate 848**

A rainforest tree uncommon in the local area, being recorded only in 2 or 3 rainforest patches near Wollongong which is the southern limit of its distribution. There it is a small tree with grey scaly bark and a dense dark crown. Leaves are alternate, ovate to lanceolate, 7 to 10 cm long, 2 to 6 cm wide and glossy green. Individual flowers are small (5 mm across), white, with 5 petals and 5 stamens and almost stalkless. They are borne in large conspicuous clusters (panicles) at the ends of the branchlets. The fruit is brown, round or egg-shaped, about 6 mm across, containing a single black shining seed.

HABITAT Rainforest, especially remnants on the coastal plain.

DISTRIBUTION Illawarra district (e.g., Mt Keira, Berkeley Hill) as far south as Minnamurra. Extending along north coast to Qld and New Guinea.

FLOWERING September–November.

Euodia

Trees or tall shrubs with opposite leaves, digitately 3-foliolate. Flowers with 4 petals, 4 sepals and 4 stamens. The generic name is derived from 'eu'—well, 'odoratus'—sweet smelling, a reference to the strong smell of the leaves.

849. *Euodia micrococca*

Euodia micrococca

White Euodia **Plate 849**

This is a fairly common tree in rainforests from the Illawarra northwards to Queensland. In the local area, it grows up to 10 m tall. The main distinguishing feature is the shape and arrangement of the leaves. These are opposite, with 3 leaflets attached to the end of a long stalk, 5 to 10 cm long. Leaflets are obovate to elliptic, 4 to 10 cm long, 2 to 4 cm wide, and bluntly pointed. The leaves are mostly clustered near the ends of the branchlets. Flowers are small, whitish and are borne in short dense clusters. The specific name 'micrococca' means 'small capsule', a reference to the small fruits about 6 mm long, which contain a single glossy black seed.

HABITAT Rainforest and sheltered gullies, including hind dune sites and rocky base of the escarpment.

DISTRIBUTION Illawarra district (e.g., Minnamurra Spit, Mangerton Park) and Gosford-Wyong district (e.g., Gap Creek Flora Reserve). Extending along north coast and northern tablelands to Qld.

FLOWERING January–February.

850. *Acronychia oblongifolia*

Acronychia

Trees or shrubs. Leaves mostly opposite and simple. Flowers with 4 petals, 4 sepals and 8 stamens. Buds with petals touching but not overlapping as in *Sarcomelicope*. Fruit a succulent berry, usually 4-angled and becoming hard with age.

Acronychia oblongifolia

Common Acronychia **Plate 850**

A small bushy rainforest tree, uncommon in the Sydney district. Leaves are obovate to lanceolate, 5 to 10 cm long and 1 to 4 cm wide, very shiny, rounded or bluntly pointed at the tip, usually slightly notched. Numerous oil dots are present in the leaves and they have a pleasant smell when crushed. Leaf stalks are pale green and up to 25 mm long. A conspicuous feature is the presence of a distinct elbow where the leaf joins the stalk. Flowers are 4-petalled, small and creamy-white; they are borne in loose clusters from the leaf axils. The fruit is a round, fleshy, lobed berry, white to green, and up to 12 mm across.

HABITAT Rainforest.

DISTRIBUTION Royal National Park (e.g., Bola Creek, Palm Jungle), Illawarra district (e.g., Minnamurra Falls, Mt Keira) and Gosford district. Extending from Gippsland (Vic) along coast to Qld.

FLOWERING January–April.

Sarcomelicope

A monotypic genus, formerly called *Bauerella* and commonly included in the genus *Acronychia* which it closely resembles. The main differences are that its flowers are dioecious (male and female flowers borne on separate plants) and its petals overlap in bud.

851. Sarcomelicope simplicifolia

852. Linum marginale

853. Geranium solanderi

Sarcomelicope simplicifolia

Yellowwood **Plate 851**

This is a shrub or small tree. Leaves are opposite, broad, glossy, elliptic to obovate, 7 to 12 cm long, 4 to 5 cm wide, with a blunt tip, distinct oil glands and an elbow where the leaf joins the stalk. The long whitish (or pale green) leaf stalks are a distinctive feature. Flowers have 4 fringed cream petals and 8 stamens. Each flower is minutely hairy, almost stalkless and borne in few-flowered panicles up to 4 cm long and from the leaf axils. Fruit is a slightly fleshy, yellow-brown, 4-angled berry.

HABITAT Rainforest.

DISTRIBUTION Coastal zone, but uncommon. Recorded from Saddleback Mountain, Royal National Park and Minnamurra Falls. Extending along north coast to Qld. Also Lord Howe Island.

FLOWERING May–September.

Linaceae

Although only a small family, the Linaceae is important because of one species, *Linum usitatissimum*, flax, cultivated in Egypt as early as 4000 BC for the making of linen cloth, ropes and nets, and still widely grown for the manufacture of linen and for linseed which on pressing provides oil and a valuable cattle feed.

Apart from 1 or 2 introduced species, the family is represented in all states of Australia by one endemic species of *Linum*, and in north-eastern Queensland by the tropical climber *Durandea*. Certain floral characteristics, such as the regular pentamerous flowers and the nature of the ovary, point to a relationship with the Geraniaceae.

Linum Flax

Linum marginale

Native Flax **Plate 852**

A slender, smooth plant with erect open branches up to 60 cm tall. It is easily identified by its 5-petalled, open flowers, which are a delicate blue with darker blue veins. There are 5 stamens and 5 styles. The styles are united above the middle; this distinguishes this species from the cultivated flax whose styles are free. Leaves are alternate, without stalks and up to 2 cm long. The fruit is a small, 5-celled, somewhat round and papery capsule. Aborigines are said to have used the fibre from Native Flax to make string and fishing nets and to have eaten the seeds.

HABITAT Grassland and open forest on shale.

DISTRIBUTION Widespread from coast to mountains, but not common. Recorded on Wianamatta shale at Menai and roadside near McMasters Beach, Gosford district. Extending to the western plains. Occurring in all States, except NT.

FLOWERING September–November.

Geraniaceae

Only 3 of about 11 genera in this family occur in Australia. Native and introduced species of all 3 occur in a variety of habitats in the Sydney area, with the introduced species being the more common.

The flowers of *Geranium* and localy occurring species of *Erodium* are regular (i.e., radially symmetrical) but those of *Pelargonium* have 2 large petals opposite 3 smaller ones, as in garden geraniums, which are in fact *Pelargonium* hybrids. The flowers are insect pollinated with those of *Pelargonium* having nectar guides in the form of purplish-red streaks radiating from the centre. They are also perfumed at night which encourages visitation by nocturnal insects. The leaves of all 3 genera are rounded and lobed like the garden geranium, or deeply dissected to the stalk.

The fruits are characteristically adapted for dispersal. On maturity they separate into 5 separate fruits called mericarps, each with a long spiral awn or beak formed from the fused styles. In *Erodium*, which is adapted to semi-desert conditions, the fruit dehisces (splits) explosively throwing the mericarps away. When rain comes the spiral beak absorbs water and straightens, driving the mericarp into the ground. In *Geranium* and *Pelargonium* the mericarps remain attached to the top of the beak, and are eventually dispersed by wind aided by feathery hairs on the inner surface of the spiral awns.

The family is related to the oxalis (Oxalidaceae) and flax (Linaceae) families. The group of families forming the order Geraniales is related to the Rutales, Sapindales and Polygalales.

Geranium Cranesbill

Small herbaceous plants with deeply divided or lobed leaves. Flowers are regular, with 5 petals and 10 fertile stamens. Awns of the fruit are hairless on the inside surface.

Geranium solanderi var. *solanderi*

Australian Cranesbill **Plate 853**

This is a decumbent to ascending softly hairy perennial. Stems up to 50 cm long shoot from a turnip-like taproot. As this species is easily confused with other *Geranium* in the area, careful examination of the nature of the hairs and the inflorescence is necessary for positive identification. Hairs on the stem and flower stalk are coarse, backward-pointing, often villous, but not appressed against the stem. Pink to white flowers are borne in pairs (*G. homeanum* is the only other species with paired flowers in this area). Sepals are 5 to 9 mm long; petals 5 to 12 mm long; flower stalk 25 to 50 mm long; inflorescence stalk 1 to 4 cm long. The bracteoles are up to 4 mm long, covered with long weak hairs and with hairy margins. This species has a larger flower than *G. homeanum*. Leaves have blades 10 to 25 mm long, deeply divided into 5 to 7 lobes; each lobe is again divided near the tip. Fruits consist of a long thin tapering beak (hence the common name), which splits, when ripe, into 5 fruitlets.

HABITAT Grassland and open woodland.

DISTRIBUTION Coast to mountains. Recordings include Richmond–Windsor, Jenolan Caves, Garie Beach, Stanwell Park, Mt Cameron (Newnes Plateau) and The Oaks. Found in most areas of NSW, except far west. Also in every other State (except NT) and NZ.

FLOWERING September–February.

Geranium homeanum **Plate 854**

This is a decumbent perennial, with much-branched stems to 70 cm long, ascending when supported by surrounding vegetation. Stems rise from a thick fleshy taproot, which is branched rather than turnip-like; they are sparsely covered with coarse, backward-pointing hairs which are appressed against the stem. Stems are sometimes almost hairless. Flowers are pink to white and borne in pairs (like those of *G. solanderi*, but smaller). Sepals are 4 to 5 mm long; petals 2 to 4 mm long; flower stalk 15 to 22 mm long; inflorescence stalk 18 to 24 mm long. The bracteoles are about 3 mm long, hairless except for the margins.

854. *Geranium homeanum*

Leaves are similar to those of *G. solanderi*, but are often purplish underneath.

HABITAT Damp sheltered sites in tall forest or rainforest margins; sometimes in grassland on basaltic soils.

DISTRIBUTION Coast to mountains. Recordings include Glenbrook, Douglas Park, Jenolan Caves, Garie Beach, Macquarie Pass and Mt Tomah. Extending along coastal zone from south coast to Qld; also northern tablelands, southern tablelands, north-west slopes and Asia.

FLOWERING November–February.

Geranium potentilloides — Plate 855

This is a small perennial with a thickened taproot. Stems are decumbent to ascending, usually less than 30 cm long, and covered with soft, white, closely appressed hairs. Stem leaves are opposite on slender stalks, 2 to 5 cm long. The leaf-blade is round in outline, 1 to 3 cm across, but deeply divided into 5 to 7 lobes which are themselves divided into secondary lobes or teeth. Leaves are covered with fine hairs; the undersurface is paler, sometimes purplish. Flowers are solitary, with pink or white petals 5 to 6 mm long. Sepals are acuminate and softly hairy. Flowers are borne on a slender stalk 2 to 4 cm long; there are 2 bracteoles about mid-way along the stalk. The 'beak' of the fruit is 9 to 10 mm long and densely covered with short stiff hairs.

HABITAT Woodlands, damp grasslands and margins of swamps.

DISTRIBUTION Higher Blue Mountains (e.g., Newnes State Forest, Clarence, Mt Wilson, Mt Irvine, Mt Tomah, Hampton). Extending from northern tablelands to southern tablelands; also Vic, Tas, SA, NZ and New Guinea.

FLOWERING November–January.

855. *Geranium potentilloides*

Geranium neglectum — Plate 856

This weak, decumbent herb is easily identified by its pink flowers which are borne singly and are larger than the flowers of other *Geranium* in the area (25 to 35 mm across). Flowering stems are often reddish, up to 120 cm long, much-branched and may bear a few appressed hairs or be almost hairless. Leaves are deeply divided into 5 to 7 lobes, 2.4 to 4 cm wide, with scattered appressed hairs on both surfaces. The solitary flowers are borne on slender stalks up to 10 cm long; two bracteoles lie about mid-way along the stalk. The petals are obovate, about 14 mm long and 6 mm wide, pink with distinctly darker veins. The rostrum is about 15 mm long.

HABITAT Banks of streams and in swamps.

DISTRIBUTION Higher Blue Mountains (e.g., Boyd Plateau, Mt Werong). A tablelands species extending from New England to ACT; also Qld and Vic.

FLOWERING December–March.

856. *Geranium neglectum*

Geranium graniticola
Granite Cranesbill — Plate 857

This is a small softly hairy perennial found in the grassy understorey of mountain forests. Flowers are solitary, 5-petalled, white and have sepals which are generally bent backwards. Leaves are lobed, but not strongly dissected, and covered with fine downy hairs.

HABITAT Open and tall forests, often in granite areas.

DISTRIBUTION Restricted to higher Blue Mountains (e.g., Jenolan Caves Reserve, Kanangra). Also near Oberon.

FLOWERING December–March.

857. *Geranium graniticola*

Erodium Crowfoot

Small herbaceous plants with deeply divided or compound leaves. Flowers are regular, blue to pink, with 5 petals and 5 stamens alternating with 5 staminodes. Awns of the fruit are hairy on the inner surface.

858. *Erodium crinitum*

Erodium crinitum
Blue Crowfoot, Blue Heronsbill — Plate 858

Blue Crowfoot is very common and widespread west of the Great Dividing Range but less common in the Sydney district. It is an annual or biennial herbaceous plant, with branched sprawling or ascending hairy stems, 30 to 60 cm long. Stems rise from a fleshy taproot, which was cooked and eaten by the Aborigines. Leaves at the base of the plant are up to 6 cm long and 3 cm wide, deeply dissected into 3 or more lobes, which are in turn jaggedly lobed; stem leaves are smaller and opposite. Flowers are deep blue and borne in umbels of 2 to 6 in the leaf axils. Fruit is a tapering beak, 4 to 7 cm long.

HABITAT Grassland and woodland.

DISTRIBUTION Cumberland Plain to Blue Mountains. Found throughout NSW, except north coast, and in all mainland States.

FLOWERING September–November.

Pelargonium Storksbill

The flowers of *Pelargonium* differ from those of *Geranium* and *Erodium* in having 2 upper petals larger than the 3 lower ones. A nectar-spur is at the base of the calyx (obsolete in one species) and the awns of the fruit are hairy on the inner surface.

859. Pelargonium australe

Pelargonium australe

Native Storksbill, Wild Geranium **Plate 859**

Native Storksbill closely resembles the cultivated Geranium, but differs from it in having flatter leaves and smaller petals. It is a slightly aromatic perennial, with sprawling or erect stems to 50 cm long, usually covered with downy hairs. Leaves are opposite, up to 9 cm long and 8 cm wide, with a number of rounded lobes. Plants on coastal dunes usually have leaves which are more hairy than those of plants growing away from the coast. Flower stems are erect and carry 4 to 12 flowers in an umbel. Each flower is pink, streaked with dark pink veins and has 6 to 8 fertile stamens. Fruit is a thin tapering beak, 8 to 12 mm long.

HABITAT Coastal sand-dunes, sea-cliffs and rocky outcrops in mountains.

DISTRIBUTION Coast to mountains. Extending to western slopes. Occurring in all States, except NT.

FLOWERING December–January.

Polygalaceae

The remarkable resemblance of the flowers of this family to the pea flowers (Fabaceae) is most readily seen in introduced species of *Polygala* which are sometimes found growing in sandy soils near the ocean. Whereas the standard in Fabaceae is formed from 1 petal, the equivalent in Polygalaceae consists of 2 enlarged purplish sepals which differ considerably from the other 3 small green sepals. A keel (fringed at the tip in *Polygala*) formed by fusion of the lower 3 petals encloses the 8 stamens which are connate in the lower half into an open sheath. As in some tribes of Fabaceae, pollination and pollen transfer is facilitated when the weight of a visiting insect depresses the petals exposing the stigma and anthers.

The flowers of *Comesperma*, the only native genus present in the Sydney area, are as described above but are much smaller than in *Polygala* and are not crested or fringed on the keel. They are, however, frequently arranged in large attractive racemes of purple or blue flowers.

860. Comesperma ericinum

Comesperma Matchheads, Milkwort

Comesperma ericinum

Pink Matchheads, Heath Milkwort **Plate 860**

An attractive erect shrub common on sandstone soils around Sydney. Its many flowers are borne on several terminal racemes. Unopened buds are rounded like match heads. The flower shape is characteristic of the genus. Two petal-like sepals stick out like wings or ears, whilst the petals are small and enclose the stamens. The tips of the yellow stamens can be seen projecting through the pink petals. Leaves are alternate, narrow, 5 to 12 mm long and with margins which are curved underneath.

HABITAT Heath and dry eucalypt forest.

DISTRIBUTION Widespread from coast to mountains. Found throughout NSW, except in drier western area; also Vic, Tas and Qld.

FLOWERING September–December.

Comesperma sphaerocarpum

Broom Milkwort **Plate 861**

Broom Milkwort has neither the erect shrub habit of Pink Matchheads nor the twining stems of Love Creeper. It has a woody stem base from which rise weak straggling leafless stems. These seldom grow taller than 20 cm. Flowers are blue and terminal with the large lateral petal-like sepals most conspicuous. In this species, the remaining outer sepals are up to two-thirds as long as the larger ones. Fruit is an orbicular capsule.

861. Comesperma sphaerocarpum

HABITAT Heath and dry forest in sandy soils.

DISTRIBUTION Widespread from coast to mountains, but not common. Recordings include Brisbane Water and Royal National Parks and Munmorah State Recreational Area. Found throughout the coast and tablelands of NSW.

FLOWERING November–February.

862. Comesperma volubile

Comesperma volubile

Love Creeper **Plate 862**

Love Creeper has delicate twining stems which twist around grasses and low shrubs, usually in shaded areas. Its leaves are few and scattered, narrow-elliptic, 5 to 30 mm long. Flowers are blue to mauve, with a darker centre and are borne on upright racemes.

HABITAT Heath, woodland and open forest.

DISTRIBUTION Widespread from coast to mountains. Extending along the coast and tablelands divisions north to Qld and south to Vic; also Tas.

FLOWERING August–November.

Tremandraceae

The Tremandraceae is a small endemic Australian family of 3 genera, with *Platytheca* and *Tremandra* (2 species each) confined to south-western Australia, and *Tetratheca* (39 species) occurring in all southern states including Tasmania. The 3 genera show a range of characters, with *Tetratheca* the most variable and occupying a middle position, frequently sharing characters with one, or the other, or both.

Although the flowers of all eastern species of *Tetratheca* are 4-merous (i.e., have 4 sepals, 4 petals and 8 stamens) the western species are 4-merous, or

5-merous (as are *Platytheca* and *Tremandra*), or both 4 and 5-merous. Despite this complication, *Tetratheca* is distinguished by all the stamens being of equal size and occurring in pairs opposite the petals, with each pair enfolded in bud by a petal.

Ten species of *Tetratheca* occur in the Sydney area, most having for many years suffered confusion in naming. Species are distinguished by leaf shape and arrangement, the nature of stem hairs and the number of ovules in each loculus of the ovary.

863. *Tetratheca juncea*

Tetratheca Black-eyed Susan

Tetratheca juncea Plate 863

T. juncea is the easiest of the local Black-eyed Susans to identify as its stems are usually leafless, with 2 to 3 narrow wings which give them an angular appearance. Stems are 30 to 60 cm long, branching above a woody root-stock but usually sprawling or even decumbent. Leaves are reduced to narrow scales about 3 mm long. The dark pink flowers are borne singly or 2 together along the stems. Flower-stalks are hairless, dark towards the top and 5 to 10 mm long.

HABITAT Sandy and swampy heath and open forest.

DISTRIBUTION Lake Macquarie–Munmorah–Wyee area. Extending northwards to Bulahdelah. Recorded in the past from the St George district of Sydney (Kogarah in 1893, Bexley in 1913), but now believed extinct in that area.

FLOWERING August–October.

Tetratheca glandulosa Plate 864

This is a low-growing spreading shrub, 20 to 50 cm tall, branching several times close to the woody root-stock. Its leaves are mostly opposite, or sometimes alternate, 5 to 10 mm long, about 1 mm wide, with a margin rolled under; small stiff hairs near the margins often give the leaf a finely toothed and rough appearance. The most conspicuous feature of this plant is its glandular hairs, hence its specific name. The flower stalks and sepals are covered with numerous dark-red, glandular hairs. Flowers are dark pink and are borne in the axils of the upper leaves. The sepals are not deciduous and persist around the fruit, a feature which separates this species from all other *Tetratheca* in the area.

864. *Tetratheca glandulosa*

HABITAT Sandy heath and scrub.

DISTRIBUTION Restricted to the central coast, north of Port Jackson. Recordings include Mangrove Mountain, Glenorie, Colo Heights, Cowan, Pennant Hills Park, Oxford Falls and Duffys Forest.

FLOWERING July–October.

865. *Tetratheca shiressii*

Tetratheca shiressii Plate 865

This is a sprawling shrub with numerous wiry stems up to 70 cm long. An unusual and distinctive feature is that the stems are of 2 kinds on the one plant. Some are hairless, with narrow opposite (sometimes whorls of 3) leaves; others have pale brown stiff hairs, with broad-elliptical to ovate leaves in whorls of 3 or 4. Flowers are dark pink, occurring in the axils of the upper leaves. Flower stalks are dark coloured, 5 to 12 mm long and usually hairless.

HABITAT Rocky or sandy heath and open forest.

DISTRIBUTION Restricted to the central coast. Occurs in 2 separate areas: Hawkesbury River–Watagan State Forest (e.g., Patonga Creek, Girrakool, Somersby, Mangrove Mountain); and Royal National Park–Helensburgh (e.g., Wallumarra Track, Uloola, Waterfall).

FLOWERING August–October.

866. *Tetratheca decora*

Tetratheca decora Plate 866

This is a very attractive small shrub with erect branches, 30 to 40 cm tall. Its stems are covered with short greyish hairs. Leaves are alternate, a feature which gives an immediate clue to its identity. These leaves are about 10 mm long, narrow (1 to 1.5 mm), with scattered hairs above and margins which curl under; the leaf stalk is short or absent. Flowers are dark pink, on short (3 to 6 mm), dark-coloured stalks which are covered with whitish hairs. The sepals are dark-coloured with a few white hairs (sometimes hairless).

HABITAT Sandy heath and dry scrub.

DISTRIBUTION Scattered on the central and south coasts, with an isolated occurrence outside the area in the Warrumbungle Mountains. Recorded from 3 areas in the Sydney district: Howes Valley–Putty; Ingleside–Mona Vale; and Coxs River–Yerranderie–Nattai River.

FLOWERING September–October.

867. *Tetratheca ericifolia*

Tetratheca ericifolia Plate 867

This diffuse shrub has numerous branching stems, 15 to 40 cm tall, and leaves in whorls of 4 to 6. Care needs to be taken to separate this species from other *Tetratheca* with whorled leaves. The main differences are that *T. ericifolia* has short, pale brown, upward-pointing, almost stem-clasping hairs on the stems (known as 'antrorse' hairs), long (10 to 15 mm), slender, hairless flower-stalks and reddish glandular hairs covering the small calyx. Leaves are up to 10 mm long and 1 mm wide, with tubercles on the upper surface and margins curved under; the apex is obtuse or with a minute recurved point. Flowers are pale pink (sometimes white) and occur singly in the leaf axils. Examination of the ovary with a lens shows that it is hairless (or with a few simple hairs), with 4 ovules, 2 in each loculus.

HABITAT Sandy heath and forest.

DISTRIBUTION Scattered along the central coast, extending to the lower north coast and isolated areas of southern highlands. Recorded from Port Jackson–Gosford (including Duffys Forest, Hornsby, Pearl Beach, Mt White, Bateau Bay); Engadine–Appin Road; and Mittagong.

FLOWERING July–November.

869. Tetratheca neglecta

868. Tetratheca rubioides

Tetratheca neglecta Plate 869

T. neglecta is a compact or diffuse shrub up to 50 cm tall with stems branching from near the base. It is often confused with *T. ericifolia* and *T. rubioides*, but unlike those species, it bears a characteristic pattern of long and short stem hairs. The longer whitish-brown hairs are neither retrorse nor antrorse but directed outwards in an irregular manner. Leaves are in whorls of 4 to 6, linear, with the margins curved under to the midrib. The upper surface has tubercles and often short stiff hairs. Flowers are dark pink, occurring singly in the axils of the upper leaves. Flower stalks are long (to 15 mm) and usually hairless. The sepals are dark-colour and hairless. Examination of the ovary with a lens shows that it has numerous minute hairs and 2 ovules, 1 in each loculus.

HABITAT Sandy heath and dry open forest.

DISTRIBUTION South of Port Jackson. Common in the Oatley–Como–Lucas Heights area. Also recorded in Royal National Park, upper Georges River and Carrington Falls, with an isolated occurrence near Yerranderie.

FLOWERING July–November.

Tetratheca rubioides Plate 868

T. rubioides is a tablelands species. It is a compact shrub with erect stems to 60 cm tall. Its leaves are in whorls of 5 to 7 and crowded along the stem. Stems are covered in shining short white hairs bent down the stem (known as 'retrorse' hairs), a feature which separates it from the closely related *T. neglecta*. Leaves are linear to 15 mm long, the margins curved under to the midrib and the apex has a short blunt point. Flowers are dark pink, occurring singly in the upper leaf axils. Flower stalks and sepals are mostly hairless or rarely with a few fine white hairs. Examination of the ovary with a lens shows that it is hairless or slightly downy, with 2 ovules, 1 in each loculus.

HABITAT Sandy heath and dry open forest.

DISTRIBUTION Blue Mountains. Recordings include Rylstone, Newnes State Forest, Blackheath, Mt Wilson, Mountain Lagoon, St Albans, Bucketty and Mt Werong. Also recorded on southern tablelands at Clyde River and Braidwood.

FLOWERING September–January.

Tetratheca rupicola Plate 870

This Blue Mountains species is a compact or diffuse shrub, seldom more than 30 cm tall, with stems branching several times from the base. Stems are covered with fine irregular spreading whitish-grey hairs. These hairs distinguish it from *T. ericifolia* and *T. rubioides* and its distribution and ovule number separate it from *T. neglecta*. Leaves are in whorls of 5 to 6; they are linear, up to 15 mm long, with margins rolled under to the midrib and the upper surface dotted with tubercles. The apex is acute and with a recurved point. Flowers are dark pink, occurring singly in the leaf axils. Flower stalks and sepals may be hairless or with fine white hairs. Examination of the ovary with a lens shows that it has dense white hairs, with 4 ovules, 2 in each loculus.

HABITAT Open forest in sandy soils.

DISTRIBUTION Confined to the Blue Mountains. Recordings include Springwood, Lawson, Kings Tableland, Leura, Govetts Leap and Mt Tomah.

FLOWERING October–December.

870. Tetratheca rupicola

Tetratheca thymifolia Plate 871

Only 2 local species of *Tetratheca* have leaves which are elliptic (broader than linear) and in whorls. They are *T. thymifolia* and *T. bauerifolia*. *T. thymifolia* is the more widespread. It is an erect undershrub with stems covered with dense spreading fawn hairs. Leaves are in whorls of 3 to 5, elliptical, 5 to 20 mm long and up to 8 mm wide, with margins curled under. In some coastal plants, the leaves are hairless, while in the mountains they may be densely hairy. Flowers are dark pink (occasionally white) on long stalks which usually are covered with whitish hairs. The sepals are dark coloured and also covered with white hairs or reddish bristles.

HABITAT Sandy heath and forest.

DISTRIBUTION Coast to tablelands. Recorded from north of Port Jackson (Oxford Falls, Brisbane Waters National Park, Gosford), north-west to Putty, Blue Mountains (Mt Irvine, Blackheath, Mt Werong) and southern highlands (Mittagong, Fitzroy Falls, Wingello). Generally absent from the coast and plateaux between Port Jackson and Nowra, Cumberland Plain and lower Blue Mountains. Extending north to Qld and south to Vic.

FLOWERING August–November.

871. Tetratheca thymifolia

Tetratheca bauerifolia **Plate 872**

This uncommon ascending undershrub somewhat resembles *T. thymifolia* as it also has whorled non-linear leaves. However, there are a number of significant differences. Its leaves are narrower (2 to 3 mm wide) and its slender flower stalks are hairless. Flower stalks are conspicuously hooked beneath the flower. Flowers are dark pink, with dark-coloured hairless sepals.

HABITAT Open forest, usually in rocky situations.

DISTRIBUTION Chiefly a species of southern tablelands and Victoria. Recorded in the Sydney district near Kanangra Walls.

FLOWERING September–November.

872. Tetratheca bauerifolia

Araliaceae

The Araliaceae is a family of trees, shrubs, climbers and herbs closely related to the larger and better known heraceous family, Apiaceae (Umbelliferae). The basic inflorescence of both families is an umbel, in which individual flowers are small with an inconspicuous calyx and 5 greenish or white deciduous petals alternating with 5 stamens attached to a disc which surmounts the ovary. The 2 families differ in their fruits, with Araliaceae forming a drupe with up to 5 seeds and Apiaceae usually forming a flattened dry 2-lobed, 2-seeded fruit. However the distinction tends to break down with a number of Australian genera such as *Astrotricha* and *Polyscias* which have flattened 2-locular fruits not unlike those of Apiaceae.

The best-known members of the family are the ivies (cultivars of *Hedera helix*), the indoor plant *Fatsia japonica*, and the Australian umbrella tree *(Schefflera actinophylla)*. The most important product is ginseng, a Chinese cure-all and supposed aphrodisiac obtained from the roots of *Panax quinquefolia.*

Astrotricha Star-hair

Shrubs, more of less covered with dense woolly star-shaped hairs. The generic name is a reference to this feature ('astro', star; 'tricha', hair). Flowers small, white, 5-petalled and borne in umbels which are arranged in large terminal panicles.

Astrotricha floccosa

Woolly Star-hair **Plate 873**

This is an erect shrub, 2 to 3 m tall, often common near estuarine waterways but also found further inland. Soft woolly hairs cover its stems, leaf stalks and undersurface of the leaves. The specific name 'floccosa' is a reference to the likeness of these hairs to tufts of wool. Leaves are large, 10 to 20 cm long, 2 to 7 cm wide, ovate to lanceolate, pointed at the tip, light green and smooth above and woolly below. It is distinguished from the closely related *A. latifolia* by the short leaf stalks (under 25 mm long). Some leaf forms appear to be intermediate between the 2 species. Flowers are small, greyish white and in umbels. The inflorescence is large and conspicuous, rising above the stem 30 to 50 cm in a much-branched panicle.

HABITAT Sheltered forested slopes, on sandy soils.

DISTRIBUTION Restricted to the coastal zone north of Port Jackson, where it is common near estuaries (e.g., Cowan Creek, West Pittwater, Mangrove Creek, Brisbane Water National Park); also recorded in the past from Casula. A narrow-leaf form occurs near Howes Valley.

FLOWERING October–January.

873. Astrotricha floccosa

874. Astrotricha latifolia

Astrotricha latifolia

Broad-leaf Star-hair **Plate 874**

At first glance, *A. latifolia* may be confused with *A. floccosa* as both are tall shrubs with broad leaves, but there are a number of differences. *A. latifolia* is more widely distributed, being found in the Blue Mountains as well as along coastal inlets. The covering of hair on the branches and stems is less loose and woolly, being shorter and more compact. Leaves are large, about 15 cm long and 5 cm wide, ovate to lanceolate at the base, glossy green above and woolly below. The leaf stalk is 4 to 8 cm long, in contrast with the short leaf stalk of *A. floccosa*. The large and conspicuous inflorescence is much more branched and diffuse than that of *A. floccosa.*

HABITAT Sheltered forest, on sandy soils.

DISTRIBUTION Widespread from coast to mountains. Recordings include Nattai Valley, West Pittwater, Pennant Hills Park, Mt Wilson and Mt Irvine. Also occurring south coast, north coast, northern tablelands and Qld.

FLOWERING October–November.

875. *Astrotricha longifolia*

Astrotricha longifolia

Long-leaf Star-hair **Plates 875, 876**

A. longifolia is a slender upright shrub, 1 to 2 m tall, with downy stems, leaf stalks and undersurface of leaves. The narrow, acute, flat leaves help distinguish this species from others in the genus. Leaves are less than 10 mm wide, 7 to 12 cm long, narrow lanceolate, dull green above and downy white underneath. The terminal inflorescence is 10 to 20 cm long and narrower than that of *A. latifolia*.

HABITAT Sheltered forest, on sandy soils.

DISTRIBUTION Widespread from coast to mountains. Also occurring north coast, northern tablelands, western slopes, north-western plains and Qld.

FLOWERING October–December.

876. *Astrotricha* sp. aff. *longifolia* (Newnes S.F.)

Astrotricha ledifolia **Plate 877**

This is a small shrub, usually about 1 m tall, with leaves which are narrow, oblong-linear, 2 to 7 cm long, obtuse at the tip and with margins curled under. The upper surface is flat and rather dull, the lower surface is covered with close, dense hairs. Flowers are borne in a loose panicle 10 to 20 cm long. The obtuse leaves help separate this species from *A. longifolia* which has acute leaves.

The rare *Astrotricha crassifolia* occurs in the Brisbane Water National Park and on parts of the Woronora Plateau. Its leaves are convex, shiny on the upper surface and margins are rolled under.

HABITAT Sheltered forest, on sandy soils.

DISTRIBUTION Blue Mountains and southern highlands. Extending to southern tablelands, south-western slopes and Vic.

FLOWERING October–December.

877. *Astrotricha ledifolia*

Polyscias

Shrubs or trees with compound leaves. Flowers small, with 5 petals and 5 stamens arranged in umbels along large panicles. Fruit more or less succulent.

Polyscias sambucifolia

Elderberry Panax **Plate 878**

Elderberry Panax is a tall shrub, up to 3 m high. Leaves are 15 to 30 cm long, compound, with about 3 pairs of leaflets and a terminal leaflet. These leaflets are very variable. In the type form, they are ovate, 5 to 7 cm long, slightly toothed, and paler underneath. There is also a bipinnate form in which the leaflets are lobed or subdivided. Flowers are greenish in many-flowered umbels which are borne in branched terminal sprays above the leaves. Fruit is a bluish compressed succulent capsule about 5 mm across. The leaves of this species resemble those of the more common *Sambucus australasica* (as the specific name suggests) and it is easy to confuse the two. However, *Sambucus* is a less robust shrub, with softer leaves and stems, yellow-white flowers in terminal clusters (not umbels) and yellow berries.

HABITAT Forest, on sheltered slopes and gullies.

DISTRIBUTION Widespread from coast to mountains. Also occurring north coast, northern tablelands and Qld. The bipinnate form extends southward along south coast and southern tablelands to Vic.

FLOWERING December–January.

878. *Polyscias sambucifolia*

Polyscias murrayi

Pencil Cedar, Umbrella Tree **Plate 879**

The common names of this rainforest tree refer to its distinctive appearance. It has a single, thin trunk, unbranched for about 6 m. Large compound leaves radiate from the upper half of the trunk and give the tree an umbrella-like appearance which clearly distinguishes it from the surrounding vegetation. Leaves are often over 1 m long, with up to 30 opposite leaflets and one terminal leaflet. Each leaflet is ovate to lanceolate, 8 to 20 cm long, pointed at the apex and unequal at the base. Flowers are numerous, creamy-yellow, in umbels on large branching sprays. Fruit is a succulent berry, pale blue, flattened on 2 sides and 2 to 3 lobed. Pencil Cedar is sometimes confused with a small Red Cedar *(Toona australis)* but Red Cedar has shorter leaves, fewer leaflets, no terminal leaflet and dry fruiting capsules.

HABITAT Rainforest, especially in partly cleared areas and along roadsides.

DISTRIBUTION Common in coastal zone. Recordings include Cambewarra, Jamberoo, Minnamurra Falls, Mt Kembla, Bola Creek, Strickland and Watagan State Forests. Extending from Gippsland (Vic) along coast to northern Qld.

FLOWERING February–March.

879. *Polyscias murrayi*

880. *Polyscias elegans*

Polyscias elegans

Celery Wood **Plate 880**

An attractive small tree with decorative dark glossy leaves and an umbrella-like mass of graceful foliage. It is sometimes grown as an ornamental. Fresh bark and broken shoots are said to have the smell of celery. Leaves are large (up to 1 m long) and bipinnate. Leaflets are numerous, ovate, 5 to 10 cm long, entire, pointed, usually opposite with a terminal leaflet. Flowers are purple, small, numerous and borne on a much-branched spray above the foliage. Fruit is a purple-black succulent berry, round but flattened at the sides.

HABITAT Rainforest.

DISTRIBUTION Illawarra district, but uncommon. Recordings include Minnamurra Falls, Kiama and Kelly's Falls. Extending from Beecroft Peninsula (south coast) to north coast and northern Qld; also New Guinea.

FLOWERING February–March.

Cephalaralia

Monotypic genus consisting of a climber with 3-foliolate leaves.

Cephalaralia cephalobotrys

Climbing Panax **Plate 881**

A tall rainforest climber, its stems are hairless but rather rough to touch. New growth and inflorescences have a sprinkling of brown hairs. Leaves are on slender stalks, 8 to 10 cm long, and are divided into 3 broad leaflets. Each leaflet is entire, oblong to lanceolate and up to 10 cm long. Juvenile leaves have a quite different appearance. They are serrated, rough to the touch and the 2 side leaflets are almost stalkless. Flowers are small, crimson, stalkless, with 5 petals and 5 stamens and are arranged in small heads along short racemes.

HABITAT Coastal rainforest and deeper sheltered gullies of Blue Mountains.

DISTRIBUTION Scattered from coast to mountains. Recordings include Burning Palms, Minnamurra Falls and Gap Creek Flora Reserve. Extending along north coast to Qld.

FLOWERING December–January.

881. *Cephalaralia cephalobotrys*

Apiaceae (Umbelliferae)

Partly because of its distribution throughout Europe and much of Asia, and partly because of the distinctive umbellate inflorescence of many species, this was one of the first plant families to be recognised and studied. The herb hemlock, remembered as the cause of death of Socrates, is one of a number of poisonous species. Other species such as parsley, celery, coriander, caraway, anise, fennel, dill and cumin have been cultivated for centuries as culinary herbs and for oils present in the seeds and fruits. Carrots and parsnips are the best known of a number of species used as vegetables. Although the common Australian representatives belong to a subfamily in which the fruits lack secretory oil canals (vittae), their identity can often be established in the absence of flowers or by the odour of the crushed leaves and stems which is sometimes offensive but more frequently 'carroty'.

The family is a large one, consisting of about 300 genera and 3000 species widely distributed in temperate areas of the world. It is subdivided into 3 subfamilies, Apioideae (e.g., *Apium, Daucus*) which is bipolar in distribution but mainly northern, Hydrocotyloideae (e.g., *Hydrocotyle, Platysace*) which is predominantly southern, particularly South American, and Saniculoideae (e.g., *Eryngium, Aciphylla*) which is distributed in both hemispheres and includes a number of spiny species quite unlike the others in appearance. All 3 subgroups are represented in the Sydney area by both native and introduced species.

Most members of the family are annual, biennial or perennial herbs with alternate or radical leaves which are often deeply divided into segments and have a broad sheathing base to the petiole. The basic flower structure is uniform, consisting of 5 petals alternating with 5 stamens, a tiny calyx, an inferior ovary with 2 carpels (*Actinotus* is an exception) and a characteristic structure called a stylopodium formed from the swollen bases of the styles and a nectar-secreting disc above the ovary. The fruits have a much greater variation in form, and may have spines, hooks, hairs or wings to aid dispersal. Basically the fruit is a dry schizocarp which splits into 2 one-seeded mericarps which are usually ridged longitudinally on the outer surface.

Although the family is characterised by the umbellate inflorescence, this is often difficult to recognise because of reduction and modification of parts or the enlargement of subtending bracts. The simple umbel (as in *Trachymene*) is a flat-topped inflorescence in which individual flower stalks arise from the same point on a common stalk (Figure 12a). The outermost flowers in an umbel open earliest. A compound umbel (as in *Platysace*) is one in which the umbels themselves are arranged in umbels.

Sometimes the umbels are organised in a manner similar to the Asteraceae, the marginal flowers irregular with enlarged outer petals (as in *Daucus*) and the inner flowers or umbels functionally unisexual, usually male. In other genera such as *Xanthosia* the umbels may be reduced to a few or even a single flower. Sometimes (as in some *Hydrocotyle* species) the flowers are sessile and appear to be in heads. Sometimes the subtending bracts and bracteoles are absent (as in *Apium*) or small (as in *Platysace*), or they may be the most noticeable part of the inflorescence. In *Actinotus* (Flannel Flower) the umbels are surrounded by large whitish woolly bracts which give the inflorescence a daisy-like appearance, while in *Eryngium* the enlarged spiny bracts give a thistle-like appearance.

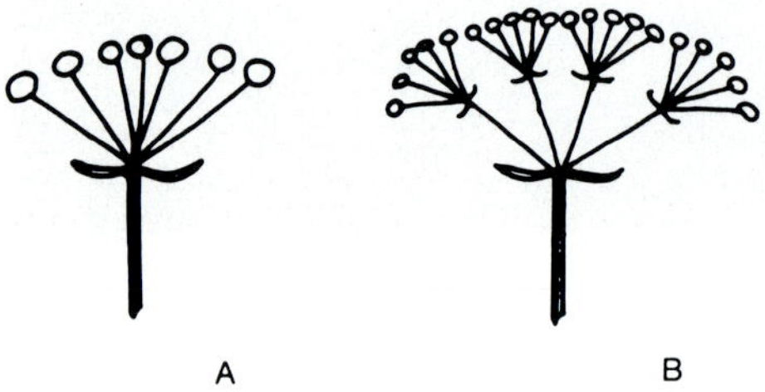

Simple umbel **Compound umbel**

Figure 12

Actinotus Flannel Flower

Herbs, with flowers crowded into an inflorescence resembling a head which is surrounded by radiating petal-like bracts (an involucre). Leaves divided into a number of narrow lobes.

Actinotus helianthi

Flannel Flower **Plate 882**

A conspicuous and well-known part of the Sydney flora, Flannel Flowers are short-lived perennials, rarely living more than 4 years. Large numbers of seed germinate after fire; some flower within one year but the most profuse flowering is in the second year. In forested situations, Flannel Flower is a slender herb, 30 to 40 cm tall, but in coastal dunes it is much more robust, with woody stems and compact foliage, and sometimes more than 1 m tall. The whole plant is covered with softy woolly-white hairs. Leaves are an attractive soft green-grey colour, deeply divided into a number of narrow lobes. What appears to be a flower with long white flannel-like petals, tipped with green, is in fact a central flower head of closely packed flowers, 3 to 4 mm long, without petals and arranged in an umbel. Surrounding the flower head, like radiating petals, are flannel-like bracts. The whole inflorescence (including the bracts) may be 5 to 8 cm across.

HABITAT Most abundant on old dunes and coastal heaths but also occurring in open forest on sandy soils.

DISTRIBUTION Widespread from coast to mountains. Occurring throughout central tablelands and western slopes divisions; also Qld.

FLOWERING Chiefly September–December, but some flowers all year round.

882. *Actinotus helianthi*

Actinotus minor

Lesser Flannel Flower **Plate 883**

A sprawling plant with wiry branches, common throughout the area. The grey-green leaves are small, divided into segments; hairless above and silky-hairy underneath. Flowers are small versions of those of *A. helianthi* with numerous white flowers in the centre surrounded by velvet-like bracts. The flower head, including the bracts, is usually only 10 mm across.

HABITAT Open forest and heath, on sandy soils.

DISTRIBUTION Widespread from coast to mountains. Also occurring south coast and southern tablelands.

FLOWERING Some flowers all year round.

883. *Actinotus minor*

Actinotus forsythii **Plate 884**

This small sprawling branched herb is rarely seen because of its restricted distribution and because it appears only sporadically, being locally common in some years and totally absent in others. After producing seed between April and May, the plant dies off and is blown away by the winds. It is believed that an important factor in the subsequent germination is fire, followed by suitable rain. Leaves are divided to the base into narrow segments. Flower heads are unmistakable as they are about 25 mm across with pink flowers in the centre and white bracts around the central flowers.

884. *Actinotus forsythii*

885. *Actinotus gibbonsii*

Actinotus gibbonsii (Plate 885) has also been recorded from wet mossy soils on a sandstone rock shelf in Wingello State Forest. It is a small tufted herb with leaf blades deeply divided into narrow lobes and long smooth leaf stalks. Flowers are pinkish buttons, about 10 mm across, surrounded by pointed green bracts. Flowers observed in November and December.

HABITAT Exposed heath, on rocky or sandy sites.

DISTRIBUTION Restricted to a few locations in the higher Blue Mountains (e.g., Narrow Neck, Mt Hay, Glow Worm Tunnel). Also recorded on south coast and southern tablelands between Nerriga and Sassafras.

FLOWERING December–February.

Hydrocotyle Pennywort

Herbs with small flowers in simple umbels; involucral bracts absent or very small. Stipules dry and mebranous, more or less fused to the leaf stalk. The most common member of this genus in the area is *Hydrocotyle bonariensis*, an introduced plant from South America which has taken root on sand dunes behind beaches all along the coast. It has round leaves, 5 to 6 cm across, and sturdy erect flower stalks.

Hydrocotyle geraniifolia

Forest Pennywort **Plate 886**

This is a delicate, lax trailing herb of sheltered situations. Stems may be 50 cm long and root at the nodes. Leaves are thin, smooth and deeply divided (almost to the base) into 5 to 7 segments. Each segment is broadly toothed, acute and may be 5 to 6 cm long. Flowers are white in a simple umbel which is borne on a slender stalk. Fruit is typical of the genus, a capsule which is 5-ribbed and laterally compressed.

886. *Hydrocotyle geraniifolia*

HABITAT Sheltered understorey of forest and margins of rainforest.

DISTRIBUTION Widespread from coast to mountains. Recordings include Patonga, Garrawarra and Colo River. Extending along coastal zone from south coast to north coast.

FLOWERING December–March.

887. *Hydrocotyle laxiflora*

Hydrocotyle laxiflora

Stinking Pennywort **Plate 887**

This small herb receives its rather unattractive common name from the strong foetid smell emitted by its flowers, especially in wet weather. Its stems are creeping and rooting at the nodes and branches and both surfaces of the leaves are covered with long spreading hairs. Leaves are round, lobed or with wavy margins, somewhat resembling those of the geranium. Flowering stems are erect, about 12 cm high, and bear numerous tiny yellowish flowers, each with a distinct stalk, arranged in an umbel.

HABITAT Open forest and woodland.

DISTRIBUTION Blue Mountains and southern highlands (e.g., Coxs River, Tallong). Extending throughout the tablelands and western slopes divisions; also occurring south coast, Qld, Vic and Tas.

FLOWERING October–December.

Trachymene

Herbs with dissected leaves and small white flowers borne in simple umbels. Flowers have 5 petals, 5 stamens and minute calyx-teeth. The involucre consists of linear bracts, shorter than the flower stalk and often shortly united at the base. The fruit consists of 2 orbicular united fruitlets which are laterally compressed.

889. *Trachymene saniculifolia*

891. *Xanthosia pilosa* (Helensburgh form)

888. *Trachymene incisa*

Trachymene incisa ssp. *incisa* — Plate 888

T. incisa is a robust erect perennial herb with a rootstock which is a reminder that the plant belongs to the same family as the cultivated carrot and parsnip. Its leaves are chiefly at the base of the stems on long stalks about 4 cm across and finely dissected into 20 to 30 lobes. Stems are up to 50 cm tall and are topped by a compact flower head which is an umbel of 40 to 80 white flowers. Umbels are 10 to 15 mm across.

Less common is the weak sprawling herb *Trachymene procumbens*, which has leaves present along the bottom of the flowering stems.

The rare *Trachymene saniculifolia* (Plate 889), last collected in 1899 from the Jenolan Caves, was noted by the authors growing near the Boyd River in Kanangra-Boyd National Park and flowering in December–January. It is a decumbent herb with small umbels of small pink/white flowers and glossy buttercup-like leaves.

HABITAT Forest on sandy soils.

DISTRIBUTION Coast, Cumberland Plain and lower Blue Mountains. Also occurring south coast, north coast, northern tablelands, north-western slopes and Qld.

FLOWERING August–December.

Xanthosia

Herbs and small shrubs with flowers in simple or compound umbels, which may be reduced to a few or even a single flower. Flowers stalkless or almost so in the ultimate umbel, surrounded by 1 to 3 bracts. Calyx lobes broad and conspicuous. Leaves usually 3-lobed.

Xanthosia pilosa — Plate 890

This small plant is striking because of its hairiness. The typical form is a shrub 30 to 60 cm tall, but habit and hairiness are very variable. Stems, leaves and flower stalks are covered with a mixture of long greyish hairs and close-matted silver-brown stellate hairs. Plants on fixed dunes tend to be more compact, erect and hairy than those in forested situations. In the Blue Mountains, there is a form with only close-matted hairs; a hairless form has been recorded near Berowra Creek (Hornsby Heights). Leaves are up to 40 mm long, with 3 to 7 broad rounded lobes. Flowers are white to creamy-green and borne in few-flowered simple or compound umbels from the leaf axils. Umbels are sometimes almost stalkless or on short stalks. Some forms (e.g., at Helensburgh (Plate 891) and Wingello State Forest) have flower stalks up to 4 cm long and the petal-like bracts surrounding the umbel are a deep pink in bud.

HABITAT Heath and open forest, on sandy soils.

DISTRIBUTION Widespread and common from coast to mountains. Also occurring north coast, south coast, southern tablelands, Qld, Vic, and Tas.

FLOWERING Chiefly August–February, but some flowers at other times.

890. *Xanthosia pilosa*

Xanthosia dissecta — Plate 892

A small perennial with tufts of leaves spreading from a woody rootstock. Leaves are hairless, with the leaf blade finely dissected, like parsley leaves. Leaf stalks are up to 12 cm long. Flowers are small, reddish and borne in simple reduced umbels from the leaf axils on stalks shorter than the leaves.

HABITAT Open forest, in sandy soils.

DISTRIBUTION Blue Mountains and southern highlands (e.g., Penrose, Newnes State Forest). Also recorded in Vic and Tas.

FLOWERING November–January.

892. *Xanthosia dissecta*

Xanthosia atkinsoniana — Plate 893

This plant is easily recognised by its general appearance. The leaves are all clustered around the base of the stems and the flower stalks are tall, thin, branched and almost leafless. Stems may be upright to almost 60 cm or more, often decumbent. The whole plant is hairless when mature. Leaves are 2 to 4 cm long, on long stalks, divided into acutely lobed segments which are thin and wiry. The stems carry terminal compound umbels of white flowers. Buds often have a pinkish tinge.

HABITAT Open forest, on sandy soils.

DISTRIBUTION Blue Mountains, Colo Plateau and southern highlands. Recordings include Mt Wilson, Bowens Creek, Wombeyan Caves, Penrose, Joadja Creek and Colo Heights. Extending along coastal and tablelands divisions; also found in WA.

FLOWERING November–January.

893. *Xanthosia atkinsoniana*

894. Xanthosia vestita

Xanthosia vestita — Plate 894

This small shrub closely resembles *X. pilosa* and some botanists regard it as a form of *X. pilosa*, rather than a separate species. It has similar hairy stems and lobed leaves with grey downy undersurfaces. There are, however, a number of significant differences. It is a smaller, more trailing plant, seldom more than 30 to 40 cm tall and often found on sheltered rock faces. Flowers are in loose compound umbels. The umbels are on thin stalks from the leaf axils and the flower head is usually longer than the leaves (unlike *X. pilosa* which usually has inflorescences shorter than the leaves). The flowers are surrounded by broad, petal-like white bracts which are longer than the flowers.

HABITAT Sheltered forest, often on sandstone cliff faces.

DISTRIBUTION Widespread from coast to mountains. Uncommon near coast. Recordings include Undercliff Walk, Wentworth Falls. Restricted to the Sydney district.

FLOWERING December–February.

Xanthosia tridentata — Plate 895

A diffuse little plant, seldom more than 20 cm tall. It is fairly common in the Sydney district, but because of its insignificant flowers it is often overlooked. Leaf shape and the degree of hairiness are very variable, but the typical form has stems and leaves sprinkled with loose hairs. The leaf undersurface is covered with silky-white hairs. Leaves are wedge-shaped, 10 to 12 cm long, with 3 teeth at the broad apex (hence the specific name 'tridentata'). Flowers are greenish yellow, in small umbels of 1 to 3 flowers, borne on short stalks amongst the leaves.

895. Xanthosia tridentata

HABITAT Heath and open forest, on sandy soils.

DISTRIBUTION Widespread from coast to mountains. Extending along south coast to Vic.

FLOWERING March–November.

Apium Celery

Herbs with hollow grooved stems. Flowers white, in compound umbels. Fruit subglobular and laterally compressed.

896. Apium prostratum

Apium prostratum
Sea Celery — Plate 896

Sea Celery and the common vegetable Celery *(Apium graveolens* var. *dulce)* belong to the same genus. This close relationship can be seen in the similarity of the leaves which are deeply divided into lobed segments. Sea Celery is a prostrate herb of the coast and estuaries. It has a thick rootstock and rather fleshy spreading branches. Flowers are small, white and borne in dense compound umbels about 2 cm across. Fruit is laterally compressed with 5 prominent ribs.

HABITAT Near or on coastal dunes and on alluvial flats of estuaries.

DISTRIBUTION Scattered along coast, but uncommon as most of its original habitat has been disturbed. Extending all along coast of NSW. Also found in all other States (but not NT) and NZ.

FLOWERING September–November.

Platysace

Shrubs with entire or lobed leaves. Flowers small, white and in terminal compound umbels. Crushed leaves have a distinct 'carroty' smell.

Platysace ericoides
Heath Platysace — Plate 897

A low shrub with short diffuse stems which are clearly rough-hairy. Leaves are crowded, spreading, up to 10 cm long, narrow-linear and finely pointed.

897. Platysace ericoides

Flowers are white, and in compound umbels, 5 to 9 mm across. This species is sometimes confused with the more common *Platysace linearifolia*, but it differs in having firmer leaves, hairy stems and being a much smaller, less erect shrub.

HABITAT Heath, shrubland and open forest, on sandy soils.

DISTRIBUTION Widespread from coast to mountains. Extending along north coast, northern tablelands, and north-western slopes to Qld.

FLOWERING November–February.

Platysace linearifolia
Narrow-leaf Platysace — Plate 898

This is a very common small shrub, growing to 1.5 m tall, with hairless branches which may be erect or somewhat spreading. Leaves are narrow-linear, up to 25 mm long, softer, less crowded and more slender than those of *P. ericoides*. Crushed leaves have a carrot-like smell, characteristic of the genus. Flowers are white, in compound umbels 5 to 7 mm across.

HABITAT Open forest, on sandy soils.

DISTRIBUTION Common from coast to mountains. Recordings include Heathcote, Brisbane Water National Park, Burragorang Plateau and Blackheath. Restricted to the Sydney district.

FLOWERING December–April.

898. Platysace linearifolia

899. Platysace lanceolata

Platysace lanceolata

Lance-leaf Platysace **Plate 899**

This is a very common undershrub in most dry sandy heaths and open forests. Branches are erect, up to 1.5 m tall, but mostly less than 1 m. Plants growing on old dunes near the sea are often much more robust and compact than those growing in open forests. Leaves are variable in size and shape, linear to narrow lanceolate or elliptic, 1 to 3 cm long, stalkless or shortly stalked and usually pointed at the apex. Crushed leaves have a carrot-like smell. Flowers are white, numerous and borne in compound terminal umbels up to 3 cm across.

HABITAT Old dunes, heath, shrubland and open forest, on sandy soils.

DISTRIBUTION Widespread from coast to mountains. Extending inland to the edge of the western plains. Found throughout the coast, tablelands and western slopes of NSW; also Qld and Vic.

FLOWERING December–April.

900. Platysace stephensonii

Platysace stephensonii **Plate 900**

A low shrub with slightly downy stems, its leaves are distinctive. They are about 10 mm long and 8 to 10 mm wide, rather rigid and deeply divided into 3 pointed lobes. The 2 side lobes are sometimes again divided, so the leaf has 5 pointed lobes. Flowers are white, numerous and borne in compound, terminal umbels up to 3 cm across.

HABITAT Coastal heath and open forest, on sandy soils.

DISTRIBUTION Restricted to the coastal zone. Recorded from Royal National Park (Bundeena) and Ku-ring-gai Chase National Park (McCowans Trig). Extending to south coast (e.g., Beecroft Peninsula, Jervis Bay).

FLOWERING December–July.

Platysace clelandii **Plate 901**

A shrub with hairy stems and small toothed leaves, somewhat similar to *P. stephensonii.* Its distribution helps in its identification as it is only found on the Hornsby and Colo Plateaux. Leaves are characteristic. They are small and more fan-like than those of *P. stephensonii*, wider than long (5 to 6 mm long, 7 to 8 mm wide), shortly stalked and with 5 to 7 acute but small teeth. Flowers are white, numerous and borne in compound, terminal umbels.

FLOWERING Open forest, on sandy soils.

DISTRIBUTION Restricted to the Hornsby Plateau and Colo-Wollemi area.

FLOWERING December–February.

901. Platysace clelandii

Eryngium

Thistle-like plants, with rigid spiny leaves and stalkless flowers in compact heads surrounded by spiny bracts. This genus includes Blue Devil *(E. rostratum)*, a conspicuous plant resembling a blue thistle. It is widespread on the western tablelands and western slopes, but generally absent from the Sydney district.

Eryngium vesiculosum

Pricklefoot **Plate 902**

A prostrate spiny plant with branches like runners spreading from a rootstock. Leaves are at the base of the stems, with a few opposite leaves along the stems. Leaves are 4 to 15 cm long, flat, tough, straplike, with spreading sharp teeth. Flowers are in heads, each a rounded cluster of needle-like bracts and pale blue flowers.

HABITAT Woodland, on dry slopes and sandy soils.

DISTRIBUTION Valleys and slopes west of the Blue Mountains (e.g., Hartley Vale). Occurring throughout the drier tablelands and western slopes; also Vic and Tas.

FLOWERING January–March.

Centella

A small herb with a few tiny flowers in simple umbels. The one species in Australia is found in damp places, often in roadside ditches.

Centella asiatica

Swamp Pennywort **Plate 903**

Swamp Pennywort is a common herb in damp sites, but is often overlooked because of its small size and lack of conspicuous flowers. Stems are creeping and rooting at the nodes. Leaves are rounded or kidney shaped, with cordate bases, 1 to 3 cm across and are borne on relatively long stalks. Margins are entire, wavy or bluntly toothed. Stems bear 3 to 4 tiny white or pinkish flowers in small heads or umbels. Each flower has 5 petals. Fruit is about 3 mm across, compressed laterally with 3 to 4 ribs on each side.

HABITAT Swamp margins, soils regularly inundated and other damp places.

DISTRIBUTION Central coast and north coast. Occurring in all States (but not NT), Asia and Pacific Islands.

FLOWERING December–January.

903. Centella asiatica

902. Eryngium vesiculosum

Olacaceae

The order Santalales consists of several families showing a transition from mostly non-parasitic trees and shrubs with 'primitive' floral characteristics to those completely dependent on a parasitic lifestyle with curious forms and 'advanced' features such as tiny simple unisexual flowers. The Olacaceae is a small family at the 'primitive' end of this series. It is represented in Australia by 13 species in 3 genera, and in the Sydney area by 1 species of *Olax*, a shrubby root-parasite.

Olax

Olax stricta

Olax **Plate 904**

Olax is a slender branched shrub, seldom more than 1 m tall, with a yellowish-green appearance. It is parasitic on the roots of nearby plants. Leaves are alternate, 5 to 10 mm long and arranged in 2 opposite rows. Flowers are solitary, white or yellowish, with 5 or 6 spreading petals, 4 to 5 mm long; there are 3 stamens and 5 or 6 staminodes opposite the petals. The fruit is a fleshy berry, about 8 mm long.

HABITAT Open forest, woodland, shrubland and coastal dunes.

DISTRIBUTION Widespread from coast to mountains. Extending from Gippsland (Vic) along coast and tablelands divisions of NSW to Qld and westwards to western slopes.

FLOWERING Most of the year.

904. Olax stricta

Santalaceae

The herbs, shrubs and trees that make up this family are hemiparasites; that is they obtain water and soluble salts from the host plant but fix carbon dioxide by photosynthesis in their own leaves. Most are root parasites but a few are parasitic on branches.

The family consists of 400 species in 30 genera, with 46 species in 10 genera in Australia. Of the 5 genera in the Sydney area, 3, *Leptomeria, Choretrum* and *Omphacomeria*, are endemic to Australia. The other 2, *Exocarpos* and *Santalum*, are centred in Malesia and the western Pacific, but are well represented in this country.

Because of the diversity of features displayed by the various genera the family is usually divided into 3 or 4 tribes. Although in *Santalum* the leaves are normally developed, in many genera they are reduced to scales which are sometimes caducous (i.e., fall off before developing fully). Most of the Santalaceae have tiny bisexual flowers with 4 or 5 perianth lobes and a floral disc above an inferior ovary. However the flowers in *Omphacomeria* are unisexual and the ovary in *Exocarpos* and *Omphacomeria* is superior. In most genera the fruit is a fleshy drupe, but in *Exocarpos* it is the receptacle below the ovary, rather than the ovary itself, that is enlarged, fleshy and coloured at the fruiting stage.

Santalum is the best-known and most sought-after genus in the family. The Indians and Chinese use the sandalwood tree *(S. album)*, which grows naturally on the islands of Indonesia, for making idols and sacred ornaments and for burning as incense. Also, sandalwood oil, distilled from the heartwood, is used in medicines, perfumery and sacred unguents. During the nineteenth century a number of Pacific islands were ruthlessly exploited for their sandalwood and large amounts were exported from Australia. Although this trade continues, the quantities involved are variable and much smaller. Two or three Australian species provide the Aborigines with edible roots and fruits known as quandongs.

Exocarpos Ballart

Shrubs or small trees with reduces leaves, persistent or falling early. The botanical name is derived from the Greek 'ex'—out, 'carpos'—fruit, a reference to the characteristic fruit which is borne on a colourful swollen succulent receptacle.

Exocarpos cupressiformis

Native Cherry, Cherry Ballart **Plate 905**

A tall shrub or small tree to 8 m, with numerous thin graceful pendulous branchlets. It is a parasite on the roots of nearby plants and may have a yellowish-green appearance. Leaves have been reduced to minute scales, 0.5 mm long. Flowers are very small, with 5 perianth segments and borne in spikes 4 to 6 mm long in the axils of the upper branchlets. Fruit is a drupe, 4 to 5 mm long, on a swollen red receptacle, 4 to 6 mm long.

905. Exocarpos cupressiformis

HABITAT Forest and woodland.

DISTRIBUTION Widespread from coast to mountains. Found throughout NSW, except in far western plains; also occurring Qld, Vic, Tas and SA.

FRUIT Ripe July–October.

Exocarpos strictus

Pale Ballart **Plate 906**

A shrub to 3.5 m, with angular, striate, erect branchlets. It is a parasite on the roots of nearby plants and may have a bronze-green appearance. Leaves are very small and pointed, 1 to 3 mm long, and fall off early to leave a tooth-like base on the stems. Flowers are very small, mostly with 4 perianth segments, and borne in small stalkless axillary clusters. Fruit is a black drupe, 2.5 to 4 mm long, on a swollen red, mauve or white receptacle, 2 to 4 mm long.

HABITAT Open forest.

DISTRIBUTION Widespread from coast to mountains. Found throughout coast, tablelands and western slopes divisions; also south-western plains, Vic, Tas and SA.

FRUIT Ripe November–January.

906. Exocarpos strictus

907. *Omphacomeria acerba*

Leptomeria Currant-bush

Shrubs, with leafless branches. Flowers are bisexual in spikes and fruit is a fleshy or dry drupe. The botanical name is derived from the Greek 'leptos'—slender, 'meros'—a part, a reference to its branches.

Leptomeria acida

Native Currant **Plate 908**

A common small to medium shrub with slender leafless green branches and numerous green fleshy fruits. The minute scale-like leaves fall off early, leaving a notch on the stem. Unlike *Omphacomeria acerba*, the flowers of Native Currant are bisexual. They are small, green-brown, and 20 or more flowers are borne in slender side spikes 1 to 4 cm long. This flowering arrangement distinguishes this species from the leafless *Choretrum* which has flowers scattered singly along the branches. Fruit is a succulent drupe, 7 mm long, which is edible, having a refreshingly tart taste.

HABITAT Open forest, mainly on sheltered slopes.

DISTRIBUTION Widespread from coast to mountains. Extending from south coast northwards to south-eastern Qld; also southern tablelands.

FLOWERING February–April.

910. *Choretrum pauciflorum*

Choretrum pauciflorum

Dwarf Sourbush **Plate 910**

A small ascending shrub, 30 to 100 cm tall, with green, terete branches, appearing leafless, but with minute scale-like persistent leaves. Flowers are creamy-white, cylindrical, arranged singly towards the ends of the branches; they are 5 lobed, almost stalkless. Fruit is globular, dry, 4 to 5 mm long, crowned by persistent perianth segments.

HABITAT Open forest, on sandy soils.

DISTRIBUTION Blue Mountains and southern highlands (e.g., Grose Valley, Bundanoon). Extending along south coast, southern tablelands to Vic; also northern tablelands.

FLOWERING September–October.

Omphacomeria Sourbush

There is only one member of this genus. Characteristic features are its leafless stems and male and female flowers on different plants. The botanical name is derived from the Greek 'omphakos'—bitter, 'meros'—a part, a reference to the sour-tasting fruit.

Omphacomeria acerba

Leafless Sourbush **Plate 907**

A leafless shrub to 1 m, with male and female flowers borne on separate plants. Branches are erect, terete and conspicuously striped. The scale-like leaves fall off, leaving scars on the stems. Male flowers are small, in axillary clusters; female flowers are solitary, axillary and stalkless. The fruit is a green ovoid drupe, 6 to 7 mm long, crowned by persistent perianth segments.

HABITAT Open forest, often on drier rocky hills.

DISTRIBUTION Gippsland (Vic) along coast to Hunter River and central and southern tablelands.

FLOWERING August–November.

908. *Leptomeria acida*

909. *Choretrum candollei*

Choretrum

Hairless shrubs, appearing leafless. Flowers are bisexual and scattered singly along the stems. The botanical name is derived from the Greek 'khoris'—separate, 'etron'—abdomen, a reference to the separation between the receptacle and the perianth.

Choretrum candollei **Plate 909**

An erect shrub, 1 to 2 m tall, with green striped branches, appearing leafless, but with minute scale-like persistent leaves. Flowers are white and very numerous, scattered singly along the branchlets; they are 5 lobed with short stalks up to 2 mm long. Flowering branches are numerous and wiry, with short raised lines beneath the scale-leaves and flowers. Fruit is globular, dry, 4 to 5 mm long, crowned by persistent perianth segments.

HABITAT Open forest, on sandy soils.

DISTRIBUTION Widespread from coast to mountains, but more common in higher areas (e.g., Blackheath, Barren Grounds). Also occurring south coast, southern tablelands, northern tablelands, western slopes and Qld.

FLOWERING June–September.

Santalum Sandalwood

Santalum is a tropical genus of about 25 species worldwide. It includes a number of trees noted for their valuable scented wood and oils. The one species in the Sydney district is easily identified by its distinctive fruit.

Santalum obtusifolium

Sandalwood **Plate 911**

An erect shrub, 1 to 2 m tall, uncommon in the Sydney district. Its leaves are opposite, dark green above, paler below, narrow lanceolate, 3 to 6 cm long, with margins curled under. Flowers are small, greenish white, about 4 mm long, borne in axillary clusters. Fruits are conspicuous, as they are blue-black, globular, 8 to 10 mm long, and have a circular scar on the summit.

HABITAT Forested slopes, especially near streams, on sandy soils.

DISTRIBUTION Scattered and uncommon. Recorded from Georges River near Ingleburn, Royal National Park, Bargo River, Douglas Park, Wattamolla and Springwood. Extending along coastal zone from Gippsland (Vic) to Lamington Plateau (Qld).

FLOWERING October–December.

911. *Santalum obtusifolium*

Loranthaceae

The majority of mistletoes fall into 2 families differing in structure, biology and geographic distribution. The larger, Loranthaceae, has a mainly southern distribution reflecting its Gondwanan development, while the other, Viscaceae, is mainly northern, reflecting a development under different conditions. Both groups are hemiparasites relying on host trees for water and mineral nutrients, but performing photosynthesis in their own leaves. All but 3 genera (these in Loranthaceae) are epiparasites growing on the branches of trees. Two of the three which retain the ancestral root-parasitic lifestyle are monotypic genera endemic in south-western Australia *(Nuytsia)* and the Blue Mountains of New South Wales *(Atkinsonia)*.

The most obvious differences between the 2 families are in their floral characteristics. Whereas the flowers of Loranthaceae are mostly bisexual and are typical 'bird flowers' with a large narrow tubular orange/red corolla, those of Viscaceae are unisexual, minute and insect-pollinated. The flowers of Loranthaceae are subtended below the calyx by a ring of tissue called a calyculus, which is absent in Viscaceae. The 2 families also differ in pollen, fruits, chromosome numbers and embryology.

Loranthaceae consists of 65 genera and 950 species widely distributed in the tropics and southern temperate regions. Half of the 12 genera and most of the 70 species which occur in Australia are endemic. The Australian group comprises both small primitive relict genera such as *Atkinsonia, Nuytsia* and *Muellerina* and more advanced genera of tropical origin such as *Amyema, Dendrophthoe* and *Amylotheca*. Many advanced genera have species that are host-specific and have leaves and a general appearance closely resembling those of the host tree.

Host-specificity is a feature of mistletoes of the drier open forests and woodland; those of rainforests are rarely host-specific. Reasons for this are apparent. Open forest and woodland are frequently dominated by such genera as *Eucalyptus, Acacia* and *Allocasuarina*, but in rainforests there is a much greater diversity of trees. Under the former conditions, adaptation to a dominant tree is selectively advantageous, and may therefore be expected, but specific adaptations to a host occurring with low frequency are not likely to be advantageous and are therefore unlikely to develop.

Host mimicry, however, cannot be as easily explained. There is good evidence that it is not due to the transfer of growth-regulating substances from the host. Neither is it simply due to evolutionary convergence resulting from similar adaptations to environmental factors. One possibility is that mimicry, which appears to be a peculiarly Australian adaptation, provides concealment and protection from browsing fauna, particularly possums. In support of this there is evidence that the prevalence of mistletoes is inversely related to the possum population. Another explanation relates host mimicry to the close association that has developed between mistletoes and certain small birds which act as pollinating and seed dispersal agents. The mistletoe bird feeds on mistletoe berries and insects. The seeds pass rapidly through the bird and, when excreted, adhere by the remains of the sticky pulp to a tree branch. Permanent attachment of the embryo to the branch can occur soon after. If, at the fruiting stage, the mistletoe is more or less hidden in the host tree the mistletoe bird has to search these trees for food, and this increases the likelihood of seed dispersal to a compatible host. On the other hand, at pollination, when high visibility is desirable, the presence of the mistletoe is adequately advertised by the brightly coloured flowers.

912. Atkinsonia ligustrina

Atkinsonia

A monotypic genus. A terrestrial root-parasitic shrub, named after Caroline Atkinson, a nineteenth century Australian botanist.

Atkinsonia ligustrina — Plate 912

An unusual and uncommon root parasite, with a very narrow distribution in the Blue Mountains. It is possibly the most primitive existing member of the family. It is an erect shrub, to 1 m tall, with opposite, lanceolate leaves, 3 to 5 cm long and 6 to 14 mm wide. Flowers are an attractive yellow and red, borne on short stalks from the leaf-axils. There are usually 6 free petals, 6 to 8 mm long. Fruit is ovoid, scarlet coloured and about 12 mm long.

Habitat Open forest and woodland, on sandy soils.

Distribution Endemic in the Blue Mountains and Colo Plateau. Recordings include Linden, Mt Tomah, Mt Wilson and Wollemi National Park.

Flowering November–December.

913. Amyema cambagei

Amyema Mistletoe

A large genus of about 100 aerial parasitic shrubs. Its characteristic features are: petals free to the base, anthers fused at their base to the filament (not attached at the back and swinging freely, as are those of *Muellerina*) and flowers in umbels of 2 to 3 flowered cymes.

Amyema cambagei

Sheoak Mistletoe — Plate 913

This Mistletoe is epiphytic on Sheoaks (*Casuarina* spp.), especially the River Oak *(C. cunninghamiana)*, and its leaves are long and terete, mimicking those of its host. It is a spreading to pendulous plant, with greyish short hairs on young stems, leaves and inflorescences. Leaves are 6 to 15 cm long and about 1 mm diameter. Flowers are slender, 15 to 21 mm long, pink to deep red under a sprinkling of white hairs, club-shaped in bud, splitting into 5 free petals. Fruit is globular, 5 to 6 mm across, and pink-red.

Habitat Epiphytic on several species of *Casuarina* in open forest and woodland.

Distribution Widespread from coast to mountains (e.g., Nepean River, Wisemans Ferry, Wombeyan Caves). Occurring throughout coast, tablelands and western slopes divisions and north-western plains, extending to northern Qld.

Flowering June–December.

Amyema gaudichaudii

Paper-bark Mistletoe — Plate 914

This erect and compact epiphyte is found on *Melaleuca* spp., especially on *M. decora* in the open swampy woodlands of Cumberland Plain. Its leaves are linear-oblong to lanceolate, 20 to 35 mm long, 2 to 5 mm wide, upright and rounded at the apex. Flowers are slender, 7 to 10 mm long, deep red, club-shaped in bud, splitting into 4 or 5 free petals. Fruit is globular, about 4 mm across, and red.

Habitat Epiphytic on *Melaleuca* spp. in woodland.

Distribution Coast and Cumberland Plain (e.g., Milperra, Berkshire Park). Extending from Hunter River to near Nowra. Also occurring northern tablelands and Darling Downs (Qld).

Flowering November–December.

914. Amyema gaudichaudii

Amyema congener ssp. *congener* **Plate 915**

The branches of this epiphyte are erect or spreading, immediately separating it from all other *Amyema* species in the area which are pendulous (except *A. gaudichaudii*). Its leaves are broadly lanceolate, 4 to 10 cm long, 1 to 4 cm wide, with a rounded apex and a short leaf-stalk. Leaves have obscure venation. Buds are green, slender and club shaped. Flowers are 16 to 35 mm long, green or yellow with pink stamens and style, splitting into 5 free petals. Fruit is elliptic to globular, about 8 mm long and green.

HABITAT Epiphytic on *Allocasuarina, Acacia* and other trees, including exotics, but not on *Eucalyptus* species, in open forest and rainforest margins.

DISTRIBUTION Widespread from coast to mountains. Extending from near Merimbula (south coast) along north coast to northern Qld; also northern tablelands and central western slopes.

FLOWERING June–November.

915. Amyema congener ssp. *congener*

Amyema quandang var. *quandang*

Grey Mistletoe **Plate 916**

A pendulous epiphyte with a general greyish appearance as the leaves are covered with short whitish matted hairs. Leaves are opposite, lanceolate to elliptic, often falcate, 3 to 12 cm long and 10 to 20 mm wide, tapering at the base to a stalk 4 to 12 mm long. Flowers are 16 to 30 mm long, green beneath grey hairs, with red stamens and style, splitting into 5 free petals. Fruit is pear-shaped to globular, 6 to 10 mm long and greyish.

HABITAT Epiphytic on *Acacia* spp., in drier woodlands.

DISTRIBUTION Western Blue Mountains. Widely distributed throughout the drier western slopes and plains; also Hunter Valley, northern tablelands and all mainland States.

FLOWERING August–December.

916. Amyema quandang var. *quandang*

Amyema miquelii **Plate 917**

A. miquelii is widely distributed throughout mainland Australia and may be parasitic on many species of *Eucalyptus* and some *Acacia* spp. In the Sydney district, however, it seems to be most common on Grey Box *(E. moluccana)*. Branches are pendulous with elliptic to narrow oblong leaves, up to 35 cm long and 30 mm wide; they are often falcate and narrowed at the base into a stalk 10 to 50 mm long. Leaves often have a pink or yellow tinge. Flowers are 15 to 28 mm long, red, with 5 to 7 free petals. All flowers have short stalks, a feature which helps distinguish this species from *A. pendulum* in which the central flower is sessile. Fruit is ovoid, 8 to 12 mm long.

HABITAT Epiphytic on *Eucalyptus* spp., in open forest and woodland.

DISTRIBUTION Widespread from coast to mountains. Common on Cumberland Plain (e.g., Wallacia, Berkshire Park). Occurring throughout NSW in all divisions, and in all mainland States,

FLOWERING Most of the year.

917. Amyema miquelii

918. Amyema pendulum ssp. *pendulum*

Amyema pendulum ssp. *pendulum*

Drooping Mistletoe **Plate 918**

A common epiphyte on many species of *Eucalyptus*. Its branches are pendulous with narrow oblong to lanceolate leaves, 10 to 20 cm long and 3 to 15 mm wide, narrowed at the base into a stalk 10 to 40 mm long. Flowers are 22 to 40 mm long, red, with 5 free petals. The central flower of each group of 3 is stalkless, a feature which separates this species from *A. miquelii*. Fruit is ovoid, 10 mm long.

HABITAT Epiphytic on *Eucalyptus* spp., in open forest and woodland.

DISTRIBUTION Widespread from coast to mountains. Recordings include Mt Kembla, Wentworth Falls and Mt Hay. Extending throughout coast, tablelands and western slopes divisions; also SA, Vic and Qld.

FLOWERING August–November.

Dendrophthoe

A genus of about 30 species of aerial parasitic shrubs from tropical Africa to Australia. Petals united into a curved tube; flowers borne in axillary clusters. The botanical name is derived from the Greek 'dendron'—a tree, 'phthienin'—to wither, a reference to the effect the mistletoe has on its host.

919. Dendrophthoe vitellina

920. *Amylotheca dictyophleba*

921. *Muellerina eucalyptoides*

Dendrophthoe vitellina **Plate 919**

This is a spreading to slightly pendulous epiphyte, with conspicuous runners along the stems of the host. Leaves are alternate, elliptic, 5 to 8 cm long and 20 to 25 mm wide, rounded at the apex and narrowing at the base into a stalk 3 to 12 mm long. Leaves are thick and often glaucous; the midvein is prominent. Flowers are an attractive yellow with a red tip, 25 to 40 mm long; petals are joined into a slightly swollen tube, with only the 5 lobes curled back. Flowers are borne in short dense clusters. Fruit is ovoid, 10 to 15 mm long, yellow to red.

HABITAT Epiphytic on many different hosts, mainly Myrtaceae, in open forest and woodland.

DISTRIBUTION Widespread from coast to mountains. Recordings include Wisemans Ferry and Lapstone Hill. Extending along coast and tablelands divisions from Mallacoota (Vic) to northern Qld; also north-western slopes.

FLOWERING July–December.

Amylotheca

A genus of 5 species of aerial parasitic shrubs from Melanesia to eastern Australia. Only one occurs in the Sydney district. Petals are united into a straight tube, which is often split on the upper side. Flowers are borne in axillary clusters of 3-flowered cymes. The botanical name is derived from the Latin 'amylum'—starch, 'theca'—case, a reference to the starchy nature of the ovary.

Amylotheca dictyophleba **Plate 920**

The stems of this rainforest epiphyte are erect, about 50 cm long; runners spread along the stems of the host. Leaves are mostly opposite, elliptical, 6 to 12 cm, rounded or acute at the apex and narrowed at the base to a distinct stalk 2 to 8 mm long. Flowers are red, becoming green then yellow near the apex, up to 35 mm long, with petals joined into a tube for most of their length; the tube is inflated in the middle and the lobes are curled back. Flowers are borne in axillary clusters. Fruit is globular, 10 to 12 mm across, red or purple.

HABITAT Epiphytic on rainforest trees (e.g., *Acmena smithii*).

DISTRIBUTION Coastal rainforest, especially Illawarra district. Extending northwards from Illawarra to north coast, northern Qld and New Guinea.

FLOWERING November–January.

Muellerina Mistletoe

A genus of only 4 species of aerial parasitic shrubs. Its characteristic features are petals united into a tube, anthers attached at the back and swinging freely and clusters of 3 flowers. Named after Ferdinand Mueller, pioneer Australian botanist.

Muellerina eucalyptoides **Plate 921**

The stems of this epiphyte are usually 1 to 2 m long. It may have a stiff bushy habit or hang in conspicuous clumps from branches of large eucalypts. Leaves are opposite, thick, pale green, oblong to lanceolate, 8 to 20 cm long, with indistinct veins. Flowers are greenish yellow, with red centres, 30 to 45 mm long, borne in loose clusters. Fruit is pear-shaped, 8 to 15 mm long, green to yellow.

HABITAT Epiphytic mainly on *Eucalyptus* and *Angophora* species, but sometimes on other trees, in open forest.

DISTRIBUTION Widespread from coast to mountains (e.g., Gerringong, Wombeyan Caves). Occurring throughout coast, tablelands and western slopes divisions; also Qld, Vic and SA.

FLOWERING December–February.

922. *Muellerina celastroides*

Muellerina celastroides **Plate 922**

M. celastroides is a more erect or spreading shrub than *M. eucalyptoides* and is epiphytic on hosts other than eucalypts. Its leaves are opposite, obovate to elliptic, darker above than below, 3 to 7 cm long, somewhat rounded at the apex. Flowers are flesh coloured, with green tips, pink bases and red styles, 22 to 35 mm long, borne in loose clusters. Fruit is pear shaped, 7 to 11 mm long, green to pink.

HABITAT Epiphytic on many hosts (including *Banksia integrifolia* and *Doryphora sassafras*), but not eucalypts, in open forest and rainforest.

DISTRIBUTION Coastal zone (e.g., Mt Keira, Manly). Extending from Gippsland (Vic) northwards along coast to Noosa (Qld).

FLOWERING December–February.

Viscaceae

In the grey European winters, the presence on deciduous trees of green fruiting branches of the hemiparasite *Viscum album* must have attracted attention to these plants which have long been regarded as objects of religious and supernatural significance. The mystique was undoubtedly heightened by a recognition of their origin; the name 'mistletoe' derives from the Anglo-Saxon 'misteltan' (mistel: dung, tan: twig) which describes the germination of seeds from the droppings of birds that feed on mistletoe fruits.

Although early attention was focussed on the Viscaceae, the family is smaller and much less diverse than the Loranthaceae. It consists of only 7 genera with about 400 species, mainly in the tropics. The 3 Australian genera are centred in Madagascar, tropical Africa to southern Asia, Malesia and some Pacific Islands. Seven of the 14 Australian species are endemic. Several species are parasitic on other mistletoes.

Korthalsella

A mistletoe with flattened, apparently leafless, stems. One species in the Sydney district. Named after the nineteenth century Dutch botanist Korthals.

Korthalsella rubra ssp. *rubra*

Jointed Mistletoe **Plate 923**

An unmistakable erect epiphyte with green flattened stems up to 20 cm long and 3 to 9 mm wide. Distinct longitudinal veins run along the stems. Each stem is jointed a number of times. Tiny flowers develop in clusters around these joints. The flowers are unisexual (male and female separate on the one plant), minute, green or yellowish, with 10 to 80 flowers per cluster in 2 to 5 rows. Rudimentary leaves, less than 1 mm long, form a border around the flower clusters.

HABITAT Epiphytic on many trees, especially *Doryphora sassafras*, in rainforest and wet eucalypt forest.

DISTRIBUTION Coastal (e.g., Mt Keira, Cambewarra Mountain) and mountains (e.g., Mt Wilson, Mt Irvine). Extending from Gippsland (Vic) along coast and lower tablelands to northern Qld; also New Guinea and Lord Howe Island.

FLOWERING September–December.

923. *Korthalsella rubra* ssp. *rubra*

Notothixos

A genus of 8 tropical mistletoes, with new growth covered with dense matted hairs. Flowers are unisexual. The botanical name is derived from the Greek 'notos'—southern, 'ixos'—mistletoe.

924. *Notothixos subaureus*

Notothixos subaureus

Golden Mistletoe **Plate 924**

A small epiphytic spreading shrub with the unusual habit of growing on other mistletoes. It is conspicuous as its young growth is a bright yellow colour because of the covering of small closely matted hairs. Leaves are opposite, rounded or elliptic, 2 to 4 cm long, 1 to 2 cm wide. Flowers are very small (2 to 3 mm long) and clustered at the ends of the branchlets in 1 terminal and 2 lateral cymes. Fruit is elliptical, 7 mm long, crowned by persistent perianth segments.

HABITAT Epiphytic on other mistletoes, such as species of *Dendrophthoe, Muellerina* and *Amyema*, in rainforest and open forest.

DISTRIBUTION Coast to mountains. Extending from Gippsland (Vic) along coat and tablelands to northern Qld; also western slopes.

FLOWERING Most of the year.

Notothixos cornifolius **Plate 925**

N. cornifolius is a spreading or pendulous epiphyte, with stems up to 1 m long. Its leaves and stems lack the bright golden hairs of *N. subaureus*. Leaves are opposite, elliptic, blunt, 5 to 7 cm long, 15 to 25 mm wide, with 3 to 5 distinct longitudinal veins. Flowers are small and are in pairs in a raceme up to 4 cm long near the ends of the branches. Fruit is elliptical, 7 mm long, crowned by persistent perianth segments.

HABITAT Epiphytic on Kurrajong *(Brachychiton populneus)*, in open forest.

DISTRIBUTION Uncommon in Sydney district because of the limited occurrence of its host. Recorded from coastal zone. Extending to north coast and northern Qld and throughout the western slopes.

FLOWERING October–February.

925. Notothixos cornifolius

Icacinaceae

The Icacinaceae and Celastraceae are 2 of 4 families forming a solid core in the order Celastrales which also includes a variable number of smaller families whose relationships are less clear. Although the major families have an extensive fossil record dating from at least the early Tertiary period, this has provided little assistance in determining the phylogeny (evolutionary development) of the group and its relationships with others.

The Icacinaceae consists of trees, shrubs and woody climbers occurring mainly in tropical rainforests. The flowers have a 5-lobed calyx and 5 petals, and are bisexual or sometimes unisexual. Occasionally bisexual and unisexual male and female flowers are found on the same tree. Male flowers have 5 stamens alternating with the petals. Female flowers have a superior ovary with usually 3 carpels surmounted by a short style with 3 stigmas. Bisexual flowers have both. The Icacinaceae are similar in many respects to the Celastraceae but the leaves always lack stipules, the flowers lack the nectar-secreting disc present in Celastraceae, and the fruit is a one-seeded fleshy drupe, never a dry capsule.

Seven species in 6 genera occur in Australia, with 2 species, *Citronella moorei* and *Pennantia cunninghamii* in the Sydney area. Both are comparatively rare in the area but can be recognised by tiny hollow glands on the undersurface of the leaves at the junction of the mid and lateral veins in *Citronella*, and in the forks of the lateral veins in *Pennantia*.

Pennantia

Pennantia cunninghamii
Brown Beech **Plate 926**

A small tree, with smooth green weak branchlets resembling a climber, distinctly zig-zagged between the leaves. Leaves are broadly elliptic, 7 to 15 cm long, shiny green above and clearly marked with small hollow glands in the forks of the lateral veins. Flowers are white and in dense branched panicles. The fruit is an egg-shaped black berry, about 12 mm long, containing a single seed.

HABITAT Rainforests, on richer soils and moist valleys, especially along watercourses.

DISTRIBUTION Illawarra escarpment and gullies (e.g., Minnamurra Falls, Macquarie Pass, Mt Keira). Also recorded from central tablelands. Extending from about Milton (south coast) to north coast and northern Qld; also southern tablelands and central western slopes.

FLOWERING November–January.

926. Pennantia cunninghamii

Citronella

Citronella moorei
Silky Beech, Churnwood **Plate 927**

A large tree, uncommon in the area with a dense dark green crown and grey bark with prominent fissures. Leaves are alternate, entire, broadly lanceolate, 5 to 10 cm long, glossy green above and paler underneath. Flowers are creamy green in narrow panicles shorter than the leaves. The fruit is an egg-shaped black berry, about 20 mm long, containing a single large seed.

HABITAT Rainforest, on sheltered valleys and slopes.

DISTRIBUTION Coastal, mainly Illawarra (e.g., Cambewarra, Macquarie Pass, Mt Keira); may still occur in rainforest gullies near Wyong. Extending from Clyde River (south coast) to north coast and northern Qld.

FLOWERING May–September.

927. Citronella moorei

Celastraceae

Celastraceae is a tropical and subtropical family of trees, shrubs and climbers. It consists of more than 800 species in 50 genera with about 35 species in 15 genera in Australia. Only 5 species in 4 genera occur in the Sydney area. Although some species are cultivated as ornamentals no Australian species is cultivated.

The family is closely related to the holly family (Aquifoliaceae) which was widespread in Australia during much of the Tertiary period but is now represented by only 3 species in the far north. Members of Celastraceae are distinguished from those in the Aquifoliaceae and Icacinaceae families chiefly by a conspicuous fleshy floral disc which surrounds the superior ovary. The flowers have 3 to 5 sepals and petals and an equal number of stamens alternating with the petals and attached to the outer edge of the disc. The fruit may be dry or drupaceous, and the seeds often have a brightly coloured pulpy covering.

Of the genera in the Sydney area, *Celastrus* (2 species) and *Maytenus* are widespread but uncommon. The other, *Cassine* (formerly *Elaeodendron*) is common in and near coastal rainforests from Batemans Bay to Queensland. The fourth genus, *Apatophyllum*, is represented by *A. constablei*, a small shrub restricted to a remote area near Glen Davis.

Celastrus Staff Vine

A genus of 32 tropical and subtropical species, with 2 occurring in Australia. They are scrambling shrubs or climbers, with conspicuous orange fruits which split into 3 valves.

Celastrus subspicatus
Large-leaf Staff Vine **Plate 928**

A woody climber, with alternate leaves. Leaves are dark ovate to lanceolate, 5 to 14 cm long, 2 to 7 cm wide, leathery, dark green above, pale below, with an acute tip. Margins are entire or minutely toothed. Flowers are yellowish green in a terminal panicle. Fruits are conspicuous as they are loose bunches of orange berries. Each berry is 5 to 10 mm long, splitting into 3 valves. The inner surface of each valve is a dull grey-brown at maturity. Some confusion has existed between this species and *C. australis*. The leaves of *C. subspicatus* are generally larger, usually acute and have fine reticulate veins. Those of *C. australis* have a blunter point and more open veins. The fruits of *C. subspicatus* are larger and in less compact bunches than the fruits of *C. australis*.

HABITAT In or near rainforests.

DISTRIBUTION Uncommon in Sydney area. Recorded from Cambewarra Mountain (in 1920) and Mt Keira (1989). Extending northwards along north coast to northern Qld and New Guinea.

FLOWERING November–December.

928. *Celastrus subspicatus*

Celastrus australis

Staff Vine **Plate 929**

A woody climber, similar in appearance to *S. subspicatus*. Its leaves are smaller, 3 to 9 cm long, 1 to 4 cm wide, tapering to a blunt point. The fruits are smaller (3 to 6.5 mm long), orange or light brown when fresh, splitting into 3 valves.

HABITAT Rainforests or sheltered slopes.

DISTRIBUTION Coast to mountains (e.g., Razorback Mountain, Mt Kembla, Mt Wilson, Mt Tomah). Extending to north coast, southern tablelands, western slopes, Darling Downs (Qld) and East Gippsland (Vic).

FLOWERING October–June.

929. *Celastrus australis*

Maytenus

A large world-wide genus of about 200 species. Only 1 species occurs in the Sydney district. It differs from *Celastris* as it is an erect shrub with a capsule which splits into 2 segments.

Maytenus silvestris **Plate 930**

A medium-sized shrub in this area, but may be a small crooked tree further north. Its leaves are alternate, narrow, 2 to 5 cm long, less than 1 cm wide, with small teeth, slightly curved downwards and paler underneath. Flowers are pale green, in short clusters and fairly inconspicuous. The fruit is an orange egg-shaped berry, which splits into two.

HABITAT Forests, especially near rainforest margins.

DISTRIBUTION Sydney suburbs (e.g., Parramatta River, Liverpool, St Ives, Ingleburn), Hornsby Plateau (e.g., Gosford, Hornsby Valley, Watagan Mountains, Colo Plateau), Razorback Mountain–Douglas Park and Wingello. Extending to north coast, northern tablelands, western slopes and Qld.

FLOWERING October–January.

930. *Maytenus silvestris*

Cassine Olive Plum

Cassine is a genus of about 80 species, mostly African, with 2 species endemic in Australia. Leaves are opposite and the fruit a drupe with 2 segments.

Cassine australis var. *australis*

Red-fruited Olive Plum **Plate 931**

A small tree with broad, elliptical, opposite leaves 4 to 10 cm long and 2 to 5 cm wide. These leaves are toothed or wavy, tapering to a blunt tip (or sometimes rounded), smooth, green and paler underneath. Flowers are pale green, in clusters shorter than the leaves; each flower has 4 small petals (2 to 4 mm) and 4 stamens. The fruit is conspicuous when it is ripe during autumn and winter. It is an orange-red berry, mostly globular but sometimes egg-shaped, 12 to 20 mm long.

HABITAT Rainforests and nearby wet forests.

DISTRIBUTION Coastal (e.g., Minnamurra, Royal National Park, Brisbane Water National Park, Terrigal). Extending from near Batemans Bay to central Qld. A narrow-leafed variety occurs on the northern tablelands, central western slopes northwards to northern Qld.

FLOWERING August–November.

931. *Cassine australis*

Stackhousiaceae

This is a small family of slender perennial herbs, without clear or close association with any other family. It is usually included in the order Celastrales on account of its 5-merous flowers with a superior multicarpellate ovary, but is the only herbaceous family in the order. The family consists of 3 genera with 16 species in Australia (1 extending to Malesia) and one endemic in New Zealand.

Macgregoria, with 1 species scattered across inland Australia, is distinguished from the other 2 by having free petals and stamens of equal length. In the others the petals are fused into a narrow tube for about half their length; then spread into 5 narrow lobes. The stamens, 3 long and 2 short, are enclosed in the corolla tube. *Tripterococcus*, with 2 species in south-western Australia, differs in minor respects from the wider-ranging *Stackhousia*. Four species of *Stackhousia* occur in a variety of habitats in the Sydney area.

Stackhousia

Stackhousia spathulata

Coast Stackhousia **Plate 932**

Dense white flower spikes on short stout stems, combined with its restricted habitat, make this species easy to identify. Leaves are thick, spoon shaped to obovate, 15 to 30 mm long. The floral tube is 6 to 7 mm long, topped by 5 obtuse spreading lobes.

HABITAT Sand dunes near the sea, or in depressions or beside lagoons behind beaches.

DISTRIBUTION Suitable coastal sites. Extending from Vic to southern Qld; also Tas and SA.

FLOWERING August–November.

932. *Stackhousia spathulata*

933. *Stackhousia monogyna*

Stackhousia monogyna

Candles **Plate 933**

The common name of this erect herb (usually about 30 cm tall) is derived from the dense spike of white flowers on top of slender, hairless stems; the spikes are 3 to 10 cm long. Leaves are thin, narrow, 1 to 3 cm long, and scattered along the stem. The floral tube is 5 to 8 mm long, topped by 5 obtuse spreading lobes.

HABITAT Open forest and heath, in sandy soils or soils derived from Narrabeen shales.

DISTRIBUTION Widespread from coast to mountains (e.g., Brisbane Water National Park, Jenolan Caves road). Occurring throughout NSW except far western plains; also Qld, Vic, Tas, SA and WA.

FLOWERING August–November.

934. *Stackhousia nuda*

Stackhousia nuda

Leafless Stackhousia **Plate 934**

Some confusion has been evident in the past over the naming of this erect leafless perennial. It has been frequently called *S. scoparia*, a species restricted to south-western Western Australia, and confused with *S. viminea*, which has well-developed leaves near the base of the stems. The stems of *S. nuda* may be 30 to 80 cm long and branched above. Leaves are reduced to scales. Flowers are in clusters of 1 to 4, on one side of the stem, sessile or shortly stalked, yellow to yellow-green, with a tube 2 to 3.5 mm long and 5 star-like pointed lobes.

HABITAT Low heath or swampy margins and coastal flats, in moist sand.

DISTRIBUTION Coastal zone. Extending from Gippsland (Vic) northwards to south-eastern Qld.

FLOWERING November–February.

935. *Stackhousia viminea*

Stackhousia viminea

Slender Stackhousia **Plate 935**

S. viminea has slender, erect, unbranched stems, 30 to 60 cm long. Leaves are sparse, but well developed, especially near the bottom of the stems, narrow to obovate, 10 to 30 mm long. Flowers are in scattered clusters of 1 to 5, almost sessile, pale green to deep yellow (sometimes reddish), with a tube 2 to 3.5 mm long and 5 star-like obtuse or pointed lobes.

HABITAT Open forest and woodland, on rocky hillsides and open sandy country.

DISTRIBUTION Widespread from coast to mountains (e.g., Heathcote National Park, Bundanoon, Mt Wilson). Occurring throughout coast and tablelands divisions; also Qld, Vic and Tas.

FLOWERING October–December.

Rhamnaceae

Rhamnaceae, the buckthorn family, and Vitaceae, the grape family, constitute the order Rhamnales. Whereas the former are mainly trees and shrubs the latter are mostly vines.

The flowers of Rhamnaceae are small, bisexual and regular with 4 or 5 sepals, 4 or 5 or sometimes no petals, and 4 or 5 stamens attached either to the rim of a disc surrounding the ovary or to the floral tube. The flowers are similar in a number of respects to those of Celastraceae, but differ in having stamens opposite the petals instead of alternating with them. In some genera the petals are hooded and partly cover the anthers.

The 900 species in 58 genera have a cosmopolitan distribution. Dyes are obtained from a few species of *Rhamnus*, and a larger number are used medicinally. The well-known purgative, cascara sagrada, or sacred bark, is obtained from a North American species. A few other genera are grown as ornamentals. Of the 16 genera and 150 species in Australia the largest and best known are *Pomaderris* (43 spp.), *Cryptandra* (40 spp., endemic), *Spyridium* (30 spp., endemic), *Trymalium* (12 spp., endemic) and *Alphitonia* (6 spp.).

Pomaderris

Shrubs, with stems, leaf undersurfaces and inflorescences often covered with a mixture of long simple hairs and close-matted stellate hairs. Flowers small and yellowish (sometimes white), with or without minute petals; stamens 5; style divided into 3 at the tip, rarely almost entire. Flowers borne in conspicuous clusters, usually terminal. Species are often difficult to identify and close examination under a lens of the nature of the hairs, the presence or absence of petals and the size and hairiness of the sepals and floral tube is usually required.

Pomaderris phylicifolia var. *ericoides*

Plate 936

This is an uncommon erect shrub to 2 m tall with densely hairy branchlets. Leaves are linear, up to 10 mm long, with edges curled under to the mid-rib, a characteristic feature. Flowers are pale creamy-yellow, without petals, and borne in short clusters from the leaf bases. The floral tube is densely covered with simple and star-like hairs.

An unnamed but closely related species (Plate 937) also occurs within the area (e.g., on the escarpment below Panorama Lookout, Bulli). It has broader lanceolate leaves to 25 mm long; it flowers in September–October.

936. Pomaderris phylicifolia var. *ericoides*

West of Berrima, there is another shrub with some relationship to *P. phylicifolia* (Plate 938). It has leaves 8 to 15 mm long, rough above with stellate hairs, grey below, and with margins curled under. Its stems are covered with close brown hairs and the floral tube and sepals are densely hairy. It flowers in October–November.

HABITAT Forest and wooded slopes, on sandy soils.

DISTRIBUTION Mt Irvine–Mt Wilson area. Extending to the southern tablelands, Vic and Tas.

FLOWERING October–November.

937. Pomaderris sp. (Bulli)

938. Pomaderris sp. (West Berrima)

Pomaderris aspera

Hazel Pomaderris **Plate 939**

A tall shrub, 2 to 3 m tall, common in sheltered gullies further south into Victoria, but uncommon and reaching its northern limit just south of Sydney. Even when not in flower, it is a conspicuous plant, with large lanceolate leaves, up to 12 cm long, dark green and rough above, whitish beneath with a covering of star-shaped hairs. Leaves are scalloped with irregular rounded teeth and appear wrinkled because of strong venation. Stems and flower stalks are covered with dense soft hairs. Flowers are yellowish and borne in a large terminal branched inflorescence.

HABITAT Moist forest gullies.

DISTRIBUTION South of Sydney (e.g., Macquarie Pass, Upper Kangaroo Valley, Tallong) and Blue Mountains. Extending to southern tablelands, central western slopes and Vic.

FLOWERING October–November.

939. Pomaderris aspera

Pomaderris prunifolia **Plate 940**

An uncommon shrub growing to 2 m tall, it can be identified by its leaves which are blunt lanceolate, 2 to 4 cm long, and wrinkled because of strong venation. The upper surface is covered with short stiff hairs (simple, not star-shaped) and the undersurface is rusty-hairy with brown star-shaped hairs. Young branches are conspicuously rusty-hairy. Flowers are in terminal clusters, without petals; the floral tube is covered with long hairs.

HABITAT Sheltered forest.

DISTRIBUTION Some old records from western suburbs (Parramatta), but probably confined to isolated areas of Blue Mountains (e.g., Mt Caley). Also found on the south coast, southern tablelands, western slopes, Vic and Qld.

FLOWERING October–November.

940. Pomaderris prunifolia

Pomaderris eriocephala **Plate 941**

This medium-sized shrub (to 2 m tall) has distinctive leaves. They are almost round, about 20 mm long, 15 mm wide, with raised corrugations between the lateral veins. The upper surface is velvety-hairy; the undersurface has a dense cover of yellow-grey simple hairs. Flowers are in short dense terminal clusters, without petals, and covered with dense tan hairs.

HABITAT Woodland and forested gullies.

DISTRIBUTION Upper Blue Mountains (e.g., Newnes). A tablelands species, extending from northern tablelands southwards to Vic. Also recorded for higher parts of central coast, north coast and south coast.

FLOWERING September–November.

941. Pomaderris eriocephala

942. Pomaderris ligustrina

Pomaderris ligustrina **Plate 942**

A shrub to 2 m tall, with rusty-hairy young shoots and lanceolate leaves, 25 to 45 mm long. The specific name is derived from the supposed resemblance of its leaves to those of Privet *(Ligustrum)*. The upper leaf surface is without hairs, the lower surface covered in dense, long, silky, rusty-grey hairs. Flowers are creamy, without petals, and borne in loose terminal panicles. The floral tube is covered with long grey hairs.

HABITAT Open and sheltered forest.

DISTRIBUTION Coast to mountains (e.g., upper Georges River, Brisbane Water National Park). Also occurring south coast, north coast, northern tablelands and Vic.

FLOWERING August–September.

Pomaderris brunnea **Plate 943**

This is an uncommon shrub, endemic in the Sydney coastal zone and with a very restricted distribution. It may grow 2 to 3 m tall. Leaves are 15 to 20 mm long, elliptic, hairless above and with a covering of dense grey-brown silky hairs underneath. Flowers are creamy, without petals and borne in dense panicles. The floral tube and sepals are covered with long grey and light brown hairs.

HABITAT Woodland and semi-cleared scrub, on clay and alluvial soils of floodplain and creek lines.

DISTRIBUTION Recorded from Nepean River floodplain at Menangle, in creek lines at Wirrimbirra Sanctuary (Bargo) and along Hawkesbury River.

FLOWERING September–October.

943. Pomaderris brunnea

944. Pomaderris discolor
945. Pomaderris lanigera

Pomaderris discolor **Plate 944**

A shrub to 3 m tall. Its leaves are ovate, about 6 cm long and with edges slightly curved downwards. The upper leaf surface is without hairs, the lower surface whitish-grey with minute hairs and often a few scattered longer hairs. Flowers are yellow, with a style deeply split into 2, and borne in numerous terminal panicles. Petals are mostly absent, but may occasionally be present.

This species is easily confused with *P. intermedia* and *P. elliptica*; all 3 species were once lumped together under *P. elliptica*. However, *P. discolor* is the only one with a deeply cleft style and slightly recurved leaf margins. Its petals are usually absent, or if present, not ear-shaped at the base as are the petals of the other two.

HABITAT Forest.

DISTRIBUTION Coast to mountains. Also occurring north coast, south coast and Vic.

FLOWERING August–September.

Pomaderris lanigera

Woolly Pomaderris **Plate 945**

Although other *Pomaderris* have leaves with both surfaces softly hairy, *P. lanigera* is the most widespread and common of that group. It is a shrub to 2 m tall. The specific name 'lanigera' means 'woolly', a reference to the simple velvety rusty hairs which cover stems, inflorescence and lower leaves. The upper leaf surface is slightly rough to touch, the lower surface paler, rusty-hairy (sometimes totally grey) and soft. Leaves are elliptic, to 10 cm long. The numerous yellow flowers have tiny petals and are borne in showy terminal panicles. The floral tube is covered with dense hairs.

HABITAT Forest.

DISTRIBUTION Coast to mountains (e.g., Menai, Bundanoon). Extending all along coast and tablelands from Qld to Vic; also north-western slopes.

FLOWERING August–September.

946. *Pomaderris velutina*

Pomaderris velutina

Velvety Pomaderris **Plate 946**

The specific name of this erect shrub refers to the velvety feel of the upper surface of the leaves. The undersurface is covered with dense grey woolly hairs. Leaves are about 3 cm long, elliptic and often notched at the apex. Flowers are pale, with petals, and borne in loose terminal panicles. The floral tube is covered with dense long grey hairs.

HABITAT Forest.

DISTRIBUTION Uncommon; restricted to Water Board Catchment area of Nattai River and Burragorang Valley. Extending to south coast, southern tablelands and Vic.

FLOWERING September–October.

947. *Pomaderris intermedia*

Pomaderris intermedia **Plate 947**

This is a shrub to 2 m tall with close tan woolly hairs on its young branches. Leaves are broadly lanceolate, 5 to 7 cm long. The upper surface is quite hairless; the undersurface has close whitish-grey hairs, not silky, with a few scattered longer simple hairs, especially on the midrib. Flowers are yellow, with petals, and borne in large branched terminal panicles. Buds and base of floral tube bear simple hairs.

This species is easily confused with *P. discolor* and *P. elliptica. P. discolor* has a deeply cleft style and petals are usually absent. *P. elliptica* has no simple hairs on the floral tube or on the undersurface of the leaves.

HABITAT Forest.

DISTRIBUTION Coast to mountains (e.g., Girrakool, Heathcote Creek). Also occurring south coast, southern tablelands, northern tablelands, central western slopes and Vic.

FLOWERING September–October.

948. *Pomaderris elliptica*

Pomaderris elliptica **Plate 948**

This is a shrub similar in appearance to *P. intermedia.* Its upper leaf surface is hairless and the undersurface is grey-white with close short hairs, lacking the longer hairs of *P. intermedia.* Flowers are yellow, with petals, and borne in large branched terminal panicles. The floral tube is also hairless.

HABITAT Forest.

DISTRIBUTION Coast to mountains (e.g., Royal and Brisbane Water National Parks, Newnes Tunnel). Extending to north coast and south coast.

FLOWERING September–October.

Pomaderris ferruginea

Rusty Pomaderris **Plate 949**

This is a tall shrub (to 2.5 m) with a distinctly rusty appearance. Young branches and the undersurface of the leaves are covered with velvety loose rusty-silky hairs. The upper leaf surface is hairless. Leaves are lanceolate, up to 10 cm long. The floral tube is also densely hairy. Flowers are with petals and are whitish, a characteristic feature distinguishing this species from similar shrubs like *P. intermedia* and *P. elliptica.* Flowers are borne in large terminal panicles up to 10 cm across.

HABITAT Open forest, usually in sandstone soils.

DISTRIBUTION Coast to mountains (e.g., Brisbane Water National Park, Woronora River). Extending along the coastal zone from Qld to Vic.

FLOWERING September–October.

949. *Pomaderris ferruginea*

Pomaderris andromedifolia **Plate 950**

This is a small shrub (to 1.5 m) with close tan woolly hairs on its young branches. Leaves are lanceolate, smaller than those of *P. ferruginea* (mostly 15 to 35 mm long), dark green and smooth above, the undersurface yellow-white with close silky hairs or somewhat rusty. A characteristic feature is the presence of a distinct network of lateral veins on the leaves. Flowers are yellow, with petals, and are borne in short dense terminal panicles up to 4 cm across.

HABITAT Forest, especially in gullies or near watercourses.

DISTRIBUTION Coast south of Sydney (e.g., Menai, Heathcote Creek) and Blue Mountains (e.g., Mt Banks). Extending all along coast and tablelands of NSW, west to Cocoparra Range and north to southern Qld.

FLOWERING September–October.

950. *Pomaderris andromedifolia*

951. *Pomaderris ledifolia*

Pomaderris ledifolia Plate 951

A small shrub (to 1.5 m), similar to *P. andromedifolia*, but lacking the distinct lateral veins on the undersurface of the leaves. The close hairs on the young branches are white, not rusty-tan. Leaves are lanceolate, 5 to 15 mm long, hairless above and close silky-white underneath. Flowers are yellow, with petals, and are borne in short dense panicles 1 to 2 cm across. The floral tube and sepals are covered with white hairs.

Habitat Forest.

Distribution Upper Blue Mountains (e.g., Kings Tableland) and southern highlands (e.g., Mittagong, West Berrima); also recorded from Appin. Extending to southern tablelands and Vic.

Flowering September–October.

952. *Cryptandra amara*

Cryptandra

Small shrubs with alternate leaves. Flowers bell-shaped or cylindrical, with 5 petals at the top of the floral tube and 5 persistent sepals. Fruit is a capsule which is enclosed in the persistent floral tube.

Cryptandra amara var. *amara* Plate 952

This is a low shrub with rigid branches and small but abundant white or pink flowers. Branches are intertwined and frequently end in a spine. Leaves are variable, obovate to linear, 2 to 4 mm long, often with edges curved under and appearing cylindrical. Flowers are scattered along short side branches. The floral tube is about 3 mm long, covered with white silky hairs, or almost hairless, surrounded at the very base by blunt, overlapping brown bracts.

Two other varieties occur in the area.

Var. *floribunda* has flowers in short spikes, 10 to 15 mm long, borne at the ends of the branchlets and is without spines. Leaves are alternate, 2 to 4 mm long, less than 1 mm wide and appearing cylindrical. It occurs in heaths and dry forests from the coast to the western plains.

Var. *longiflora* has larger flowers (floral tube 4 to 6 mm long), branches which are often spinescent and flat, oblong to oblanceolate leaves, 4 to 6 mm long. It occurs in woodlands from the coast to the western plains.

Habitat Heath and dry forest communities.

Distribution Widespread from coast to mountains. Extending to south coast; also SA.

Flowering June–August.

Cryptandra propinqua

Silky Cryptandra Plate 953

An open shrub with many small branches. Its leaves are linear, 3 to 5 mm long, clustered, with edges curved down and appearing cylindrical. Flowers are white, bell-shaped, 4 to 7 mm long, and clustered at the ends of the branches. A characteristic feature of the flowers is that the floral tube is silky-hairy and almost covered by conspicuous overlapping brown bracts.

Habitat Dry rocky forest and woodland.

Distribution Coast to mountains, but uncommon. Occurring throughout coast and tablelands of NSW, extending to north-western slopes and south-western plains; also Qld, Vic and WA.

Flowering July–September, but also observed flowering in April.

953. *Cryptandra propinqua*

954. *Cryptandra ericoides*

Cryptandra ericoides Plate 954

A small shrub, with linear leaves, 4 to 8 mm long, curled under so that they are almost terete, and scattered along the twiggy stems in small clusters. Flowers are white, borne in dense terminal heads. The leaves and terminal flowers help distinguish this species from others in the genus, but close examination of the flower is necessary for positive identification. The floral tube is narrow, bell-shaped, about 4 mm long, covered with silky hairs and surrounded by pointed, overlapping, fringed bracts, about half as long as the floral tube.

Habitat Heath and scrubland, on sandy soils.

Distribution Coastal (e.g., Brisbane Water and Royal National Parks). Extending to the south coast.

Flowering March–April.

955. *Cryptandra spinescens*

Cryptandra spinescens

Spiny Cryptandra **Plate 955**

A spreading shrub with many small spiny-tipped side branches, its leaves are small (3 to 4 mm long) and obovate. The small white flowers are crowded together on the side branches. The floral tube is 3 to 4 mm long, hairless at the base and silky-hairy above. Blunt brown bracts surround the base of the flowers.

HABITAT Open forest, mainly on soils derived from Wianamatta shales.

DISTRIBUTION Coast to mountains, but uncommon. Recordings include Douglas Park and Howes Valley. Extending to central western slopes.

FLOWERING June–October.

956. Alphitonia excelsa

Alphitonia

Alphitonia excelsa

Red Ash **Plate 956**

A medium-sized tree (6 to 10 m tall), with conspicuous rusty new branchlets and leaves with a silvery-white undersurface. The contrast between the dark green glossy top and whitish undersurface of the leaves is very noticeable, especially when the foliage is being blown by the wind. Leaves are alternate, broad-elliptic, 8 to 15 cm long, with rusty leaf stalks. Flowers are numerous, small, creamy, in branched clusters near the ends of the branchlets. Fruit is a round, black berry, about 10 mm across, which sits on the ring-like scar of the calyx rim.

HABITAT Rainforest margins and protected drier gullies.

DISTRIBUTION Coast and lower Blue Mountains (e.g., Jamberoo, Strickland State Forest, Erskine Creek, Kurrajong). Extending from Tilba Tilba northwards to Qld, NT and WA; also north-western slopes (Mt Kaputar).

FLOWERING December–March.

957. Emmenosperma alphitonioides

Emmenosperma

Emmenosperma alphitonioides

Yellow Ash, Bonewood **Plate 957**

Yellow Ash is a rainforest tree, uncommon in the area, but the occasional tree is conspicuous in autumn to winter when it bears abundant orange-yellow berries. Leaves are opposite, elliptic, 4 to 8 cm long, dark green, with short (4 to 10 mm) smooth leaf stalks. Flowers are numerous, small, yellow-white, in branched clusters at the ends of the branchlets, mostly shorter than the leaves. Fruit is a small round berry, 4 to 6 mm across, splitting into 2 parts.

HABITAT Rainforest.

DISTRIBUTION Mainly coastal (e.g., Upper Hacking River, Illawarra escarpment, Minnamurra Falls), but also recorded from Kurrajong, Colo River and Gap Creek Flora Reserve. Extending from Nowra to north coast and northern Qld.

FLOWERING September–October.

Vitaceae

The Vitaceae is a tropical and subtropical group of woody or herbaceous climbers related to the Rhamnaceae. The flowers are small, regular and bisexual with a small 4 to 5 lobed calyx, 4 to 5 petals and 4 to 5 stamens opposite the petals and inserted on the rim of a thick disc. The leaves are arranged alternately on the stems, sometimes with a climbing tendril or an inflorescence attached opposite the leaf stalk. The largest genera are *Cissus* with 350 species (10 in Australia), *Tetrastigma* with 90 species (one in Australia), *Vitis* with 50 species (none in Australia) and *Cayratia* with 45 species (8 in Australia). Six native species occur in the Sydney area.

The economic and sociological importance of this family rests almost entirely with 1 species, *Vitis vinifera*, the European grape, which was known to stone age man, and of which various varieties have been cultivated since the earliest times. Winemaking from grapes began in Egypt and Mesopotamia about 4000 BC, and from there spread to Rome and Greece, and later to France and Germany. Winemaking was encouraged by the Church because of the religious significance of wine. The census ordered by William the Conqueror in 1085, and recorded in the Domesday Book, showed a number of vineyards and a flourishing wine industry in England in areas much too cool to ripen grapes today. The industry has been periodically afflicted by disease and parasites and it is usual today for vines to be grafted to stocks of American species having roots that are resistant to the Phylloxera aphis.

Cissus Native Grape, Water Vine

A genus of climbers, with 4 species in the Sydney area, 2 common and 2 uncommon.

Cissus antarctica

Water Vine **Plate 958**

A common tall climber, with rusty-coloured matted woolly hairs on young shoots and leaf-stalks. Its leaves are simple and alternate, distinguishing it from the other 3 species of *Cissus* in the area which have compound leaves. Leaves are ovate, 7 to 10 cm long, mostly prominently toothed, but some almost entire. Glands are obvious in the axils of some of the main undersurface veins. Tendrils or inflorescences are produced opposite the leaves. Flowers are small and borne in short dense clusters. The fruit is a round black berry, about 10 mm across.

HABITAT Rainforest.

DISTRIBUTION Coastal zone. Extending from south coast to north coast and Qld; also lower parts of northern tablelands.

FLOWERING December–January.

958. Cissus antarctica

Cissus hypoglauca

Five-leaf Water Vine **Plate 959**

A common tall climber easily identified by its clusters of blue-black berries and by its 5 digitate leaflets. Leaflets are 5 to 10 cm long, entire or toothed towards the apex and pale grey to ash-grey below; they are rounded at the base and borne on stalks about 25 mm long (a feature which separates this species from *C. sterculiifolia* which has leaflets narrowing at the base to very short stalks). Tendrils or inflorescences are produced in short dense clusters. Fruit is 10 to 20 mm across.

HABITAT Rainforest and sheltered gullies.

DISTRIBUTION Widespread from coast to mountains. Also occurring south coast, north coast, northern tablelands, central western slopes, Vic and Qld.

FLOWERING November–December.

959. Cissus hypoglauca

960. Cissus sterculiifolia

Cissus sterculiifolia

Long-leaf Water Vine **Plate 960**

An uncommon to rare climber, somewhat similar to the more common *C. hypoglauca*. Its leaves are compound, consisting of 5 digitate leaflets. Each leaflet is 6 to 16 cm long, entire (or sometimes slightly toothed), green on both surfaces, with glands present in the axils of the main veins. Leaflets narrow at the base into a short stalk. The fruit is an ovoid berry.

HABITAT Rainforest.

DISTRIBUTION Scattered along coast (e.g., Palm Jungle, Minnamurra Falls). Extending along north coast to Qld.

FLOWERING January–February.

961. Cissus opaca

Cissus opaca

Small-leaf Water Vine **Plate 961**

A small weak climber, usually found twining over low shrubs, rare in the Sydney district. Its leaves are compound, consisting of 5 digitate leaflets. Each leaflet is 1 to 6 cm long, narrow (5 to 15 mm wide), toothed, sessile or almost so, and paler on the undersurface. Some forms have entire margins and very short leaf stalks. Flowers are borne in short clusters opposite the leaves. The fruit is a dark blue berry, about 12 mm across.

HABITAT Dry rainforest.

DISTRIBUTION Razorback Mountain (between Camden and Picton). Also occurring north coast, western slopes and Qld.

FLOWERING January–February.

Cayratia

Cayratia clematidea

Slender Grape **Plate 962**

A weak climber, often with stems a few metres long, which twine over low shrubs in more open sunny areas of rainforest. Leaves are distinctive. They are

962. Cayratia clematidea

thin, soft, light green, coarsely toothed and compound. Each leaf is divided into 3 leaflets, but the 2 side leaflets are further divided into 2, giving the impression of 5 leaflets. Flower clusters are dense, upright and on long stalks. The fruit is a depressed berry, about 6 mm across.

HABITAT In and near rainforest and gullies, usually in more open situations.

DISTRIBUTION Coastal zone (e.g., Bouddi National Park, Palm Jungle, Razorback Mountain). Extending from south coast to north coast and Qld; also western slopes.

FLOWERING January–February.

Oleaceae

The relationships of this family are uncertain and it is placed in different orders in different classifications. The family consists of about 30 genera of trees, shrubs and woody climbers widely distributed in both tropical and temperate regions. It is represented in Australia by 5 genera with about 30 native species. The 4 species which occur in the Sydney area are from the genus *Notelaea*. They are known as native olives because of the resemblance of their fruits to those of the European olive, *Olea europaea*, also from this family.

Other well-known members of this family are jasmine (*Jasminum* with more than 200 species, 9 native to northern Australia), lilac *(Syringia)*, ash *(Fraxinus)*, privet *(Ligustrum)* and *Forsythia.*

Notelaea Native Olive, Mock Olive

Shrubs or small trees with opposite leaves. Flowers small, yellowish, with 4 free petals joined at the base by the stamens. Fruit an olive-like drupe.

963. A. Notelaea longifolia (flowers)
B. Notelaea longifolia (fruit)

Notelaea longifolia

Large Mock Olive **Plate 963**

A tall shrub or small tree, common on the margins of rainforest. Its leaves are opposite, entire, narrow lanceolate and tapering to a pointed tip and gradually into a leaf stalk. Leaves are very variable in their hairiness, ranging from velvety to quite smooth. Three forms have been distinguished. Forma *longifolia* has velvety new growth, young stems and flower stalks. The leaves often become hairless and leathery with age. In forma *intermedia*, these features are shortly downy or almost smooth. Forma *glabra* has smooth leaves, stems and flower stalks. Leaf size is also variable, from 5 to 15 cm long and 1 to 3 cm wide. Leaf veins are prominent above and fainter beneath; the net veins are irregularly spaced and of varying thickness, not close and evenly reticulated as are those of *N. venosa*. Flowers are small, yellow and borne on short racemes from the leaf axils. Fruit is a globular bluish-black drupe.

HABITAT Rainforest margins and nearby sheltered eucalypt forest.

DISTRIBUTION Forma *longifolia* occurs along the coast from the south coast to about Gloucester (north coast), extending to central and southern tablelands. Forma *intermedia* occurs only at the northern limit of the Sydney district (Watagan State Forest), extending to far north coast and central western slopes. Forma *glabra* occurs along the north coast and in Qld.

FLOWERING April–July.

964. Notelaea venosa

Notelaea ovata

This is a shrub or small tree with hairless or softly hairy branches. Leaves are entire, opposite, leathery, mostly ovate, 4 to 10 cm long, 2 to 5 cm wide (about twice as long as wide), hairless or softly downy, paler underneath, with margins which are often wavy. Small veins are prominent above and below. The leaf base may be obtuse, heart-shaped or taper into a short stalk. Flowers are small and borne in short clusters from the leaf axils. Fruit is a round or ovoid purple-black drupe, 8 to 12 mm long. This species is often difficult to separate from the more common *N. longifolia* and *N. venosa* as leaf characteristics tend to intergrade.

HABITAT Sheltered eucalypt forest and rainforest margins.

DISTRIBUTION Coastal zone. Extending from south coast (Jervis Bay) to north coast, central western slopes and Qld.

FLOWERING January–March.

965. Notelaea neglecta

Notelaea venosa

Smooth Mock Olive **Plate 964**

A tall shrub or small tree, leaves are opposite, 6 to 14 cm long, 2 to 5 cm wide, leathery, thick, pale beneath, hairless or softly hairy when young and becoming hairless with age. They may be broad or narrow, entire or with wavy margins, and have a rounded base or taper to a short thick stalk. Veins are distinct, with the smaller net veins fine and evenly reticulated and the main lateral veins slightly raised below. Flowers are small, yellow and borne in short racemes from the leaf axils. Fruit is an ovoid-ellipsoid, dark purple drupe, 10 to 15 mm long.

HABITAT Sheltered eucalypt forest and rainforest margins.

DISTRIBUTION Widespread from coast to mountains. Extending from East Gippsland (Vic) along coast and tablelands, and central western slopes to Qld.

FLOWERING November–January.

Notelaea neglecta — Plate 965

A small tree, leaves are entire, opposite, narrow lanceolate, up to 10 cm long, 5 to 7 mm wide, and acute at the apex and the base. The smaller veins of the leaves are indistinct above and below with only the primary veins clearly visible. The absence of these smaller veins is a characteristic feature of this species, separating it from the other *Notelaea* in the area. Leaves are softly hairy when young, but more or less hairless when mature. Flowers are small (about 2 mm long) and borne in short clusters. The olive-like fruit is about 5 mm across.

HABITAT Open forest, on sandy soils.

DISTRIBUTION Higher Blue Mountains, Capertee Valley and western part of the southern highlands (e.g., West Berrima). Endemic in the central tablelands.

FLOWERING September–October.

966. Sambucus australasica

967. Logania albiflora

Caprifoliaceae

The only native Australian representatives of this fairly small family are 2 species of *Sambucus* which are widespread in the eastern states. *Sambucus* is atypical of the family and is sometimes given a family of its own, the Sambucaceae. The genus consists of small trees and shrubs with pinnate leaves, unlike the rest of the family which have simple leaves. Also, the tubular flowers of *Sambucus* have stamens with anther openings facing outwards, away from the style, instead of towards it like the other members of the family.

Sambucus

Sambucus australasica

Yellow Elderberry — **Plate 966**

This is a tall shrub from sheltered forests and rainforest margins. Its leaves are compound, usually with 5 broad leaflets on a stalk. Leaflets are soft, lanceolate, up to 10 cm long, entire or slightly toothed, and pointed at the tip; the lowest pair of leaflets is distant from the stem. Flowers are pale yellow (or sometimes creamy) in clusters which are borne above the leaves, often hang on long (8 to 10 cm) thin stalks, and develop into yellow fleshy berries.

Less common is *S. gaudichaudiana* which has herbaceous spreading branches; the lowest pair of leaflets close to the stem resemble leafy stipules. Its flowers and berries are white.

HABITAT Rainforest margins and nearby sheltered forest.

DISTRIBUTION Widespread from coast to mountains, but not common. Also occurring north coast, south coast, southern tablelands, central western slopes, Qld and Vic.

FLOWERING December–February.

Loganiaceae

The Loganiaceae is a heterogeneous family widely distributed in tropical, subtropical and temperate regions. Although the family contains about 30 genera, most of the Australian species belong to *Mitrasacme* (40 species, 30 endemic) and *Logania* (30 endemic species). The former genus consists of herbs, the latter mainly of shrubs. Both have small white tubular flowers with 4 or 5 lobes. The stamens are attached to the inside of the tube and alternate with the corolla lobes.

Members of the family have opposite leaves with the leaf bases joined by raised lines on the stem representing rudimentary or reduced stipules. Usually the stem is slightly expanded below the leaf insertions.

Some extra-Australian genera such as *Buddelia* and *Fagraea* are much more showy than their Australian counterparts and are widely cultivated. Poisonous alkaloids and glycosides occur in a number of genera, the best known being strychnine and curare from species of *Strychnos*.

Logania

Shrubs with opposite leaves, the bases connected by a raised line or ridge of tissue. Flowers small and white.

Logania albiflora — Plate 967

An attractive erect shrub up to 2 m tall. In late winter and early spring its numerous white flowers and sweet scent demand attention. Its leaves are opposite, narrow lanceolate, 2 to 5 cm long, olive green above and pale on the undersurface; the midvein is very prominent. Flowers are small, tubular and are in branched clusters shorter than the leaves. These clusters are borne from the leaf axils all along the upper stems.

HABITAT Forest.

DISTRIBUTION Widespread from coast to mountains. Recordings include Munmorah State Recreation Area, Brisbane Water National Park, Pennant Hills Park, Bundanoon and Wombeyan Caves. Occurring throughout coast and tablelands divisions; also western slopes, Qld and Vic.

FLOWERING August–October.

Logania pusilla — Plate 968

L. pusilla is easily overlooked as it is a low plant, erect or spreading, seldom more than 10 cm tall. Its preference for soils derived from Wianamatta shales means that it is absent from the coastal sandstone areas. Its leaves are opposite, elliptic, 5 to 10 mm long and rather blunt at the tip. Flowers are bell-shaped, with a tube less than 6 mm long, white, solitary and borne in the axils of the leaves. The specific name 'pusilla' is a reference to the insignificant nature of the plant: 'pusillus'—petty, unimportant.

HABITAT Open forest, usually in shady situations on shales and clayey soils.

DISTRIBUTION Cumberland Plain (e.g., Castlereagh State Forest) and Hornsby Plateau. Also occurring north coast, southern tablelands, Qld and Vic.

FLOWERING September–October.

968. Logania pusilla

Mitrasacme

Small herbs, with opposite leaves without a raised connecting ridge (as in *Logania*). Flowers white, bell-shaped with 4 spreading petals and 4 stamens. Five species occur in the area, but only 1 is widespread and common. Three species *(M. serpyllifolia, M. paludosa* and *M. pilosa)* are prostrate plants with creeping stems found in wet soils or on wet rocks. *M. alsinoides* is a very small (6 to 10 cm tall), rare annual found in wet soils.

969. *Mitrasacme polymorpha*

Mitrasacme polymorpha

Mitre Weed **Plate 969**

Mitre Weed is a common, small, branched herb, seldom more than 25 cm tall. The specific name is a reference to its variability ('poly-morpha'—many forms). Sometimes stems and leaves are covered with greyish hairs, sometimes it is hairless or rough to touch. Leaf shape is also variable from linear to elliptical and 4 to 10 mm long. Leaves are opposite. White, 4-lobed, bell-shaped flowers are borne on long, thin stalks, often in an umbel of 3 to 5 flowers; the flower tube is bearded with white hairs.

HABITAT Heath and woodland, on both dry open sandy soils and on wet swampy ground.

DISTRIBUTION Widespread from coast to mountains. Also occurring north coast, south coast, southern tablelands, Qld, NT and Vic.

FLOWERING September–December.

Apocynaceae

The Apocynaceae is a large tropical family of trees, shrubs and climbers with simple opposite leaves and a milky or clear sap. The flowers are bisexual and regular with 5 sepals and 5 petals which are united into a tube in the lower part and spreading in the upper. Frequently the inward-facing anthers of the 5 stamens are joined. The superior ovary consists of 2 usually free carpels united by a common style with a club-shaped stigma. The fruits are berries or paired follicles containing seeds bearing a tuft of long silky hairs. Many of the members contain poisonous glycosides or alkaloids.

The family is represented in the Sydney area by *Parsonsia*, a common climber in rainforest and moist open forest, and by *Melodinus*, another rainforest climber, but uncommon in the area. Unlike some members of the family which have large colourful fragrant flowers these 2 have yellowish insignificant flowers. The 3 local species of *Parsonsia* have clear yellowish or colourless sap which, in the absence of flowers and fruit, provides a convenient distinction between this genus and similar climbers belonging to the closely related family Asclepiadaceae, which have milky sap in local species.

Parsonsia

Woody climbers with 5-lobed tubular flowers and 5 exserted stamens. Fruit long and splitting into 2 sections to release seeds covered with silky hairs.

Parsonsia straminea

Common Silkpod **Plate 970**

Silkpod is one of the most common climbers in local rainforest. Adult plants have woody stems which twine over branches and through the canopy for 20 m or more. Adult leaves are opposite, thick, glossy above, paler below, with a fine network of veins. Leaves are variable in shape and size but are mostly broadly lanceolate, 7 to 12 cm long and up to 5 cm wide. A watery yellowish sap is exuded when stems or leaves are broken. Juvenile plants are often found climbing up tree trunks with the aid of a special root structure. Juvenile leaves are smaller (1 to 4 cm long), with cordate bases and are bright purple below. Flowers are yellowish, 5-petalled and bearded inside at the base. Stamens are united into a cone. The fruit is conspicuous as it is pendulous, and up to 20 cm long; it splits into 2 follicles to release numerous seeds, each covered with long silky hairs.

HABITAT Rainforest, gullies and sheltered eucaylpt forest.

DISTRIBUTION Widespread from coast to mountains. Also occurring south coast, north coast, central western slopes and Qld.

FLOWERING November–January, but some flowers at other times.

970. *Parsonsia straminea*

Parsonsia brownii

Mountain Silkpod **Plate 971**

P. brownii is a tall climber. Adult leaves are opposite, thick, glossy above, paler below, 4 to 10 cm long, 1 to 4 cm wide, lanceolate, tapering to a tip at both ends; 9 to 15 pairs of main lateral veins are prominent and make an angle of 30° to 45° with the midvein. A clear sap is exuded from broken leaves and stems. Young stems are often covered with downy hairs. Flowers are minutely hairy and have white hairs inside near the base of the tube; they are borne in

971. *Parsonsia brownii*

loose clusters. The inflorescence is usually rusty-hairy, but is occasionally hairless. The fruit is 5 to 8 cm long.

Another *Parsonsia* (formerly *P. leichhardtii*) occurs in gullies of the Blue Mountains. It has rounded leaf bases, leaves which are green tinged with purple, only 3 to 8 pairs of main lateral veins more widely spaced than those of *P. brownii* and a hairless inflorescence.

HABITAT Mountain rainforest.

DISTRIBUTION Blue Mountains, with a record from Saddleback Mountain (Kiama). Extending from northern tablelands to southern tablelands; also north coast, south coast, Vic and Tas.

FLOWERING September–December.

Melodinus

972. *Melodinus australis*

Melodinus australis

Southern Melodinus **Plate 972**

A common but rather insignificant woody climber reaching its southern limit in the coastal rainforests of the Illawarra district. Its leaves are opposite, hairless, oblong to elliptic, 5 to 10 cm long and almost stalkless. A milky sap is exuded when stems or leaves are broken. Flowers are yellow and small, with a floral tube about 4 mm long and 5 spreading lobes. A number of flowers are borne on a stalk from the leaf axil. Stamens are enclosed in the floral tube. The fruit is an ovoid, red-orange berry, 4 to 5 cm long.

HABITAT Rainforest.

DISTRIBUTION Scattered along coast from about Kiama northwards. Extending along north coast to Qld.

FLOWERING December–April.

Asclepiadaceae

The Asclepiadaceae is closely related to the Apocynaceae but has a more specialised floral structure necessary for the elaborate pollination mechanisms that characterise the family. Although the family contains trees and shrubs there is a high proportion of woody and herbaceous climbers with twining twems. Both the introduced and native species in the Sydney area exude milky sap from the petioles and stems when cut. In the absence of flowers this family can be distinguished by the tiny raised glands on the leaves at the base of the lamina (leaf blade) on the midvein.

The flowers of Asclepiadaceae comprise a 5-lobed calyx, a 5-lobed tubular corolla and 5 stamens which are fused to each other and to the top of the style to form a central column called a gynostegium. In addition, the stamens or corolla, or both, bear appendages which form a single or double corona having a variety of forms which apparently serve to direct insects into certain positions to receive pollen. The anthers contain 2 locules or compartments and the pollen in each locule sticks together in a waxy ball called a pollinium. Each pollinium is attached by a thin threadlike arm called a translator to an adjacent pollinium. The connecting arm also carries a sticky gland by which the 2 pollinia and the arm are transferred to insect visitors. In a minority of species the pollinia are transferred to insects by means of pollen spoons. The net result of this complexity seems to be that many insect visitations do not result in pollen transfer and fertilisation. The fruit of asclepiads is commonly a pair of follicles with only 1 developing fully. The flattened seeds bear a tuft of long silky hairs to aid dispersal.

Marsdenia Milk Vine

Twiners with milky sap. Leaves opposite with minute glands at the base of the leaf-blade on the midvein. Flowers fragrant and urn shaped.

973. *Marsdenia flavescens*

Marsdenia flavescens

Hairy Milk Vine **Plate 973**

Twining plant of rainforests, its leaves are opposite. Stems and leaf stalks are often covered with soft down, but this may be absent. Leaves are soft, oblong-elliptic, with a short point, about 6 cm long and 3 cm wide. The upper surface is glossy green, the undersurface pale yellowish green and downy. A milky sap is exuded when stems or leaves are broken. Flowers are small, pale yellow and on compound umbels which often exceed the leaves. This clearly distinguishes this species from the other 2 *Marsdenia* in the area which have flowers in short simple umbels. Fruit is a narrow follicle, 4 to 6 cm long, containing seeds covered with silky hairs.

HABITAT Coastal rainforests and gullies of the Blue Mountains.

DISTRIBUTION Coast to mountains. Also occurring along north and south coast.

FLOWERING December–January.

974. *Marsdenia rostrata*

Marsdenia rostrata

Common Milk Vine **Plate 974**

A climber with stems up to 10 m long. Leaves are 4 to 10 cm long, rounded or slightly cordate at the base, ovate, narrowing to a blunt point, dark green above and somewhat paler underneath. A cluster of minute glands is visible at the base of the leaf on the upper midvein. A milky sap is exuded when stems or leaves are broken. Flowers are small, yellowish and borne in simple umbels in the leaf axils. The stigma is long and beaked, a key feature which separates this species from *M. suaveolens*. Fruit is a broad, pointed follicle, about 5 cm long, containing seeds covered with silky hairs.

HABITAT Coastal rainforest and mountain gullies.

DISTRIBUTION Most common along coast (e.g., Macquarie Pass, Palm Jungle), but also found in mountains. Occurring throughout coast and tablelands divisions; also Vic, Qld and Lord Howe Island.

FLOWERING October–January.

Marsdenia suaveolens

Scented Marsdenia **Plate 975**

This species occurs in sheltered forest sites as well as in rainforests. It is usually a slender climber, but in drier conditions it may be a small low shrub.

975. *Marsdenia suaveolens*

Leaves are opposite, lanceolate, shorter and narrower than those of *M. rostrata*, short stalked, dark green above and paler underneath. A milky sap is exuded when stems or leaves are broken. Flowers are white or creamy, and are borne in simple umbels in the leaf axils. The stigma is narrow and covered by anther appendages, not long and beaked as in *M. rostrata*. Fruit is a narrow, pointed follicle, 5 to 10 cm long, containing seeds covered with silky hairs.

HABITAT Rainforest and sheltered siutations in forest and heath.

DISTRIBUTION Widespread from coast to mountains (e.g., Oatley Park, Wentworth Falls). Also occurring along north and south coasts.

FLOWERING November–December.

Tylophora

A small climber with opposite leaves. A cluster of minute glands is located at the base of the midvein. Flowers purplish, 5 to 6 mm across, with 5 spreading lobes.

Tylophora barbata

Bearded Tylophora **Plate 976**

A common small twiner with opposite leaves. It is found twisting over shrubs or trailing along the ground in sheltered forests. Its leaves are ovate to elliptic, 2 to 5 cm long, 1 to 3 cm wide, obtuse or sometimes slightly cordate at the base. A cluster of minute glands is visible at the base of the leaf on the upper midvein. A milky or watery sap is exuded when stems or leaves are broken. Flowers are small, about 5 mm across, but very distinctive. They have 5 purple star-like petals, shortly hairy, a yellow stigma, and a ring of 5 knobs around the staminal column, known as a 'corona'. The generic name is a reference to this corona ('tylo', knobs; 'phora', bearing). Flowers are borne in umbels of 3 to 5 flowers in the leaf axils.

Less common is *Tylophora woollsii* which has flowers in loose clusters (not umbels) and broad-ovate leaves, up to 5 cm wide and cordate at the base. It occurs in open forest on the Cumberland Plain.

HABITAT Rainforest and tall eucalypt forest.

DISTRIBUTION Widespread from coast to mountains. Also occurring north coast, south coast, southern tablelands and Vic.

FLOWERING August–December, but some flowers at other times.

976. Tylophora barbata

977. Villarsia exaltata

Menyanthaceae

This small family of 5 genera of aquatic and marsh plants has an almost worldwide distribution. Two of the genera occur in the northern hemisphere, 2 in the southern, and 1, *Nymphoides*, in both. Of the 3 genera in Australia, *Liparophyllum* is confined to mountain bogs in Tasmania, and the others, *Villarsia* and *Nymphoides*, occur in all states and in the Sydney area. Both have yellow flowers (local species) 15 mm to 35 mm across. Those of *Villarsia* are conspicuously bearded inside the corolla tube and those of *Nymphoides* have fringed corolla lobes.

The family is related to the Gentianaceae, and is sometimes regarded as a subfamily, but differs in having radical and alternate leaves, and petals which do not overlap.

Villarsia

Erect swamp plant with yellow flowers.

Villarsia exaltata

Yellow Marsh Flower **Plate 977**

This is an erect, hairless herb, which even when not in flower is easily identified by its basal clump of glossy-green leaves. These have broad ovate blades up to 12 cm long and 8 cm wide, on long stalks. The yellow flowers are borne on a tall, branching, almost leafless stem, well above the leaves. Each flower has 5 free spreading petals, which are bearded inside at the base. Fruit is a capsule opening through 4 valves at the top.

HABITAT Swampy ground, especially around coastal lagoons.

DISTRIBUTION Chiefly along coast and Cumberland Plain, but also recorded in mountains. Recordings include Kurnell, Maddens Plains and Penrose. Also occurring north coast, south coast, southern tablelands, Qld, Vic and Tas.

FLOWERING November–December.

Nymphoides

Aquatic plant with round floating leaves and yellow flowers with fringed petals. (The white-flowered *N. indica* also occurs in the coastal zone, but is very uncommon.)

Nymphoides geminata

Marshwort **Plate 978**

An aquatic herb with circular floating leaves, similar to small water lily *(Nymphaea)* leaves. These leaves are entire, 3 to 8 cm across and V-shaped at the junction of the leaf stalk and blade. Flowers are yellow, 5-lobed, up to 25 mm across, with distinctly fringed margins and are borne in pairs or clusters on long stems. Stems, leaves and flowers are often found on mud as standing water dries up. The 5 spreading petals are without the fringed ridge down the centre which is a feature of Wavy-leaved Marshwort *(N. crenata)*, a species from the western slopes and western plains.

HABITAT Lagoons, dams and still margins of rivers.

DISTRIBUTION Coast, Cumberland Plain and lower Blue Mountains (e.g., Castlereagh State Forest, Shane Park, Mountain Lagoon). Also occurring southern tablelands, north coast, northern tablelands and Qld.

FLOWERING October–March.

978. Nymphoides geminata

Rubiaceae

Rubiaceae is named after *Rubia*, a genus of largely Mediterranean herbs (not after *Rubus*, which is in Rosaceae). It has a chiefly tropical and subtropical distribution, but also has some temperate and a few Arctic species.

The family is one of the largest of the dicotyledonous families and comprises trees, shrubs and herbs with opposite entire leaves bearing stipules. Often the stipules are relatively large and give the appearance of a whorl of leaves. In some cases the stipules are fused and form a sheath between the bases of the leaf pair, while in others they are small or deciduous. The inflorescence, flowers and fruit show such variation that it is difficult to describe any feature that does not have exceptions. Despite this, the Rubiaceae is a well-defined natural group having affinities with both the Gentianales and the Dipsacales. It is usually either placed in the former or given an order of its own, the Rubiales.

The family contains a number of important, interesting and attractive plants. These include *Coffea* (coffee), *Cinchona* (quinine) and *Cephaelis* (ipecacuanha). *Myrmecodia* and *Hydnophytum*, known as ant plants because their swollen bases contain cavities inhabited by ants, are quaint epiphytes from tropical Asia and Australia. The best known of the ornamentals is *Gardenia*, with more than 100 African and Asian species. The best-known European genera are *Galium* (bedstraw) and *Asperula* (woodruff), both consisting of small herbs and both represented in the Sydney area by native and introduced species.

979. Psychotria loniceroides

Psychotria

Psychotria loniceroides

Hairy Psychotria **Plate 979**

A shrub from rainforest margins, its stems, leaf stalks and leaves are softly downy to rusty-hairy, an easy feature to help identification. Leaves are opposite, soft, ovate to elliptic and 6 to 10 cm long. The stipules between the pairs of leaf stalks fall off before maturity. Flowers are tiny, white and in terminal branched clusters. The fruits are conspicuous as they are yellow fleshly berries about 6 mm across in clusters at the ends of the branches.

HABITAT Rainforest margins and sheltered forested slopes.

DISTRIBUTION Scattered along coastal zone. Recordings include Bola Creek, Gosford district and Ourimbah State Forest. Extending along coast from near Bega (south coast) to north coast and northern Qld. Also recorded from central tablelands and northern tablelands.

FLOWERING January–March.

Coprosma Currant Bush

Shrubs with opposite leaves. Male and female flowers often borne on separate plants. Male flowers with long filaments and 4 to 6 exserted anthers. Female flowers with a style divided to the base into 2 branches. Fruit a currant-like drupe.

Coprosma quadrifida

Prickly Coprosma, Prickly Currant Bush **Plate 980**

Although common in the sheltered forests of Victoria and Tasmania, Prickly Coprosma is less common in the Sydney district. It is a shrub to 3 m tall, with slender branches and numerous short side branches ending in spines. Leaves are opposite, narrow to broad lanceolate, 5 to 12 mm long. Male and female flowers are borne on separate plants. Fruits are red, rounded and fleshy.

980. Coprosma quadrifida

HABITAT Sheltered forests, often on the margins of rainforest.

DISTRIBUTION Illawarra Escarpment (e.g., Bulli, Mt Keira, Mt Kembla) and upper Blue Mountains (e.g., Mt Wilson–Mt Tomah). Scattered throughout coast and tablelands; also Vic and Tas.

FLOWERING September–December.

981. Coprosma hirtella

Coprosma hirtella

Rough Coprosma **Plate 981**

A shrub of mountain forests, Rough Coprosma has opposite, broad-ovate leaves, about 30 mm long, rather thick, dark green and scabrous above, light green underneath. Flowers are clustered together in short heads. Fruits are red to coffee coloured, rounded and fleshy.

HABITAT Forest, especially on rocky slopes.

DISTRIBUTION Higher Blue Mountains (e.g., Blackheath, Hassans Walls). Extending from northern tablelands to southern tablelands, Vic and Tas.

FLOWERING October–January.

Nertera

Prostrate herbs with opposite leaves and minute 4-lobed flowers.

Nertera granadensis **Plate 982**

This is a soft creeping herb which roots along the prostrate stems and forms mats up to 10 cm across. Leaves are opposite, round to ovate, 3 to 8 mm long, with very short stalks. Flowers are small (corolla about 2 mm long), solitary, stalkless, terminal, and cream-coloured. Fruit is an oval drupe.

HABITAT Mossy rocks and wet areas often below cliff-lines.

DISTRIBUTION A tablelands species, recorded from Govetts Leap, Centennial Glen at Blackheath and Newnes States Forest. Occurring from northern tablelands to southern tablelands, Vic and Tas; also NZ and South America.

FLOWERING November–February.

982. *Nertera granadensis*

984. *Pomax umbellata*

986. *Opercularia diphylla*

Morinda

Woody climbers with 4 to 12 flowers in stalkless heads, the heads in pairs on terminal stalks.

Morinda jasminoides

Morinda **Plate 983**

Morinda is a climber with stems 1 to 2 m long, often found scrambling over rocks or through the undergrowth of sheltered eucalypt forests. It is easily identified by either its leaves, flowers or fruit. Leaves are opposite, with stipules between the pairs of leaf-bases, lanceolate, 3 to 7 cm long, with a narrow point and up to 4 conspicuous raised bumps (domatia) at the junction of the midrib and lateral veins. Flowers are tubular, yellow-white and are united at their bases to form simple heads. The heads are on stalks (15 to 25 mm long) in pairs at the ends of the branches. Fruit is a compound berry, usually orange-red.

HABITAT In and near rainforest and sheltered eucalypt forest.

DISTRIBUTION Widespread from coast to mountains. Occurring throughout coast and tablelands; also Qld.

FLOWERING November–December.

983. *Morinda jasminoides*

Pomax

A monotypic genus consisting of a small herb with minute flowers arranged in umbels.

Pomax umbellata

Pomax **Plate 984**

Pomax is one of those common small plants frequently overlooked because it lacks attractive flowers or foliage. It is erect, much branched, seldom more than 20 cm tall. Most plants are hairy, but a few are hairless. Leaves are opposite, lanceolate, up to 15 mm long. Individual flowers are minute, but are arranged in simple terminal umbels, which are the most noticeable feature of the species.

HABITAT Heath and open forest, on sandy soils.

DISTRIBUTION Widespread from coast to mountains. Found throughout most of NSW, and in every mainland State.

FLOWERING September–November.

Opercularia Stinkweed

Sprawling herbs with ridged stems. Leaves opposite with an unpleasant smell when crushed. Stipules form a sheath around the leaf base. Flowers are very small, mauve-coloured and clustered into a globular head. Styles and stamens are long and exserted. Fruit is a compound capsule.

985. *Opercularia aspera*

Opercularia aspera

Common Stinkweed **Plate 985**

A widespread and common small perennial herb. Like other members of the genus, it is characterised by a strong unpleasant foetid smell when crushed. Stems are usually about 30 cm long (but may be much longer) sprawling from a woody root-stock and are usually covered with stiff, sandpapery hairs but may be almost hairless. Leaves are opposite, ovate to lanceolate, 1 to 4 cm long, and may be rough or almost smooth. Flowers are in dense globular heads on stalks. The fruit is a compound capsule with the appearance of a spiky globe. The other members of the genus occur in the area, but are less common.

O. hispida has long soft hairs on its stems and leaves.

O. varia has leaves 8 to 12 mm long and less than 5 mm wide; stems are numerous and the whole plant is usually quite hairless.

O. diphylla (Plate 986), is a rare erect plant to 20 cm tall, with hairless stems and leaves. It has been recorded from Castlereagh State Forest and clay flood plain of Eastern Creek, Doonside.

HABITAT Heath, woodland and open forest, on sandy soils or among rocks.

DISTRIBUTION Widespread from coast to mountains. Especially common on rocky headlands near the sea. Also occurring throughout coast and tablelands divisions, central western slopes, Qld and Vic.

FLOWERING September–November.

Asperula Woodruff

Herbs with 4-angular stems. Leaves appearing whorled because of the presence of leaf-like stipules. Flowers white, with a 4-lobed tubular corolla and exserted anthers. Male and female flowers borne on separate plants.

Asperula scoparia

Prickly Woodruff **Plate 987**

A tablelands species, found scattered through grassy understorey, it grows to 15 cm tall. Stems are 4-angled. Leaves appear to be in whorls, with 6 leaves in each whorl. Leaves are linear, up to 10 mm long, 1 mm wide, and have a long acuminate tip. Lower leaves often point downwards. Flowers are small (2 mm long), white, tubular and 4-lobed; the anthers protrude from the male flowers; a 2-lobed style protrudes from the female flowers. Flowers are borne in terminal clusters.

Asperula gunnii, another Blue Mountains species, has 4 leaves in each whorl. Its leaves are wider (up to 3 mm) than those of *A. scoparia*.

HABITAT Forest, among grassy understorey.

DISTRIBUTION Higher Blue Mountains (e.g., Hampton, Jenolan Caves, Yerranderie). Extending to southern tablelands. Also occurring Vic, Tas and SA.

FLOWERING October–December.

987. *Asperula scoparia*

Asperula conferta

Common Woodruff **Plate 988**

A weak herbaceous plant with numerous thin-ridged straggling stems, up to 30 cm tall. These stems are often tangled. Leaves are narrow, 3 to 9 cm long, about 1 mm wide, pointed but without stipules. Leaves are often strongly reflexed (point downwards). Flowers are white (often with some pink), shortly tubular and terminal.

HABITAT Grassland and woodland, often on shale soils.

DISTRIBUTION Widespread from coast to mountains. Occurring throughout the State and in Qld, Vic, Tas and SA.

FLOWERING September–November.

988. *Asperula conferta*

Galium Bedstraw

Herbs with 4-angular stems. Leaves appearing whorled because of the presence of leaf-like stipules. Flowers small, with a minute floral tube which is scarcely visible and 4 spreading lobes. The apparent absence of the floral tube distinguishes this genus from *Asperula*. Fruit is dry, with 2 segments, often covered with bristles or hooks.

989. *Galium propinquum*

Galium propinquum

Maori Bedstraw **Plate 989**

This is a trailing perennial with stems up to 30 cm long rising from an extensive creeping underground branched rhizome system. This feature helps distinguish *G. propinquum* from other *Galium* in the area which have stems rising from a central rootstock. Leaves and stipules are 4 in each whorl, equal in size, thin, elliptic to ovate, 3 to 12 mm long and 1 to 2.5 mm wide. Flowers are white to cream, each on a short stalk and borne in clusters in the leaf axils. Fruits are about 1 mm across, rough and warty or covered with short curved bristles.

An undescribed species, closely related to *G. propinquum*, occurs between Berambing and Mt Tomah.

G. binifolium is very similar to *G. propinquum* except that in each whorl it has 2 leaves larger than the 2 leaf-like stipules, the fruits are smooth and the stems rise from a rootstock not a branched rhizome.

HABITAT Sheltered forested sites, especially in damp soils, and often on shaded rocks.

DISTRIBUTION Widespread from coast to mountains. Recordings include Mt Tomah and Gap Creek Flora Reserve. Occurring throughout coast and tablelands divisions; also Qld, Vic and NZ.

FLOWERING December–February.

990. *Galium gaudichaudii*

Galium gaudichaudii

Rough Bedstraw **Plate 990**

This is a perennial herb with tufted stems up to 30 cm tall rising from a stout rootstock. The edges of the square stems feel rough when rubbed, a characteristic feature useful for identification. Leaves and stipules are 4 in each whorl, equal in size, linear to narrow lanceolate, 4 to 6 mm long, almost stalkless, rough to the touch, with down-curved margins. Flowers are very small, white or yellow, each on a short stalk and borne in clusters in the leaf axils. Fruits are 1 to 2 mm across, mostly smooth and hairless, but sometimes warty.

G. liratum is much less common. It has leaves and stipules up to 18 mm long and net-veined; fruit is wrinkled with irregular furrows.

HABITAT Open dry forest, especially at the base of trees and rocks.

DISTRIBUTION Widespread from coast to mountains, but most common west of the central Blue Mountains (e.g., Glen Davis, Jenolan Caves). Occurring throughout tablelands, western slopes and western plains; also Vic, Tas, SA and WA.

FLOWERING August–October.

Convolvulaceae

The Convolvulaceae consists mainly of herbaceous and woody climbing plants seemingly well represented in lush tropical and subtropical areas, but in fact occupying a wide range of habitats. The family includes the sweet potato which is arguably the most important crop of tropical regions.

The funnel-shaped flowers of the native species illustrated are typical of the family, although somewhat smaller than those of the introduced species of *Ipomoea* known as Morning Glory. The flowers of this family often remain closed in bad weather; in fine weather they open in the morning and begin to wilt before evening. The close relationship of this family with the Solanaceae is more immediately apparent in some introduced Solanaceous genera such as *Datura* than in the common Australian representatives such as *Solanum*.

Convolvulus Bindweed

Twining plants, with funnel-shaped flowers which have a stigma divided into 2 elongated lobes. Bracteoles are minute and placed well away from the calyx.

Convolvulus erubescens

Australian Bindweed, Pink Bindweed **Plate 991**

Australian Bindweed has prostrate trailing or twining stems up to 80 cm long. Leaves are very variable. They are usually softly hairy, but may be almost hairless, and are on short stalks or almost stalkless. Leaf blades are 10 to 30 mm long and up to 15 mm wide, usually shaped like a spear-head (sagittate), with entire margins or sometimes lobed. Flowers are pink to white, 6 to 20 mm across when fully open

991. *Convolvulus erubescens*

993. *Dichondra repens*

and borne in groups of 1 to 4 from the leaf axils. Sepals are ovate, pointed and softly hairy. Fruit is a capsule with 2 chambers. *C. erubescens* is often mistaken for the introduced *C. arvensis*, a common roadside weed, but that species has hairless leaves and obtuse hairless sepals.

HABITAT Grassland, woodland and open forest, especially in drier shale areas.

DISTRIBUTION Widespread, but uncommon in Sydney district. Found throughout NSW and in all other States.

FLOWERING September–January.

Polymeria Bindweed

Trailing or slightly erect plants, with funnel-shaped flowers which have a stigma divided into 4 to 8 lobes. Bracteoles are minute and placed well away from the calyx. Sepals are unequal, with 2 larger than the other 3.

Polymeria calycina

Polymeria **Plate 992**

A trailing plant with stems to 40 cm long. Leaves are 15 to 30 mm long, 8 to 15 mm wide, with stalks

992. *Polymeria calycina*

up to 25 mm long; they are round-oblong, with entire margins and cordate at the base. Flowers are pink, solitary, trumpet-shaped, 10 to 12 mm long. The stigma is divided into 6 lobes. The sepals are unequal in size and overlapping. Fruit is a capsule, about 5 mm across, with 1 or 2 seeds. *Polymeria* differs from *Convolvulus* in having leaves on longer stalks, solitary flowers, a 6-lobed stigma and unequal sepals.

HABITAT Grassland and woodland, often in damp places.

DISTRIBUTION Coast and lower Blue Mountains. Recordings include Burning Palms, Pymble, Eastern Creek and Castlereagh State Forest. Extending along coastal zone from south coast to Qld.

FLOWERING October–April.

Dichondra

Dichondra repens

Kidney Weed **Plate 993**

An inconspicuous herb, with hairless creeping stems which root at the nodes. Leaves are alternate, on stalks 2 to 6 cm long. Leaf blades are kidney-shaped or orbicular, 8 to 12 mm wide, with entire or angular margins, green on both surfaces, but often with a few hairs on the undersurface when young. Flowers are small, yellowish-white, 5-lobed, with 2 free styles, and are borne singly on stalks shorter than the leaf stalks. Fruit is a capsule with 1 or 2 cells.

HABITAT Grassland and forest, in damp and shaded locations.

DISTRIBUTION Widespread and common from coast to mountains. Found throughout coast and tablelands divisions; also Qld, Vic, Tas, SA and NZ.

FLOWERING September–December.

Wilsonia

Wilsonia backhousei

Wilsonia **Plate 994**

An uncommon small hairless perennial which forms dense mats on the landward side of estuarine salt marshes. Its stems are usually about 15 to 20 cm tall, bearing thick, fleshy, narrow-oblong leaves, 6 to 15 mm long. The flower is erect, white and tubular, 5 to 8 mm long, borne singly near the ends of the stems. Stamens and style project beyond the flower tube.

HABITAT Margins of salt-marshes.

DISTRIBUTION Restricted to a few scattered salt-marshes which have not been damaged by urban development. Recordings include Salt Pan Creek and Mill Creek along Georges River, and the Parramatta River. Also occurring Jervis Bay, Vic, Tas, SA and WA.

FLOWERING October–November.

994. *Wilsonia backhousei*

Cuscuta Dodder

Leafless annual parasitic plants, with yellowish or reddish stems which drape over host shrubs, attaching themselves by special absorbing organs known as haustorium. Often confused with *Cassytha* (see below for differences).

Cuscuta australis

Australian Dodder **Plate 995**

Dodder has thread-like twining leafless stems. Its flowers are small (3 to 4 mm across), bell-shaped, with 5 whitish obtuse lobes, 5 stamens and 2 styles. Each flower is stalkless or on stalks 1 to 2 mm long, and borne in small globular clusters of 2 to 7. The

995. *Cuscuta australis*

fruit is a capsule, sometimes semi-succulent. *Cuscuta* may be confused with *Cassytha*, but that genus has 6 perianth segments in 2 whorls (flowers appear to have 3 petals), 9 stamens and its fruit is a fleshy berry.

HABITAT Parasitic on many herbs (e.g., *Persicaria*).

DISTRIBUTION Coastal zone. Also recorded north coast, north-western slopes, Qld, Vic, SA, WA, Europe, Africa and Asia.

FLOWERING November–January.

Boraginaceae

Although this family includes *Heliotropium* (Heliotrope) and *Myosotis* (Forget-me-not), the best known representative in Australia is *Echium plantagineum* (Paterson's Curse or Salvation Jane), a major weed of agricultural land, but also a valued bee plant. Native members in the Sydney district are uncommon erect or straggling herbs with rough hairs on stems, leaves and inflorescences. Flowers are small and tubular with a 5-lobed blue, white, yellow or pink corolla. The family is closed related to the Ehretiaceae from which it is distinguished by differences in the ovary and fruit.

Cynoglossum Hound's Tongue

Perennial herbs with minute tubular flowers and 5 spreading lobes; scales in the throat of the tube opposite the lobes. The dry fruit has 4 segments covered with short hooked spines. The name is derived from the Greek 'kyon'—a dog, 'glossa'—a tongue, a probable reference to the shape of the leaves in some species.

996. *Cynoglossum latifolium*

Cynoglossum latifolium

Forest Hound's Tongue **Plate 996**

C. latifolium is a straggling plant with stems covered with rough hairs which often lengthen into prickles. Leaves are alternate, ovate, 3 to 4 cm long, dark green and sandpapery above, with 5 distinct veins spreading from the central vein. Flowers are white, with a tube divided almost to the base, solitary or in loose clusters near the ends of the stems.

HABITAT Sheltered moist forest and rainforest margins.

DISTRIBUTION Uncommon in Sydney district. Recorded from Illawarra Escarpment (e.g., Bulli, Macquarie Pass) and Blue Mountains (e.g., Mt Wilson, Mt Tomah). Found throughout the coast and tablelands; also Qld and Vic.

FLOWERING January–May.

Ehretiaceae

The Ehretiaceae consists of a group of about 400 species of largely trees and shrubs in 13 genera which are sometimes separated from the predominantly herbaceous family, Boraginaceae, and sometimes regarded as a subfamily. The group occurs widely in the subtropics and is well represented in Central and South America where a number of species provide valuable timber.

Ehretia

Some classifications place this genus in the family Boraginaceae. Trees or shrubs with alternate leaves and tubular flowers with 5 spreading lobes and 5 stamens alternating with the lobes. Fruit is a succulent berry.

Ehretia acuminata

Koda **Plate 997**

A rainforest tree with hairless alternate, elliptic to lanceolate leaves, up to 15 cm long and 5 cm wide, with toothed margins. Flowers are small, white and borne in large terminal clusters. Fruits are round, succulent and about 5 mm across. New growth and

997. *Ehretia acuminata*

flowers are often covered in a web, constructed by the caterpillar of the moth *Ethmia heliomela*.

HABITAT In or near rainforest, especially in creek valleys.

DISTRIBUTION Scattered throughout Illawarra Escarpment (e.g., Minnamurra Falls, Macquarie Pass, Mt Keira). Also recorded at Kurrajong and Palm Grove near Gosford. Extending from south coast to north coast and Qld.

FLOWERING December.

Solanaceae

This large, diverse cosmopolitan family contains many plants such as potatoes, tomatoes, tobacco, chillis, peppers (capsicum) and paprika that are important to man. The more notorious members such as belladonna, stramonium, mandrake and hyoscyamus contain highly poisonous alkaloids. These and others such as *Duboisia* have been used medicinally and ceremonially since the earliest times. Some, such as *Petunia, Browallia* and *Capsicum*, are grown for their colourful flowers or fruits. It is not surprising therefore that many Solanaceous plants in the more fertile areas of the Australian countryside are plants which have escaped from cultivation and become naturalised. Nonetheless, Australia has about 140 widely distributed native species, mostly from the genus *Solanum*. Two of the 3 species of *Duboisia* are endemic, the third extends to New Caledonia.

Formerly, great importance was attached by the Australian Aboriginals to *Duboisia hopwoodii*, a shrub or small tree of the arid inland known as pituri, the dried leaves and twigs of which were smoked or chewed with the ashes of certain Acacia species for their narcotic effect, or used to poison waterholes for the capture of game. Some species of *Nicotiana* were similarly used as a narcotic. The leaves of *D. myoporoides* contain hyoscyamine and related alkaloids which are used in ophthalmology as a pupil dilator.

The flowers of this family have an essentially bell-shaped (campanulate) corolla, but with considerable variation. Those of *Solanum* generally appear rotate because of their short tube and large rounded lobes. Those of *Duboisia* have a longer tube with short rounded or narrow acute lobes. In *Anthocercis* and *Cyphanthera* the flowers are more funnel-shaped with long, narrow, spreading lobes, while those of *Nicotiana* have a long, narrow tube with short, broadly rounded lobes.

Solanum Kangaroo Apple, Nightshade

Shrubs and herbs with purple or blue-lilac flowers, with a short tube, 5 spreading lobes and contrasting orange stamens. Flowers borne in lateral clusters. Fruit a globular berry.

998. Solanum aviculare

Solanum aviculare

Kangaroo Apple **Plate 998**

An erect shrub, 2 m or more tall, with hairless green or purple stems. Leaves are large, thin, up to 30 cm long, variable in shape, mostly divided with deep pointed lobes, although the upper leaves are often entire. Flowers are 3 to 4 cm across, blue-violet, with spreading acute lobes. Fruit is egg-shaped, 25 mm long, smooth and yellow, turning to orange and red.

HABITAT Sheltered forests and rainforest margins.

DISTRIBUTION Blue Mountains (e.g., Mt Tomah and Mt Wilson) and Gosford area (e.g., Palm Grove). Occurring throughout NSW, except far western plains; also Qld, Vic, New Guinea, Lord Howe Island and NZ.

FLOWERING November–December.

999. Solanum vescum

Solanum vescum

Gunyang **Plate 999**

A hairless shrub to about 2 m tall. Leaves are lanceolate, to 15 cm long, usually entire (occasionally with a few long lobes), stalkless, and with the leaf blades extending down the stem forming raised lines. Flowers are 3 to 4 cm across, violet, with spreading lobes. Fruit is globular, 20 to 25 mm across, smooth and green to ivory.

HABITAT Sheltered forest, old coastal dunes and along streams.

DISTRIBUTION Widespread but uncommon. Recordings include Mittagong and Jenolan Caves. Extending throughout NSW, except drier western slopes and plains; also Qld, Vic, Tas and SA.

FLOWERING August–October near coast, but as late as December in mountains.

1000. Solanum stelligerum

Solanum stelligerum

Devil's Needles **Plate 1000**

A shrub to 2 m tall with lanceolate leaves, 3 to 7 cm long, 1 to 2.5 cm wide, hairless or with a few star-shaped hairs above, and with soft woolly hairs below. Prickles up to 12 mm long are usually present on the stems, but are absent from the leaves. Flowers are lilac, about 25 mm across, with spreading narrow petals, which are often bent backwards. The flower stalk and calyx are covered with conspicuous hairs. Fruit is globular, about 10 mm across, and bright red.

HABITAT Sheltered forest, rainforest margins and coastal sand dunes.

DISTRIBUTION Coastal zone (e.g., Macquarie Pass, Howes Valley). Extending from south coast to north coast and Qld; also southern tablelands.

FLOWERING June–November.

1001. Solanum brownii

Solanum brownii

Violet Nightshade **Plate 1001**

S. brownii is easily confused with *S. stelligerum*, but has larger flowers and yellow or greenish-white berries. It is a shrub, usually up to 2 m tall, with lanceolate leaves, up to 12 cm long and 1 to 3 cm wide, with sparse hairs above and dense woolly hairs below. Prickles up to 10 mm long are usually present on the stems (sometimes absent), but are absent from the leaves. Flowers are pale blue to purple, 25 to 40 mm across, with a short tube and spreading broad petals. The flower stalk and calyx are covered with conspicuous hairs. Fruit is globular, 15 to 20 mm across, yellow to greenish-white.

HABITAT Open forest and woodland.

DISTRIBUTION Scattered from coast to mountains (e.g., Glen Davis), but not common. Occurring throughout coast, tablelands and western slopes from south-eastern Qld to north-eastern Vic.

FLOWERING July–October.

1002. Solanum cinereum

Solanum cinereum

Narrawa Burr **Plate 1002**

A prickly undershrub, about 1 m tall. Branches, leaves and calyces all have spreading yellow spines. Branches and flower stalks are covered in dense, close woolly white hairs. Leaves are broad ovate, up to 12 cm long, with wavy lobes, green and hairless above, and with close woolly white hairs below. Flowers are purple, in groups of 2 to 7, about 35 mm across. Fruit is globular, 15 to 20 mm across, mottled green to yellowish with a parchment-like texture.

HABITAT Forests, especially in rainforest margins and on rocky slopes.

DISTRIBUTION Scattered throughout higher elevations, but not common. Occurring along tablelands and western slopes from south-eastern Qld to central Vic.

FLOWERING January–April, but some flowers at other times.

1003. *Solanum campanulatum*

Solanum campanulatum — Plate 1003

This is a very prickly spreading undershrub, about 1 m tall. Branches, leaves and calyces all have spreading yellow spines. Branches and flower stalks are covered in short woolly hairs and longer loose hairs. Leaves are ovate-elliptical, up to 12 cm long, with wavy irregular lobes; the upper surface often has matted brown hairs, and the lower surface has dense woolly and long hairs. Flowers are violet, in groups of 4 to 10, about 20 to 30 mm across, and distinctly funnel or bell-shaped, a characteristic feature of this species. Fruit is globular, 20 to 25 mm across and yellowish-green, turning pale brown then black.

HABITAT Open forest.

DISTRIBUTION Coastal zone, but uncommon. Recordings include Brisbane Waters National Park, Culoul Range (Wollemi National Park) and Nepean River. Also occurring north coast, northern tablelands, central western slopes and south-eastern Qld.

FLOWERING September–March.

Solanum pungetium

Eastern Nightshade — Plate 1004

This low prickly plant is very similar in appearance to *S. prinophyllum*. It has similar spines on its stems, leaves and calyces. Its elliptical leaves are thin, bluntly lobed, hairless or with stellate hairs on both surfaces. The main difference is that the flowers are solitary or in pairs, each flower on its own stalk 2 to 4 cm long, not deeply campanulate but more spreading. Fruit is globular, 25 to 30 mm across, and yellow.

HABITAT Forest, especially in volcanic soils on hillsides, or along watercourses.

DISTRIBUTION Widespread from coast to mountains. Recordings include Brisbane Water National Park, Ourimbah State Forest, Mt Kembla, Macquarie Pass and Bundanoon. Extending to north and south coasts.

FLOWERING September–May.

1004. *Solanum pungetium*

Solanum prinophyllum

Forest Nightshade

Forest Nightshade is a low prickly plant with stems seldom more than 30 cm long. The stems, leaves and calyces have slender erect spines. Leaves are broad-oblong, up to 8 cm long, with rounded or toothed lobes, and green on both surfaces. Flowers are lilac-blue, 20 to 25 mm across, deeply campanulate, borne 2 to 3 together in a loose cluster with a common stalk. Fruit is globular, about 15 mm across, mottled green or purple when ripe. This species is similar to *S. pungetium*, but that species has 1 or 2 flowers, each on a separate stalk, and is usually more hairy.

HABITAT Forest and rainforest margins, often in disturbed areas.

DISTRIBUTION Widespread from coast to mountains. Recordings include Brisbane Water National Park, Burning Palms, Macquarie Pass and Doonside. Extending throughout coast and tablelands to Qld and Vic.

FLOWERING October–April.

1005. *Duboisia myoporoides*

Duboisia

Duboisia myoporoides

Duboisia, Corkwood — Plate 1005

Duboisia is a tall shrub or small tree, with light-coloured corky bark. Leaves are alternate, oblanceolate, 5 to 10 cm long, entire, green above, slightly paler underneath, gradually tapering to the leaf stalk. Leaves are often damaged from beetle attack. Flowers are white, with a tube 3 to 4 mm long, and 5 broad lobes and 4 stamens; they are borne in a loose, branched terminal cluster. The fruit is a globular, purple-black berry, 5 to 8 mm across. The leaves contain alkaloids used for medical purposes and are regarded as poisonous. Aborigines are said to have used its branches to stupefy eels in pools.

HABITAT Forested gullies and rainforest margins, especially in more open areas.

DISTRIBUTION Coastal zone. Recordings include Macquarie Pass, Nepean Dam, Mooney Mooney Creek, Bouddi National Park, St Albans and Agnes Banks. Extending from Clyde River (south coast) to northern Qld; also New Caledonia.

FLOWERING May–November.

1006. *Cyphanthera scabrella*

Cyphanthera Ray-flower

Shrubs, with tubular flowers, 5 spreading lobes and 4 stamens. Fruit a small, dry capsule.

Cyphanthera scabrella — Plate 1006

The stems and leaves of this small uncommon shrub are covered with short star-shaped hairs giving the whole plant a rough appearance. Leaves are small, oval to elliptic, 6 to 12 mm long, and scattered along the stems. Flowers are white, or yellowish green, striped with indigo, with pointed lobes and are borne on hairlike stalks.

HABITAT Forest, in gullies.

DISTRIBUTION Endemic in valleys of lower Blue Mountains (e.g., Bowen Creek, Erskine Creek, Euroka Creek).

FLOWERING October–November.

Cyphanthera albicans ssp. *albicans*

Grey Ray-flower — Plate 1007

A rare plant in the Sydney district, it may grow to 2 m tall. Branches and leaves appear greyish because of a covering of grey-white hairs. Leaves are elliptic to oblong, 6 to 15 mm long, 2 to 7 mm wide and almost stalkless. Flowers are creamy-white, striped with indigo, almost stalkless, with few or many flowers borne along the stems. In the Nepean River district, the shrub known sometimes as *Cyphanthera frondosa* appears to be a natural hybrid of *C. albicans* and *Duboisia myoporoides.*

HABITAT Rocky slopes in open forest.

DISTRIBUTION Restricted in the Sydey district to Nepean River gully at Douglas Park and valleys of western Blue Mountains. Scattered along the Great Dividing Range from Qld to Vic. Other subspecies occur on the western slopes and plains.

FLOWERING August–November.

1007. *Cyphanthera albicans*

1009. *Parahebe derwentiana*

Scrophulariaceae

At one time about 30 families having flowers with the petals fused into a tube around a superior ovary were grouped in the order Tubiflorae. It is possible to arrange these families into a series showing progression from radially symmetrical (rotate) or shortly tubular flowers through long tubular flowers with 5 small lobes at the end (actinomorphic) to 2-lipped tubular flowers with bilateral symmetry (zygomorphic). The series also shows changes from a calyx of free sepals to a calyx of fused sepals with 5 lobes or 2 lips, and progressive reduction in the number of stamens from 5 to 4 with 1 staminode, or 4, or 2 with 2 staminodes, or 2 only. Although it is tempting to regard this as evidence of an evolutionary series, or as radiation from a single central group, modern research shows that these families have diverse origins and consequently they are now divided into 3 or more orders. The largest of these, the Scrophulariales, consisting of a dozen closely related families, here includes the families Scrophulariaceae, Bignoniaceae, Gesneriaceae, Lentibulariaceae, Myoporaceae and Acanthaceae. The relationship between Scrophulariaceae and Solanaceae is not as close as once believed, and the latter is now usually placed in a separate order, the Solanales.

The Scrophulariaceae is a large cosmopolitan herbaceous family, well represented in temperate regions of both hemispheres, particularly the northern. The great variation in floral structure within the family is due to the develpment of specialised mechanisms favouring cross-pollination in newly opened flowers, but later allowing self-pollination if this fails. Because of this diversity the family has been divided in several different ways into subfamilies and tribes.

More than 40 genera occur in Australia, nearly half of them consisting of introduced weeds. Most of the genera contain, or are represented by, only a few species, with the largest number occurring in *Stemodia, Euphrasia* and *Veronica*. Of the 20 genera found in the Sydney area, the native species from 9 genera are uncommon and scattered or easily overlooked.

The family contains a number of genera used horticulturally and 1, *Digitalis*, used medicinally. Some genera are parasitic or semi-parasitic on the roots of other plants, and a few are serious pests in cereal crops in other countries.

Parahebe

Shrubs or stout herbs with opposite leaves. Flowers with 4 segments and 2 stamens projecting beyond the tube, borne in long racemes.

Parahebe perfoliata

Digger's Speedwell **Plate 1008**

An upright leafy perennial to about 1 m tall, easily identified by its stalkless, opposite, ovate leaves, which are often joined at the base to encircle the stem. Leaves are 3 to 5 cm long, blue-green, with entire margins. Flowers are deep blue and borne on slender racemes from the axils of the upper leaves.

HABITAT Forested slopes, especially in rocky areas.

DISTRIBUTION Higher Blue Mountains (e.g., Mt Victoria–Clarence). Extending to southern tablelands, south coast, western slopes and Vic.

FLOWERING November–January.

1008. *Parahebe perfoliata*

Parahebe derwentiana

Derwent Speedwell **Plate 1009**

Derwent Speedwell has the habit of *P. perfoliata*, but its leaves are toothed and do not encircle the stem. Leaves are stalkless, opposite, lanceolate, and about 10 cm long (sometimes longer). Flowers are lavender to white, with a purple tinge, and are borne in terminal racemes.

HABITAT Forested slopes.

DISTRIBUTION Higher Blue Mountains (e.g., Jenolan Caves Reserve), extending along southern tablelands. Two other subspecies occur on northern tablelands, southern tablelands, south coast, Qld, Vic, Tas and SA.

FLOWERING December–January.

1010. *Parahebe* sp. (Clarence)

Parahebe sp. Plate 1010

This is an attractive plant with numerous hairless stems spreading from the rootstock, giving a clumped appearance. Stems are about 40 cm tall. Leaves are opposite, with each pair 2 to 3 cm apart and at a right-angle to the pair above and below. Leaves are ovate, about 4 cm long and 15 to 18 mm wide, stalkless and have many small teeth along the margins. The most conspicuous feature of the leaves is that they are downward pointing and strongly concave, so that the base of the leaf encloses the base of the flower stem. Flowers are deep blue, about 10 mm across, with one broad upper petal, 3 smaller lower petals and 2 exserted stamens with yellow anthers. Flowers are borne singly on short stalks in a raceme 10 to 22 cm long from the upper leaf axils.

HABITAT Roadside margins and sandy slopes above swamps.

DISTRIBUTION Restricted to one swampy site near Clarence in higher Blue Mountains.

FLOWERING November–December.

1011. Parahebe sp. (Mt Colong)

Parahebe sp. Plate 1011

This is a rare species, atypical in habit, first collected last century by the explorer Cayley from a small area near Mt Colong (southern Blue Mountains), and rediscovered in recent years in a few isolated mountain locations. It has delicate hairless stems up to 30 cm long which hang from shaded cliff-faces. Leaves are opposite, about 3 cm long, shortly stalked and toothed. Flowers are pale lavender, with the upper petals streaked with violet; they are borne in long clusters from the upper leaf axils.

HABITAT Shaded, somewhat damp, sites on cliff-faces.

DISTRIBUTION Confined to a few scattered locations in the Blue Mountains (e.g., McMahons Lookout, Kanangra Walls, Pierce's Pass).

FLOWERING August–December.

1012. Veronica notabilis

Veronica

Herbs with oposite leaves. Flowers with 4 segments and 2 stamens projecting beyond the tube, borne singly or in racemes from the leaf axils.

Veronica notabilis

Forest Speedwell Plate 1012

A partly erect or spreading plant with softly hairy stems to 35 cm long rising from a creeping root. Leaves are opposite, about 3 cm long, 1 to 2 cm wide, stalked, ovate to lanceolate and coarsely toothed. Flowers are pale lilac, streaked with violet and are borne in racemes 10 to 20 cm long from the axils of the upper leaves.

HABITAT Sheltered forest, usually in valleys.

DISTRIBUTION Blue Mountains and higher areas of Woronora Plateau, but uncommon in Sydney district. Extending along coast and tablelands divisions; also Qld, Vic and Tas.

FLOWERING December–January.

1013. Veronica calycina

Veronica calycina Plate 1013

A small plant with hairy stems usually less than 25 cm tall; these hairs are long and weak or softly downy. Leaves are firm, broadly ovate, 12 to 25 mm long, with obtuse teeth, heart-shaped or truncate at the base and shortly stalked. Flowers are violet, in loose clusters of 2 to 4 flowers from the upper leaf axils. Calyx segments are blunt, unlike those of *V. plebeia* which are acute. Fruit is a compressed capsule, heart-shaped or truncate at the summit.

HABITAT Sheltered forest.

DISTRIBUTION Widespread from coast to mountains. Chiefly a tablelands species, extending to central and north western slopes; also Vic, Tas, SA and WA.

FLOWERING December–January.

Veronica plebeia

Creeping Speedwell Plate 1014

A small prostrate plant with stems to about 20 cm tall. Stems are covered in minute downy hairs. Leaves are ovate, 10 to 25 mm long, with coarse teeth, heart-shaped or truncate at the base and on stalks about 10 mm long. Flowers are violet to pink, in loose clusters of 2 to 8 flowers from the upper leaf axils. Calyx segments are acute. Fruit is an orbicular capsule, rounded at the summit.

HABITAT Sheltered situations, especially moist forests, usually on heavier soils.

DISTRIBUTION Widespread from coast to mountains. Occurring all along coast and tablelands, extending to western slopes and north-western plains; also Qld Vic, SA and WA.

FLOWERING September–December.

1014. Veronica plebeia

1015. Euphrasia collina ssp. *paludosa*

Euphrasia

Herbs with opposite leaves. Flowers with 4 segments, 2 lipped; the upper lip is 2-lobed and the lower lip spreading and 3-lobed. Stamens 4, in pairs.

Euphrasia collina ssp. *paludosa*

Eye-bright Plate 1015

Eye-bright is an erect perennial herb with several stems, 30 to 40 cm tall, branching from the base. Leaves are opposite, 5 to 15 mm long, oblong to wedge-shaped, stalkless, with 1 to 3 pairs of teeth towards the tip. Flowers are violet, pink or white, stalkless and borne in terminal spikes. Fruit is an oblong, compressed capsule.

HABITAT Heath and open forest, especially in damp sandy soil.

DISTRIBUTION Chiefly tablelands (e.g., Newnes State Forest, Jenolan Caves Reserve) but also occurring along coast (e.g., Munmorah State Recreation Area). Extending to north coast, western slopes and from northern tablelands to Vic.

FLOWERING September–December.

1016. *Gratiola latifolia*

Gratiola

Creeping herbs from wet places. Leaves opposite; flowers irregular, 2-lipped, trumpet-shaped, with 2 fertile stamens.

Gratiola latifolia

Brooklime **Plate 1016**

A hairless spreading plant with a creeping rootstock and stems ascending to about 20 cm. Leaves are opposite, stalkless, ovate to elliptic, up to 25 mm long and 8 to 15 mm wide; margins may be entire, wavy or toothed. Flowers are pale pink, solitary from the leaf axils. Flowers are tubular, with spreading lobes; there are 3 lower lobes and an entire or notched upper lobe. The calyx is hairless and the flowers are stalkless or on short stalks less than 5 mm long. Two other species of *Gratiola* occur in the Sydney district. *G. pubescens* has a glandular-downy calyx and flowers on stalks less than 5 mm long. *G. pedunculata* has a glandular-downy calyx and flowers on stalks more than 10 mm long.

HABITAT Wet places.

DISTRIBUTION Widespread from coast to mountains, but most common at higher altitudes (e.g., Boyd Plateau, Nattai River valley). Found all along coast and tablelands divisions; also central western slopes, Qld, Vic, Tas and SA.

FLOWERING December–January.

Mimulus Monkey-flower

Creeping herbs, rooting at the nodes. Leaves opposite; flowers tubular, 5-lobed and 2-lipped, with 4 fertile stamens.

Mimulus repens

Creeping Monkey-flower **Plate 1017**

A hairless prostrate plant with soft slightly succulent stems which root at the nodes. Leaves are opposite, stalkless, crowded, elliptic and small (2 to 6 mm long). Flowers are violet with a yellow centre, like small Snapdragons, with a tubular calyx, expanding into 5 broad spreading lobes. Flowers are borne singly on short stalks from the axils of the upper leaves.

HABITAT Wet places, around drying swamps and brackish coastal lagoons.

DISTRIBUTION Coastal zone. Extending from north coast to south coast, Vic, Tas and SA. Another subspecies is found in the western plains division.

FLOWERING February–March.

Bignoniaceae

The Bignoniaceae is a family of mainly woody climbers showing greatest development in the forests of tropical South America. The 5 genera with species native to Australia are centred to the north, in Malesia or New Guinea. *Pandorea*, the best known of the 5, occurs in all states. One species, *Pandorea pandorana*, occurs in the Sydney area and is typical of the family in having opposite glabrous pinnate leaves without stipules. The flowers are tubular, and showy with 5 irregular lobes and 4 stamens in 2 unequal pairs attached to the base of the tube.

1018. *Pandorea pandorana*

Pandorea

Pandorea pandorana ssp. pandorana

Wonga Vine **Plate 1018**

A woody climber with branches several metres long. Leaves are opposite, compound, with 2 to 4 pairs of leaflets and one terminal leaflet. Leaflets are ovate to lanceolate, 2 to 7 cm long and 10 to 15 mm wide. Juvenile leaflets are much smaller, bluntly toothed, with 9 to 17 leaflets per leaf. Flowers are tubular, up to 2 cm long, whitish with red spots in the throat; they are borne in attractive loose clusters.

HABITAT Sheltered forest, often on rocky slopes.

DISTRIBUTION Widespread from coast to mountains. Recordings include Royal National Park, Bouddi National Park, Oatley Park, St Albans, Wombeyan Caves and Bateau Bay. Occurring throughout NSW except for drier western plains where its place is taken by another subspecies. Also Qld, Vic, Lord Howe Island, New Guinea and Indonesian islands.

FLOWERING August–September.

1017. *Mimulus repens*

Gesneriaceae

This is a fairly large family of tropical herbs and shrubs which include many popular garden and indoor plants. Some from Central and South America have striking red 2-lipped tubular flowers adapted for pollination by humming birds. In other countries also, birds appear to be the main pollinators and dissemination of seed. The Australian representatives, numbering 6 species in 6 genera, have large softly hairy leaves and colourful tubular flowers not unlike those of *St Paulia* (African Violet), a member of this family.

In Australia the family is confined to the moist east coast. Only 1 species, *Fieldia australis*, occurs in the Sydney area.

1019. *Fieldia australis*

Fieldia

Fieldia australis

Fieldia **Plate 1019**

Fieldia is a scrambling semi-epiphytic plant, mostly found climbing over shaded rocks and mossy tree trunks in sheltered rainforest gullies. Young branches and leaves are covered with long spreading hairs. Leaves are opposite, often unequal in size; larger leaves are up to 7 cm long and 2.5 cm wide, the smaller leaves less than 1 cm long. Leaves are toothed and paler underneath. Flowers are tubular, greenish-yellow bells, about 3.5 cm long, solitary and pendulous. The fruit is an oblong berry to 2 cm long, whitish with red spots and with recurved calyx-lobes.

HABITAT Rainforest, especially on mossy rocks and tree trunks.

DISTRIBUTION Coast to mountains. Recordings include Carrington Falls (The Gorge), Mt Tomah and Gap Creek Flora Reserve. Occurring throughout the coast and tablelands divisions; also Vic and Qld.

FLOWERING April–May.

Lentibulariaceae

This small family of carnivorous herbs inhabiting sandy or muddy seepage areas, or growing in water, is virtually cosmopolitan, being absent from only the icebound and desert areas of the world.

The best known of the 4 genera comprising the family are *Utricularia* (Bladderworts) and *Pinquicula* (Butterworts). Australia has about 50 species of *Utricularia* and 3 species of the very similar endemic, *Polypompholyx*. Seven species of *Utricularia* occur in the Sydney area.

Although all family members have adaptations for capturing tiny creatures, most interest centres on the traps attached to the roots of *Utricularia* and *Polypompholyx*. These consist of tiny hollow bladders having a small entrance closed by a trapdoor which is activated by a trigger of hairs or glands. The trap is under negative pressure so that, when triggered, the door opens quickly and the triggering organism is sucked in with the rush of water. The door then shuts and the victim is digested.

Utricularia Bladderwort

Delicate herbs growing in wet soils and mud, or floating aquatic plants. Flowers borne on thin erect leafless stems. Flowers mauve (sometimes yellow), with a tubular corolla, 2-lipped.

1020. Utricularia exoleta

Utricularia exoleta — Plate 1020

U. exoleta is an aquatic plant. Its leaves and stems are submerged, but its yellow flowers are borne on fine leafless stalks 2 to 7 cm long above the water. Its stems are fine and often interwoven and its leaves are divided into hair-like segments interspersed with numerous small bladders about 1 mm across. Flower stalks bear 1 to 3 flowers; each flower is 4 to 6 mm across. A similar species, *U. aurea*, also occurs along the central and north coasts. Its submerged stems are up to 60 cm long, straight, not interwoven. Its leaves are much more divided than those of *U. exoleta* and its yellow flowers are 9 to 10 mm across, with 3 to 6 flowers on each stalk.

HABITAT Shallow water.

DISTRIBUTION Coastal zone, extending to lower north coast. Also Qld, NT and tropical Africa and Asia.

FLOWERING January–February.

1021. Utricularia dichotoma

Utricularia dichotoma

Fairies' Aprons — Plate 1021

Fairies' Aprons is a common plant in wet places, especially in the peaty edges of swampy heaths. Flowers are on thin, leafless stems, 10 to 20 cm tall. Each stem may bear a solitary, terminal flower, a pair of flowers, or occasionally 3 to 4 flowers in whorls. Flowers are purple or lilac with a yellow centre; the upper lip is small and the lower lip is broad, semicircular, 1 to 2 cm across.

HABITAT Swampy areas.

DISTRIBUTION Widespread from coast to mountains. Recordings include Ku-ring-gai Chase and Royal National Parks, Bundanoon and Blackheath. Occurring throughout coast, tablelands and western slopes. Also Qld, Vic, Tas and SA.

FLOWERING August–March.

1022. Utricularia uliginosa

1023. Utricularia lateriflora

Utricularia uliginosa — Plate 1022

A small swamp plant with leafless stems only 5 to 8 cm tall. A few tiny pale blue or white flowers are scattered alternately along the stem, on stalks 4 mm long. Leaves, when present, are at the base of the stem; they are pale green, narrow to spoon-shaped, and 6 to 12 mm long. The upper lip of the flower is obovate, 3 to 5 mm long, shorter than the calyx; the lower lip is longer, broader (6 to 8 mm wide) and convex. A similar species, *U. lateriflora* (Plate 1023), has purple flowers on stems bearing 1 to 4 stalkless flowers, with the upper lip oblong-linear, slightly exceeding the calyx. It occurs along the coast and in the central tablelands.

HABITAT Swampy areas.

DISTRIBUTION Coastal zone. Extending from south coast to north coast, Qld and NT. Also New Guinea and tropical Asia.

FLOWERING February–March.

Myoporaceae

The Myoporaceae is a small family of woody trees and shrubs with sticky or resinous leaves. The family of only 4 genera is centred in Australia and the western Pacific. *Myoporum* (32 species, 19 Australian), occurs in eastern Asia, New Guinea, New Zealand and throughout Australia and is the most widespread of the 2 Australian genera, but *Eremophila* (112 endemic species), which is confined to the arid Australian inland, is the larger.

Whereas the flowers of *Myoporum* are small (less than 1 cm), and white with purple dots inside a 4 or 5-lobed campanulate tube, those of *Eremophila* are usually larger (2 to 3 cm long) with clustered exsert stamens and the corolla tube variously coloured and lobed. Prostrate and upright species of both genera are widely available in Australia for garden culture. Since *Myoporum* are commonly coastal plants they are tolerant of light sandy soils, but *Eremophila* do best in a warm well-drained position in heavier soils in medium to low rainfall areas.

1024. *Myoporum insulare*

1026. *Myoporum montanum*

Myoporum

Myoporum insulare

Boobialla **Plate 1024**

M. insulare is a tall shrub, usually about 2 m tall, but may grow to a small tree. Its leaves are alternate, light green, obovate, 3 to 8 cm long and up to 3 cm wide. The tip has a short point and the margins are either entire or with a few small teeth near the tip. Flowers are white, with purple dots, hairy inside, tubular and with 5 spreading nearly equal lobes. The 4 stamens do not project beyond the floral tube. Flowers are borne from the leaf axils, in clusters of 2 to 7, but occasionally solitary. Ripe fruit is globular, purplish blue and 6 to 8 mm across.

HABITAT Sea coasts and estuaries.

DISTRIBUTION Scattered along coast (e.g., Pearl Beach, Minnamurra Spit). Also occurring north coast, south coast, Lord Howe Island and in all other States, except NT.

FLOWERING October–March.

Myoporum acuminatum

Northern Boobialla **Plate 1025**

This species is often confused with *M. insulare* as it is also a tall shrub or small tree and is found in a similar habitat. Its leaves are, however, longer and lanceolate. They are alternate, 5 to 10 cm long, up to 3.5 cm wide, hairless, glossy, light green, tapering to a long point at the tip and to a stalk 1 to 2 cm long at the base; the margins may be entire or toothed. Flowers are white, with purple dots, hairy inside, tubular and with 5 spreading nearly equal lobes. The 4 stamens project slightly beyond the floral tube. Flowers are borne from the leaf axils, in clusters of 1 to 8. Ripe fruit is globular, purplish-blue, 4 to 6 mm across.

HABITAT Chiefly coast and estuaries, but also on margins of wet forests or littoral rainforests.

DISTRIBUTION Scattered along coast (e.g., Gerringong, Macquarie Pass, Port Hacking, Patonga Creek). Occurring along coastal zone from near Bega (south coast) to northern Qld. Also NT and WA.

FLOWERING August–October.

1025. *Myoporum acuminatum*

Myoporum montanum

Western Boobialla, Water Bush **Plate 1026**

A common plant in drier western regions of the State, but occurs in only a few isolated localities closer to the coast. It is an erect, bushy shrub to 3 m tall. Leaves are alternate, narrow lanceolate, 3 to 10 cm long and less than 10 mm wide; they are pointed at the tip, shortly stalked and have entire margins. Flowers are white, with purple dots, tubular, hairy inside, with 5 spreading lobes. The 4 stamens project beyond the floral tube. Flowers are borne from the leaf axils, in clusters of 1 to 6. Ripe fruit is globular, purplish, 6 to 8 mm across.

HABITAT Forest, often along creek banks and on alluvial flats.

DISTRIBUTION Blue Mountains and southern highlands (e.g., Wombeyan Caves, Wollondilly River). Found throughout most of NSW, except southern tablelands and southern far western plains; also occurring Qld, Vic, SA and NT.

FLOWERING July–September.

Myoporum debile

Amulla, Winter Apple **Plate 1027**

An attractive ground cover plant, with prostrate stems up to 1 m long spreading from a thick rootstock. Leaves are alternate, dark green, lanceolate to elliptic, 3 to 6 cm long, stalkless or almost so; the margins are usually entire but may have a few scattered teeth. Flowers are pink to mauve, with 1 or 2 borne from the leaf axils. Fruit is ovoid, 6 to 8 mm long, reddish and slightly succulent, hence the common name 'Winter Apple'.

HABITAT Open forest, usually on clay soils.

DISTRIBUTION Cumberland Plain (e.g., Doonside) and Glen Davis. Extending to north coast and northern tablelands. Also scattered throughout western slopes, western plains and Qld.

FLOWERING November–February.

1027. *Myoporum debile*

Acanthaceae

The Acanthaceae is a large tropical family of mainly herbs and small shrubs. The flowers, which are often showy, have a short or long slender corolla tube spreading at the top into 5 lobes or 2 lips (1 in *Acanthus*) which may be entire or lobed. Four stamens are arranged in pairs inside the tube, although sometimes 1 pair is reduced to staminodes or absent entirely.

The family is represented in Australia by about 20 genera with a few species each; 3 species from 2 genera occur in the Sydney area.

Brunoniella

Small perennials with opposite leaves. Flowers are tubular, mauve-blue, gradually enlarging towards the throat, with 5 almost equal lobes and 4 enclosed stamens.

1028. *Brunoniella australis*

Brunoniella australis

Blue Trumpet, Blue Yam **Plate 1028**

A small perennial with erect or spreading stems, usually 10 to 20 cm tall, rising from a fleshy white root. Leaves are opposite, 1 to 4 cm long, obovate to lanceolate and stalked; leaf-blades have scattered long hairs. Flowers are mauve-blue, about 15 mm across, tubular at the base, with 5 spreading broad lobes. They are borne in the leaf axils. Sepals are narrow (about 0.5 mm wide) and covered with long spreading hairs, a feature which separates this species from *B. pumilio* (Plate 1029), which has narrow-

1029. *Brunoniella pumilio*

oblong to lanceolate sepals (about 1 mm wide) covered with shorty downy hairs. *B. pumilio* is a coastal species, with smaller flowers (about 10 mm across) on stems only about 5 cm tall, and with hairless leaves. Fruit for both species is an oblong-linear capsule.

Habitat Open forest and woodland, chiefly on shales and clays.

Distribution Coastal areas, especially Cumberland Plain. Also occurring on western slopes and western plains (where it is uncommon), and Qld.

Flowering October–December.

Pseuderanthemum

Small perennial with opposite leaves. Differs from *Brunoniella* in having a slender cylindrical flower tube, spreading lobes and only 2 stamens.

1030. Pseuderanthemum variabile

Pseuderanthemum variabile — Plate 1030

A common small perennial with a creeping root and erect stems, 7 to 15 cm tall. Leaves are opposite, narrow to broad lanceolate, 3 to 5 cm long. Flowers are pale blue, pink or white, with a slender tube and 5 broad spreading oblong lobes. The 2 stamens project partially beyond the floral tube. Flowers are borne in clusters near the ends of the stems. Fruit is a narrow capsule, 10 to 14 mm long.

Habitat Sheltered forest.

Distribution Coastal areas (e.g., Royal and Brisbane Water National Parks, Macquarie Pass). Extending from south coast to north coast, Qld and NT.

Flowering December–January.

Plantaginaceae

Most of the 260 species in this family belong to the genus *Plantago*, of which Australia has about 24 native species and 8 naturalised species. Four native and 5 naturalised species occur in the Sydney region. Some in the latter category are common weeds on roadsides and disturbed land.

The family consists of annual and perennial herbs with entire or lobed parallel-veined leaves in a basal rosette surrounding one to several erect leafless stems bearing tiny flowers in dense terminal spikes. Being mostly wind pollinated, the flowers have an insignificant perianth and achieve cross-fertilisation by exserting their long stigmas before the stamens elongate to shed their pollen.

The family is not closely related to any other, although it has some similarities with the Scrophulariaceae.

1031. Plantago gaudichaudii

Plantago Plantain

Plantago gaudichaudii

Narrow-leaf Plantain — **Plate 1031**

This is a perennial herb with a thick fleshy taproot. Leaves are in basal tufts, linear to narrow-oblong, 7 to 24 cm long, 8 to 12 mm wide, entire or with a few scattered teeth, hairless or with long hairs, and with long parallel veins. Flowers are small (3 to 4 mm long) and are borne along upright stems; stems are about as long as the leaves. Flower spikes are 4 to 12 cm long. Fruit capsules are ellipsoid, 3 to 4 mm long.

Habitat Grassland and open forest, usually on clay or heavier soils.

Distribution Cumberland Plain (e.g., Doonside, Casula, Camden). Also recorded in the past from Moss Vale. Extending to near Singleton (north coast), Huskisson (south coast), and throughout tablelands and western slopes divisions; also Qld, Vic, Tas and SA.

Flowering Chiefly September–April, but some flowers at other times.

1032. Plantago debilis

Plantago debilis

Slender Plantain — **Plate 1032**

This is a perennial (sometimes an annual), usually with a short fleshy taproot. Leaves are in basal tufts, thin, oblanceolate to obovate, 3 to 15 cm long, 1 to 4 cm wide, blunt at the tip, with or without small irregular teeth and with long parallel veins. Leaves are usually softly hairy, but hairs are sometimes confined to the leaf stalks. Flowers are small (1 to 2 mm long) and are borne along upright stems, 12 to 20 cm long. Flower spikes are 4 to 10 cm long. Fruit capsules are ovoid, 2 to 3 mm long.

Habitat Moist forest.

Distribution Widespread from coast to mountains (e.g., Brisbane Water National Park, Colo-Putty, Mt Hay, Royal National Park). Occurring throughout most of NSW (uncommon on drier western areas) and in all other States.

Flowering Chiefly December–February, but some flowers at other times.

Verbenaceae

At present there is no consensus as to the delimitation of this family since a number of groups once regarded as tribes or subfamilies have been raised to family rank. In some modern classifications *Avicennia*, the Grey Mangrove, is given a family of its own, Avicenniaceae, and *Chloanthes* with about 10 other genera form the Chloanthaceae. The three groups are here treated as separate families.

The Verbenaceae are more woody (the herbaceous species have a tendency to brittle woodiness in the lower/older stems) than the closely related Lamiaceae which are mostly herbaceous. Many members of both families can be recognised by the opposite leaves without stipules and the quadrangular glabrous or scabrous/hairy stems. The flowers of Verbenaceae usually have a 4 or 5-toothed calyx, a 4 or 5-lobed corolla and 4 stamens. The superior ovary has 2 fused carpels, each containing 2 axile ovules separated by a false septum (partition). This is similar to, but distinct from, the arrangement in Lamiaceae which is a 4-locular 2-carpellate ovary with 1 basal ovule in each locule.

The Verbenaceae is represented in the Sydney district by a number of common introduced weeds such as lantana and verbena (Purple Top) and 3 native species.

Clerodendrum

Tall shrubs with broad, opposite leaves. Flowers with a long floral tube, 5 spreading lobes and 4 stamens projecting well beyond the tube.

1033. A. Clerodendrum tomentosum (flowers)

Clerodendrum tomentosum

Hairy Clerodendrum **Plate 1033**

This is a tall shrub or small tree. Leaves are opposite, entire, broadly elliptic, 7 to 12 cm long, with a stalk 2 to 3 cm long. The undersurfaces of the younger leaves and inflorescences are covered with dense velvety hairs; older leaves are often hairless. Flowers are white and borne in dense heads. The floral tube is narrow, about 2 cm long; 4 conspicuous stamens protrude 1 to 2 cm beyond the tube. The fruit is also conspicuous as it consists of a large, black, shining berry, 8 mm across, sitting on a red 5-lobed enlarged calyx. Fruits remain on the shrub throughout summer and autumn.

HABITAT Protected forests and rainforest margins.

DISTRIBUTION Widespread from coast to mountains. Most common along coastal zone (e.g., Patonga, Maitland Bay, Oatley Park, Mt Kembla, Razorback Mt, Barren Grounds), but also Blue Mountains (e.g., Kurrajong, Kanangra, Howes Valley). Occurring along the coast from near Batemans Bay (south coast) to northern Qld; also northern tablelands, central western slopes and WA.

FLOWERING October–November.

1033. B. Clerondendrum tomentosum (fruit)

Gmelina

An uncommon large rainforest tree, with opposite leaves and white bell-shaped flowers.

Gmelina leichhardtii

White Beech **Plate 1034**

White Beech is an uncommon tree (to 40 m tall), considered 'at risk' because of the reduction of its range. It is rare in the Sydney district. Leaves are opposite, entire, ovate, 7 to 14 cm long, hairless above; the undersurface is paler, densely hairy with prominent raised veins. Young shoots and leaf stalks are covered with dense velvety hairs. Flowers are borne in large open terminal clusters; individual flowers are white, with purple and yellow markings, bell-shaped, with 5 spreading petals and 4 stamens, 2 long and 2 short. Fruit is succulent, globular, blue when ripe, about 25 mm across, with enlarged and persistent calyx.

HABITAT Rainforest.

DISTRIBUTION Scattered along coastal zone. Recordings include Minnamurra Falls, Hacking River (Royal National Park), upper Patonga Creek and Gap Creek Flora Reserve (Watagan Mountains). Extending along north coast to north Qld.

FLOWERING November–January.

1034. Gmelina leichhardtii

Avicenniaceae

This monogeneric family of mangroves colonises the intertidal zone along sheltered tropical and southern hemisphere subtropical coastlines.

In Australia, mangroves fringe a total of about 6000 km of coast in sites from Bunbury, south of Perth, around the northern coast and down the east coast to Corner Inlet in Victoria, with isolated occurrences in Westernport and Port Phillip Bays and the Spencers Gulf region. Mangroves are absent from Tasmania and the Bass Strait islands although they were probably present in warmer periods during the Tertiary period.

The most widely distributed mangrove in Australia is *Avicennia marina*, the Grey Mangrove. Wherever mangroves approach their southern limits, this is the sole species. On tropical coastlines where a dozen or more mangrove species form distinct zones relative to the shoreline, this species usually dominates the seaward zone and, because of its tolerance of high salinities, may also form an extensive woodland in the landward zone where tidal inundation is infrequent. *Avicennia marina* is associated with *Aegiceras corniculatum* in the Sydney area where it occurs in muddy and sandy estuaries, in salty lagoons and along the banks of tidal rivers and creeks. *Avicennia* is the taller and is readily identified by its pointed opposite leaves, clusters of tiny orange flowers, and the broad flat embryo which germinates before dropping from the tree.

Avicennia

Avicennia marina var. *aus alasica*

Grey Mangrove **Plate 1035**

A small tree of intertidal mud flats, usually surrounded by conspicuous aerating roots, or pneumatophores, which project above the

1035. Avicennia marina

surrounding mud. Leaves are opposite, broad lanceolate, 5 to 8 cm long, hairless, leathery dark green and shining above, lighter below. Flowers are orange, small, with 5 spreading lobes and 4 stamens; they are in small dense clusters either at the ends of the branchlets or in the upper leaf axils. Fruit is a large, compressed oval capsule, about 3 cm long, containing a single seed which germinates on the tree before dropping into the mud.

HABITAT Intertidal mud flats and estuaries.

DISTRIBUTION Widespread and common along coast (e.g., Port Hacking, Botany Bay, Parramatta River, Broken Bay). Found all around coastline of Australia, except cooler areas of Bass Strait, Tasmania and southern ocean. Also New Guinea, Malesia and Pacific Islands.

FLOWERING February–April.

Chloanthaceae

This is a small endemic family of about 10 genera and 100 species well represented in Western Australia. The famiy consists of small shrubs with simple opposite leaves and branches frequently clothed in soft woolly hairs. It is divided into 2 tribes, Physopsideae with a symmetrically lobed corolla, and Chloantheae with a 2-lipped or unequally 5-lobed corolla.

Chloanthes

1036. Chloanthes stoechadis

1037. Chloanthes glandulosa

Chloanthes stoechadis

Common Chloanthes **Plate 1036**

This small undershrub is unmistakable. Its stems ae covered with woolly white hairs and its opposite leaves are covered with small bubble-like wrinkles. These leaves are narrow, 3 to 5 cm long, with margins curved under, stalkless and continuing as ridges down the stems, and woolly underneath. Flowers are tubular, yellowish green, up to 3 cm long, irregularly 5-lobed, with the upper lip 2-lobed and erect, the lower lip 3-lobed and spreading. The style protrudes beyond the floral tube; the 4 stamens slightly protrude. Flowers are borne singly in the upper leaf axils. The dry fruit splits into 2 sections. An allied species, *C. glandulosa* (Plate 1037), occurs in the Blue Mountains (e.g., Burralow, Kurrajong). Its leaves are only slightly recurved and have yellow glandular hairs as well as rough sandpapery hairs scattered over the surface; they are not white woolly underneath.

HABITAT Heath and open forest, on sandy soils and near rocky outcrops.

DISTRIBUTION Widespread from coast to mountains. Extending from south coast to north coast and Qld; also to southern tablelands.

FLOWERING July–October, with some flowers as late as April.

1038. Spartothamnella juncea

Spartothamnella

Spartothamnella juncea

Bead Bush **Plate 1038**

A hairless upright shrub (sometimes scrambling) to 2 m tall. Stems are 4-angled and appear leafless as the small opposite, narrow-elliptic leaves (3 to 18 mm long) fall off early, leaving only small scales on the stems. Flowers are whitish, 3 to 4 mm long, shortly tubular, but divided almost to the base into 5 narrow spreading acute lobes. Fruit is a conspicuous, round orange bead-like drupe, about 3 mm across.

HABITAT Forested hillsides, often in rocky situations and dry rainforests.

DISTRIBUTION Rare in the Sydney district. Recorded from Razorback Mt (Camden) and Mt Wilson. Also recorded north coast, western slopes and Qld.

FLOWERING December–January.

Lamiaceae

The mints are a large cosmopolitan group, frequently cultivated for perfumery, as culinary and medicinal herbs, and as ornamental shrubs. Most of the better known members such as lavender, rosemary, marjoram, sage, thyme, mint and basil are from Mediterranean and central European countries.

Although none of the Australian representatives find application in the food or pharmaceutical areas, many of the 90 or so *Prostanthera* species are ideal garden subjects. They are easy to propagate from cuttings, and easy to grow in sun or shade provided they are in cool and moist but well-drained soil. In spring many flower spectacularly in shades of mauve, lilac and violet. A few species or forms have red, pink, orange, green or white flowers. *Westringia*, with white or mauve flowers, are less showy, but are tolerant of the most adverse conditions, and this has led to the use, or overuse, of *Westringia fruticosa* for the landscaping of public areas. Some 'touchy' mints are now grown by grafting to stocks of this species.

After *Prostanthera*, the main Australian genera are *Hemigenia* (about 40 species, mainly in the south-west of WA), *Westringia* (27 species, predominantly eastern states), *Microcorys* (15 species, south-west WA) and *Plectranthus* (15 species, chiefly in Queensland). Unlike the above which are Australian endemics, genera such as *Prunella, Mentha, Salvia, Scutellaria* and *Ajuga* are typically European, but have 1 or 2 Australian species.

Members of the family have a number of easily recognised characteristics. They are herbs and shrubs with square stems and simple whorled or opposite decussate leaves. Many secrete aromatic oils from glands or glandular hairs that cover the leaves and stems. The flowers are tubular with 5 petals which are more or less fused and arranged to form 2 lips. In the major Australian genera the upper lip is composed of 2 segments, and the lower of three. The ovary has 4 compartments, each with one ovule, and the style arises centrally from between the compartments. Flowers may have either 4 stamens, 2 stamens and 2 staminodes, or 2 stamens. The calyx has either 5 narrow teeth or 2 lips.

The four common local genera can be readily distinguished by the following features.

Feature	*Westringia*	*Hemigenia*	*Plectranthus*	*Prostanthera*
Plant form	shrub	shrub	herb	shrub
Leaf arrangement	whorled	whorled	opposite	opposite
Stamens	2 (+2 staminodes)	4	4	4
Calyx	5-toothed	5-toothed	2-lipped (lower lip with 4 acute teeth)	2-lipped (lips entire)

Prostanthera Mint-bush

Shrubs, often strongly scented, with square stems and opposite leaves. Flowers with upper lip 2-lobed and erect, lower lip 3-lobed and spreading; calyx divided into 2 lips, each usually entire; 4 stamens in pairs. The anthers are 2-celled, often with 1 or 2 appendages; the presence or absence of these appendages are useful identification aids. This genus is currently undergoing revision. It is expected that some species will have name changes and a number of new species will be named.

1039. Prostanthera lasianthos

Prostanthera lasianthos

Victorian Christmas-bush **Plate 1039**

This tall shrub or small tree (to 5 m tall) derives its popular name from the fact that it is common in the sheltered gullies of Victoria where its abundant flowers are very conspicuous in late December. It is less common in the Sydney area. Leaves are lanceolate, 3 to 10 cm long, flat, finely toothed (sometimes entire), dark green above and paler below. Flowers are pale mauve or white, spotted inside with purple. The lobes are very broad. One anther-appendage is long, the other is very short. Flowers are borne in large showy clusters.

HABITAT Sheltered forested slopes and mountain gullies, in richer soils.

DISTRIBUTION Scattered from coast to mountains. Recordings include Macquarie Pass, Nattai River, Fitzroy Falls, Grose River, Mt Tomah, Saddleback Mountain and Bowen Creek. Found throughout coast, tablelands and western slopes, south to Vic and Tas.

FLOWERING December–January.

1040. Prostanthera prunelloides

Prostanthera prunelloides

Prunella Mint-bush **Plate 1040**

This is an aromatic bushy shrub, 1 to 2 m tall, with leaves which resemble those of Self-heal *(Prunella vulgaris)*. They are ovate, 2 to 5 cm long, entire or with a few rounded teeth, and with stalks up to 20 mm long. Flowers are mauve to nearly white, with faint yellow dots in the throat, and softly hairy on the outer surface. One anther-appendage is long, the other is very short. Flowers are borne in simple terminal clusters. The leaves below the flowers are reduced to broad bracts, which enclose the calyx when in bud, but fall off when flowers develop.

HABITAT Sheltered forested gullies and rocky slopes.

DISTRIBUTION Lower Blue Mountains (e.g., Grose Lookout, North Richmond) and Colo Plateau (e.g., Putty road). Also found central western slopes (Goulburn River valley) and south coast.

FLOWERING September–November.

1041. Prostanthera caerulea

Prostanthera caerulea **Plate 1041**

This is an attractive shrub or small tree, 2 to 4 m tall. Leaves are lanceolate, 2.5 to 5 cm long, with irregular teeth. Flowers are deep lilac, sometimes bluish (hence 'caerulea') and are hairless on the outside. Both anther appendages are shorter than the anthers. Numerous flowers are borne in simple or branched clusters. The leaves below the flowers are reduced to ovate bracts which fall off when the plant is in bud. The flower colour and the short anther-appendages separate this species from *P. lasianthos*.

HABITAT Sheltered gullies and tall wet forests, mainly on sandstone just below basaltic soils.

DISTRIBUTION Scattered from coast to mountains but uncommon. Recordings include Mt Irvine, Mt Banks, Mt Tomah and Pierces Pass (Grose River). Also recorded north coast and northern tablelands.

FLOWERING October–November.

1042. Prostanthera ovalifolia

Prostanthera ovalifolia **Plate 1042**

Oval-leaf Mint-bush, Purple Mint-bush

This is a showy shrub to 2 m tall, often cultivated for its attractive masses of flowers. Branchlets are softly hairy. Despite its name, its leaves are mostly lanceolate to oblanceolate, 1 to 3 cm long, flat, tapering into a stalk, strongly aromatic, with entire or slightly lobed margins. Flowers are purple or mauve. Both anther-appendages are short. Flowers are borne abundantly in short, slender clusters. The leaves below the flowers are reduced to broad bracts which fall off when the plant is in bud.

HABITAT Forested hillsides and ridges, especially amongst rocks.

DISTRIBUTION Isolated occurrences in Hawkesbury River district (e.g., Mooney Mooney, Wiseman Ferry), Blue Mountains (e.g., Mt Banks) and Colo Plateau (e.g., Culoul Range). Recorded throughout coast, tablelands and western slopes; also Vic and Qld.

FLOWERING August–November.

1043. Prostanthera sieberi

Prostanthera sieberi **Plate 1043**

P. sieberi is a tall shrub with downy stems. Leaves are hairless, flat, broad-ovate to diamond-shaped, 5 to 15 mm long, coarsely and irregularly toothed, narrowing to a long stalk, with a strong aroma when crushed. The lower leaves are larger than leaves near the tips of the branches. Flowers are purple to mauve. Both anther appendages are short, but one is slightly longer than the other. Calyx lobes are blunt, hairless and nearly equal in size. Flowers are borne in short loose clusters. The leaves below the flowers are similar to the lower leaves but slightly smaller. This species is easily confused with *P. incisa* var. *incisa* as the leaves are similar in shape. *P. incisa* has hoary minutely hairy leaves and a lower calyx lobe longer and slightly narrower than the upper lobe.

HABITAT Sheltered gullies.

DISTRIBUTION Scattered along central coast (e.g., Royal National Park, Helensburgh). Extending to central tablelands and lower north coast.

FLOWERING September–October.

Prostanthera incisa var. *incisa*

Cut-leaf Mint-bush **Plate 1044**

This is a variable spreading shrub, 1 to 2 m tall. Leaves are hoary with minute hairs, flat, broad-ovate, 10 to 25 mm long, with 2 to 4 pairs of obtuse teeth, green above, paler below, narrowing to a stalk about 5 mm long, with a strong aroma when crushed. Flowers are pale mauve. Both anther appendages are shorter than the anther cell. The lower calyx lobe is longer and slightly narrower than the upper lobe. Flowers are borne in short clusters. Leaves below the flowers are much smaller than other leaves. This species is easily confused with *P. sieberi* (see that species for differences).

HABITAT Moist, sheltered forest.

DISTRIBUTION Hawkesbury River to Newcastle (e.g., Mooney Mooney Creek, Watagan State Forest). Also recorded south coast and northern tablelands.

FLOWERING September–October.

1044. Prostanthera incisa var. incisa

Prostanthera incisa var. *pubescens* **Plate 1045**

This is a medium-sized shrub. The whole plant is covered with conspicuous soft downy hairs. Leaves are about 3 cm long, coarsely toothed, paler below and have a short stalk. Flowers are mauve. Both anther appendages are short. The lower calyx lobe is longer and narrower than the upper lobe. Flowers are borne in small clusters at the ends of short side branches.

HABITAT Sheltered gullies, often on the edge of rainforest.

DISTRIBUTION Restricted to small areas near Gosford, Ourimbah and Strickland State Forest. Also Port Macquarie (north coast) and south coast. A small-leaved form is recorded from Norton's Basin (Nepean River).

FLOWERING September–October.

1045. Prostanthera incisa var. pubescens

Prostanthera violacea

Violet Mint-bush **Plate 1046**

This spreading, much-branched shrub (to 2 m tall) has slender branches which may be smooth or covered with soft hairs; a sprinkling of short stiff hairs covers its leaves. Leaves are small, 2 to 6 mm long, ovate, dark green above, wrinkled and with

1046. Prostanthera violacea

margins curved under; they are very aromatic. Flowers are mauve. Both anther appendages are short. The lower calyx lobe is slightly longer and narrower than the upper lobe. This species may be confused with *P. incana*, which occurs in similar habitats. *P. incana* is more robust, has larger leaves, and usually has darker flowers.

HABITAT Sheltered forest and gullies.

DISTRIBUTION Nepean River and Nattai River valleys and Blue Mountains (e.g., Pierces Pass, Mt Wilson). Also occurring south coast and southern tablelands.

FLOWERING September–November.

1047. Prostanthera incana

Prostanthera incana — Plate 1047

The specific name 'incana' is a reference to the grey, hoary hairs which cover the stems and leaves of this spreading shrub (1 to 2 m tall). Leaves are ovate, 5 to 20 mm long, wrinkled, bluntly toothed, with margins curved downwards. Flowers are mauve, often with a dark purple centre (a Blue Mountains form has pink flowers). Both anther appendages are short (or absent). The lower and upper calyx lobes are both broad and, unlike those of *P. violacea*, the lower calyx lobe is scarcely longer than the upper lobe. Flowers are borne in small crowded clusters.

HABITAT Sheltered forest.

DISTRIBUTION Nepean River and Nattai river valleys, Bundanoon, Wingello State Forest, Blue Mountains and Gosford area. Also occurring south coast, southern tablelands and lower north coast.

FLOWERING September–October.

Prostanthera hirtula — Plate 1048

P. hirtula is a shrub to 2 m tall, with branchlets and leaves covered with short rough hairs. Leaves are narrow lanceolate, about 15 mm long and 4 mm wide, blunt, entire and with margins curved downwards; they are green and scabrous above, paler underneath. Flowers are mauve with a purple throat. Both anther appendages are short. Calyx lobes are

1048. Prostanthera hirtula

almost equal. Flowers are borne on short stalks in loose clusters at the ends of branchlets.

HABITAT Sheltered forest.

DISTRIBUTION Woronora Plateau (e.g., Carrington Falls, Belmore Falls). Also recorded south coast, southern tablelands, central tablelands and Vic.

FLOWERING September–October.

Prostanthera denticulata

Rough Mint-bush — Plate 1049

Rough Mint-bush is a variable small shrub with a number of forms which merge into forms of *P. hirtula*. The typical form in the Sydney district is a low scrambling shrub, seldom 1 m tall. Leaves are ovate to lanceolate, 5 to 12 mm long, stalkless or almost so, rough above, scattered distantly along the stems and with margins rolled under. Flowers are violet to lilac. One anther appendage is longer than the other. The calyx lobes are reddish, hairy, broad and nearly equal. Buds are enclosed by small pointed bracts which fall off. Flowers are borne in small clusters at the ends of the branches.

An unnamed upright shrub (to 1 m tall) with affinities to *P. denticulata–P. hirtula* (Plate 1050) occurs on the rocky escarpment above the Kowmung River on Scotts Range. It has small, crowded, scabrous, almost triangular leaves with down-turned margins. Flowers are violet, small (8 mm long) and borne in the upper axils. The calyx and outside of the flower are covered with short hairs.

HABITAT Sheltered forest, especially on Narrabeen shales.

DISTRIBUTION Restricted in the Sydney district to a few sites in and around Mona Vale (Katandra Reserve), West Head (Ku-ring-gai Chase National Park), French's Forest and Cowan. Also recorded Goulburn River Valley, Pilliga (north-western slopes), central tablelands, southern tablelands, north coast, south coast and Vic.

FLOWERING September–October.

1049. Prostanthera denticulata

1050. Prostanthera denticulata-hirtula

Prostanthera linearis

Narrow-leaf Mint-bush — Plate 1051

This is a slender, erect shrub, sometimes to 3 m tall. Leaves are narrow, 12 to 25 mm long, oblong to linear, blunt at the tip, mainly hairless, with entire margins which are curled under. Flowers are a paler colour, whitish to mauve, often with yellow spots in the throat. One anther appendage is longer than the other. Flowers are borne either from the leaf axils or in clusters at the ends of branches.

HABITAT Sheltered forest, often along streams.

DISTRIBUTION Widespread from coast to mountains. Recordings include Woronora River, Mill Creek (Dharug National Park), Berowra, Pearl Beach–Patonga, Tallowa Dam, Wollangambe Wilderness and Mt Hay road. Also occurring south coast, north coast and Qld.

FLOWERING August–March.

1051. Prostanthera linearis

1052. *Prostanthera scutellarioides*

Prostanthera scutellarioides Plate 1052

This is a medium-sized shrub up to 1 m tall. Leaves are narrow, about 10 mm long, oblong to linear, hairless, acute at the tip, with margins curled under. Flowers are purple, or a deep lilac. The calyx is 5 to 6 mm long; the upper lobe is slightly longer than the lower lobe. A feature of this species is the presence of a prominent line of fine white hairs inside the base of the upper calyx lobe. Anther appendages are absent (or very rudimentary). Flowers are borne on short stalks from the leaf axils.

A form with darker and more abundant flowers from the Bundanoon–Wingello area may be an unnamed species. Its stems, leaves and calyx lobes are glandular-hairy and leaves are rough to the touch. Flowers have purple spots in the throat and purple anthers contrasting with a deep mauve corolla. It flowers from September to November.

HABITAT Forest, often in open situations on sandy soils.

DISTRIBUTION Type form occurs around Castlereagh (Cumberland Plain), with reports from Woronora Plateau, southern highlands and scattered locations in Blue Mountains (e.g., Bell). Also occurring north coast, northern tablelands and Qld.

FLOWERING July–December.

Prostanthera howelliae Plate 1053

A small erect shrub, seldom reaching 1 m tall. Its stems and leaves are rough to touch. Leaves are lanceolate, 3 to 5 mm long, blunt at the tip, with margins curled under, stalkless and hairy underneath. Flowers are dark mauve. Anther appendages are absent. The calyx has broad lobes covered with short stiff hairs; the upper calyx lobe is larger than the lower lobe. Flowers are borne from the leaf axils near the ends of the branches.

HABITAT Open forest.

DISTRIBUTION Uncommon. Recorded from Sackville–Maroota area, with old records closer to Sydney (e.g., Lane Cove River). Also occurring near Orange (central tablelands) and Bylong (central western slopes).

FLOWERING October–November.

1053. *Prostanthera howelliae*

1054. *Prostanthera rhombea*

Prostanthera rhombea Plate 1054

A small shrub with straggling branches which are usually covered with soft downy hairs. Leaves are round to almost diamond-shaped (rhomboidal), only 4 to 6 mm long, wrinkled and sparsely hairy, especially around the margins. Flowers are small and mauve. Anther appendages are short and of unequal length. The calyx is slightly downy and marked with glandular dots; calyx lobes are nearly equal, with the upper lobe broad and the lower lobe narrower. Flowers are borne in the upper leaf axils on very short stalks.

HABITAT Forest.

DISTRIBUTION Scattered from coast to mountains. Recordings include Douglas Park, Brisbane Water National Park and Bell. Extending to the lower north coast.

FLOWERING August–October.

1055. *Prostanthera densa*

Prostanthera densa Plate 1055

This compact shrub (to 1 m tall) is a rare and endangered plant, being restricted to a few scattered sites along rocky sea-coasts. Its stems are densely hairy. Leaves are ovate to heart-shaped, 5 to 10 mm long, soft, somewhat succulent, shortly stalked, dark green and rough above, paler below, with margins curled under and covered with short stiff white hairs. Leaves are usually densely crowded along the smaller branches, especially near the ends. Flowers are light mauve with white and orange markings in the throat; they are borne in the leaf axils on short stalks (2 mm long) and subtended by 2 narrow bracts. One anther appendage is twice as long as the anther cell. The calyx lobes are almost equal, green and covered in short stiff hairs.

HABITAT Rocky slopes near the sea.

DISTRIBUTION Rare. Recorded from Cronulla and Marley (Royal National Park); also Port Stephens area (north coast).

FLOWERING May–December.

1056. *Prostanthera marifolia*

Prostanthera marifolia Plate 1056

This is a small rare shrub, restricted to just a few locations along the coastal zone. Branches are up to 1 m tall, covered with soft down or short rigid hairs. Leaves are triangular-ovate, 6 to 10 mm long, about 6 mm wide, obtuse, stalkless or very shortly stalked, with rough glandular hairs on the upper surface and lighter below. These leaves are somewhat similar to those of the closely related *P. densa*, but are less thick, less recurved and usually do not have a cordate base. Leaves are scattered along the branches, in pairs about 2 to 3 cm apart, except at the ends of the branchlets where they are bunched. Flowers are mauve and borne from the leaf axils; they are smaller than those of *P. densa* (the lower centre corolla lobe is about 7 mm long, compared to about 10 mm in *P. densa*). One anther appendage is 1 to 1½ times as long as the anther cell (shorter than that of *P. densa*). The calyx lobes are both broad, almost equal and more or less covered in long weak hairs.

HABITAT Open forest on protected slopes.

DISTRIBUTION Restricted to a few sites near Helensburgh. Also recorded on south coast (Beecroft Peninsula) and north coast.

FLOWERING October–November.

Prostanthera saxicola var. *saxicola*

Slender Mint-bush Plate 1057

Slender Mint-bush is a low shrub, with straggling branches seldom more than 60 cm tall. It is very variable in the size and shape of its leaves, the hairiness of its stems and leaves and the size and colour of its flowers. At least 2 of the named varieties occur in the Sydney district. Variety *saxicola* has flat linear to elliptic leaves, up to 10 mm long and 4 mm wide, entire, obtuse and somewhat thick. Flowers are mauve (occasionally white), with a calyx 4 mm long and corolla 8 mm long. One anther appendage is about twice as long as the anther cell. Flowers are few and borne in the upper leaf axils.

Another form occurs along Heathcote Road between Lucas Heights and Sandy Point. It has flat,

1057. *Prostanthera saxicola* var. *saxicola*

elliptic leaves up to 8 mm long, appressed white hairs on the stems, loose scattered hairs on the leaves and calyx, and the stems and leaves are rough to touch. Flowers are lilac. It may be a new variety or be classified as a separate species when the genus is revised.

HABITAT Forest and heath.

DISTRIBUTION Coast to mountains. Also occurring south coast, southern tablelands, northern tablelands and north-western slopes.

FLOWERING May–September.

1058. Prostanthera saxicola var. *montana*

1059. Prostanthera rotundifolia

Prostanthera saxicola var. *montana*

Plate 1058

This variety is an undershrub about 30 cm tall. It occurs at higher altitudes. Its leaves are flat, elliptic, 7 to 9 mm long and 2 to 4 mm wide, entire, obtuse and somewhat thick. Flowers are white, with a purple throat and yellow dots, larger than those of variety *saxicola*, with a calyx about 8 mm long and corolla 12 mm long.

HABITAT Forest and heath.

DISTRIBUTION Blue Mountains (e.g., Narrow Neck, Kanangra Walls). Extending to northern tablelands and south coast.

FLOWERING December.

Prostanthera rotundifolia

Round-leaf Mint-bush **Plate 1059**

A tall, compact shrub, often with hoary-downy stems. Its leaves are flat, rounded to spoon-shaped, about 10 mm long, with stalks about 4 mm long; they are paler below and emit a strong aroma when crushed. Flowers are lilac to purple and anther appendages are both very short. The middle lobe of the lower lip of the flower is larger than those on either side. The calyx is dull green, with 2 equal broad, rounded lobes. Flowers are borne in short clusters either in the leaf axils or at the ends of branchlets.

HABITAT Forest, often in rocky areas.

DISTRIBUTION Uncommon in Sydney district, occurring mainly on western fringe (e.g., Newnes, Glen Davis). Recorded throughout coast and western slopes divisions; also southern tablelands, Vic, Tas and SA.

FLOWERING September–October.

Prostanthera rugosa **Plate 1060**

A slender shrub to 2 m tall, with branches covered with short stiff white hairs. Leaves are small, 3 to 6 mm long, stalkless, very wrinkled, with crinkled margins which are curved under. The upper surface often has a few short tooth-like hairs on the margins. The lower surface is hairless. Flowers are purple, and borne abundantly along the underside of branchlets. One anther appendage is longer than the other but shorter than the anther cell. Calyx is purple-green, hairless inside, covered with glands and short stiff white hairs outside; the lower lip is slightly longer than the upper lip.

1060. Prostanthera rugosa

HABITAT Sheltered slopes and gullies.

DISTRIBUTION Uncommon in Sydney district; recorded from above Wingecarribee River at Berrima West. Extending along coastal zone south from Bulahdelah area (north coast) to south coast.

FLOWERING September–October.

1061. Prostanthera cryptandroides

Prostanthera cryptandroides **Plate 1061**

This small rare shrub (to 60 cm tall) is regarded as endangered because of its restricted distribution. It has sticky-downy stems and glossy sticky-varnished foliage which has a strong aroma when crushed. Leaves are about 10 mm long and 3 mm wide, deeply divided into rounded lobes. Flowers are mauve, with whitish markings and yellow dots in the throat; they are borne singly or in opposite pairs near the ends of the branches. Flowers are relatively large (up to 18 mm long). The anther appendage exceeds the anther cell on all 4 stamens. The calyx is about 6 mm long, with a broad upper lobe and narrow lower lobe.

HABITAT Forested slopes and gullies, in rocky areas, especially at the base of scree slopes.

DISTRIBUTION Glen Davis–Capertee area, extending to Goulburn River valley. Also recorded near Yetman (north-western plains).

FLOWERING May–December.

Prostanthera sp. **Plate 1062**

This small unnamed hairless shrub is found in the crevices of sandstone 'pagoda' rock formations in the Clarence–Wolgan area. Leaves are flat, ovate to lanceolate, 12 to 15 mm long, 5 to 8 mm wide, thick, paler underneath, often with a purplish tint and with a stalk about 1.5 mm long. Leaves just below flowers are much smaller than the lower leaves. Flowers are deep purple, large (15 mm long) and borne singly in the upper leaf axils. Anther appendages are equal. The corolla is hairless in the throat but short hairs cover the lobes and the outside. The lower calyx lobe is slightly shorter than the upper lobe, or similar; the upper part of the calyx is purplish, the inside and underneath is green. The calyx is subtended by 2 linear bracts, 2.5 mm long.

HABITAT Crevices of sandstone rock formations.

DISTRIBUTION Clarence–Wolgan area, possibly extending to Khyber Pass and Upper Cudgegong River, east of Rylstone.

FLOWERING September–October.

1062. Prostanthera sp. (Clarence-Wolgan)

Plectranthus

Herbs, with opposite leaves. Flowers tubular, with upper lip 4-lobed and lower lip entire; calyx 2-lipped, the upper lip entire, the lower lip with 4 acute teeth; 4 stamens.

1063. Plectranthus parviflorus

Plectranthus parviflorus

Cockspur Flower **Plate 1063**

A perennial herb with semi-succulent stems, which may be sprawling or upright to 40 cm tall. Leaves are opposite, ovate to orbicular, 2 to 7 cm long, toothed and on long stalks. Leaves and stems may be covered with soft hairs or almost hairless. Flowers are purple to blue-mauve, and borne on erect leafless stems. Each flower is 6 to 10 cm long and curved below the middle. The inflorescence has an unpleasant odour when crushed. The plant dies back to its fleshy tuberous base after flowering. An allied species, *P. graveolens*, occurs on rocky outcrops of the Blue Mountains. It is a robust plant with a mixture of glandular and long loose hairs on its stems and leaves.

HABITAT Rocky outcrops in open forest and rainforest margins.

DISTRIBUTION Widespread from coast to mountains. Also occurring south coast, north coast, northern tablelands, western slopes, Qld and Vic.

FLOWERING August–March.

Prunella

Herb, with opposite leaves. Flowers tubular, with upper lip entire and erect, and lower lip 3-lobed and spreading; calyx with 3-toothed upper lip and 2-toothed lower lip; 4 stamens.

1064. Prunella vulgaris

Prunella vulgaris

Self-heal **Plate 1064**

Self-heal is a perennial herb, often mistaken for a weed especially when it occurs in gardens and damp roadsides. Although a native species, it is also abundant in damp pastures and along the roads of Britain and Europe. It has a creeping rootstock, with ascending flowering branches, up to 30 cm tall. The stem is square. Leaves are opposite, ovate to lanceolate, 2 to 5 cm long, stalked and usually with entire margins. Both stem and leaves are more or less hairy. Flowers are purple, 7 to 9 mm long and arranged in dense terminal clusters; flowers are in whorls within these clusters, each whorl surrounded by 2 broad bracts. Prior to modern medicine, Self-heal was considered one of the most useful medicines for internal and external use.

HABITAT Moist places, especially cleared and disturbed areas, and rainforest margins.

DISTRIBUTION Widespread from coast to mountains. Also occurring south coast, southern tablelands, northern tablelands, western slopes, Qld, Vic, Tas, SA and Europe.

FLOWERING November–December.

Scutellaria Skullcap

Herbs with erect stems. Flowers with a blue-mauve corolla tube and small lobes; upper lobe concave, lower lobe divided into 3 spreading segments. Stamens 4, in pairs, enclosed in the upper lobe. Calyx divided into 2 lobes, the upper lobe having a hump or protuberance from which the common name is derived.

1065. Scutellaria humilis

Scutellaria humilis

Dwarf Skullcap **Plate 1065**

This is a small perennial with a creeping rootstock and stems ascending to about 15 cm tall. Stems may be covered with minute appressed white hairs or be almost hairless. Leaves are opposite, triangular-ovate, 8 to 12 mm long, bluntly toothed, often heart shaped at the base and with a short stalk. Flowers are small (about 5 mm long) and mauve. A characteristic feature is that from the pairs of upper leaves there is a pair of flowers; each flower is on a fine stalk and both stick out in the same direction.

HABITAT Open forest, often on flood-plains or clay soils.

DISTRIBUTION Widespread but uncommon. Recorded on old river sediments at Menangle and rainforest margins in Olney State Forest. Occurring throughout coast, tablelands and western slopes; also Qld, Vic and Tas.

FLOWERING September–December.

Hemigenia

Shrubs with lilac flowers; upper lip 2-lobed, lower lip longer and 3-lobed. Differs from *Prostanthera* in having 5-toothed calyx (not 2-lipped) and from *Westringia* in having 4 perfect stamens. Leaves in whorls of 3 to 4.

1066. Hemigenia purpurea

Hemigenia purpurea

Common Hemigenia **Plate 1066**

This is a common slender undershrub on sandy heaths around Sydney. Leaves are terete, channelled above, crowded, up to 15 mm long. Flowers are lilac and borne singly in the upper leaf axils. Less common is *H. cuneifolia*, a shrub of open forests from the central and north coasts. It has flat, oblong to wedge-shaped leaves.

HABITAT Heath, on sandy soils.

DISTRIBUTION Restricted to the central coast and Blue Mountains. Common in coastal heaths.

FLOWERING June–January.

Westringia

Shrubs with leaves in whorls of 3 or 4. Flowers with upper lip erect and broadly 2-lobed, lower lip spreading and 3-lobed. There are 2 fertile stamens and 2 lower stamens reduced to staminodes, separating this genus from *Hemigenia*. The calyx is 5-toothed, separating it from *Prostanthera*.

1069. Westringia eremicola

1067. Westringia fruticosa

Westringia longifolia

Long-leaved Westringia **Plate 1068**

This is a slender shrub, 1 to 2 m tall, with hairless leaves in whorls of 3. Leaves are narrow, linear, 2 to 5 cm long, 2 to 3 mm wide, flat, with the midvein depressed on the upper surface, shining green above, duller below. Flowers are borne singly from the leaf axils. The corolla is about 8 mm long, softly downy outside, with the floral tube as long as or slightly exceeding the calyx teeth. Flowers are mostly white, but some plants have pale mauve flowers (such as those at Colo River).The calyx is 2 to 3 mm long, green and often fringed with hairs.

HABITAT Rocky slopes and gullies.

DISTRIBUTION Scattered throughout Sydney area (e.g., upper Georges River, Appin, Colo River, Cumberland Plain, Gosford district). Also recorded central tablelands and central western slopes.

FLOWERING July–December.

Westringia eremicola

Slender Westringia **Plate 1069**

This is a widespread species found from coast to drier inland. It is a slender bushy shrub, 1 to 2 m tall, with leaves in whorls of 3. Leaves are narrow, linear, 5 to 25 mm long, 1 to 2 mm wide, spreading, often hoary-grey below, rough above and with margins rolled under. Flowers are borne singly, or 2 to 3 together from the leaf axils, usually towards the ends of the branches. The corolla is 5 to 7 mm long, with a tube exceeding the calyx. Flowers vary from pale mauve to white, with purple spots in the throat. The calyx is 2 to 3 mm long, hairless or sometimes fringed with hairs.

HABITAT Rocky areas, often in drier gullies. Occurring in red sands and mallee communities further inland.

DISTRIBUTION Restricted in Sydney district to Nattai valley and southern Blue Mountains. Also occurring south coast, southern tablelands, western slopes, western plains, Qld, Vic and SA.

FLOWERING August–October.

1068. Westringia longifolia

Westringia fruticosa

Coast Rosemary **Plate 1067**

This is a common compact shrub along sea-cliffs and rocky foreshores. In sheltered locations, it may be up to 1.5 m tall, but it is often dwarfed and gnarled as it clings tenaciously in the cracks of bare rocks right on the edge of the cliff line. Leaves are in whorls of 4, narrow lanceolate, 1 to 2 cm long, tough, spreading, hairless above and with margins curled underneath. The branches, undersurface of leaves and calyx are hoary-white with appressed hairs. Flowers are white, with pink or orange spots on the lower lip and are borne in the upper leaf axils.

HABITAT Heath and cliffs near the sea and along harbour foreshores.

DISTRIBUTION Coastal zone from far south coast to lower north coast. Also recorded central tablelands. Commonly planted in home gardens and used in landscape gardening.

FLOWERING Some flowers all year round.

Teucrium

Erect herbs with opposite leaves. Flowers white with 5 spreading lobes; the lower middle lobe is much larger than the others. Stamens 4, long and projecting beyond the floral tube.

Teucrium corymbosum

Forest Germander **Plate 1070**

A much-branched erect herb, mostly 50 to 60 cm tall. Leaves are thin, ovate to narrow lanceolate, 2 to 4 cm long, deeply toothed or lobed, stalked, strongly aromatic, the upper surface hairless, the lower surface covered with greyish hairs. Leaves beneath flowers are smaller than the lower leaves. Flowers are white and borne in small loose clusters. Corolla is divided into 5 lobes, 4 almost equal; the lower middle is longer and broader. Calyx is tubular, hoary, with 5 equal teeth. Stamens project conspicuously above and beyond the flower.

HABITAT Sheltered forests, chiefly on richer soils. Occurring in red sands and mallee communities further inland.

DISTRIBUTION Widespread from coast to mountains, but not common. Recordings include Macquarie Pass, Mt Tomah and Wombeyan Caves. Found throughout all areas of NSW, except far western plains; also Qld, Vic, SA and Tas.

FLOWERING December–January.

1070. Teucrium corymbosum

Ajuga

Erect herb, with opposite leaves. Flowers purple with long spreading lower lip and small upper lip. Stamens 4, projecting beyond the upper lip.

Ajuga australis

Austral Bugle **Plate 1071**

A softly hairy perennial, with leafy ascending stems 4 to 30 cm tall. Leaves at the base of the stems are obovate, up to 15 cm long and 4 cm wide, stalked, with irregular broad, rounded lobes. Stem leaves are smaller, stalkless and often entire. Flowers are purple and borne in whorls of 6 to 20 flowers in the axils of the stem leaves. Calyx is 5-toothed and 4 to 8 mm long. The flower may be 20 mm long, dominated by the spreading 3-lobed lower lip. Coastal forms are less robust than inland forms.

1071. Ajuga australis

HABITAT Open forests, especially on Wianamatta shales; occasionally weeds in waste places.

DISTRIBUTION Scattered, but uncommon in Sydney district (e.g., Tallong). Occurring throughout NSW; common on western slopes and plains; also Qld, Vic, SA and Tas.

FLOWERING October–January.

1072. Wahlenbergia gracilis

Campanulaceae

Campanulaceae is a family of mostly annual and perennial herbs centred in the temperate regions of the northern hemisphere and represented in Australia by 15 to 20 species of *Wahlenbergia*, a widespread genus with a mainly southern distribution.

The name of the type genus, *Campanula*, refers to the campanulate (bell-shaped) flowers that are characteristic of the family. Since the predominant flower colour is blue a number of species in various countries are known as bluebells (English Bluebell is in Liliaceae).

The flowers of *Wahlenbergia*, and the family generally, are protandrous, i.e., the stamens mature before the gynoecium to increase the probability of cross-pollination. In this case the style is sticky or hairy and catches pollen released before the flowers open. Some time after opening, when most of the pollen has been carried away, the stigma surfaces open and become receptive.

The Campanulaceae has a number of features in common with the Asteraceae, the daisy family, which suggests that it is a fairly advanced family.

Wahlenbergia Bluebells

Wahlenbergia gracilis

Australian Bluebell **Plate 1072**

This is a many-stemmed annual or perennial, growing after favourable rains and flowering in late spring and early summer. Stems are erect or sprawling, 10 to 50 cm tall and leaves are small, entire, lanceolate, to 15 mm long and 1 to 2 mm wide. The flowers are the smallest of all local Bluebell, only 5 to 6 mm across; this separates this species from *W. communis* which has flowers twice as large.

HABITAT Open forest and woodland.

DISTRIBUTION Widespread from coast to mountains. Also occurring north coast, south coast, southern tablelands, western slopes, Qld, Vic, Tas and Lord Howe Island.

FLOWERING November–January.

1073. Wahlenbergia communis

Wahlenbergia communis

Tufted Bluebell **Plate 1073**

Usually a many-stemmed sprawling herb, with an extensive rhizome system. Flowering stems are mostly branched, 15 to 45 cm tall. The lower leaves are opposite or alternate, narrow and about 3 cm long. Flowers are blue, bell shaped, 10 to 15 mm across. The capsule is an elongated cone, surmounted by persistent calyx lobes. Tufted Bluebell is sometimes confused with *W. gracilis* but that species has much smaller flowers (5 to 6 mm across).

HABITAT Forest and woodland, especially among grassy understorey.

DISTRIBUTION Widespread from coast to mountains. Recorded throughout the State; also Qld, Vic, NT and SA.

FLOWERING Chiefly September–February, but some flowers at other times.

Wahlenbergia graniticola

Granite Bluebell **Plate 1074**

An upright herb with just a few stems, it is found in cooler parts of the tablelands especially in rocky granite areas. Its flowers are blue to almost mauve, 20 to 25 mm across, bell shaped but with 5 spreading star-like lobes. The style is thickened around the middle, an important feature distinguishing this Bluebell from others in the area. The upper stems are bare; lower leaves are alternate, small and narrow.

HABITAT Open forest in rocky situations.

DISTRIBUTION Higher Blue Mountains and southern highlands.

FLOWERING October–November.

1074. Wahlenbergia graniticola

Wahlenbergia stricta

Tall Bluebell **Plate 1075**

A perennial herb with a few stems, usually about 30 cm tall. The flowers are solitary on long stalks and are larger than those of any other Bluebell in the area. The corolla is bell shaped, about 30 mm across, blue with a white centre. Basal leaves are opposite, narrow oblong, about 10 mm long, 5 to 8 mm wide, often wavy edged and hairy. The fruiting capsule is globular to egg shaped, and surmounted by persistent calyx lobes.

HABITAT Forest and woodland.

DISTRIBUTION Widespread from coast to mountains. Occurring throughout the State, except far western plains, and in all other States, except NT.

FLOWERING October–November.

1075. Wahlenbergia stricta

Lobeliaceae

The Lobeliaceae differs structurally from the Campanulaceae in having irregular flowers with anthers fused into a tube. It also has a different distribution, being centred in the tropical Americas and much better represented in the southern hemisphere. Although the family is herbaceous, some giant *Lobelia* species from mountainous parts of Africa attain heights of up to 8 m and inflorescences up to 4 m long.

In Lobeliaceae the 5 fused petals form 2 lips which may be equal or unequal. The corolla tube is entire or notched, or slit part way, or to the base. The staminal filaments are fused to the tube at their bases, and to each other at the top. The fused anthers usually have bristles at the apex which assist in pollen release. The flowers are mostly bisexual but the 2 common species of *Pratia* in the Sydney area are unisexual.

As with Campanulaceae the flowers are protandrous. After the flowers open, the style emerges through the anther tube pushing the pollen out the top where it can be collected by insects. The delayed opening of the stigmatic lobes ensures that cross-fertilisation is favoured over self-fertilisation.

Many of the Lobeliaceae contain alkaloids, and some local species of *Pratia* and *Isotoma* have an irritant milky sap and are considered poisonous.

1076. Pratia pedunculata

Pratia

Small herbs, with 5 spreading irregular petals. Floral tube split open to the base. Anthers and filaments are united in a column, the 2 front anthers bearing short bristly hairs. Flower colour ranges from white to pale blue and mauve. Fruit somewhat fleshy. Although 4 species occur in the Sydney district, only one *(P. purpurascens)* is widespread and common.

Pratia pedunculata **Plate 1076**

This is a ground-hugging plant (occasionally ascending when supported), with weak stems which creep over the mud. Leaves are alternate and spread out on either side of the stem; they are stalkless, ovate-elliptic, 6 to 14 mm long and unevenly toothed. Male and female flowers are borne on different plants; each is on a stalk 4 to 10 mm long. Flowers may be pinkish-white or purplish. Fruit is a capsule about 5 mm long.

HABITAT Swampy situations, often on damp peaty soils.

DISTRIBUTION Coast to mountains. Also occurring south coast, northern tablelands, southern tablelands, Vic and Tas.

FLOWERING December–March.

1077. Pratia purpurascens

Pratia purpurascens

White Root **Plate 1077**

A common small herb of damp and sheltered places. Stems rise from a rootstock and may be ground-hugging or ascending, seldom more than 12 cm tall. Leaves are alternate, elliptic, 10 to 25 mm long, toothed and usually purplish on the undersurface. Flowers are solitary on long stalks, white to pale purple, with 5 irregular lobes. Flowers are unisexual, with male and female flowers borne on separate plants. *Pratia* is often confused with *Lobelia* and *Isotoma*, but *Pratia*, unlike the others, has fleshy fruits and hairs on the 2 front anthers.

HABITAT Forest and woodland, usually in damp or sheltered situations.

DISTRIBUTION Widespread from coast to mountains. Also occurring south coast, north coast, northern tablelands, north-western slopes, Qld and Vic.

FLOWERING December–May.

1078. Lobelia trigonocaulis

Lobelia

Erect or semi-prostrate herbs, with 5 spreading irregular petals. The 3 front petals are broad and spreading, the 2 back petals are much reduced. Floral tube slit open to the base. Anthers and filaments united into a column; anthers all hairy at the apex, or at the apex and on the back of the anthers (unlike the anthers of *Pratia*). Flower colour ranges from white to deep blue, pink or mauve. Fruit a capsule.

Lobelia trigonocaulis Plate 1078

This is a low herb with weak 3-angled stems up to 30 cm long. It is easily identified by its heart-shaped leaves, 20 to 30 mm long, which are pointed, bluntly toothed and are on a stalk about as long as the leaf. Blue-mauve flowers are in loose terminal racemes of 3 to 6 flowers.

HABITAT In and near rainforest and sheltered eucalypt forest.

DISTRIBUTION Wyong–Gosford (e.g., Strickland State Forest), and Watagan Mountains (e.g., Gap Creek Flora Reserve, Olney State Forest). Extending along north coast and northern tablelands to Qld.

FLOWERING November–February.

1079. Lobelia gracilis

Lobelia gracilis Plate 1079

This is a trailing or procumbent plant, with stems up to 30 cm tall, often branched, with more or less erect flower stems. Flowers are deep blue, borne singly on slender stalks. The main feature of this species is that the 3 front petals are broad-obovate (rounded and broader near the tip), unlike the narrow-oblong petals of *L. dentata* and *L. gibbosa.*

HABITAT Open forest, on sandy soils.

DISTRIBUTION Coastal zone. Extending along north coast and northern tablelands to Qld.

FLOWERING November–May.

1080. Lobelia dentata

Lobelia dentata Plate 1080

An erect annual herb with weak stems up to 40 cm tall. The lower leaves are ovate, 15 to 30 mm long, 4 mm wide, stalked and deeply toothed. Flowers are 15 to 25 mm long, deep to bright blue, on stalks about 20 mm long and are borne in racemes of 8 to 10 flowers. The 3 front petals are narrow-oblong (unlike those of *L. gracilis* which are obovate); the middle petal is longer than the others. *L. dentata* is often confused with *L. gibbosa* which also has 3 narrow-oblong front petals. The leaves of *L. gibbosa* are mostly linear, entire (sometimes with a few teeth), the flowers are on stalks about 10 mm long, and the capsule is gibbose (lobed or swollen) on the upper side.

HABITAT Forest and woodland, on sandy soils.

DISTRIBUTION Widespread from coast to mountains. Also occurring south coast, north coast, southern tablelands and Qld.

FLOWERING July–February.

1081. Lobelia gibbosa

Lobelia gibbosa Plate 1081

An erect annual herb with weak stems up to 40 cm tall. The lower leaves are mostly linear, up to 50 mm long, 4 mm wide, with entire margins (occasionally with a few teeth). Flowers are 10 to 25 mm long, bright blue, on stalks about 10 mm long and are borne in many-flowered racemes. The 3 front petals are narrow-oblong, the middle petal usually longer. The capsule is gibbose on the upper side.

HABITAT Forest, often in shaded spots.

DISTRIBUTION Widespread from coast to mountains. Occurring throughout coast, tablelands and western slopes and in all other States.

FLOWERING November–February.

1082. Lobelia alata

Lobelia alata Plate 1082

A small variable herb of damp situations, its stems are trailing or erect and somewhat angular. Leaves are variable, 15 to 50 mm long, 6 to 10 mm wide, obovate to almost linear, entire or with a few teeth. Flowers are white to pale blue, fan-shaped and are solitary in the leaf axils.

HABITAT Damp places in various communities from heath to rainforest. Common on wet rocks and drainage lines near the sea.

DISTRIBUTION Widespread from coast to mountains. Also occurring north coast, south coast, in all other States (except NT), NZ, Lord Howe Island, South Africa and South America.

FLOWERING November–January.

Isotoma

Prostrate herbs with a tubular corolla and 5 spreading, almost equal lobes. Stamens fused to the corolla tube. Anthers united; the 2 front anthers each bear a long bristle. Flower colour white to pale blue and mauve. Fruit a capsule.

1083. Isotoma axillaris

Isotoma axillaris

Rock Isotoma **Plate 1083**

This is a low bushy plant to 30 cm tall, widely distributed throughout south-eastern Australia, but most commonly found around boulders on the cooler tablelands. Its leaves are narrow, up to 7 cm long, deeply divided into jagged lobes. The flowers are easily identified as they are pale blue to mauve, with a long tube (25 mm) and 5 equal spreading lobes, 15 to 18 mm long. Flowers are on stalks up to 12 cm long. Fruit is a dry brown capsule.

HABITAT Rock crevices, especially granite.

DISTRIBUTION Widespread from coast to mountains, but generally absent from coastal sandstone areas. Typically found in rocky situations such as on the volcanic Mt Gibraltar between Mittagong and Bowral. Occurring throughout the State in all divisions; also Qld and Vic.

FLOWERING October–May.

1084. Isotoma fluviatilis ssp. *fluviatilis*

Isotoma fluviatilis ssp. *fluviatilis*

Swamp Isotoma **Plate 1084**

This is a prostrate creeping plant with ovate-elliptic, alternate leaves, 5 to 10 mm long and 3 to 5 mm wide. They are slightly toothed and spread out on one plane. Flowers are pale blue; the base of the petals is yellow, surrounded by a deep mauve edging. The inside of the corolla is hairless. In this subspecies, flowers are unisexual, with male and female flowers borne on separate plants. Each flower is on a stalk, 5 to 60 mm long. This is a variable species, forming a low compact mat in exposed sites, but having longer scrambling stems and larger leaves in sheltered sites.

Ssp. *borealis* occurs in wet bogs along the Great Dividing Range in northern NSW and southern Qld (e.g., Ben Bullen, Rylstone). The inner surface of the corolla is hairy; the corolla is 6 to 10 mm long.

Ssp. *australis* occurs from Bathurst southwards (e.g., Jenolan Caves, Paddys River, Goulburn); also Vic, SA and Tas. The inner surface of the corolla is hairy; the corolla is 7 to 15 mm long.

HABITAT Damp places, especially around the edges of dams and swamps.

DISTRIBUTION Coastal zone around Sydney (e.g., Wilberforce, Narrabeen Lake, Richmond, Wallacia) extending to lower Blue Mountains (e.g., Springwood) and north coast (Gloucester).

FLOWERING October–April.

Stylidiaceae

This small family is confined to Australia except for 8 species of *Stylidium* (trigger plants) in India, Sri Lanka, China and South East Asia, and 9 alpine species from 3 genera in New Zealand and South America. The largest genus, *Stylidium*, with about 140 species has its greatest development in south-western Australia where there are about 90 endemic species. Eight to 10 species of *Levenhookia* (styleworts) occur in southern Australia, again with major development in the south west. Six species of *Stylidium* and 1 of *Levenhookia* occur in the Sydney area where 4 of the *Stylidium* species are common and the others rare.

The family is characterised by a column formed by fusion of 2 stamens with the style. The anthers are sessile at the top of the column and conceal the stigma in newly opened flowers. Later, after the flower has released its pollen, the stigma enlarges and pushes the empty anthers apart enabling it to receive pollen from a visiting insect.

The column is sensitive in *Stylidium* and can be triggered and reset numerous times. The flowers have a tubular corolla with 5 deeply divided lobes. Usually the labellum (the lobe behind the column) is smaller than the others, and points downward. The other lobes, which are commonly pink or yellow, spread out in pairs. At the base of each lobe 2 or 3 tiny upright appendages form a trigger around the top of the tube. When the trigger is touched by an insect the reflexed column swings upwards then down onto the insect to deposit or pick up pollen.

Although *Levenhookia* are smaller than *Stylidium*, and are scarcely noticeable individually, they are sometimes massed in great numbers. Unlike *Stylidium, Levenhookia* can only be triggered once. In newly opened plants the labellum forms a sheath which covers the short erect column. As the flower ages the tip of the labellum becomes sensitive, and, if touched, causes the labellum to spring back exposing the column which moves forward spraying pollen onto the insect intruder.

The perennial species of *Stylidium* should be easy to cultivate since they are generally found in poor sandy soils and seem to withstand both dryness and periods of inundation.

Stylidium

Stylidium graminifolium

Grass-leaf Trigger Plant **Plate 1085**

A common plant throughout south-eastern Australia, extending from the coast to alpine areas and the drier western slopes. One or more long stalks, 30 to 40 cm tall, rise from a tuft of grass-like leaves and bear a number of bright pink flowers. The leaves are linear, up to 20 cm long and 4 mm wide, and rise in a single tuft from ground level, unlike those of *S. productum* which are scattered at intervals along the stems.

HABITAT Heath and open forest, on sandy soils.

DISTRIBUTION Widespread from coast to mountains. Found throughout coast, tablelands and western slopes; also Qld, Vic, Tas and SA.

FLOWERING October–December.

1085. Stylidium graminifolium

Stylidium productum

Plate 1086

This Trigger Plant closely resembles *S. graminifolium*, and was formerly regarded as a variety of that species. It is endemic in the Sydney district and less common than *S. graminifolium*. It has sprawling and elongated stems which may be 50 cm or more tall. Tufts of grass-like leaves are scattered at intervals along the stems, not rising from the base, as in *S. graminifolium*. Deep pink flowers are borne at the ends of the long stalks.

HABITAT Heath and open forest, on sandy skeletal soils.

DISTRIBUTION Scattered from coast to mountains. Recordings include Gosford, Cataract Dam, Pennant Hills Park, Woronora River, Oatley Park, Gunderman, Katoomba and Wentworth Falls.

FLOWERING November–January.

1086. Stylidium productum

1087. *Stylidium lineare*

Stylidium lineare

Narrow-leaf Trigger Plant **Plate 1087**

A small plant with a slender upright stem seldom more than 20 cm tall. Leaves are narrow linear, 2 to 5 cm long and about 1 mm wide and are in a small basal tuft. Flowers are smaller than those of *S. graminifolium*, with a corolla 5 to 8 mm long; they vary from light to deep pink.

Habitat Heath and open forest, on sandy soils.

Distribution Widespread from coast to mountains. Extending to south coast and southern tablelands.

Flowering September–December near coast, as late as April in mountains.

1088. *Stylidium laricifolium*

Stylidium laricifolium

Larch-leaf Trigger Plant **Plate 1088**

The common name of this Trigger Plant comes from the similarity of its leaves to those of the coniferous tree, the larch. These are linear, 1 to 4 cm long and about 1 mm wide, and are spirally arranged along the stem, a feature clearly distinguishing this Trigger Plant from others in the area. Stems are up to 50 cm tall and bear numerous pale-pink flowers.

Habitat Forests, often on hillsides, in sandy soils.

Distribution Widespread from coast to mountains. Recordings include Thirlmere Lakes, Erskine Creek, The Basin (upper Georges River) and Winifred Falls (Royal National Park). Found throughout coast and tablelands divisions; also central western slopes, Qld and Vic.

Flowering August–November.

Goodeniaceae

Goodeniaceae, with about 16 genera and 400 species, is largely confined to Australia, particularly to the south-western part of the continent. The majority of species are in the genera *Goodenia* (170 species), *Scaevola* (110 species) and *Dampiera* (70 to 80 species). Half the genera have only 1 or 2 species each and are exclusively Western Australian. Possibly the most attractive and commonly cultivated genus is *Lechenaultia* (about 20 species) which is widespread but absent from south-east Australia.

In the Goodeniaceae the corolla tube is slit almost to the base on the upper side. The 5 corolla lobes may spread more or less evenly as in *Scaevola* and *Selliera* or form an upper lip of 2 lobes and a lower lip of 3 as in *Goodenia, Velleia,* and *Dampiera*. The lobes have a thickened tapering central portion, which is usually hairy on the back, and thin wings which may be narrow and entire or broad and undulate. In *Dampiera*, and to a lesser extent, *Goodenia*, the outer wings of the upper lobes are infolded into a rounded auricle which surrounds the indusium. The indusium, which serves to identify the family, is a 2-lipped pollen cup surrounding the stigma. It may be rounded and glabrous *(Dampiera)* or flattened and densely hairy on the ends and inside *(Goodenia, Velleia, Scaevola)*. Pollen is shed from the anthers and falls into the indusium before the flowers open but does not cause self-fertilisation because the stigma is undeveloped and unreceptive. Later, after the pollen is released, the stigma enlarges and becomes receptive.

Velleia

Herbs with short thick rootstocks and leaves in basal tufts. Flowers yellow, with 2 upper and 3 lower petals. Indusium with stiff hairs; ovary superior. Characteristic opposite bracts occur around each fork of the branched inflorescence.

1089. *Velleia lyrata*

Velleia lyrata

Plate 1089

An erect herb of moist sandy soils. Leaves are in basal tufts, up to 10 cm long, blunt and entire at the tip, deeply toothed near the base. The name 'lyrata' refers to the lyrate leaves (leaves with a large terminal lobe and smaller side lobes). Flowers are yellow and borne on branched, leafless, hairless stems up to 30 cm tall. Each flower is terminal above 3 conspicuous leafy green heart-shaped or round sepals.

Habitat Heath or woodland, on open moist sandy soils or mossy sandstone rocks.

Distribution Gosford district and lower Blue Mountains. Recordings include Mt White, Mangrove Creek, Patonga, Glenbrook and Bowens Creek. Endemic in the Sydney district.

Flowering June–July.

1090. *Velleia spathulata*

Velleia spathulata

Plate 1090

V. spathulata is similar to *V. lyrata* as it has leaves in basal tufts, leafless flowering stems and yellow flowers above broad green sepals. The main differences are that *V. spathulata* has weak spreading flowering stems (up to 20 cm long) which are rather hairy (when viewed under a lens), 3 sepals which are lanceolate to ovate, and obovate-spathulate leaves, up to 12 cm long which are usually bluntly toothed.

Habitat Heath and woodland, on open moist sandy soils.

Distribution Coastal zone north of Sydney. Extending along north coast to Qld and New Guinea.

Flowering October–January.

Coopernookia

Herbs with mauve-pink flowers; petals unequal and anthers free. Differs from *Goodenia* in having 2 ovules and 2 seeds in each ovary. Fruit a capsule.

1091. Coopernookia barbata

Coopernookia barbata

Coopernookia **Plate 1091**

Formerly regarded as a mauve-pink flowered *Goodenia, Coopernookia* has been established as a separate genus on the basis of different ovary and seed structure. It is easily identified by its flower colour. Although locally common in parts of the tablelands, it is not often seen around Sydney. It is an erect plant, 50 to 60 cm tall, with short narrow leaves, about 25 mm long, glandular-hairy and with margins curled under. Flowers are on stalks from the leaf axils and have conspicuous pale hairs in the throat.

HABITAT Open forest, usually in sheltered situations.

DISTRIBUTION Lower Blue Mountains, Nattai and Wollemi. Also occurring south coast, southern tablelands, northern tablelands, western slopes and Vic.

FLOWERING Chiefly September–October, but as late as April in some areas.

Goodenia

Herbs or small shrubs with yellow flowers. Petals uneven with 3 spreading lower lobes and 2 erect back lobes. Stamens free; indusium covered with stiff hairs. Ovary under the flower, not superior as in *Velleia*. Fruit a capsule. Species are separated into 2 groups on the basis of the presence or absence of stellate hairs on the outside of the corolla.

1092. Goodenia stelligera

Goodenia stelligera **Plate 1092**

This species is characterised by an erect stem, 30 to 40 cm tall, rising from a rosette of basal leaves. These leaves are hairless, thick, narrow lanceolate or oblanceolate, up to 15 cm long and about 5 mm wide, sometimes entire, but usually toothed. Stem leaves are reduced or absent. Flowers are in stalkless clusters and have star-shaped hairs on the outside of the yellow petals. The combination of basal leaves and star-shaped hairs on the flowers distinguishes this species from all other *Goodenia* in the area, except for *G. dimorpha* which has flowers with slender stalks on branching inflorescences.

HABITAT Wet heath.

DISTRIBUTION Coastal zone (e.g., La Perouse). Extending from south coast to north coast and Qld; also southern tablelands.

FLOWERING July–December.

1093. Goodenia dimorpha var. *angustifolia*

Goodenia dimorpha

G. dimorpha is distinguished by a combination of 3 features. Its yellow flowers have star-like hairs on the outside of the petals; its leaves are in a distinct basal rosette; and its flowers are arranged in a spreading inflorescence, each flower with a short stalk. The erect stems are almost leafless. Two varieties are recognised, based on different-shaped basal leaves.

Var. *dimorpha* has oblong to spoon-shaped leaves, 2 to 5 cm long; var. *angustifolia* (Plate 1093) has narrow linear to lanceolate leaves entire or slightly toothed, 2 to 4 cm long. Both varieties are restricted to the Sydney district.

HABITAT Heath and scrubland, on wet sandy soils.

DISTRIBUTION Var. *dimorpha* occurs north of Sydney to Gosford and in the Blue Mountains. Var. *angustifolia* occurs south of Sydney at Kurnell, Royal National Park and on the Woronora Plateau.

FLOWERING November–March.

1094. Goodenia decurrens

Goodenia decurrens **Plate 1094**

G. decurrens is chiefly found in the Blue Mountains, where it is an erect leafy plant, 50 to 60 cm tall. It is easily recognised by its decurrent leaves; these are stalkless, toothed and have edges which continue down the stem as raised lines. Leaves are 5 to 10 cm long. The yellow flowers are borne near the ends of leafless stalks; star-like hairs are present on the outside of the petals.

HABITAT Sheltered forested slopes, often near cliffs.

DISTRIBUTION Blue Mountains (e.g., Mt Tomah, Bowens Creek, Wentworth Falls, Blackheath) and scattered along lower Hawkesbury River. Also occurring central western slopes.

FLOWERING October–February.

Goodenia rostrivalvis **Plate 1095**

This is regarded as a species at risk because of its very restricted distribution and its narrow habitat preference. Its lower stems and leaves are decumbent amongst the vegetation, but the upper stems are ascending, and bear loose sprays of yellow flowers. Leaves are dark green, glossy, oblanceolate to obovate, toothed near the tip, 6 to 10 cm long. The leaves are somewhat clumped at the lower part of the stem but scattered along the ascending stem. The

1095. Goodenia rostrivalvis

whole plant has a distinctive smooth glossy appearance. Flowers have star-like hairs on the outside of the petals.

HABITAT Damp benches along and below cliff line.

DISTRIBUTION Endemic in the central Blue Mountains at Lawson, Wentworth Falls and Leura.

FLOWERING December–January.

1096. Goodenia glomerata

Goodenia glomerata Plate 1096

This is an erect herb to 40 cm tall, with stems and young leaves conspicuously covered with long weak greyish hairs. Leaves are narrow obovate, 2 to 7 cm long, pointed at the tip and distinctly toothed. Flowers are fairly large (corolla about 2 cm long) and are arranged in compact hairy clusters at the ends of the stems. The ovary is covered with long hairs and star-like hairs are present on the outside of the petals.

HABITAT Damp soil pockets in flat rock outcrops.

DISTRIBUTION A plant of the south coast and southern tablelands with its northern limit on the southern edge of the Sydney district (e.g., Bundanoon, Wingello State Forest).

FLOWERING November–January.

1097. Goodenia bellidifolia

Goodenia bellidifolia Plate 1097

This is a common small plant with an erect, mostly undivided stem about 30 cm tall, which is covered with long weak hairs. The stems rise from a basal rosette of leaves. These leaves are thick, hairless, obovate, 5 to 10 cm long, irregularly toothed and taper into a distinct leaf-stalk. Flowers are yellow, stalkless or almost so, and are borne singly or in small clusters.

HABITAT Open forest and woodlands, especially around swamps or in damp open places.

DISTRIBUTION Widespread from coast to mountains. Occurring throughout coast and tablelands divisions.

FLOWERING Chiefly November–February, but some flowers at other times.

1098. Goodenia paniculata

Goodenia paniculata Plate 1098

Like *G. bellidifolia*, this species has a distinct basal rosette of leaves and a flower stem which is almost leafless. However, *G. paniculata* differs in having glandular-hairy stems which are up to 50 cm tall, spreading and branched towards the top. Basal leaves are softly downy on both surfaces, obovate to narrow spoon-shaped, up to 10 cm long and irregularly toothed. The yellow flowers are solitary at the ends of the spreading branches.

HABITAT Forest, woodland and shrubland, in damp soils.

DISTRIBUTION Widespread from coast to mountains. Extending along coastal zone from south coast to north coast and Qld.

FLOWERING Chiefly December–February, but some flowers at other times.

Goodenia hederacea
Ivy Goodenia Plate 1099

A common small plant with a cluster of basal leaves from which rise a number of trailing stems, up to 25 cm long. The basal leaves are spoon shaped, 4 to 9 cm long, dark green above, entire or broadly

1099. Goodenia hederacea

toothed, with stalks about as long as the leaf blades. The leaves on the trailing stems are smaller, 1 to 3 cm long, more rounded, and irregularly toothed. The flowers are on slender stalks along the trailing stems. This species is very variable and 4 subspecies are recognised in New South Wales; only 1 occurs in the Sydney district.

HABITAT Open forest, woodland and heath.

DISTRIBUTION Widespread from coast to mountains. Also occurring north coast, north-western plains and Qld.

FLOWERING October–April.

Goodenia ovata
Hop Goodenia Plate 1100

The most distinctive feature of this erect shrub is the glossy appearance of the leaves, especially the new growth, which at times seem to be varnished and sticky. Crushed leaves are also somewhat aromatic. Hop Goodenia may be 1 m or more tall. Its leaves are light green, hairless, ovate, 2 to 8 cm long, clearly stalked and toothed. The yellow flowers are borne in clusters from the leaf axils.

HABITAT Sheltered forest or protected sites, often near the sea.

DISTRIBUTION Widespread from coast to mountains. Also occurring north coast, south coast, southern tablelands, western slopes and the southern division of western plains; also Qld, Vic, Tas and SA.

FLOWERING October–December.

1100. Goodenia ovata

1101. Goodenia stephensonii

Goodenia stephensonii — Plate 1101

This is an erect herb with stems bearing scattered white hairs. Leaves are narrow-ovate, 3 to 5 cm long, irregularly toothed, with a short stalk or almost stalkless. Flowers are solitary from the leaf axils on long stalks or arranged in leafy racemes near the ends of the stems. A feature of this species is that the bracteoles are located about half-way along the flower stalk, or a little below halfway, not close to the base.

HABITAT Sheltered forest.

DISTRIBUTION An uncommon plant with distribution restricted to the Gosford district and Goulburn River area (central western slopes).

FLOWERING September–October.

Goodenia heterophylla — Plate 1102

A variable plant, its stems are usually erect, 30 to 40 cm tall, and leaves and stems distinctly hairy (rarely hairless). Leaves are borne along the stem except young plants which often have a basal rosette of leaves. Leaves are 2 to 3 cm long, deeply toothed or almost entire, mostly stalkless. Flowers are mostly solitary, sometimes in groups of 3, on slender stalks from the leaf axils.

HABITAT Open forest and scrubland, mainly in sheltered situations.

DISTRIBUTION Widespread from coast to mountains. Also occurring north coast, south coast and southern tablelands.

FLOWERING Chiefly October–February, with some flowers at other times.

1102. Goodenia heterophylla

1103. Selliera radicans

Selliera

A creeping herb of salt-marsh edges. Flowers with 5 petals spreading digitately. Fruit an ovoid berry.

Selliera radicans — Plate 1103

This is a hairless creeping plant with stems which root at the nodes. Leaves are shiny green, thick, entire and rounded at the tip. They are very variable in size and shape, ranging from 2 to 8 cm long and from linear to ovate or spoon shaped. They may form thick mats or be quite scattered. Flowers are small, 6 to 8 mm long, very shortly stalked and fairly inconspicuous; they are white with mauve and brown markings. Fruit is a berry 4 to 6 mm long.

HABITAT Margins of mangrove and saline marshes, and wet rocky areas within spray range of the ocean.

DISTRIBUTION Coastal range. Recordings include Kurnell and Salt Pan Creek. Extending along south coast to Vic; also recorded southern tablelands, Tas, SA, NZ and South America.

FLOWERING October–March.

Scaevola Fan-flower

Low spreading plants. Flowers with 5 equal petals, spreading digitately. Indusium covered with stiff hairs. Fruit a drupe. Differs from *Selliera* in having a 1 or 2 celled ovary with 2 ovules and 2 seeds, whereas *Selliera* has numerous ovules and seeds.

Scaevola ramosissima
Purple Fan-flower — Plate 1104

The 5 purple spreading petals of this straggling plant make it a conspicuous part of the understorey in coastal heath and dry open forest. Purple Fan-flower has wiry, spreading branches, 30 to 50 cm long, covered with short stiff hairs. Leaves are narrow lanceolate, 2 to 5 cm long, entire or toothed, and without stalks. The purple flowers are borne singly or in clusters of 3, on a slender stalk, from the leaf axils.

HABITAT Heath and open forest; on sandy soils.

DISTRIBUTION Chiefly coastal; less common in mountains. Recorded throughout coast and tablelands divisions; also Qld and Vic.

FLOWERING Chiefly November–January, but some flowers at other times.

1104. Scaevola ramosissima

1105. Scaevola hookeri

Scaevola hookeri
Creeping Fan-flower — Plate 1105

This prostrate or creeping herb is a plant of the high sub-alpine country where it often forms mats in swampy areas. It is rare in the Sydney district. Stems are covered with long loose whitish hairs. Leaves are variable, 10 to 20 mm long, 5 to 14 mm wide, broadly toothed at the tip, with short stalks; they may be covered with long hairs or glandular-hairy with longer hairs along the margins. Flowers are light pink, with a yellow-brown centre, reddish underneath and borne on short stalks from the leaf axils.

HABITAT Wet rock ledges.

DISTRIBUTION Higher Blue Mountains (e.g., Govetts Leap, Blackheath). Extending from northern tablelands to southern tablelands, Vic and Tas; also higher parts of south coast.

FLOWERING December–February.

1106. Scaevola aemula

1108. Scaevola calendulacea

1109. Dampiera stricta

Scaevola aemula

Fairy Fan-flower **Plate 1106**

A sprawling herb with a few slightly hairy ascending stems, 10 to 30 cm tall. Leaves are ovate to obovate, 2 to 7 cm long, 10 to 25 mm wide, coarsely toothed, with scattered hairs and a short stalk. Flowers are mauve to blue with yellow at the base, stalkless and borne singly in the axils of the upper leaves. A characteristic feature is the presence of a tuft of long purple hairs on the back of the indusium, a feature which separates this species from *S. albida.*

HABITAT Coastal dunes and sands on sea-cliffs. Also deeper sands in forests.

DISTRIBUTION Coastal zone from Qld to Vic and SA. Also recorded southern tablelands, western slopes and western plains.

FLOWERING February–March.

Scaevola calendulacea

Dune Fan-flower **Plate 1108**

A common plant on coastal foredunes, where it forms extensive mats, helping to bind the shifting sands. Its leaves are usually thick, semi-succulent, obovate to spoon shaped, 25 to 40 mm long and up to 20 mm wide. Most leaves are entire, but sometimes there are a few teeth near the apex. Flowers have blue spreading petals with yellow centres. Fruit a berry, 6 to 7 mm across, creamy coloured at first, turning purple.

HABITAT Dunes near the sea.

DISTRIBUTION Coast. Extending all along NSW coast from Vic to Qld.

FLOWERING November–May, but some flowers at other times.

Dampiera purpurea

Purple Dampiera **Plate 1110**

A many stemmed shrub which may grow to 1 m or more tall. Stems, leaf undersurfaces, buds and the backs of flowers are covered with dense woolly hairs, making it easy to distinguish from *D. stricta.* Leaves are oval or almost round, 25 to 30 cm across, entire, rough above and with brown hairs below. Flowers are purple with yellow centres and are clustered in the leaf axils near the ends of the branches.

HABITAT Open forest, on sandy soils.

DISTRIBUTION Widespread from coast to mountains. Occurring throughout the coast and tablelands divisions; also Vic.

FLOWERING October–November.

1107. Scaevola albida

Scaevola albida

Pale Fan-flower **Plate 1107**

A soft, low, spreading plant with a few ascending branches. Leaves are pale, 1 to 3 cm long, ovate to narrow lanceolate and coarsely toothed. Flowers are pale blue to mauve, stalkless and borne in the upper leaf axils. The indusium is covered with silvery hairs. These silvery hairs help separate this species from *S. aemula* which has a tuft of long purple hairs on the indusium.

HABITAT Open forest and woodland.

DISTRIBUTION Widespread from coast to mountains. Also occurring north coast, northern tablelands, southern tablelands, western slopes, Qld, Vic and SA.

FLOWERING November–February.

Dampiera

Small shrubs with blue or purple flowers. Flowers with 3 upper petals and 2 lower petals. Anthers are united into a tube around the style; indusium is hairless (unlike *Goodenia* and *Scaevola*). Fruit is a small nut.

Dampiera stricta

Blue Dampiera **Plate 1109**

A small shrub with straggling, angular, hairless stems up to 60 cm long. Leaves are smooth, rather thick, ovate to lanceolate, 1 to 3 cm long, often with marginal lobes or broad teeth. Flowers are light to dark blue (sometimes mauve or pink), 10 to 14 mm long, with conspicuous brown hairs on the back and on the buds. Flowers are borne singly or clustered in the axils of the upper leaves.

HABITAT Open forest and heath.

DISTRIBUTION Widespread and common from coast to mountains. Occurring throughout the coast and tablelands divisions.

FLOWERING August–December, but some flowers at other times.

1110. Dampiera purpurea

Brunoniaceae

Brunoniaceae, 1 of only 4 endemic Australian vascular families, consists of 1 genus with 1 species, *Brunonia australis*. It is closely related to, and sometimes included in, Goodeniaceae, although in its inflorescence it resembles the Asteraceae.

Brunonia

The flowers of *Brunonia* are similar to those of the Goodeniaceae in having an indusium (a 2-lobed pollen cup around the stigma), but differ in being regular (symmetrical) instead of the corolla tube being split and the flowers irregular (1-sided). They also have a superior ovary, which is unusual in Goodeniaceae.

1111. Brunonia australis

Brunonia australis

Blue Pincushion **Plate 1111**

A common plant over large areas of inland Australia, it reaches its eastern limit in drier areas of the western Blue Mountains. It is absent from the coastal zone. Leaves are in a large basal rosette; each leaf is obovate to oblanceolate, entire, 4 to 10 cm long and clothed with long silky hairs. Stems are up to 30 cm tall, topped with a bright blue pincushion of flowers. Each head of flowers is 15 to 20 mm across. Individual flowers are tubular, hairy, with 5 hairless unequal lips.

HABITAT Open forest, on sandy soils.

DISTRIBUTION Higher Blue Mountains, such as between Clarence and Bell and in Newnes State Forest. Also occurring southern tablelands, western slopes, western plains and far western plains; also found in all other States.

FLOWERING November–December.

Asteraceae

In contrast to the imaginative and occasionally bizarre contrivances of the Orchidaceae, this advanced group of plants relies for pollination on a few variations of a simple economical plan. The secret is revealed in the former name, Compositae, which refers to the fact that what at first appears to be a conspicuous single flower is actually a composite of many small flowers, called florets, arranged in a head or capitulum. The general impression of a single flower is increased by the presence of a protective cluster of bracts which are sometimes soft and green, and resemble a calyx, and sometimes dry and papery and resemble brightly coloured petals. Individual florets may be male, female, bisexual or sterile; that is, they have respectively anthers, stigmas, both anthers and stigmas, or neither.

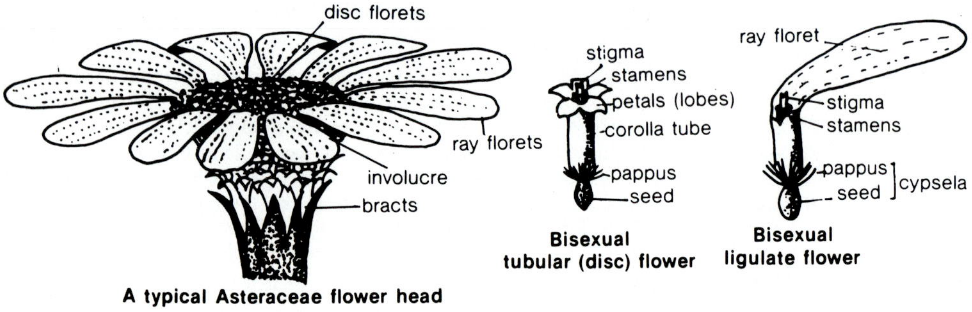

Figure 13

The commonest type of floret is the disc floret, which is usually bisexual, and has a narrow tubular corolla with 5 lobes and an emergent style separating at the top into 2 style arms. In many cases these are surrounded by male or sterile ray florets or by bisexual ligulate florets. The ray and ligulate florets have a flat fused corolla, resembling a large petal, with respectively 3 and 5 tiny teeth at the end. For convenience, the central tubular florets are often referred to collectively as the 'disc', and the outer petal-like florets are called 'rays'. The sole function of sterile rays is to attract and direct insects to the closely packed fertile flowers. In some cases the flower heads are composed of only one type of floret, either disc florets or ligulate florets.

After fertilisation, the calyx is pushed upward by the enlarged inferior ovary, or sometimes by a stipe or stalk which develops at the top of the ovary. The calyx consists of a parachute of bristles or scales, called a pappus, which enables the seed to be carried long distances by the wind. Some genera form burrs or develop spiny ornamentation on the fruit as a protection and as an aid to dispersal by animals. The mechanisms of insect attraction, fertilisation and seed dispersal employed by these plants are so successful that they are found worldwide and are common everywhere except in aquatic and tropical rainforest habitats.

Many of the Asteraceae such as *Chrysanthemum*, *Dahlia* and *Zinnia*, have been developed into popular garden plants, but others such as Dandelion, Crofton Weed, Boneseed and thistles, are troublesome weeds. For a brief period after rain vast areas of inland Australia are covered with brightly coloured ephemeral species of *Helipterum* and *Myriocephalus*. Under quite different conditions, the family is also well represented in the alpine herbfields of Australia and other countries. The presence of well-known Australian genera such as *Brachycome, Olearia, Senecio, Helichrysum, Cassinia, Craspedia* and *Celmisia* in temperate and montane areas of New Zealand, and in some cases, New Guinea, attests to the ability of these plants to cross large areas of ocean and unsuitable country, become established, survive the adversities of climatic change, and differentiate into new species. As a proportion of the estimated 20 000 or more species worldwide, the 110 native species in the Sydney region are insignificant in terms of numbers. Unfortunately, the 70 odd weed species have a much greater impact on disturbed and agricultural land.

Asteraceae has traditionally been divided into 12 tribes and about 1000 genera, but some of the groupings have been questioned, and revision seems to be warranted. Individual genera are defined according to the nature of the florets and bracts, the pappus, ornamentation of the fruit, the presence of chaffy scales between florets, and the arrangement of heads into a compound inflorescence. Individual species are often difficult to tell apart because of the numerous forms and subspecies, the small differences between many species, and the small size of the floral parts.

Lagenifera

Small perennials with solitary flower-heads, basal rosette of leaves and unbranched stems. Ray flowers strap-like and female; disc flowers tubular and male. Pappus absent. Differs from *Brachycome* in the cypselas; those of *Lagenifera* have a neck or beak at the summit, and never have wings on the side (not blunt at the summit, like those of *Brachycome*).

Lagenifera stipitata

Plate 1112

This small daisy has an erect stem 30 to 40 cm tall, rising from a rosette of leaves. Leaves are oblanceolate, toothed, up to 15 cm long, with a short stalk. Flowers are solitary, with mauve, pink or white rays and yellow disc. Rays are small (3 to 4 mm long) and the involucre less than 10 mm across. Cypselas are beaked.

HABITAT Open forest and grassland.

DISTRIBUTION Widespread from coast to mountains (e.g., Royal National Park, Castlereagh State Forest, Brisbane Water National Park). Extending from Vic to Qld along coast and tablelands; also Tas, SA and New Zealand.

FLOWERING October–December.

1112. Lagenifera stipitata

1113. Brachycome multifida var. *dilatata*

Brachycome

A genus of over 50 species Australia-wide. Mostly small herbs with rosette or basal leaves (sometimes also stem leaves) and solitary, often mauve, flower heads. Ray flowers strap-like and female; disc flowers tubular and bisexual. Pappus either absent or minute ('brachys', 'comus' is Greek for 'short hair'). Cypselas are blunt at the summit and often have wings on the side.

Brachycome multifida

Cut-leaf Daisy **Plate 1113**

A small sprawling or erect annual (to 40 cm tall) with fine divided leaves and branched stems. Flowers are mauve (sometimes pink-white) with yellow centres. Rays are 7 to 10 mm long. A characteristic feature is that the cypselas are rough and warty, without wings, and the pappus is white and minute. Two varieties occur in the area, separated by their leaf shapes.

Var. *multifida* has leaves which are narrow-linear, about 1 mm wide, divided into segments about 2 cm long and narrowing to a point at the ends.

Var. *dilatata* (Plate 1113) has leaves which are broad-linear, about 2 mm wide, more divided, ascending the lower part of the stem, with segments 2 to 2.5 cm long and oblanceolate at the ends.

HABITAT Open forest and grassland.

DISTRIBUTION Var. *multifida* occurs from coast to mountains. It extends along the north coast and northern tablelands and westwards to far western plains; also Vic and Qld. Var. *dilatata* is a coastal form from the central and north coasts; also Vic.

FLOWERING September–February, but some flowers at other times.

Brachycome graminea

A weak upright to sprawling plant, 30 to 50 cm tall, with stems which spread along the ground, producing roots at the nodes. Leaves are scattered along the upright stems; they are entire, linear to oblanceolate, 6 to 12 cm long and 7 mm wide. Flowers are terminal, with 20 to 30 mauve-pink or purple rays each about 9 mm long, around a yellow disc. Cypselas are smooth and the pappus minute.

HABITAT Grassland, often in damp areas.

DISTRIBUTION Mountains and higher areas of the coastal zone (e.g., Thirlmere). Extending throughout the tablelands and to south coast; also Qld, Vic, Tas and SA.

FLOWERING October–February.

Brachycome ptychocarpa **Plate 1114**

This small daisy is unmistakable, as it has a small dense cluster of finely divided leaves with short narrow lobes. The flower stalk is 10 to 12 cm long, erect and leafless. Flowers are pink to white, with rays about 3 mm long. Cypselas are smooth and the pappus short.

HABITAT Wet mossy rocks.

DISTRIBUTION Higher Blue Mountains (e.g., Boyd River, Kanangra-Boyd National Park). Extending throughout the tablelands division and central western slopes; also Vic.

FLOWERING December–January.

Brachycome angustifolia **Plate 1115**

A sprawling perennial, up to 35 cm tall, with stems which spread along the ground, producing roots at the nodes. Leaves are scattered along the upright stems; they vary in shape, but are mostly about 5 cm long. Flowers are pink to mauve on leafless stalks. Rays are 6 to 10 mm long. Cypselas are winged and warty; the pappus is minute. Two varieties occur in the area.

1114. Brachycome ptychocarpa

Var. *angustifolia* (Plate 1115) has narrow lanceolate leaves, which are mostly entire.

Var. *heterophylla* has broader, more elliptical leaves (about 9 mm wide), which are deeply divided or toothed.

HABITAT Grassland, woodland and hind dune forest.

DISTRIBUTION Var. *angustifolia* occurs from coast to mountains. It extends throughout tablelands and along south coast to Vic, Tas and SA. Var. *heterophylla* occurs from coast to mountains. It extends throughout tablelands and to western slopes and north coast; also Vic.

FLOWERING September–December.

1115. Brachycome angustifolia var. *angustifolia*

1116. *Brachycome scapiformis*

Brachycome scapiformis **Plate 1116**

This tablelands species has a rosette of leaves which are spoon shaped, 5 to 10 cm long, with blunt teeth and narrowed at the base to a short stalk. The flower is terminal on a long leafless stem which is up to 50 cm tall. Flowers are mauve, with an involucre 10 to 15 mm across and rays 5 to 6 mm long. Cypselas are winged and smooth; the pappus is small.

HABITAT Grassland, especially in damp soil.

DISTRIBUTION Blue Mountains and southern highlands (e.g., Hampton, Wombeyan Caves road). Extending throughout tablelands to western slopes and to south coast, Vic and Tas.

FLOWERING October–January.

Calotis Burr-daisy

The Burr-daisies are sprawling herbs or small shrubs. The pappus consists of rigid thistles with recurved barbs which develop into a burr. Flowers consist of female rays in 1 or more rows, with yellow tubular disc flowers, usually male.

Calotis cuneifolia

Bindi-eye, Blue Burr-daisy **Plate 1117**

A very common small perennial on the western slopes and plains, but less common in coastal areas. It is a sprawling plant, 25 to 60 cm tall, covered with stiff hairs. Leaves are up to 4 cm long, wedge to spoon shaped, with 3 to 6 lobes near the apex. Flowers are abundant, 1 to 2 cm across, on the ends of slender stems; ray flowers are blue-mauve to white, numerous, about 10 mm long, around a yellow centre. Cypselas are flattened and wedge shaped, topped by 2 (sometimes 3 or 4) barbed spines.

HABITAT Grassland and open forest, often near road edges and in heavier soils.

DISTRIBUTION Widespread, but most common west of the coastal plain (e.g., Lapstone Hill, Hartley Valley). Found throughout NSW (except southern tablelands), Qld, Vic, NT and SA.

FLOWERING Chiefly September–December, but some flowers at other times.

1118. *Calotis scapigera*

Calotis scapigera

Tufted Burr-daisy **Plate 1118**

C. scapigera occurs as small clumps in moist places. These clumps are formed as the result of this plant's habit of sending out stems along the ground which root at the nodes. It is a hairless perennial, usually 25 to 30 cm tall, with a basal cluster of leaves. These leaves are narrow, linear-lanceolate, 10 to 20 cm long, entire or slightly toothed. Flowers are white to light mauve with yellow centres, and are borne on erect stems which exceed the leaves and bear a few bract-like leaves. Flowers are about 15 mm across, with rays 3 to 5 mm long. Cypselas are flattened and wedged shaped, topped by 4 to 6 rigid spines of equal length (1 to 4 mm long).

HABITAT Grassland and woodland, in damp places on clayey soils.

DISTRIBUTION Coastal zone. Absent from central tablelands but occurring throughout western plains and along Murray–Darling system; also north coast, northern tablelands, Qld, Vic and SA.

FLOWERING October–March.

Calotis dentex

White Burr-daisy **Plate 1119**

This is an upright undershrub to 80 cm tall, with smooth or slightly hairy stems, which are often brownish in colour. Leaves are stalkless, with stem-clasping bases, up to 8 cm long, deeply and irregularly lobed; some upper leaves may be entire. Flower heads are numerous and borne on stalks up to 5 cm long from the leaf axils. Flowers have white (sometimes mauve) rays about 10 mm long and yellow centres, 5 to 10 mm across. Cypselas are reddish-brown and flattened, topped by 2 (sometimes 1 or 3) rigid spines.

HABITAT Grassland and open forest, chiefly on clay soils.

DISTRIBUTION Coast to mountains. Also occurring north coast, northern tablelands, north-western slopes and Qld.

FLOWERING Some flowers all year round.

1117. *Calotis cuneifolia*

1119. *Calotis dentex*

1120. *Calotis lappulacea*

Calotis lappulacea

Yellow Burr-daisy **Plate 1120**

Yellow Burr-daisy has a scattered distribution in the Sydney district because of its preference for heavier clay soils. It is a much-branched perennial, to 50 cm tall, with numerous small yellow flower heads. Younger plants have toothed, wedged-shaped basal leaves; stem leaves are linear, 25 mm long and 4 mm wide, sessile, entire or toothed. Flowers are about 18 mm across; ray flowers are numerous, yellow, about 3 mm long. Cypselas are flattened and slightly warty, topped by 2 rigid spines and several smaller ones.

HABITAT Grassland and open forest, in heavier clay soils.

DISTRIBUTION Widespread (e.g., Narellan, Yerranderie). Occurs throughout NSW, except on far western plains; also Qld, Vic, SA and WA.

FLOWERING Most of the year.

1121. Olearia microphylla

Olearia Daisy-bush

A genus of over 100 species, with about 80 species in Australia. They are shrubs, often quite tall, with large attractive flower heads. Ray flowers are strap-like, female and in a single row; disc flowers are tubular and bisexual. Cypselas are terete and marked with longitudinal lines; pappus consists of numerous soft hair-like bristles.

Olearia microphylla

Small-leaved Daisy-Bush **Plate 1121**

A common shrub, about 1 m tall, with the smallest leaves of any *Olearia* in the Sydney district (2 to 4 mm long). Flowers are small and white, but borne abundantly along the branches making the plant very conspicuous. Each flower has 7 to 9 rays, each 3 to 5 mm long.

HABITAT Forest, on both shales and sandy soils.

DISTRIBUTION Widespread from coast to mountains. Occurring throughout coast and tablelands, extending to western slopes and Qld.

FLOWERING August–November.

1122. Olearia viscidula

Olearia viscidula **Plate 1122**

An erect shrub, 1 to 1.5 m tall, with opposite leaves more or less coated with a sticky gloss above and whitish below. Leaves are flat, 6 to 10 cm long, 4 to 7 mm wide, with entire margins. Flowers are white, with 8 to 10 rays, each about 3 mm long. Flowers are borne in small clusters along short lateral branches.

HABITAT Sheltered forest and rainforest margins.

DISTRIBUTION Scattered throughout coast and tablelands divisions. Recordings include Loftus, Thirlmere–Nattai and West Berrima. Extending to western slopes.

FLOWERING August–October.

1123. Olearia argophylla

Olearia argophylla

Musk Daisy-bush **Plate 1123**

A small tree, common in the gullies of Vic and Tas, and in cooler mountain forests of the southern tablelands, but less common around Sydney. Its leaves are alternate, lanceolate, about 8 to 13 cm long, 3 to 4 cm wide, green above and silvery below. Some leaves (like those on plants near Yerranderie) have toothed margins, others (like those along Illawarra Escarpment) are entire. The foliage has a musk-like scent. Flowers are small, creamy-white, with 3 to 5 ray florets; numerous flowers are borne in conspicuous terminal clusters.

HABITAT Sheltered forested gullies and slopes.

DISTRIBUTION Blue Mountains, southern highlands and Illawarra Escarpment. Extending throughout coast and tablelands southwards to Vic and Tas.

FLOWERING October–December.

Olearia quercifolia

Oak-leaved Daisy-bush **Plate 1124**

This small shrub (to 50 cm tall) is regarded as a species 'at risk' because of its limited range of

1124. Olearia quercifolia

distribution. Its name is derived from the similarity of its leaves to the Oak (*Quercus* sp.). Leaves are about 35 mm long, 15 to 20 mm wide, with 3 deep wavy teeth along each edge and blunt tips; they are alternate, dark green and rough above, with star-shaped hairs underneath. Flowers are pale mauve to white, with 10 to 12 ray florets and borne singly on the woolly stems from the leaf axils.

HABITAT Sheltered forest, on hillsides, often in drainage lines.

DISTRIBUTION Endemic in central Blue Mountains between Wentworth Falls and Clarence.

FLOWERING November–December.

Olearia asterotricha **Plate 1125**

A small shrub to 80 cm, with small, greyish alternate leaves which are densely stellate-hairy on both surfaces. Leaves are about 20 mm long, 2 to 4 mm wide and deeply lobed near the tip. Flowers are pale mauve to white, borne singly on short axillary branches, or sometimes terminal.

HABITAT Sheltered forest, especially in rocky areas.

DISTRIBUTION Higher Blue Mountains and southern highlands; also Vic.

FLOWERING October–December.

1125. Olearia asteroticha

1126. Olearia nernstii

Olearia nernstii — Plate 1126

This is a shrub to 2 m tall which reaches the southern-most limit of its range north of Broken Bay. Stems are covered with close rusty-white hairs. Leaves are alternate, 5 to 9 cm long, up to 2 cm wide, coarsely toothed or lobed, with a fine mucro at the tip. The teeth along the leaf edge often have a fine point. The upper surface of the leaf is dark green and sometimes sticky; the undersurface is covered with rusty-white close short hairs. Flower-heads are large (to 25 mm across) and showy in terminal clusters. The yellow centre is surrounded by numerous narrow white rays. Bracts and flower stalks are rusty-hairy.

HABITAT Coastal forest, in sheltered sites on sandy soils.

DISTRIBUTION Coastal zone, north of Broken Bay. Also recorded central tablelands. Extending along north coast and northern tablelands to Qld.

FLOWERING August–October.

Olearia phlogopappa

Alpine Daisy-bush — Plate 1127

This medium shrub (to 1.5 m) is common in the cooler forests of the sub-alpine zone and in Victoria, but is uncommon in the Sydney district. It shows considerable variation in habit, leaf characteristics and inflorescence. Branches are spreading and the flowers and leaves tend to face upwards. Leaves are 3 to 5 cm long, 5 to 9 mm wide, with a wavy margin (sometimes entire). The upper margin is green with a few scattered hairs; the undersurface has soft white stellate hairs. Flower-heads are numerous, with 10 to 13 white to pink rays and yellow centres. Flowers are either solitary on slender axillary stalks or in small lateral clusters. The involucre is 4 to 5 mm long and cypselas are 2.5 mm long. Flower stalks may be softly woolly or quite hairless. This species may be confused with *O. erubescens*, but that species has simple, not stellate hairs beneath the leaves, a larger involucre (6 to 10 mm across) and larger cypselas (4 mm long).

1127. Olearia phlogopappa

HABITAT Forest.

DISTRIBUTION Higher Blue Mountains and southern highlands. Extending along tablelands to Vic and Tas.

FLOWERING October–December.

1128. Olearia tomentosa

Olearia tomentosa — Plate 1128

A small shrub (to 60 cm tall), with distinctive leaves. They are broadly ovate, about 5 cm long, 10 to 25 mm wide, with rounded lobes, rough above and woolly white-rusty below. Flower heads are 2 to 3 cm across, white, pink or bluish, on stalks about 4 cm long. The involucre is 6 to 10 mm across and rays 6 to 8 mm long.

HABITAT Forests and sheltered sites.

DISTRIBUTION Widespread from coastal headlands to mountain gullies. Occurring all along coast and southern tablelands.

FLOWERING October–November.

Olearia erubescens — Plate 1129

A medium shrub to 1.5 m with oblong to lanceolate leaves, 3 to 4 cm long and 7 to 15 mm wide. These leaves are alternate, toothed, green above and white below because of a covering of short simple hairs; new growth is often a distinct red colour. Flower heads are white to pink-mauve, with 3 to 5 flowers

1129. Olearia erubescens

in axillary clusters. Flower stalks are up to 4 cm long. Ray flowers are few (3 to 5), involucre 6 to 10 mm across and cypselas about 4 mm long. The length of the involucre and cypselas, and the simple hairs beneath the leaves separate this species from *O. phlogopappa.*

HABITAT Forest.

DISTRIBUTION Blue Mountains, southern Illawarra Escarpment and southern highlands (e.g., Barren Grounds, Wombeyan Caves road). Extending along south coast and southern tablelands to Vic and Tas.

FLOWERING September–November.

1130. Olearia myrsinoides

Olearia myrsinoides

Silky Daisy-bush — Plate 1130

This small shrub (to 60 cm high) is closely related to *O. erubescens*. It is a spreading bush with silky young branches. Leaves are ovate, up to 15 mm long and 4 to 8 mm wide, alternate, minutely toothed (sometimes entire), green above and white below because of a covering of simple hairs. Flower heads are white, with 2 to 5 flowers in short axillary clusters; flowers stalks are less than 10 mm long. Ray flowers are few (2 to 3) and the involucre is 2 to 3 mm across.

HABITAT Open forest and sheltered gullies.

DISTRIBUTION Chiefly Blue Mountains, but also recorded in higher parts of the central coast (Nattai Valley). Extending throughout tablelands and south coast to Vic and Tas.

FLOWERING September–November, but some flowers at other times.

1131. Olearia elliptica

Olearia elliptica

Sticky Daisy-bush **Plate 1131**

O. elliptica is easily identified as it is a shrub to 2 m tall, with alternate glabrous leaves, elliptical to lanceolate, shiny or slightly viscid, 3 to 7 cm long and 2 to 3 cm wide. It is the only Daisy-bush in the area with such broad hairless leaves. Flower heads are numerous and borne in terminal clusters longer than the leaves. Rays are white, 6 to 9 per head and the involucre is 6 to 8 mm across.

HABITAT Forest, especially on protected slopes and wet situations.

DISTRIBUTION Widespread from coast to mountains. Recordings include Mt Keira, Fitzroy Falls, Wentworth Falls and Medlow Bath. Extending along north coast and northern tablelands to Qld; also Lord Howe Island.

FLOWERING October–December.

Celmisia Snow Daisy

Snow Daisies are abundant in the alpine areas where their silver-grey foliage and large white flower heads are conspicuous during summer months. They are herbs with basal leaves and large solitary flower heads. The strap-like ray flowers are female, the disc flowers are tubular and bisexual. The cypselas are slightly compressed; the pappus consists of numerous soft hair-like bristles.

1132. Celmisia sp. aff. *longifolia*

Celmisia sp. aff. *longifolia*

Snow Daisy **Plate 1132**

It was once thought that the *Celmisia* from the higher areas of the Blue Mountains was *C. longifolia*, a common species from Kosciusko National Park. It is now regarded as a local disjunct population of a new species. Its leaves are basal, silvery, entire, up to 20 cm long, with a sheathing base. Flower heads are on erect stalks longer than the leaves. Ray flowers are white, numerous, 1 to 2 cm long, and in a single spreading row; the centre is yellow. The involucre is woolly, 2 to 3 cm across, with several rows of overlapping bracts.

HABITAT Grassland and open woodland, in wet soils or swampy margins.

DISTRIBUTION Wentworth Falls to Blackheath and Newnes Plateau.

FLOWERING November–January.

Vittadinia

Small branched herbs with alternate leaves along stems and solitary flower heads. Involucre is cylindrical or bell shaped; ray flowers strap like and mauve, disc flowers tubular; pappus consisting of numerous soft hair-like bristles.

1133. Vittadinia triloba

Vittadinia triloba

Fuzzweed **Plate 1133**

V. triloba is part of a complex which includes *V. cuneata* and which has several varieties and forms. Recent revisions have divided this group into a number of new species. The most common form of Fuzzweed is a small plant with a woody base and branched, somewhat rough (scabrous) stems to 30 cm tall. Leaves are small, 5 to 25 mm long, oblanceolate or wedge-shaped, usually 3 lobed at the apex. Flower heads are narrow, with mauve rays about 5 mm long and involucre about 5 mm across. Cypselas are 5 to 6 mm long, ribbed and topped by spreading pappus hairs about as long as the cypsela. The fluffy seeds are produced in abundance. Those plants with stems covered in soft woolly hairs rather than short scabrous hairs have been known as *V. cuneata.*

HABITAT Woodland, in both clay and sandy soils, often along roadsides.

DISTRIBUTION Widespread from coast to mountains (e.g., Narellan, Hartley). Varieties found throughout the State; also Qld, Vic, Tas, SA and NT.

FLOWERING October–February, but some flowers at other times.

Cotula

Small sprawling herbs with leaves along the stalks and distinctive button-like flowers on stalks. Petal-like ray flowers are absent and the flower-head is made up of tubular bisexual disc flowers with an outer ring of female flowers. The seeds do not have any hairy or spiny pappus.

1134. Cotula coronopifolia

Cotula coronopifolia

Waterbuttons **Plate 1134**

A weak perennial with succulent stems commonly found rooted in the mud in or around wet saline pools. Leaves are entire (sometimes lobed or jagged), oblong, up to 5 cm long and stem-clasping. The yellow button-like flowers are 6 to 10 mm across and borne at the end of long leafless stalks, making them quite conspicuous.

HABITAT In shallow water or mud, in both fresh and brackish water.

DISTRIBUTION Coastal zone around Sydney. Extending to north and south coasts; also scattered through southern tablelands, and south-western slopes and plains. Found in all States, except NT.

FLOWERING July–October.

Senecio Fireweed, Groundsel

A large cosmopolitan genus of about 3000 species with more than 40 species in Australia. Shrubs or herbs with alternate leaves and yellow flower-heads. Some species have petal-like ray flowers, others have all tubular disc flowers. Pappus of soft hairs as long as the cypsela, forming conspicuous white heads at fruiting time.

1135. Senecio lautus ssp. *maritimus*

1138. Senecio macranthus

Senecio lautus ssp. *maritimus*

Coast Groundsel **Plate 1135**

A sprawling or creeping perennial commonly found on deeper sands near the sea. Its leaves are narrow-oblong, 2 to 4 cm long, toothed (sometimes almost entire) and often somewhat fleshy and stem-clasping. Flowers are yellow, with 12 to 20 ray flowers which are 1 to 2 cm long and a bell-shaped involucre, 5 to 8 mm across. In protected sites, Coast Groundsel may form large low clumps, but stems, leaves and flowers are much more sparse on open dunes.

HABITAT Sand dunes near the sea.

DISTRIBUTION Coastal (e.g., Marley Beach, Kurnell). Extending along north and south coasts. Found in all States, except NT.

FLOWERING Some flowers most of the year.

1136. Senecio lautus ssp. *dissectifolius*

Senecio lautus ssp. *dissectifolius*

Variable Groundsel **Plate 1136**

This subspecies is a more upright plant than ssp. *maritimus*, growing to 70 cm tall, and it has leaves which are deeply dissected into short narrow lobes. Flower heads are on the ends of branched stems. There are 10 to 15 yellow ray flowers, 5 to 10 mm long, and a bell-shaped involucre, 5 to 6 mm across.

HABITAT Open forest and woodland.

DISTRIBUTION Widespread from coast to mountains. Recordings include Mt Gibraltar (Bowral) and Jenolan Caves road. Also occurring north coast, northern tablelands, western plains, and in all other States.

FLOWERING May–October.

Senecio velleioides

Plate 1137

A hairless erect perennial to 1 m tall, easily identified by its large stem-clasping leaves. These are entire or coarsely toothed, up to 12 cm long and have rounded ear-like bases. Flowers are large, 2 to 3 cm across, with 8 to 10 ray flowers, each 1 cm long and borne in loose terminal clusters.

HABITAT Sheltered forested slopes, often near watercourses.

DISTRIBUTION Blue Mountains, southern highlands and higher areas of coast (e.g., Macquarie Pass, Fitzroy Falls, Bundanoon). Also occurring north coast, south coast, southern tablelands, central western slopes, Vic and Tas.

FLOWERING October–May.

1137. Senecio velleioides

Senecio macranthus

Plate 1138

This erect perennial (to 1 m tall) is conspicuous because of its large flower heads. Leaves are hairless, linear to narrow lanceolate, to 8 cm long and 10 mm wide, with scattered irregular teeth and sessile to stem-clasping bases. Flowers are terminal, with ray flowers 1 to 3 cm long and the involucre about 1 cm across.

HABITAT Forest, on sheltered slopes and gullies.

DISTRIBUTION Higher parts of western Blue Mountains (e.g., Jenolan Caves road). Extending to north coast, northern tablelands and western slopes.

FLOWERING October–November.

1139. Senecio linearifolius

Senecio linearifolius

Fireweed Groundsel **Plate 1139**

This is a common tall perennial (to 1 m tall) of sheltered forests. Leaves are very variable, narrow or broad lanceolate, 6 to 10 cm long and 8 to 15 mm wide, sessile or slightly stem-clasping and with either toothed, wavy or entire margins. Flower heads are numerous in terminal clusters. Ray flowers are 5 to 8 mm long and the involucre cylindrical, about 3 mm across.

HABITAT Sheltered forest and rainforest margins.

DISTRIBUTION Widespread from coast to mountains. Recordings include Watagan State Forest, Bouddi National Park and Bundanoon. Occurs throughout coast and tablelands and less commonly on western slopes; also Vic and Tas.

FLOWERING Chiefly September–November, but some flowers at other times.

1140. Senecio vagus ssp. *eglandulosus*

Senecio vagus ssp. *eglandulosus* Plate 1140

This is a tall hairless perennial shrub, 1 m or more tall, with distinctive stalked leaves, 8 to 15 cm long, which are deeply dissected by irregular lobes or teeth. Flower heads are in loose terminal clusters. There are 10 to 15 ray flowers, about 10 mm long, and a bell-shaped involucre, about 7 mm across.

HABITAT Sheltered forests, gullies and rainforest margins.

DISTRIBUTION Widespread from coast to mountains. Recordings include Kellys Falls, Bouddi National Park and Pierces Pass, Grose River. Extending to north coast and northern tablelands.

FLOWERING Chiefly October–December, but some flowers at other times.

Pseudognaphalium Cudweed

Cudweeds are small woody herbs with flower heads crowded into dense clusters. Petal-like ray flowers are absent; disc flowers are tubular, surrounded by 2 or more rows of tubular filiform female flowers. The involucre is bell shaped, with several rows of overlapping bracts. Pappus consists of soft fine white hair-like bristles.

Pseudognaphalium luteoalbum

Jersey Cudweed

This is an erect annual herb, with stems to about 30 cm long. Stems and leaves are covered in grey-white woolly hairs. Leaves vary from spoon shaped to linear and are 3 to 6 cm long. Flower heads are in dense clusters at the ends of the stems, with an involucre about 3 mm across, surrounded by obtuse, straw-coloured bracts. The cypselas are minute, less than 1 mm long, topped by small pappus hairs, 3 to 4 mm long.

HABITAT Grassland and disturbed places, such as roadsides and cultivated land.

DISTRIBUTION Widespread from coast to mountains. Occurring throughout NSW and in all other States.

FLOWERING Chiefly October–January, but some flowers at other times.

Cassinia

Cassinias are shrubs with alternate and entire leaves. Numerous small flower heads are borne in terminal flat-topped clusters or loose panicles. Flowers are all tubular, without any petal-like ray flowers. Involucral bracts are dry, white or straw coloured. Small chaffy scales are between the tubular flowers, a feature which separates this genus from *Helichrysum*.

1141. Cassinia arcuata

Cassinia arcuata

Chinese-shrub, Drooping Cassinia **Plate 1141**

A shrub more typical of the western slopes and western plains where it spread quickly in overgrazed pastures and in disturbed ground such as around old mines. It is uncommon in the Sydney district and usually occurs along roadsides where it may have been introduced by seeds falling from passing stock. Its common name is probably derived from the gold-rush days when Chinese miners used this species for thatch on their roofs. It grows to 2 m tall. Branches are woolly-white and leaves linear, blunt, 5 to 10 mm long, with margins rolled under. Crushed foliage has a strong curry-like aroma. Flower heads are pale brown and numerous; they are borne in loose, drooping terminal panicles.

HABITAT Roadsides and disturbed clay or loamy soils.

DISTRIBUTION Scattered west of Blue Mountains (e.g., Hartley Valley), but also recorded from coast; also southern tablelands, western slopes and western plains.

FLOWERING December–June.

1142. Cassinia quinquefaria

Cassinia quinquefaria

Cough-bush **Plate 1142**

Cough-bush, like *C. arcuata*, is a shrub to 2 m tall, more common in drier areas of the western slopes. It differs from *C. arcuata* in having hairless stems, longer leaves and a more open panicle of flowers. Branchlets and leaves are often sticky (viscid); crushed leaves are strongly aromatic. Leaves are linear, up to 4 cm long, with margins rolled under. Flower heads are numerous, white or pale brown in loose terminal panicles.

HABITAT Open forest.

DISTRIBUTION Scattered west of the Blue Mountains, but also recorded at lower Kangaroo Valley. Occurring throughout coast, tablelands and western slopes; also Qld.

FLOWERING January–February.

Cassinia aculeata

Dogwood, Dolly Bush **Plate 1143**

A shrub to 2 m tall, its branchlets are covered with short whitish matted glandular hairs. Leaves are 1 to 5 cm long, 0.5 to 2 mm wide, dark green above and slightly rough or sticky. Margins are curled under, sometimes almost obscuring the pale undersurface. Leaves of various lengths are usually present together. Flower heads are white to straw coloured and borne in dense clusters to 10 cm across. Individual flowers have 2 types of bracts; the inner bracts are white and shiny, the outer bracts are dull, papery and brownish.

HABITAT Open forest, often in disturbed areas.

DISTRIBUTION Widespread from coast to mountains (e.g., Brisbane Water National Park, Agnes Banks, Tallong and Clarence). Occurring throughout coast and tablelands; also Tas, SA and Vic.

FLOWERING December–March.

1143. Cassinia aculeata

1144. Cassinia uncata

1147. Cassinia trinervia

Cassinia uncata

Sticky Cassinia, Bent Cassinia **Plate 1144**

This is a shrub to 2 m tall. It is easily confused with *C. aculeata*. Branchlets are covered with short hairs which are glandular but not softly woolly. Leaves are 15 to 25 mm long (up to 50 mm in mountain forms), very narrow (about 0.5 mm), appearing terete as the margins are rolled under to form a groove beneath. Leaves may be smooth, rough or somewhat sticky. A characteristic feature is that most leaves are curved down at the tip. Flower heads are white to straw coloured and borne in dense almost flat clusters 3 to 8 cm across.

HABITAT Open forest, often in rocky areas.

DISTRIBUTION Widespread from coast to mountains (e.g., Agnes Banks, Clarence). Extending from northern tablelands to southern tablelands and to western slopes; also Vic and SA.

FLOWERING December–January.

Cassinia cunninghamii **Plate 1145**

An erect shrub, 1 to 2 m tall, with hairy white stems and crowded linear leaves. These leaves are 20 to 35 mm long, with margins rolled under to the woolly undersurface. Flower heads are yellow or straw coloured and borne in compact terminal clusters, about 12 cm across, which tend to be flat or domed. The involucral bracts are straw coloured throughout, unlike those of *C. aculeata*, with which this species could be confused.

HABITAT Open forest.

DISTRIBUTION Widespread but not common. Recorded from mountains and higher areas of coastal zone (e.g., Bucketty). Also occurring north coast, south coast and central western slopes.

FLOWERING August–September, but some flowers at other times.

Cassinia denticulata **Plate 1146**

This is an erect shrub, to 2 m tall, with broad lanceolate stalkless leaves and creamy flower heads. Its leaves are characteristic of the species. They are smooth above, shiny green, 10 to 25 mm long, stem-clasping, entire or occasionally toothed, with a single longitudinal vein, and are often bent downwards. Flower clusters are terminal, flat or oval on top.

HABITAT Open forest.

DISTRIBUTION Widespread from coast to mountains. Recordings include Brisbane Water, Royal and Ku-ring-gai Chase National Parks and Mt Wilson. Endemic in the Sydney district.

FLOWERING October–November.

1145. Cassinia cunninghamii

1146. Cassinia denticulata

Cassinia longifolia

Shiny Cassinia

A large spreading shrub (to 3.5 m) of rainforest margins, especially abundant in regrowth areas, where its masses of white flower heads make it very conspicuous. Its leaves are narrow, 3 to 8 cm long, 2 to 5 mm wide, flat, green above and covered with matted whitish hairs below. Flower heads are in large clusters 6 to 12 cm across. Involucre is oblong, about 1 mm across and involucral bracts are white.

HABITAT Sheltered forest and rainforest margins, especially in partly cleared areas.

DISTRIBUTION Widespread from coast to mountains (e.g., Mooney Mooney Creek). Extending along south coast and southern tablelands to Vic.

FLOWERING November–January.

Cassinia trinervia

Three-veined Cassinia **Plate 1147**

A large shrub, usually 2 to 3 m tall, although odd plants may be much taller and have thick tree-like trunks. It closely resembles *C. longifolia*, except the leaves have a midvein and two longitudinal veins running close to the margin. Leaves are flat, 3 to 10 cm long, 3 to 8 mm wide, green and hairless above and paler below. The undersurface is covered with minute glandular hairs, not the matted cottony hairs of *C. longifolia*. Flower heads are in large whitish flat-topped clusters up to 12 cm across.

HABITAT Sheltered forests and slopes, especially near rainforest.

DISTRIBUTION Widespread from coast to mountains (e.g., Macquarie Pass, Mt Kembla). Also occurring north coast, northern tablelands, south coast and Vic.

HABITAT December–February.

Cassinia aureonitens

Yellow Cassinia **Plate 1148**

This erect shrub is easy to identify as it is the only *Cassinia* in the Sydney district with bright yellow flower heads. Leaves are narrow lanceolate, 3 to 5 cm long, flat, green above and lighter below. Flower clusters are dense and terminal, rounded on top. The whole plant is generally hairless, or with a few short yellowish stem hairs.

HABITAT Open forest, usually in sheltered sites.

DISTRIBUTION Widespread from coast to mountains. Recordings include Georges River, Waterfall, Nattai Valley and Mt Irvine. Also occurring north and south coasts.

FLOWERING October–December.

1148. *Cassinia aureonitens*

Helichrysum Everlasting

Everlastings are herbs or shrubs with flower heads usually containing all tubular flowers, without any petal-like flowers. Involucral bracts are scarious and in several rows; in the Everlasting or Paper Daisy group, these bracts are enlarged like petals. There are no chaffy scales between the flowers as there are in *Cassinia*. Pappus hairs are simple or feathery near the tips, unlike those of *Helipterum* which are feathery from the base.

1149. *Helichrysum diosmifolium*

Helichrysum diosmifolium

Pill Flower, Ball Everlasting **Plate 1149**

This is a common large branched shrub, often 2 m or more tall, with narrow crowded leaves and large terminal white flower clusters. It is sometimes mistaken for a *Cassinia*, but lacks the chaffy scales between the flowers. Leaves are 1 to 2 cm long with margins rolled under. Flower-heads are borne in dense cauliflower-like clusters up to 7 cm across. The involucre is 3 to 6 mm across, with bracts white or tinged with pink, and there are 20 or more flowers in each head.

HABITAT Open forest, on both shale and sandy soils.

DISTRIBUTION Widespread from coast to mountains. Recordings include Brisbane Water National Park, Menai, Nattai, Macquarie Pass, Tallong, Agnes Banks and Thirlmere Lakes. Extending through most of NSW, except far western plains; also Qld.

FLOWERING October–December.

1150. *Helichrysum dendroideum*

Helichrysum dendroideum

Tree Everlasting **Plate 1150**

Although widespread and common in southern NSW and in Victoria, Tree Everlasting is seldom seen in the Sydney district. Here it is usually not a tree, but an untidy shrub 2 to 3 m tall with grey-brown fibrous bark and rather open foliage. Leaves are flat, lanceolate, 3 to 5 cm long, 2 to 3 mm wide, green above and paler below. Flowers are small and white, with only 5 or 6 in a head; outer bracts are yellow-green. Flower heads are numerous and borne in broad round-topped clusters. Inflorescence suggests *Cassinia*, but flower heads have short spreading rays.

HABITAT Open forest and scrub, on moist or richer soils.

DISTRIBUTION Scattered and uncommon from coast to mountains. Recordings include Mt Tomah. Extending along coast and tablelands to Vic, Tas and SA (Port Augusta).

FLOWERING December–January.

1151. *Helichrysum adnatum*

Helichrysum adnatum **Plate 1151**

An open shrub, about 1 m tall, with upright white stems. Leaves are small, 2 to 8 mm long, linear, without stalks and decurrent (margins continue down along the stems). The lines of small leaves flattened along the stems with outturned tips are characteristic of this species. Flower heads are in terminal clusters, not as dense as those of *H. diosmifolium*. The involucre is 3 to 4 mm across, with bracts white or straw coloured, and there are 10 to 12 flowers in each head.

HABITAT Forest.

DISTRIBUTION Blue Mountains (e.g., Scotts Range, Yerranderie), with a rare recording for southern Sydney. Extending along tablelands south to Vic.

FLOWERING November–December.

Helichrysum argophyllum **Plate 1152**

This is a bushy shrub to 2 m tall. Leaves are strongly aromatic, lanceolate, 4 to 7 cm long, 6 to 10 mm wide, green above and grey-white below. Flower heads are in large terminal clusters. The involucre is about 2 mm across. Flower heads contain about 14 flowers, yellowish in the centre, surrounded by small white

1152. *Helichrysum argophyllum*

1153. Helichrysum apiculatum

rays. *H., argophyllum* may be confused with *Cassinia longifolia*, but the *Cassinia* does not have the small ray flowers.

HABITAT Tall sheltered forest.

DISTRIBUTION Coastal zone (e.g., Illawarra Escarpment). Extending to south coast and southern tablelands; also Tas.

FLOWERING October–November.

Helichrysum apiculatum

Yellow Buttons **Plate 1153**

A perennial herb, erect (to 50 cm tall) or a loose mat, often with a woody base. It is a very variable plant and some specimens seems to grade into forms of *H. semipapposum*. The typical form has stems and leaves covered with dense felt-like grey hairs. Leaves are lanceolate to oblong wedge shaped, 2 to 5 cm long. Flower heads are bright yellow, in terminal clusters, with the broad centre disc surrounded by radiating petal-like bracts, unlike the bell-shaped flower heads of *H. semipapposum.*

HABITAT Grassland and forest, mainly on clay soils.

DISTRIBUTION Widespread, especially in mountain valleys. Occurring throughout the State and in all other States.

FLOWERING October–December, with a few flowers at other times.

1154. Helichrysum semipapposum

Helichrysum semipapposum

Clustered Everlasting **Plate 1154**

An erect perennial herb to 50 cm tall, somewhat woody near the base. It is a variable plant, with stems and leaves often covered with loose white hairs when young; older plants tend to be hairless and sometimes sticky. Leaves are linear, 1 to 5 cm long, longer at the base of the stem than on the upper stems; margins are rolled under to the woolly undersurface. Flower heads are bright yellow, in terminal clusters, bell shaped, 4 to 5 mm across, with the central disc surrounded by small petal-like bracts.

HABITAT Grassland and woodland, often in rocky areas on hillsides.

DISTRIBUTION Widespread from coast to mountains. Occurring throughout tablelands, western slopes and north coast; also Qld, Vic and Tas.

FLOWERING November–January, with some flowers at other times.

1155. Helichrysum elatum

Helichrysum elatum

White Paper Daisy, Tall Everlasting **Plate 1155**

A common erect shrub in sheltered situations, it is usually about 1 m tall with robust branches covered with white woolly hairs. Leaves are broad lanceolate, up to 10 cm long and 2 to 4 cm wide, mostly smooth above and grey below. Flower heads are large, about 3 cm across, with yellow centres surrounded by numerous crisp papery white petal-like bracts.

HABITAT Sheltered forests and richer soils of gullies.

DISTRIBUTION Widespread from coast to mountains. Recordings include Hacking River, Brisbane Water National Park, Macquarie Pass and Thirlmere Lakes. Extending throughout coast and tablelands; also Qld.

FLOWERING August–October.

1156. Helichrysum leucopsideum

Helichrysum leucopsideum

Satin Everlasting **Plate 1156**

A small erect perennial, with stems 30 to 50 cm tall. Stems and lower surfaces of leaves are covered with grey woolly hairs, sometimes only slightly so. Leaves are variable in shape; they may be oblong or linear, 2 to 4 cm long, rough above, with jagged margins. Flower heads are solitary and terminal, about 3 cm across, yellow centred with white papery petal-like pointed bracts.

HABITAT Open forest.

DISTRIBUTION Mainly higher Blue Mountains (e.g., Clarence, Wombeyan Caves road) but also reported from central coast. Extending along south coast and southern tablelands, with an isolated occurrence in the south-west of the State around Wentworth; also Vic, Tas, SA and WA.

FLOWERING December–January.

Helichrysum adenophorum var. *waddelliae*

Waddell Everlasting **Plate 1157**

This small decumbent to erect perennial lacks the grey woolly hairs of *H. leucopsideum.* Its stems and upper leaf surfaces are almost hairless or with a few glandular hairs. Leaves are linear, 3 to 5 cm long, green above, whitish below. Flower heads are solitary and terminal, about 3 cm across, yellow centred with white papery petal-like bracts. The outer bracts are usually distinctly pink.

HABITAT Open forest.

DISTRIBUTION Mainly a mountain and alpine species, but found in the coastal area from Windsor to Putty. Extending along central and southern tablelands to the Victorian high country (e.g., Mt Buffalo National Park).

FLOWERING December–April.

1157. *Helichrysum adenophorum* var. *waddelliae*

1159. *Helichrysum rutidolepis*

1161. *Helichrysum calvertianum*

Helichrysum bracteatum Plate 1158

Yellow Paper Daisy, Golden Everlasting

A common plant along roadsides in tall forests, it is often grown for use in dried arrangements. Its erect stems may reach 1 m high, but it is usually half that height. Leaves are linear to lanceolate, 2 to 8 cm long, dark green, hairless or almost so, with bases which are somewhat stem-clasping. Flower heads are bright yellow, about 3 cm across, with the yellow centre surrounded by several rows of stiff shiny petal-like yellow bracts.

HABITAT Sheltered forests, especially on better soils.

DISTRIBUTION Widespread from coast to mountains. Found throughout NSW, and in all other mainland States.

FLOWERING October–March.

1158. *Helichrysum bracteatum*

Helichrysum rutidolepsis

Pale Everlasting **Plate 1159**

A decumbent to erect perennial, with hairy white stems, 20 to 50 cm tall. Leaves are oblong, linear or lanceolate, to 5 cm long, stem-clasping and with margins slightly turned under. Flower heads are terminal on stalks bearing a few small leaves. The involucre is about 1 cm across. The yellow disc is surrounded by short white or yellow spreading bracts. Pale Everlasting may be confused with *H. scorpioides*, but that species has more woolly stems, the longer leaves are mainly around the base of the stems and the flower head is larger.

HABITAT Grassland, sheltered forest and rainforest margins.

DISTRIBUTION Widespread from coast to mountains. Occurring throughout coast and tablelands and western plains; also Vic and SA.

FLOWERING January–May.

Helichrysum scorpioides Plate 1160

H. scorpioides has an erect woolly flower stalk, up to 30 cm tall, rising from tufts of light green hairy leaves. These leaves are oblong to linear, up to 6 cm long, woolly or hairless above and woolly below, stalkless and somewhat stem-clasping. The flower stalk has long leaves near its base, but reduced leaves along the upper part. Flower heads are terminal. The involucre is 1 to 2 cm across. The yellow disc is surrounded by short white or yellow papery bracts.

HABITAT Grassland and forest, often on clay soils.

DISTRIBUTION Widespread from coast to mountains (e.g., Thirlmere, Castlereagh State Forest). Occurring throughout coast, tablelands and western plains; also Vic, Tas and SA.

FLOWERING September–November.

1160. *Helichrysum scorpioides*

Helichrysum calvertianum Plate 1161

This small branched Everlasting has a very restricted distribution and has only recently been recorded for the Sydney district. Stems are wiry, 10 to 12 cm tall, with crowded, narrow, blunt leaves 6 to 8 mm long. Flower heads are solitary, small (10 to 12 mm across), with yellow centres and numerous rows of white papery petal-like bracts. Outer bracts are pinkish or brown and slightly hairy.

HABITAT Rocky sandstone escarpments, often in moist sites.

DISTRIBUTION Cliff-tops near Fitzroy Falls, Carrington Falls and near Mittagong. Extending to southern tablelands.

FLOWERING November–January.

1162. *Helichrysum viscosum*

Helichrysum viscosum

Sticky Everlasting **Plate 1162**

A multi-stemmed, sticky perennial, erect or sprawling, 40 to 80 cm tall, with stems and leaves which have a distinctly varnished appearance. Leaves are narrow linear, 3 to 9 cm long and dark green. Flower heads are terminal, bright yellow, with several rows of stiff shining petal-like yellow or brownish bracts. This species may be confused with *H. bracteatum* which also has yellow flower heads, but the latter species has broader leaves which are not so conspicuously varnished. Sticky Everlasting usually dies off in summer and re-shoots in autumn.

HABITAT Open forest, especially on shallow stony soils.

DISTRIBUTION Widespread throughout NSW, except north coast and far western plains; also Qld and Vic.

FLOWERING September–December.

1165. Helipterum anthemoides

1163. Helipterum albicans ssp. *albicans*

Helipterum Sunray

Sunrays are herbs with flower heads containing all tubular flowers, without any petal-like ray flowers. The genus includes a number of Paper Daisies or Everlastings, in which the involucral bracts are dry, spreading and enlarged like petals; these are absent in some species. The genus is often confused with *Helichrysum*, the other genus containing Everlastings. The pappus hairs of *Helipterum* are plumose (feathery) from the base, whilst those of *Helichrysum* are simple or feathery only near the tips.

Helipterum albicans ssp. *albicans*

Hoary Sunray **Plate 1163**

A variable perennial, 12 to 45 cm tall, with a branched and often woody base. Young stems are erect and woolly; older stems are often covered with persistent brown leaf bases. Leaves are linear to narrow oblong, 3 to 10 cm long, 1 to 10 mm wide. Flower heads are solitary on woolly stalks. The yellow centre is surrounded by spreading, dry, petal-like bracts which may be pale yellow, brown, purple or colourless. Inner bracts are yellow or white. Four varieties of this subspecies have been recognised; three of these occur in the Sydney district.

Var. *albicans* has leaves which are filiform to linear, woolly on both surfaces; the outer involucral bracts are brown, the inner bracts are yellow.

Var. *incanum* forma *purpureo-album* has leaves which are filiform to narrow oblanceolate, woolly on both surfaces; outer bracts are pale brown to purple, the inner bracts are white.

1164. Helipterum albicans ssp. *graminifolium*

Var. *graminifolium* (Plate 1164) has leaves which are narrow, closely revolute, grasslike and hairless above, woolly below, except for the hairless midrib.

HABITAT Forest, grassland and rocky outcrops.

DISTRIBUTION Var. *albicans* occurs at higher altitudes (e.g., Hartley, Lithgow). Although absent from the central and south coasts, it occurs throughout most of eastern NSW and in Qld and Vic. Var. *incanum* is restricted to the central and southern tablelands, extending to Vic. Recordings include Berrima, Bundanoon, Wingello and Tallong. Var. *graminifolium* is known only from the higher Blue Mountains between Clarence and Wolgan (e.g., Newnes State Forest).

FLOWERING Chiefly November–December, but some flowers at other times.

Helipterum anthemoides

Chamomile Sunray **Plate 1165**

This is a perennial bushy herb to 30 cm tall, with hairless, linear leaves, about 10 mm long, crowded along the stems. Flower heads are solitary on smooth unbranched stems. Flower heads are about 25 mm across, with a small yellow centre surrounded by spreading, dry, white, petal-like bracts.

HABITAT Forest, grassland and rocky outcrops.

DISTRIBUTION Western parts of Sydney district, but also recorded from higher areas of the central coast. Extending throughout tablelands and western slopes; also Qld and Tas.

FLOWERING October–December.

Ammobium

A genus of 2 species, 1 in the Sydney district. Perennial herbs with winged stems. Flowers all tubular, with involucral bracts papery white and pappus a membranous crown on a 4-angled cypsela.

Ammobium alatum **Plate 1166**

This tall herb has silver-white stems 60 to 80 cm high, with 4 distinct wings running from top to bottom. Lower leaves are in a rosette; they are oblanceolate, 12 to 18 cm long, with silver-white hairs below. Stem leaves are entire, lanceolate, with leaf edges which continue onto the stem to form wings. Flower heads are about 2 cm across, with a yellow centre and surrounded by several rows of white papery bracts.

HABITAT Woodland and grassland.

DISTRIBUTION Higher mountains and western valleys. Extending along tablelands and western slopes; also south and north coasts, but not recorded from central coast.

FLOWERING December–January.

1166. *Ammobium alatum*

1168. *Podolepis jaceoides*

Calomeria

Tall plants with striking large drooping terminal panicles of tiny flower heads. Involucre small, 1 to 2 mm across; flower heads with 3 to 4 tubular flowers; cypselas hairless and without any pappus.

Calomeria amaranthoides

Plume Bush **Plate 1167**

A conspicuous plant, tall (2 m or sometimes taller), strongly scented, with large stem-clasping, entire, alternate, lanceolate leaves, 12 to 25 cm long. The erect stems end in a large, much-branched, loose panicle, from which tiny red-brown flower heads hang on their dropping stalklets. Plume Bush appears to be short lived, seldom lasting more than 2 seasons.

HABITAT Sheltered forested slopes and rainforest margins, mainly on volcanic soils.

DISTRIBUTION Blue Mountains (e.g., Mt Hay, Mt Tomah). Extending along coast southwards to Vic.

FLOWERING January–February.

1167. *Calomeria amaranthoides*

Podolepis

Erect herbs with terminal flower heads, usually yellow. Disc flowers tubular; outer flowers are spreading rays with 2 to 4 lobes at the tip, giving the appearance of a fringed flower head. Involucral bracts are in several rows, of unequal length; the inner bracts are stalked, hence the botanical name (from Greek, 'podos'—a foot, 'leptos'—slender).

Podolepis jaceoides

Showy Podolepis **Plate 1168**

An attractive branched perennial with large yellow flower heads on stems 30 to 80 cm tall. Basal leaves are lanceolate, green, 10 to 15 cm long and up to 2 cm wide; stem leaves are smaller and stem-clasping. Flower heads are usually solitary. Spreading rays are cut into narrow lobes; there are 30 to 40 rays, each up to 25 mm long.

HABITAT Grassland and woodland, especially on heavier clay soils.

DISTRIBUTION Widespread from coast to mountains. Found throughout NSW; also Qld, Vic, Tas and SA.

FLOWERING September–December.

Podolepis hieracioides **Plate 1169**

This is an erect plant to 70 cm tall with showy yellow flower heads. It differs from *P. jaceoides* in having smaller flowers which are in clusters of 3 or more (not solitary). Basal leaves are oblanceolate, 9 to 10 cm long and 15 to 20 mm wide; stem leaves are narrow, about 7 cm long and 7 mm wide. Flowers have 15 to 20 spreading fringed rays, each 15 to 18 mm long. The bracts surrounding the flower heads are shiny.

HABITAT Woodland with grassy understorey.

DISTRIBUTION Higher Blue Mountains (e.g., Hartley Valley). Extending from northern tablelands to southern tablelands and Vic.

FLOWERING November–December.

1169. *Podolepis hieracioides*

Monocotyledons

Alismataceae

The Alismataceae is a small cosmopolitan family of aquatic or amphibious plants. It is 1 of 3 families in the order Alismatales which includes the most primitive of the monocotyledons. Some botanists believe that the dicotyledon group from which the monocotyledons are derived was an aquatic plant similar to the Nymphaeales (a primitive dicot order of water lilies) and the Alismatales.

Like several primitive dicot families, the Alismataceae has trimerous flowers (i.e., with parts in threes). They have 3 free sepals, 3 free petals, 3, 6, 9 or many stamens, and a superior ovary with 3 to many free carpels. *Sagittaria* has unisexual flowers, the others are bisexual. Some members of the family produce nectar to attract insects, but others appear to be wind pollinated. Fruit walls contain air spaces which provide buoyancy needed for water dispersal. Although the star-shaped fruits develop above water, young plantlets are often produced asexually when the plants are submerged. Some species show more than one leaf form, with narrow submerged leaves differing in structure and shape from the aerial leaves.

One species each of *Alisma* and *Damasonium* and one introduced species of *Sagittaria* occur in the Sydney area. None is common in the area. A few species are cultivated as ornamentals in ponds and aquaria. Others are an important wildlife food source.

Damasonium

Damasonium minus

Star-fruit **Plate 1170**

A hairless aquatic annual with a perennial roostock, occurring in shallow water or around wet and swampy margins of ponds. Leaves rise from the base on long stalks; leaf blades are ovate to lanceolate, often heart shaped at the base, up to 10 cm long, with 3 or 5 prominent veins. Stems are up to 50 cm tall. Flowers are small, white, with 3 petals and 6 stamens. Each flower is solitary on long slender stalks which emerge in whorls from the branched flower stems. Fruit is star like, about 10 mm wide, and formed by 6 to 9 triangular fruitlets which are joined at their bases.

1170. Damasonium minus

HABITAT Shallow water and muddy margins of swamps and lagoons.

DISTRIBUTION Coastal zone around Sydney (e.g., Castlereagh State Forest, Shane Park, Campbelltown, Leumeah), but becoming uncommon because of drainage works, filling of swamps and water pollution. Also recorded Nattai valley. Widely distributed through north coast, northern tablelands, western slopes, western plains and in all other States.

FLOWERING Chiefly November–December, but varying according to rainfall and flooding.

1171. Alisma plantago-aquatica

Alisma

Alisma plantago-aquatica

Water Plantain **Plate 1171**

An erect aquatic plant with leaves on long stalks to 1 m tall which rise from a fleshy root. Leaves are 10 to 30 cm long and have 5 to 9 conspicuous longitudinal veins. Flowers are borne on erect branched stems up to 1.5 m tall. Stems are branched into ray-like whorls and the white flowers are in whorls on short stalks. Each flower is about 5 mm across, with 3 broad petals which fall off early. Fruit consists of about 20 small rounded fruitlets.

HABITAT Pools, swamps and along streams.

DISTRIBUTION Widespread from coast to mountains, but not common. Also occurring north coast, northern tablelands, southern tablelands, south-western slopes, Vic and in temperate regions of the northern hemisphere.

FLOWERING November–December.

Hydrocharitaceae

The Hydrocharitaceae is mainly tropical and includes marine as well as freshwater plants. It appears to be closely related to the Butomaceae, but differs from that family and others in the Alismatales in having flowers with an inferior ovary with fused carpels. Like the Alismatales, the Hydrocharitaceae has both bisexual and unisexual flowers with the perianth segments in threes. Some members are cultivated as ornamentals, others are problem weeds which clog waterways.

The most interesting feature of the family is the variety of specialised pollination mechanisms. Some genera with showy emergent flowers are insect pollinated but others undergo self-fertilisation below water. In some cases male flowers or buds break off and rise to the surface and upon opening, sail like tiny boats to the female flowers. In other cases they remain attached by long stalks and float in a wide circle. Sometimes the stamens hang overboard and brush pollen onto passing objects, or the anthers may open explosively scattering pollen in the air and on the surface of the water. Occasionally pollen is released underwater and carried by currents. In *Vallisneria*, the coiled stems of the female flowers contract after fertilisation, dragging the developing fruit to the bottom.

Five native species from 4 genera occur in the Sydney area where none is common.

Vallisneria

Submerged fresh-water herbs with ribbon-like leaves. One species in Australia.

1172. Vallisneria gigantea

Vallisneria gigantea

Eel-weed, Ribbon-weed **Plate 1172**

A submerged freshwater perennial, with a stoloniferous rootstock. Leaves are long, ribbon-like, 30 to 40 cm long (sometimes much longer), up to 2 cm wide, usually obtuse, with minutely toothed margins near the apex, and 5 or more parallel veins. Leaves emerge in tufts from along the stolons and are often covered with brown silt. Flowers are inconspicuous; male and female flowers are borne on separate plants. Male flowers are produced near the base of the plant in a sheath; this opens to allow the flowers to float to the surface of the water. Female flowers are borne at the end of a slender spirally coiled stalk which elongates to carry the flowers to the surface where pollination takes place with the free-floating male flowers. The coiled stalk pulls the developing fruit underwater. Reproduction occurs mainly from new shoots and roots along the stolons.

Habitat Larger streams, swamps and lakes.

Distribution Coastal zone (e.g., Hacking River near Audley, McCarrs Creek) but uncommon because of pollution and destruction of suitable habitat. Formerly recorded from Nepean, Cooks and Wollondilly Rivers. A common and troublesome weed in inland irrigation areas. Recorded from the coast, southern tablelands, western slopes, western and far western plains, in all other States and in warmer parts of Asia.

Flowering February–April.

Ottelia

Freshwater herbs with floating oblong leaves. Two species in Australia, one in NSW.

1173. Ottelia ovalifolia

Ottelia ovalifolia

Swamp Lily **Plate 1173**

Swamp Lily is a perennial freshwater herb, rooted in the mud. Its leaves rise from the base of the plant on long stalks and the leaf blades usually float on the surface. Adult leaves are ovate to elliptic, up to 15 cm long, obtuse at the apex and often heart shaped at the base, with 5 to 7 longitudinal veins. Flowers are 3 to 4 cm across, with 3 whitish-yellow petals, and 9 to 15 conspicuous yellow stamens spreading from a reddish base. Flowers are solitary and borne above the surface of the water at the end of a stout stem. Seeds develop and are released underwater.

Habitat Lagoons, dams and shallow slow water.

Distribution Widespread, but not common around Sydney because of lack of suitable habitat. Found throughout coast, central tablelands, western slopes and western plains; also all mainland States.

Flowering December–March.

Halophila

Submerged marine herbs.

Halophila ovalis

Sea Wrack **Plate 1174**

This species is widely distributed around the coast, especially in bays and inlets where it is often found in conjunction with *Zostera* and *Posidonia*. It has creeping rhizomes which anchor the plant to the sediment and submerged paired elliptic leaves on long stalks. Leaf blades are 2 to 5 cm long, almost transparent, hairless, with a conspicuous central vein and 10 to 14 pairs of lateral veins. Male and female flowers are borne on separate plants and pollination takes place beneath the surface of the water. Flowers are minute and borne in a small axillary spathe.

1174. Halophila ovalis

Less common is *H. decipiens*, which is found on muddy substrates. Its leaves are smaller than those of *H. ovalis* (10 to 25 mm long), with fine hairs on the lower surface and 6 to 9 pairs of lateral veins. It has been recorded from Quibray Bay (Botany Bay).

Habitat Bays and inlets, where it is a fast coloniser of unstable sediments.

Distribution Recorded in Lake Macquarie, Pittwater, Port Jackson, Botany Bay (Towra Point) and Port Hacking. Occurring all along the NSW coast, in every other State, tropical Africa, Asia and western Pacific.

Flowering Flowers rarely seen.

Juncaginaceae

This small family of aquatic and marsh plants occurs in temperate and cold regions in both the northern and southern hemispheres. The largest and most widespread genus is *Triglochin*, with 12 species (7 in NSW). *Maundia*, with 1 species, is confined to eastern Australia.

The reddish-green flowers occur in racemes or spikes. They are regular and unisexual or bisexual with a perianth in 2 series of 3 segments. The superior ovary consists of 4 to 6 free or partly united carpels, each with a single ovule.

Two species of *Triglochin* are fairly common in the Sydney area in quite different habitats. *Maundia triglochinoides* is rare in the area, and easily mistaken for *Triglochin procera*.

Triglochin

Aquatic plants of freshwater streams or salt marsh. Leaves narrow, rising from the base of the plant. Flowers small and borne on an erect spike.

Triglochin procera

Water Ribbons **Plate 1175**

A robust freshwater plant with strap-like leaves up to 1 m long and 3 cm wide, which spread from a thick rhizome. In permanent streams and ponds, the upper parts of these leaves float on the surface of the water; in muddy swampy soils the leaves are often reduced in size. Underwater leaves and stems often have a reddish tint. Flowers are numerous, small, almost stalkless and are crowded along an erect spike to 30 cm long which is at the end of a stout cylindrical stem. Fruit is rounded, with 6 fertile segments, 4 to 10 mm long and twisted.

Habitat Ponds and slow-floating water to about 1 m deep; sometimes in swampy margins.

Distribution Widespread. Recordings include Royal National Park, Bents Basin (Nepean River) and Paddys River. Found throughout most of NSW, except drier western plains; also all other States, New Guinea and Malaysia.

Flowering August–September.

1175. Triglochin procera

Triglochin striata

Streaked Arrowgrass **Plate 1176**

A small, inconspicuous, grass-like component of salt marsh and wet saline areas near the sea. Tufts of narrow linear leaves, up to 30 cm long and 1 to 3 mm wide, rise from small stoloniferous root stocks. Flowers are small, greenish and borne on short stalks on a slender erect spike which is about as long as the leaves. Fruit is rounded, with the 3 fertile segments winged or streaked.

HABITAT Saline swamps, along coast and estuarine salt marsh, but occasionally in wet places away from coast.

DISTRIBUTION Coastal zone. Recordings include Kurnell, Curl Curl, Mill Creek (Georges River), Mooney Mooney Creek and Port Hacking. Found in all other States, South Africa and the Americas.

FLOWERING November–December.

1176. Triglochin striata

Najadaceae

The Najadaceae is a small family of submerged aquatic plants found throughout the world in fresh and brackish waters. It consists of 1 genus, *Najas*, divided into about 50 species which are often difficult to tell apart. *Najas* is a source of food for fish. A few species are serious weeds in rice fields and irrigation ditches.

Two species are found in the Sydney area in fresh or brackish water.

1177. Najas tenuifolia

Najas tenuifolia

Water Nymph, Thin-leaved Naiad **Plate 1177**

Water Nymph is a submerged freshwater plant, with slender, much-branched stems, up to 50 cm long, and often appearing as a black mass of foliage just beneath the water surface. The creeping stems root at the nodes. Leaves are narrow-linear, 12 to 35 mm long, edged with minute teeth, emerging in bunches from sheaths at intervals along the stems. Flowers are tiny, with male and female flowers separate on the one plant. Pollination takes place underwater. Fruit is 2 mm long and covered by the leaf sheath. Less common in the area is *N. marina*, which has leaves prominently toothed, the teeth stout and spine like.

HABITAT Still waters of lagoons and lakes, and backwaters of slow-moving streams.

DISTRIBUTION Widespread, but uncommon in Sydney district because of lack of suitable habitat. Recorded north coast, central coast, central tablelands, Lachlan, Murrumbidgee and Murray Rivers; also all mainland States and Malesia.

FLOWERING Probably summer, but flowers seldom seen.

Potamogetonaceae

This is a family of freshwater submerged or floating plants known as Pondweeds. Its closest relationships are to the marine families Zosteraceae (Eel-grasses) and Posidoniaceae. The family is worldwide but is most common in the temperate northern hemisphere. All except one of the 100 or so species are in the genus *Potamogeton*.

The leaves of some species have 2 forms—broad floating leaves and narrow linear submerged leaves. The inconspicuous brownish flowers are carried in spikes above the water, and are probably wind pollinated. After pollination the spike becomes submerged and the fruits develop beneath the surface.

Seven species of *Potamogeton* occur in the Sydney area; all but 2 are uncommon.

Potamogeton

1178. Potamogeton tricarinatus

Potamogeton tricarinatus

Floating Pondweed **Plate 1178**

This freshwater perennial grows rooted in the mud of still and slow-moving water. Rounded stems rise from a creeping rhizome. Leaves are of 2 types, some floating, some submerged. Those floating are ovate to broad elliptic, 2 to 10 cm long, green, leathery, with 15 to 25 parallel veins, and on long stalks. Lower leaves are stalkless, thin, translucent, narrow lanceolate and up to 20 cm long. Flowers are small and borne in short dense spikes 2 to 4 cm long at the ends of stout erect stems. Fruit is nut-like, about 3 mm long, with 3 ridges across its back.

HABITAT Rivers, creeks, lagoons and other permanent fresh water.

DISTRIBUTION Widespread, although suitable habitats in the Sydney district are limited. Occurring in most districts of NSW, in all other States and Asia.

FLOWERING November–January.

Posidoniaceae

This is a small family of submerged marine plants closely related to the Zosteraceae (Eel-grasses). The family consists of 3 species of *Posidonia*, one restricted to the Mediterranean, and the other 2 to Australia's southern shores. One of these, *P. ostenfeldii*, with narrow leaves, occurs only in Western and South Australia, but the other, with broad leaves, is found in all southern states.

1179. Zostera capricorni (left) and *Posidonia australis* (right)

1180. Zostera capricorni

Posidonia australis

Strapweed, Marine-fibre Plant **Plate 1179**

Strapweed grows on the sandy bottom of protected areas in bays and inlets from just below low tide mark to a depth of about 10 m. It prefers slightly deeper water than *Zostera* and is rarely exposed at low tide. Leaves are attached to a creeping rhizome which, aided by the roots, helps anchor the plant to the substrate. The smooth brownish strap-like leaves are sheathed at the bases. The leaves of the local species are up to 1 m long, 1 cm wide, and with 10 to 20 fine nerves running from the base to the smooth rounded-truncate apex. Leaf sheaths are persistent and fibrous with ear-like expansions at the top; they are open at the top (the margins do not meet or overlap). The nature of the leaf sheaths helps to distinguish this species from *Zostera*. The small flowers are borne in terminal spikes subtended by leaf-like bracts. Flowers are bisexual with 3 stamens. Fruits are egg shaped and fleshy.

HABITAT Protected bays and inlets, anchored on bottom and growing underwater.

DISTRIBUTION Central and south coast. Recordings include Pittwater, Botany Bay (Quibray Bay, Woolooware Bay) and Port Hacking. Also Vic, Tas, SA and WA.

FLOWERING Flowers rarely seen.

Zostera capricorni

Eel-grass **Plate 1180**

A seagrass which often forms large stands below high tide level in estuaries and bays. It is tolerant of a variety of environmental conditions, being found in a range of salinities and on soft substrates. The best specimens are usually found growing in mud or in a mixture of mud and sand on the landward side of *Posidonia* beds. Its leaves are strap like, rising from a creeping rhizome, 10 to 50 cm long, 2 to 4 mm wide, with conspicuous sheathing bases, 3 to 5 main veins and the ends rounded or truncate (occasionally with small teeth). The margins of the leaf sheaths overlap. Small unisexual flowers are borne from a sheath near the base of a floral leaf. Male flowers have a single stamen and female flowers have a short style with 2 relatively long stigmas. The sexes are arranged alternately in a flat spike with as many as 10 flowers of each sex. The fruit is a single egg-shaped seed, about 2 mm long.

HABITAT Protected bays and inlets, anchored on bottom and growing underwater.

DISTRIBUTION All along NSW coast. Recordings include Pittwater, Sydney Harbour, Botany Bay and Port Hacking. Extending from Mallacoota (Vic) northwards to Qld. Also Lord Howe Island.

FLOWERING Flowers rarely seen.

Palms are characteristically woody, an unusual feature in the monocotyledons which are generally herbaceous. They usually have a single unbranched trunk with the apical bud well protected by leaf bases and spines, since damage to the bud frequently causes death to the tree. As palms grow the trunks do not thicken by production of new cells at a cambium layer beneath the bark in the same way as dicotyledons. Instead, the palm seedling attains its full girth below ground before it begins to grow upward. Consequently palms do not have wood with annual rings although the trunks are sometimes ringed due to the clean detachment of old leaf bases.

The structure of the palm trunk also differs from that of dicotyledonous trees in having bundles of conducting vessels enclosed in fibrous sheaths (fibro-vascular bundles) distributed throughout a soft ground tissue in the trunk or concentrated in the outer layers. This structure, analogous to that of a fibre-reinforced fishing rod, provides the flexibility and strength which enable palms to withstand lashing by winds that break other trees.

Although further subdivision is possible, palm leaves have 2 basic shapes, pinnate (feather shaped) and palmate (fan shaped). Usually, the leaves are crowded into a crown at the top of the trunk, but a few like *Rhapis*, and climbing palms such as *Calamus*, have them spaced at intervals along the stem. The inflorescence is borne in the crown or just below it, and ranges from a simple spike with a few flowers to huge panicles with many thousands. The flowers are bisexual or unisexual and are pollinated by either wind or insects. Floral parts are normally in threes and, although the ovary is basically 3-carpellate, usually only 1 ovule matures into a seed.

The fruits are berries, or more commonly, drupes, with a 3-layered pericarp surrounding a small to very large seed containing oily or ivory-hard endosperm in which a small embryo is embedded near either end or to one side. The middle layer of the pericarp, called the mesocarp, may be fleshy (as in *Livistona*) or fibrous (as in *Cocos*, the coconut palm). The former fruit type is adapted for dispersal by birds and animals, the latter is dispersed by water, often by ocean currents.

The palm family has more than 200 genera and about 2800 species. Fifty-two species in 21 genera occur in Australia, mostly in the tropical north. Two species in 2 genera occur in the Sydney area.

Zosteraceae

The Zosteraceae, or Eel-grass family, is larger and much more widely distributed than the Posidoniaceae to which it is related. It consists of about 18 species in 3 genera found on many shores in temperate regions of the northern and southern hemispheres. Its range in Australia is from Cape York southwards to Geraldton in Western Australia. The 4 Australian species were until recently included in the genus *Zostera*, but 1 southern species is now separated into the monotypic *Heterozostera* due to differences in the rhizome. One species of Zostera occurs in the Sydney area.

Eel-grasses grow on the sandy or muddy bottom in protected seas from the intertidal zone to depths of about 6 m. Extensive beds, uncovered by the receding tide, are areas of great interest to numerous birds in search of food.

Both Zosteraceae and Posidoniaceae are an important part of a marine food chain which extends to fish and sea birds.

Arecaceae

The Arecaceae, also known as Palmae, the palm family, is second to the grasses in economic importance but first in the variety of products provided. This distinctive group is largely restricted to the tropics and subtropics where it is often a prominent part of the vegetation. The richest areas are Malesia, Central America, the West Indies, the Amazon region of South America, and numerous tropical islands in the Pacific and Indian Oceans. In warmer and moister periods of the late Cretaceous and early Tertiary periods the family was much more widespread.

The palms are not closely related to any other extant group, and there are few clues as to their evolutionary development. Their antiquity (relative to other angiosperms) and some unique features of the palm trunk lend support for the view that palms, and possibly all monocotyledons, are derived from some unknown proto-angiospermous stock, and not from a herbaceous aquatic dicotyledon. The group is usually placed alone in the order Arecales.

Archontophoenix

Tree with leaves pinnately divided; flowers unisexual and borne on the same plant.

Archontophoenix cunninghamiana

Bangalow Palm **Plate 1181**

A medium to tall tree easily recognised by its long pinnately divided leaves which are 3 to 4 m long, and its large sheathing leaf bases. Flowers are pale lilac and borne on a large branched inflorescence which emerges from the base of the leaf sheaths. Fruits are red, fleshy, globular and about 12 mm wide.

HABITAT Rainforest, on sheltered slopes and alluvial flats.

DISTRIBUTION Coastal zone, from south coast to northern Qld. Common in Gosford–Watagan area (e.g., Strickland State Forest, Palm Grove, Bouddi National Park and Olney State Forest).

FLOWERING December–February.

1181. Archontophoenix cunninghamiana

1182. Livistona australis

Livistona

Trees with fan-shaped leaves; flowers bisexual.

Livistona australis

Cabbage Tree Palm **Plate 1182**

A medium to tall tree with palmately divided leaves (leaflets spreading from the same point) on long stalks which often have stout prickles along their margins. Flowers are small, yellowish and numerous. They are borne on a much-branched hanging panicle about 1 m long which emerges from among the leaf bases. Fruits are black, hard, globular and 12 to 20 mm wide. The common name was given by early settlers because the apical bud or heart of the palm is edible.

HABITAT Rainforest and sheltered eucalypt forests.

DISTRIBUTION Coastal zone, from south coast to northern Qld, with an isolated occurrence near Orbost, Vic. Common in Illawarra district, but also widespread north of Sydney.

FLOWERING August–October.

Araceae

The Araceae, although essentially tropical, occurs in most countries. It is a large family of about 2000 species characterised by an inflorescence consisting of a spadix, or spike of closely packed tiny bisexual or unisexual flowers, subtended or partly enclosed by a large bract called a spathe. The once-popular arum 'lily' *(Zantedeschia aethiopica)*, a typical member of the family, is perhaps less well known now than other aroids such as *Philodendron, Anthurium, Dieffenbachia* and *Spathiphyllum*. Many members contain a poisonous watery or milky sap and needle-like crystals of calcium oxalate. Leaves may be long, narrow and parallel veined (as in *Gymnostachys* and most monocotyledons) or with large heart-shaped pinnately or palmately veined blades (as in *Alocasia* and the plants mentioned above).

The Araceae contains a number of plants with edible corms cultivated throughout the tropics. The most important of these are *Colocasia* (taro), *Alocasia* and *Amorphophallus*. These and species of *Typhonium* were also eaten by Aborigines after preparation by baking and pounding.

1183. Gymnostachys anceps

Gymnostachys

Gymnostachys anceps

Settlers Flax **Plate 1183**

A tough upright perennial with rigid, linear, fibrous leaves, 1 to 2 m long and up to 25 mm wide, rising from the base of the stem. The flowering stem is flattened (sometimes angular), with sharp sandpapery margins and is about as long as the leaves. Flowers are stalkless, very small, with 4 scale-like perianth segments. They are borne in narrow spikes up to 15 cm long, which are grouped in clusters along the upper stem. Fruit is a blue-black berry, globular to egg shaped, 5 to 10 mm long, with thin white flesh enclosing a large brown seed.

HABITAT Rainforest and sheltered gullies.

DISTRIBUTION Chiefly coastal, but also recorded in central and northern tablelands. Recordings include Macquarie Pass, Royal National Park and Gosford district. Extending from near Milton (south coast) to Qld.

FLOWERING October–December.

1184. Alocasia macrorrhizos

Alocasia

Alocasia macrorrhizos

Cunjevoi, Spoon Lily **Plate 1184**

A common plant in the rainforests of the north coast, but uncommon in the Sydney district. It is a large clumped perennial with a tuberous rhizome and broad leaf blades on a thick, spongy stalk which is 30 to 100 cm long. Leaf blades are up to 90 cm long, light green and have strong lateral veins. Flowers are borne at the end of upright fleshy stems about as long as the leaves. Individual flowers are in an erect yellow spike ('spadix'), with male flowers above and female flowers below. The spike is surrounded by a large greenish-yellow sheathing bract ('spathe'), which is 15 to 20 cm long. Fruit is an egg-shaped berry, 4 to 8 mm long.

HABITAT Rainforest, especially along more open margins.

DISTRIBUTION Coastal zone (e.g., Macquarie Pass). Extending to north coast and Qld.

FLOWERING December–March.

Lemnaceae

The Lemnaceae is a small family of small to minute aquatic herbs known as Duckweeds which sometimes form a green carpet or scum on the surface of still water. The family is generally considered to be derived by reduction from the Araceae; the two families constitute the order Arales which has affinities with the palm order, Arecales.

Reduction in the Lemnaceae has been carried to the extent that leaves and stems cannot be distinguished. Each plant consists of a small flat or subglobular disc with or without roots. Inconspicuous unisexual flowers develop infrequently, hence propagation is generally vegetative by lateral branches which become detached and form new plants.

Three genera occur in the Sydney area. One of these, *Wolffia*, is less than 1 mm long and is the smallest known flowering plant.

Spirodela Duckweed

Small floating plants with an ovate to elliptic leaf-like body (cladode) and 2 to 5 suspended roots (sometimes 1 or up to 11). Flowers very small and borne in a pocket along the edge of the cladode. There are two species in the Sydney district. They are very similar to *Lemna* (1 species in the area), but more common, as *Lemna* is chiefly a plant of still waters west of the Great Dividing Range where it occurs in such abundance that it becomes a nuisance. *Wolffia* is also a small floating herb, but without roots and with ellipsoid cladodes less than 1.5 mm long. Three species occur in the Sydney district, usually in dams or ponds enriched by drainage from grazing land.

Spirodela pusilla

Small Duckweed **Plate 1185**

A free-floating plant which often forms a mat of green on the surface of backwaters and lagoons. It consists of a green elliptical cladode up to 4 mm long and less than 2 mm wide, from which are suspended thread-like roots. In this species there are usually 2 to 3 roots, but specimens with 1 root or 4 to 5 roots have been recorded, making it difficult to distinguish between it and *Lemna minor* (with 1 root) and *Spirodela oligorrhiza* var. *oligorrhiza* (cladode 5 mm long, 4 mm wide, and 2 to 4 roots). Flowers are minute and not often found. New plants usually develop by vegetative means from the sides of parent plants where they may remain attached for some time.

HABITAT Ponds, lagoons and slow-moving steams.

DISTRIBUTION Widespread from coast to mountains. Common in Georges River south of Liverpool. Also occurring north coast, south coast, southern tablelands, western slopes, south-western plains and Vic.

FLOWERING October–February, but rarely observed.

Commelinaceae

The Commelinaceae contains about 600 species widely distributed in tropical, subtropical and warm temperate regions. It consists of soft-leaved creeping or ascending herbs, usually inhabiting moist places. The flowers are mostly regular and bisexual with 3 sepals and 3 blue or white free petals. Usually there are 6 stamens in 2 series of 3, or 3 stamens and 3 staminodes. The superior ovary consists of 3 fused carpels, each with 1 to several ovules. The fruit is usually a thin-walled capsule.

Several genera such as *Tradescantia* and *Zebrina* are popular indoor plants. Five native and 2 introduced species occur in the Sydney area.

Commelina

Trailing herb with blue flowers borne in axillary inflorescences.

Commelina cyanea

Scurvy Weed **Plate 1186**

This weak herb has a similar appearance to the introduced troublesome weed, Wandering Jew (*Tradescantia)* but has blue axillary flowers, not white and terminal like those of Wandering Jew. Stems are trailing or ascending, hairless and rooting at the nodes. Leaves are lanceolate, 3 to 8 cm long and stem-sheathing. Fine white hairs are on the margins of the sheath. Flowers are blue, with 3 petals, each about 12 mm long, and yellow stamens. Flowers are borne along the stems from a folded green sheath; each sheath contains 2 flower stems, one with a single flower, the other with 2 or 3. Fruit is a small capsule or nut. The common name is derived from the fact that this plant was cooked as a vegetable by early settlers to combat scurvy.

HABITAT Shaded forest; sometimes a weed in gardens.

DISTRIBUTION Coast to mountains. Also recorded north coast, south coast, northern tablelands, central western slopes, north-western plains and Qld.

FLOWERING January–February.

1186. Commelina cyanea

1185. Spirodela pusilla and *Spirodela oligorrhiza*

1187. Aneilema acuminatum

Aneilema

Weak herbs with white flowers in open terminal clusters. Fruit a capsule, splitting on maturity to release seeds.

Aneilema acuminatum

Pointed Aneilema **Plate 1187**

This is a weak herb with trailing stems, with flower stems ascending to about 30 cm tall. Leaves are lanceolate, 3 to 8 cm long, drawn out to a long pointed tip and stem sheathing. Fine hairs are on the top margin of the sheath. Flowers are white, with 3 petals, each about 4 to 6 mm long. The 6 stamens are conspicuous, as some are short and others long and project beyond the petals; there are 2 to 3 perfect stamens and 2 to 4 staminodes. Flowers are borne on thread-like stalks in an open terminal cluster, much exceeding the upper leaves. The fruit capsule is oblong, flattened, about 4 mm long and splits to release the seeds.

Aneilema biflorum (Plate 1188), a smaller plant with 2 short terminal stalks each bearing 2 white flowers, has been recorded at Macquarie Pass and Royal National Park.

HABITAT In or near rainforest.

DISTRIBUTION Coastal zone, but uncommon. Recorded at Mt Keira. Also found north coast, northern tabelands, central western slopes and Qld.

FLOWERING December–January.

1188. Aneilema biflorum

1189. Murdannia graminea

Murdannia

Weak herb, with erect stems and mauve flowers borne in a branched terminal inflorescence.

Murdannia graminea

Murdannia **Plate 1189**

This is a rare plant, with a scattered distribution, seldom seen in flower and often mistaken for a member of the Iris family. Its stems are erect, up to 50 cm tall. Leaves are grass like, 6 to 10 cm long (sometimes longer). Flowers are mauve to blue, with 3 petals, each about 10 mm long, 3 perfect stamens, 3 staminodes and bearded filaments. Flowers are borne in a branched terminal inflorescence. Fruit is a capsule, splitting to release its seeds.

HABITAT Open forest, often in damp soil.

DISTRIBUTION Scattered along coastal zone (e.g., Shane Park). Also recorded north coast, northern tablelands, western slopes, Qld, NT and WA.

FLOWERING December–January.

Pollia

Weak herb with white to blue flowers in short branched terminal clusters. Fruit a shining nut, not splitting as does the fruit of *Aneilema*.

1190. Pollia crispata

Pollia crispata

Pollia **Plate 1190**

A low hairless perennial with creeping stems, rooting at the nodes, with ascending flower stems usually 30 to 50 cm tall. Leaves are lanceolate, 10 to 15 cm long, drawn out to a long pointed tip and stem-sheathing. The margins of the leaf near the sheath are often wavy (crisped, hence the specific name). Flowers are white to blue, with 3 petals, each 4 to 6 mm long, 3 stamens and 3 staminodes. They are borne in short terminal clusters, not elongated and open as is the inflorescence of *Aneilema*. Fruit is an egg-shaped nut about 6 mm long.

HABITAT In and near rainforest.

DISTRIBUTION Coast and central tablelands, but uncommon. Recorded from Illawarra rainforests (e.g., Macquarie Pass). Common along north coast to Qld.

FLOWERING December–January.

1191. Xyris gracilis ssp. gracilis

Xyridaceae

Most of the 270 species in this family of tufted rush-like plants belong to the widespread genus *Xyris* found in tropical and subtropical regions of the Americas, Africa, Asia and Australia. Nineteen species occur in Australia, with 5 in the Sydney area.

The leaves are basal, linear, flat or terete, distichous and sheathing at the base. The yellow flowers are borne at the top of erect flattened or terete stems and arranged in globular or cylindrical heads, with each flower in the axil of a stiff dark bract. Each flower has a calyx consisting of 2 narrow lateral bract-like brown sepals with a large inner membranous sepal which envelops the corolla in bud. The corolla of 3 fragile petals is fused into a tube at the bottom and spreads into 3 large lobes at the top. Three hairy staminodes alternate with the petals and the stamens which are often attached to the petals. The fruit is a capsule, usually enclosed in the withered corolla tube.

The Xyridaceae is related to the Commelinaceae and the Eriocaulaceae.

Xyris gracilis ssp. *gracilis*

Slender Yellow-eye **Plate 1191**

Two subspecies of *X. gracilis* occur in the Sydney district and careful examination is necessary to distinguish one from the other. Intermediate forms can also be found exhibiting different combinations of characteristics. Subspecies *gracilis* has leaves which are tufted, flat, 12 to 18 cm long, 1 to 2 mm wide, erect, rigid, often twisted, with margins pale and often scabrous. Flowers are on terete stalks up to 60 cm long. Flower heads are egg shaped, 6 to 8 mm long, 3 to 5 mm wide, with 6 to 14 bracts; the lower bracts are pressed closely against the head. There are 1 to 4 yellow flowers in each head. The style is branched into 3 for half its length (not lobed). This subspecies is characterised by its narrow leaves, narrow flower heads, closely appressed bracts and fewer flowers. These features separate it from ssp. *laxa*.

HABITAT Sedge swamps and in deep sandy soils near margins of coastal swamps.

DISTRIBUTION Coast and tablelands. Recordings include Cowan, Waterfall to Darkes Forest, Thirlmere to Buxton, Clarence to Lithgow, Blackheath, Mt Irvine and Kanangra Walls. Also north coast, south coast, northern tablelands, southern tablelands and Vic.

FLOWERING November–January.

1192. Xyris gracilis ssp. laxa

Xyris gracilis ssp. *laxa* **Plate 1192**

This subspecies differs from the type form in a number of ways. It has wider leaves, larger flower heads, more numerous bracts (with lower ones lax and spreading) and more flowers. Leaves are flat, 12 to 18 cm long, 2 to 5 mm wide, erect, with margins pale and often scabrous. Flower heads are almost globular, about 10 mm long, 7 to 8 mm wide, with 18 to 29 bracts; the basal bracts are lax, somewhat spreading, often pale and conspicuous. There are 8 to 12 yellow flowers in each head. Intermediate forms betwen ssp. *gracilis* and ssp. *laxa* are found between Sydney Harbour and Broken Bay, near Linden and Kings Tableland and from Putty road near Howes Valley.

HABITAT Sedge swamps and in deep sandy soils near margins of coastal swamps.

DISTRIBUTION South of Botany Bay, through Royal National Park to Bulli district (e.g., Kurnell, Picnic Point, Bundeena, Waterfall, Stanwell Park). Also recorded Mt Colah and Jervis Bay.

FLOWERING November–January.

1193. *Xyris complanata*

1195. *Xyris operculata*

Xyris complanata — Plate 1193

X. complanata is distinguished from other members of the genus by a combination of its flat leaves, flattened flower stalk, long cylindrical flower heads and 3-lobed style. Leaves are tufted, flat, 5 to 30 cm long, 1 to 4 mm wide, with margins pale and often scabrous. Flowers are on flattened stalks, 20 to 60 cm long. Flower heads are cylindrical to egg shaped, 10 to 25 mm long, 7 to 10 mm wide. The yellow flowers have a 3-lobed style, not branched for half its length like that of *X. gracilis.*

HABITAT Swampy margins and other damp sandy places.

DISTRIBUTION Coastal zone and Cumberland Plain (e.g., Kurnell, Agnes Banks, Castlereagh State Forest). Also found north coast, western slopes, Qld, NT, WA and Asia.

FLOWERING November–March.

Xyris juncea

Dwarf Yellow-eye — Plate 1194

A small tufted plant distinguished from other members of the genus by a combination of its smaller leaves and flower stalk, its terete leaves and flower stalk and its rounded flower heads. Leaves are 2 to 12 cm long, about 1 mm wide, terete or slightly flattened. Flowers are on terete stalks, 7 to 30 cm long and about 1 mm wide. Flower heads are almost globular, 4 to 8 mm across.

HABITAT Swampy margins and other damp sandy places.

DISTRIBUTION Coast to mountains (e.g., Bundanoon, Kariong). Extending along the coastal zone from Vic to Qld.

FLOWERING November–December.

Xyris operculata

Tall Yellow-eye — Plate 1195

Tall Yellow-eye is characterised by its terete leaves and flower stalks and its arrangement of lower bracts on the egg-shaped flower head. Leaves are tufted, 20 to 60 cm long, 1 mm wide, terete and shiny brown-red at the base. The lower bracts are arranged in 5 closely overlapping rows (a feature separating this species from the closely related *X. ustulata*).

HABITAT Swampy margins and other damp sandy places.

DISTRIBUTION Coast to mountains. Recordings include Gosford, Ku-ring-gai Chase National Park and Maddens Plains. Found throughout coast and tablelands. Also Qld, Vic, Tas and SA.

FLOWERING August–November.

Xyris ustulata — Plate 1196

This species is closely related to *X. operculata*, differing in having lax and spreading lower bracts which are not arranged in 5 rows. Leaves are tufted, up to 60 cm long, 1 mm wide, terete and shiny brown-red at the base. Flowers are on terete stalks (sometimes slightly angular) which may be more than 1 m tall. Flower heads are egg shaped, 8 to 20 mm long, 5 to 15 mm across, and rounded at the base.

HABITAT Swamp margins and other damp sandy places.

DISTRIBUTION Restricted to Woronora Plateau and Blue Mountains. Recordings include Kurnell, Maddens Plains, Lawson and Blackheath.

FLOWERING November–December.

1194. *Xyris juncea*

1196. *Xyris ustulata*

Eriocaulaceae

The Eriocaulaceae is a family of about 1200 species of grass-like tropical and subtropical marsh plants, being particularly well developed in tropical America. *Eriocaulon* is the only genus of the family represented in Australia. One species occurs in the Sydney area.

The flowers of Eriocaulaceae are inconspicuous and are arranged in crowded heads subtended and surrounded by bracts in a manner similar to the Xyridaceae. Unlike Xyridaceae, however, the flowers are unisexual. Usually both males and females occur on the one plant. The perianth has 6 or fewer segments in 2 series which are not clearly differentiated into petals and sepals.

The family is somewhat isolated, and is usually placed alone in the order Eriocaulales. Its closest relationships appear to be with the Xyridaceae and the Rapateaceae, a small family from tropical South America. The family is of no economic or horticultural significance.

Flagellaria

Flagellaria indica

Whip Vine **Plate 1198**

A robust climber with solid cane-like stems to 10 m long supported by coiled tendrils formed by the long, spirally twisted leaf tip. Leaves are narrow lanceolate, 10 to 30 cm long, with many parallel veins and a sheathing base which completely surrounds the stem. The remains of these sheathing bases cover the upper part of the stems. Flowers are small, hermaphrodite, creamy-white, numerous, stalkless and borne in a dense terminal branched inflorescence. Fruit is small, smooth, round, red and fleshy.

HABITAT Rainforest.

DISTRIBUTION Uncommon in Sydney district where it reaches its southern limit. Recorded in Palm Jungle and along Hacking River (Royal National Park) and Strickland State Forest. Extending to north coast; also Qld, NT, WA, tropical Africa, Asia and Lord Howe Island.

FLOWERING December.

1198. Flagellaria indica

Eriocaulon

1197. Eriocaulon scariosum

Eriocaulon scariosum

Common Pipewort **Plate 1197**

This small tufted glabrous herb has short grass-like leaves 2 to 6 cm long, sheathing at the base and tapering gradually to the apex. The tiny white flowers are in dense globular heads about 5 mm across at the top of erect stalks of variable length, but always much longer than the leaves.

HABITAT In open swampy ground.

DISTRIBUTION Widespread but uncommon. Extending throughout coast and tablelands; also NT, Qld and Vic.

FLOWERING December–January, possibly also at other times.

Flagellariaceae

There is some disagreement amongst botanists as to whether the Flagellariaceae consists of 3 rather disparate genera, or 2, or 1. No matter which view is held, the family is represented in Australia by 1 species, *Flagellaria indica*, a wide-ranging climber found in forests in Africa, Asia, Malesia and north-western and eastern Australia.

Flagellariaceae has affinities with the Restionaceae.

Restionaceae

The Restionaceae is a southern hemisphere family with centres of distribution and diversity in southern Africa, south-western Australia and, to a lesser extent, eastern Australia. The family is also represented in Madagascar, Malesia, New Zealand and South America. It consists of sedge-like or grass-like plants with aerial stems (culms) in tufts or rising at intervals from a scaly creeping rhizome. The leaves are reduced to open sheaths on the aerial stems.

Most genera are dioecious, with male and female flowers on different plants and often very different in appearance. The tiny wind-pollinated flowers are usually arranged in spikelets with overlapping rigid bracts (glumes). The perianth consists of 3 to 6 thin dry segments. The male flowers have 3 stamens, the females a 2 or 3 locular ovary, a stigma with 1 to 3 styles, and sometimes 1 to 3 staminodes. The fruit is a 2 or 3 angled capsule, or in advanced forms, a small nut. Restionaceae plants are usually found in acid nutrient-poor soils that are periodically waterlogged.

Although individual species show some variability, the family as a whole is reasonably uniform in appearance. This not only makes identification of species difficult, but also masks a considerable diversity in anatomy which has been used with other evidence to define genera and species. As a result of recent studies 3 genera formerly considered to be common to both Africa and Australia are now regarded as distinct in each country, and a revision of the family, now proceeding (1989), will result in generic and specific name changes for some Australian taxa.

The Restionaceae is related to the Centrolepidaceae and the Flagellariaceae which are included with it in the order Restionales. It consists of some 30 genera and possibly more than 400 species. Five genera and 15 species occur in the Sydney area.

Figure 14

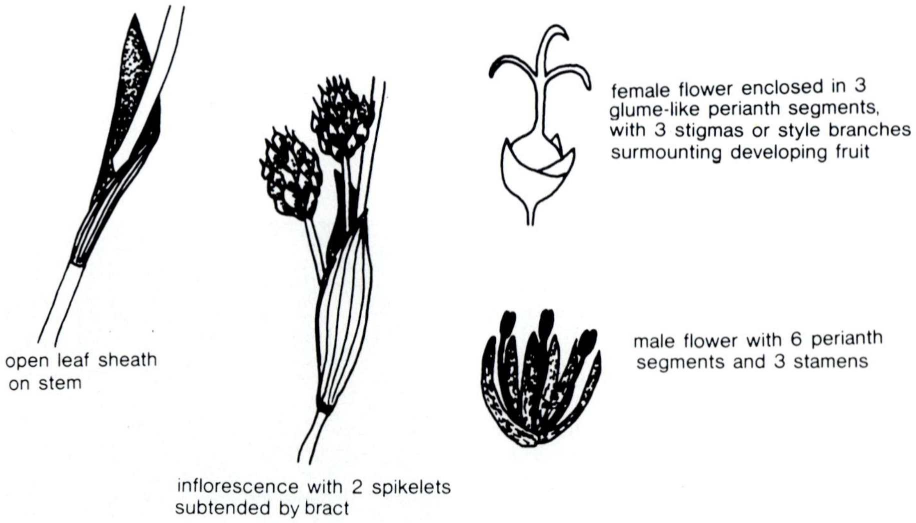

Comparison of 4 families of grass-like plants

Character	*Juncaceae (rushes)*	*Restionaceae (rushes)*	*Cyperaceae (rushes and sedges)*	*Poaceae (grasses)*
leaves	at base of stems, terete to flat or reduced to a sheath	reduced to open sheaths on stems, overlapping on lower stem, distant on upper	3-ranked, basal or on lower stem, terete or with grasslike blade or reduced to sheathing scales	2-ranked, alternate, at nodes; consisting of sheath, ligule and blade
leaf sheaths	open or closed	open	closed, tubular	open, with overlapping margins
stems	terete, solid	solid or hollow, unbranched to many branched	triangular or terete, solid, without nodes	terete, hollow at internodes, solid at nodes
basic unit of inflorescence	cyme	usually a spikelet	spikelet	spikelet
spikelets	absent	1 to many flowered, solitary or numerous, subtended by bracts; male and female spikelets often different	glumes in 2 ranks or imbricate and spirally arranged, most glumes bearing flowers, one to several lower glumes empty	of 1 or more florets subtended by 2 lower empty glumes; florets consisting of 2 bracts (palea, lemma) and 1 reduced flower; glumes and bracts often awned
flowers	mostly bisexual	mostly unisexual (dioecious)	bisexual or unisexual (mostly monoecious)	mostly bisexual
perianth segments	6, in 2 whorls of 3, well developed	6 or less, or absent, dry, glumaceous	bristles, hairs, scales, or absent	usually 2, reduced (lodicules)
stamens	usually 6, sometimes 3	3 (male), 0 to 3 staminodes	3, sometimes less	3
styles/stigmas	styles 1 or 3, stigmas 3	styles 1 to 3, stigmas 1 or 3 (female)	style 1, stigmas 2 or 3 (female)	stigmas 2, occasionally 1 or 3
bracts per flower	1, or absent	1 (glume)	1 (glume)	2 (palea, lemma)
fruit	capsule with 1 to many seeds	3-sided capsule or nutlet	nutlet, flat or 3-angled	caryopsis (grain)
habitat	in water and water-logged ground	infertile soils periodically wet	mostly damp, wet or marshy areas	mostly dry land in a variety of soils

Lepyrodia

A genus of about 20 species confined to Australia. Flowers in panicles which are sometimes small and spike like, loosely arranged, not in spikelets, each flower with 1 to 2 tiny bracteoles beneath it. Male and female inflorescences similar. Male flowers with 3 stamens, females with a 3-branched style. Fruit a capsule with 2 to 3 compartments.

1199. Lepyrodia scariosa

1200. Lepyrodia gracilis

Lepyrodia scariosa

Scale-rush **Plate 1199**

The commonest of the 4 species of *Lepyrodia* in the Sydney area, *L. scariosa* is most noticeable during flowering because of the large pale stigmas of the flowers which are loosely arranged on short axes in the axils of large straw-coloured long-pointed bracts. The plant is usually 30 to 100 cm high, with smooth glabrous unbranched erect stems. It is immediately recognisable as a member of the Restionaceae by the loose open leaf sheaths at intervals on the stems. Male and female flowers occur on separate plants.

HABITAT Widespread and common on damp sandy or peaty ground in heath and open forest.

DISTRIBUTION Coast and tablelands of NSW and Qld.

FLOWERING January–February.

Lepyrodia gracilis

Slender Scale-rush **Plate 1200**

This species has fine erect stems with few to many branches and tight, closely appressed leaf sheaths. The flowers are arranged in narrow panicles about 3 cm long, each 1 with 6 scaly brown perianth segments, the outer 3 shorter and narrower than the inner 3 which are reddish when fresh. The plant is not often seen since it is both uncommon and inconspicuous.

HABITAT Open wet sandy ground.

DISTRIBUTION Central coast and central and southern tablelands of NSW.

FLOWERING Chiefly January–February.

Restio

About 30 endemic Australian species currently included in this South African genus; 11 species in NSW, the remainder mostly Western Australian. Male and female flowers in spikelets of 1 to many flowers, the spikelets either similar or dissimilar. Bracteoles absent. Male flowers with 3 stamens, females with a 2-branched style (NSW sp.). Fruit a capsule with 2 compartments, or 1 by abortion.

Restio tetraphyllus ssp. *meiostachyus*

Tassel-rush, Tassel Cord-rush **Plate 1201**

With stems often exceeding 1 m in height, this is the tallest of the local Restios. It is an attractive plant, easily recognised by the large tassel-like clusters of slender, bright-green divided branchlets in the axils of the sheaths on the main stems. Male and female flowers occur on separate plants in spikelets arranged in long loose narrow panicles above the clustered branchlets. In this subspecies the spikelets are less than 5 mm long, with the females narrower and longer than the males.

HABITAT Damp sandy ground on the coast in heath and open forest.

DISTRIBUTION Coastal zone. Extending from south coast to north coast and Qld.

FLOWERING September–November.

1202. Restio fastigiatus

1201. Restio tetraphyllus ssp. *meiostachyus*

Restio fastigiatus **Plate 1202**

This species and the next have branched stems but are without the cluster of barren branchlets that characterises *R. tetraphyllus*. In *R. fastigiatus* the stems have tight leaf sheaths over most of their length and are divided near the top into numerous fine erect straight branchlets bearing (on separate plants) either male or female spikelets about 4 mm long. The male spikelets have several flowers; the females are stalked with several empty glumes and 1 terminal flower.

HABITAT Damp sandy ground in heath and scrub.

DISTRIBUTION Coast to mountains. Extending to south coast and southern tablelands; also Qld.

FLOWERING September–December.

1203. Restio dimorphus

Restio dimorphus **Plate 1203**

In this species the flowering stems are divided from near the base into long weak flexuose branchlets, all bearing broad loose leaf sheaths. The male spikelets are numerous, about 2 mm long and almost sessile within the axil of a short floral bract; the females have several barren glumes and 1 fertile glume. The rhizome is longer, creeping and less densely clothed in woolly brown hairs than in *R. fastigiatus*.

HABITAT Damp sandy ground in heath and scrub.

DISTRIBUTION Central coast of NSW.

FLOWERING September–December.

1204. Restio gracilis

Restio gracilis

Slender Cord-rush **Plate 1204**

This is one of 4 species in the Sydney area having unbranched terete stems and a terminal inflorescence of spikelets in a raceme or narrow panicle. It is distinguished by the male and female spiklets being of different appearance. The purplish, narrow-cylindrical female spikelets 6 to 16 mm long differ from the brown ellipsoid male spikelets 5 to 10 mm long. The female spikelets are sessile or nearly so; the males are on slender stalks. The subtending bracts taper gradually to the apex and usually exceed the lowest spikelet.

R. pallens, a north coast species, also occurs on the central coast. It occurs at Richmond, Agnes Banks and Castlereagh, and differs in having crowded nearly-sessile ellipsoid male and female spikelets of similar appearance subtended by short, abruptly pointed bracts.

HABITAT Wet or damp sandy soils.

DISTRIBUTION Widespread on the central and south coast from Gosford to Jervis Bay and on the southern highlands. Absent from the Blue Mountains.

FLOWERING September–December.

1205. Restio fimbriatus

Restio fimbriatus

Fringed Cord-rush **Plate 1205**

Two members of the *Restio gracilis* complex, *R. fimbriatus* and *R. australis*, occur on the Blue Mountains and are joined on the southern highlands by *R. gracilis*. *R. fibriatus* has close-fitting leaf sheaths, the upper ones bearing a tuft of hairs 1 to 4 mm long at the apex. The glumes of the spikelets are also fringed with long fine hairs. Male and female spikelets are similar in appearance, purplish when fresh, 4 to 7 mm long, pointed or rounded at the top and carried on slender stalks. *R. australis* is easily distinguished from this species by its stouter stems, 1.5 to 3 mm in diameter, and its loose, rough wrinkled sheaths on the upper part of the stem. The sheaths and glumes also lack fine hairs.

HABITAT Damp sandy soils, often near swamps.

DISTRIBUTION Tablelands and south coast of NSW.

FLOWERING October–December.

1206. Restio complanatus

Restio complanatus

Flat Cord-rush **Plate 1206**

The stems of *R. complanatus* are pale green, flat, unbranched and 30 to 70 cm high with close-fitting blunt leaf sheaths spaced along them below an erect terminal brown inflorescence. The flowers are arranged in dense spikelets, the females usually longer and narrower than the males.

HABITAT Wet sandy or peaty soils in heath and scrub.

DISTRIBUTION Mainly coastal, but also recorded from central tablelands. Extending from Qld to Vic and Tas.

FLOWERING November–December.

Hypolaena

A genus of 2 species endemic in Australia. Stems wiry, branched, striate, often flexuose. Male and female spikelets terminal, dissimilar. Male flowers with 3 stamens, females with a 3-branched style. Fruit a nut.

1207. Hypolaena fastigiata

Hypolaena fastigiata

Tassel Rope-rush **Plate 1207**

This species is closely related to *Leptocarpus tenax*, but is much less common. It differs from the latter in having branched, flexuose, striate wiry stems often covered in a dense greyish tomentum. Leaves are reduced to dark-brown, closely appressed sheathing bracts with a small erect point. The inflorescence is terminal as in *Leptocarpus*, not axillary as in *Empodisma* which the plant otherwise resembles. The male flowers are in narrow elliptical brown spikelets on slender stalks forming nodding or erect panicles; the females in narrow cylindrical erect sessile spikelets which are solitary or 2 to 3 together.

HABITAT Uncommon in damp sandy heath.

DISTRIBUTION All states (but not NT). Chiefly coastal in NSW, but also recorded from central tablelands.

FLOWERING September–December.

1208. Leptocarpus tenax (male)

Leptocarpus

A genus of about 15 species, 12 endemic in Australia, 1 in NSW. Creeping or tufted perennials with mostly unbranched erect stems, smooth, not obviously striate, with closely appressed leaf sheaths. Male and female spikelets terminal, dissimilar (NSW sp.). Male flowers with 3 (rarely 2) stamens, females with a 3-branched (rarely 2) style. Fruit a nut.

Leptocarpus tenax

Slender Twine-rush **Plate 1208**

At times very abundant on sandy soils in open country with impeded drainage, *Leptocarpus* has smooth, greyish-green, mostly unbranched erect stems 60 to 100 cm tall, with the leaves reduced to closely appressed dark brown leaf sheaths. It is readily recognised by the purplish-brown narrow male spikelets about 4 mm long which are numerous on long slender stalks in a drooping panicle at the top of the stems. The female spikelets by contrast are reddish, about 10 mm long, few, nearly sessile and erect in narrow panicles.

HABITAT On damp sandy soils, often in heath.

DISTRIBUTION All states (but not NT). Mainly coastal in NSW, with a few records from central and northern tablelands.

FLOWERING October–December.

Empodisma

A genus of 2 species. Stems thin, wiry, green, flexuose bearing distant sheaths. Upper leaf sheaths green or pale with a reflexed pointed tip. Spikelets axillary, the females solitary and 1-flowered with 1 to 3 barren glumes, the males solitary or several together with 1 or several flowers. Fruit a nut.

Empodisma minus

Spreading Rope-rush **Plate 1209**

The fine smooth green stems of this plant may be short and erect when in open sandy country, but can also be long, flexuose and highly branched, forming a tangled mass with other weak herbs in wet shady places. The tiny green leaf sheaths have a narrow pointed tip, 1 to 3 mm long, bent at right angles to the stems, and a light tuft of woolly hairs in the throat of the overlapping sheath margins. Male and female spikelets are axillary, solitary and sessile, or sometimes the males are paired with 1 sessile and the other shortly stalked. The fruit is a small nut about 1.5 mm in diameter, sometimes retaining the 3 or 2-branched coiled style.

HABITAT Common on wet or boggy soil on creek banks, cliff faces and heath or in periodically wet sandy soil in open forest.

DISTRIBUTION Coast and tablelands; also Qld, Vic, SA, Tas and New Zealand.

FLOWERING Most of the year, particularly September–November.

1209. Empodisma minus

Centrolepidaceae

The distribution of the Centrolepidaceae, and that of Restionaceae to which it is related, suggests that the families developed on the former southern supercontinent, Gondwanaland, and were dispersed by the process popularly known as continental drift. The present centre of distribution of the Centrolepidaceae is Australia which has about 30 species in 3 genera. One species occurs in southern South America, another in New Zealand, and 4 are found in New Guinea and South-east Asia as a result of recent northward extension.

The family consists of small grasslike herbs found throughout Australia, often on wet ground, and small moss-like cushion plants in alpine areas. The inflorescence in *Centrolepis* and *Gaimardia* is reduced to a structure called a pseudanthium which consists of 1 or 2 male flowers with 1 stamen each and 2 or more female flowers with 1 carpel each fused together at their bases and subtended by 2 sheathing bracts. The family has no particular uses although some species make interesting and hardy plants for bog gardens.

1210. Centrolepis fascicularis

Centrolepis

Two species of *Centrolepis* are widespread and fairly common in the Sydney area, though rarely noticed.

Centrolepis fascicularis

Tufted Centrolepis **Plate 1210**

These tiny tufted annuals form rounded clumps up to 8 cm high and across in damp or wet sandy ground. The crowded narrow linear leaves at the base are exceeded by the flowering stems which terminate in a flower cluster subtended by 2 hairy bracts. In *C. fascicularis* the bracts each have a conspicuous erect awn 3 to 4 mm long; those of *C. strigosa* have only a short blunt tip.

HABITAT Widespread and common in damp to wet seepage areas.

DISTRIBUTION Coast to mountains. Extending along north coast, south coast, northern tablelands and to all other States (except NT); also Malesia.

FLOWERING July–December.

Juncaceae

The Juncaceae, or rush family, is found mostly in the cold temperate areas of the world. The family consists of 2 widespread genera, *Juncus* and *Luzula*, with about 300 and 80 species respectively, and 7 small genera with restricted distributions in South America and Africa.

Juncus and *Luzula* are represented in Australia by 40 or 50 native species and about 20 foreign species. They are similar in many respects but differ in the leaves (glabrous in *Juncus*, hairy in *Luzula*), the leaf sheaths open (i.e., split) in *Juncus*, closed in *Luzula* and the ovary 3- or 1-locular with many ovules in *Juncus*, 1-locular with 3 ovules in *Luzula*.

The Juncaceae are wind-pollinated, hence the flowers are insignificant, though not reduced to the same extent as the Restionaceae (rushes), the Cyperaceae (rushes and sedges) and the Poaceae (grasses). The flowers have 6 scaly perianth segments in 2 whorls of 3 and 6 or 3 stamens, usually opposite the perianth segments. The leaves of *Juncus* are variable. They may be flat like a blade of grass, stiff and terete and pointed like a knitting needle, soft and sometimes compressed with internal septa, or reduced to sheaths at the base of the aerial stems.

Five native species of *Luzula* and more than 30 species of *Juncus* occur in the Sydney area. Many of the species are similar in appearance and weed-like, and in most cases require expert knowledge for identification. Consequently we have not attempted to show more than a few representative species.

Juncus Rush

Plants growing in or near water. Leaves glabrous, cylindrical, channeled or flat, often reduced to basal sheaths, the sheaths slit to the base. Flowers small, bisexual, usually in clusters. Perianth of 6 glume-like segments, stamens 6 or 3, style 3-branched. Fruit a capsule with numerous seeds.

1211. Juncus planifolius

Juncus planifolius

Broad Rush **Plate 1211**

This is one of the rushes with thin, flat, grass-like leaves about 5 mm wide rising from the bases of flowering stems which are 30 to 40 cm high with numerous small globular clusters of tiny brown flowers in an umbel-like arrangement. The clusters contain 8 or more flowers, each with 6 scaly segments and 3 stamens surrounding an ovary or, at the fruiting stage, a developing capsule surmounted by a 3-branched style.

HABITAT Widespread and common on wet sandy ground.

DISTRIBUTION Found throughout coast, tablelands and western slopes. Also in other States (but not NT) and the Pacific region.

FLOWERING November–March.

1212. Juncus prismatocarpus

Juncus prismatocarpus

Branching Rush **Plate 1212**

This rush belongs to a group with soft terete or slightly flattened septate leaves (i.e., hollow with internal cross partitions). In *J. prismatocarpus* the leaves are flattened and the cross-septa are often incompletely formed. The flowering stems are firmer than the leaves and 30 cm or more high with flowers in clusters arranged in a loose panicle, the panicle subtended by a short leafy bract which is also septate. The flowers have 6 acute perianth segments and 3 stamens.

HABITAT In or near water.

DISTRIBUTION Coast and tablelands. Also in all other States (but not NT), NZ, New Guinea and Mauritius.

FLOWERING November–January.

1213. Juncus krausii

Juncus krausii

Sea Rush **Plate 1213**

J. krausii is a common rush covering large areas of tidal marshland in estuaries. It has dull-green, stiff terete leaves about 3 mm in diameter and a metre high. The flowers are in small pale brown clusters arranged in a loose panicle which appears lateral due to the erect rigid floral bract which subtends the inflorescence. The leaves and bracts are pungent pointed, though less so than in the naturalised *J. acutus*, which is found mostly in freshwater habitats.

HABITAT Saline and brackish marshes near the coast.

DISTRIBUTION Coastal zone. Found in all other States, NZ and South Africa.

FLOWERING November–February.

Juncus continuus **Plate 1214**

This is one of a large number of species in which leaves are reduced to sheaths at the base of the flowering stems. Each sheath has a small upright acute or mucronate tip which constitutes the leaf blade. *J. continuus* is a common rush on damp sandy soils, forming clumps 50 to 150 cm high. It can usually be recognised by the stems which are light green, 2 to 3.5 mm in diameter, with continuous even pith. The sheaths are loose, dull, straw-coloured at the top, shiny brown near the base. Flowers are usually clustered towards the ends of the inflorescence branchlets.

HABITAT Damp sandy soils.

DISTRIBUTION Widespread throughout coast, tablelands and western slopes. Also Qld, Vic, NT, SA, NZ and Norfolk Island.

FLOWERING January–February.

1214. Juncus continuus

Juncus usitatus

Common Rush, Tussock Rush **Plate 1215**

In *J. usitatus* the leaves are reduced to sheaths at the base of the stems; the sheaths are up to 15 cm long, straw coloured near the top and shiny reddish brown near the base. Stems are nearly a metre high and 1 to 2 mm dia with internal pith regularly interrupted by air spaces. The pale flowers are fairly evenly spaced along the ultimate (finest) branches of open inflorescences 2.5 to 6 cm long. The capsule is ellipsoid and much longer than the perianth segments of the flowers. This is not the only rush in the area having spaced, rather than clustered, flowers, but it is the commonest.

HABITAT In damp soil, often on the banks of creeks and ponds.

DISTRIBUTION Widespread in NSW. All mainland States (but not NT), NZ and New Caledonia.

FLOWERING September–January.

1215. Juncus usitatus

Luzula

Luzulas are small perennial herbs found mostly in damp cool habitats. Five species, formerly regarded as variants of the one widespread species, are presently recognised in the Sydney area. Since differences between them are generally small and indistinct, identification at the generic level is probably sufficient in most cases. Luzulas can be recognised by their flat linear leaves with closed leaf sheaths and slightly hairy margins. Flowers are borne in 1 or more small clusters arranged irregularly and subtended by a large erect leafy bract and a few smaller ones at the top of stems about 20 cm high. The fruit is a capsule containing up to 3 seeds.

1216. Luzula meridionalis

Luzula meridionalis — Plate 1216

This species is characterised by the plant having a slight onion-like bulb at the base, and the flowers being in 4 or more stalked clusters on each stem subtended by obtusely pointed bracts.

HABITAT Damp, often swampy conditions, particularly on the highlands.

DISTRIBUTION Central coast, tablelands and slopes; also Vic, Tas, SA and WA.

FLOWERING October–December.

Cyperaceae

The Cyperaceae (sedge family) and the Poaceae (grass family) both exhibit extreme adaptations to habitat and to wind pollination giving a superficial resemblance which suggests a closer relationship than actually exists. Whereas grasses are found in habitats ranging from fresh or brackish water to very dry, and on soils ranging from pure sand to heavy clay, sedges are generally found in permanently or seasonally moist habitats on sandy soils.

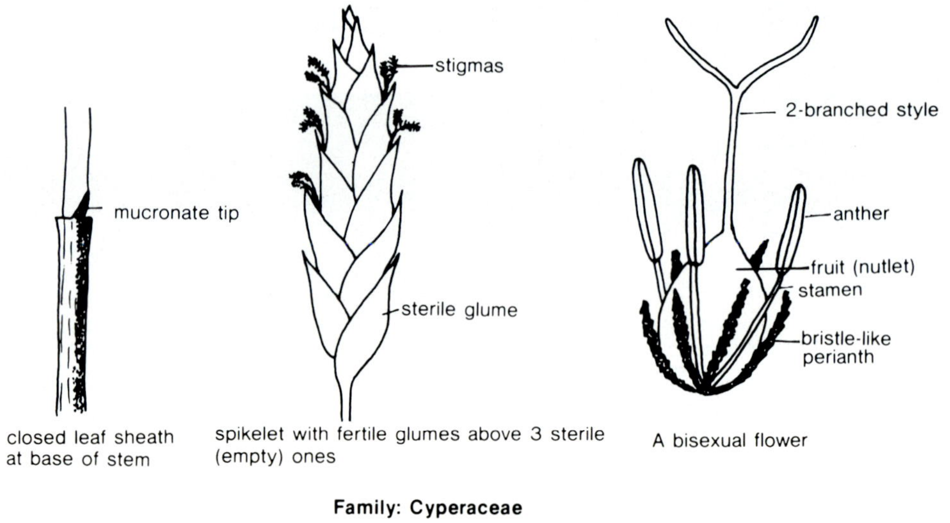

Figure 15

Apart from this, the 2 families are readily distinguished on a number of anatomical features. Sedges differ from grasses in having solid terete or triangular aerial stems (culms) containing pith or occasionally transverse septa and leaves which are 3-ranked and either basal or less often on the lower part of the culms. Leaves may be terete, or reduced to a small tip at the top of the leaf sheath, or may have a grass-like blade. The leaf sheaths which surround the culms are usually closed (i.e., tubular and not split down one side as in Poaceae and Restionaceae) and there is usually no ligule (collar) at the junction of the sheath and blade (except in *Gahnia*).

The inflorescence varies considerably, but individual flowers are always arranged in spikelets composed of light to dark brown or black bracts which may be imbricate (all around the axis of the spikelet) or distichous (in 2 rows along the axis). Although 1 to several of the lower glumes are often empty, each of the upper glumes contains a tiny reduced flower which is usually bisexual or occasionally unisexual. The perianth may be absent or represented by bristles or scales surrounding the stamens, ovary and style. There are 1 to 3 stamens with large basifixed anthers, and a style which is usually divided into 2 to 3 branches or stigmas. The fruit is a small angular or flattened nut with one seed. The main points of difference between the families Cyperaceae, Juncaceae, Restionaceae and Poaceae are set out on page 329.

The Cyperaceae is a large cosmopolitan family of about 90 genera and 4000 species, with familiar genera such as *Carex, Isolepis (Scirpus), Eleocharis* and *Schoenus* distributed abundantly in temperate areas of both hemispheres, and others such as *Cyperus, Fimbristylis, Bulbostylis* and *Gahnia* in tropical countries as well as in Australia.

The family is well represented in Australia by about 650 species and nearly 50 genera, of which 150 species in 26 genera occur in the Sydney area. These plants are a common component of heath, particularly in swampy or seepage areas, also the understorey of forests of different kinds, with a few even present in rainforest. The species illustrated in the following pages provide a fairly good representation of those found in the area, and of the family generally.

Cyperus

A cosmopolitan genus of more than 500 species; about 30 in the Sydney area. Stems 3-angular or terete. Leaves at base of stems, long, linear or rarely reduced to sheaths. Inflorescence terminal with few to many flat spikelets in a cluster, or in spikes or clusters on the rays of a simple or compound umbel; subtended by 1 to about 9 leafy bracts. Spikelets 2 to many flowered with glumes in 2 opposite rows along the axis. Nut 3-angular or biconvex.

Cyperus exaltatus

Tall Flat-sedge

This native perennial grows to about 1 m high from a short rhizome. It has triangular stems 2 to 3 mm wide near the top with a large inflorescence subtended by 3 to 6 long, finely serrated bracts. The leaves are about as long as the stems and have a prominent lateral vein on each side of the midvein near the scabrous margins. Each branch of the compound inflorescence ends in a cluster of brown spikes 2 to 6 cm long; each spike with many spikelets 2 to 10 mm long and 1 to 2 mm broad along its axis. The flowers develop into a small 3-angled nut topped by a deciduous 3-branched style.

HABITAT In damp depressions and near running or still water.

DISTRIBUTION Mainly the western parts of the central coast. Absent from the central tablelands, but recorded from western slopes and western plains. All mainland States, Africa, Asia.

FLOWERING December–February.

Cyperus difformis

Variable Flat-sedge **Plate 1217**

This is a widespread and common annual weed of damp waste places. It has numerous dark spikelets in small dense globular clusters at the ends of the long rays of a simple umbel which is subtended by 1 to 3 spreading bracts, the lowest longer than the inflorescence and nearly erect so that the inflorescence is pushed to one side. The soft-celled stems are sharply triangular with concave surfaces. They grow to about 50 cm tall and are often reddish at the base. The leaves at the base of the stems are 2 to 4 mm broad, grasslike and usually shorter than the stems. The fruit, which is produced abundantly, is a tiny dark 3-sided nut with a deciduous 3-branched style.

HABITAT In shallow water or on boggy ground.

DISTRIBUTION Widespread on the central coast but absent from the central tablelands. Also recorded northern tablelands, southern tablelands, western slopes and western plains; also all mainland States, pantropical.

FLOWERING December–March.

1217. Cyperus difformis

1218. Cyperus polystachyos

1219. Cyperus sanguinolentus

1220. Cyperus enervis

Cyperus polystachyos **Plate 1218**

This is a fairly common tufted perennial about 30 to 60 cm high. The broad flat spikelets are green when immature, later turning yellowish to pale brown, and are crowded into a dense sessile head subtended by 2 to 4 horizontal bracts exceeding the inflorescence. Flowering stems are firm, shiny green, and 3-angular with convex surfaces; the leaves flat, 2 to 3 mm wide and much shorter than the stems. The fruit is a flat lens-shaped (biconvex) nut surmounted by a deciduous 2-branched style.

HABITAT Damp open ground.

DISTRIBUTION In coastal areas of NSW but not on the tablelands. Widespread in Australia and the tropics.

FLOWERING Mainly March–April.

Cyperus sanguinolentus **Plate 1219**

This species is readily recognised by its long flat spikelets of regularly arranged broad green glumes marked with reddish purple or black near the margins and tip. Careful removal of the glumes shows the style at the top of each developing fruit to be 2-branched. The plant forms a small tuft with 1 or a few short stems bearing simple umbels of 3 to 5 short rays. Inflorescences are subtended by about 3 long ribbony bracts. The fruit is a brown 2-sided nut.

HABITAT Damp grassy areas and margins of swamps and lakes.

DISTRIBUTION Throughout NSW, except far western plains. Widespread in eastern Australia and tropical countries.

FLOWERING December–April.

Cyperus enervis **Plate 1220**

This is one of 3 or 4 species in the area sometimes known as Slender Sedge. As suggested by this name, they are all small tufted plants with relatively few short erect or long trailing slender stems with narrow leaves. Inflorescences are small, sometimes with only 1 or 2 short narrow clustered spikelets. This species has greenish spikelets which eventually dry out to a pale colour. The glumes are keeled and have 1, or sometimes 2, slightly raised darker green nerves on each side of the keel. Flowers have a 3-branched style and develop into a tiny smooth 3-sided brown nut with a short point at the blunt apex.

HABITAT Moist sandy soils in sheltered places on the coast.

DISTRIBUTION Coastal districts from about Kiama northward to Qld (e.g., Minnamurra Spit).

FLOWERING November–May.

Eleocharis

A genus of about 200 species; 10 in the Sydney area. Leafless annuals or perennials with terete or flattened stems usually produced along linear rhizomes or occasionally tufted. Inflorescence a single terminal spikelet without bracts, the glumes spirally arranged, overlapping, often deciduous. Flowers bisexual, with 0 to 10 perianth bristles (mostly 6), 1 to 3 stamens and a 2 or 3 branched style. Fruit a nut topped by the persistent style base.

1221. *Eleocharis sphacelata*

Eleocharis sphacelata

Tall Spike-rush **Plate 1221**

Often forming extensive stands in freshwater lakes, ponds, lagoons, and slow-moving streams, this robust leafless plant is one of the few truly aquatic members of Cyperaceae. It is readily recognised by its stout erect stems 50 to 200 cm high and 5 to 15 mm in diameter rising from a thick horizontal rhizome and terminated by a tapering flowering spike 3 to 6 cm long or by a short, pointed, scaly tip on an otherwise obtuse apex. The stems are hollow with white cross-septa spaced along their length. Flowering spikes have overlapping, spirally arranged glumes from which the 3-branched white stigmas, and later, the 3 yellowish anthers, emerge. At maturity the hard round fruit and the glumes fall away from the spike leaving the central axis at the top of the stem.

HABITAT Rooted in shallow freshwater.

DISTRIBUTION Widespread in NSW. Also Qld, Tas, SA, WA, New Guinea and NZ.

FLOWERING November–April.

1222. *Eleocharis cylindrostachys*

Eleocharis cylindrostachys **Plate 1222**

Most spike-rushes are smaller than *E. sphacelata*, and this one is typical of several in the area. It resembles *E. acuta* but differs in having only a very short rhizome; the stems therefore are more densely tufted. Stems are usually 30 to 50 cm high and 1 to 2 mm in diameter with a pale brown more or less cylindrical flowering spike 10 to 20 mm long and 2.5 to 3 mm wide. Both the spike and the individual glumes are obtuse at the tip, features which distinguish this species from *E. pallens*. The small shiny brown biconvex nut has prominently ribbed margins.

HABITAT In or near water on boggy ground.

DISTRIBUTION Central coast. Scattered in NSW and Qld; also New Caledonia.

FLOWERING November–December.

1223. *Eleocharis acuta*

Eleocharis acuta

Common Spike-rush **Plate 1223**

E. acuta is more common and widespread than *E. cylindrostachys* and differs in a number of respects. It has a long-creeping rhizome with erect stems irregularly spaced along it. Stems are sometimes taller than those *E. cylindrostachys* and are triangular immediately below the flower spike. Flower spikes are dark brown or variegated, 15 to 30 mm long and 3 to 7 mm broad and, as the name suggests, usually taper to an acute tip. Unlike those of *E. cylindrostachys* and *E. pallens*, the glumes do not readily fall away after flowering. The biconvex nut is nearly 2 mm long (including the persistent dilated style base) and has narrow margins.

HABITAT In or near water on boggy ground, sometimes in dense stands.

DISTRIBUTION Widespread throughout NSW and in all other States (but not NT); also NZ.

FLOWERING October–April.

1224. *Fimbristylis dichotoma*

Fimbristylis

A genus of 300 species; about 80 in Australia, 4 in the Sydney area. Tufted annuals or perennials with narrow leaves at base of stems, sometimes short or reduced to sheaths. Spikelets many-flowered, in clusters or irregular umbels subtended by several unequal bracts resembling the leaves. The name refers to the fringe of hairs often present on part or all of the style.

Fimbristylis dichotoma

Common Fringe-rush **Plate 1224**

F. dichotoma is a slender perennial 30 to 40 cm high with long narrow linear leaves having a membranous ligule at the top of the leaf sheath. The spikelets are on long rays in open umbels subtended by 2 to 5 narrow leaf-like bracts often longer than the inflorescence. They are about 5 mm long, ovoid and brown with prominently keeled glumes all round the axis. The fruit is a pale shiny biconvex nut with numerous longitudinal and transverse ribs on either side.

HABITAT On periodically damp sandy or clayey soils.

DISTRIBUTION Uncommon in the Sydney area on the coast, absent from the tablelands. Various forms scattered throughout NSW, northern Australia and warmer parts of the world.

FLOWERING December–April.

Fimbristylis ferruginea **Plate 1225**

In this species the stems are 30 to 60 cm high, rigid, striate and usually compressed (slightly flattened), and the leaves are few, narrow and quite short. The light to dark brown spikelets are 8 to 14 mm long and 3 to 5 mm broad with numerous glumes. Spikelets are arranged in umbels with short rays subtended by 1 or 2 short linear bracts. The nut is brown with a smooth or finely striate surface.

HABITAT Saline and brackish swamps near the sea (e.g., at Towra Point).

DISTRIBUTION Central and north coast of NSW, northern Australia and tropics.

FLOWERING December–April.

1225. *Fimbristylis ferruginea*

1226. *Scirpus polystachyus*

1227. *Isolepis fluitans*

1228. *Isolepis nodosa*

Scirpus

The genus *Scirpus* has long been an unsatisfactory residue of heterogeneous elements. Recently, embryological, physiological and other evidence has been used to place the 40 or so Australian species into 5 segregate genera. Following this revision, only one Australian species is retained in *Scirpus*.

Scirpus polystachyus

Large-headed Club-rush **Plate 1226**

This species grows in clumps and resembles local *Bolboschoenus* species with stout, pithy, noded stems a metre or more high and 6 to 7 mm in diameter bearing leaves at the nodes up to 60 cm long and 15 mm wide. The leaves are fairly thick, striate on the back, scabrous on the keel and margins, and taper gradually to a fine point. Inflorescence bracts are similar but smaller. Spikelets are 4 to 7 mm long with numerous brown glumes 4 to 5 mm long. The flower and rather large nut are surrounded by bristles. Identification can be reliably confirmed by the scabrous hairs covering the inflorescence branches.

HABITAT Creek margins and boggy or wet sandy ground.

DISTRIBUTION Sparsely distributed on the tablelands from southern Qld to eastern Vic.

FLOWERING October–November.

Isolepis

A genus of about 70 cosmopolitan species; 28 in Australia, 13 in the Sydney area, formerly included in *Scirpus*. Mostly small tufted annuals with 3-angular or terete stems. Leaves few, narrow linear or reduced to a sheath with a short blade or tip. Inflorescence terminal but often appearing lateral by the growth of 1 erect bract apparently continuing the stem. Spikelets in clusters or umbels. Perianth bristles absent. Fruit a smooth flattened or 3-angular nut.

Isolepis fluitans

Floating Club-rush **Plate 1227**

This species forms dense masses of long branching leafy stems floating just beneath the surface of freshwater ponds and streams. The fine leaf blades, about 1 mm wide and up to 15 cm long, are attached without auricles or a ligule to the summit of open, pale-green sheaths at nodes along the stem. Flowering spikelets are emergent on long erect peduncles. Each spikelet is about 5 mm long with 4 to 10 glumes, the lowest barren and occasionally elongated, the others with a bisexual flower in the axil. The tiny flowers have 1 or 2 stamens and a 2-branched style. *I. producta* is very similar and is possibly only a variant of *I. fluitans*.

HABITAT In freshwater or on the muddy bank of streams and ponds.

HABITAT Central coast, south coast, southern tablelands, northern tablelands, Vic, Tas, SA, WA, NZ, Europe, Africa, Asia.

FLOWERING October–April.

Isolepis nodosa

Knobby Club-rush **Plate 1228**

This is an easily recognised species with densely tufted stiff hard stems, 40 to 90 cm high, terminating in a globular cluster of numerous sessile brown spikelets about 5 mm long. The cluster is subtended by an erect bract, 2 to 4 cm long, appearing to be a continuation of the stem, thereby making the inflorescence appear lateral. Flowers have 3 stamens and a 3-branched style, and develop into a smooth, shiny plano-convex nut. The perianth is reduced to a disc at the base of the nut. Leaves are reduced to sheaths at the base of the stems.

HABITAT Damp sandy places, especially near saltwater.

DISTRIBUTION Common on the coastal margin of NSW. All States (but not NT), and temperate southern hemisphere.

FLOWERING October–February.

Isolepis inundata

Swamp Club-rush **Plate 1229**

This species is typical of many in the genus. It has tufted erect stems 10 to 30 cm high terminating in a cluster of 4 to about 6 brown spikelets about 4 mm long subtended by several bracts. Usually one bract is erect and longer than the cluster, and appears to be a continuation of the stem. Frequently, as in the more robust, common and introduced *I. prolifera*, the inflorescence is proliferous (i.e., capable of rooting and forming separate plants). Flowers have 1 stamen (*I. prolifera* has 3) and usually a 3-branched style. As is common in this genus, the leaf blades are short and slender or reduced to a small point on the sheath.

HABITAT Damp sandy or muddy ground.

DISTRIBUTION Widespread on the coast and tablelands of NSW. All states (except NT), NZ, South America, Malesia.

FLOWERING August–December.

1229. *Isolepis inundata*

1230. Schoenoplectus mucronatus

1231. Schoenoplectus validus

Schoenoplectus

A genus of about 60 species, 11 in Australia, 4 or 5 in the Sydney district, formerly included in *Scirpus*. Rhizomatous perennials with terete or 3-angular stems. Leaves basal, reduced to sheaths or with very short blades. Inflorescence terminal, sometimes appearing lateral due to continuation of the stem by an erect bract. Spikelets in a sessile cluster or an irregular umbel. Perianth bristles present. Fruit a nut.

Schoenoplectus mucronatus — Plate 1230

In *S. mucronatus* the stems are 30 to 70 cm high, acutely 3-angled and leafless except for sheathing scales at the base. The pale brown ovoid spikelets, 8 to 12 mm long, are numerous in a dense lateral sessile cluster subtended by an erect bract 3 to 5 cm long exceeding the inflorescence and continuous with the stem. Hypogynous perianth bristles (i.e., bristles surrounding and attached below the ovary) are usually longer than the dark-coloured plano-convex nut.

HABITAT In shallow water and on creek and river banks.

DISTRIBUTION Central coast, north-eastern NSW and northern Australia; also Malesia, Asia, Europe.

FLOWERING November–March.

Schoenoplectus validus
River Club-rush — Plate 1231

This species has soft, easily compressible cylindrical stems 1 to 2 m high and 4 mm or more in diameter. Leaves are reduced to sheaths, or the uppermost sheath may have a short rudimentary blade. The brown spikelets, up to 13 mm long and 5 mm wide, are in a loose compound irregular umbel which is usually longer than the erect subtending bract. The glumes are notched with the keel drawn out into a short awn in the notch. Flowers have a 2-branched style, 3 stamens and 6 hypogynous bristles with downturned hairs.

HABITAT Margins of rivers, creeks and swamps, sometimes forming dense stands.

DISTRIBUTION Widespread in NSW but absent from the central tablelands. All States (but not NT), NZ, New Caledonia, and Americas.

FLOWERING December–April.

Bolboschoenus

A genus of about 16 widespread species; 3 in Australia, 3 in the Sydney district, formerly included in *Scirpus*. Stout rhizomatous perennials with 3-angular stems to 1.5 m high. Leaves well developed, long, linear, at the base or from nodes on the stems. Inflorescence terminal, umbel-like with spreading rays, subtended by 2 to 3 unequal leafy bracts longer than the inflorescence. Spikelets large, ovoid, golden brown. Perianth bristles present. Fruit a nut.

Bolboschoenus caldwellii — Plate 1232

All 3 Australian species of *Bolboschoenus* occur on the central coast near Sydney and are similar in appearance. They have erect triangular stems 30 to 100 cm high with long keeled leaves (V-shaped in cross-section) at nodes on the stems. Spikelets are solitary or up to 3 together on the unequal rays of an umbel subtended by 2 or more long narrow leafy bracts. In *B. caldwellii* the leaves are 6 to 8 mm wide with finely scabrous margins and keel, and the flowers have a 2-branched style, 2 or 3 stamens and 2 to 6 short hypogynous bristles. The nut is pale or golden brown, compressed and often depressed on both faces, differing from that of *B. medianus* which is glossy black at maturity and obtusely 3-angled.

1232. Bolboschoenus caldwellii

HABITAT Boggy ground.

DISTRIBUTION Widespread but uncommon. Occurring along the coastal zone, and scattered further inland; also Vic, Tas, SA, WA and NZ.

FLOWERING October–December.

1233. Bolboschoenus fluviatilis

Bolboschoenus fluviatilis
Marsh Club-rush — Plate 1233

This species resembles *B. caldwellii* but is more robust and usually grows in water. It has tall 3-angled stems with long linear leaves 7 to 11 mm across and an inflorescence of reddish-brown ovoid spikelets 10 to 25 mm long. The flowers have a 3-branched style (2-branched in *B. caldwellii*). They mature into acutely 3-angled light to dark brown nuts which are longer than the hypogynous bristles.

HABITAT Usually in shallow fresh or brackish water.

DISTRIBUTION Coastal NSW and south-eastern Qld; also NZ, Asia and North America.

FLOWERING October–December.

Lepironia

Lepironia articulata — Plate 1234

Widespread in the Indian/Pacific region, this is the only species of a genus characterised by having 8 stamens and numerous (about 16) hypogynous scales on the flowers. It has rigid, cylindrical grey-green stems, 4 to 7 mm diameter, with cross partitions. The stems rise 2 to 3 m high from a creeping rhizome and terminate in a single ovoid spike 2 to 3 cm long covered in numerous overlapping scaly glumes which conceal the flowers. The flower spike appears lateral due to continuation of the stem by an erect bract at the base of the spike. The fruit is a smooth flat or rounded nut lacking the prominent longitudinal ribs of *Chorizandra*.

HABITAT In mud on the margins of freshwater lakes and lagoons.

DISTRIBUTION NSW central coast to Moreton Bay (Qld) (e.g., Thirlmere Lakes and Glenbrook Lagoon). Also NT, Malesia, Madagascar.

FLOWERING March–May.

1234. *Lepironia articulata*

Chorizandra

Chorizandra is a small genus of 6 species closely related to *Lepironia* but having a greater number of stamens (up to 12), a spikelet with more exserted perianth scales and a ribbed fruit.

Chorizandra cymbaria

Bristle-rush **Plate 1235**

The bright-green stems of *C. cymbaria* are 60 to 100 cm high and 3 to 4 mm in diameter, rising from a thick creeping rhizome. They have transverse septa and an erect ovoid spikelet partly enclosed by the sheathing base of an erect bract which continues the stem. The leaves are few, terete and stem-like, sometimes longer than the stems. Each flower is subtended by a broad obtuse glume and has about 15 overlapping perianth scales longer than the glume. The nearly globular nut is about 3 mm across with about 8 prominent longitudinal ridges.

HABITAT Damp marshy ground.

DISTRIBUTION Central coast and tablelands. Extending along north coast, south coast and northern tablelands; also Qld, Tas, Vic, WA and New Caledonia.

FLOWERING September–November.

1235. *Chorizandra cymbaria*

1236. *Chorizandra sphaerocephala*

Chorizandra sphaerocephala

Round-headed Bristle-rush **Plate 1236**

This species resembles *C. cymbaria* but differs in having a globular spikelet which is not sheathed by the base of the subtending bract. Leaves and stems are stiff, erect, septate and 50 to 80 cm high, the leaves usually shorter than the stems. Each flower is subtended by a narrow-pointed glume and has 14 to 18 hypogynous perianth scales shorter than the glume. The 2 outer lateral scales are folded and have fringed keels. The slightly flattened nut has about 8 prominent longitudinal ridges.

HABITAT Damp marshy ground.

DISTRIBUTION Central coast and tablelands. Extending along the coastal zone from Qld to Vic.

FLOWERING September–November.

1237. *Cyathochaeta diandra*

Cyathochaeta

Cyathochaeta diandra

Plate 1237

This species is not often seen, although sometimes it is very common in the early stages of regrowth after fire in sandy coastal heath and scrub. It has slender stems 50 to 90 cm high, and stiff, channelled, pointed, twisted leaves shorter than the stems and chiefly at the base forming a tuft with many white fibres at ground level. The inflorescence is a long narrow panicle of brown spikelets. Each spikelet is about 15 mm long with 4 narrow-pointed glumes, the 2 lower ones empty, the topmost containing a fertile bisexual flower, the remaining 1 a male or sterile flower. Each flower has 2 stamens and usually 4 long stiff hypogynous perianth bristles. Fertile flowers have a 2-branched style.

HABITAT Sandy soil, usually in heath.

DISTRIBUTION Coast to mountains. Extending along south coast to Vic; also Qld.

FLOWERING Irregular, usually following fire.

1238. *Schoenus villosus*

Schoenus

Perennials, mostly with leafless terete stems less than 1 m high. Leaves basal, narrow linear or reduced to a short blade or sheath. Spikelets in terminal clusters, panicles or solitary, with 2 to 8 usually dark-coloured glumes in 2 opposite rows on the axis, the axis short and straight between the lower empty glumes, often distinctly zig-zagged between the upper fertile glumes. Perianth of bristles, scales or absent. Stamens usually 3, style 3-branched, fruit a nut.

Schoenus villosus

Hairy Bog-rush **Plate 1238**

This is one of the few eastern species of *Schoenus* with long leaves. These are linear, about 20 cm long and 1 to 2 mm wide with 2 or 3 raised ribs, rough margins and woolly dark sheaths at the base. The spikelets are in dense clusters in an interrupted panicle on stems taller than the leaves. Unlike those of *S. moorei*, a much smaller plant also with basal long leaves, the glumes are bearded with long hairs on the margins and the lowest 3 or 4 are empty.

HABITAT Sandy soils in heath and open forest.

DISTRIBUTION Central coast and tablelands. Extending along south coast and southern tablelands; also Qld.

FLOWERING May–June.

1239. Schoenus ericetorum

Schoenus ericetorum

Heath Bog-rush **Plate 1239**

This species and *S. imberbis* form small clumps of erect stems about 30 cm high with the spikelets clustered at the ends in heads. The leaves are very short or reduced to sheaths at the base of the stems, the sheaths brown with a short point and bearded at the opening with long hairs. Spikelets have about 6 to 9 black glumes with bearded margins, the upper 2 enclosing flowers, the lower ones empty. The flowers have 3 stamens, no hypogynous bristles and develop into a small wrinkled 3-angled nut.

HABITAT Sandy and lateritic soils.

DISTRIBUTION Widespread throughout NSW; also Qld, Vic and Tas.

FLOWERING October–November.

1240. Schoenus imberbis

Schoenus imberbis

Beardless Bog-rush **Plate 1240**

Schoenus imberbis is very similar to *S. ericetorum* but has glabrous leaf sheaths with a subulate or linear point representing the leaf blade. The flower heads are subtended by several dark bracts which are usually shorter than the head, but sometimes one is longer. The heads are about 10 mm across with up to 20 tightly clustered black spikelets about 6 mm long, each with 2 fertile glumes above 3 or 4 empty ones. The glumes are sometimes lightly fringed but not hairy as in *S. ericetorum*. Flowers and fruit are similar in the 2 species.

HABITAT Sandy and lateritic soils.

DISTRIBUTION Central coast and tablelands southward on the coast to East Gippsland (Vic).

FLOWERING October–November.

1241. Schoenus melanostachys

Schoenus melanostachys

Black Bog-rush **Plate 1241**

This species is variable in form, depending upon conditions. It sometimes has only a few stems 30 to 40 cm tall with narrow spike-like few-flowered inflorescences, but often forms large spreading clumps on sandy creek banks, and in other damp places, and may have large loose panicles of black spikelets on stems a metre or more tall. The panicles are about 8 to 15 cm long with the short sheathing bracts subtending the branches slightly woolly at the top and with a short reflexed obtuse point. This species can be distinguished from other sedges of similar appearance by the dark tubular sheaths on the inflorescence and at the base of the stems. The spikelets are about 5 to 8 mm long and numerous on slender stalks about 5 mm long. Each has 1 to 3 fertile glumes and about 4 empty ones.

HABITAT Widespread and fairly common in damp places.

DISTRIBUTION Coast to mountains. Occurring throughout coast and tablelands from Qld to Vic; also Malesia.

FLOWERING August–February.

1242. Schoenus apogon

Schoenus apogon

Common Bog-rush, Fluke Bog-rush **Plate 1242**

Schoenus apogon is a small tufted plant with slender weak striate stems from 5 to 25 cm high bearing several clusters of small brown or black spikelets both terminal and in the axils of the long leafy bracts. The soft filiform or linear leaves, as long as the stems on small plants or shorter on tall plants, are both basal and on the stems below the inflorescence. The spikelets are 4 to 6 mm long with 2 to 4 empty acute glumes below 2 or 3 fertile obtuse glumes subtending flowers with 3 stamens, a 3-branched style and 6 perianth bristles. The fruit is a small white 3-angled nut.

HABITAT Damp to wet sandy soil.

DISTRIBUTION Widespread and common in NSW, except in drier western plains; also Qld, Vic, Tas, SA and NZ.

FLOWERING November–December.

1243. Schoenus brevifolius

Schoenus brevifolius

Zig-zag Bog-rush **Plate 1243**

This species forms dense clumps of stout stiff erect stems rarely exceeding 70 cm high, with leaves reduced to glabrous brown sheaths at the base. The stems have brown sheathing bracts at the inflorescence branches tapering to a short erect point. The panicle is narrow and erect with clusters of cylindrical pointed brown spikelets about 10 mm long on short stalks. Spikelets have 3 to 5 fertile glumes on a distinctly zig-zag rachis above 2 to 4 shorter empty glumes.

HABITAT Damp places.

DISTRIBUTION Widespread and fairly common on coastal zone and central tablelands; also Qld, Vic, Tas, WA, NZ and New Caledonia.

FLOWERING December–March.

1244. Schoenus paludosus

Schoenus paludosus **Plate 1244**

This species is atypical of the genus in several respects and was formerly included in *Tricostularia*. It has a loose panicle of reddish-brown spikelets on slender weak stems 20 to 40 cm long with narrow-linear leaves up to 10 cm long, chiefly at the base but with a few on the stem passing into floral bracts on the upper part. The spikelets are 4 to 5 mm long, stalked with 7 or 8 distichous glumes, the lower 4 and the uppermost empty. Usually only the upper flower is fertile and develops into a brown nut with 3 ribs.

HABITAT Wet sandy soil.

DISTRIBUTION Central coast and tablelands. Extending from south coast to Qld.

FLOWERING May–June.

Schoenus turbinatus **Plate 1245**

This species has the unusual combination of well-developed leaves with spikelets in a terminal head. It is a tufted perennial with slender erect striate stems

1245. Schoenus turbinatus

30 to 40 cm tall and filiform leaves 10 to 15 cm long which are inrolled and appear to be terete. Each stem has a dense top-shaped head of 1-flowered spikelets surrounded by 3 to 4 scaly bracts with minutely scabrous margins, the longest about 6 cm long. The flowers have a perianth consisting of 6 fringed bristles which, at the fruiting stage, are longer than the 3-ribbed brown nut.

HABITAT Moist sandy soil in heath and open forest.

DISTRIBUTION Uncommon on the coast and central tablelands; also Vic (rare), Tas.

FLOWERING October–February.

Gymnoschoenus

Gymnoschoenus sphaerocephalus

Button Grass, Button Bog-rush **Plate 1246**

Very striking when in flower, this tussocky perennial has many small spikelets in a dense globular head about 15 mm in diameter at the top of rigid leafless stems a metre or more tall. The stems greatly exceed the stiff flat or channelled basal leaves. Each spikelet has 3 short broad bracts at the base and about 6 to 8 glumes, the lower ones empty, the upper 2 containing a male flower and a bisexual flower above it. The bisexual flower has 3 yellow stamens, a white 3-branched style and 3 small stiff hypogynous bristles. It develops into a smooth shiny obovoid nut 3 to 4 mm long crowned by the narrow pubescent style base.

HABITAT In cool wet places, often in heath and along streams.

DISTRIBUTION Scattered on the central coast and tablelands of NSW (e.g., Maddens Plains, Carrington Falls swamp, above Minne Ha-Ha Falls, Katoomba, Govetts Leap, Blackheath). Extending along northern tablelands, southern tablelands and south coast. Common in Tas and Vic; also SA.

FLOWERING October–December.

1246. Gymnoschoenus sphaerocephalus

1247. Ptilanthelium deustum

Ptilanthelium

Ptilanthelium deustum **Plate 1247**

This small species is common on damp sandy soils throughout the area, though probably rarely noticed. It has the same floral structure as *Gymnoschoenus* but differs in the inflorescence. The leaves at the base of the plant are similar to the stems, though shorter. The leafless stems rise 30 to 40 cm (rarely to 80 cm) from a horizontal rhizome to a single narrow-ovoid flower head, 12 to 18 mm long, subtended by 2 erect pointed bracts with black-brown sheaths having pale membranous margins, the outer bract with a longer point than the inner, the sheaths enlarged and enclosing the spikelets. The spikelets are numerous, 10 to 14 mm long with about 6 glumes, the lower 4 or 5 and the uppermost empty, the remaining 1 containing a bisexual flower. The fruit is an angular ovoid nut.

HABITAT Infertile soils in heath, open forest and woodland.

DISTRIBUTION Coast and tablelands; also Qld.

FLOWERING September–November.

Lepidosperma

A genus of about 50 species, 40 (mostly endemic) in Australia, 13 (sometimes indistinct) in the Sydney area. Tufted perennials with cylindrical, flat, concave or convex stems, often with sharp edges. Leaves basal, similar to the stems or reduced to sheaths. Inflorescence a panicle of sessile spikelets, sometimes reduced to a spike. Spikelets with 1 fertile bisexual flower, 1 or more male flowers below it and several outer empty glumes. The name comes from the Greek *'lepidos'*—a scale; *'sperma'*—seed; alluding to the hypogynous perianth scales which become thickened at the fruit stage forming a small spongy white base with 6 short or long lobes under and around the nut.

1248. *Lepidosperma longitudinale*

1249. *Lepidosperma squamatum*

1250. *Lepidosperma viscidum*

Lepidosperma longitudinale

Pithy Sword-sedge **Plate 1248**

The different species of *Lepidosperma* are known as sword sedges when the stems are broad and flattened, or rapier sedges when they are slender and terete. *L. longitudinale*, one of the former type, is readily recognised by the soft loose pith inside the stems and leaves, making them easily identified. Stems are 1 to 2 m high, 5 to 7 mm wide and distinctly convex on both faces. Leaves are similar but shorter, 9 to 10 mm wide, and taper at the apex to a long acute point. The brown spikelets are crowded on the erect branches of a narrow panicle 15 to 25 cm long. Each branch has a very short subtending bract. The nut is 2.5 mm by 1.5 mm, smooth, shiny, pale with dark brown markings.

HABITAT Swamps and the margins of lakes.

DISTRIBUTION Widespread along coastal zone (e.g., Agnes Banks, Thirlmere Lakes). Occurring in all States (but not NT).

FLOWERING November–January.

Lepidosperma squamatum

Sword-sedge, Hill Sword-sedge,
Sticky Sword-sedge **Plate 1249**

This species form an intergrading complex in which the stems are solid, flat, plano-convex or concave-convex, up to 70 cm high and 3 to 10 mm wide, with spikelets crowded on the secondary branches of the inflorescence. The leaves are thinner and usually shorter than the stems, though often similar in width. They are brown at the base, becoming noticeably wider where they overlap in a flat fan-like arrangement. The stems and leaves of *L. squamatum* tend to be narrow (less than 6 mm) and the panicle rather short (about 4 to 7 cm long) with the lowest bract about as long or occasionally longer than the panicle. The glumes of the spikelets are minutely pubescent and acutely pointed, and the nut is shiny brown and about 3 mm long. In *L. viscidum* (Plate 1250) the edges of the stems are rough with minute resinous particles and the stems and leaves are sticky where they overlap at the base. Stems are similar in width or broader than in *L. squamatum*, and the panicle is longer (about 7 to 15 cm).

HABITAT Widespread and common on sandy soils in heath and open forest.

DISTRIBUTION Central coast and tablelands. Also occurring north coast, south coast, southern tablelands, Vic, Tas, SA and WA.

FLOWERING Mostly December–February, also at other times, particularly after fire.

1251. *Lepidosperma limicola*

Lepidosperma limicola

Razor Sedge **Plate 1251**

This species has dense solid biconvex stems a metre or more tall, and 4 to 5 mm wide. The leaves are shorter but of similar shape and width. Both have sharp cutting edges. The panicle is compact, about 8 to 12 cm long with the lower bract 5 to 10 cm long and the spikelets very crowded on the secondary branches. Spikelets are narrow and sessile with acutely pointed glumes all round the rhachis. The shiny pale brownish-grey ovoid nut has 3 raised ridges and is 2.5 to 3 mm long.

HABITAT Swamp or wet ground.

DISTRIBUTION Widespread from coast to mountains (e.g., Dancefloor Cave at Kanangra Walls, Braeside Walk at Blackheath). Also southern tablelands, northern tablelands, north coast and north-eastern Vic.

FLOWERING December–February.

1252. *Lepidosperma elatius*

Lepidosperma elatius

Tall Sword-sedge **Plate 1252**

L. elatius is a distinctive though uncommon species. Its arching stems are about 2 m long, 6 to 11 mm wide, convex on both sides, rather thick in the middle but with narrow, acute, sharp, slightly scabrous margins. Leaves are shorter and thinner, concave/convex and up to 13 mm wide. The panicle is large, loose and drooping. It is 30 to 50 cm long with the lower branches distant from the upper ones; the lowest subtended by a large erect bract up to 12 cm long, the upper branches with shorter bracts. Spikelets are brown, 4 to 5 mm long and more crowded than in *L. laterale*, but do not completely obscure the axis, and are not clustered. The nut is greyish brown or pale brown, 2.5 to 3 mm long, smooth and shiny with a very obtuse apex. The scales surrounding the nut are long and acuminate, but are not tipped by fine hairs as in *L. laterale.*

HABITAT Forming large clumps in shady gullies, usually near rainforest.

DISTRIBUTION Central coast and tablelands of NSW (e.g., Katandra Reserve, Mona Vale). Extending throughout coast, tablelands and western slopes and in all other States (but not NT).

FLOWERING December–February.

1253. *Lepidosperma laterale*

1254. *Lepidosperma urophorum*

1255. *Lepidosperma flexuosum*

Lepidosperma laterale

Variable Sword-sedge **Plate 1253**

L. laterale is part of a complex which includes Tall Sword-sedge, *L. elatius*, in which the spikelets are not so dense on the secondary axes as to completely hide them. It is smaller than *L. elatius*, with a shorter, less-branched inflorescence. The stems are 50 to 100 cm high (rarely to 150 cm) and 3 to 5 mm wide (occasionally to 10 mm), thin and shiny, more or less flat, or slightly convex on one side, or concave/convex. The leaves are thinner and shorter than the stems. Both stems and leaves have sharp, minutely scabrous cutting edges. The panicle is erect, open and 6 to 25 cm long, the lowest bract up to 18 cm long, but usually about 6 cm, the upper bracts much smaller. Spikelets are evenly spaced, not clustered, with the glumes drawn out to a short point, the lower empty glumes shorter and broader than the upper. The wrinkled green nut turns pale brown, finally grey. It is 2.5 to 3 mm long with about 6 narrow hypogynous scales sometimes tipped with fine long hair-like points about as long as the nut.

HABITAT Widespread and common on moist sandy soils.

DISTRIBUTION Coast, tablelands and western slopes; also all other States (but not NT).

FLOWERING September–March.

Lepidosperma urophorum **Plate 1254**

L. urophorum, L. filiforme and *L. flexuosum* all have slender terete stems and leaf blades reduced to filiform tips on sheaths at the base of the stems. Because the inflorescences are variable, the different species at times resemble each other. *L. urophorum* has stems 1 mm in diameter and 80 to 150 cm tall. The inflorescence is 5 to 8 cm long and is either a spike of closely appressed dark narrow spikelets on a straight axis or, more often, a panicle with several short spicate branches in the lower part. The lowest bract and the bracts subtending inflorescence branches are about twice as long as the spikelets; the others are only slightly longer than the spikelets they subtend. All have filiform tips 1 to 8 mm long. Inner spikelet bracts are brown with 2 scabrous keels. The dark sheathing bracts at the base of the stems are open on one side and rather loose. They are 10 to 15 cm long and have a blunt apex with a filiform tip about 10 mm long. The nut is whitish, wrinkled and rounded at the top.

HABITAT Sandy soil, mainly on the coast, favouring shady gully slopes near rainforest.

DISTRIBUTION Coastal zone and central tablelands; also western Vic.

FLOWERING July–August.

Lepidosperma flexuosum **Plate 1255**

The inflorescence of *L. flexuosum* is a spike or simple panicle, 4 to 8 cm long, of brown spikelets arranged on a more-or-less zig-zag (flexuose) axis. The spikelets are 8 to 10 mm long, shortly stalked and well spaced, at first erect but later spreading and accentuating the flexuose rhachis. Each spikelet is solitary within a narrow rigid obtuse or shortly pointed bract of similar length. Stems are slender, striate, terete and about 80 to 100 cm long. The brown or grey leaf sheaths at the base of the stems are long and closely appressed with an erect subulate tip usually about 60 mm long but sometimes much longer. The nut is shiny white mottled with brown, 3 mm long and obtuse at the apex.

L. filiforme resembles *L. flexuosum*, and at times intergrades with it. Generally, it has a shorter inflorescence of about 3 to 6 brown spikelets arranged in a spike with a straight axis. Stems are less than 1 mm in diameter and usually about 40 cm long, though sometimes longer.

HABITAT Wet or damp sandy soil in heath or open forest.

DISTRIBUTION Coastal NSW.

FLOWERING April–May.

Lepidosperma forsythii

Stout Rapier-sedge **Plate 1256**

Rather than forming compact upright clumps like most other sword and rapier sedges, this species is likely to have stout arching stems 3 m long rising singly, or a few together, through shorter thick undergrowth. The stems are terete or obtusely angular, 1.5 to 2 mm across, and finely striate with a loose pith in the centre. Leaves are reduced to loose, broad sheaths at the base of the stems, the sheaths with a short fine tip less than 1 cm long. The panicle is often a single spreading cluster of very flexuose branchlets subtended by a pointed bract 2 to 3 cm long, shorter than the cluster. The keel of the glumes and short upper floral bracts forms a fine reflexed point at maturity. The slightly angled, pointed nut is 3.5 mm long and pale brown with a finely dotted surface.

HABITAT Wet or swampy ground near the coast.

DISTRIBUTION North and central coast of NSW and eastern Vic. Uncommon near Sydney. Locations include La Perouse. Originally described from a specimen at Centennial Park.

FLOWERING May–June.

1256. *Lepidosperma forsythii*

1257. Lepidosperma neesii

Lepidosperma neesii

Stiff Rapier-sedge **Plate 1257**

L. neesii is the only *Lepidosperma* in the Sydney district with terete or slightly angular stems and long leaves. It forms rather thin clumps with stems 40 to 80 cm high and leaves 30 to 50 cm. The stems and leaves are striate, 1 to 2 mm across, and often flattened or slightly grooved on one side. The panicle, 4 to 8 cm long, has an almost straight axis with short erect branches giving a spike-like appearance. The spikelets are in clusters or short dense spikes within each bract. The lowest bract has a rigid subulate tip several centimetres long; the others are shorter and acutely pointed. Spikelets are numerous, about 7 mm long, with dark narrow acute or acuminate slightly pubescent glumes. The pale nut is thick at the apex and surrounded at the base by short, pointed or fringed perianth scales. The distinction, if any, between this species and *L. quadrangulatum* is not clear at present.

HABITAT Damp heath.

DISTRIBUTION Coast and central tablelands. Recordings include Maddens Plains. Extending from north coast to south coast and Vic.

FLOWERING May–June.

Cladium

1258. Cladium procerum

Cladium procerum

Leafy Twig-rush, Tall Twig-rush **Plate 1258**

This is a very widespread but locally uncommon plant with stout terete hollow stems 1 to 2 m high, and long narrow flat tapering leaves with finely toothed margins and keel. The leaves are upright and continuous with the smooth green sheaths on the stem, or are tufted at nodes on the stem. The inflorescence is 20 to 40 cm long, consisting of dense corymbose clusters 4 or 5 cm apart on the upper stem subtended by long leafy bracts. Spikelets are brown with overlapping glumes all around the axis, the lowest 4 glumes empty, the upper 2 containing flowers, only 1 of which is bisexual and fertile. Flowers have usually 2 stamens and a 2-branched style.

HABITAT Swampy ground near the coast.

DISTRIBUTION Coastal areas in all mainland States. Recorded locally at Scarborough Park, Ramsgate.

FLOWERING November–January.

Baumea

A genus of about 30 species; 15 in Australia, 11 in the Sydney district; formerly included in *Cladium*. Rhizomatous perennials with terete or somewhat flattened pithy or septate stems, the leaves similar or occasionally reduced to sheaths. Inflorescence paniculate, sometimes large and open, more often contracted. Spikelets with 1 to 3 flowers, the lowest or middle one perfect (bisexual). Glumes 2-ranked, the lowest and uppermost empty. Fruit a nut.

1259. Baumea tetragona

Baumea tetragona

Square Twig-rush **Plate 1259**

B. tetragona is readily recognised by the square leaves and stems 30 to 100 cm high and about 2 mm wide. The inflorescence is a dense, erect, reddish-brown panicle 5 to 10 cm long bearing 1-flowered spikelets 4 mm long. The nut is indistinctly 3-ribbed, irregularly wrinkled and crowned by a pointed style base.

HABITAT Swampy ground in cool areas.

DISTRIBUTION Uncommon in the central coast and tablelands (e.g., Carrington Falls swamp). Extending along south coast to Vic; also Tas and SA.

FLOWERING September–November.

1260. Baumea juncea

Baumea juncea

Bare Twig-rush **Plate 1260**

A common plant in coastal marshes where it forms clumps or erect, stiff, greyish stems 40 to 70 cm high rising from long creeping rhizomes covered with large, shiny, pale-brown acuminate scales. Leaves are reduced to light brown or reddish mucronate sheaths at the base, with another 2 or 3 on the stem below the inflorescence. The inflorescence, 3 to 10 cm long, is narrow and spike-like; the spikelets, 4 to 5 mm long, are few and 1 to 2 flowered with only the lowest one fertile. The dull dark ovoid nut is 2 to 5 mm long, lightly dimpled and crowned by a small pubescent style base.

HABITAT Swamps, salt marshes, estuaries, lakes and damp heaths.

DISTRIBUTION Coastal. All states except Qld and NT; also NZ, Lord Howe Island and New Caledonia. Locations include Kurnell (beside refinery), La Perouse (near golf links) and Manly Dam.

FLOWERING October–December.

1261. Baumea articulata

Baumea articulata

Jointed Twig-rush **Plate 1261**

This species is very distinctive owing to the prominent white cross-partitions (septa) in its hollow stems and leaves. The stems are terete, 1 to 2 m tall and about 10 mm in diameter, with a large, loose, drooping inflorescence 25 to 50 cm long of brown spikelets. The leaves are similar in size but taper to a long pointed apex. Spikelets, about 4 cm long, have 3 to 5 bisexual flowers, usually with only the lower or middle flower fertile. Glumes have minutely scabrous keels. The obovoid or 3-sided brown nut, about 5 mm long, is topped by a pointed style base.

HABITAT Coastal lagoons and swamps in water less than 1 m deep.

DISTRIBUTION Coastal zone and northern tablelands. Recordings include Marley Lagoon (Royal National Park), Kurnell and La Perouse. Also found Qld, Vic, SA, WA and south-west Pacific.

FLOWERING October–December.

1262. *Baumea rubiginosa*

Baumea rubiginosa

Soft Twig-rush **Plate 1262**

The terete, angular or slightly flattened stems of *B. rubiginosa* rise a metre or so from a scaly creeping rhizome and bear narrow, irregular inflorescences of densely clustered red-brown spikelets. The clusters occur several together on long and short curved or ascending peduncles in the axils of brown sheathing bracts. Each spikelet has 5 or 6 membranous, fringed, acuminate glumes with the upper 2 or 3 enclosing flowers of which usually only the lowest one is fertile. The basal leaves have a loose pith and are shorter and broader than the stems, flattened and sharply pointed at the apex; the 1 or 2 leaves on the stems have long sheaths and short channelled blades. The shiny reddish nut, about 3 mm long, is topped by a pubescent style base.

HABITAT Wet or swampy ground.

DISTRIBUTION Widespread on coast and tablelands throughout NSW. Recordings include Darkes Forest turn off and Wentworth Falls dam. Also NT, Qld, SA, WA, NZ, New Caledonia, Malesia.

FLOWERING October–December.

1263. *Baumea teretifolia*

Baumea teretifolia

Wrinkle-nut Twig-rush **Plate 1263**

This species has a compact inflorescence 5 to 15 cm long with dense clusters of brown spikelets borne on short erect peduncles in the axils of brown sheathing bracts. The spikelets, about 5 cm long, have one fertile bisexual flower and sometimes a male flower above it. Glumes are membranous, keeled, and acuminate. Stems and leaves are terete, or nearly so, the stems striate and 30 to 100 cm tall, the leaves shorter, erect, acutely pointed and with long loose sheaths at the base. The dull brown nut is about 3 mm long with raised longitudinal ribs and is crowned by an indistinct style base.

HABITAT Wet or swampy ground near the coast.

DISTRIBUTION Coastal zone and northern tablelands. Absent from Blue Mountains. Also Qld, NZ and New Guinea.

FLOWERING August–December.

1264. *Gahnia aspera*

Gahnia

A genus of about 30 species, 20 in Australia, 11 in the Sydney district. Tufted perennials with leafy terete stems, the leaves ligulate, scabrous, long-linear, terete or inrolled, passing on the upper stems into fine long bracts usually exceeding the inflorescence. Inflorescence large, paniculate with numerous spikelets. Spikelets usually with 2 flowers in the upper glumes, only the uppermost flower bisexual and fertile. Glumes overlapping, all round the axis, usually dark brown to black at maturity. Perianth absent. Stamens 3 to 6. Style 3 or more branched. Fruit a nut.

Gahnia aspera

Rough Saw-sedge **Plate 1264**

The stems of this species are robust but usually only 40 to 60 cm high, and much shorter than the leaves. The leaves, like those of most *Gahnia* species, are flat with inrolled margins with tiny sharp upward-pointing teeth. Spikelets are crowded in dense clusters in a narrow interrupted inflorescence. They are about 8 mm long with 7 or more narrow pointed empty glumes passing into a few shorter more obtuse inner glumes and one short round glume enclosing a single fertile bisexual flower with usually 6 stamens and a 3-branched style. The bright red or brown-red nuts are smooth and shiny with a dark acutely pointed apex. They are 4 to 6 mm long and 3 mm across, which is larger than in most other species.

HABITAT Widespread, often in or near rainforest.

DISTRIBUTION Coast, tablelands and western slopes divisions. Also Qld, Malesia and Polynesia.

FLOWERING Irregular.

Gahnia melanocarpa

Black-fruit Saw-sedge **Plate 1265**

When in fruit this species can be recognised by the small shiny black nuts, 2 to 4 mm long, hanging from the spikelets by the elongated staminal filaments. It is a taller species than *G. aspera* to which it is closely

1265. *Gahnia melanocarpa*

related, with stems 1 to 2 m high bearing a dense narrow panicle of spikelets. Each spikelet is about 3 to 4 mm long with 3 or 4 outer pointed empty glumes and 2 or 3 shorter obtuse inner ones, the uppermost quite rounded and enclosing a fertile bisexual flower with usually 3 stamens.

HABITAT Often in or near rainforest.

DISTRIBUTION Coast and central tablelands of NSW. Also Vic and Qld.

FLOWERING June–July.

1266. Gahnia erythrocarpa

Gahnia erythrocarpa — Plate 1266

Ths species resembles *G. sieberiana*, but is much less common, and grows on damp hillsides rather than in swampy situations like the latter species. The stems are about 2 m high with the inflorescence a compound panicle 30 to 60 cm long. Spikelets are about 8 mm long with 10 to 12 dark brown to black acuminate glumes. The lower empty glumes are very small but become longer towards the top. The 2 upper (inner) glumes are smaller and obtuse, the uppermost enclosing a fertile bisexual flower, the other a male or sterile one. The flowers have 4 stamens, and the upper one, a 4-branched style. The nut is smooth shiny and red-brown, about 5 mm long, ovoid or slightly flat on 1 side, and acutely pointed.

HABITAT Damp hillsides on sandy soil.

DISTRIBUTION Restricted to central coast of NSW. Recordings include Brisbane Water National Park.

FLOWERING Irregular.

Gahnia clarkei

Tall Saw-sedge — Plate 1267

With stout leafy stems rising 2 m or more from bases clothed in the tattered grey remains of the lower leaves, or from bare woody clumps marked with horizontal leaf scars, this species sometimes forms impenetrable thickets in damp shady gullies. The leaves are about as long as the stems, scabrous on the back and margins, broad and flat near the base,

1267. Gahnia clarkei

becoming narrow and inrolled towards the apex. The inflorescence is a 1-sided compound panicle about 1 m long with the primary branchlets drooping and 20 to 30 cm long. Each cluster of branchlets is subtended by a long leafy bract. The dark brown spikelets are well spaced along the branchlets. They are about 6 mm long with 12 to 15 closely overlapping glumes, the lower glumes short and acute, the upper longer and obtuse. The uppermost glume encloses a fertile bisexual flower, the one below it a male flower. The fertile flower has 6 stamens and a 3-branched style. It develops into a shiny red-brown slightly angled ovoid nut 3 mm long which at maturity is held, or hangs from the spikelet, by the short staminal filaments.

HABITAT Damp shady places, usually along the banks of streams or near rainforest.

DISTRIBUTION Coast and central tablelands of NSW. Extending all along coast from Qld to Vic; also SA.

FLOWERING Irregular.

1268. Gahnia filifolia

Gahnia filifolia — Plate 1268

This is a slender species, fairly common in some parts of the Blue Mountains. Stems are about 1 m high; the leaves of similar length and only about 1 to 2 mm wide. The outer leaves have short dark open sheaths, the upper ones have long sheaths. The dark narrow panicle is very compound with numerous erect branches in clusters. The species can be recognised by the long subulate points on the bracts subtending the clusters; the lowest bract is much longer than the cluster, the upper bracts only gradually become shorter. Spikelets are dense on the branchlets, dark grey or black, 6 to 9 mm long, with about 5 empty pointed glumes outside 2 shorter thin obtuse glumes enclosing flowers. The upper flower has 4 stamens and a 3-branched style, and develops into a smooth 3-angled oblong nut.

HABITAT Damp sandy soil in open forest.

DISTRIBUTION Central and southern tablelands. Locations include Wentworth Falls Lookout and Fairy Falls.

FLOWERING Irregular.

1269. Gahnia microstachya

Gahnia microstachya

Slender Saw-sedge — Plate 1269

This is a slender species, 60 to 100 cm high, superficially resembling *G. filifolia* but immediately distinguishable by its brown, rather than black, spikelets at both the flowering and fruiting stages. Leaves are shorter than the stems, striate or ridged, about 3 mm wide with inrolled scabrous margins, becoming filiform towards the apex. The spikelets are shortly stalked and only 2 to 3 mm long with about 5 acuminate glumes. They are evenly spaced and only slightly spreading on the erect branches of a narrow panicle 20 to 60 cm long. The lowest cluster of branchlets is subtended by a sheathing bract with a subulate point longer than the cluster; the higher clusters are closer together with sheathing bracts shorter than the clusters. The dull dark-brown oblong nuts are slightly flattened and about 2 mm long.

HABITAT Drier sandy soils.

DISTRIBUTION Chiefly a tablelands species, but also recorded from central coast and central western slopes. Also Vic where it is very rare. Locations include track west of Carrington Falls swamp and near Evans Lookout, Blackheath.

FLOWERING Irregular.

1270. Gahnia sieberiana

Gahnia sieberiana

Red-fruited Saw-sedge **Plate 1270**

This widespread species is the commonest of the Gahnias in the Sydney district. Stems are hollow and usually about 2 m high, though occasionally much taller; the leaves 8 to 20 mm wide with scabrous margins and back. The panicle is held erect above the leaves and is 25 to 60 cm long with numerous dividing spreading branches up to 15 cm long. Spikelets are numerous and dense along each branch. They are at first yellowish brown, becoming dark brown to black at the fruiting stage. Each spikelet has about 5 or 6 glumes of which the lowest and the upper 2, which enclose flowers, are shorter than the middle ones. The flowers have 4 stamens, and the upper one a 3-branched style. At maturity the bluntly 3-angled shiny brownish-red nuts, about 3 mm long, hang from the spikelets by short staminal filaments.

HABITAT Damp sandy ground and waterlogged areas with bottlebrush, tea-tree and other swamp-loving plants.

DISTRIBUTION Widespread throughout coast and tablelands divisions; also Vic, Tas, SA, Qld, New Guinea and New Caledonia.

FLOWERING September–December.

1271. Gahnia subaequiglumis

Gahnia subaequiglumis **Plate 1271**

This coarse erect perennial grows in clumps a metre across and a metre or more high. The scabrous leaves have loose, dark-brown sheaths at the base and are usually inrolled towards the apex, ending in fine long points. The leaves pass into scabrous bracts on the upper stems, with both leaves and bracts often exceeding the inflorescence. Spikelets are densely arranged on the ascending branches of a long narrow panicle. They are yellowish brown at first, turning black at maturity, each with 4 to 6 rough scabrous acuminate empty glumes below 2 nearly acute fertile glumes (i.e., enclosing flowers), the lower and upper ones only slightly shorter than the middle ones. The nuts are 4 to 5 mm long, obtusely 3-angled, shiny and reddish at maturity, and remain suspended from the ends of the spikelets by the short staminal filaments.

HABITAT Open ground on the margins of swampy flats.

DISTRIBUTION Chiefly a tablelands species, but also recorded from central coast; also Vic, Tas, Qld.

FLOWERING Irregular.

Gahnia filum

Chaffy Saw-sedge **Plate 1272**

This is a southern species recently discovered in the Sydney district. It forms clumps about 1 m high of fine erect stems with leaves which taper to long slightly scabrous points and often exceed the stems. The leaves have long striate brown sheaths and are about 5 mm wide, but are often inrolled and appear terete. They are smooth on the back and margins (except sometimes the finest leaves) and have 8 to 10 prominent ridges on the upper or inner surface. The panicle is 30 to 50 cm long, very narrow with clusters of erect branchlets subtended by long leaf-like bracts which become shorter and relatively broad towards the top. The spikelets are 4 to 6 mm long, numerous, crowded and straw coloured, usually with only 1 flower. The 3-sided narrow oblong pale-brown nut is about 5 mm long.

HABITAT On the margins of salt marshes.

DISTRIBUTION Known in the Sydney district only from Mill Creek, and Salt Pan Creek on the Georges River. Occurring on south coast and widespread and often abundant in Vic, Tas and SA.

FLOWERING September–October.

1272. Gahnia filum

Caustis

A genus of about 9 species, 4 in the Sydney district. Rhizomatous perennials with terete striate stems, the leaves reduced to sheaths at the base and along the stems. Spikelets with 3 to 5 overlapping glumes all around the axis, the lower ones short and empty, the upper 2 enclosing flowers. Perianth absent. Stamens 3 to 6. Style 3-branched. Fruit a nut.

Caustis flexuosa

Curly Sedge, Old Mans Whiskers **Plate 1273**

This is a common plant about 30 to 80 cm high bearing brown sheaths at the base and at intervals along the stems. The upper sheaths contain clusters of curved branchlets, with each branchlet dividing several times from sheaths and the ultimate divisions becoming fine and flexuose or coiled. Spikelets occur singly on fine stalks resembling the ultimate branchlets. The upper, bisexual flower has 3 stamens which hang from the end of the spikelet to release pollen. The nut is topped by a large persistent brown style base.

HABITAT Widespread on sandy soil in heath and open forest.

DISTRIBUTION Coast to mountains. Occurs throughout coast, tablelands and western slopes; also Qld and Vic.

FLOWERING September–November.

1273. Caustis flexuosa

1274. Caustis recurvata

Caustis recurvata — Plate 1274

In this species the functionally male and female spikelets are on separate stems of quite different appearance. The male spikelets are in narrow panicles with erect branches on stems 30 to 60 cm high. They are solitary or in short spike-like clusters subtended by broad striate sheathing bracts. The female spikelets are solitary on short stalks on stems having numerous crowded curved or flexuose branchlets. Flowers usually have 5 or 6 stamens (the females without anthers) and a 3-branched style (rudimentary in the males). The ovoid nut has a small conical style base on top. An uncommon form, once known as variety *hirsuta*, with pubescent glumes and rhachis, and larger spikelets, occurs in the Royal National Park. It is now regarded as a new species.

Habitat On sandy soil in heath and dry open forest.

Distribution Coast and central tablelands. Extending from south coast to Qld. Relatively common in the Royal National Park from Audley to Wattamolla. Uncommon in the Blue Mountains.

Flowering September–October.

1275. Caustis pentandra

Caustis pentandra — Plate 1275

This widespread species has erect stems with straight flowering branches rising to about 2 m from a stout rhizome. The stems are terete below the branches, but in the branches are flattened or angular on the inner side, fitting together to form a terete cluster. As in other species, the leaves are reduced to brown sheaths with an erect subulate tip, the sheaths becoming smaller and more bract-like on the upper stems. The spikelets are 12 to 16 mm long in pairs, 1 sessile, the other shortly stalked, in the upper sheaths or bracts. They contain 1 bisexual flower with 4 to 6 yellow or orange stamens, and a male flower below it. The nut has a large persistent pubescent style base on top.

Habitat Widespread and common on sandy soils in heath and dry open forest.

Distribution Coast and central tablelands. All States (but not NT).

Flowering April–May, also spring–summer.

Carex

A cosmopolitan genus of about 200 species, 45 in Australia, 15 in the Sydney district. Perennial herbs with flat leaves, often with scabrous margins. Stems usually 3-angled. Inflorescence of 1-flowered unisexual spikelets, the sexes in the same or separate spikes. Spikes in panicles, umbels or solitary. Female spikelets consisting of a tiny subtending bract and a modified glume forming a flask-shaped utricle around the ovary or nut. Male flowers with 3 stamens, females with a 2 or 3-branched style. Fruit a flat or 3-angular nut.

1276. Carex pumila

Carex pumila — Plate 1276

A small plant with short stems rising from a creeping rhizome, and leaves much longer than the stems, this species occurs along the southern and eastern coasts of Australia, but is locally uncommon. Each plant has about 3 to 6 spikes, the narrow terminal one male, the lower ones entirely female, or female with a few male spikelets at the top. At the fruiting stage the female spikes are erect, ovoid and 2 to 3 cm long. The utricle surrounding each nut is about 5 mm long, greenish to golden brown, thick and corky in texture to aid dispersal in water, and tapers to a notched beak 2 mm long. The flower and nut are topped by a 3-branched style.

Habitat Moist depressions in coastal sand dunes.

Distribution Coastal zone from SA to Qld; also Tas, Lord Howe Island, NZ, Asia and South America.

Flowering September–December.

Carex longebrachiata

Drooping Sedge — Plate 1277

The narrow stems and leaves of this species form grassy clumps 20 to 50 cm high. Each stem has 5 to 20 loosely flowered slender spikes 3 to 5 cm long on fine drooping stalks several times longer than the spikes. The spikes are in bundles of 1 to 5 at nodes on the stem and each node is subtended by a long erect leafy bract. One or a few of the upper spikes are male, the others female, sometimes with a few male spikelets at the end. The utricles on the female spikes are 3-angular with prominent veins and narrow entire or slightly notched beaks with fine bristles. The female flowers have a 3-branched style in, or emerging from, the neck of the utricle. A similar species from the central and southern tablelands, *C. iynx*, has 1 to 4 upper male spikes on short stalks, and the female spikes more densely flowered.

Habitat Moist open forest, rainforest, margins of swamps and creek banks.

Distribution Coast and tablelands divisions; also Vic and NZ.

Flowering October–December.

1277. Carex longebrachiata

Carex hattoriana

The slender stems and leaves of this species are 30 to 80 cm high. Male and female flowers occur together in erect or slightly spreading narrow oblong greenish spikes 1 to 3 cm long with males in the upper part of the spike, the females with a 2-branched style. Each flower is subtended by a tiny acute or acuminate bract. The spikes are on short stalks in clusters of 2 or 3 subtended by short subulate bracts on the stem, the whole forming a loose panicle. Often the lowest spike is much lower on the stem and subtended by a longer leaf-like bract. The utricle on the female flowers is flattened with many prominent veins, pubescent, brown and tapers to a long beak with a 2-pointed apex.

HABITAT Moist soil, usually in rainforest.

DISTRIBUTION Coast and tablelands divisions; also Qld, south-west Pacific and Asia.

FLOWERING October–November.

1278. Carex gaudichaudiana

Carex gaudichaudiana

Tufted Sedge **Plate 1278**

The leaves and stems of this sedge form tufts 30 cm or more high rising at intervals from a short horizontal rhizome. The leaves are about 4 mm wide and are often longer than the stems. Stems are triangular with raspy edges and bear inflorescences consisting of 3 to about 6 dark-coloured spikes, with 1 or 2 narrow spikes of male flowers at the top and 1 or more broader oblong-elliptic spikes, 2 to 4 cm long, of female flowers below them. Each inflorescence is subtended by 1 or more leafy bracts longer than the spikes. The style of the female flowers is 2-branched and the surrounding utricle has 4 or 5 strongly marked nerves.

HABITAT Swamps and the sandy or muddy banks of streams, preferring the cooler conditions of the tablelands.

DISTRIBUTION Coast tablelands and slopes; also Qld, Vic and Tas.

FLOWERING September–December.

1279. Carex appressa

Carex appressa

Tall Sedge **Plate 1279**

This common sedge forms clumps a metre tall and sometimes covers fairly large areas near shady streams and swamps. Stems are acutely triangular, hard and finely scabrous on the edges towards the top. The long grass-like leaves are about 4 or 5 mm broad and also scabrous on the margins. The inflorescence is a narrow spike-like panicle 5 to 20 cm long of numerous short closely appressed erect brownish spikes, the lowest spike distant, the upper ones crowded. The floral bracts subtending the spikes are small or absent, or sometimes the lowest bract is long and thread-like. Spikes carry both male and female flowers, the males sometimes few, in the upper part of the spike. The utricle surrounding the female flowers or nut is flattened with fringed or toothed margins and tapers abruptly to a short 2-pointed beak. Female flowers and nut have a 2-branched style.

HABITAT Margins of swamps, creek banks and damp ground in open forest and rainforest.

DISTRIBUTION Widespread in NSW except the far western plain. All States (but not NT), New Guinea, NZ and New Caledonia.

FLOWERING October–December.

Poaceae

The Poaceae, or grass family, contains about 650 to 700 genera and 9000 species, and although fourth in terms of number of species (behind Asteraceae, Orchidaceae and Fabaceae) is first amongst plant families in many other respects. It is the most abundant family of plants; it has been estimated that there are probably more individual grass plants than there are of all other vascular plants combined. It is the most widespread, being found from the Arctic to the Antarctic (2 species on the Antarctic Peninsula) in the most diverse habitats from the very edge of the sea to the highest elevations where vascular plants can exist. It is the dominant form of vegetation over vast areas of the earth (steppes, prairies, pampas, savannah, tundra) giving sustenance to both grazing livestock and wild herbivores. It surpasses all other families in agricultural and economic importance, providing cereal grains (rice, wheat, maize, barley, oats, rye), pasture and hay for farm animals, sugar and also bamboo. Finally, from the botanical point of view, it is the most highly developed and diverse of the wind-pollinated plant groups.

Grasses have stems with solid nodes and hollow internodes. Each node on the aerial stems (culms) bears a single leaf consisting of a sheath which surrounds the stem, an expanded blade with parallel veins and a ligule or collar at the junction of the sheath and blade. The leaves are distichous (in 2 ranks with each leaf turned 180 degrees from those immediately above and below). Grass plants may be tufted with roots at the basal nodes of the aerial stems, or may have creeping leafy stems (stolons) with roots at the nodes, or leafless scaly stems beneath the soil surface (rhizomes). Stolons and rhizomes enable the plant to reproduce and spread vegetatively. Grasses without stolons and rhizomes spread by basal branching (tillering) to form larger clumps. The culms bear the flowers in a terminal inflorescence above the leafy portion of the culm.

Grasses have inflorescences of many types ranging from a small spike to large, many-branched panicles. The basic unit of the inflorescence is the spikelet. Each spikelet consists of a central axis (rachilla) with 2 empty bracts (glumes) at the base and 1 or more florets above. Each floret consists of 2 bracts (the lemma and the palea) surrounding a grass flower. The flowers are tiny and much reduced or simplified. They are usually bisexual with 3 stamens having large versatile anthers, a superior ovary topped by 2 styles with large feathery stigmas, and a perianth represented by 2 bodies called lodicules. At flowering the lodicules swell with water forcing the lemma outwards thereby allowing the anthers and stigmas to emerge. The anthers produce copious quantities of light fluffy pollen which floats in the air and is collected by the stigmas of other flowers. After pollination the florets close and the 'seed' (a caryopsis or grain) develops.

In many grasses, including some cereals such as wheat, oats and barley, the florets do not open, consequently the flowers are self-pollinated. In other grasses seed is produced asexually from a vegetative cell in the ovule by a process called apomixis. At maturity the spikelets either break up to release individual grains enclosed in the subtending bracts or fall whole and are disseminated as a unit. Occasionally the entire inflorescence breaks off forming tumbleweeds which disperse seed as they are blown along by the wind.

The earliest reliable records of grasses date from Eocene times (about 40 million years ago), which is about the time of final separation of the southern continents, but considerably earlier than the period of great expansion of grasslands in various countries during the Miocene–Pliocene Epochs (26 to 2 million years ago) brought about by climatic change and the coevolution of herds of grazing animals on these open lands.

In the absence of significant pressure from herbivores, grasses remained uncommon in Australia until the Pliocene, and expansion was due mainly to a cooler and more seasonal climate and the increasing occurrence of drought and fire. Extensive grasslands developed in areas with clay soils and in the driest parts of the northern and western inland. Elsewhere saltbush, heath and scrub developed on saline, shallow and infertile soils. For the last 60 000 years extensive burning by the Aborigines has assisted the spread of grasses. Comments by many of the early explorers attest to the effectiveness of this practice in maintaining areas of woodland and forest open and free of shrubby undergrowth. In recent times the native grasses have felt the onslaught of urbanisation, agriculture and the introduction of species more palatable to sheep, cattle and horses.

Thus in Australia today about 210 genera and more than 1200 species of grass are recognised. Only 17% of the genera are endemic, a further 53% contain indigenous species, and the remaining 30% are entirely introduced. Approximately 25% of species are introduced. In the Sydney area, which has about 100 genera and 300 species, only 6% of the genera are endemic to Australia, 57% contain indigenous species and 37% are introduced. Nearly half the species are introduced. As a rule, the endemic genera contain a few species restricted to a few plant communities. Most of the common native species are in large non-endemic genera with distributions beyond Australia that owe more to geological events than to long-distance dispersal of grass seed.

A few common or striking native species are illustrated in the following pages. The identification of grass species is not readily achieved from photographs; books devoted exclusively to this subject which illustrate the detail of grass morphology should be consulted.

Figure 16

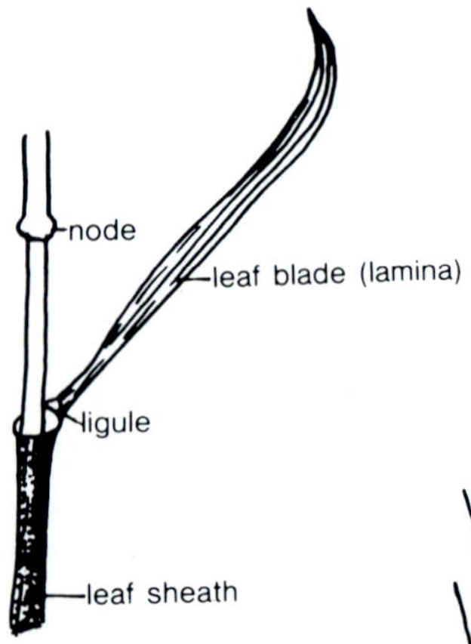

A grass stem (culm) with leaf

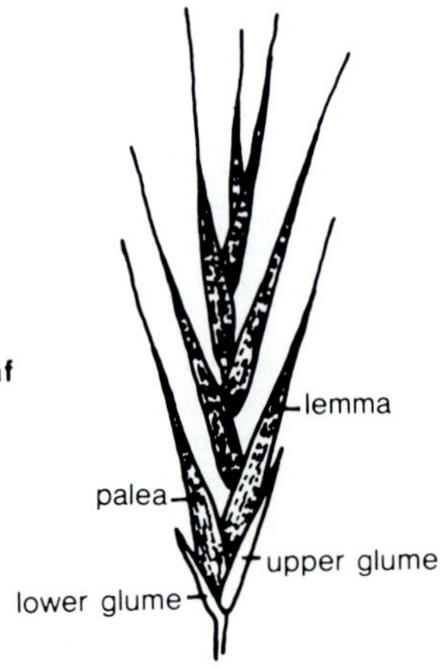

A grass spikelet

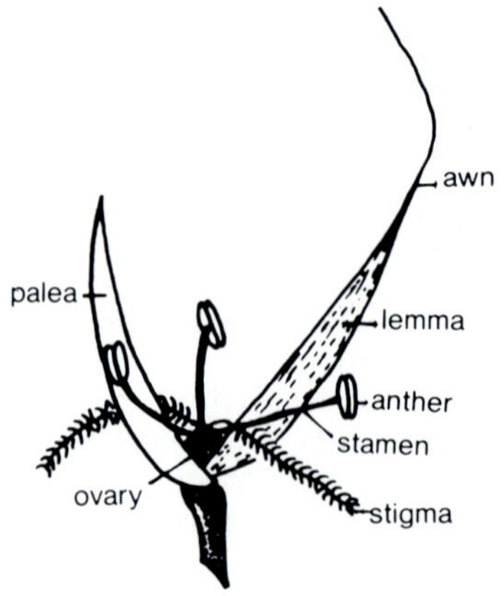

A fertile floret

Agrostis

Agrostis avenacea

Blown Grass **Plate 1280**

This is a common and widespread tufted annual grass usually found in dense patches in places with poor drainage or around ponds and waterholes. It grows through winter and spring, maturing in early summer at which time the seedhead becomes dry and brittle causing it to be easily broken off and blown away by the wind to pile up against fences and other obstructions. The slender stems are up to 70 cm high, the leaves 10 to 25 cm long and 3 mm wide. The stems are hairless but rough below the inflorescence, the leaves are also rough and have a large membranous ligule. The developing inflorescence is at first narrow, erect and partly enclosed in the uppermost leaf sheath, but it later expands to a light open spreading or drooping panicle with fine whorled branches. Spikelets are borne on slender stalks at the ends of the branches and consist of 2 outer glumes enclosing a single flower with a softly hairy lemma bearing a fine bent awn about 4 mm long.

HABITAT Usually in damp conditions.

DISTRIBUTION Throughout NSW. All states (including NT); also NZ, Polynesia, America.

FLOWERING October–December.

1280. Agrostis avenacea

Anisopogon

Anisopogon avenaceus

Oat Spear Grass **Plate 1281**

This grass grows commonly on coastal sandy soils and usually consists of 1 or 2 arching stems about 1 m high with a few fine-pointed leaves and a few dry dangling husks (the persistent outer glumes) resembling oat husks. The leaves are hairless and ribbed on the upper surface which is usually hidden by the inrolled margins. The inflorescence is sparse and pendulous with a few large single-flowered spikelets on long fine stalks. At flowering each spikelet has 2 narrow, linear outer glumes 4 to 5 cm long with a prominent stalked lemma between them. The lemma is densely hairy and has 3 awns, the central one twisted and bent and 5 to 9 cm long, the outer 2 finer and much shorter.

HABITAT Light sandy soils in open forest and heathy scrub.

DISTRIBUTION Coast and tablelands of central and southern NSW; also south-eastern Qld and eastern Vic.

FLOWERING October–November.

Aristida

Although thin and harsh and generally lacking in feed value, Wire Grasses are nevertheless invaluable in the drier parts of Australia because of their drought resistance and widespread occurrence. They resemble Spear Grasses in having long narrow sharply pointed 'seeds' capable of penetrating the flesh of grazing animals, but differ in having 3 awns on the 'seed' instead of one. They are tufted perennials branching from the base or on the stems. The leaf blades are usually narrow and inrolled to restrict moisture loss; the seed head is either narrow or loose and open.

1281. *Anisopogon avenaceus*

1283. *Aristida personata*

Aristida species

Wire Grasses or Three-awned Spear Grasses

The lemmas of the 3 species illustrated have overlapping margins and the 'seed' is therefore circular in cross-section, but other common species such as *A. calycina* have an involute lemma with an obvious longitudinal furrow.

A. ramosa (Plate 1282) has a narrow flowerhead with erect appressed brachlets, a lemma that is clearly longer than the 2 surrounding glumes, and awns all of similar length.

1282. *Aristida ramosa*

The flowerhead of *A. personata* (Plate 1283) is also narrow with erect branchlets but the lemma is similar in length to the glumes and 1 awn is longer than the other 2.

The flowerhead of *A. vagans* (Plate 1284) is more open with spreading branchlets, the lemma is longer than the surrounding glumes and 1 awn is longer than the other 2.

HABITAT Poor shallow shaly and sandy soils, usually in dry sites on the coast, tablelands and interior.

DISTRIBUTION Widely distributed throughout Australia, particularly in Queensland.

FLOWERING December–April.

1284. *Aristida vagans*

Capillipedium

Two species of this small genus occur in tropical Australia and extend down the east coast as far as Sydney. Both have an ornamental flower head with an aromatic fragrance when crushed.

1285. *Capillipedium spicigerum*

Capillipedium spicigerum

Scented-top Grass **Plate 1285**

C. spicigerum, the larger of the two Australian species, is a robust, densely tufted perennial with leafy stems growing to almost 2 m. The lower leaf sheaths are silky hairy and the ligule at the junction of the sheath and blade is tufted with long hairs. Leaf blades are 30 to 40 cm long and 5 to 8 mm wide at the base and are keeled on the midrib. The purplish inflorescence is large and either narrow or loose and spreading with many fine spikelets borne in groups of 3 to 8 pairs at the ends of long threadlike branches. Each spikelet pair consists of a lower sessile spikelet containing 2 florets of which one has a long awn, and a smaller stalked upper spikelet without an awn.

HABITAT Open forest on a variety of soils.

DISTRIBUTION Chiefly coastal from Sydney northward; also NT, Qld and New Caledonia.

FLOWERING January–April.

Danthonia

Owing to their wide distribution and many desirable features, Wallaby Grasses are among the most important native grasses. Although as a group they are easy to recognise by their fine-leaved tussocky habit and the presence of glistening white hairs in the flowering spikelets, they are by no means easy to identify individually. While the various species are characterised by such differences as size, leaf length, and number of flowers in each spikelet, the most useful identifying feature is the nature and pattern of hairs on the back of the lemma of each flower. Generally, *Danthonia* spikelets consist of 2 large outer glumes subtending 3 to 10 florets (flowers), each with a 2-lobed lemma having a long, bent, and sometimes twisted, bristle between the lobes. Although 15 or more of the 35 Australian species occur in the Sydney district, the 4 discussed are sufficient to show the main features.

1286. Danthonia longifolia

1287. Danthonia tenuior

1288. Danthonia pallida

Danthonia species

Wallaby Grasses

D. longifolia (Long-leaf Wallaby Grass) (Plate 1286) is a tufted plant up to 80 cm high with long slender wavy leaves and stems with about 7 nodes. Flowers are in spikelets of 4 to 6 in a narrow panicle about 10 cm long. Uniformly short hairs cover the body of the lemma except near the top, which has much longer hairs.

D. linkii var. *linkii* is a common species with slightly shorter leaves and 6-flowered spikelets. In this species the hairs are short near the base of the lemma and gradually lengthen towards the top.

In *D. tenuior* (Plate 1287) the leaves are less than 20 cm long and relatively broad and flat, at least in the lower part. Unlike the other species, the hairs on the lemma are in a transverse row just beneath the cleft between the lobes.

D. pallida (Plate 1288) (Silver Top Wallaby Grass, White Topped Wallaby Grass, Red-anther Wallaby Grass) is a robust species common on light sandy soils on the tablelands. It has long curved leaves and a large loose spreading panicle of spikelets containing only about 2 or 3 florets each. As in other species, the inflorescence is often purplish.

HABITAT In a wide variety of habitats, but mainly in pastures and open scrubland.

DISTRIBUTION Common on the coast and tablelands of NSW, widespread throughout Australia; also NZ, Africa, Europe, America.

FLOWERING September–April depending on species and location.

Dichelachne

Dichelachne is a small, mainly Australian genus characterised by spikelets consisting of 1 floret (flower) with an inner glume (the lemma) having a long bent dorsal awn attached just below the apex.

1289. Dichelachne crinita

Dichelachne crinita

Long-hair Plume Grass **Plate 1289**

A slender tufted perennial grass 60 to 100 cm tall. The erect stems are unbranched with hairless nodes. Long narrow leaves occur at the base and on the stems. The inflorescence is a dense spike-like panicle 10 to 20 cm long, slender before maturity due to the erect spikelets, but later spreading slightly. The fine silky awns are tinged with purple and are sufficiently long and dense to conceal the spikelets.

A similar species, *D. micrantha*, Short-hair Plume Grass, is smaller with a fairly dense panicle but with shorter awns (15 to 20 mm compared with 20 to 40 mm for *D. crinita*) that do not quite hide the spikelets.

HABITAT Open forest and woodland, often on heavier soils.

DISTRIBUTION Central coast, tablelands and slopes; also Qld, Vic, SA, Tas, WA, NZ, Norfolk Island and Lord Howe Island.

FLOWERING October–April

1290. Dichelachne rara

Dichelachne rara

Rare Plume Grass **Plate 1290**

Three of the 4 or 5 Australian species of *Dichelachne* occur in the Sydney district. Two of these have dense narrow silky inflorescences which earn them the common name plume grass; the third, *D. rara*, has a much looser inflorescence with the spikelets relatively few and not at all hidden by the silky awns. It can, however, be recognised as a plume grass by the attachment of the awn on the dorsal surface of the lemma below the 2-lobed apex. As in the other species, the 'seed' separates above the outer glumes leaving them behind.

HABITAT Open forest and scrub on both heavy and sandy soils.

DISTRIBUTION Coast and tablelands; also Vic and Tas.

FLOWERING October–January.

1291. *Echinopogon caespitosus*

Echinopogon

Echinopogon caespitosus

Tufted Hedgehog Grass **Plate 1291**

This tufted perennial is widespread and common on poorer soils and in places such as roadsides. It has slender rough stems about 60 cm high with a few relatively short flat tapering leaves and a short cylindrical inflorescence about 4 cm long made up of tiny spikelets consisting of 1 floret. Each spikelet has 2 scabrous outer glumes and a 3-lobed inner glume (the lemma) in which the central lobe extends into a fine awn about 5 mm long.

A related species, *E. ovatus*, Forest Hedgehog Grass, is also present in the area. It is similarly rough on the stem immediately below the inflorescence but has a shorter ovoid inflorescence with awns up to about 15 mm long.

HABITAT Dry sandy soil in open forest and scrub.

DISTRIBUTION Coast and tablelands; also Qld and New Guinea.

FLOWERING November–January.

1292. *Entolasia stricta*

Entolasia

Entolasia stricta

Wiry Panic **Plate 1292**

This loosely tufted perennial is fairly common on sandy and stony ground in the Sydney district. It has wiry, somewhat straggling stems which branch at the nodes among spreading clusters of short flat or incurved rough leaves. The inflorescence is a very narrow panicle less than 10 cm long with the spikelets on short erect branches. Each spikelet has a bisexual floret and a sterile floret reduced to a lemma below (i.e., outside) it. The outer glume is very short; the inner glume is the length of the spikelet and is matched in size by the sterile lemma. The lemma of the upper (inner) fertile floret is of similar length and is covered with fine silky white hairs.

HABITAT In scrub and open forest on sandy and poor stony clayey ground.

DISTRIBUTION Coast and tablelands of NSW and Qld.

FLOWERING October–March.

1293. *Imperata cylindrica*

Imperata

Imperata cylindrica

Blady Grass **Plate 1293**

This widespread species occurs commonly in large patches in forest clearings, low fertility pastures, on roadsides and in waste land, especially in sandy areas that are subject to periodic fires. It is a perennial grass which forms tufts of harsh erect leaves 50 to 80 cm long and 5 to 20 mm wide. Leaves have rounded sheaths and silky white hairs at the ligule. The blade is glabrous with a prominent stiff midrib, has margins with tiny cutting teeth, and tapers gradually to a point. The inflorescence is narrow, cylindrical and silky white due to numerous long fine hairs at the base and on the outer glumes (husks) of the spikelets. At seeding the spikelets and hairs fall together leaving a bare axis, the hairs assisting in dispersal of the seed by wind.

HABITAT In a variety of habitats, usually on low fertility soils.

DISTRIBUTION Coast and central tablelands. All States; widespread in tropical and temperate regions of Africa, Asia and Europe.

FLOWERING November–February and after fire.

1294. *Pennisetum alopecuroides*

Pennisetum

Pennisetum is a mainly subtropical and tropical genus of well over 100 ornamental and pasture grasses. The genus is characerised by the spikelets being surrounded at the base by a ring of fine hairs or bristles, with the spikelets and bristles falling together at maturity. Only 4 species occur in Australia, 3 endemic in the north-west, the fourth, *P. alopecuroides*, widespread in the east and in countries to the north.

Pennisetum alopecuroides

Swamp Foxtail Grass **Plate 1294**

A coarse tufted perennial growing to about 1 m high. Its leaves are basal and on the lower part of the stems. They have a large loose striate sheath and a long harsh keeled blade with prominent longitudinal veins and rough margins. The stem is hairless except just below the inflorescence which is a dense spike-like panicle 10 to 15 cm long and 4 to 5 cm in diameter with small 2-flowered spikelets more or less concealed by the numerous rough pale or purplish bristles.

HABITAT Grassy open areas along stream banks, swamp margins and other damp places.

DISTRIBUTION Coast, tablelands and slopes of NSW; also Qld, Vic (localised), WA, Malesia, Asia and Pacific Islands.

FLOWERING December–March.

1295. *Phragmites australis*

1297. *Spinifex sericeus*

Spinifex

Spinifex sericeus

Hairy Spinifex **Plate 1297**

This familiar grass colonises large areas of sand dunes on the beaches of Australia and nearby Pacific islands where its vigorous stoloniferous habit and the ability to survive when partly buried contribute to its important role as a sand binder, stabilising the dunes and enabling other pioneer herbs and shrubs to gain a foothold. Plants are of 2 types which are vegetatively similar but differ in their flower heads. The stolons, or creeping stems, produce roots, branches and tufts of softly hairy grey inrolled leaves at the thickened nodes. One type of plant produces a narrow male inflorescence consisting of 2-flowered spikelets in racemes or spikes clustered in groups of 4 to 6, the other produces large spiny spherical bisexual heads with the spikelets clustered at the centre and greatly exceeded by long hairy bristle-like rhachises. At maturity the stalk of the bisexual head breaks at the stem node allowing the head to be bowled along by the wind, dispersing seed as it goes.

HABITAT Coastal sand dunes above the high-water mark.

DISTRIBUTION All states of Australia (except NT), NZ, New Caledonia and Lord Howe Island.

FLOWERING October–March.

Phragmites

Phragmites australis

Common Reed **Plate 1295**

This tall reed-like or bamboo-like perennial is cosmopolitan in temperate regions of the world, occurring in freshwater and brackish streams, rivers, lakes, ponds and swamps on the margins, or sometimes covering extensive areas where the water depth is less than about a metre. The stout stems grow to 3 m high and bear long leaves at the nodes. Leaf bases tightly sheath the stems; the blades are 1 to 3 cm wide at the base and taper to a fine tip. The inflorescence is a large, dense, branched panicle up to about 30 cm long, though sometimes it is much shorter and sparse. It is initially green, becoming purplish-brown as the bracts colour, then finally fluffy and whitish at maturity due to innumerable long silky hairs growing along the axis of each spikelet. Spikelets have 2 outer sterile glumes and 2 to 6 sessile florets.

HABITAT Shallow fresh and brackish water.

DISTRIBUTION Widespread in all states of Australia, cosmopolitan.

FLOWERING January–April.

1296. *Poa labillardieri*

Poa

Poa labillardieri

Tussock Grass **Plate 1296**

This species and the very similar *P. sieberiana* are common native grasses throughout most of New South Wales. They form dense perennial tussocks with stems up to a metre or more high and long slender greyish-green leaves, the former species usually taller and preferring moist alluvial soils, the latter growing more commonly on drier poorer sites in open forest and woodland. The leaves, which are produced mostly from the base, are inrolled and scabrous on the back (especially noticeable when rubbed downward). The inflorescence is purplish grey, narrow at first, but opening at maturity into a large spreading panicle about 20 cm long with spikelets occurring along the upper parts of the branches. Each spikelet contains about 4 florets with hairy lemmas (husks).

HABITAT In a variety of habitats in open forest, woodland and grassland.

DISTRIBUTION Throughout NSW except the far west; also Qld, Vic, SA and Tas.

FLOWERING March–April.

Stipa

Spear grasses are characterised by their single-flowered spikelets in which one of the inner glumes (the lemma) has a long fine awn or bristle which is spirally twisted and bent when mature. Spikelets separate below the lemma leaving the outer glumes behind and releasing a long, shar-pointed 'seed'. These spear-like 'seeds' are particularly troublesome to sheep, becoming embedded in the fleece and piercing the skin by the cork-screw action of the awn in response to moisture changes.

1298. *Stipa pubescens*

Stipa pubescens
Tall Spear Grass **Plate 1298**

Although spear grasses are frequently the dominant pasture plant over wide areas of the inland, this species is confined to the coast and tablelands where it is relatively common on light sandy soils. It is a thin open plant about 1 m tall with pubescent nodes on the stems and an elongated loose panicle up to 30 cm long. The leaves are 20 to 30 cm long, narrow and inrolled with a short fringed ligule. Outer glumes of the spikelets are slightly unequal in length with one or both 3-pointed at the end. Inner glumes (lemma and palea) are usually hairy.

HABITAT Widespread in open forest and scrub on sandy and stony soils.

DISTRIBUTION Coast and tablelands; also Qld.

FLOWERING October–January.

1299. Stipa ramosissima

Stipa ramosissima
Stout Bamboo Grass **Plate 1299**

This grass immediately attracts attention due to its growth in large 2 m tall cane-like clumps. The stems are stiff and erect but are often weeping near the apex owing to their length. They are branched at the nodes where there is usually a cluster of both branches and short slender leafy shoots. Nodes are thickened, hairless and well separated on the stems. The inflorescence is a lax narrow panicle 40 to 50 cm long with fine whorled branches. Single-flowered spikelets are attached by slender stalks along the branches. The lemma of each spikelet has a long fine awn or bristle. At maturity the spikelets separate above the outer glumes releasing fine, sharply pointed 'seeds' characteristic of the genus.

S. verticillata, Slender Bamboo Grass, is similar, but smaller, and is found in drier habitats. It differs principally in its awned lemmas which are covered on the outside with fine hairs.

HABITAT Rainforest margins and moist protected hillsides.

DISTRIBUTION Eastern NSW extending to the western plain except in the south; also Qld and Lord Howe Island.

FLOWERING October–March.

1300. Themeda australis

Themeda

Themeda australis
Kangaroo Grass **Plate 1300**

Found in all States from the coast to the dry inland, this is one of the most widespread species in the Australian flora. It grows mainly in the spring and summer and is a major fodder grass for kangaroos—hence the common name. It is a tufted perennial growing to 1 m high, often bluish green with purple and red when young, but generally of a brownish colour at maturity. Leaves occur at the base and on the culms (stems) and have a fairly short loose sheath and a blade up to about 30 cm long and 5 mm wide, keeled near the base but flat towards the finely pointed tip. At flowering the upper leaf sheaths bear 2 or 3 compound clusters of spikelets on long slender stalks. Each cluster consists of a complex arrangement of bisexual, male and sterile spikelets enclosed by a pair of long pointed bracts. Several fertile flowers in each cluster have a fine twisted awn about 5 cm long. When ripe, the fertile spikelets separate with their hard outer glumes to form a sharp-pointed barb tipped with stiff hairs, an arrangement, presumably, which facilitates both dispersal on the bodies of animals and the penetration of hard ground.

HABITAT A common ground cover in grassy open forest and woodland, particularly on heavier clay soils.

DISTRIBUTION Widespread throughout Australia.

FLOWERING September–March.

Sparganiaceae

The Sparganiaceae is a distinctive family of 1 genus, *Sparganium*, divided into about 15 species of upright aquatic plants known generally as bur-reeds.

Sparganium

Sparganium occurs mainly in the temperate and arctic regions of the northern hemisphere but also has a discontinuous distribution through South-east Asia to Australia and New Zealand. It was probably introduced into the southern hemisphere from the north by way of seeds in the gut or on the muddy feet of migrating waterfowl. Two species occur sparsely in eastern Australia, one in the Sydney area.

In some countries stands of *Sparganium* provide shelter, nesting and food for waterfowl. *Sparganium* is not common enough in the Sydney area to be of significance in this respect.

1301. Sparganium antipodum

Sparganium antipodum
Bur-reed **Plate 1301**

Bur-reed is an aquatic rush-like plant with stout leaves rising from a creeping rhizome. Leaves are erect, 60 to 80 cm long, 3 to 6 mm wide, often 3-angled, acute at the apex, sheathing at the base and with the midvein prominent on the undersurface. The flowering stem is thick at the base, thinning towards the top and zig-zagged. Male and female flowers are borne separately along the stem in numerous white globular clusters. The upper clusters consist of male flowers, the lower clusters of female flowers. Flowers are wind-pollinated. Fruits are crowded, stalkless and pointed like a burr.

HABITAT Freshwater ponds and streams, either in shallow water or mud.

DISTRIBUTION Coast to mountains, but uncommon. Recordings include Castlereagh and Watagan Mountains. Extending from Vic throughout coast and tablelands to Qld; also NZ.

FLOWERING November–December.

1302. *Typha orientalis*

Typhaceae

The Typhaceae consists of a single genus, *Typha*, with about 15 species of plants known as bulrushes found commonly in dense stands in shallow freshwater habitats in most countries. It is generally agreed that the Typhaceae and Sparganiaceae are closely related; there is, however, much less agreement concerning their derivation and relationships with other monocotyledons.

In both families the thick spongy leaves are 2-ranked around the flowering stems which arise at intervals from horizontal rhizomes in the mud. Both have tiny wind-pollinated unisexual flowers with the males above the females in the inflorescence. In *Sparganium* the 2 sexes are in separate globose heads, in *Typha* they are in separate cylindrical spikes. The flowers of *Sparganium* are protogynous (the females mature first) while those of *Typha* are protandrous (the males mature first). In either case cross-fertilisation is assured. At maturity the male spikes of *Typha* disintegrate and the females turn velvety brown.

Typha

Two native species of *Typha* and one introduced species occur in Australia. The native species are found in all states and are difficult to tell apart. The introduced species, distinguishable by its dark brown female spike, is present in only a few locations from New South Wales to Tasmania.

Typha orientalis

Broad-leaf Cumbungi, Bulrush **Plate 1302**

A tall reed-like perennial, widely distributed and a major weed in some areas where its masses of leaves block water courses and channels. Erect stems rise up to 2 to 3 m from extensive branched rhizomes. Leaves are 1 to 2 m long, 10 to 25 mm wide, blue-green, flat or convex on both surfaces, stem-sheathing and arranged in the same plane on each side of the stem. Flowers are borne in cylindrical chestnut brown spikes near the top of the stem. Male flowers are above, female flowers below; they are usually separated by about 2 cm of bare axis (sometimes up to 5 cm). The mature female spike is 1 to 4 cm wide. The female spike produces masses of fluffy seeds which are distributed by the wind.

T. domingensis also occurs in the Sydney district, but is less common. It is a smaller plant, 1 to 2 m tall with a thinner paler brown female spike (usually less than 2 cm wide) and greener leaves. These characteristics are not very reliable and distinguishing between the 2 species in the field is difficult. The main botanical differences are the shape and abundance of bracts in the female spike and the shape of the stigmas. *T. orientalis* has few bracts (sometimes none), narrow spoon-shaped, and narrow-obovate stigmas. *T. domingensis* has numerous broadly spoon-shaped bracts and linear stigmas.

HABITAT Lagoons, backwaters and other freshwater sites.

DISTRIBUTION Widespread throughout NSW, except drier far western plains. Also Vic, SA, WA, NZ and north to Philippines.

FLOWERING December–February.

Zingiberaceae

The Zingiberaceae is the largest of the 4 to 8 families in the tropical order Zingiberales which includes some interesting plants such as *Musa* (banana, manila hemp), *Ravenala* (travellers palm), *Strelitzia* (bird of paradise flower), *Canna* (arrowroot), *Maranta* (arrowroot) and *Zingiber* (ginger).

The family is a fairly large one, with more than 1000 species in about 50 genera. About 20 species in 10 genera occur in Australia, mostly in Queensland rainforests. One species extends to the Hawkesbury River in New South Wales

The Zingiberaceae have large simple leaves arising spirally, or more usually in 2 ranks, from an underground rhizome. The leaves have sheathing bases and a broad blade with pinnate venation (veins diverging from the midrib), unusual in monocotyledons which are generally parallel veined (veins running lengthwise, parallel with the midrib). The flowers are bisexual with one functional stamen in the inner whorl of stamens. The other 2 are either reduced to staminodes or are absent altogether. One stamen in the outer whorl is absent, the other 2 are fused into a large petaloid lip called the labellum. A slender style is fused to the back of the anther with only the stigma protruding above. The ovary is inferior and 1 or 3 locular.

1303. *Alpinia caerulea* var. *caerulea*

Alpinia caerulea

Native Ginger **Plate 1303**

Native Ginger has leafy stems to 1.5 m tall, erect or arching, produced from a stout rhizome. Leaves are oblong-elliptic, up to 50 cm long and about 5 cm wide, with a prominent midrib; the lower leaves sheath the stem, but the upper leaves are shortly stalked. Flowers are borne in short terminal clusters. Flowers are white in var. *caerulea*. Fruit is globular, about 10 mm across, with a brittle blue outer covering when mature (hence the specific name 'caerulea' which means 'blue').

Var. *arundelliana* is a smaller plant with wavy leaf margins. Its flowers are rosy red and its fruits are smaller.

HABITAT Rainforest, on moist but well-drained soil, usually in shaded sites with some filtered sunlight.

DISTRIBUTION Chiefly north coast and Qld, with its southern limit in Gosford district. Var. *arundelliana* recorded at Gap Creek, Olney State Forest. Var. *caerulea* recorded in gully behind Maitland Bay, Bouddi National Park.

FLOWERING November–December.

Iridaceae

This is a large family of herbaceous perennials with underground corms, rhizomes or tubers and narrow erect parallel-veined leaves which are often arranged in 2 ranks on opposite sides of the stem giving a flat fan-like appearance.

The family is distinguished from related families in the Liliales by the flowers which have 6 perianth segments, 3 stamens and an inferior ovary. The perianth is in 2 similar or dissimilar whorls. Either all 6 perianth segments are fused in the lower part into a narrow tube which expands in the upper part into a regular or irregular flower or the 3 outer segments are fused and the 3 inner ones are erect in the centre of the flower. The style is usually 3-branched, and in some genera the branches are enlarged with showy coloured crests. Some well-known members include *Iris, Ixia, Gladiolus, Freesia, Crocus* and *Watsonia*.

Of the 5 Australian genera, only *Patersonia* is widespread and well known. It contains about 18 species in south-western and eastern Australia and 1 or 2 in New Guinea, Borneo and the Philippines. The others, *Isophysis, Libertia, Orthrosanthus* and *Diplarrhena*, consist of only 1 or a few species which are occasionally cultivated.

Three species of *Patersonia* and 2 of *Libertia* occur in the Sydney area.

Patersonia Native iris, Purple-flag

Small herbs with rhizomes. Flowers blue to purple with 3 broad conspicuous petals which are united at the base into a narrow tube, 3 smaller petals, 3 yellow stamens and a 3-branched style. Flowers are fragile and short-lived, emerging in succession on sunny days. They are in terminal clusters enclosed by 2 large bracts known as spathes. Leaves linear, overlapping and arranged in the same plane on both sides of the flower stem.

1304. Patersonia fragilis

Patersonia fragilis
Short Purple-flag **Plate 1304**

This is the only *Patersonia* in the Sydney area which has hairless leaves, flower stems and spathes. Leaves are 15 to 45 cm long, flat to almost cylindrical, 1 to 6 mm wide, green to blue-green and sometimes twisted. Flower stems are leafless and usually much shorter than the leaves. Spathes are green to pale brown, similar to the leaves in texture. Flowers are a delicate mauve-blue. The larger petals are about 10 mm long and the tube of the flower is hairless and extends well beyond the spathe.

HABITAT Heath and woodland especially in damp sandy soils.

DISTRIBUTION Widespread from coast to mountains. Also recorded north coast, south coast, southern tablelands, Vic, Tas, SA and Qld.

FLOWERING September–December.

Patersonia sericea var. *longifolia*
Dwarf Purple-flag **Plate 1305**

Dwarf Purple-flag has lax trailing leaves to 30 cm tall and 1 to 3 mm wide. They are mostly blue-green and often twisted. The lower parts of the leaves are

1305. Patersonia sericea var. *longifolia*

fringed with a single row of hairs. Flower stems are up to 15 cm tall, leafless and silky-woolly when young. The outer bracts of the spathe are 25 to 30 mm long, dark coloured, silky-woolly when young. Flowers are purple-mauve. This plant is very similar to var. *sericea* but has narrower leaves, a shorter flower stalk, a shorter and narrower spathe and has few leaves in each clump.

HABITAT Heath, woodland and open forest, on sandy and rocky soils.

DISTRIBUTION Widespread from coast to mountains. Extending from Hunter River to south coast, southern tablelands and Genoa River (Vic).

FLOWERING October–December.

1306. Patersonia sericea var. *sericea*

Patersonia sericea var. *sericea*
Silky Purple-flag **Plate 1306**

A densely tufted plant with erect to somewhat lax leaves to 30 cm long and 2 to 6 mm wide. Leaves are usually flat, but may be almost terete. The lower margins of the leaves are woolly. Flower stems are up to 40 cm tall, leafless and silky-woolly at the top when young. The outer bracts of the spathe are about 30 mm long, dark coloured, silky-woolly when young. Flowers are deep violet to mauve. This variety is similar to var. *longifolia* but larger in all its parts.

HABITAT Heath, woodland and open forest in sandy and rocky soils, or on swampy margins.

DISTRIBUTION Widespread from coast to mountains. Occurring from eastern Vic along coast, tablelands and western slopes of NSW to Qld.

FLOWERING July–December.

1307. Patersonia glabrata

Patersonia glabrata
Leafy Purple-flag **Plate 1307**

Leafy Purple-flag differs from the other *Patersonia* in the Sydney area as its leaves are not tufted from the root but emerge from the bases of the flower stems. The lower part of the stem usually has the remains of old leaves around it. Leaves are up to 30 cm long and 2 to 5 mm wide, and have silky hairs along the margins near their bases. Flower stems are slender, 10 to 30 cm tall, hairless or with short silky downy hairs at the base. The outer bracts of the spathe are brown and silky-hoary but not covered with woolly hairs. Flowers are violet.

HABITAT Heath and woodland, in sandy soils.

DISTRIBUTION Widespread from coast to mountains. Extending from Wilsons Promontory (Vic) along coast and tablelands to south-eastern Qld.

FLOWERING September–November.

Libertia Grass-flag

Small herbs with rhizomes. Flowers white with 6 free perianth segments, 3 stamens and style divided into 3. Fruit is a 3-valved capsule. Some botanists place *Libertia* in the genus *Sisyrinchium*.

1308. Libertia pulchella

Libertia pulchella

Pretty Grass-flag **Plate 1308**

This small grass-like plant is usually no more than 25 cm tall. Narrow, flat, overlapping leaves are arranged in the same plane on both sides of the flower stem; they are about 20 cm long and 5 mm wide. Flower stems slightly exceed the leaves and bear a few irregular loose clusters of white flowers. Inner petals are 3 to 5 mm long and elliptic. The 3 stamens are upright with conspicuous yellow anthers.

HABITAT Mossy and wet rock ledges, usually beneath waterfalls, or along shaded rocky creeks.

DISTRIBUTION Blue Mountains (e.g., Dantes Glen, Lawson). Recorded from Barrington Tops, southwards along Great Dividing Range to Vic; also Tas, NZ and New Guinea.

FLOWERING November–December.

1309. Libertia paniculata

Libertia paniculata

Branching Grass-flag **Plate 1309**

L. paniculata is larger in all its parts than *L. pulchella* and has a flower stem which is much more branched. Leaves are flat, overlapping and arranged in the same plane on both sides of the flower stem; they are up to 60 cm long and 12 mm wide. Flower stems bear numerous loose spreading clusters of white flowers. Three obovate petals, 8 to 15 mm long, alternate with 3 smaller narrower petals. The 3 stamens are upright with conspicuous yellow anthers.

HABITAT Damp shaded slopes and rainforest margins.

DISTRIBUTION Widespread from coast to mountains. Extending from eastern Vic along coast to south-eastern Qld; also northern tablelands.

FLOWERING September–October.

1310. Blandfordia cunninghamii

Liliaceae

The Liliaceae is the central family in the order Liliales which consists of mostly terrestrial plants with swollen underground storage organs, linear leaves clustered at the base, and usually showy flowers.

A number of families in the Liliales contain saponins and alkaloids, and the occurrence of these biochemicals, together with chromosomal evidence, is used to help elucidate complex relationships within the group. Closely related monocotyledon orders, Iridales, Zingiberales and Orchidales, and some more distantly related orders such as Commelinales, Juncales and Restionales are probably all derived from an ancestral Liliales stock.

The flowers of the family Liliaceae are usually borne in racemes or cymose umbels. They are bisexual and regular with 6 petaloid perianth segments which are either free or fused into a short or long tube. The 6 segments are usually similar but occasionally the outer whorl (sepals) and the inner whorl (petals) are different. Six stamens are opposite the perianth segments and surround a superior 3-carpellate ovary. In more advanced forms the number of parts is reduced and the ovary is occasionally inferior. The fruit is a capsule or berry.

With over 3500 species the Liliaceae is one of the largest families of flowering plants. It is found throughout the tropical and temperate zones of the world and is represented in Australia by approximately 200 species in 40 genera, many of which are either endemic or have only limited representation outside Australia. A few Australian genera such as *Blandfordia, Dianella, Stypandra* and *Thysanotus* are cultivated as ornamentals. Twenty-eight species in 15 genera occur in the Sydney area where most are reasonably common.

Blandfordia Christmas Bell

Erect perennials with showy red bell-shaped flowers with yellow lobes. Fruit is a 3 sided capsule.

Blandfordia cunninghamii

Mountain Christmas Bell **Plate 1310**

This is a stout erect plant with a basal tuft of linear leaves, 4 to 10 mm wide (sometimes up to 15 mm) and up to 1 m long. Leaves have quite smooth margins, unlike the other 2 species of Christmas Bells in the Sydney area which have rough wavy margins. The bell-shaped flowers are borne in a cluster of 7 to 18 at the end of a stout stem. The floral tube may be up to 45 mm long and 18 to 30 mm across; it narrows near the base. The 6 stamens are attached to the tube below its middle.

HABITAT Damp places, in sandy soils.

DISTRIBUTION Blue Mountains and ranges west of Illawarra.

FLOWERING December–January.

1311. Blandfordia nobilis

Blandfordia nobilis

Christmas Bell **Plate 1311**

Once a common sight in wet heaths and swamps along roadsides around Sydney, its numbers have been reduced by indiscriminate picking. It is still common in less accessible areas. It is an erect plant with a basal tuft of linear leaves, about 5 mm wide and up to 60 cm long. Leaves have a rough wavy edge. The flowers are roughly cylindrical, narrowing only 5 to 10 mm above the base (not broadening conspicuously from the base to the tip as do the flowers of *B. grandiflora*), and borne in a cluster of 3 to 10 at the end of a stout stalk. The floral tube is 20 to 35 mm long, 5 to 10 mm across. The 6 stamens are attached to the middle of the tube.

Intermediates between *B. nobilis* and *B. grandiflora* occur where their distributions overlap (e.g., between Hawkesbury River and Gosford).

HABITAT Damp places, in heath or swampy margins.

DISTRIBUTION Coastal zone, south of Hawkesbury River. Extending to south coast and southern tablelands.

FLOWERING December–February.

1313. *Bulbine bulbosa*

1312. *Blandfordia grandiflora*

Blandfordia grandiflora

Northern Christmas Bell **Plate 1312**

This species is similar to *B. nobilis*, differing mainly in the shape of its flowers. The floral tube is slightly larger (35 to 45 mm long, 25 to 40 mm across) and is funnel shaped. Intermediates between *B. grandiflora* and *B. nobilis* occur where their distributions overlap.

HABITAT Damp places, in heath or swampy margins.

DISTRIBUTION Coastal zone north of Gosford (e.g., Munmorah State Recreation Area) and north-western part of Sydney district. Extending along north coast and northern tablelands to Qld.

FLOWERING December–January.

Bulbine Leek Lily

Yellow-flowered, erect perennials, with tufts of succulent leaves. Stamens 6, some or all bearded above the middle.

Bulbine bulbosa

Native Leek, Golden Lily **Plate 1313**

An erect herb which dies back in summer after its seeds have shed. It is sometimes called 'Wild Onion' because of its tuberous roots from which the leaves shoot annually after autumn or winter rains. Leaves are in basal tufts, narrow, succulent, hollow and up to 40 cm long. Flowers are numerous near the end of an erect stem up to 50 cm tall. The lower flowers open first. Flowers are yellow, with 6 petals, 10 to 12 mm long and 6 bearded stamens. Fruit is a round 3-celled capsule about 6 mm across.

The allied *B. semibarbata* has also been recorded in rocky areas near Kiama and in the central tablelands. It has smaller petals (5 mm long) and only 3 of the 6 stamens are bearded.

HABITAT Grasslands and woodlands, on shale or clay soils, and on rocky outcrops in higher country.

DISTRIBUTION Western and southern Blue Mountains (e.g., Wallerawang, Jenolan Caves, Wombeyan Caves road) and Cumberland Plain. Occurring throughout most of NSW (except south coast) and Qld, Vic, SA and Tas.

FLOWERING October–November.

Thysanotus Fringe Lily

Erect herbs (1 species twining) with 3 broad purple petals which are conspicuously fringed. Fruit a 3-valved globular capsule containing several black seeds.

1314. *Thysanotus tuberosus*

Thysanotus tuberosus

Common Fringe Lily **Plate 1314**

This is an erect perennial to 40 cm tall with much-branched upper stems. Slender linear leaves, 10 to 20 cm long, rise from the base of the flowering stem. These leaves shoot annually (usually in spring) from the tuberous roots and die off at flowering time. Stem leaves may be present, but they are small, few, dry and usually subtending a branch or flower stalk. Flowers are purple with 3 broad fringed petals interspersed with 3 narrower sepals, all 12 to 15 mm long. Flowers are borne in groups of 1 to 4 at the ends of short branchlets. Each flower is short-lived, opening on sunny days and lasting only 1 day, but the plant produces flowers over several months.

HABITAT Heath and woodland, on dry sandy soils.

DISTRIBUTION Common from coast to mountains. Found throughout NSW, except drier far western plains; also Qld, Vic and SA.

FLOWERING October–December.

1315. *Thysanotus juncifolius*

Thysanotus juncifolius — Plate 1315

This Fringe Lily is often mistaken for the more common *T. tuberosus*, but there are a number of significant differences. Its roots are not tuberous but are horizontal fibrous rhizomes. Its basal leaves are short or often completely absent; stem leaves, when present, are small and green. Stems are stiff, hairless (lower parts sometimes hairy) and often flexuose (bent from side to side in zig-zag pattern). The inflorescence is branched only once or twice, unlike the much-branched *T. tuberosus*. Flowers are conspicuously fringed and usually a lighter mauve than those of *T. tuberosus*. The sepals are 5-nerved and less than 2 mm wide.

A closely related species, *T. virgatus* (Plate 1316) occurs on lateritic and gravel outcrops between Loftus Heights and Heathcote (Royal National Park). It has stout ridged stems, covered with short dense hairs, and 7-nerved sepals, 3 to 4 mm wide.

HABITAT Heath and woodland, on dry sandy soils.

DISTRIBUTION Chiefly coastal (e.g., Belrose, Brisbane Water National Park) but also recorded from Blue Mountains (e.g., Kings Tableland). Extending to north coast, south coast, Vic and SA.

FLOWERING October–December.

1316. *Thysanotus virgatus*

1317. *Thysanotus patersonii*

Thysanotus patersonii

Twining Fringe Lily — Plate 1317

This delicate leafless climbing plant is rare in the Sydney district, but has been recorded twisting over low shrubs in the vicinity of Tallong. Its stems are much branched and rise annually from tuberous roots. Basal leaves are slender and wither before the flowering period. Flowers are solitary at the ends of short branchlets. The 3 broad fringed petals are 7 to 8 mm long.

HABITAT Open forest, on sandy soils.

DISTRIBUTION Tallong area of central tablelands. More commonly found on western slopes, southern tablelands and south-western plains. Recorded from every State.

FLOWERING October–November.

Arthropodium Vanilla Lily

Erect herbs; flowers purplish-white with 6 bearded filaments.

Arthropodium milleflorum

Pale Vanilla Lily — Plate 1318

A branching slender herb, 20 to 50 cm tall, with narrow, flat, grass-like leaves about 20 cm long and 4 to 5 mm wide rising from the base. Flowers are pale lilac on stalks 8 to 15 mm long and arranged in clusters of 2 or 3 along the erect flowering stem. Petals are 6 to 7 mm long and filaments have a yellow 'beard' above the middle. The closely related Small Vanilla Lily *(A. minus)* also occurs in the area. It is smaller in all its parts than *A. milleflorum* and has rich purple flowers, borne singly on shorter stalks (less than 8 mm long); filaments are bearded almost to the base.

HABITAT Damp grassy areas and open woodland.

DISTRIBUTION Widespread, but not in sandstone zones. Common on grassy slopes near Jenolan Caves and Wombeyan Caves. Found throughout most of NSW, except far western plains; also Qld, Vic, SA and Tas.

FLOWERING November–December.

Dichopogon Chocolate Lily

Erect herbs. Flowers lilac or mauve, chocolate or vanilla scented, with 3 broad wavy-edged petals alternating with 3 narrower sepals.

1318. *Arthropodim milleflorum*

Dichopogon fimbriatus

Nodding Chocolate Lily — Plate 1319

This is an erect perennial herb, occasionally 1 m tall but usually 50 to 60 cm, with short, flat, grass-like basal leaves 1 to 3 cm long, which often wither before flowering commences. Flowers are lilac with darker anthers and hairy yellow centres, borne on downward-curving stalks in groups of 2 to 3 along the erect unbranched flower stem. The plant dies off after flowering, and shoots in winter from its fibrous roots.

HABITAT Grassland and open forest.

DISTRIBUTION Mainly at higher elevations (e.g., Tallong), but also recorded in central coast division (e.g., Doonside). Extending throughout tablelands, western slopes and western plains; also Qld, Vic and SA.

FLOWERING November–December.

1319. *Dichopogon fimbriatus*

1320. Caesia vittata

Caesia Grass Lily

Small erect herbs. Flowers blue-mauve to white, with hairless filaments (unlike *Arthropodium*); perianth spirally twisted after flowering.

Caesia vittata

Blue Grass Lily **Plate 1320**

Blue Grass Lily has a tuft of crowded linear, grass-like leaves rising from fibrous, often tuberous roots. Flowers have 3 narrow blue-mauve petals alternating with 3 narrow sepals and 6 blue stamens tipped with yellow. Flowers are 5 to 8 mm across and are arranged in irregular clusters along the erect flower stalk. Fruit is a 3-angular capsule.

HABITAT Grassland and woodland.

DISTRIBUTION Widespread from coast to mountains. Recordings include Belrose, Picnic Point, Brisbane Water National Park and Illawarra district. Found throughout coast and tablelands divisions, western slopes and in all other States.

FLOWERING November–December.

Caesia parviflora

Pale Grass Lily **Plate 1321**

This small tufted herb is smaller in all its parts than *C. vittata* and has white to pale flowers less than 5 mm across. The small flowers are borne along an erect thin flower stalk (to 30 cm tall), with flowers and the twisted perianth of finished flowers on the same stalk. Stamens are white, hairless and with yellow anthers.

HABITAT Grassland and woodland.

DISTRIBUTION Coastal districts and Blue Mountains. Also recorded north coast, south coast, south-western slopes, Qld, Vic, Tas and WA.

FLOWERING November–December.

1321. Caesia parviflora

Dianella Flax Lily

Erect hairless herbs. Leaves distichous (arranged in 2 rows on opposite sides of the stem in the same plane). Flowers blue-mauve with 6 yellow hairless thickened stamens. Fruit a berry, usually blue. Its tough leaves produce a flax-like fibre which was used by Aborigines.

Dianella caerulea

Paroo Lily **Plate 1322**

An upright plant with branched stems up to 50 cm long which bear reduced, dry membranous leaves. Leaves at the base of the stem are distichous, sheathing, and vary in size from 10 to 60 cm long and 1 to 20 mm wide. An important identifying feature is that the leaf-sheaths of young leaves are closed at the base (these split open with age), with the leaf blade folded tightly just above the sheath. Leaves are flat for most of their length and usually have rough, sandpapery edges and keel. Flowers are blue, with yellow anthers which are 2 to 3 times as long as the thickened part of the orange filament. Flowers are borne on spreading branched flower stems. Fruit is a blue-purple berry.

HABITAT Coastal heath and forest, in sandy soils.

DISTRIBUTION Widespread from coast to mountains. Extending from Qld to Vic along coastal zone and to northern tablelands.

FLOWERING October–February.

1322. Dianella caerulea

Dianella laevis

Smooth Flax Lily **Plate 1323**

D. laevis has distichous, sheathing leaves from the base of the plant. They are flat above, with margins often curving downwards, green or bluish-green, and varying in size from 20 to 60 cm long and 6 to 10 mm wide. The leaf sheaths are open at the base, smooth and concave, and the leaf edges and keel are usually smooth. Flowers are pale blue, with yellow anthers which are twice as long as the thickened part of the filament. Flowers are borne on loose spreading branched flower stems, 20 to 50 cm long. Fruit is a blue berry.

HABITAT Open forest, woodland and grassland in sandy or clayey soils.

DISTRIBUTION Widespread from coast to mountains. Recordings include Brisbane Water National Park, Macquarie Pass, Camden and Agnes Banks. Found throughout most of NSW and in all States (but not NT).

FLOWERING October–December.

1323. Dianella laevis

1324. Dianella revoluta

Dianella revoluta **Plate 1324**

Spreading Flax Lily, Black-anther Flax Lily

Spreading Flax Lily resembles *D. laevis*, but its leaves are greener and more revolute (margins rolled downwards to the midrib). The leaf-sheaths are open at the base and the keel is prominent. Leaf-margins are usually rough. Flowers are dark blue, and the anthers are yellow-brown or black (not yellow, as are those of *D. laevis*). Anthers are 3 or more times as long as the thickened part of the yellow filament. Flowers are borne on loose spreading flower stems, 20 to 60 cm long. Fruit is a blue berry. A form with broader glaucous leaves is presently known as *Dianella* sp. aff. *revoluta*.

HABITAT Open forest, on sandy soils.

DISTRIBUTION Widespread from coast to mountains. Recordings include Brisbane Water National Park, Macquarie Pass, Picnic Point and Agnes Banks. Found throughout most of NSW (except north coast and far western plains) and in all States (but not NT).

FLOWERING October–January.

Stypandra Blue Lily

Erect herbs with leafy stems. Flowers with 6 stamens and yellow bearded filaments; anthers rolled back after pollen is shed. Fruit a 3-valved capsule containing tiny black seeds.

1325. Stypandra glauca

Stypandra glauca

Nodding Blue Lily **Plate 1325**

This attractive shrubby perennial may have stems up to 1 m tall. Leaves are stem-clasping, 5 to 20 cm long, up to 10 mm wide, smooth, often slightly bluish, and arranged alternately up the stem in the same plane. Flowers are blue, about 20 mm across, with the upper part of the filaments yellow woolly. Flowers are borne on nodding stalks in branching clusters at the ends of the stems.

HABITAT Forest, often in rocky situations.

DISTRIBUTION Widespread from coast to mountains. Recordings include Thirlmere Lakes, Upper Georges River and Wisemans Ferry. Found throughout NSW and in Qld, Vic, SA and WA.

FLOWERING Chiefly July–October, but some flowers at other times.

1326. Stypandra caespitosa

Stypandra caespitosa

Tufted Blue Lily **Plate 1326**

This tufted herb has grass-like erect leaves growing in the one plane from the base of the creeping rootstock. These are 10 to 35 cm long and 5 to 8 mm wide. Flower stems are usually longer than the leaves (to 60 cm tall) and leafy only at the base. They are branched 2 to 3 times near the tip. Flowers are blue (sometimes yellowish-white) with filaments yellow woolly almost to the base and petals 8 to 12 mm long.

HABITAT Open forest, in sandy soils or on rocky outcrops.

DISTRIBUTION Widespread from coast to mountains. Extending to north coast, south coast and southern tablelands; also Qld, Vic and Tas.

FLOWERING October–December.

1327. Stypandra umbellata

Stypandra umbellata **Plate 1327**

This tufted herb resembles *S. caespitosa* and intermediates between the 2 species occur in the southern Blue Mountains. Its leaves are grass-like, 10 to 30 cm long and about 5 mm wide. The flower stems are shorter than those of *S. caespitosa*, up to 30 cm long, and just exceed the leaves. The ends are more branched and they bear 3 to 10 creamy-white flowers, with yellow filaments, and petals 5 to 7 mm long.

HABITAT Open forest, in sandy soils or on rocky outcrops.

DISTRIBUTION Restricted to central coast, south coast and Blue Mountains. Recordings include Gosford, Kurnell, Dobroyd Head and Carrington Falls.

FLOWERING August–October.

Tricoryne Rush Lily

Small herbs. Flowers yellow, with yellow woolly filaments and borne in terminal umbels. Perianth persistent and spirally twisted after flowering. Fruit splitting into 3 segments.

1328. Tricoryne elatior

Tricoryne elatior

Yellow Rush Lily **Plate 1328**

A small perennial with wiry much-branched stems, 10 to 50 cm tall. Leaves are grass-like, short (seldom more than 10 cm long) and wither before the plant flowers. Flowers are yellow and borne in umbels of 2 to 6 flowers at the ends of the stems. Filaments have woolly yellow hairs and the petals and sepals twist spirally after flowering.

HABITAT Grassland, woodland and open forest.

DISTRIBUTION Widespread from coast to mountains. Recordings include Brisbane Water National Park, Doonside and Megalong Valley. Found throughout NSW, except in far western plains; also Qld, Vic, SA, NT and WA.

FLOWERING December–February.

1329. Tricoryne simplex

Tricoryne simplex Plate 1329

T. simplex is smaller than *T. elatior*, usually less than 30 cm tall, and the flowering stems are unbranched or almost so. Leaves are grass-like and about as long as the flower stems. Flowers are yellow and borne in umbels at the ends of the stems. Filaments have woolly yellow hairs and the petals and sepals twist spirally after flowering.

HABITAT Grassland, woodland and open forest.

DISTRIBUTION Coastal zone, extending from north coast to south coast.

FLOWERING September–December.

Laxmannia Wire Lily

Wiry herbs with flowers in compact terminal heads, surrounded by dry and membranous bracts.

1330. *Laxmannia gracilis*

Laxmannia gracilis

Slender Wire Lily **Plate 1330**

A wiry plant with fibrous roots and erect stems to 30 cm tall. Leaves are grass-like, narrow, linear, 2 to 4 cm long, in basal tufts and tufted in intervals along the stems. Individual flowers are stalkless and are clustered together in heads, which are borne at the ends of long leafless stalks. Petals and sepals are arranged in 2 whorls of 3; the outer whorl is dry and membranous and the inner whorl has a pinkish tinge. The fruit is a small capsule.

HABITAT Open forest, in sandy soils.

DISTRIBUTION Widespread from coast to mountains. Recordings include Brisbane Water National Park, Howes Valley, Oatley Park, Royal National Park, Castlereagh and Tallong. Found throughout coast and tablelands, western slopes and north-western plains; also Qld and Vic.

FLOWERING November–December.

1331. *Allania endlicheri*

Allania

Herbs with clumps of grass-like leaves and umbels of white flowers. Grows on sheltered wet rock ledges.

Allania endlicheri

Allania **Plate 1331**

When not in flower, Allania appears to be a tuft of grass clinging to wet cliff-faces. Its leaves are dense, linear, up to 12 cm long and less than 1 mm wide; they have broader, dry and papery bases. Stems are prostrate or erect, short and covered with leaves. White rounded clusters of flowers are borne on slender stalks. Each flower has 6 segments, about 3 mm long, and 6 free stamens. Flowers are surrounded by dry, membranous brown bracts.

HABITAT Wet rock-faces, in sheltered gullies.

DISTRIBUTION Restricted to the Blue Mountains (e.g., Wentworth Falls, Kings Tableland, Linden, Mt Irvine, Mt Coricudgy) with an unusual occurrence along Piles Creek, Brisbane Water National Park.

FLOWERING November–December.

Sowerbaea Rush Lily

Herbs with onion-like leaves and terminal umbels of pink-lilac flowers. Grows in damp soils.

1332. *Sowerbaea juncea*

Sowerbaea juncea

Rush Lily, Vanilla Plant **Plate 1332**

Rush Lily has soft cylindrical leaves, 20 to 40 cm long, growing from fibrous roots. Flower stems are erect and smooth, up to 45 cm long each with a dense terminal cluster of pink-lilac flowers with yellow stamens; 3 perfect stamens alternate with 3 staminodes. Fruit is a capsule. Rush Lily is often locally abundant and its colourful flower heads make it very conspicuous.

HABITAT Wet soil or intermittently water-logged sites.

DISTRIBUTION Widespread from coast to mountains. Recordings include Brisbane Water National Park, Royal National Park, Macquarie Pass National Park, Narrow Neck and Bowens Creek. Extending along the coast from Qld to Vic; also southern tablelands.

FLOWERING October–November.

Schelhammera Lilac Lily

Small soft herb with lilac flowers which have a style divided into 3.

1333. *Schelhammera undulata*

Schelhammera undulata

Lilac Lily **Plate 1333**

Lilac Lily has soft, almost succulent stems, which are diffuse or erect to about 18 cm tall. Leaves are stalkless, 3 to 4 cm long, broadly lanceolate and have wavy margins; they are scattered along the stems. Flowers are lilac, with 3 petals and 3 similar sepals, about 15 mm across, and 6 purple stamens. Flowers are solitary on slender stalks from the upper leaf axils. Fruit is a globular, almost fleshy capsule.

HABITAT Shaded forest sites, on sandy soils.

DISTRIBUTION Coastal zone. Extending from Qld to Vic.

FLOWERING Chiefly October–November, but some flowers at other times.

1334. Burchardia umbellata

Burchardia

Erect herb with grassy basal leaves and white flowers arranged in a terminal umbel; style divided into 3.

Burchardia umbellata

Milkmaids **Plate 1334**

Milkmaids has erect flowering stems rising 15 to 35 cm from surrounding sheaths of 1 or 2 grass-like, limp, basal leaves. White flowers are borne in terminal clusters. The 6 'petals' are spreading and star-like. The ovary is 3-angular and red, and carries a white 3-fid style. Fruit is a capsule.

HABITAT Heath, grassland and woodland.

DISTRIBUTION Widespread from coast to mountains. Also recorded north coast, south coast, southern tablelands, western slopes and all other States (but not NT).

FLOWERING September–November.

Wurmbea

Erect herbs with male and female flowers borne on separate plants or sometimes together on the same plant or occasionally bisexual. Leaves stem-sheathing. Flowers cream-white with a style divided into 3.

1335. Wurmbea dioica

Wurmbea dioica ssp. *dioica*

Early Nancy **Plate 1335**

This is a common species of the tablelands and western slopes, but uncommon in the Sydney district. It dies off after seeding, but shoots from its swollen rootstock soon after autumn rains and is one of the first flowers to appear of the new winter-spring season, hence its common name. Leaves are stem-sheathing, about 10 to 20 cm long and 5 to 10 mm wide, and tapering towards the tip. There are only about 3 leaves on each plant. Flowers are white, with 6 star-like 'petals', some with a circle of purple. Female flowers have a 3-segment pink-purple ovary and a 3-fid style. There are up to 8 flowers on each stem. They are borne in a characteristic manner, with each flower above the other in an arrangement known as a monochasial cyme. Fruit is a 3-celled capsule.

HABITAT Grassland and woodland.

DISTRIBUTION Higher altitudes of Sydney region. Recorded from central coast and central tablelands. Extending throughout tablelands, western slopes and western plains and in all States (but not NT).

FLOWERING August–October.

Xanthorrhoeaceae

Except for 1 or 2 species of *Lomandra* in New Guinea and New Caledonia, the Xanthorrhoeaceae is entirely Australian and contributes much to the unique character of Australian scenery.

This small group of 8 or 9 genera and about 70 species was formerly included in the Juncaceae or in the Liliaceae which it closely resembles in floral structure. It is a heterogeneous group which may warrant further division into subfamilies or families.

The best known and most striking members of the family are the arborescent genera *Xanthorrhoea* (Grass Tree, Blackboy), *Kingia* (Black Gin, Drum Head) and *Dasypogon* (Pineapple Bush). These are stout perennials with a thick caudex or stem covered by the persistent remains of leaf bases and topped by a dense tuft of long arching narrow-angular or flattened leaves above a grey skirt of dead leaves. The inflorescence is a spike up to 2 m long on a smooth scape of similar length *(Xanthorrhoea)* or a spherical head 3 to 5 cm in diameter on peduncles about 1 m long or shorter *(Kingia, Dasypogon)*. Although *Xanthorrhoea* is widespread, the others are restricted to the extreme south-west between Perth and the Stirling Ranges near Albany. In New South Wales, *Xanthorrhoea* is more or less restricted to infertile parts of the coast, tablelands and western slopes; apart from an occurrence near Narrabri, it is not seen on the western plains.

Some *Xanthorrhoea* and the smaller members of the family have leaves arising from a tuft or clump at ground level. Except for *Lomandra*, which has about 40 widespread species, the smaller members consist of genera with 1 to 3 species with restricted distribution. *Lomandra* differs from all the others in being dioecious (in having male and female flowers which are borne on separate plants). Frequently the floral arrangement of the sexes is dissimilar.

Lomandra and *Xanthorrhoea* are the only genera which occur in all states. Twenty species of *Lomandra* and 7 of *Xanthorrhoea* occur in New South Wales, 12 and 6 respectively in the Sydney area.

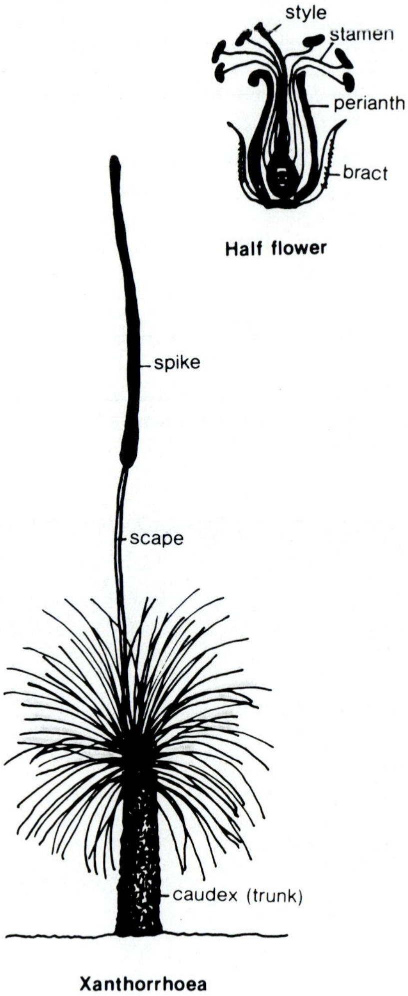

Figure 17

Xanthorrhoea Grass Tree, Blackboy

Stout perennials, with thick woody underground or above-ground trunks. Leaves long, narrow and tough, spreading from the top of the trunk, often appearing like a skirt around the trunk. Flowers creamy-white, numerous, surrounded by floral bracts and borne on a long spike at the end of a smooth erect scape. Fruits are shiny brown hard capsules which protrude prominently from the spike. Trunks and leaf bases are strongly impregnated with resin.

Identification of species is often difficult because of considerable variation within species correlated with distribution and because of a marked tendency of species to hybridise.

The following characterisics are useful for identification purposes: (i) the habit of the plant—whether it has an above-ground trunk more than 75 cm high; (ii) the shape and hairiness of the floral bracts; (iii) the relationship between the lengths of the spike and the scape; and (iv) the shape of the leaf in section, when cut through the middle.

Xanthorrhoea australis

Austral Grass Tree **Plate 1336**

X. australis is uncommon in the Sydney district and there is some disagreement as to whether the form found in the area is its typical form. It is easily recognised by its large dark trunk, distinctly quadrangular leaves, long acute floral bracts and a stout flower spike which is longer than the relatively short scape. The trunk is usually 1 to 3 m tall, sometimes taller, and is often branched. Leaves are narrow (usually 1 to 2 mm wide) and up to 1 m long, quadrangular or flattened quadrangular in section and often have a greyish or silvery appearance. A skirt of old leaves hangs from the top of the trunk, distinct from the more erect younger leaves. The spike is stout, usually 90 to 200 cm long and 3 to 6 cm across. The scape is also stout, and may be 35 to 150 cm long. Floral bracts are light brown, long and with slender points. Austral Grass Tree hybridises with *X. media* in the region south of the Hunter Valley towards Putty and with *X. latifolia* in the coastal ranges north of Wyong.

Another arborescent species with quadrangular leaves *(X. johnsonii)* occurs to the north and north-west of the Sydney district (e.g., Old Darkey Creek, Howes Valley). It is closely related to *X. australis* and its features are variable, probably due to hybridisation. It has an undivided trunk to 2 m tall, leaves about 1.5 mm wide, green not greyish in colour, and a more slender spike and scape.

HABITAT Forests, especially on rocky hillsides, often on limestone and volcanic rock.

DISTRIBUTION Southern Highlands (e.g., Bundanoon) and southern Blue Mountains (e.g., Wombeyan Caves). Occurring south of Nowra into Vic, SA and Tas.

FLOWERING November–December; dependent on previous bushfires.

1336. Xanthorrhoea australis

1337. Xanthorrhoea resinosa

Xanthorrhoea resinosa

Plate 1337

This is a common Grass Tree in the Sydney area, especially in coastal sites with swampy-sandy conditions. Its flower spikes dominate large areas of wet heath in summer months; flowering is at its best following fire. The trunk is underground or, rarely, a few centimetres above ground. Leaves are rather thick, mostly 3 to 3.5 mm wide and quadrangular or convex-triangular in section, often blue-green. The spike is dark brown, up to 1.8 m long. The scape is often over 2 m long and 8 to 30 mm wide. The scapes are nearly always longer than the spikes. Trunks and leaf bases exude yellow resin (most other species have dark red-brown resin). A characteristic feature is the presence of dense matted hairs on the tips of the floral bracts, giving the spike a dark brown velvety appearance. *X. resinosa* is often confused with *X. media*, but *X. resinosa* has hairy bracts and is more tolerant of swampy conditions. When found in the same area, *X. resinosa* grows in the wetter parts and *X. media* in the drier parts.

HABITAT Heaths and wet places on sandy soils.

DISTRIBUTION Coast and Blue Mountains. Occurring from Gippsland (Vic) along south coast to central coast of NSW.

FLOWERING June–January; dependent on bushfires.

1338. Xanthorrhoea concava

Xanthorrhoea concava

Plate 1338

This species was formerly regarded as a subspecies of *X. resinosa*. It differs mainly in its leaves. These are broader (3 to 6 mm wide), flattish or concave above and often blue-green. The spike is up to 1.5 m long and shorter than the scapes which are up to 2.5 m tall and about 10 mm wide. The spike appears distinctly light brown and velvety because of long pale hairs on the bracts and sepals.

HABITAT Heath, swamps and open forest, on sandy soils and shale.

DISTRIBUTION South of Sydney to southern highlands (e.g., Buxton, Menai, Nepean Dam, Hilltop, Bundanoon, Wingello). Extending to southern tablelands and as far south as Eden.

FLOWERING Chiefly October–December; dependent on previous bushfires.

Xanthorrhoea fulva

This species was formerly regarded as a subspecies of *X. resinosa*. It differs mainly in its leaves. These are slender, depressed triangular in cross-section, 2 to 3 mm wide and usually blue-green. The size of the spike is very variable; in young plants it may be slender and less than 30 cm long, but in older plants may be stout and up to 90 cm long. The scape is up to 1.6 m tall and 5 to 20 mm wide.

HABITAT Heaths, low dunes and wet places on sandy soils.

DISTRIBUTION North of Gosford (e.g., Wyong to Warnervale). Extending along north coast to Qld.

FLOWERING Chiefly September–November; dependent on previous bushfires.

1339. Xanthorrhoea media

1340. Xanthorrhoea latifolia

Xanthorrhoea media **Plate 1339**

X. media closely resembles *X. resinosa*, but has floral bracts which are short-acute, almost hairless, with only a few scattered short hairs on the tips, not densely woolly like those of *X. resinosa*. The trunk is underground or above ground to 80 cm tall. Leaves are glossy green (not blue-green) and the spike is shorter than the scape by the ratio of 1:2 or 1:3. Leaves are less than 3 mm wide and triangular to quadrangular in section. The spike is usually 30 to 80 cm long (up to 1.4 m in more southern specimens). The scape is up to 2 m tall and 8 to 12 mm wide. Hybrids of this species and *X. australis* occur in the north-west of the Sydney district and show a tendency to arborescence. Hybrid derivatives from *X. minor* occur in the western and south-western outskirts of Sydney.

HABITAT Dry rocky ridges on Hawkesbury sandstone; occasionally on drier coastal heath.

DISTRIBUTION From near Hilltop to the Blue Mountains and along the coast as far north as Goulburn Valley (e.g., Putty road, Central Mangrove, Kariong, Warrimoo, Hornsby).

FLOWERING Chiefly August–March; dependent on previous bushfires.

Xanthorrhoea latifolia **Plate 1340**

This species was formerly regarded as a subspecies of *X. media*. It differs from that species by its broader, nearly flat leaves which are 3 to 4 mm wide and by its tendency to arborescence. Trunks may be 1 m high, especially at higher altitudes, or completely underground. The spike is 20 to 100 cm long and much shorter than the scape which is usually 1.5 to 2 m tall and 7 to 11 mm wide. This species may be confused with *X. arborea* but differs in its glossy green leaves and the high scape-to-spike ratio. In some areas along the north coast, there are large populations of hybrids from *X. latifolia* and *X. australis*.

HABITAT Forests and coastal ranges and coastal heaths.

DISTRIBUTION North of Gosford (e.g., Doyalson, Dora Creek, Munmorah State Recreation Area). Extending along north coast to about Townsville (Qld).

FLOWERING Chiefly March–October; dependent on previous bushfires.

1341. Xanthorrhoea macronema

Xanthorrhoea macronema **Plate 1341**

The only Grass Tree in the Sydney district which completely maintains its distinctness as a species. All others have hybrid derivatives. It is easily identified by its leaf section, short flower spike and, when in flower, by its large flowers and long stamens. The trunk is underground. Leaves are glossy, up to 1 m long, 2 to 4 mm wide, triangular in section, ribbed with several deep grooves between the ribs and sandpapery on the margins. The spike is short, 3 to 12 cm long, on long slender scapes 50 to 200 cm long and 2 to 5 mm wide. Flowers are large, yellow, with petals larger than the floral bracts; stamens stick out about 10 mm from the petals. Flowers on the lower part of the spike open before those on the upper part.

HABITAT Forest, often in shaded areas.

DISTRIBUTION Coastal zone. Recordings include Helensburgh, Garrawarra, Wyong, Tuggerah, Bateau Bay and Munmorah. Extending along north coast as far as Fraser Island (Qld).

FLOWERING July–August; dependent on previous bushfires.

1342. Xanthorrhoea arborea

Xanthorrhoea arborea **Plate 1342**

This is the common arborescent Grass Tree in the Sydney district. It usually occurs on cooler slopes and sheltered rocky hills. Its trunk may be up to 2 m tall. Leaves are flattish, up to 1.5 m long and up to 8 mm wide. Near the coast, leaves are often greenish-grey in colour, but are mid-green in the Colo and Nepean districts. The stout spike is nearly equal in length to the scape (from 1 to 2 m). The spike-to-scape ratio helps separate this species from *X. latifolia*.

HABITAT Sheltered sites, such as south-facing slopes or cooler lower slopes of north-facing hills.

DISTRIBUTION Restricted to sandstone areas of central coast and central tablelands, from Port Hacking to Wollemi National Park, east of Rylstone.

FLOWERING Chiefly November–January; dependent on previous bushfires.

Xanthorrhoea minor ssp. *minor*

Small Grass Tree **Plate 1343**

X. minor is the smallest of the local Grass Trees, with a scape and spike only about 60 cm tall. It has a branched underground trunk which produces several

1343. *Xanthorrhoea minor*

1344. *Lomandra micrantha*

scapes on the one plant. Leaves are triangular in section, often concave above, about 2 to 3 mm wide. The spike is only 5 to 12 cm long on slender scapes 25 to 50 cm long and 2 to 5 mm wide.

HABITAT Open forest, usually on clay soils which are seasonally water-logged.

DISTRIBUTION Coastal zone and lower Blue Mountains. Recordings include Holsworthy, Picnic Point, Agnes Banks and Glenbrook.

FLOWERING October–November; dependent on previous bushfires.

Lomandra Mat-rush

Perennials with narrow wiry leaves. Male and female flowers borne on separate plants in semi-globular or branched clusters or panicles. Inflorescence of each sex is often dissimilar. Flowers small, with 6 perianth segments in 2 whorls. Fruit a shiny globular capsule. It is often difficult to distinguish between some species when they are not in flower.

Lomandra micrantha ssp. *tuberculata*

Small-flowered Mat-rush **Plate 1344**

Despite the specific name, the flowers on this species are not significantly smaller than those of other Mat-rushes. Male flowers are about 2 mm across, but female flowers are about 5 mm. The flowers are the most useful feature for identification purposes. They are cream, purplish or reddish. Male flowers have 6 similar, spreading, open petals (not joined and like a closed bell as most other species). Flowers are borne in short branched panicles, singly or 2 to 3 together, the males on distinct stalks, the females on very short stalks. Leaves are in wiry tufts from the base of the plant, about 30 cm long, 1 to 2 mm wide and quadrangular to terete in section. The tips are blunt and the bases are often purplish-brown and without the characteristic shreds of some species.

HABITAT Open forest on sandy soils.

DISTRIBUTION Coast (south of Gosford) to Blue Mountains. Recordings include Minnamurra Falls, Maddens Plains, Blackheath and Hornsby. Extending to eastern part of southern tablelands; also Vic and SA.

FLOWERING April–June.

1345. *Lomandra cylindrica*

Lomandra cylindrica

Needle Mat-rush **Plate 1345**

This species is characterised by its tufts of cylindrical to semi-cylindrical leaves (often only 2 leaves per shoot) and its many small yellowish flowers borne singly on erect branched panicles to 15 cm tall. Leaves are 15 to 60 cm long, 1 to 2 mm wide, usually twisted above, pointed at the tip, with a few shreds at the base and prominent erect scale leaves below. Flowers are 1 to 3 mm long. Male flowers are like closed bells, hanging on distinct stalks; the females are on very short stalks.

HABITAT Open forest, on sandy soils.

DISTRIBUTION Coast to mountains. Recordings include Brisbane Water National Park, Hornsby, Glenbrook, Clarence, Lithgow, Agnes Banks and Hilltop. Extending to northern part of southern tablelands and south coast (e.g., Jervis Bay); also Vic.

FLOWERING September–December.

1346. *Lomandra gracilis*

Lomandra gracilis **Plate 1346**

This species is characterised by its tufts of narrow leaves which are flat above and convex below, with numerous purplish or dark brown fine shreds at the base, and its many small yellowish flowers borne singly on erect branched panicles 20 cm or more tall. Leaves are up to 60 cm long, 1 to 2 mm wide, only slightly twisted above; there are several leaves per shoot. Flowers are 1 to 3 mm long. Male flowers are like closed bells hanging on distinct stalks; the females are on very short stalks.

HABITAT Open forest, on sandy soils, often in rocky situations.

DISTRIBUTION Coast to central Blue Mountains. Recordings include Campbelltown, Blackheath, Springwood, Audley, Woodford, Hornsby, Mangrove Creek and Colo Heights. Extending to north coast, northern tablelands and Qld.

FLOWERING September–October.

Lomandra filiformis ssp. *filiformis*

Wattle Mat-rush, Iron Grass **Plate 1347**

L. filiformis is a variable species, with 3 subspecies recognised in NSW, 2 occurring in the Sydney district. Ssp. *filiformis* is a tufted plant characterised by narrow incurved or inrolled leaves, up to 30 cm

1347. Lomandra filiformis ssp. *filiformis*

long and 2 mm wide, with numerous dark brown shreds at the base and a ragged rounded tip. It is taller than *L. brevis*, less densely tufted and grows in dry rather than damp sites. Flowers are cream coloured, 2 to 3 mm long. Male flowers are rounded or bell-shaped and hang on distinct stalks; the females are on very short stalks. Male flowers are borne on erect branched panicles up to 20 cm long; the female inflorescence is slightly shorter.

HABITAT Open forest, on sandy soils.

DISTRIBUTION Widespread from coast to mountains. Recordings include Brisbane Water National Park, Berowra, Cumberland State Forest, Hornsby, Macquarie Pass and Erskine Creek. Occurring mainly in the north from Qld to central NSW, occasional in southern NSW and Vic. Intergrading with ssp. *coriacea* to the south and ssp. *flavior* to the west.

FLOWERING October–December.

Lomandra brevis

Tufted Mat-rush

L. brevis is regarded as a species 'at risk' because its known populations are limited in range. It is characterised by dense tufts up to 15 cm tall, narrow leaves with inrolled margins and a fondness for sheltered situations. Leaves are about 1 mm wide, with an entire and acute tip. Leaf bases are purplish above the whitish sheath. Flowers are small, yellowish and borne singly on erect narrow branched panicles up to 9 cm tall. Male flowers are like closed bells hanging on distinct stalks; the females are on very short stalks.

HABITAT Damp sandy areas or sheltered situations in dry open forest.

DISTRIBUTION Restricted to the Sydney region, mainly north of Sydney. Recordings include Hornsby, Mt Ku-ring-gai, Berowra, Girrakool, Sutherland and Heathcote.

FLOWERING September–December, with some flowers March–May.

Lomandra filiformis ssp. *coriacea*

Wattle Mat-rush **Plate 1348**

This subspecies differs from the type form in having more or less flat leaves (sometimes slightly incurved), firm and leathery, up to 30 cm long and about 3 mm wide. There are only a few pale brown shreds at the base. Flowers and inflorescences are similar to those of ssp. *filiformis.*

HABITAT Open forest, on sandy soils.

DISTRIBUTION Widespread from coast to mountains, chiefly to the south. Recordings include Brisbane Water National Park, Yarramundi, Woronora Dam, Hilltop and Colo Vale. Also occurring north coast, southern tablelands, western slopes, south-western plains, Qld and Vic.

FLOWERING October–December.

1348. Lomandra filiformis ssp. *coriacea*

Lomandra longifolia ssp. *longifolia*

Spiny-headed Mat-rush **Plate 1349**

This is a common tufted plant, found in a variety of situations from exposed sand dunes to rainforest. It is a very variable species but characterised by its tough, strap-like leaves and its large inflorescence with spiny bracts at the base of each flower cluster. Leaves are 40 to 80 cm long and mostly 8 to 12 mm wide, flat or slightly inrolled. Flowers are almost cylindrical, creamy, often with purplish centres, stalkless and borne in paired clusters 1 to 2 cm long along a flattened stem which is usually about half the length of the leaves. The spiny bracts beneath the flower clusters are 1 to 2 cm long.

HABITAT Many situations, including sand dunes, exposed lateritic ridges, open forest, creek banks and more open areas of rainforests.

DISTRIBUTION Widespread from coast to mountains. Occurring throughout coast, tablelands and western slopes; also Qld, Vic, Tas and SA.

FLOWERING August–December.

1349. Lomandra longifolia

1350. Lomandra obliqua

Lomandra obliqua

Twisted Mat-rush **Plate 1350**

This small plant (to 50 cm tall) is the easiest of the local *Lomandra* to identify as it is not tufted and has short twisted leaves all on one plane and evenly spaced up the spreading stems. Leaves are usually 2 to 4 cm long, green or greyish, flat or almost so, and have a few white shreds at the base (sometimes bare). Flowers are yellow, stalkless and grouped in small clusters along a spike.

HABITAT Open forest, on sandy, often shallow, soils.

DISTRIBUTION Common from coast to mountains. Extending from Jervis Bay (south coast) and southern tablelands to north coast and Qld.

FLOWERING September–November.

Lomandra confertifolia ssp. *rubiginosa*

Plate 1351

L. confertifolia is a variable species with 4 subspecies recognised in NSW. Two of these occur in the Sydney district, ssp. *rubiginosa* and ssp. *pallida.* Ssp. *rubiginosa* has tufts of spreading channelled leaves up to 40 cm long and to 2 mm wide. Leaf margins are somewhat incurved, the tips have irregular teeth and the leaf bases are surrounded by reddish or yellow-brown shreds. Flowers are yellowish, stalkless and grouped in small well-spaced clusters along a spike. Small spiny bracts are at the base of each flower cluster.

Ssp. *pallida* has whitish shreds at the base of the leaves. It occurs from Queensland south to near Sydney and is less widespread than ssp. *rubiginosa.* It has been recorded in rocky sandstone areas near Wolgan Gap, central tablelands.

HABITAT Open forest, on dry sandy soils.

DISTRIBUTION Widespread from coast to mountains. Recordings include Brisbane Water National Park, Berowra, Bents Basin and Nepean Dam. Extending from near Gloucester (north coast) to south coast and central western slopes.

FLOWERING July–November.

1351. Lomandra confertifolia ssp. *rubiginosa*

Lomandra glauca ssp. *glauca*

Pale Mat-rush **Plate 1352**

L. glauca has short leafy or tufted ascending stems with characteristic blue-grey leaves, which are 8 to 20 cm long, 1 to 2 mm wide, flat to concave, often slightly twisted, acute or obtuse at the tip and with white or brownish shreds at the base. Flowers are yellowish and stalkless. Male flowers are clustered along a small panicle (to 12 cm long); female flowers are clustered in a solitary head on a stalk up to 3 cm long.

HABITAT Open forest, on dry sandy soils.

DISTRIBUTION Widespread and common from coast to mountains. Recordings include Bell, Mt Victoria, Agnes Banks, Hilltop, Girrakool, Royal and Macquarie Pass National Parks. Extending from about Tea Gardens (north coast) to south coast and southern tablelands.

FLOWERING July–October.

1352. Lomandra glauca (male)

Lomandra multiflora

Many-flowered Mat-rush **Plate 1353**

L. multiflora is characterised by its abundant yellow male flowers which are borne in clusters on a branched flower stem up to 30 cm tall, each flower on a stalk about 10 mm long. Female flowers are stalkless and the inflorescence is smaller. Firm, rigid, rather thick leaves emerge in thick tufts from the rhizome. They are 40 to 70 cm long, 2 to 4 mm wide, rounded at the tip and with few, if any, shreds at the base.

HABITAT Open forest, on dry sandy soils.

DISTRIBUTION Widespread from coast to mountains. Recordings include Clarence, Agnes Banks, Menai, Brisbane Water and Ku-ring-gai Chase National Parks. Occurring throughout NSW, except far western plains; also NT, Qld, Vic and SA.

FLOWERING September–January.

1353. Lomandra multiflora

Lomandra fluviatilis

Plate 1354

This plant is regarded as rare and at risk because of its small populations and its restricted habitat preference. It grows in or near the rocky beds of larger freshwater streams. Plants usually form large tussocks in the cracks of rocks. Leaves are up to 50 cm long, 1 to 3 mm wide, with edges slightly inrolled, toothed at the tip and with shiny brown shreds at the base. Flowers are yellow, with the male flowers stalkless and borne in clusters in a narrow panicle much shorter than the leaves (5 to 17 cm long). The female inflorescence is smaller. Fine spiny bracts are at the base of each flower cluster.

HABITAT Creek beds, creek banks and shaded lower hillsides.

DISTRIBUTION Restricted to coastal zone around Sydney. Recordings include Heathcote Creek, Woronora River, Nepean River (e.g., Bents Basin), Flat Rock Creek (Royal National Park) and Colo River.

FLOWERING January–March, but irregular.

1354. Lomandra fluviatilis

Lomandra montana

Plate 1355

This is a localised plant from rocky creeks in the Blue Mountains. Its leaves are up to 50 cm long, 2 to 4 mm wide, glossy green, flat or slightly concave, with irregular teeth at the tip and a few dull shreds at the base. Male flowers are stalkless and borne in clusters along a compact spike 5 to 8 cm long; the spike is at the end of a flattened stalk. The female inflorescence is a little smaller.

HABITAT Rocky creeks, especially near waterfalls.

DISTRIBUTION Restricted to the Blue Mountains. Recordings include Lawson, Wentworth Falls, Grose River, Mt Wilson and Bowens Creek.

FLOWERING November–December.

1355. Lomandra montana

Philesiaceae

The Philesiaceae consists of 6 genera with only a dozen species, formerly included in the Liliaceae. In one modern classification the group is included in the Smilacaceae and in another it is divided into two families, the Philesiaceae and the Geitonoplesiaceae. It sometimes includes the Australian genus *Petermannia* which otherwise is separated into the monotypic family, Petermanniaceae.

The Philesiaceae occurs in the southern hemisphere in temperate South America, southern Africa, Australia, New Guinea and the Pacific Islands, and consists of shrubs and wiry climbers with a rhizomatous rootstock and oblong or ovate leaves with well-marked parallel veins. The flowers occur in clusters or cymes and consist of 6 perianth segments in 2 whorls of 3, 6 stamens and a superior ovary of 3 fused carpels surmounted by a capitate or 3-lobed style.

Two genera with 1 species each occur in eastern Australia and islands to the north and east.

Eustrephus

Wiry climber, with alternate leaves. Flowers with 6 perianth segments, the inner 3 fringed, and yellow stamens united into a tube. Fruit an orange berry.

Eustrephus latifolius

Wombat Berry **Plate 1356**

A hairless, branching climber, having no prickles or tendrils. It may climb to several metres high. Leaves are 5 to 10 cm long, glossy green, without a distinctive midvein, but with several closely spaced longitudinal veins. Leaves are spaced well apart along the stem and taper to a twisted base without a distinct stalk. Leaf shape is very variable, with 2 varieties recognised—the broad lanceolate form (var. *latifolius*) and the narrow linear form (var. *angustifolius*). Flowers are white (sometimes pinkish or mauve), with the inner 3 petals fringed and 6 yellow stamens. Flowers hang in groups of 4 to 6 on slender stalks. Fruit is a pendulous, globular, bright-orange berry.

HABITAT Shaded, moist forests.

DISTRIBUTION Widespread from coast to mountains. Also recorded south coast, north coast, northern tablelands, western slopes, Qld, Vic, New Guinea and New Caledonia.

FLOWERING October–November.

Geitonoplesium

Wiry climber, with alternate leaves. Flowers lacking the fringed petals of *Eustrephus*. Stamens yellow and free. Fruit a blue-black berry.

1356. Eustrephus latifolius

1357. Geitonoplesium cymosum

Geitonoplesium cymosum

Scrambling Lily **Plate 1357**

A hairless, branching climber, without prickles or tendrils. It may climb to several metres high. Leaves are alternate, narrow, 5 to 8 cm long, shiny above, with a distinct midvein and several closely spaced longitudinal veins. Leaf bases are asymmetrical. Leaf stalks are very short and twisted so that the surfaces of the leaves in a group are all in the one plane. Flowers are white, often with red-brown sepals. Berries are blue-black, about 6 mm across.

HABITAT Shaded, moist forests.

DISTRIBUTION Widespread from coast to mountains. Also recorded south coast, north coast, northern tablelands, Qld, Vic, Malesia and New Caledonia.

FLOWERING October–December.

Smilacaceae

This is another small group of plants at one time included in the Liliaceae but now regarded as a separate family. It differs from the Liliaceae in its woody climbing habit, its leathery 3 or 5-nerved leaves with broad blades and fine reticulate venation, its stipular tendrils and its unisexual flowers (bisexual in *Ripogonum*) with the sexes borne on separate plants. The family is widely distributed in the tropics and subtropics and consists of 4 genera with about 370 species, most of which belong to the genus *Smilax*. The flavouring and medicinal sarsaparilla is obtained from several species of *Smilax* native to the West Indies and Central America.

Two genera occur in Australia, *Smilax* with 3 Australian species and *Ripogonum* with five. Two species of each occur in the Sydney area in moist forest or rainforest.

Smilax Sarsaparilla

Climbers with alternate, hard, strongly veined leaves; a pair of tendrils present at the base of many leaves. Flowers unisexual (with male and female flowers on different plants) in an axillary umbel. Fruit consists of globular black berries, which hang like grapes on thin stalks.

1358. Smilax glyciphylla

Smilax glyciphylla
Sweet Sarsaparilla **Plate 1358**

A common climber in sheltered situations. Its leaves are alternate, ovate to lanceolate, 4 to 6 cm long, green above, ash-grey below and have 3 prominent longitudinal veins. Unlike *S. australis*, its stems have no prickles. Flowers are borne in umbels which hang on slender stalks. Petals are greenish-white and spreading. The fruit is a hanging cluster of black berries, each about 7 mm across. Young leaves and the flesh of the berries have a sweet taste.

HABITAT Common in sheltered forested gullies, but also found in more open, less protected sites.

DISTRIBUTION Coastal zone and nearby plateau areas (e.g., Colo, Wollemi). Extending from south coast to north coast and Qld; also northern tablelands.

FLOWERING October–December.

1359. Smilax australis

Smilax australis
Lawyer Vine, Austral Sarsaparilla **Plate 1359**

A common climber, with tough stems bearing small prickles. Its leaves are alternate, leathery, oblong to broad-elliptic, 5 to 10 cm long, green on both surfaces and have 5 prominent longitudinal veins with conspicuous finer cross-veins. Flowers are borne in umbels which hang on slender stalks. Petals are greenish white and spreading. The fruit is a hanging cluster of black berries each about 7 mm across.

HABITAT In and near rainforest or shaded gullies.

DISTRIBUTION Widespread from coast to mountains. Occurring all along coast and tablelands divisions; also central western slopes, Lord Howe Island, Qld, Vic, NT and WA.

FLOWERING September.

Ripogonum Supplejack

Climbers with leathery leaves, opposite or nearly so; lacking the pair of tendrils present at the base of the leaves of *Smilax*. Flowers bisexual, borne in racemes or spikes. Fruit a red globular berry.

1360. Ripogonum fawcettianum

Ripogonum fawcettianum
Small Supplejack **Plate 1360**

This is a shrubby climber with branches and leaf stalks covered with short soft pale or brown hairs. Young shoots are often covered with short soft pale yellow hairs. Branches are usually without prickles, but may be sparsely prickly. Leaves are opposite, narrow-ovate, 5 to 9 cm long, rounded or heart shaped at the base, and tapering to a point at the tip. Leaf stalks are stout and rather short. Flowers are white and arranged in racemes or spikes from the leaf axils.

HABITAT Rainforest.

DISTRIBUTION Coastal zone north of Sydney (e.g., Bouddi National Park, Strickland State Forest). Extending along north coast to McPherson Range (Qld).

FLOWERING October–November.

Ripogonum album
White Supplejack **Plate 1361**

The distribution of *R. album* overlaps that of *R. fawcettianum* and, as they are both shrubby rainforest climbers with leathery opposite leaves, one could be confused with the other. However, *R. album* has hairless branches (sometimes young shoots softly hairy), and the larger branches usually bear prickles. Leaves are ovate, 8 to 12 cm long, tapering at both ends (not heart shaped at the base as are the leaves of *R. fawcettianum*). Leaf stalks are short, stout and

1361. Ripogonum album

often twisted. Flowers are greenish white and arranged in racemes or spikes from the leaf axils.

HABITAT Rainforest.

DISTRIBUTION Coastal zone (e.g., Hacking River and Burning Palms in Royal National Park). Extending along coast from East Gippsland (Vic) to Cape York (Qld). Also New Guinea.

FLOWERING September–October.

Dioscoreaceae

The Dioscoreaceae consists of about 6 genera with about 630 species of which about 600 belong to the genus *Dioscorea*. It is a family of mostly tropical climbers resembling the Smilacaceae but differing from it in a number of respects. In both families the thin aerial stems develop from a tuber or rhizome and carry leaves with several conspicuous longitudinal nerves or veins with fine reticulate venation between. Both bear tiny unisexual flowers in axillary inflorescences with the sexes on different plants. The Dioscoreaceae has fine hastate to cordate leaves unlike the generally thick leathery leaves of Smilacaceae and lacks the stipular tendrils and climbing hooks present in the latter. The 2 differ also in flowers and fruit. The Dioscoreaceae have flowers with an inferior ovary developing (in *Dioscorea*) into a 3-angled capsule whereas the Smilacaceae have flowers with a superior ovary developing into a berry with 1 to 3 seeds.

Dioscorea is the source of yams which are the tuberous roots. About 60 species are cultivated as a staple food in South-east Asia, Africa and Central and South America. The starchy tubers usually contain bitter or poisonous alkaloids which must be destroyed or removed by boiling, baking or washing. Four species occurring in Australia were eaten by Aborigines, sometimes with ceremonial preparation, but in the case of *D. transversa*, mostly without cooking or special preparation.

One species, *D. transversa* extends from the north in rainforests to Stanwell Park, south of Sydney.

1362. Dioscorea transversa

Dioscorea

Dioscorea transversa

Native Yam **Plate 1362**

Native Yam is a slender hairless climber with stems several metres long. Leaves are alternate, 6 to 10 cm long, spear shaped, heart shaped at the base, with 5 to 7 prominent longitudinal veins. Leaf stalks are long, slender and have a swelling where they join the branch. Flowers are very small, with male flowers arranged along a spike 3 to 5 cm long and female flowers in interrupted racemes 5 to 10 cm long. Fruit is a conspicuous 3-winged capsule, about 3 cm wide, brown and rigid when mature.

HABITAT Rainforest.

DISTRIBUTION Coastal zone, north from near Stanwell Park. Extending along north coast and northern tablelands to Qld and NT.

FLOWERING September–October.

Agavaceae

The Agavaceae is a heterogeneous group of plants made up of genera formerly included in both the Liliaceae and the Amaryllidaceae. It consists of mostly robust plants with large fleshy fibrous leaves arising directly from a rhizomatous base or at the top of a short straight or branched trunk. The inflorescence is a raceme or panicle and is often large, rising well above the leaves on a stout scape.

The group has a mainly tropical and subtropical distribution, with many of the plants adapted to arid conditions. A number such as *Agave, Yucca, Dracaena, Sansevieria* and *Cordyline* with a striking appearance are frequently used as ornamentals. Others are economically important as a source of fibre (sisal) for ropes, cords and matting.

Three genera are native in northern and eastern Australia, *Doryanthes* (3 species), *Cordyline* (3 species) and *Dracaena* (1 species). One species of *Doryanthes* is common in parts of the Sydney area, and 1 species of *Cordyline* occurs in rainforests from Wheeny Creek northwards.

1363. Cordyline stricta

Cordyline

Cordyline stricta

Narrow-leaf Palm Lily **Plate 1363**

This is an erect palm-like shrub to 2 m tall (in Sydney area), with a stem 2.4 cm thick. Leaves are crowded near the top of the stem; they are strap-like and up to 80 cm long. Flowers are purplish blue, with petals about 5 mm long and arranged in a hanging panicle which may be 1 m long. Fruit is a blue-black berry, somewhat succulent when young, but becoming dry when mature.

HABITAT Rainforest and sheltered gullies.

DISTRIBUTION Gullies near Mountain Lagoon and Kurrajong (e.g., Cabbage Tree Creek and Wheeny Creek) and north of the Colo and Hawkesbury Rivers. Extending along north coast to Qld.

FLOWERING November–January.

Doryanthes

1364. Doryanthes excelsa

Doryanthes excelsa

Gymea Lily, Flame Lily, Spear Lily **Plate 1364**

Dense clusters of dark red flowers on the top of a long stem and its clumps of broad pale-green leaves make Gymea Lily one of the most conspicuous of our native plants. Leaves are fibrous, about 1 m long and up to 10 cm wide. The stem may rise to 4 m and bears the compact head-like panicle of red flowers. The 6 perianth segments ('petals') are 10 to 12 cm long, spreading when fully in flower and surrounded by deep-red bracts. The style is 10 cm long and red; filaments are 5 to 6 cm long, topped by anthers 3.5 to 4 cm long, opening in 2 slits which expand turning the anthers inside out. They bear green pollen. The fruit is a brown-black capsule.

HABITAT Open forest, on deep sandy soils.

DISTRIBUTION Coastal zone. Common south of Georges River to Wollongong (e.g., Heathcote and Royal National Park). Absent between Georges River and Broken Bay, but occurring north of Broken Bay to about Wyong (e.g., Calga, Brisbane Water National Park). Also near Port Stephens.

FLOWERING August–November.

Philydraceae

This is a small, somewhat isolated family of only 6 species in 4 genera distributed in south-east Asia, New Guinea, and Australia. The 1 species from the genus *Philydrum* is fairly common in the Sydney area in shallow ponds and waterways with restricted flow. It occurs from Victoria to northern Australia. Two other genera, *Helmholtzia* and *Orthothylax* occur in south-eastern Queensland, and the fourth, *Philydrella*, in south-western Western Australia.

The closest relationships of the Philydraceae are with the water hyacinth family, Pontederiaceae.

Philydrum

Philydrum lanuginosum

Woolly Frogmouth **Plate 1365**

This upright plant is often a conspicuous feature of coastal freshwater pools and wet areas beside roads. Leaves are 20 to 60 cm long, with sheathing bases, mostly crowded at the base of the flower stem but with smaller leaves higher on the stem. Leaves often have a reddish appearance. Flower stems are up to 1 m tall, woolly-hairy. Flowers are yellow, stalkless and surrounded by a prominent pointed bract. Flowers have 4 perianth segments, with 2 large petal-like segments resembling the open mouth of a Frogmouth. There is one perfect stamen and 2 staminodes. Fruit is a capsule, about 12 mm long, opening into 3 valves.

HABITAT Shallow ponds and other freshwater sites.

DISTRIBUTION Chiefly coastal from Vic to Qld and NT, but also recorded from central and northern tablelands. Extending to Malaysia and south China.

FLOWERING November–May.

1365. *Philydrum lanuginosum*

Amaryllidaceae

The Amaryllidaceae is one of the larger families in the order Liliales. Because relationships within the order are incompletely understood, and there is disagreement as to the relative significance of characters used to divide the order into families, there are different opinions regarding its size and composition. If, as here, the family is limited to plants with a bulb and flowers with an inferior ovary borne in an umbel (sometimes reduced to a single flower) subtended by 1 or more bracts, the family consists of about 65 genera and 800 species. It includes many fine ornamentals such as daffodils and jonquils *(Narcissus)*, snowdrops *(Galanthus)* and lilies from various genera.

The family occurs in most of the warmer countries of the world and is widespread in Australia, though never common. There are 3 native genera, *Crinum* (11 species), *Calostemma* (4 species) and *Eurycles* (2 species). One species of *Crinum* occurs infrequently in moist situations in the Sydney area.

Crinum

Crinum pedunculatum

Swamp Lily **Plate 1366**

A robust plant, with broad leaves, 70 cm or more tall. Thick stems grow from an underground bulb and numerous leaves up to 15 cm wide grow from near the base. The leafless flowering stem rises from the lower leaves and bears 8 to 25 white flowers in terminal umbels. Flowers have 6 free narrow segments and 6 stamens and each flower is stalked. The fruit is a capsule which splits to release many large seeds.

HABITAT Swampy aras.

DISTRIBUTION Coastal zone, but uncommon. Extending from south coast to north coast and Qld; also SA and Lord Howe Island.

FLOWERING November–December.

1366. *Crinum pedunculatum*

Haemodoraceae

The Haemodoraceae consists of 2 quite distinct tribes, the Haemodoreae with flowers having 6 separate perianth segments in 2 whorls and either 3 or 6 stamens, and Conostyleae in which the 6 segments are fused into a woolly tube with 6 stamens attached to the inner wall.

The Haemodoreae, with about 60 species, occurs in Africa, the Americas, Malesia and Australia (23 species, 2 genera) while the Conostyleae, with about 40 species, is largely confined to south-western Australia. The Conostyleae contains several genera such as *Anigosanthos* (kangaroo paw), *Macropidia* (black kangaroo paw), *Blancoa* (red bugle) and *Conostylis* (cone flower) of considerable horticultural interest. The sole eastern Australian representative of this family, *Haemodorum*, shares a family resemblance but is much less attractive.

Of the 20 or so families in the order Liliales, the Haemodoraceae appears to be most closely related to the Amaryllidaceae, Alstoemeriaceae and the Hypoxidaceae.

1367. *Haemodorum planifolium*

Haemodorum Bloodroot, Blood-lily

Haemodorum planifolium

Strap-leaf Bloodroot **Plate 1367**

An erect perennial with tufts of strap-like, flat, mostly basal leaves, 30 to 40 cm long and 2 to 6 mm wide. The flat leaves separate this species from the closely related *H. corymbosum*, which has rounded leaves. Flowers are borne in crowded, branched clusters at the ends of erect, brown-black flower stalks about 60 cm tall. Flowers are black, with 6 segments arranged in 2 whorls, the outer slightly shorter than the inner. There are 3 stamens, with orange-yellow anthers. Fruit is a capsule, 8 to 12 mm across.

HABITAT Heath and woodland, in damp peaty to sandy soils.

DISTRIBUTION Widespread from coast to mountains. Occurring throughout coast and tablelands and north-western slopes; also Qld.

FLOWERING November–January.

1368. *Haemodorum corymbosum*

Haemodorum corymbosum

Rush-leaf Bloodroot **Plate 1368**

H. corymbosum is an erect perennial similar to *H. planifolium* but having cylindrical leaves sheathing at the base, up to 55 cm long and less than 2 mm wide. The flower structure, inflorescence and fruit are all as described for *H. planifolium.*

HABITAT Heath and woodland, in damp peaty to sandy soils.

DISTRIBUTION Widespread from coast to mountains. Extending to south coast and southern tablelands.

FLOWERING November–January.

Hypoxidaceae

The Hypoxidaceae consists of 7 genera formerly included in the Amaryllidaceae, but now generally regarded as separate. Whereas the former develop from a corm (a swollen underground stem covered with thin scaly leaves) the latter have a bulb (an onion-like structure consisting of layers of thick fleshy leaf bases). The 2 groups also differ in the inflorescence; in Hypoxidaceae the flowers are solitary or in short racemes of 2 or 3, in Amaryllidaceae they are in terminal umbels subtended by 2 or more membranous bracts or spathes. The 2 have a similar floral structure differing from the Liliaceae in the position of the ovary—inferior rather than superior as in Liliaceae. They differ from the Iridaceae in having 6 stamens instead of three.

Hypoxis

The genus *Hypoxis*, with about 100 species, is the largest in the family. Five species occur in Australia, 1 in the Sydney area.

1369. Hypoxis hygrometrica

Hypoxis hygrometrica

Golden Star, Golden Weather-glass **Plate 1369**

Golden Star is a small herb seldom more than 20 cm tall. Grass-like leaves rise from a short tuber-like rhizome (a corm). These leaves are 10 to 18 cm long, about 1 mm wide, grooved above and sprinkled with long silky hairs. Flowers are yellow, solitary or in short racemes. The 3 petals and 3 sepals are similar, 4 to 7 mm long, and spread like a golden star. The backs of the sepals are green; small tufts of white hairs are usually at the tip of the sepals. There are 6 stamens with yellow anthers shaped like an arrowhead. Fruit is a globular capsule. This species may be confused with *Tricoryne simplex*, but that plant has flowers in umbels, filaments with a tuft of hairs and its perianth spirally twists after flowering.

HABITAT Grassland and swampy areas.

DISTRIBUTION Widespread from coast to mountains. Occurring in all areas of NSW except far western plains; also Qld, Vic and Tas.

FLOWERING November–March.

1370. Thelymitra ixioides

Orchidaceae

The Orchidaceae is the culmination of the insect-pollinated monocotyledons, justly prized for its inexhaustible variety of floral forms developed in many cases to ensure cross-pollination by specific insects. It is considered to be the largest family of flowering plants, variously estimated to contain 600 to 800 genera and 15 000 to 30 000 species, with possibly a further 60 000 hybrid cultivars produced by orchid growers.

Orchid flowers are often spectacularly beautiful, occasionally comical, and sometimes quite bizarre. They may be 40 cm from tip to tip or so small as to require magnification to be recognised. Some flowers last for months, others less than a day. They may be borne singly, or in huge sprays, or even underground. To attract particular pollinators the colour may be light or dark, vivid or dull, delicately shaded or blotchy, and they may be pleasantly perfumed, stinking or odourless (or apparently so). Some emit attractants to lure male bees or wasps then use deception to arouse them into sexual activity with the flower. Others employ traps or blast the intruder with pollen. Orchid seed is so fine that a single capsule can contain up to several million seeds, yet because of very specific growth requirements few may germinate and develop beyond the protocorm stage.

Orchid plants may be terrestrial, epiphytic, or saprophytic (obtaining food from dead or decaying organic matter), but never parasitic. They are found in all parts of the world except the very driest, coldest and highest places. Their greatest profusion is in the continually wet cloud forests in mountainous areas of the tropics. Often the range of species and genera is quite restricted in these areas, and consequently the orchid population is very different from one region to another. Numbers are far fewer in flatter country or areas with alternate rainy and dry seasons. In lowland rainforests orchids mostly live high in the trees or on the edges of the forest, particularly along river banks. The number of terrestrials on the dark forest floor is small, although some species may adopt either habit according to the conditions. The proportion of epiphytes declines in subtropical and temperate areas, and the terrestrial habit is characteristic of temperate zones. By spending cold winters in a dormant state underground, terrestrial orchids can withstand high altitudes and latitudes. A few are even found inside the Arctic Circle. Because requirements for moisture and temperature are less stringent on the ground, terrestrial orchids occupy a wide variety of habitats, and individual species sometimes have an astonishing range.

The orchid family is distinguished from other families mainly by its flowers. These are nearly always bisexual and have 3 sepals and 3 petals with the stamens, stigma and style reduced in number and fused into a structure called the column. The sepals are often similar to each other, or 2 lateral sepals may be similar but different to the third or dorsal sepal. Usually the petals are dissimilar, with 2 lateral petals differing from the labellum situated opposite the column. The labellum may be smaller than the other petals, but usually it is larger, often strikingly coloured, highly modified and ornamented with calli, hairs and fringed margins. In bud, the labellum is uppermost and points upward, but subsequently the pedicel (stalk) twists so that at flowering the labellum points downward to provide a landing platform for insect pollinators. Nectaries are present in various positions on sepals and petals.

In the majority of orchids there is a single anther at the top of the column separated from the stigma below it by a sterile lobe, the rostellum. Usually the pollen grains are bound together by sticky threads into masses called pollinia. At its simplest, pollen transfer occurs by accidental removal of pollinia by an insect searching for nectar at the base of the flower. Often the insect is partially trapped by the flower and forced to exit via a path wherein sticky fluid is brushed onto its back by the rostellum. The pollinia then become attached to the sticky material and are carried to the next flower. On entering the next flower the insect transfers the pollinia to the stigmatic surface. A variety of other ingenious devices is employed by different species to achieve the same end. In a small percentage of species self-pollination is the rule, or may occur in the absence of pollination by an external agent. Deposition of pollen on the stigma triggers changes in the flower, often within hours, but fertilisation may take anything from a day or 2 to 6 months or more. Ripening of the seed may then take another 2 to 18 months.

Mature seed consists of relatively few undifferentiated cells enclosed in a thin seed coat. If on dispersal it lands on a suitable surface, germination begins, the seed coat is broken and a tiny green protocorm is formed. However, because the seed lacks endosperm (food storage tissue), development is arrested until certain essential nutrients are supplied. This is achieved in nature with the aid of a fungus which provides enzymes to break down starches and other material, and possibly also vitamins, hormones and other growth factors. Growth resumes after the protocorm is infected by the fungus, and an apical meristem is formed which produces leaves. Finally roots develop and a fungus becomes localised in the roots, establishing a mycorrhizal association.

The relationships of the Orchidaceae with other families are not at all clear, although the most likely appear to be with the Burmanniaceae or the Liliaceae. Relationships within the family are also difficult to decipher owing to the rapid diversification and adaptive radiation of the family and numerous examples of parallel development from groups now extinct. In these circumstances intrafamily classification is based almost solely on the morphology of present-day genera, and although there is general agreement on the major subdivisions there is some disagreement as to the levels at which various groupings should be recognised.

The most primitive group, regarded either as a subfamily or as a separate family, Apostasiaceae, has regular flowers with a column consisting of 2 or 3 fertile anthers and a slender terminal style. It is restricted to the Indo-Malaysian region with 1 species in northern Queensland. Another small group, sometimes regarded as a separate family, Cypripediaceae, includes the slipper orchids with irregular flowers and a column of 2 fertile anthers, a conspicuous staminode and a thick terminal style. It is widespread but absent from Australia. Separation of these groups leaves over 99% of species which are regarded either as the subfamily Orchidoideae with 4 tribes and numerous subtribes, or as the family Orchidaceae with 2 or 3 subfamilies, or variously divided into tribes and subtribes. All the major subgroups of this subfamily or family are represented in Australia.

Considering its size, Australia is only modestly endowed with orchids. Approximately 90 genera and 550 to 750 species occur in the country, mainly distributed around the coastal fringe. Epiphytes are common in northern Queensland rainforests, consisting of a few large widespread genera such as *Dendrobium, Bulbophyllum* and *Liparis* and numerous relatively unknown tropical or endemic genera represented by only 1 or 2 species each. The number of epiphytes declines towards the south, with 23 species in the Sydney area, 5 in Victoria and 2 in Tasmania.

Approximately half the genera and three-quarters of the species in Australia are terrestrial and saprophytic. The largest terrestrial genera *(Pterostylis, Prasophyllum, Caladenia, Diuris* and *Thelymitra)* are more or less confined to Australia and New Zealand, with representation in the latter country almost certainly the result of dispersal of seed by wind from eastern Australia. These genera are widespread throughout the continent in woodland, open forest and heath communities, and relatively abundant in south-western and south-eastern Australia. As with the epiphytes, many of the northern Queensland species belong to tropical genera with restricted Australian representation. Approximately 35 genera and 190 species occur in the Sydney area; of these, 28 genera and 165 species are terrestrial.

Thelymitra

Terrestrial glabrous herbs. Stems with 1 long narrow channelled stem-clasping leaf at the base and 1 or 2 smaller bracts above. Flowers in a terminal raceme, rarely solitary, opening only in sunny weather. Perianth almost regular. Column short and broad with prominent wings, the wings bearing 2 lateral lobes often terminating in a tuft of hairs.

Thelymitra ixioides

Dotted Sun-orchid **Plate 1370**

The commonest and most widespread sun orchid in the area, this species varies from an erect robust plant 50 cm or more tall with numerous large flowers 30 to 50 cm in diameter to a much shorter slender plant with a few smaller flowers. The flowers are blue or pinkish blue and are usually marked with small dark blue spots on the upper 3 segments. The lateral lobes of the column extend upward and forward forming 2 arms ending in white hair tufts. Between these are 3 shorter mainly yellow lobes forming a hood above the anther. The short central lobe has 2 or 3 rows of calli which form a crest at the back of the column.

HABITAT Common in coastal heath and open forest, favouring damp sandy ground.

DISTRIBUTION Coast, tablelands and western slopes; also Qld, Vic, Tas, SA, WA, NZ and New Caledonia.

FLOWERING August–October on the coast, as late as December on the highlands.

1371. Thelymitra pauciflora

Thelymitra pauciflora

Slender Sun-orchid **Plate 1371**

T. pauciflora varies from a slender plant with 2 or 3 small flowers to a larger plant about 30 cm high with numerous flowers. These vary in colour from whitish to pink, blue or purple. It is the commonest of 3 species in the area in which the hooded column is dark towards the top except for a cleft at the front which is yellow. Differences between the 3 are by no means clear-cut. The column of *T. pauciflora* has a deeply cleft mid-lobe and erect lateral lobes with terminal whitish hair tufts pointing upward.

1372. Thelymitra nuda

T. nuda (Plate 1372) is similar with light-blue flowers having a deeply cleft column and lateral lobes bearing terminal whitish hair tufts pointing horizontally.

The third species, *T. aristata*, is very variable in habit, and is distinguished by having fragrant, usually mauve, flowers, a shallow cleft or notch on the column mid-lobe and more or less horizontal lateral lobes with whitish hair tufts on the upper side. *T. pauciflora* and *T. nuda* are both shy flowerers, opening only in warm sunny weather.

HABITAT In sheltered moist open forest and heath.

DISTRIBUTION Widespread on coast and tablelands (except northern tablelands); all States (but not NT) and NZ.

FLOWERING October–November.

1373. Thelymitra venosa

Thelymitra venosa

Veined Sun-orchid **Plate 1373**

This is a most attractive species found in very wet habitats, mostly on the tablelands. It has a number of characteristic features which allow easy identification. It has 1 to about 10 large blue or mauve flowers in a large loose terminal raceme on a long erect or curved stem. The flowers are about 25 mm in diameter with conspicuous dark-blue veins on the segments. The yellow column is winged, but not hooded over the anther; the mid-lobe short with small crowded calli, the 2 lateral lobes forming coiled wings on each side above the anther. Unlike those of other blue-flowered species in the area, the wings lack hair tufts.

HABITAT Prefers wet but sunny habitats such as mountain bogs and seepage areas or wet mossy banks and walls below dripping cliff faces.

DISTRIBUTION Chiefly a tablelands species, but also recorded on central coast; occurring Vic, Tas, SA (where restricted) and NZ. Locations include Porters Pass and Centennial Glen, Blackheath.

FLOWERING November–December.

1374. Thelymitra carnea

Thelymitra carnea

Salmon Sun-orchid, Pink Sun-orchid **Plate 1374**

Although the commonest of 5 species in the area with small pinkish flowers, this orchid is rarely noticed owing to its small size and its habit of opening only around midday in warm sunny still weather. It is a slender plant, usually about 10 to 20 cm tall with a flexuose or straight stem bearing 2 small stem bracts and having an erect, narrow, channelled, terete leaf at the base. It bears 1 to 3 or 4 salmon-pink flowers about 10 mm in diameter, each with a pinkish column having a yellow 3-lobed apex. The lobes are often shorter than the anther, the 2 lateral ones smooth on the outside, finely toothed in front and devoid of hair tufts. This species is easily confused with the rather rare (in the area) *T. rubra* which has densely tuberculate lateral lobes. It appears to hybridise with *T. ixioides* when growing in association.

HABITAT Damp clayey or sandy soil in open scrub, mainly on the coast, sometimes forming small colonies.

DISTRIBUTION Coast and tablelands (except northern tablelands); also central western slopes, Tas, SA, Vic and NZ.

FLOWERING September–November.

Thelymitra circumsepta **Plate 1375**

This is a rare orchid found in only a few places on the tablelands of New South Wales. It is, however, easily recognised. The leaf is erect, thick, leathery, channelled and very broad at the base. It encloses a robust stem 50 or 60 cm in height which bears 1 or 2 stem bracts and a narrow spike of 3 to 10, or more, relatively small flowers. These vary from pale to dark purple and open very shyly. Usually only 1 or 2 open at a time, and then never widely. A little probing with a twig reveals 2 upward-pointing tufts of yellow hairs on the lateral lobes of the column together with a smaller third tuft lower down in front of the anther. The column wings form an open hood above with a pinkish-yellow fringe.

HABITAT Grassy open scrub and periodically swampy ground.

DISTRIBUTION Recorded only from the Mt Irvine–Clarence district, the Fitzroy Falls–Robertson district and the New England National Park.

FLOWERING December.

1375. Thelymitra circumsepta

Calochilus

A genus of 11 species, 9 in Australia, 4 in the Sydney district. Terrestrial glabrous herbs with erect stems and a single long narrow channelled basal leaf. Flowers few to many in a terminal raceme, each subtended by a small bract. Flowers green, reddish or purplish. Dorsal sepal hooded. Labellum large and conspicuously bearded with metallic-lustrous hairs, frequently ending in a ribbony filament. Column short, broad, winged.

1376. Calochilus campestris

Calochilus campestris

Copper Beard-orchid **Plate 1376**

This is a robust plant, usually 30 to 50 cm high, with a thick deeply channelled leaf and 2 green or coppery stem bracts. Flowers are in a raceme of about 5 to 15. They are yellowish green with red-brown or purplish markings and a labellum that is shorter than in other local species. It is about 15 mm long, triangular, sometimes with a short strap-like tip, rather sparsely beset with coppery red hairs except for the upper basal part which has smooth raised metallic blue or purplish lines. The column is short and very wide with dark glands in both lower corners. The glands are not connected by a coloured ridge as in *C. robertsonii.*

Habitat Moist sandy soil in open forest and grassland.

Distribution Coast, tablelands and western slopes; also Qld, Vic, Tas, SA and NZ.

Flowering September–October.

1377. Calochilus robertsonii

Calochilus robertsonii

Purplish Beard-orchid **Plate 1377**

This is a fairly common species with 2 to 10 flowers in a terminal raceme. It is mostly 20 to 40 cm tall with a long narrow channelled leaf and 2 to 4 stem bracts. With the labellum often exceeding 20 mm in length, the flowers are larger than those of *C. campestris.* The labellum hairs are long, lustrous and purple, becoming shorter, thicker and gland-like at the base and sometimes bronze or greyish towards the bare tip which is relatively short, crooked and strap-like. Two dark glands at the base of the column are connected by a transverse raised reddish ridge.

Habitat Moist sandy soil in open forest and grassland.

Distribution Coast, tablelands and western slopes. Occurring in all other States (but not NT) and NZ. Localised in Qld, SA and WA.

Flowering September–October.

1378. Calochilus paludosus

Calochilus paludosus

Red Beard-orchid **Plate 1378**

This species resembles *C. robertsonii* but is more variable in size and habit. It has about 2 to 8 flowers, occasionally more, a relatively short slender erect leaf and 1 to 2 stem bracts. The flowers are green with red stripes. The labellum is broad at the base and tapers to a long, narrow, often straight, bare apex. It is densely beset with lustrous red hairs which become progressively shorter towards the base, eventually shortening to crowded calli at the base. The column is short, broadly winged and without basal glands. The stigma, however, is prominent and marked with red or purple along its lower margin.

Habitat Moist sandy soil in heath, open forest and grassland.

Distribution Coastal zone, central tablelands and western slopes. Occurring in all States (except WA) and NZ.

Flowering September–November.

Diuris

A genus of about 45 species, all but 1 in Australia, 10 in the Sydney district. Terrestrial glabrous herbs with erect stems and few to many linear basal leaves passing into sheathing bracts on the stem. Flowers 1 to about 10 in a terminal raceme. Perianth yellow, rarely mauve, lilac or white, often blotched with brown or purple. Dorsal sepal broad at the base, erect above the column; lateral sepals narrow linear, downward pointing. Petals broad on narrow linear claws. Labellum deeply 3-lobed, the mid-lobe similar in size or larger than the others. Column short and broad with narrow wings.

Diuris maculata

Leopard Orchid, Spotted Doubletails **Plate 1379**

This species is common throughout the area, but variable in size and number of flowers. It is usually about 30 cm high with 2 or 3 basal leaves. The dark-stemmed flexuose raceme of 2 to 8 yellow flowers is blotched with dark brown, chiefly on the underside of the perianth. The dorsal sepal is short with a broad obtuse apex; the lateral sepals 14 to 16 mm long, are greenish brown and often crossed. The petals are of similar length and have an ovate yellow lamina on a narrow brown claw. The labellum is 3-lobed from above the base with the 2 spreading lateral lobes nearly as long as the central lobe which has a prominent raised ridge down the middle extending almost to the broad blunt apex.

Habitat Heath, scrub and open forest in a variety of conditions.

Distribution Widespread and common (except northern tablelands and drier western plains); also Qld, Vic, Tas, SA.

Flowering July–October.

1379. Diuris maculata

Diuris aurea

Golden Doubletails **Plate 1380**

This species is fairly common in the area and, although similar in shape to several others, can be recognised by its dull golden-yellow colour, differing from the clear light yellow of the others. It is usually a tall robust plant, 20 to 50 cm high with 2 erect, narrow, channelled leaves reaching about half the plant height. It has 2 to 4 flowers in a loose raceme, usually lightly marked with brown. The dorsal sepal is about 12 to 16 mm and is larger and more pointed than in *D. maculata.* The petals are broad lanceolate to almost round with rather short brown claws. The labellum is about as long as the dorsal sepal and is 3-lobed from near the base with the lateral lobes less than half as long as the mid-lobe. The central part of the mid-lobe has 2 short green or brown ridges with the space between them flexed into a keel and continuing to the broadly pointed apex.

Habitat Grassy slopes of forest and heathland or in the understorey of dry open forest.

Distribution Central and north coasts and central tablelands, extending to the western slopes in places; also Qld.

Flowering Mainly September–October.

1380. Diuris aurea

1382. Diuris punctata var. *alba*

1383. Orthoceras strictum

Diuris sulphurea

Tiger Orchid **Plate 1381**

The flowers of this species are a bright sulphur-yellow colour with dark brown blotches on the labellum and on either side of the dorsal sepal near the base. The dorsal sepal is about 20 mm long and tapers to a pointed apex. It may be either hooded over the column or bent back to expose it. Lateral petals have a short brown claw and an elliptical somewhat pointed yellow blade. Lateral sepals are stout and mostly dark brown. The labellum is 3-lobed from near the base with the lateral lobes nearly half as long as the mid-lobe. The labellum is diamond shaped with a central ridge from the base to the middle, continuing as a fold to the apex.

HABITAT Open forest and grassland.

DISTRIBUTION Coast, tablelands and western slopes; also Qld, Vic, Tas and SA.

FLOWERING August–November.

1381. Diuris sulphurea

Diuris punctata var. *alba* **Plate 1382**

D. punctata is widespread in eastern Australia and varies considerably in colour and form throughout its range. The standard form, known as var. *punctata*, has large mauve, lilac or purple flowers and is widespread, although many recordings for the Sydney district date from the last century or the early 1900s in localities where it is now unlikely to be found. A smaller former, known as var. *minor*, is recorded from the central coast. The white form, often streaked with purple and known as var. *alba*, is occasionally common in coastal heath and scrub between Gosford and Newcastle. It is readily recognised by its colour and by the straight green lateral sepals which are about twice the length of the other perianth segments.

HABITAT Moist sandy coastal heath and grassy areas in scrub.

DISTRIBUTION Central coast of NSW to Qld. Relatively common in the Munmorah State Recreation Area.

FLOWERING September.

Orthoceras

Orthoceras is a monotypic genus related to *Diuris*.

Orthoceras strictum

Horned Orchid **Plate 1383**

An erect herb 20 to 60 cm high with 2 to 5 narrow-linear erect sheathing leaves at the base passing into 1 or 2 stem bracts. The inflorescence consists of 2 to about 10 almost sessile green or brown flowers in a terminal raceme with each flower subtended by a large erect floral bract. The dorsal sepal is 10 to 12 mm long, acutely pointed and hooded over the column, the lateral sepals more than twice as long and narrow-linear to filiform. Petals are tiny and more or less hidden under the dorsal sepal. The labellum has 2 short lateral lobes and a longer mid-lobe about 10 mm long. It is generally brown with a yellow centre. This orchid is sometimes confused with *Lyperanthus suaveolens* which is similar in some respects but has long narrow sepals and petals and a relatively short labellum.

HABITAT Moist sandy soils in heath and scrub.

DISTRIBUTION Coast, tablelands and western slopes; also Qld, Vic, Tas, SA, New Guinea, NZ and New Caledonia.

FLOWERING December–January.

Microtis

1384. Microtis unifolia

Microtis unifolia

Onion-orchid **Plate 1384**

Although a number of local variants of this homely little orchid are sometimes afforded specific rank, the differences between them are slight and difficult to detect in the tiny flowers, hence one is drawn to the view that they are best regarded as representatives of a single variable species. Plants are entirely green with a soft stem loosely sheathed for some distance above the base by a solitary fleshy leaf which is free and terete in its upper part. The size varies from 10 to about 40 cm and the number of flowers in the spike from few to many. Flowers vary in size, but are about 5 mm long with a hooded dorsal sepal and short spreading lateral sepals. Petals are partly hidden by the dorsal sepal. The short sessile obtuse labellum has a number of raised callosities on the upper surface. The column is short and broad with 2 wings.

HABITAT In a variety of situations in open scrub and grassland, 1 or 2 together or in small colonies.

DISTRIBUTION Coast and tablelands, extending to the slopes and plains in the south. Found in all States (but not NT), NZ, Malesia and Asia as far as China.

FLOWERING September–December.

1385. Prasophyllum elatum

Prasophyllum

Terrestrial glabrous herbs with an erect stem partly sheathed by a solitary terete leaf. Flowers few to many in a spike, not inverted. Dorsal sepal lanceolate, concave. Lateral sepals similar in length, free or joined for part or most of their length. Petals shorter, narrower, forward-pointing. Labellum oblong to lanceolate, the margins crisped, toothed, fringed or smooth. Column short.

Prasophyllum orchids are often small with a narrow dull-coloured inflorescence and are therefore difficult to detect when growing among grass and other low vegetation. Also, many of the 40 or so species recorded from the Sydney district are rare, or possibly extinct in the area, consequently few are normally encountered. Flowers appear to be upside-down, with the dorsal sepal lowermost and 2 partly joined lateral sepals above and behind the labellum. The flowers are actually the right way up; most other orchid flowers are resupinate, i.e., turned through 180°.

Prasophyllum elatum

Tall Leek-orchid **Plate 1385**

Often growing in colonies, especially following a bush fire, this is easily the commonest and most noticeable of the leek orchids in the area. It is a robust plant, frequently reaching 100 cm in height. The stem varies in colour from pale green to greenish black and is enclosed on the lower part by a thick erect tapering leaf which is usually shorter than the inflorescence. The inflorescence is 10 to 25 cm long with numerous large sessile white, green and brown flowers densely arranged in a spike. The dorsal sepal is about 10 mm long. The lateral sepals are broad and joined for most of their length. They are erect and slightly recurved above the labellum which is also slightly recurved from about the middle. The upper labellum has a white undulate margin surrounding a prominent raised green callus plate. The column is readily seen between the labellum and the dorsal sepal.

HABITAT In a variety of habitats from moist heath to open forest.

DISTRIBUTION Mainly coastal in NSW, extending to central tablelands, slopes and parts of the western plains. All States (but not NT), though restricted to southern coastal areas in Qld and WA.

FLOWERING August–November, mainly September–October.

1386. Prasophyllum flavum

Prasophyllum flavum

Yellow Leek-orchid **Plate 1386**

This moderately robust species is widely spread in eastern Australia, though apparently uncommon throughout its range. It has a dark stem growing to about 60 cm high, enclosed near the base by a dark leaf with a very short free blade. The flower spike is crowded with numerous dull yellow or yellowish green flowers of moderate size. Lateral sepals are joined for most of their length and are 7 to 10 mm long, similar in length to the other perianth segments. This species is unusual in producing a fleshy tuberous rhizome rather than ovoid tubers. It is saprophytic and the rhizome harbours mycorrhizal fungi which assist in the assimilation of food.

HABITAT Mostly in dry forest understorey.

DISTRIBUTION Scattered on coast and tablelands from south-eastern Qld to Vic and Tas.

FLOWERING November–January.

1387. Prasophyllum patens

Prasophyllum patens

Broad-lip Leek-orchid **Plate 1387**

Although widespread, this orchid is infrequently seen owing to its sparse occurrence and its predominantly brownish-green colour which blends with a grassy background. It is a slender plant growing to about 35 cm high. The leaf, which is terete and of variable length, encloses the lower stem. The flowers are more or less sessile, often numerous and about 15 mm long. Their most characteristic features are the narrow, widely diverging lateral sepals at the top of the flower and the white-bordered labellum which is bent back about the middle, but not sharply so, and not to the extent that the upper part protrudes between the lateral sepals. The labellum has a raised green callus plate extending to about the bend. Petals are narrow, green with a brown band and shorter than the sepals.

HABITAT In a variety of habitats ranging from coastal heaths to drier open forest.

DISTRIBUTION Coast and Blue Mountains; also north coast, northern tablelands, Qld, Vic, Tas and SA.

FLOWERING October–December.

1388. Prasophyllum striatum

Prasophyllum striatum — Plate 1388

One of the smaller type of *Prasophyllum* orchids known as midge-orchids, this species is occasionally seen in autumn in damp sandy habitats. Although few-flowered and only about 10 to 15 cm high, it is readily recognised by the brown-striped petals and dorsal sepal on a green and white background. The lateral sepals are erect and are usually joined together. The white labellum is recurved with wavy margins. A short subulate leaf sheaths the stem to well above the middle.

HABITAT Damp sandy soils.

DISTRIBUTION Widespread on coast and central tablelands of NSW.

FLOWERING April–June.

1389. Prasophyllum brevilabre

Prasophyllum brevilabre

Short-lip Leek-orchid — **Plate 1389**

This is one of the more common, or more noticeable, of the leek-orchids. It is a slender plant, usually less than 20 cm high, with the stem sheathed below the inflorescence by a leaf with a short free blade. The inflorescence is fairly open with about 6 to 12 small flowers, though large specimens with up to 30 flowers may occur. The flowers are dark purplish brown, made conspicuous by a white labellum which is folded back upon itself giving the appearance of shortness. The margins of the upper part are undulate and crisped. Lateral sepals are usually joined for most of their length but are occasionally free. The petals are narrow and taper to a fine point.

HABITAT In a variety of habitats ranging from heath to woodland and forest.

DISTRIBUTION Coast to western slopes, more common on the tablelands. Widespread in eastern Australia from south-eastern Qld to Tas.

FLOWERING September–November.

1390. Prasophyllum archeri

Prasophyllum archeri

Variable Midge-orchid — **Plate 1390**

Over its extensive geographic range this slender orchid varies considerably in size as well as in colour and number of flowers. It is usually less than 20 cm high and leafless except for a floral bract just below the short inflorescence. This usually consists of 2 to 6 greenish flowers with purple or red markings on the petals and dorsal sepal and a purple or reddish labellum fringed with short hairs. The dorsal sepal below the flower is acutely pointed and about 5 mm long; the green lateral sepals are longer and narrower and widely spreading. In some Victorian locations the plant is much taller with 10 to 15 wholly dark purple flowers.

HABITAT Moist sandy soil in open eucalypt forest.

DISTRIBUTION Scattered on central coast, north coast and central tablelands. All states except WA and NT. Localised in Qld and SA.

FLOWERING December–April.

1391. Prasophyllum fimbriatum

Prasophyllum fimbriatum

Fringed Midge-orchid — **Plate 1391**

This is one of the midge-orchids with a ciliate or hairy labellum which flutters in the breeze, mimicking an insect hovering above the flower. This species reputedly also emits an attractant scent which entices and arouses male insects to attempt pseudo-copulation with the flower and thereby transfer pollen from one flower to another. This plant is usually about 15 to 30 cm high, sometimes more, with few to many flowers in a narrow open spike. Flowers are green on the lateral sepals, reddish and ciliate on the narrow dorsal sepal and on the small striped petals.

HABITAT Sandy soils.

DISTRIBUTION Coast and tablelands of NSW. Common near Sydney, but easily overlooked.

FLOWERING January–May.

1392. Prasophyllum woollsii

Prasophyllum woollsii **Plate 1392**

This species resembles several others in the area which are difficult to tell apart owing to their small size. It is a slender plant, 15 to 25 cm high, with a narrow leafy bract some distance below the dense spike of dark brownish red flowers. Flowers have a broad, acuminate curved dorsal sepal about 2 mm long. The lateral sepals are longer, narrower, acutely pointed and divergent. Unlike the other segments, they do not have ciliate margins. Petals are narrow, pointed and shorter than the other segments. The labellum is obovate with a raised dark fleshy central portion and a shortly pointed tip.

HABITAT Damp sandy or clayey soils in open scrub.

DISTRIBUTION Coastal districts and central tablelands; also south-eastern Qld. Uncommon and localised.

FLOWERING January–March.

Caleana

The Flying Duck-orchids belong to 2 genera, *Caleana* with 1 species and *Paracaleana* with 2. *Caleana* is widespread and sometimes common, though usually inconspicuous unless massed.

1393. Caleana major

Caleana major

Flying Duck-orchid, Large Duck-orchid **Plate 1393**

A slender plant 15 to 50 cm high with a solitary narrow stem-clasping brownish or reddish leaf at its base and an inflorescence of 1 to 4 flowers on slender stalks with a rudimentary bud at the top. The flowers are about 20 mm long with the reddish or reddish-purple perianth segments curiously resembling a duck in flight. The labellum forms the duck's head which is attached to the enlarged column by a sensitive claw. Two narrow reflexed sepals form the wings. The dorsal sepal lies behind the column, the petals on either side. The column which forms the duck's body is incurved, broadly winged and terminated by a conspicuous anther. When triggered by movement of an insect, the claw snaps downward, forcing the insect to escape past the anther, thereby increasing the likelihood of pollination.

HABITAT Heath, woodland and open grassy forest.

DISTRIBUTION Coast, tablelands and central western slopes; also Qld, Vic, SA and Tas.

FLOWERING September–January.

Paracaleana

Paracaleana minor

Small Duck-orchid

Paracaleana minor resembles *Caleana major* but is smaller in all respects. It is a slender wiry plant 5 to 20 cm high with a short linear leaf at the base and 1 to 4 tiny green and red-brown flowers. The labellum is ovate, bluntly pointed at one end and with a narrow bifid appendage at the other. The upper surface is covered with raised dark-coloured calli. Other segments are narrow linear and surround the concave green column. The labellum claw is sensitive and snaps downward when triggered.

HABITAT In a variety of habitats, usually sandy and dry.

DISTRIBUTION More or less coextensive with *C. major*, but much less common.

FLOWERING October–December.

Chiloglottis

Small terrestrial herbs with 2 leaves at the base. Flowers usually solitary on a short stem. Dorsal sepal narrow erect, petals and lateral sepals spreading or deflexed. Labellum broad variously shaped with a characteristic arrangement of calli on the upper surface. Column slender, hooded by dorsal sepal.

1394. Chiloglottis reflexa

Chiloglottis reflexa

Autumn Bird-orchid **Plate 1394**

This species resembles 2 other small bird-orchids present in the area. Although it flowers in autumn (the others in spring) and its labellum is usually broad at the end with a small pointed tip (unlike the others), this species is most reliably distinguished by the shiny dark calli on the labellum surface. As the specific name suggests, the large stalked callus at the base of the labellum near the claw is reflexed or bent back. In front of this, a bushy outgrowth of red glandular hairs and dark calli extends towards the labellum tip, occasionally covering much of the surface.

HABITAT Cool shady moist places such as mossy clay banks.

DISTRIBUTION Coast and tablelands; also Qld, Vic, SA and Tas.

FLOWERING Chiefly March–May, but under favourable conditions may flower at any time.

1395. Chiloglottis formicifera

Chiloglottis formicifera

Ant-orchid **Plate 1395**

Relatively common in the Sydney district, this species is similar to the somewhat rarer *Chiloglottis trapeziformis*. The former species has leaves with short stem-clasping stalks, the latter has leaves with longer non-stem-clasping stalks. The labellum in both cases is rhomboid or trapezoid in shape (occasionally rounded at the end in *C. formicifera*), hence they are difficult to distinguish on this feature. However *C. formicifera* has a cluster of small calli near the apex of the labellum in addition to the main group of calli which has a conspicuous double-headed callus at the rear near the claw. *C. trapeziformis* has a single group of calli in the centre of the labellum. The specific epithet and the common name of *C. formicifera* refer to the resemblance of its calli to ants. This is incidental; the calli actually resemble a wingless female thynnid wasp and emit sexual pheromones which attract male wasps for pollination. The 2 above-mentioned species reportedly attract 2 different wasp species.

HABITAT Sandy soils in protected conditions.

DISTRIBUTION Chiefly in the central eastern part of NSW, extending to the western slopes. Also NZ. *C. trapeziformis* is more widespread, occurring also in Qld, Vic and Tas.

FLOWERING August–October.

Chiloglottis gunnii

Bird-orchid **Plate 1396**

Known popularly as the Common Bird-orchid in Victoria and Tasmania, this species is comparatively rare in the Sydney district, being confined mostly to cool shady places on the ranges. Plants have 2 large ovate or lanceolate leaves at the base of a short stem about 5 cm high. The flower is solitary, green or brownish and larger than in other species. it is usually 2 to 4 cm in diameter with widely spreading

1396. Chiloglottis gunnii

segments. The labellum is large, broadly ovate and acute with a large erect callus in the throat and a few smaller ones arranged on the labellum surface. Two relatively broad petals are spreading or erect on either side of the dorsal sepal. The calli on the labellum are sites of pheromone production, attracting several species of thynnid wasps which act as pollinators.

HABITAT Cool shady forested areas, usually in small colonies, mostly in elevated regions.

DISTRIBUTION Tablelands and central coast; also Vic and Tas.

FLOWERING September–January.

Acianthus

Small terrestrial herbs with a solitary rounded to heart-shaped leaf at the base. Flowers in a terminal raceme, small, the segments often acutely pointed. Dorsal sepal erect or bent forward. Lateral sepals narrow, forward-pointing. Labellum variously shaped with 2 basal calli. Petals usually short, reflexed or spreading. Column narrow, usually bent forward.

Acianthus fornicatus

Pixie Caps **Plate 1397**

Both the specific and common names of this ubiquitous little orchid refer to the broad dorsal sepal which is arched over the column, projecting forward to a fine point. The plant has a basal stem-clasping heart-shaped leaf 2 to 4 cm long. It is usually less than 20 cm high with about 4 to 6, sometimes more, pale reddish-brown and green flowers. Finely pointed lateral sepals project forward beneath and beyond the other segments. The petals are comparatively short. The pointed ovate labellum is slightly longer with a papillose upper surface.

HABITAT Common on sandy soils in heath, scrub and open forest.

DISTRIBUTION Coast, tablelands and western slopes; extending to Qld.

FLOWERING April–July, sometimes as late as September.

1397. Acianthus fornicatus

Acianthus exsertus

Gnat Orchid, Mosquito Orchid

This plant is similar to *A. fornicatus* but is smaller, more slender and much less common. It can be distinguished by its more erect narrow dorsal sepal and its conspicuously exserted (bent forward) column. The sepals project forward beneath the labellum and are sometimes tipped with a small knobby gland. The labellum tapers abruptly to a short point which is often deflexed (bent under). The petals are short and usually reflexed (bent back).

HABITAT Sandy soils in a variety of habitats.

DISTRIBUTION Coast, tablelands and western slopes; also south-eastern Qld, Vic, SA and Tas.

FLOWERING April–July, some as late as September.

1398. Acianthus caudatus

Acianthus caudatus

Mayfly Orchid **Plate 1398**

A slender little plant, usually no more than 10 to 15 cm tall, this species is inconspicuous in undergrowth owing to its reddish stem and narrow dark perianth segments. Once discovered, it is readily recognised by its long filiform sepals which may reach 4 or 5 cm in length. The labellum and petals are only one-tenth as long. The narrow column is bent forward above the labellum.

HABITAT Moist sheltered sites on sandy soils.

DISTRIBUTION Central coast and central tablelands; also north coast, Vic, SA and Tas.

FLOWERING July–September.

Rimacola

Rimacola is a monotypic genus restricted to protected wet clayey ledges and crevices of sandstone cliffs in the Blue Mountains.

1399. Rimacola elliptica

Rimacola elliptica **Plate 1399**

It produces characteristic radical and stem leaves. The radical leaves are dark green, glabrous and lanceolate-elliptical, tapering gradually to a narrow base. They are often about 5 cm long but may exceed 15 cm in long competing vegetation. Flowering specimens also have 1 to 3 short, broad, loosely stem-clasping leaves below a rather crowded drooping raceme of greenish-yellow flowers. Perianth segments are 2 to 3 cm long, narrow and acuminate, the dorsal sepal more or less hooded over the column and labellum, but often recurved towards the apex. The labellum is shorter than the other segments and marked with dark brownish red.

HABITAT Dripping ledges and faces on sandstone cliffs and caves, often with other growth, particularly coral fern (*Gleichenia* sp.).

DISTRIBUTION Central Blue Mountains (e.g., Linden, Wentworth Falls).

FLOWERING November–December.

1400. Eriochilus cucullatus

Eriochilus

Eriochilus cucullatus

Parson's Bands **Plate 1400**

These little orchids range widely and abundantly in eastern Australia, attracting attention by their deflexed white sepals which show some resemblance to a parson's neck-piece. Plants are usually 10 to 20 cm high with a short broad-ovate radical leaf and 1 to 3 white or pink flowers in a terminal raceme. The dorsal sepal is cucullate (hooded) over the column, roughly in the same plane as 2 small spreading petals. The labellum is green with red hairs and is distinctly recurved in the broad upper part.

HABITAT In a variety of open habitats.

DISTRIBUTION Coast, tablelands and slopes; also south-eastern Qld, Vic, SA and Tas.

FLOWERING January–May.

Lyperanthus

Lyperanthus is a small genus of about 8 species occurring in Australia, New Zealand and New Caledonia. Two species occur in the Sydney district; one *(L. nigricans)* is rarley seen and probably flowers only after fire; the other *(L. suaveolens)* is relatively common and widespread.

1401. Lyperanthus suaveolens

Lyperanthus suaveolens

Brown Beaks **Plate 1401**

This species has a stem 15 to 30 cm high sheathed at the base by an erect linear-lanceolate leaf and bearing a narrow raceme of 2 to 6 purplish-brown flowers, each almost sessile within a large subtending bract. The flowers have a narrow hooded dorsal sepal, 2 erect linear sepals and 2 similar deflexed petals, all about 20 mm long. The labellum is about 10 mm long, recurved, yellow towards the apex and with 2 dark lateral lobes near the base.

HABITAT In sandy soils in open forest and scrub.

DISTRIBUTION Coast, central tablelands and central western slopes; also Qld, Vic and Tas.

FLOWERING August–September.

1402. Caladenia alba

Caladenia

Terrestrial herbs with small round tubers, more or less hairy, with a solitary radical leaf and flowers solitary or in a small terminal raceme. Perianth segments narrow, rarely produced into long tails, the dorsal sepal erect behind the column, other segments spreading or deflexed. Labellum usually 3-lobed, the mid-lobe furnished with stalked calli in 2 to 6 rows, often recurved, fringed or toothed on the margin. Column incurved, winged.

Caladenia alba

White Caladenia **Plate 1402**

C. alba and *C. catenata* are common closely related species with medium to large pink or white flowers. Except for an orange-yellow tip on the labellum, the perianth of the former is usually wholly white, occasionally pale pink, with a white, green or tri-coloured green, red and white column, but not a white and red barred column as in the latter species. The labellum is 3-lobed, the lateral lobes rounded and often lightly stained with purple, the mid-lobe short and recurved with a fringed margin and 2 rows of white calli.

HABITAT In sandy soils in a variety of habitats.

DISTRIBUTION Widespread on coast, central tablelands and central western slopes; also south-eastern Qld, Vic, SA and Malesia.

FLOWERING June–October.

1403. Caladenia catenata

Caladenia catenata

Pink Fingers **Plate 1403**

This is a variable species with flowers in different shades of pink, or occasionally white. It is typically 10 to 20 cm high (though smaller and larger forms occur, some having varietal rank) with an erect hairy narrow linear leaf and 1 or 2 short stem bracts. It has 1 to 3 flowers which are a light dusky green on the back due to glandular hairs. Sepals and petals are narrow lanceolate and 10 to 20 mm long, the labellum much shorter. It is 3-lobed with erect rounded lateral lobes and a recurved mid-lobe with fringing calli. The upper surface has 2, or sometimes 4, rows of white, yellow or red calli. Both the labellum and the column are conspicuously marked with transverse red lines or bands.

HABITAT In sandy soils in a variety of habitats ranging from heath to open forest.

DISTRIBUTION Widespread in eastern NSW; also Qld, Vic, SA, Tas, New Caledonia and Malesia.

FLOWERING August–October.

1404. Caladenia dimorpha

Caladenia dimorpha — **Plate 1404**

Originally described by Fitzgerald from specimens gathered at Bowenfels near Lithgow, this species is part of a complex of a half dozen closely related and confusing eastern species with white flowers and more or less cucullate (hooded) dorsal sepals. It is a sparsely hairy plant 15 to 30 cm high with a long linear leaf, a stem bract and 1 to 3 large flowers subtended by short floral bracts. Perianth segments are often tinged with pink or purple towards the ends and are brownish with glandular hairs on the back. The labellum is broad, undivided and fringed along its margins with purple or yellow clubbed calli which extend almost to the recurved purple-blotched tip. Four rows of stout purple, yellow or white-headed calli run from the base almost to the tip.

HABITAT In open swampy, grassy or dry rocky ground.

DISTRIBUTION Higher parts of the central tablelands. Reasonably common in the Bell–Wolgan–Lithgow area.

FLOWERING October–November.

Caladenia testacea

Honey Caladenia — **Plate 1405**

This slim little orchid has 1 to 6 (usually 2 to 3) small honey-scented flowers with narrow bronze-tipped, glandular-pubescent pale or golden green segments.

1405. Caladenia testacea

As with several closely related white-flowered species, the dorsal sepal is closely hooded over the column revealing the darker brown outside surface. The labellum is undivided, or at most, obscurely lobed, the margins prominently fringed with erect calli or teeth which become shorter towards the dark purple recurved tip. Four crowded rows of small calli cover the labellum surface. The column is broadly winged and marked with red or pink.

HABITAT In rather dry places amongst undergrowth in scrub and open forest.

DISTRIBUTION Chiefly coastal from about Bullahdelah southward, extending to the central tablelands and southern slopes; also Vic and Tas.

FLOWERING August–September.

Caladenia angustata

Musky Caladenia — **Plate 1406**

One of a group of basically white-flowered caladenias exhibiting some variation in the colouring of the labellum and the outside of the perianth, and differing slightly in the lobes and fringe of the labellum, this species has long been a source of difficulty and confusion. The common name refers to its rather overpowering musky odour which does not appear to be an infallible identifying feature since others such as *C. cucullata* and *C. testacea* may also have a pungent or sweet odour. The flowers, numbering 1 to about 4, are generally white on the inside, sometimes tinged with pink or purple near the tips of the segments, and dusky, purplish brown or green and red on the outside. As with several related species, the dorsal sepal is cucullate or hooded over the column and labellum. The labellum is 3-lobed, sometimes obscurely so. The lateral lobes are entire, the mid-lobe coarsely toothed near the lateral lobes then very finely toothed to the tip which is usually purple or blotched with purple. The labellum is furnished with 4 rows of calli which are white or yellow headed near the base, becoming purple headed towards the apex. In the closely related *C. cucullata* the column is abruptly bent forward, the labellum is densely covered and fringed with minutely scabrous purple-headed calli and the mid-lobe is almost entirely coloured purple, not just the tip.

1406. Caladenia angustata

HABITAT Fairly dry hillsides with eucalypt forest or scrub cover, in seasonally cold elevated regions.

DISTRIBUTION Higher parts of the Blue Mountains (e.g., Bell–Lithgow). Scattered from Qld to Vic and Tas, mainly on the tablelands.

FLOWERING September–November.

Caladenia caerulea

Blue Caladenia — **Plate 1407**

Only 2 blue-flowered caledenias occur in eastern Australia and only one of these, *C. caerulea*, is common in the Sydney district. It is a slender, sparsely hairy plant about 10 cm tall with a short leaf which tends to lie on the ground. The solitary flower is small to medium in size, sky blue or purplish, and covered with minute dark blue glands on the outside. The labellum is 3-lobed with 2 large erect rounded lateral lobes with entire margins and a short recurved mid-lobe. Apart from the white or yellow tip the labellum is marked with transverse dark bands. Two rows of stalked yellow-headed calli extend from the claw to the tip.

The second blue species, *C. deformis*, has a solitary dark-blue flower with an obscurely lobed dark blue or purple labellum densely covered with stalked calli in 4 to 6 irregular rows.

HABITAT In open sandy or stony ground.

DISTRIBUTION Widespread in NSW; also Qld, Vic, SA and WA.

FLOWERING August–September.

1407. Caladenia caerulea

Adenochilus

Adenochilus is a small genus of 2 species, one endemic in New South Wales, the other in New Zealand.

Adenochilus nortonii **Plate 1408**

Rare and restricted in both distribution and habitat, this small terrestrial plant is 10 to 20 cm high with a solitary glabrous ovate to cordate leaf at, or near, the base of the stem which grows from a fleshy rhizome instead of a tuber. The solitary flower is white resembling a *Caladenia*. Perianth segments are 7 to 10 mm long, relatively broad but tapering to a fine point. The labellum is marked with red and has a band of yellow and white calli down the centre.

HABITAT Restricted to permanently moist mossy clay banks and rock crevices.

DISTRIBUTION Endemic in the Blue Mountains above about Wentworth Falls and in Barrington Tops.

FLOWERING November–December.

1408. Adenochilus nortonii

Glossodia

A genus of 2 species confined to eastern and southern Australia. Terrestrial herbs with a solitary radical, sheathing, prostrate, hairy leaf. Flowers 1 to 2, purple to blue, on an erect stem. Perianth segments nearly equal, spreading. Labellum small, almost sessile, undivided, margin entire, with 2 clubbed calli between the base and the column. Column short, erect, winged.

1409. Glossodia major

Glossodia major

Wax-lip Orchid **Plate 1409**

The 2 species of *Glossodia* are readily distinguished by size and the colour of the labellum and small calli at its base. This species grows to 20 cm or more in height with a leaf reaching 10 cm in length. Flowers are mostly 30 to 40 mm in diameter but may reach 60 mm. Floral segments are purple to mauve, becoming white near their bases. The short ovate labellum is white and pubescent in the basal half, purple to mauve and glabrous in the upper half, its base crossed by a 2-lobed yellow projection formed by the fused heads of 2 purple calli.

HABITAT In a variety of habitats ranging from heath to open forest, sometimes in large numbers.

DISTRIBUTION Widespread in eastern NSW; also Qld, Vic, SA and Tas.

FLOWERING August–September.

1410. Glossodia minor

Glossodia minor

Small Wax-lip Orchid **Plate 1410**

This smaller species is occasionally found with the larger one, but is more often seen in scattered colonies of its own kind. It grows to no more than 15 cm in height with a small prostrate hairy lanceolate leaf 2 to 4 cm long. The flowers are usually solitary, deep violet and about 20 mm in diameter. The tiny labellum is acutely ovate but is not white in the basal half as in *G. major*. The dark heads of 2 basal calli project forward above the labellum.

HABITAT In a variety of habitats, often in sandy or clayey lateritic soils.

DISTRIBUTION Coastal regions of NSW; also south-eastern Qld and Vic.

FLOWERING August–September.

Corybas

A genus of about 50 species ranging from New Zealand, Australia and the Pacific through Malesia to the Himalayas. Very small terrestrial herbs with a solitary ovate, cordate or orbicular horizontal leaf. Flower solitary, reddish purple, hooded, sessile or nearly so, the dorsal sepal and labellum well developed, the other segments insignificant.

1411. Corybas pruinosus

Corybas pruinosus **Plate 1411**

This species has long been confused with *C. fimbriatus* and *C. dilatatus*, which does not occur in the Sydney region. It has slightly smaller flowers which stand higher off the leaf than those of *C. fimbriatus* due to a longer labellum tube and ovary. The dorsal sepal is narrower, tending to greyish green with darker spots. The expanded part of the labellum has a well-developed boss and is usually shorter than the tubular basal part. The margin is fringed, but the fimbriae are not inturned near the apex as in *C. fimbriatus*. The ear-like appendages on the labellum are very small. Lateral sepals are filiform and sometimes sufficiently long to project beyond the labellum.

HABITAT In ground litter in cool, damp protected sandy sites.

DISTRIBUTION Coast and central tablelands of NSW.

FLOWERING April–July.

Corybas fimbriatus

Fringed Helmet-orchid

With a dorsal sepal 20 to 30 mm long when straightened, *C. fimbriatus* has the largest flowers of the Australian species. These are a deep reddish purple and sit close to the broad rounded heart-shaped leaf. The dorsal sepal is somewhat translucent with purplish spots. It is broad with a blunt apex and hooded over the rather long fringed labellum. The fimbriae of the fringe are spreading on the expanded part of the labellum but are usually inturned towards the apex. The boss or raised central part of the labellum is often obscure and varies from a pale to a dark colour. The base of the labellum is tubular and has 2 large ear-like appendages. Surrounding the base are the tiny petals and sepals. These are linear and white with the petals shorter than the sepals and often forked at the tip.

HABITAT In ground litter in cool damp protected sandy sites.

DISTRIBUTION Coast and tablelands of NSW, extending along the coast from Qld to Vic and eastern Tas.

FLOWERING May–July.

1412. Corybas aconitiflorus

Corybas aconitiflorus

Spurred Helmet-orchid **Plate 1412**

This species is readily identified since the dorsal sepal is large and so hooded over the labellum as to almost hide it from view. Its leaf is usually heart-shaped with a reddish undersurface. Flowers are fairly large, up to 24 mm in length, and a dull reddish purple or slaty colour. The dorsal sepal is large, deeply arched and very convex. The labellum is pale, marked with red. It has a long tubular portion and a very short lamina with recurved entire margins. The base of the labellum extends into 2 short hollow spurs which sometimes are the only part visible below the dorsal sepal. Sepals and petals are diminutive and located near the spurs.

HABITAT In ground litter in cool, damp protected sandy sites.

DISTRIBUTION Uncommon on the coast and tablelands. Extends along coast and coastal ranges from northern Qld to Vic and Tas; also NZ.

FLOWERING April–June.

Cryptostylis

A genus of about 20 species of which 5 are endemic in Australia, the others in Malesia and the Pacific. Terrestrial glabrous herbs with 1 to 3 radical lanceolate leaves on long stalks, rarely leafless. Flowers several in a terminal raceme, not inverted, the labellum large and red-brown or purple, the other segments small and green. Labellum sessile, undivided, concave or convex, the base enclosing the column. The name is derived from 'kryptos'—hidden; 'stylos'—style or column.

The Australian species of *Cryptostylis* are pollinated by ichneumon wasps in a remarkable process known as pseudocopulation. This occurs through the attraction of males by a pheromonal scent and subsequent arousal to sexual activity by the resemblance of the labellum to the female insect. In the act of mating, pollinia are transferred from flower to insect.

At least 4 of the 5 Australian species (the pollinator of the rare saprophyte, *C. hunteriana* is not known) are pollinated by the same species of wasp. The concurrent flowering of the 5 species is apparently timed to coincide with emergence of the male wasp. However, a promiscuous system such as this requires fertility barriers betwen the different species, otherwise mixed populations would eventually be reduced to a single hybrid or dominant species. Several other orchid genera are pollinated by pseudocopulation in different forms.

1413. Cryptostylis erecta

Cryptostylis erecta

Bonnet Orchid, Tartan Tongue-orchid **Plate 1413**

In this species the expanded part of the labellum is broad and concave, forming an erect hood or bonnet above the column, which is more or less enclosed in the basal part. A narrow green dorsal sepal hangs below the column; the other sepals and petals are erect or spreading behind the labellum. The plant is eye-catching, usually 15 to 40 cm tall, with 1 to 3 broad lanceolate leaves at the base and a raceme of 2 to 10 flowers marked with purple or red-brown.

HABITAT Sandstone gullies, often on mossy rock ledges but also growing in moist or fairly dry situations on the gully slopes.

DISTRIBUTION Coast and central tablelands; also south-eastern Qld and eastern Vic. Reasonably common in the northern suburbs of Sydney and in the Blue Mountains.

FLOWERING December–February.

1414. Cryptostylis subulata

Cryptostylis subulata

Large Tongue-orchid, Cow Orchid **Plate 1414**

The flowers of this species have a conspicuous red-brown labellum 25 to 35 mm long with thick upturned margins together with narrow spreading sepals and short reflexed petals at the base, the arrangement suggesting a cow's head and prompting the name Cow Orchid. Plants are usually 30 to 50 cm high with 1 to 3 erect narrow-lanceolate leaves. The inflorescence is an open raceme of 2 to 8 flowers, sometimes more. The labellum is oblong, contracted around the column at the base and lined on the undersurface by 2 dark beaded ridges ending in a 2-lobed knob near the apex.

HABITAT In dry sandy soil in open forest; also in west marshy habitats.

DISTRIBUTION Coast and tablelands of NSW; also Qld, Vic, SA and Tas.

FLOWERING October–January.

Cryptostylis leptochila

Small Tongue-orchid **Plate 1415**

Growing to about 40 cm in height, this species is smaller than *C. subulata* and has smaller flowers. The leaves at the base are broader, nearly ovate and purplish or red on the undersurface. The labellum has a short broad base and then abruptly narrows

1415. *Cryptostylis leptochila*

to an oblong-linear channel. It is usually bent back near the base and coiled or recurved at the apex. The surface is pubescent, mainly deep red with 2 or 3 rows of raised shiny dark calli. The other perianth segments are green, narrow and spreading or reflexed.

Habitat Mainly in scrub and open forest, chiefly on the tablelands.

Distribution Tablelands and coast from the New England district of NSW to eastern Vic.

Flowering December–March.

Pterostylis

Small terrestrial herbs with leaves in a radical rosette or tuft and ovate to lanceolate, or cauline and narrow. Flowers solitary or in a terminal raceme. Dorsal sepal arched, the lateral petals appressed to it and forming a hood or galea which conceals the column. Lateral sepals erect or deflexed, joined in their lower part and forming a V-shaped cleft or sinus, their upper parts free and often tapering to fine points. Labellum small, attached to the foot of the column by a moveable, often irritable, claw. A genus of over 70 species, chiefly Australian, but also in New Zealand, New Caledonia and New Guinea.

Pterostylis nutans

Nodding Greenhood **Plate 1416**

With its large nodding head, this is a common and favourite orchid likely to be seen alone or in colonies in moist sheltered sites from winter to early spring. It is variable in height, up to 40 cm, but is more often about 15 to 20 cm. It has 4 or 5 stalked ovate leaves with undulate margins in a basal rosette. The flower is usually solitary, 20 to 30 mm long, green and bent forward from the floral bract so that it faces the ground. The galea is shortly pointed, often with an upturned tip; lateral sepals are erect on either side. The labellum is linear, strongly curved downward and with a pubescent upper surface.

Habitat Favours sandy soils in undergrowth in sheltered positions near the coast, sometimes on damp mossy rocks.

Distribution Widespread on the coast, tablelands and slopes in NSW; also south-eastern Qld, Vic, SA, Tas and NZ.

Flowering June–September.

1416. *Pterostylis nutans*

1417. *Pterostylis acuminata*

Pterostylis acuminata

Sharp Greenhood **Plate 1417**

Resembling *P. nutans*, but more delicate and without a nodding flower, this species is relatively common in coastal areas of NSW and Qld. It is usually about 15 cm high with a basal rosette of short ovate to broad lanceolate leaves. The stem has 1 or 2 very small bracts and a solitary green flower 25 to 30 mm long, the flower erect or only slightly bent forward. The dorsal sepal is rather long and conspicuously acuminate. The lower sepals are narrow at the base, slightly bulging at the V-shaped sinus then produced into long erect filiform points. The labellum is linear with an acuminate tip and strongly curved downward.

Habitat Sandy soils in light coastal scrub and open forest.

Distribution Coastal districts from central NSW to Qld.

Flowering April–May.

Pterostylis baptistii

King of the Greenhoods **Plate 1418**

Popular in cultivation owing to its impressive size, this species is like a large version of *P. acuminata* but differs in having an acute rather than acuminate tip on the dorsal sepal, and lateral sepals which are broad and bulging at the V-shaped sinus then swept back almost behind the galea rather than beside it. It is a tall orchid, usually 30 to 40 cm high, with a basal rosette of reticulate elliptical to oblong leaves 5 to 8 cm long. Flowers are up to 60 mm long, reputedly the largest in the genus. They are translucent green and white with brown tonings towards the tip of the galea. The labellum is linear with a suddenly constricted narrow tip. It is only slightly recurved and points forward.

Habitat Sandy soils in coastal scrub, usually near swampy ground, sometimes forming extensive colonies.

Distribution Coastal districts from Qld to eastern Vic. Uncommon and scattered near Sydney.

Flowering August–October.

1418. *Pterostylis baptistii*

Pterostylis curta

Blunt Greenhood **Plate 1419**

This is a sturdy squat little orchid whose most distinctive feature is its twisted oblong labellum. It is usually 15 to 25 cm tall with a basal rosette of ovate leaves, 1 or 2 stem bracts and a solitary flower subtended by a loosely sheathing floral bract. The flower is usually green with brownish tints towards the apex of the galea and has a brown labellum. The galea is broad with a short acute tip scarcely exceeding the petals. The lower sepals are very wide at the sinus then bent back slightly with short acuminate tips rarely rising above the galea. The twisted labellum is easily visible in the sinus, but snaps back when the flower is shaken.

Habitat In cool shady gullies, often in grassy areas beside walking tracks and forming extensive colonies.

Distribution Widespread and fairly abundant in eastern NSW; also south-eastern Qld, Vic, SA, Tas and New Caledonia.

Flowering June–September.

1419. *Pterostylis curta*

Pterostylis hildae

Hilda's Greenhood **Plate 1420**

This species resembles *P. curta* but has smaller flowers with a narrower sinus and a shorter non-twisted labellum. Its usual habitat in or near rainforest is also distinctive. It is a slender glabrous plant about 15 cm high with 2 or 3 fairly large leaves at the base. The stem bracts are small and sheath the stem tightly. The solitary flower is entirely green or tinged with brown at the apex. The labellum is oblong-elliptic with an obtuse tip.

HABITAT In small grassy open areas in or near subtropical rainforest, usually in small colonies.

DISTRIBUTION Coastal NSW and Qld. Uncommon to rare in the Sydney ditrict. Observed at Mt Kembla.

FLOWERING August–October.

Pterostylis concinna

Trim Greenhood **Plate 1421**

The most distinctive features of this greenhood are the double-tipped labellum and the long erect lateral sepals. The plant is very variable in height, depending on the conditions, but is usually of a neat trim appearance with 1 flower 10 to 20 mm long. The flower is green with darker green stipes and brown suffusions at the tip of the galea and on the lateral sepals. The sepals form a very wide V-shaped sinus through which the labellum is visible, then sweep upwards into long filiform points well above the galea. The labellum is oblong, less than 10 mm long, almost straight and vertical, but on an irritable claw which snaps it back against the column when triggered by the movement of an insect.

HABITAT In open forest in both sandy and clayey soils in conditions ranging from moist to dry.

DISTRIBUTION From Qld to Vic and Tas, generally coastal but extending to the tablelands and slopes in central NSW. Fairly common near Sydney.

FLOWERING June–September.

Pterostylis grandiflora

Superb Greenhood **Plate 1422**

Truly a superb species combining elegance of form with contrasting translucent green and rich brown colouring, this orchid is widely acknowledged as the most beautiful of the genus. It is a slender plant 15 to 25 cm high with both radical and cauline (stem) leaves. The radical leaves form a rosette separate from the stem and are often absent at flowering time. The stem-clasping cauline leaves are about 5 cm long and taper to an acute apex from a broad base. The flower is fairly large, up to 40 mm long, but often less. It is translucent green with darker green lines near the base, becoming darker green towards the tip of the dorsal sepal with the petals forming broad bronze-brown wings on either side. The lower sepals are erect, united in the lower third then abruptly contracting at the broad flat sinus to 2 long filiform points which extend high above the galea. The labellum is oblong-linear with a long filiform tip.

HABITAT Amongst undergrowth on sandy soil in cool moist gullies.

DISTRIBUTION Coast and central tablelands; also Qld, Vic and Tas.

FLOWERING May–July.

1420. Pterostylis hildae

1421. Pterostylis concinna

1422. Pterostylis grandiflora

Pterostylis pedunculata

Little Red Riding Hood, Maroon Hood **Plate 1423**

Usually a slender plant with a distinctive narrow flower atop a long stem, this species is well known in coastal districts of south-eastern Australia. It grows to about 25 cm high from a basal rosette of ovate to oblong leaves. It has 2 to 3 narrow stem bracts, the upper one some distance below the flower. Flowers may reach 20 mm in length but are often considerably smaller. They are green with brown suffusions on the galea and lower sepals, or sometimes almost wholly reddish-brown. The galea is erect in the lower two-thirds then becomes abruptly horizontal, ending in an acutely pointed tip. The lower sepals are narrow and united for less than half their length then spread upwards or backwards into long linear points extending well above the galea. The labellum is red-brown, elliptic-oblong, about 5 mm long with a central ridge above and a corresponding groove on the underside.

HABITAT Scrub and open forest in open fairly dry situations.

DISTRIBUTION South-eastern Qld to Vic, SA and Tas, extending from the coast to the western slopes in central NSW; reasonably common.

FLOWERING August–September.

1423. Pterostylis pedunculata

Pterostylis mutica

Midget Greenhood **Plate 1424**

P. mutica varies from short and robust to about 30 cm tall and rather slender. It has small ovate to lanceolate leaves which are sometimes numerous in a basal rosette. It also has 3 to about 8 closely sheathing stem bracts and a similar number of very small, pale-green flowers in a spike-like raceme. The galea is about 7 mm long with a high curved apex contracting suddenly to a short blunt point. The lower sepals are short and broad and joined for most of their length. They point downward exposing the short irritable labellum which has a thick dark basal appendage turned backward towards the claw.

HABITAT Fairly dry open country.

DISTRIBUTION Extending from the coast to the western plains in NSW. Distributed in all States. Also NZ.

FLOWERING July–October.

1424. Pterostylis mutica

Pterostylis parviflora

Baby Greenhood, Tiny Greenhood **Plate 1425**

Although reportedly common in some areas, this plant is difficult to detect in undergrowth. It is very slender and of variable height with numerous dark-tipped little flowers well spaced along the stem. Each flower is subtended by a leaf-like bract and faces inward towards the stem. If basal leaves are present at flowering time they usually form a rosette separate from the stem but connected to it underground. The galea is about 10 mm long and white with green stripes, becoming green tinged with brown towards the tip. The lower sepals are erect and united for two-thirds of their length, forming a wide sinus where they separate into 2 short acuminate points which rarely exceed the galea. The labellum is oblong and quite short.

HABITAT In a variety of habitats.

DISTRIBUTION From Qld to Vic, SA and Tas, extending from the coast to the western slopes in NSW.

FLOWERING February–April, possibly also at other times.

1425. Pterostylis parviflora

Pterostylis longifolia

Tall Greenhood **Plate 1426**

P. longifolia is readily distinguished from its 2 closest allies, *P. parviflora* and *P. daintreana* by its long stem leaves. These are narrow lanceolate with a stem-clasping base and vary in length from 3 to about 8 cm. Basal leaves are absent at flowering time. Plants are usually 25 to 50 cm in height, but may be shorter or occasionally much taller. Large specimens may have as many as 12 flowers. These are green and about 12 to 20 mm from the base of the galea to its short-pointed tip. Lateral sepals are narrow and point downward. The labellum is 6 to 8 mm long with a raised appendage at the base and a narrow upturned bifid apex. The labellum claw is short and extremely irritable, snapping the labellum up against the column at the slightest disturbance.

HABITAT Usually in moist shady gullies.

DISTRIBUTION Widespread and fairly common in temperate parts of eastern Australia, extending to the central western slopes in NSW.

FLOWERING June–September.

1426. Pterostylis longifolia

Pterostylis daintreana

Plate 1427

The flowers of *P. daintreana* resemble those of *P. longifolia* but it is a slender smaller species, usually less than 25 cm in height. Plants have one, sometimes 2, basal rosettes of small ovate leaves separate from the flowering stem which has 3 to 5 small bracts resembling narrow stem-clasping leaves. At the top of the stem are 3 to 7 dainty green and white flowers on slender stalks. The galea, about 10 mm long, forms a high curved hood then suddenly constricts to a fine point 5 to 7 mm long. The lower sepals are produced forward and downward into long filiform points. They are united only at the base, forming a narrow V-shaped sinus below the short narrow obtuse labellum.

HABITAT Usually on moist mossy sandstone outcrops.

DISTRIBUTION Mainly central coast and central tablelands, with isolated occurrence elsewhere on the coast and ranges; also Qld.

FLOWERING April–July.

1427. Pterostylis daintreana

Pterostylis sp. aff. *gibbosa*

Plate 1428

This is one of the *rufa* group of greenhoods, a confusing complex with predominantly reddish flowers found mostly in the slightly drier areas of both eastern and western Australia but uncommon in the Sydney district. As with many others in this complex, the basal rosette of leaves is often withered at flowering time. It makes up for this by having as many as 5 or 6 stem bracts and floral bracts subtending each of the 2 to 6 flowers. Flowers are reddish brown on the petals and lower sepals but greenish with translucent spaces between the markings on the dorsal sepal. The galea, which consists of the dorsal sepal and petals, forms a high arch over the column then tapers abruptly to a fine slightly upturned tip. The lower sepals are deflexed (bent downwards) forming a backdrop for the labellum. These sepals are 18 to 20 mm long with narrow tails 8 to 9 mm long. The labellum is attached to the base of the column by a sensitive narrow green claw. It has a brown oblong-elliptical lamina with a few long hairs on the margin.

HABITAT In dry sandy soil in open eucalypt forest.

DISTRIBUTION Central coast (e.g., Picnic Point, Georges River).

FLOWERING September–October.

1428. Pterostylis sp. aff. *gibbosa*

Galeola

Galeola is a genus of about 70 species ranging from Japan through Malaysia to Australia and New Caledonia. One of two endemic Australian species extends to the Sydney region.

1429. Galeola cassythoides

Galeola cassythoides

Climbing Orchid **Plate 1429**

A leafless terrestrial saprophyte with 1 or more flexuose branched stems rising from a thick fleshy rhizome. By attaching itself by sucker-like aerial rootlets, it can climb to 5 or 6 metres on tree trunks and rock surfaces, but is more often low and scrambling. Being a saprophyte, it lacks the green photosynthetic pigment, chlorophyll. Stems are dark brown, the flowers pale yellow and white with greenish-brown tints. The flowers occur in clusters at the ends of the stem branches and have spreading sepals and petals, 12 to 15 mm long, and a somewhat shorter whitish labellum with undulate-crisped margins. Although tall northern specimens often develop hundreds of flowers, local examples are more likely to have 20 to 50, with only a few open at the time.

HABITAT Open forest at the base of trees or rocks, usually sandy soil.

DISTRIBUTION South-eastern Qld and north-eastern NSW, extending along the coast to the Illawarra.

FLOWERING October–November.

Gastrodia

A small genus of about 17 mainly Indo-Malaysian species, *Gastrodia* is represented in Australia by 2 species, one restricted to northern Queensland, the other, *G. sesamoides*, occurring in all States and extending to New Zealand.

1430. Gastrodia sesamoides

Gastrodia sesamoides

Potato Orchid, Cinnamon Bells **Plate 1430**

Like *Galeola cassythoides, G. sesamoides* is a leafless saprophyte with brown, white and yellow tonings, but is readily distinguished by its erect unbranched straight stems and tubular flowers in a terminal raceme. At the fruiting stage the seed capsules are broadly ovoid rather than long, narrow and terete like those of *G. cassythoides*. Stems are 30 to 60 cm tall and dark to medium brown with a few short scaly bracts. The narrow bell-shaped flowers are unusual for an orchid since the sepals and petals are fused into an inflated tube with a brown warty exterior. The tube is constricted near the end before opening into 5 short pale lobes and exposing the tip of the yellow labellum in the short split between the lateral sepals. The common names refer respectively to the tuberous rhizomes and the cinnamon-like odour of the flowers.

HABITAT Cool moist forests, often in sandy soil but preferring well-drained humus-rich fertile soils.

DISTRIBUTION All states and NZ. Scattered from the coast to the western parts of the tablelands in NSW.

FLOWERING November–January.

1431. Spiranthes sinensis

Spiranthes

The genus *Spiranthes* contains about 50 very widely distributed species of which one extends to Sydney.

Spiranthes sinensis ssp. *australis*

Austral Ladies' Tresses **Plate 1431**

A quaint little orchid that almost requires magnification for one to appreciate its intricate beauty. The slender stem rising 15 to 35 cm from a cluster of linear leaves has 3 or 4 short bracts below a long spiral plait of closely packed tiny bright pink and white flowers. In each flower the petals and dorsal sepal form a protective hood over the column which is partly submerged in the white fringed labellum. The lateral sepals point forward on either side of the labellum.

HABITAT In a variety of open situations, mostly on wet sandy or swampy ground, sometimes forming sparse colonies.

DISTRIBUTION Coast and tablelands of NSW; also south-eastern Qld, Vic, SA, Tas, NZ, New Guinea and Asia.

FLOWERING January–March.

Liparis

1432. Liparis reflexa

Liparis reflexa

Yellow Rock Orchid **Plate 1432**

Facetiously called the wet dog orchid because of its unpleasant odour, this species grows densely on sandstone outcrops in shady forested gullies deriving nourishment from soil and humus trapped in the root mass. As with other members of the genus, the stems are swollen into short ovoid pseudobulbs from which rather thick long channelled leaves emerge almost distichously on opposite sides of the flowering scape. Scapes are usually 10 to 20 cm long with about 6 to 20 or more small greenish-yellow flowers in a loose raceme. The narrow sepals and petals are 10 to 15 mm long, at first spreading, but later reflexed. The labellum, broader than the other segments, is sharply reflexed about the middle and has a blunt irregularly shaped apex.

HABITAT Rock faces, ledges and crevices.

DISTRIBUTION Widespread and common along the NSW coast, extending to the western slopes in the central part; also south-eastern Qld.

FLOWERING March–May.

Dendrobium

A genus of about 1000 species ranging from eastern Asia and Japan through Malesia, Australia and the Pacific Islands. About 45 Australian species, mostly endemic, 9 in the Sydney district growing on rocks and trees. Stems, leaves and general form diverse. Flowers in racemes, the lateral sepals forming a short spur at their base, often similar in length to the petals but usually broader. Labellum similar in length to the other segments or shorter, usually 3-lobed, the lateral lobes embracing the column, the mid-lobe recurved or spreading.

Dendrobium speciosum var. *speciosum*

Rock Orchid **Plate 1433**

Easily the showiest of the local species, *D. speciosum* forms large clumps on rock faces and rocky outcrops, but unfortunately those in accessible places are often grazed by wallabies and deer or 'rescued' by acquisitive orchid fanciers. Mature plants have numerous swollen yellowish and brownish green ribbed stems 10 to 40 cm high, encircled by scars left by membranous bracts and topped by 2 to 5 thick leathery dark green ovate leaves. Racemes in the upper axils are up to 30 or 40 cm long with many white, cream or yellow fragrant flowers on long stalks. The flowers, which usually do not open widely, are 20 to 25 mm long with narrow segments. The labellum is shorter than the other segments and usually marked with purple or red. The common name, Rock Lily, should be discouraged; calling an orchid a rock lily is akin to calling a horse a short-necked giraffe.

HABITAT Sandstone cliffs and outcrops in sheltered but often sunny and rather dry places.

DISTRIBUTION Northern Qld to eastern Vic, mainly coastal in NSW but extending to a few places on the central western slopes.

FLOWERING August–September.

1433. *Dendrobium speciosum*

1434. *Dendrobium aemulum*

Dendrobium aemulum

Ironbark Orchid **Plate 1434**

This species grows on the trunks of Myrtaceous rainforest trees and on eucalypts with persistent bark, particularly the ironbark, *E. paniculata*. In rainforest it usually has long slender cylindrical pseudobulbs or stems, but in drier open forest it has short, stout, rather crowded pseudobulbs and smaller leaves. Pseudobulbs have usually 2 or 3 ovate leaves and 1 to 4 short racemes each carrying 4 to 10 fragrant flowers. These are wholly white or tipped with pale yellow; the labellum and the spur at the base of the lateral sepals are often tinged with reddish purple.

HABITAT On the trunks and branches of Myrtaceous trees in rainforest and open forest.

DISTRIBUTION Coastal districts from northern Qld to the Clyde River in southern NSW.

FLOWERING September–October.

1435. *Dendrobium tetragonum*

Dendrobium tetragonum

Tree Spider Orchid **Plate 1435**

Although variable in flower colour and size over its extensive ranges, *D. tetragonum* is easily recognised by its distinctive 4-sided pseudobulbs (stems) which taper to both the base and the apex. It forms small pendulous clusters of about 3 to 8 stems on the trunks of rainforest trees, or occasionally on rocks. The stems have long narrow bases, become thickened above the middle then taper shortly to the apex which bears 2 or 3 broad oblong-lanceolate leaves 3 to 7 cm long. Short racemes of fragrant flowers develop from the leaves or from nodes on the bare stems. Local forms have rather small flowers with the sepals rarely exceeding 25 mm in length. They are yellowish green with brownish red markings. The petals are yellow, narrow and shorter than the sepals, the labellum short and broad with brownish or reddish markings.

HABITAT Subtropical rainforest, usually on the trunks of trees (especially *Backhousia* adjacent to shady streams), less often on rocks.

DISTRIBUTION Northern Qld to near Nowra in NSW, chiefly coastal.

FLOWERING September–November.

1436. Dendrobium teretifolium

Dendrobium teretifolium var. *teretifolium*

Pencil Orchid, Rat's-tail Orchid **Plate 1436**

In favourable locations in northern New South Wales and Queensland this is often a very spectacular orchid bearing massive sprays of white to yellow flowers. However, local forms are usually much less showy and sometimes have only a few flowers in the raceme. The stems do not form swollen pseudobulbs; instead they are long, wiry and extensively branched and bear numerous slender, terete, pendulous leaves 10 to 50 cm long. The racemes have few to many white or creamy flowers with long narrow-linear perianth segments. A form from the Blue Mountains, known as var. *fairfaxii*, has longer, more slender leaves and small racemes of yellowish flowers with red-brown markings at the base of the perianth segments.

HABITAT Epiphytic on trees or less commonly on rocks, the coastal form usually in quiet estuaries on *Casuarina glauca*, the Blue Mountains form in cool moist rainforest.

DISTRIBUTION Northern Qld to south coast of NSW; also New Guinea, New Caledonia and Pacific Islands.

FLOWERING Chiefly August–September.

1437. Dendrobium linguiforme

Dendrobium linguiforme

Tongue Orchid **Plate 1437**

This small orchid is often seen clinging to rock surfaces in sheltered moist gullies, or occasionally in exposed dry places where its leaves become dessicated and purplish. It is easily recognised by its thick leathery ovate leaves, 2 to 3 cm long, which have 3 or 4 shallow longitudinal grooves or ridges on the upper surface. In spring it bears attractive sprays of white flowers with narrow perianth segments. The labellum is faintly marked with purple and has 3 narrow ridges along its upper surface. It is 3-lobed, the lateral lobes broad, the mid-lobe narrow, recurved and with undulate margins.

HABITAT On rock surfaces, usually in eucalypt forest.

DISTRIBUTION Coast and tablelands of NSW, southern Qld and New Caledonia.

FLOWERING Late September–October.

Dendrobium pugioniforme

Dagger Orchid **Plate 1438**

This species grows in large pendulous masses high on the trunks and branches of tall rainforest trees, its presence often only becoming obvious on fallen trees. The long, branched stems form a tangled network and produce numerous erect or hanging rigid ovate to lanceolate leaves 2 to 5 cm long with a sharp dagger-like point. The flowers are solitary, or 2 or 3 together, on rather short stalks. They are greenish white with purple markings on the labellum and brown lines at the base of the other segments. Lateral sepals are broad and spreading, the labellum recurved, narrow at the base then expanding towards the tip. The upper surface of the labellum has 3 narrow ridges down the centre.

HABITAT Epiphytic on rainforest trees, favouring cool rainforests, rarely found on rocks.

DISTRIBUTION Coast and tablelands of NSW; also southern Qld.

FLOWERING October–November.

1438. Dendrobium pugioniforme

1439. Dendrobium striolatum

Dendrobium striolatum

Streaked Rock Orchid **Plate 1439**

Often locally abundant, the Streaked Rock Orchid forms dense mats on rocks and cliff faces. These mats have many pointed terete leaves about 4 to 7 cm long and 2 to 3 mm in diameter arising from branched stems on a creeping rhizome. The leaves taper gradually to both ends, are usually slightly curved (falcate) and have several shallow grooves running from the base to the tip. Numerous small yellow flowers streaked with red or brown on the sepals and petals are produced in spring. The flowers are solitary, or 2 together, on long stalks rising from the leaf bases. Sepals are 10 to 13 mm long and relatively broad; the petals are shorter and narrower. The labellum is 3-lobed, the mid-lobe recurved with white undulate-crisped margins.

Habitat On rock surfaces and cliff faces in rainforest and in the more protected cool parts of open eucalypt forest.

Distribution Hunter River in NSW to eastern Vic and north-eastern Tas, extending to the tablelands and slopes in central NSW.

Flowering September–November.

Bulbophyllum

A large mainly tropical genus of over 1000 species with about 20 in Australia and 3 in the Sydney district. Epiphytic herbs forming an extensive network of creeping stems bearing small to minute pseudobulbs and fleshy leaves. Flowers usually very small, solitary or in racemes. Sepals erect, free and forming a short blunt spur at the base. Petals small, the labellum often attached to the column by a short claw.

1440. Bulbophyllum crassulifolium

Bulbophyllum crassulifolium **Plate 1440**

This species forms an extensive network of creeping stems often covering large areas of rock surfaces. The pseudobulbs are minute but the thick, fleshy, grooved leaves are very noticeable along the stems. Although the flowers are often numerous they are small, on rather short stalks and occur singly, hence they are usually inconspicuous. The 3 sepals are white and about 3 mm long but the petals are very tiny and difficult to see. The labellum is yellowish and visible between the sepals.

Habitat Sandstone outcrops in moist cool gullies in open forest and rainforest, frequently near the ocean or along estuarine creeks.

Distribution Coastal districts of NSW, extending to the central and northern tablelands; also Qld.

Flowering Chiefly August–October.

1441. Bulbophyllum exiguum

Bulbophyllum exiguum

Autumn Bulbophyllum **Plate 1441**

Both the pseudobulbs and the leaves of this species are readily seen. The pseudobulbs are about 5 mm in diameter and usually deeply furrowed, the leaves solitary on the pseudobulbs, oblong-linear, pointed and varying in length from 10 to 40 mm. Flowers are small, whitish or slightly green, in small racemes of only a few flowers. As with other species, the sepals are larger and more noticeable than the petals, and the labellum is yellowish.

Habitat Usually in rainforest on the trunks of trees and on rocks, sometimes covering large areas.

Distribution Chiefly coastal, but extending to the Blue Mountains; also Qld.

Flowering March–May.

Dipodium

1442. Dipodium punctatum

Dipodium punctatum

Hyacinth Orchid **Plate 1442**

Hyacinth Orchid grows to about 50 cm high displaying numerous attractive pink-to-red-spotted flowers in a terminal raceme. Being a saprophyte dependent on decaying vegetable matter for nutrition, it lacks chlorophyll. Sometimes the stem is greenish, although more often it is dark brown with several short scaly bracts below the inflorescence. Sepals and petals are free, spreading, about 15 mm long and similar to each other. The labellum, attached directly to the base of the column, is 3-lobed with small lateral lobes and a pubescent mid-lobe.

Habitat In sandy soils in scrub and open forest.

Distribution Occurs widely and commonly on the coast, tablelands and slopes of NSW; also Qld, Vic, Tas and SA.

Flowering Chiefly November–January.

Cymbidium

Cymbidium is a genus of 40 to 50 species distributed from Africa through Asia to Australia where there are 3 endemic species, only 1 of which occurs in the Sydney district.

1443. *Cymbidium suave*

1444. *Sarcochilus falcatus*

Cymbidium suave

Snake Flower **Plate 1443**

This species is most frequently seen growing in the decayed heart-wood of tree stumps and logs, forming large clumps of strap-shaped leaves 20 to 40 cm long and 1 to 2 cm wide. Less often it grows in the hollow branches of large trees or in the papery bark of some *Melaleuca* species. Its extensive fleshy root system usually penetrates deeply inside its host. In spring and summer it bears large, dense racemes of fragrant golden or greenish flowers which are sometimes indistinctly blotched with red. Sepals and petals are 10 to 12 mm long, rather broad and spreading. The labellum is slightly shorter and reddish brown near the base but yellow near the tip.

HABITAT In stumps, logs and tree branches and on paperbarks, usually in open forest.

DISTRIBUTION Eastern NSW and Qld, chiefly coastal but extending at times to the tablelands and slopes.

FLOWERING October–December.

Sarcochilus

A genus of about a dozen species, almost exclusively Australian, 9 species in NSW, 4 in the Sydney district. Herbs with broad lanceolate to linear leaves crowded along short stems. Flowers often small, few to many in axillary racemes. Sepals and petals free, spreading, nearly equal. Labellum articulate, 3-lobed, the lateral lobes large and erect, the mid-lobe small. Column short and erect.

Sarcochilus falcatus

Orange-blossom Orchid **Plate 1444**

An attractive orchid, this species is now uncommon in many parts of its range due to collecting and destruction of its habitat. It has thick, falcate oblong-lanceolate leaves 5 to 15 cm long crowded along the short stem and, when not in flower, resembles *S. olivaceus*. The flowers, however, are predominantly white compared to green or golden for the latter species. Flowers are in short racemes which hang downward below the leaves. They are 20 to 35 mm across, larger than in some species, with broad spreading segments nearly equal in size. The labellum is distinctly marked with dark red lines over an orange background on the broad erect lateral lobes.

HABITAT On trunks and branches of rainforest trees, often in association with lichens and Rock-felt Fern *(Pyrrosia)*.

DISTRIBUTION Northern Qld to eastern Vic, mainly coastal but extending to the tablelands, e.g., at Robertson.

FLOWERING September–November.

1445. Sarcochilus australis

1446. Sarcochilus hillii

Sarcochilus australis

Butterfly Orchid **Plate 1445**

This plant resembles *S. falcatus*, although, in local forms at least, it is less robust in all respects. It has 3 to 6 lanceolate leaves about 5 to 10 cm long, crowded on short stems and shorter than the rather sparse racemes. These are up to 10 cm long with 3 to about 10 small fragrant greenish-brown flowers with a mainly white labellum. The dorsal sepal is slightly larger and broader than the other segments. The labellum is prominently spurred at the base and has 2 erect lateral lobes blotched with yellow and streaked with red or purple. The mid-lobe is small.

HABITAT Epiphytic on rainforest trees and shrubs, often on small branches.

DISTRIBUTION Eastern states of Australia, mainly coastal in NSW but extending to the southern tablelands.

FLOWERING October–December.

Sarcochilus hillii

Little Gem Sarcochilus **Plate 1446**

This is a small epiphyte with an extensive root system connecting small clusters of narrow linear, channelled, spotted leaves about 3 to 6 cm long. Racemes are usually shorter than the leaves and consist of 2 to 6 tiny white to pink fragrant flowers. Perianth segments are nearly equal, about 3 to 4 mm long and relatively broad. The labellum is short and broad with a prominent spur at the base beneath and a densely hairy white mid-lobe.

HABITAT Epiphytic on the trunks and small branches of trees in moist open forest and marginal rainforest, frequently near steams growing on *Backhousia myrtifolia.*

DISTRIBUTION Coastal districts from central Qld to Bega, NSW.

FLOWERING October–February.

Plectorrhiza

Plectorrhiza tridentata

Tangle Orchid **Plate 1447**

This orchid belongs to a small genus of 3 species of which 2 occur in eastern Australia, the other on Lord Howe Island. Tangle Orchid is uncommon in the Sydney district, occurring along well-shaded streams in coastal rainforests where it is occasionally seen hanging in the cool humidity by 1 or a few slender threads. It usually consists of a small tangled mass of roots and a few narrow-lanceolate leaves 5 to 10 cm long produced along slender aerial stems. Flowers are tiny in small racemes of about 3 to 10. The perianth segments are spreading, similar in size and brownish or greenish with a pale labellum. The dorsal sepal is hooded over the column and slightly broader than the lateral sepals; petals are slightly narrower. The labellum has 2 prominent acute lateral lobes and a short obtuse mid-lobe.

HABITAT Along shady rainforest streams, hanging in the air and attached by a few slender roots to twigs and branches.

DISTRIBUTION Northern Qld to eastern Vic, mainly coastal but is reported from the Blue Mountains.

FLOWERING September–December.

1447. Plectorrhiza tridentata

Glossary

acroscopic: pointing towards the apex of a frond.
acuminate: tapering gradually to a prolonged point.
acute: pointed, sharp.
adnate: fused to a different part, e.g., stamens to petals, anthers to the filament by their whole length..
anadromous: with the first branch towards the apex; applied to veins and pinnules of ferns.
anther: the part of a flower bearing the pollen.
antheridium: the sperm-producing structure on the prothallus, especially on ferns.
anthesis: the time of flowering.
antrorse: directed forwards or upwards.
appressed: pressed closely against; lying flat against.
arborescent: resembling a tree in habit of growth.
archegonium: the egg-producing structure on the prothallus, especially on ferns.
aril: the swollen part of the stalk attaching seed to pod, often covering the testa of a seed.
articulated: jointed; having joints where separation occurs.
awn: a bristle-like projection.
axillary: occurring in the axil or angle between stem and leaf.

barren: lacking sporangia, sterile.
basiscopic: pointing towards the base of a frond.
bipinnate: twice pinnately divided, i.e., with the primary divisions (pinnae) again divided into separate leaflets (pinnules).
bipinnatifid: twice pinnatifid, i.e., with the primary lobes again lobed.
bract: small scale-like modified leaf, associated with a flower or inflorescence.
bracteole: small bract-like structure on the stalk or calyx of a flower.

callus: a thickened mass or lump found on the labellum of some orchids and on the axis of the spikelet of some grasses.
calyx: collective term for the sepals of a flower.
campanulate: bell-shaped.
capsule: a dry fruit composed of more than one carpel and breaking open at maturity to release the seeds.
carpel: a unit of the female part of a flower in which the ovules are enclosed; usually divisible into ovary, stigma and style.
catadromous: with the first branch towards the base; applied to veins and pinnules of ferns.
caudex: a thick, erect trunk.
cauline: borne on the stem or old wood (hence 'cauliflorous').
ciliate: having a margin fringed with fine hairs.
circinnate: coiled from the apex; the coiled arrangement of a young fern frond.
clathrate: forming a lattice; often applied to markings on scales.
cleistogamous: of flowers which never open but are self-pollinating and set fertile seeds.
concolorous: of leaves, the same colour on both surfaces.
confluent: flowing together, merging.
connate: referring to organs of the same kind which are fused together.
cordate: heart-shaped.
coriaceous: leathery.
corolla: the collective term for the petals of a flower.
corymbose: an inflorescence arrangement in which the flower stalks become shorter towards the top so that all flowers are borne at about the same level.
cotyledon: the seed-leaf of the embryo of a seed plant.
crenate: scalloped, with small rounded teeth.
crenulate: slightly or minutely crenate.
crisped: curled, very wavy.
culm: the aerial stem of grasses, sedges, rushes.
cyme: an inflorescence arrangement in which a flower is at the end of the main axis and further flowers occur on lateral stems arising from below it.
cypsela: a dry, one-seeded fruit of Asteraceae.

decompound: divided several times.
decumbent: branches which spread along the ground but whose tips grow upwards.
decurrent: of leaves, extending downwards to form a wing or raised flange on the stem.
dehiscent: opening to shed seeds or spores at maturity.
dentate: toothed.
denticulate: finely toothed.
dichotomous: dividing equally into two, forking.
dicotyledon: a flowering plant whose embryo has two seed-leaves.
digitate: radiating from a single point like the fingers of a hand.
dimorphic: having two forms, e.g., in ferns where sterile and fertile fronds differ.
dioecious: having the male and female flowers borne on different plants.
discolorous: of leaves, a different colour on each surface.
distichous: in two opposite rows in the same plane.
drupe: a succulent one-seeded fruit with the seed enclosed in an inner stony layer.

emarginate: having a shallow notch at the apex.
endemic: natural distribution confined to a particular geographic region.
epicalyx: a whorl of bracts below the flower resembling a second calyx.
epiphyte: a plant growing on another plant, but not parasitic.
exsert: protruding from, as stamens projecting from floral tube.

falcate: sickle-shaped.
fertile: (in ferns) bearing sporangia.
-fid: cleft, e.g., 3-fid.
filament: the stalk of a stamen.
filiform: thread-like.
flexuose: bent from side to side in a zig-zag pattern.
floret: a grass flower; includes the lemma and palea which enclose it.
follicle: a dry fruit formed from one carpel which splits along one of its edges.
free venation: not joining other parts; applied when the veins pass directly from the midrib to the margin without forming a network.
frond: the leaf of a fern, consisting of stipe (stalk) and lamina (blade).
funicle: the stalk of an ovule.

galea: perianth segment of an orchid shaped like a helmet.
gametophyte: a plant producing eggs and sperm, not spores.
gibbose: lobed with a slight swelling to be more convex on one plane than another.
glabrous: without hairs.
glaucous: blue-green in colour, with a waxy whitish bloom which may rub off.
glume: a small chaffy bract at the base of most grass spikelets.

hastate: having the shape of an arrow-head, i.e., triangular, with spreading basal lobes.
hirsute: with long, coarse, spreading hairs.
hispid: with stiff bristly hairs.
hypogynous: referring to stamens which are attached below the ovary.

imbricate: having the edges overlapping.

incurved: bent inwards.
indehiscent: not opening at maturity.
indusium: the pollen cup of Goodeniaceae; also, the tissue covering the sorus of a fern.
inflexed: bent inwards.
inflorescence: the arrangement in which flowers are borne on a plant.
involucre: a ring of bracts surrounding an inflorescence or several flowers.
involute: rolled upwards and inwards.

labellum: a lip; the middle petal of an orchid on which insects land.
lamina: the blade of a leaf.
lanceolate: shaped like a lance-head; long and narrow, tapering at both ends with the broadest part below the middle.
lemma: the lower of the two bracts enclosing a grass flower.
lignotuber: a woody swelling partly or wholly below ground.
ligulate: strap-shaped.
ligule: a membranous or hairy appendage at the junction of a leaf-sheath and blade of grass; the corolla limb of a ray flower of Asteraceae.
linear: long and narrow with parallel sides.
lithophyte: growing exclusively on rocks.
loculus: a compartment within an ovary.
lomentum: a dry fruit, a legume, having distinct constrictions between the seeds and which breaks into one-seeded sections when mature.

Malesia: a botanic–geographic region consisting of Malaysia, Indonesia, Philippines and New Guinea.
marginal: on or along the edge.
membranous: thin and translucent.
monocotyledon: a flowering plant whose embryo has only one seed-leaf.
monoecious: having unisexual flowers with the male and female borne on the same plant.
monopodial: growing on a single main stem and producing lateral branches successively.

obcordate: heart-shaped leaf, attached at the pointed end and notched at the tip.
oblanceolate: roughly lance-shaped but broader above the middle nearer the tip.
oblong: longer than broad, with the sides parallel.
obovate: similar in shape to ovate, but attached by the narrow end.
obtuse: blunt or rounded at the apex.
operculum: the cap covering the bud in *Eucalyptus.*
orbicular: circular.
ovary: the base of a carpel enclosing the ovules.
ovate: shaped like a section through the long axis of an egg and attached by the wider end.
ovoid: an egg-shaped solid.
ovule: a structure in a seed plant which develops into a seed after fertilisation.

palmate: divided like an outstretched hand, with several leaflets arising from the same point.
panicle: a branched inflorescence.
pappus: tuft (or ring) of hairs or scales at the top of the cypsela of the Asteraceae.
pedicel: the stalk of a flower.
peduncle: the stalk of an inflorescence.
penniveined: leaf veins which diverge from the midrib and run parallel to each other to the margin.
perianth: the calyx and corolla of a flower.
petiole: the stalk of a leaf (hence 'petiolate').
phyllode: a flattened and enlarged leaf stalk which looks like and functions like a leaf (as in *Acacia*).
phyllopodium: the base of a leaf or frond.
pinna: the first division of a compound leaf, consisting of stalk and leaflets (plural : pinnae).
pinnate: a compound leaf with leaflets on a common stalk.
pinnatifid: leaves or leaflets cut deeply into lobes but not to the midrib or rhachis.
pinnule: a leaflet of a bipinnate leaf.
plano-convex: flat on one side, convex on the other.
procumbent: lying or spreading along the ground.
proliferous: capable of developing into a separate plant, or having adventitious buds with this capability.
prothallus: the gametophyte of a fern.
pubescent: covered with short downy hairs.
pungent: sharp and stiffly pointed or having an acrid taste or smell.

raceme: an inflorescence with a central axis and a series of lateral flowers on stalks of nearly equal length.
rachilla: the axis of a grass spikelet above the glumes.
rachis: the axis of an inflorescence or of a pinnate leaf.
radical: leaves clustered at the base of the stem.
recurved: curved downwards.
reflexed: bent sharply downwards or backwards.
reniform: kidney-shaped.
reticulate: marked with a network of veins.
retrorse: directed backwards or downwards.
revolute: rolled backwards to the undersurface.
rhachis: the axis of an inflorescence or of a pinnate leaf.
rhizome: an underground stem, usually horizontal.
rhomboid: somewhat diamond-shaped.

saprophyte: a plant deriving its nutrition from dead organic matter; usually lacking chlorophyll.
scabrous: rough to the touch.
scale: (of ferns) a flat papery outgrowth on the surface of rhizomes, stipes and rhachises.
scarious: dry and membranous, not green.
sepal: one of the outer segments surrounding a flower; collectively known as the calyx.
septate: having cross partitions.
septum: a partition.
serrate: toothed like a saw.
serrulate: finely toothed.
sessile: without a stalk.
sorus: a cluster of sporangia in a fern.
spathulate: spoon-shaped.
spike: an unbranched inflorescence with stalkless flowers.
spore: a microscopic cell, produced in a sporangium, which germinates to form a gametophyte.
sporangium: a spore-producing organ.
sporophyll: a leaf-like structure which bears one or more sporangia.
sporophyte: a spore-producing plant.
stamen: the male organ of a flower, consisting of a stalk and anther.
staminode: a sterile stamen.
sterile: barren; (in ferns) lacking sporangia.
stigma: the part of the carpel which receives the pollen, usually at the tip of the style.
stipe: the stalk of a fern frond.
stipule: one of the two appendages found at the base of the leaves in many dicotyledons.
stolon: a prostrate stem that roots at the nodes.
stoloniferous: having stolons.
striate: marked with lines or ridges.
strobilus: a cluster of sporophylls.
subulate: gradually tapering to a fine point.
sympodial: without a single growing point; changing the direction of growth by lateral growth below the apex.
synangium: a group of united sporangia.

tendril: part of a plant modified into a climbing organ.
terete: cylindrical.
terminal: at the apex.
terrestrial: growing in the ground.
testa: the seed coat.
tomentum: a covering of dense matted hairs, like felt.
tripinnate: thrice divided pinnately; the first two divisions referred to as pinnae (primary and secondary), the third as pinnules or ultimate segments.
truncate: ending abruptly as if cut off.
tuber: a swollen underground part of a stem or root.
tubercle: a small swelling or protuberance.
tuberculate: having swellings making the surface rough or raspy.
turgid: swollen, bloated.

ultimate segments: the finest divisions (excluding lobes).
umbel: an inflorescence in which all the flower stalks arise in a cluster and are of about equal length so that the flowers are borne at the same level.
undulate: wavy.
urceolate: urn-shaped.

valvate: meeting edge to edge, not overlapping.
venation: the arrangement of veins in a leaf or frond.
villous: with dense, long, weak hairs.
viscid: sticky.

Bibliography

Books

Armitage, I. *Acacias of New South Wales*. Sydney: Surrey Beatty, 1978.

Baker, M., Corringham, R. and Dark, J. *Native Plants of the Sydney Region*. Katoomba: Three Sisters Publishing, 1986.

Beadle, N.C.W., Evans, O.D. and Carolin, R.C. *Flora of the Sydney Region*. 3rd edn. Sydney: A.H. & A.W. Reed, 1982.

Beadle, N.C.W. *Students Flora of North-eastern New South Wales*. Parts I–V. Armidale: University of New England, 1971–84.

Hall, N., Johnston, R.D. and Chippendale, G.M. *Forest Trees of Australia*. Melbourne: Nelson-CSIRO, 1975.

Brooker, M.I.H. and Kleinig, D.A. *Field Guide to Eucalypts, Vol 1, South-eastern Australia*. Melbourne and Sydney: Incata, 1983.

Burbidge, N.T. and Gray, M. *Flora of the A.C.T.* Canberra: Australian National University Press, 1970.

Bureau of Flora and Fauna. *Flora of Australia*. Vols 1, 3, 4, 8, 19, 22, 25, 29, 46. (other volumes to follow) Canberra: Australian Government Publishing Service, 1981–89.

Costermans, L. *Native Trees and Shrubs of South-eastern Australia*. Adelaide: Rigby, 1983.

Cunningham, G.M., Mulham, W.E., Milthorpe, P.L. and Leigh, J.H. *Plants of Western New South Wales*. Sydney: Soil Conservation Service of NSW, 1981.

Floyd, A.G. *N.S.W. Rainforest Trees*. Parts I–XII. Sydney: Forestry Commission of NSW, 1979–83.

Fuller, L. *Wollongong's Native Trees*. 2nd ed, Wollongong, 1982.

Hartley, W. and Leigh, J. *Plants at Risk in Australia*. Occasional paper No. 3, Canberra: National Parks and Wildlife Service, 1979.

Jacobs, S.W.L. and Pickard, J. *Plants of New South Wales. A Census*. Sydney: Royal Botanic Gardens, 1981.

Morley, B.D. and Toelken, H.R. (eds). *Flowering Plants in Australia*. Adelaide: Rigby, 1983.

Nicholls, W.H. *Orchids of Australia*. Melbourne, Sydney: Nelson, 1969.

Rotherham, E.R., Briggs, B.G., Blaxell, D.F. and Carolin, R.C. *Flowers and Plants of New South Wales and Southern Queensland*. Sydney: A.H. & A.W. Reed, 1975.

Rupp, H.M.R. *The Orchids of New South Wales*. Facsimile, Sydney: Government Printer of NSW, 1969.

Williams, J.B. and Harden, G.J. *Rainforest Trees and Shrubs—a field guide to trees and shrubs of rainforests in NSW and SE Qld using leaf features*. Armidale: University of New England, 1979.

Journals

The following articles dealing with aspects of Sydney's vegetation and plant communities occur in *Cunninghamia*, an occasional publication of National Herbarium of New South Wales, Royal Botanic Gardens, Sydney.

Benson, D.H. The vegetation of the Gosford and Lake Macquarie 1:100 000 map sheet. *Cunninghamia*, 1(4), 1986.

Benson, D.H. Vegetation of Upper Mangrove Creek, Wyong, New South Wales. *Cunninghamia*, 1(1), 1981.

Benson, D.H. Vegetation of the Agnes Banks sand deposit, Richmond, New South Wales. *Cunninghamia*, 1(1), 1981.

Benson, J.S. and Fallding, H. Vegetation survey of Brisbane Water National Park and environs. *Cunninghamia*, 1(1), 1981.

Fallding, H. and Benson, J.S. Natural vegetation and settlement at Macquarie Pass, Illawarra region, New South Wales. *Cunninghamia*, 1(3), 1985.

Keith, D.A. Floristic lists of New South Wales. *Cunninghamia*, 2(1), 1988.

Keith, D.A. The natural vegetation of the Katoomba 1:100 000 map sheet. *Cunninghamia*, 2(1), 1988.

Outhred, R., *et al.* A floristic survey of Ku-ring-gai Chase National Park. *Cunninghamia*, 1(3), 1985.

Guide to Identification

Use of the following tables should make identification easier.

Select the correct table from the choices below then work through the table by choosing the statement in the first column that best describes the specimen in question. Work across the page choosing only from statements that follow the one previously selected. The table will indicate the most likely family. The specimen can then be compared with illustrations of members of that family. If a match is not obtained work through the table again trying different alternatives. Most species in the area should be identifiable by this means.

Botanical terms are avoided wherever possible so that a minimum of botanical knowledge is required. Terms such as leaf, berry, head, toothed, etc., which have precise and restricted botanical meanings are used here in their broadest sense.

The plants are divided into 5 major groups; the first 4 consist of terrestrial plants, the fifth contains aquatic plants. The major groups are as follows:

Group 1: Herbaceous soft-stemmed plants less than 50 cm high.
Group 2: Climbers and scramblers, mostly with wiry or fibrous stems greater than 50 cm long.
Group 3: Shrubs with 1 to many woody stems, 10 cm to 3 m high.
Group 4: Trees with 1 or several trunks, greater than 2 m high.
Group 5: Submerged, floating or emergent plants growing in salt or fresh water.

The first 4 groups are further divided on the basis of leaf characteristics such as size, shape and arrangement. Each subgroup is treated in a separate table as set out below.

Group 1 Herbs

Table 1.1 Leaves simple, opposite or whorled.
Table 1.2 Leaves simple, alternate, stipules absent.
Table 1.3 Leaves simple, alternate, stipules present.
Table 1.4 Leaves narrow, linear with parallel longitudinal veins, mostly at the base of erect flowering stems.
Table 1.5 Leaves basal, tufted or flat on the ground, not narrow linear, without parallel veins.
Table 1.6 Leaves absent or reduced to sheaths or scales on stems.
Table 1.7 Leaves compound (divided into 3 or more separate leaflets).

Group 2 Climbers and Scramblers

Table 2.1 Leaves simple, opposite or whorled.
Table 2.2 Leaves simple, alternate or sheathing the stem.
Table 2.3 Leaves compound (divided into 3 or more separate leaflets).

Group 3 Shrubs

Table 3.1 Leaves mostly less than 1 cm long, any shape.
Table 3.2 Leaves more than 1 cm long, lobed.
Table 3.3 Leaves 1 to 10 cm long, less than 5 mm wide.
Table 3.4 Leaves 1 to 10 cm long, 5 to 50 mm wide, margins entire.
Table 3.5 Leaves 1 to 10 cm long, 5 to 50 mm wide, margins toothed.
Table 3.6 Leaves mostly greater than 10 cm long or 5 cm wide.
Table 3.7 Leaves compound (divided into 3 or more separate leaflets).

Group 4 Trees

Table 4.1 Leaves simple, opposite, whorled or clustered.
Table 4.2 Leaves simple, alternate, margins entire.
Table 4.3 Leaves simple, alternate, margins toothed.
Table 4.4 Leaves compound (divided into 3 or more separate leaflets).

Group 5 Aquatic Plants

Table 1.1

Herbs. Leaves simple, opposite or whorled.

			Family
Leaves whorled and/or succulent		Leaves narrow, about 5 mm long, often reddish. Small succulent plant with tiny white or pink flowers among the upper leaves. On rocks or damp ground.	CRASSULACEAE (p.113)
		Leaves thick, fleshy ovate-rhomboidal, 5 to 25 mm long, opposite or in whorls of 4. Flowers in erect terminal spikes 3 cm long. On rocks in rainforest.	PEPEROMIACEAE (p.57)
		Leaves linear, 3 to 10 mm long, in whorls of 4 to 6 on 4-angled stems. Small moss-like plants.	RUBIACEAE (p.275)
		Leaves linear, 3 to 10 cm long, thick, fleshy angular. Flowers purple, showy, about 5 cm diameter. Prostrate succulent near the sea.	AIZOACEAE (p.66)
Leaves opposite, less than 3 cm long	Flowers in terminal spikes or panicles	Leaves 5 to 15 mm long, obtuse, wrinkled, toothed. Flowers purple, 2 lipped. Small herbs with erect stems.	SCROPHULARIACEAE (p.282)
		Leaves 10 to 25 mm, irregularly shaped, dull green, greyish or mealy. Fruit succulent red berries 2 to 3 mm diameter. Herbaceous shrub, often coastal.	CHENOPODIACEAE (p.69)
	Flowers in upper leaf axils	Leaves 2 to 20 mm long, lanceolate and soft or narrow and pungent. Flowers white, star shaped, with 5 deeply notched petals appearing as 10.	CARYOPHYLLACEAE (p.68)
		Leaves 5 to 10 mm long, linear. Flowers white, about 5 mm diameter, 4 lobed, usually on long stalks. Slender herb.	LOGANIACEAE (p.271)
		Leaves 2 to 6 mm long, elliptic, slightly succulent. Flowers blue with yellow centre, solitary. Prostrate herb rooting at nodes.	SCROPHULARIACEAE (p.282)
		Leaves 1 to 3 cm long, linear. Flowers white, papery, in clusters 4 to 10 mm diameter in the leaf axils. Shrubby.	AMARANTHACEAE (p.68)
		Leaves 1 to 3 cm long, narrow or lanceolate, with stipules. Flowers in globular heads, greenish. Stems ridged, smooth or rough hairy. Plant often with foetid smell.	RUBIACEAE (p.275)
Leaves opposite, mostly more than 3 cm long		Leaves with stinging hairs, irregularly toothed, 2 to 10 cm long, thin, soft. Flowers small, greenish. Weed.	URTICACEAE (p.63)
		Leaves broad ovate, 2 to 7 cm long, stalked, toothed, softly hairy. Flowers bluish, 2-lipped, the lower lip larger, in whorls forming a long terminal inflorescence. On rocks in gullies.	LAMIACEAE (p.289)
		Leaves ovate, 2 to 5 cm long. Flowers purple, 2-lipped, in whorls of 6 forming a dense terminal spike. Small erect herb.	LAMIACEAE (p.289)
		Leaves 1 to 5 cm long. Flowers pale blue or pinkish, tubular with 5 spreading lobes, the lowest pink dotted, in a terminal raceme. Small upright herb.	ACANTHACEAE (p.286)
		Leaves 1 to 5 cm long. Flowers blue, tubular with 5 equal rounded lobes, 10 to 20 mm across, axillary. Small sprawling herb.	ACANTHACEAE (p.286)
		Leaves very unequal in size in each pair, the larger 5 to 8 cm long. Young branches and leaves softly hairy. Scrambling semi-epiphyte on rocks in rainforest.	GESNERIACEAE (p.284)

Table 1.2

Herbs. Leaves simple, alternate, stipules absent.

			Family
Flowers brightly coloured	Flowers white or pink	Flowers with a sensitive (moveable) column and 4 spreading pink lobes, in a narrow erect arrangement.	STYLIDIACEAE (p.300)
		Flowers 5 to 25 mm diameter, solitary or several on erect stems. Leaves with sensitive dewy glandular hairs.	DROSERACEAE (p.106)
		Flowers 5 to 15 mm diameter with 2 sepals and 4 or 5 petals, in racemes. Small spreading herb with thick fleshy leaves 1 to 5 cm long, usually on mossy rock faces.	PORTULACACEAE (p.67)
		Flowers 10 to 30 mm diameter, with 4 notched petals above a long slender 4-sided ovary.	ONAGRACEAE (p.223)
		Flowers 10 mm diameter, white, open bell shaped with 5 lobes and 5 stamens opposite the lobes alternating with 5 slender staminodes. Swampy ground.	PRIMULACEAE (p.98)
		Flowers 10 to 60 mm diameter, white, woolly. Shrubby or wiry herb with deeply dissected leaves.	APIACEAE (p.250)
		Flowers tiny, white, in umbels. Leaves with 3 to 5 segments, the segments themselves often lobed or dissected.	APIACEAE (p.250)
		Flowers daisy-like with radiating petal-like rays or stiff papery bracts.	ASTERACEAE (p.306)
		Flowers pink, bell or funnel shaped. Twining herbs with linear to heart-shaped leaves.	CONVOLVULACEAE (p.277)
		Flowers small, white, tubular with 5 narrow spreading lobes, in erect terminal spikes.	STACKHOUSIACEAE (p.262)
	Flowers white to mauve, purple or blue	Flowers about 20 mm diameter with 5 rounded pale blue petals with dark blue streaks.	LINACEAE (p.243)
		Flowers 10 to 30 mm diameter, bell or star-shaped with 3 to 5 blue lobes.	CAMPANULACEAE (p.297)
		Flowers daisy-like with radiating petal-like rays.	ASTERACEAE (p.306)
		Flowers blue with 3 petals, enclosed by a folded green spathe.	COMMELINACEAE (p.325)
		Flowers tubular, 10 to 20 mm long, with a 3-lobed lower lip and 2 smaller upper lobes. Stamens forming a tube surrounding the style.	LOBELIACEAE (p.298)
		Flowers with a 3-lobed lower lip and a 2-lobed upper lip of similar size. Stamens forming a 2-lipped indusium surrounding the stigma.	GOODENIACEAE (p.301)
		Flowers yellow centred with 5 upward-facing spreading lobes.	GOODENIACEAE (p.301)
	Flowers yellow	Flowers greenish yellow, tiny, tubular, with 5 narrow triangular lobes, in interrupted erect spikes.	STACKHOUSIACEAE (p.262)
		Flowers daisy-like, with radiating soft petal-like rays or stiff papery bracts.	ASTERACEAE (p.306)
		Flowers with a 3-lobed lower lip and a 2-lobed upper lip. Stamens forming a 2-lipped hairy indusium surrounding the stigma.	GOODENIACEAE (p.301)
Flowers insignificant or surrounded by scales or husks	Flowers tiny, green or mealy, solitary or clustered in leaf axils, sessile. Fruit a green, yellow or red berry about 5 mm diameter. Leaves succulent.		CHENOPODIACEAE (p.69)
	Flowers enclosed by dark scales, in large, erect plume-like inflorescences above narrow scabrid leaves. Sedge in damp ground.		CYPERACEAE (p.334)
	Flowers with 2 husks. Tufted herbs with erect hollow stems and leaves at the nodes.		POACEAE (p.349)

Table 1.3

Herbs. Leaves simple, alternate, stipules present.

			Family
Leaf blades about as long as broad		Flowers tiny, white, pink or yellowish, in umbels.	APIACEAE (p.250)
		Flowers pansy-like, white to mauve, on long stalks.	VIOLACEAE (p.84)
		Flowers with 5 petals and 5 sepals, white or pink, axillary, single or paired.	GERANIACEAE (p.243)
Leaves much longer than broad	Leaves narrow linear, less than 5 cm long	Flowers blue, in slender arrangement, the lowest petal largest, about 12 mm long.	VIOLACEAE (p.84)
		Flowers white, tiny, in flat or rounded small or large heads.	EUPHORBIACEAE (p.98)
	Leaves otherwise	Leaves 3 to 15 cm long, ovate to lanceolate with entire margins. Stipules forming a membranous sheath around the stem. Flowers white or pink, mostly in spikes.	POLYGONACEAE (p.70)
		Leaves 10 to 17 cm long, elliptical, toothed, upper surface dull. Flowers tiny, in dense heads 1 to 2 cm diameter. In rainforest.	URTICACEAE (p.63)

Table 1.4

Herbs. Leaves narrow linear with parallel longitudinal veins, mostly at the base of erect flowering stems.

			Family
Flowers with 3 rounded lobes		Flowers yellow, 2 to 3 cm diameter, subtended by dark scaly bracts in heads on leafless stalks.	XYRIDACEAE (p.326)
		Flowers blue-purple, 2 to 3 cm diameter, enclosed in 2 large smooth or hairy bracts.	IRIDACEAE (p.356)
Flowers with 4 lobes		Flowers pink, with 4 lobes in 2 pairs and a slender trigger-like column. Leaves 2 to 20 cm long, 1 to 10 mm wide, tufted at base, sometimes also on stems.	STYLIDIACEAE (p.300)
Flowers with 6 free segments	Stamens 3	Flowers white, 6 to 25 mm diameter, in clusters. On margins of rainforest or on damp rocks in Blue Mountains.	IRIDACEAE (p.356)
	Stamens 6	Flowers yellow, solitary or a few, segments nearly equal. Leaves 10 to 18 cm long, about 1 mm wide, with long white hairs. Small herb.	HYPOXIDACEAE (p.374)
		Flowers blue, purple, white or yellow, segments all alike or inner whorl broader or fringed.	LILIACEAE (p.358)
		Flowers white, large, narrow tubular with 6 narrow spreading lobes, in a terminal umbel subtended by 2 scaly brown bracts. Leaves large. Swamps.	AMARYLLIDACEAE (p.373)
	Stamens not free	Stamens fused into an erect column. Flowers regular or the segments highly modified, on stems bearing 1 to 3 bracts. Leaves 1 to 5, usually thick, chanelled above.	ORCHIDACEAE (p.374)
			Family
Flowers bell-shaped		Flowers orange-red, 25 to 45 mm long, pendulous, 3 to 15 at the top of erect stems. In wet ground.	LILIACEAE (p.358)
Flowers irregularly shaped		Flowers 1 to many in a terminal spike or raceme. Leaves 1 to 5, often thick, channelled above and clasping the lower part of the flowering stem. Stems with 1 to 3 bracts.	ORCHIDACEAE (p.374)

Table 1.4

Herbs. Leaves narrow linear with parallel longitudinal veins, mostly at the base of erect flowering stems (continued).

		Family
Flowers tiny and lacking colourful segments or lobes	Leaves 1 to 5 cm long, crowded, fine, forming a round tuft 2 to 6 cm diameter. Flowering stems bearing a tiny terminal spikelet with 2 sheathing bracts. Wet sandy soil.	CENTROLEPIDACEAE (p.332)
	Leaves 2 to 8 cm long with broad sheathing bases, tapering to apex. Flowers in dense globular greyish heads 6 to 8 mm diameter. Wet ground.	ERIOCAULACEAE (p.328)
	Leaves 1 to 2 mm long, 10 to 25 mm wide, tough, fibrous. Stems flat or angular with scabrous edges. Flowers in narrow green spikes. Rainforests.	ARACEAE (p.324)
	Leaves flat or cylindrical, tufted, or reduced to a sheath at the base of flowering stems. Flowers with 6 scaly segments in 2 whorls, arranged in loose cymes or heads. In or near water. Rushes.	JUNCACEAE (p.332)
	Leaves at base of stems or reduced to sheaths. Sheaths closed (i.e. not split on one side). Stems solid without nodes. Sedges and rushes.	CYPERACEAE (p.334)
	Leaves at the base and at nodes on flowering stems. Leaves with a flat blade and an open sheath (i.e. split on one side). Stems hollow with solid nodes. Grasses.	POACEAE (p.349)

Table 1.5

Herbs. Leaves basal, tufted or flat on the ground, not narrow linear, without parallel veins.

			Family
Leaves solitary		Leaves heart-shaped, 2 to 5 cm long. Flowers solitary above the leaf, or in a few flowered raceme 5 to 18 cm high.	ORCHIDACEAE (p.374)
Leaves 2		Leaves stalked or stem clasping, 1 to 6 cm long, opposite each other. Flowers solitary on stem 10 to 12 cm high.	ORCHIDACEAE (p.374)
Leaves forming a flat basal rosette		Leaves ovate to broad lanceolate, soft, green, glabrous. Flowers green or tinged with brown, solitary or a few on upright soft stems.	ORCHIDACEAE (p.374)
		Leaves spoon shaped and reddish with dewy glandular sensitive hairs. Insectivorous plant growing in damp soils.	DROSERACEAE (p.106)
Leaves forming a basal rosette or tuft	Flowering stems leafless	Leaves large, hastate-cordate, up to 90 cm long on long spongy stalks. Flowers tiny, in a spike subtended by a green bract (spathe).	ARACEAE (p.324)
		Leaf blades broad lanceolate, 2 to 7 cm long on stalks 2 to 12 cm, arising from the base. Flowers violet with 5 petals, on long erect stalks.	VIOLACEAE (p.84)
		Leaves obovate, 5 to 10 cm long, silky hairy. Stems 15 to 40 cm high with many small flowers forming a blue cushion 10 to 30 cm diameter at the top. Higher mountains.	BRUNONIACEAE (p.306)
		Leaves oblanceolate or linear, up to 20 cm long. Flowers daisy-like, white, blue to pink, yellow, 15 to 40 mm diameter, solitary on unbranched stems.	ASTERACEAE (p.306)
		Leaves spathulate, 3 to 10 cm long, with entire or irregularly toothed margins, tapering gradually to the base. Flowers yellow, in a spike or an open arrangement.	GOODENIACEAE (p.301)
	Stems leafy	Leaves succulent, 7 to 30 mm long, mostly basal. Spreading stems with smaller leaves, terminating in racemes of pink flowers 5 mm diameter. On rocks.	PORTULACACEAE (p.67)

Table 1.6

Herbs. Leaves absent or reduced to sheaths or scales on stems.

		Family
Slender twining parasites	Flowers 2 mm diameter, with 6 perianth segments in 2 whorls.	CASSYTHACEAE (p.56)
	Flowers clustered, 3 mm diameter, 4 or 5 lobed with stamens alternating with the lobes.	CONVOLVULACEAE (p.277)
Plants not parasitic	Saprophyte with dark brown stems, erect or flexuous and climbing. Flowers brownish white or yellowish, in racemes or pendulous terminal clusters.	ORCHIDACEAE (p.384)
	Small succulent herb with erect jointed stems growing in coastal salt marshes.	CHENOPODIACEAE (p.69)
	Bog carnivorous herb with 2-lipped white to purplish blue flowers on slender erect stems.	LENTIBULARIACEAE (p.285)
	Rush with leaves reduced to open sheaths, overlapping on lower part of stem, distant on upper. Flowers in spikelets.	RESTIONACEAE (p.328)
	Rush with leaves reduced to open sheaths at base of stem. Flowers with 6 scaly perianth segments, in clusters.	JUNCACEAE (p.332)
	Sedge with leaves reduced to closed tubular sheaths at base of stem. Flowers in spikelets.	CYPERACEAE (p.334)

Table 1.7

Herbs. Leaves compound (divided into 3 or more separate leaflets).

		Family
Flowers less than 5 mm diameter	Flowers reddish, in dense globular heads or long interrupted spikes. Leaves pinnate with bases sheathing the stem.	ROSACEAE (p.113)
	Flowers white, in umbels. Leaves deeply dissected into 3 to 5 segments, the segments themselves often lobed or dissected.	APIACEAE (p.250)
Flowers more than 5 mm diameter	Flowers shiny yellow with usually 5 petals. Leaves pinnate, lobed or dissected to the base.	RANUNCULACEAE (p.58)

Table 2.1

Climbers and scramblers. Leaves simple, opposite or whorled.

			Family
Clear or milky sap exuded from cut stems and leaf stalks	Sap clear	Leaves 5 to 15 cm long, tapering to acute apex, rounded at base, leathery, rather dull, veins distinct. Tall climber with wiry stems. Common.	APOCYNACEAE (p.272)
	Sap milky	Leaves 4 to 14 cm long, several tiny raised glands on leaf blade above stalks. Flowers white, yellow or purple, with corona formed from fused anther appendages.	ASCLEPIADACEAE (p.273)
		Leaves 5 to 10 cm long, elliptic/oblong, stalk short or absent. Flowers yellow. Fruit an ovoid orange berry 5 cm long. Uncommon climber in rainforest.	APOCYNACEAE (p.272)
Sap not exuded from stems and leaf stalks	Leaves thick	Leaves 8 to 15 cm long with 3 to 5 distinct veins from base to apex, leathery, opposite, sub-opposite or in whorls of 3. Stems often prickly. In rainforests.	SMILACACEAE (p.370)
	Leaves not thick	Leaves 5 to 7 cm long, lanceolate, acuminate, shiny dark green above. Stipules present on stem. Flowers tiny, white. Fruit an irregular orange berry.	RUBIACEAE (p.275)
		Leaves 8 to 15 cm long, broad elliptic, dull, raspy to touch, branchlets and veins with stellate hairs. Scrambler in rainforest.	MONIMIACEAE (p.53)

Table 2.2

Climbers and scramblers. Leaves simple, alternate or sheathing the stem.

			Family
Leaves with 3 to 7 major veins radiating from the base	Leaves 3 lobed	Climber with axillary tendrils. Leaf stalks 5 to 10 cm long, often with 2 glands near top. Flowers white to red.	PASSIFLORACEAE (p.85)
		Trailing or climbing prickly shrub. Leaf stalks prickly. Leaves 5 to 10 cm long, dark green above, pale or rusty hairy beneath, margins finely toothed.	ROSACEAE (p.113)
	Leaves not lobed	Leaves up to 20 cm long on stalks 4 to 10 cm long, broad ovate, heart shaped or rounded. Rainforest.	MENISPERMACEAE (p.60)
		Leaves 7 to 10 cm long, broad ovate, dark green and shiny above, paler beneath, rather soft, with a sharp peppery taste. Tall climber with adventitious roots. Rainforest.	PIPERACEAE (p.55)
		Leaves 4 to 10 cm long, broad, leathery, with strongly marked reticulate veins. Prickles present or absent. Climber with stipular tendrils at base of leaf stalks.	SMILACACEAE (p.370)
		Leaves 6 to 10 cm long, hastate-cordate, thin, drawn out to a point at apex. Stalks 20 to 30 mm. Fruit a 3-angled capsule. Coastal rainforest.	DIOSCOREACEAE (p.371)
Leaves with numerous longitudinal veins		Leaves 12 to 25 mm long, cordate, stem clasping, fringed with white hairs. A wiry undershrub with red stems and small white tubular flowers in clusters. Gullies.	EPACRIDACEAE (p.86)
		Leaves 5 to 10 cm long, narrow or broad, apex pointed, scattered or regularly arranged. Fruit an orange or purplish black berry. Moist shady places.	PHILESIACEAE (p.370)
		Leaves 10 to 30 cm long, narrow, sheathing base surrounding stem, apex drawn out to a fine coiled tendril. Tall climber in rainforest.	FLAGELLARIACEAE (p.328)
Leaves pinnately veined, or veins obscure	Rain-forest plants	Leaves 4 to 10 cm long, entire or with a few teeth. Bark tough or stout thorns present. Straggling or climbing shrubs with milky sap.	MORACEAE (p.61)
		Leaves 4 to 12 cm long, elliptic-ovate, margins finely toothed. Stems with numerous raised pale outgrowths. Fruit 3 to 10 mm long, orange.	CELASTRACEAE (p.261)
		Leaves 7 to 10 cm long, 6 to 7 cm wide, ovate to heartshaped, margins finely toothed. Tendrils opposite the leaves. Common woody climber with black berries 10 mm diameter.	VITACEAE (p.269)
	Plants not in rain-forest	Leaves 5 to 20 mm long, linear, sparse. Flowers small, blue, in racemes 5 cm long. Slender twiner.	POLYGALACEAE (p.245)
		Leaves 1.5 to 5 cm long, narrow ovate, margins undulate, undersurface hairy. Flowers pale yellow, 2 cm long, on long stalks. Wiry twiner.	PITTOSPORACEAE (p.111)
		Leaves 3 to 9 cm long, lanceolate, leathery, strongly veined. Twiner or shrub with purple pea flowers with a green centre.	FABACEAE (p.131)
		Leaves 4 to 9 cm long, toothed or entire, sheathing or clasping the stem. Flowers yellow 25 to 50 mm diameter. Trailing or climbing plants.	DILLENIACEAE (p.72)

Table 2.3

Climbers and scramblers. Leaves compound (divided into 3 or more separate leaflets).

			Family
Leaves opposite	Leaves with 3 ovate leaflets on stalks up to 10 cm long. Leaflets entire or toothed. Flowers 25 to 50 mm diameter, white with 4 narrow sepals.		RANUNCULACEAE (p.58)
	Leaves with usually 5 to 7 leaflets, pinnate with a terminal leaflet. Leaflets 2 to 6 cm long, ovate, venation faint. Flowers tubular, 20 mm long, white with a purplish throat.		BIGNONIACEAE (p.284)
Leaves alternate	Tendrils present	Leaves with 5 digitate or pedate leaflets. Leaflets 3 to 12 cm long, entire or toothed. Tendrils opposite the leaves.	VITACEAE (p.269)
	Stipules present	Leaves 3-foliolate. Leaflets 3 to 10 cm long, ovate. Stipules 3 to 5 mm, triangular, ribbed. Flowers 4 cm long, dusky red. Fruit a hairy pod 5 to 10 cm long.	FABACEAE (p.131)
		Leaves 3-foliolate. Leaflets 1 to 5 cm long, linear to lanceolate, thin, soft. Stipules narrow. Flowers 8 mm long, mauve. Small twiner.	FABACEAE (p.131)
	Prickles present	Leaves with 3 or 5 toothed leaflets, pinnate or digitate. Flowers white, pink or red. Fruit a red raspberry. Trailing or climbing shrub.	RUBIACEAE (p.275)
	Tendrils stipules and prickles absent	Leaves 3-foliolate on stalks 7 cm long with bristly hairs. Leaflets 5 to 10 cm long, lanceolate, acuminate, thin, glossy, margins and midribs minutely scabrous.	ARALIACEAE (p.248)

Table 3.1

Shrubs. Leaves mostly less than 1 cm long, any shape.

		Family
Leaves reduced to tiny scales	Shrub less than 0.5 m.	EUPHORBIACEAE (p.98)
	Shrub or tree greater than 0.5 m.	SANTALACEAE (p.255)
	Parasitic shrub entwining other plants.	CASSYTHACEAE (p.56)
Leaves sharply pointed	Flowers yellow, in globular heads.	MIMOSACEAE (p.115)
	Flowers tubular, mostly white.	EPACRIDACEAE (p.86)
Flowers white or pink	Flowers with 4 petals, nodding, centre dark.	TREMANDRACEAE (p.245)
	Flowers with 4 petals, facing upward, centre pale.	RUTACEAE (p.231)
	Flowers with 5 small petals and/or numerous stamens.	MYRTACEAE (p.176)
	Flowers with 5 spreading petals and 10 stamens.	RUTACEAE (p.231)
	Flowers with 4 to 10 petals, on slender stalks.	BAUERACEAE (p.110)
	Flowers with 6 or more narrow segments.	EUPHORBIACEAE (p.98)
	Flowers tiny, shortly tubular, crowded at ends of branches.	RHAMNACEAE (p.263)
	Flowers tiny, in small or large corymbs.	EUPHORBIACEAE (p.98)
	Leaves warty due to oil glands, strong aromatic odour.	RUTACEAE (p.231)
	Leaves with longitudinal veins on lower surface.	EPACRIDACEAE (p.86)
	Leaves densely hairy.	SOLANACEAE (p.279)
	Stems and leaves covered in hairy tomentum.	RUTACEAE (p.231)
	Plant with fine thorns, orange berries.	PITTOSPORACEAE (p.111)
	Plant prostrate, leaves wrinkled.	STERCULIACEAE (p.78)
Flowers yellow to orange	Flowers with 5 spreading petals.	DILLENIACEAE (p.72)
	Flower narrow tubular with 4 spreading tips.	PROTEACEAE (p.157)
	Flowers with many stamens, in globular heads.	MIMOSACEAE (p.115)
	Flowers red centred, pea-like.	FABACEAE (p.131)
Flowers mauve to purple	Flowers tiny, in large racemes.	POLYGALACEAE (p.245)
	Flowers 2-lipped. Plant with strong smell.	LAMIACEAE (p.289)
	Flowers 1 to 2 cm diameter, with 5 petals, 10 stamens.	RUTACEAE (p.231)

Table 3.2

Shrubs. Leaves more than 1 cm long, lobed.

		Family
Lobes sharply pointed	Leaves and stems with spines. Flowers bluish.	SOLANACEAE (p.279)
	Shrub 1 to 2 m high with yellow or orange pea flowers.	FABACEAE (p.131)
	Large shrub or tree in coastal rainforests. Rare.	EUPHORBIACEAE (p.98)
	Leaves 1 cm long, 3 to 5 lobed. Flowers tiny, white. Small shrub in heath.	APIACEAE (p.250)
Lobes not sharply pointed	Leaves large, thin, lax, irregularly lobed. Flowers yellow, daisy-like.	ASTERACEAE (p.306)
	Leaves linear, pinnately lobed to the midrib. Flowers red, in one-sided racemes.	PROTEACEAE (p.157)
	Leaves large, rough, 3-lobed, dark green above. Flowers large, white to pink.	MALVACEAE (p.82)
	Leaves soft, thick, hairy. Flowers small, whitish. Small shrub.	APIACEAE (p.250)

Table 3.3

Shrubs. Leaves 1 to 10 cm long, less than 5 mm wide.

		Family
Shrub with thorny branches	Leaves entire. Flowers white, small, numerous in terminal panicles.	PITTOSPORACEAE (p.111)
	Leaves entire or irregularly toothed. Flowers yellowish, axillary, 1 or 2 together on slender stalks.	VIOLACEAE (p.84)
Leaves sharply pointed	Leaf veins parallel with midrib. Flowers with a long or short tube. Wiry shrub.	EPACRIDACEAE (p.86)
	Leaves needle-like, 2 to 8 cm long, 1 to 2 mm diameter. Flowers white or yellowish. Fruit large, woody.	PROTEACEAE (p.157)
	Leaves needle-like, 0.5 to 3 cm long. Flowers yellow/red, pea-like. Fruit a small pod.	FABACEAE (p.131)
Leaves opposite	Stems quadrangular. Flowers 2-lipped, white, mauve or purple.	LAMIACEAE (p.289)
	Stems hidden by leaf bases. Flowers yellow, tubular, 2 to 5 cm long. ...	CHLOANTHACEAE (p.289)
Leaves in whorls of 3	Leaves rigid, 3 to 5 cm long, entire, recurved, dark green above, pale beneath. Flowers terminal, red.	PROTEACEAE (p.157)
Branchlets long, cylindrical	Branches flexible, cylindrical, not pungent pointed. Leaves represented by whorls of tiny triangular teeth every 1 to 2 cm.	CASUARINACEAE (p.64)
Flowers in cylindrical spikes	Spikes red, purple, white or creamy bottle-brushes 1 to 10 cm long. ...	MYRTACEAE (p.176)
	Spikes yellow to orange, erect 5 to 20 cm long, 5 to 8 cm diameter. Leaves entire or with recurved toothed margins, 3-toothed at apex.	PROTEACEAE (p.157)
Flowers small, mostly white, in terminal flat or rounded heads 2 to 20 cm across.	Flowers tubular with 4 lobes and 2 orange stamens. Heads 5 to 25 mm wide.	THYMELAEACEAE (p.104)
	Flowers sometimes with bluish tinge, tubular with a broad upper lobe and 3 spreading narrow lower lobes.	PROTEACEAE (p.157)
	Leaves limp, 1 to 3 veined. Flowers in large heads. Brittle shrub to 4 m.	ASTERACEAE (p.306)
	Leaves very narrow. Flowers in rounded heads 2 to 4 cm diameter.	APIACEAE (p.250)
	Leaves with stipules. Flowers in small or large corymbs.	EUPHORBIACEAE (p.98)
Flowers white, pinkish or pink, mostly axillary, not in heads or spikes	Flowers unisexual (i.e. with either stamens or carpels), sepals and petals 4 to 6.	EUPHORBIACEAE (p.98)
	Flowers 5 to 25 mm diameter, with 4 or 5 petals and few to many stamens.	MYRTACEAE (p.176)
	Flowers in spider-like clusters. Flowers and leaf undersurface silky.	PROTEACEAE (p.157)
	Leaves flat or needle-like. Fruit a narrow beaked or rough woody capsule 10 to 40 mm diameter.	PROTEACEAE (p.157)
	Leaves terete, or flat with recurved toothed margins, often warty due to oil glands.	RUTACEAE (p.231)
	Leaves with veins on undersurface parallel to midrib.	EPACRIDACEAE (p.86)
	Leaves in 2 opposite ranks. Slender greyish or yellowish shrub 1 to 2 m high.	OLACACEAE (p.255)
Flowers yellow to orange	Flowers in globular heads, stamens prominent. Leaves thick, usually with raised gland on upper margin.	MIMOSACEAE (p.115)
	Flowers pea-like with reddish centres.	FABACEAE (p.131)
	Flowers 1 to 2 cm long, narrow tubular, 4 lobed at tip.	PROTEACEAE (p.157)
	Flowers with 5 petals, 10 spreading stamens. Pubescent or scurfy shrubs.	RUTACEAE (p.231)
Flowers greenish	Leaves simple, entire or toothed near apex. Fruit orange.	CELASTRACEAE (p.261)
Flowers mauve	Pea flowers. Leaves dark green above, brown hairy beneath. ..	FABACEAE (p.131)

Table 3.4

Shrubs. Leaves 1 to 10 cm long, 5 to 50 mm wide, margins entire.

			Family
Plants parasitic on the branches of other trees.			LORANTHACEAE (p.257)
Straggling shrub with milky sap in twigs and leaf stalks. Uncommon.			MORACEAE (p.61)
Mangrove growing in mud. Leaves with rounded tips. Flowers white.			MYRSINACEAE (p.97)
Shrubs with opposite leaves		Flowers white 5 to 15 mm diameter, with ring of stamens. Petals 4 or 5 or absent.	MYRTACEAE (p.176)
		Shrub covered with stellate tomentum. Flowers tubular with 4 lobes.	RUTACEAE (p.231)
		Flowers white, tubular with 4 lobes, in terminal rounded heads.	THYMELAEACEAE (p.104)
		Leaves thick, stiff, dull, venation visible. Flowers small, creamy.	OLEACEAE (p.270)
		Leaves 5 to 7 mm wide, sticky above, greyish below. Flowers white daisy-like.	ASTERACEAE (p.306)
		Leaf bases connected by a raised line. Flowers small, white, tubular.	LOGANIACEAE (p.271)
		Leaves dark green, on long stalks. Flowers white tubular.	VERBENACEAE (p.288)
Hairy shrubs with alternate leaves		Leaves oblong or heart shaped. Flowers pale or rusty, in drooping clusters.	STERCULIACEAE (p.78)
		Leaves narrow heart shaped. Flowers deep red or violet, on long stalks.	MALVACEAE (p.82)
		Flowers tiny, yellow, in small to large terminal sprays.	RHAMNACEAE (p.263)
		Branches with thick white woolly hairs. Flowers tiny, creamy or greenish.	ARALIACEAE (p.248)
		Leaves thick, irregular in shape. Flowers 20 to 30 mm diameter, white, daisy-like.	ASTERACEAE (p.306)
Glabrous shrubs with alternate leaves (new growth sometimes with fine hairs, or leaf undersurface hairy)	Flowers white, creamy, straw, yellowish, pink or pinkish.	Flowers in cylindrical spikes, with prominent stamens.	MYRTACEAE (p.176)
		Flowers in spider-like clusters. Fruit narrow or round and woody.	PROTEACEAE (p.157)
		Flowers with 5 petals and 10 stamens, 15 to 20 mm diameter. Leaves with oil glands.	RUTACEAE (p.231)
		Flowers stalked, in terminal clusters. Leaves glossy, acute. Stems red. Berry black.	WINTERACEAE (p.52)
		Leaves regular on either side of stems, soft, glossy, broad. In rainforest.	EUPOMATIACEAE (p.53)
		Flowers 1 to 2 mm diameter, with 5 white petals, in terminal heads 2 to 3 cm diameter.	APIACEAE (p.250)
		Flowers daisy-like 1 to 2 mm diameter, in large inflorescences.	ASTERACEAE (p.306)
		Flowers 12 to 15 mm diameter, white, purple-dotted inside.	MYOPORACEAE (p.285)
		Flowers tiny, axillary. Leaves in 2 ranks, thin, very obtuse.	EUPHORBIACEAE (p.98)
	Flowers yellow	Flowers 1 to 2 cm long, tubular, with fine rusty or grey hairs.	PROTEACEAE (p.157)
		Flowers 12 mm long, tubular with 5 petals. Fruit orange with red seeds.	PITTOSPORACEAE (p.111)
		Flowers 15 to 30 mm diameter, with 5 large petals and 12 or more stamens.	DILLENIACEAE (p.72)
		Leaves with 2 to 5 longitudinal veins and a raised gland on upper margin.	MIMOSACEAE (p.115)
		Leaves thick, with prominent veins, often broad with obtuse tip.	FABACEAE (p.131)
		Brittle shrub with soft leaves. Flowers tiny, in heads 2 to 7 cm diameter.	ASTERACEAE (p.306)

Table 3.4

Shrubs. Leaves 1 to 10 cm long, 5 to 50 mm wide, margins entire (continued).

			Family
Glabrous shrubs with alternate leaves (new growth sometimes with fine hairs, or leaf undersurface hairy)	Flowers red	Prostrate shrub on higher Blue Mountains. Leaves broad, rounded.	PROTEACEAE (p.157)
		Leaves usually thick with an acute apex. Flowers in spikes, stamens prominent.	MYRTACEAE (p.176)
	Flowers mauve to bluish purple	Leaves large, thin, sometimes lobed. Flowers 20 to 50 mm diameter.	SOLANACEAE (p.279)
		Leaves soft, greyish, broad at end, tapering to stalk. Bark corky.	SOLANACEAE (p.279)
	Flowers grey or green	Flowers in spider-like clusters. Leaves silky on undersurface.	PROTEACEAE (p.157)
		Flowers greenish, inconspicuous. Fruit a green winged capsule.	SAPINDACEAE (p.226)
		Flowers greenish, on older stems. Leaves glossy. In or near rainforest.	MYRSINACEAE (p.97)

Table 3.5

Shrubs. Leaves 1 to 10 cm long, 5 to 50 mm wide, margins toothed.

			Family
Shrubs with opposite or whorled leaves		Leaves in whorls of 3. Small glabrous shrub from lower Blue Mountains. Flowers white to pink.	CUNONIACEAE (p.108)
		Leaves shiny above, white to rusty beneath. Flowers pale, like large wattle balls.	CUNONIACEAE (p.108)
		Leaves thick, broad, shiny, pale beneath, margins wavy. Fruit orange.	CELASTRACEAE (p.261)
		Leaves with oil glands. Stems square. Flowers 2-lipped, white to purple.	LAMIACEAE (p.289)
		Leaves sessile or stem clasping, tapering to apex. Flowers white or blue. Higher Blue Mountains.	SCROPHULARIACEAE (p.282)
Hairy shrubs with alternate leaves		Plant with grey or rusty tomentum. Flowers pale. Fruit with bristles.	STERCULIACEAE (p.78)
		Leaves thick, irregularly toothed or lobed. Flowers white, daisy-like.	ASTERACEAE (p.306)
Glabrous shrubs with alternate leaves	Leaves thick or stiff	Leaves silvery grey or rusty beneath. Flowers in spikes 5 to 15 cm long.	PROTEACEAE (p.157)
		Leaves glabrous, glossy, sometimes entire or toothed at apex only.	MYRSINACEAE (p.97)
	Leaves rough or dull on upper surface	Milky sap in twigs and leaf stalks. Fruit 5 to 15 mm diameter, yellow-orange.	MORACEAE (p.61)
		Leaves resembling peach leaves, regularly arranged. Fruit 3 mm diameter, black.	ULMACEAE (p.60)
		Leaves narrow hastate, dark green above, paler below, veins raised on lower surface. Thin shrub to 1 m.	MALVACEAE (p.82)
	Leaves smooth on upper surface	Leaves narrow lanceolate, sometimes entire, tapering gradually to stalk. Flowers greenish yellow. Fruit long, narrow, seeds in 2 rows.	PROTEACEAE (p.157)
		Leaves with 2 narrow glands at junction of blade and stalk. In or near rainforest.	EUPHORBIACEAE (p.98)
		Leaves often sticky. Fruit a green or purplish winged capsule.	SAPINDACEAE (p.226)
		Leaves more or less clasping stem. Flowers yellow, daisy-like.	ASTERACEAE (p.306)
		Leaves thin, glossy, entire or toothed at apex. In or near rainforest.	MYRSINACEAE (p.97)

Table 3.6

Shrubs. Leaves mostly greater than 10 cm long or 5 cm wide.

			Family
Shrubs or small trees in or near rainforest	Leaves 10 to 20 cm long, 20 to 40 mm wide, entire, lobed or square at base, pointed at apex. Flowers white, stalked, in terminal clusters.		WINTERACEAE (p.53)
	Leaves 10 to 15 cm long, 40 to 50 mm wide, entire, thick, glabrous, glossy, arranged regularly on either side of stem. Flowers creamy.		EUPOMATIACEAE (p.53)
	Leaves 5 to 15 cm long, elliptic, entire, rough and sandpapery.		MORACEAE (p.61)
	Leaves up to 15 cm long, 70 mm wide, toothed, heart shaped, dull, thin. Flowers small, white, in clusters opposite leaves. Bark tough, fibrous. ..		STERCULIACEAE (p.78)
	Leaves up to 25 cm long, 30 to 40 mm wide, or 3-lobed, toothed, rough. Flowers large, white to pink. Fruit with stiff hairs.		MALVACEAE (p.82)
	Leaves broad at base tapering to fine apex, on stalks 20 to 80 mm long, entire, thin, dull, pale on undersurface, older leaves often red.		EUPHORBIACEAE (p.98)
	Leaves 10 to 25 cm long, 40 to 80 mm wide, on stalks 15 to 45 mm long, elliptical, tapering at both ends, soft, glabrous, marginal teeth small.		ESCALLONIACEAE (p.110)
	Leaves up to 22 cm long and 25 mm wide, toothed or entire, thin, tapering gradually to stalk. Flowers greenish yellow.		PROTEACEAE (p.157)
	Leaves opposite, oval with wavy margins, thick, shiny, pale on undersurface. Fruit an orange berry. Coastal.		CELASTRACEAE (p.261)
Shrubs or small trees not in rainforest	Leaves less than 25 mm wide	Branches narrow cylindrical, flexible, pendulous with whorls of tiny triangular teeth.	CASUARINACEAE (p.64)
		Branches long, wiry, pendulous, green or grey, with yellow or orange pea flowers.	FABACEAE (p.131)
		Leaves up to 25 cm long, linear or narrow lanceolate, usually toothed. Flowers red or greenish yellow.	PROTEACEAE (p.157)
	Leaves greater than 25 mm wide	Leaves stiff, toothed, rusty or grey beneath. Flowers in large spikes.	PROTEACEAE (p.157)
		Leaves leathery, toothed or entire, veins strongly marked. Flowers red or pale. Fruit narrow and leathery or a woody pear.	PROTEACEAE (p.157)
		Leaves rusty on undersurface. Milky sap in stems and leaf stalks.	MORACEAE (p.61)
		Leaves in vertical plane with raised gland on upper margin near stalk. Often sickle shaped with 2 to 5 nerves.	MIMOSACEAE (p.115)
		Leaves tapering to a point or rounded with eucalyptus smell.	MYRTACEAE (p.176)
		Leaves 10 to 20 cm long, 30 to 50 mm wide, on stalks 20 to 80 mm long, tapering to apex. Flowers white. Shrub clothed in woolly hairs.	ARALIACEAE (p.248)

Table 3.7

Shrubs. Leaves compound (divided into 3 or more separate leaflets).

			Family
Leaves divided into narrow segments	Segments again divided into irregular flat segments or narrow pointed segments.		PROTEACEAE (p.157)
Leaves with 3 or 5 leaflets spreading from the one point	Leaflets 3	Leaves opposite. Leaflets entire, narrow or elliptic, smooth or with rough glands, strong odour.	RUTACEAE (p.231)
		Leaflets toothed. Flowers small, white.	CUNONIACEAE (p.108)
		Leaflets soft, thick, irregularly lobed.	APIACEAE (p.250)
	Leaflets 3 or 5	Shrub with yellow pea flowers. Pod inflated.	FABACEAE (p.131)
Leaves divided pinnately into more than 3 leaflets, the leaflets entire, toothed or again divided	Leaflets less than 2 cm long	Leaves with oil glands and strong aromatic smell.	RUTACEAE (p.231)
		Leaves with stiff or soft hairs, without smell.	SAPINDACEAE (p.226)
	Leaflets 2 to 5 cm	Raised glands present on axis between some or all pairs of leaflets. Flowers yellow-orange.	MIMOSACEAE (p.115)
		Shrub to 1.5 m with mauve pea flowers.	FABACEAE (p.131)
	Leaflets 5 to 12 cm long	Leaflets thick, entire, oblanceolate.	MELIACEAE (p.230)
		Leaflets thin, toothed, on long main axis. Leaves opposite.	CAPRIFOLIACEAE (p.271)
		Leaflets entire, toothed towards apex or pinnately divided. Leaves large, alternate.	ARALIACEAE (p.248)

Table 4.1

Trees. Leaves simple, opposite, whorled or clustered.

			Family
Leaves with entire margins	Trees in or near rain-forest	Large palm with broad fan-shaped leaves 1 m across.	ARECACEAE (p.323)
		Small tree with red-brown flaky/stringy bark and 3-veined oval leaves drawn out to a point.	MYRTACEAE (p.176)
		Large emergent trees with red-brown fibrous bark. Flowers creamy or white, in globular clusters on long stalks.	MYRTACEAE (p.176)
		Trees often along creeks. Bark soft, pale brown. Flowers yellow.	MYRTACEAE (p.176)
		Leaves with prominent oil glands, tapering gradually or sharply to a point. Flowers white with 4 or 5 petals. Fruit a fleshy berry.	MYRTACEAE (p.176)
		Leaves with oil dots and aromatic smell, apex blunt. Leaf stalks with swollen elbow at top. Fruit 4-angled.	RUTACEAE (p.231)
		Leaves in clusters at ends of branches, 10 to 30 cm long, elliptic, on glabrous stalks 3 to 7 cm long, dark green and glossy above, pale green beneath. Fruit sticky.	NYCTAGINACEAE (p.66)
		Leaves opposite, 5 to 15 cm long, ovate, on hairy stalks up to 5 cm long. New growth and undersurface of leaves softly hairy.	VERBENACEAE (p.288)
		Leaves shortly stalked, glabrous, thick, glossy, dark green above, broadest near apex, narrowing abruptly to a short point.	EUPHORBIACEAE (p.98)
		Leaves rather stiff with a dry leathery feel, 6 to 14 cm long, tapering to both ends, apex acute. Fruit a blackish olive 12 mm long.	OLEACEAE (p.270)
	Trees not in rain-forest	Leaves up to 1 m long, less than 10 mm wide, clumped at top of rough caudex of old leaf bases. Flowers in a tall spike.	XANTHORRHOEACEAE (p.364)
		Small to large tree with narrow flexible cylindrical branches with whorls of tiny triangular teeth.	CASUARINACEAE (p.64)
		Mangrove growing in mud. Leaves dark green, whitish below, glossy, pointed at apex. Flowers orange.	AVICENNIACEAE (p.288)
		Tree resembling a eucalypt. Flowers 10 to 20 mm diameter, white or creamy, with 5 petals and numerous stamens, in large clusters at ends of branches. Fruit a ribbed capsule.	MYRTACEAE (p.176)
Leaves with toothed margins		Leaves stiff, dotted with oil glands, aromatic, veins strongly marked, raised on undersurface. Rainforest trees.	MONIMIACEAE (p.53)
		Leaves 8 to 16 cm long, narrow ovate, pointed at apex, dark green, with saw-tooth margins. Common tree with smooth grey bark growing in rainforest and along shaded creeks.	CUNONIACEAE (p.108)
		Leaves 5 to 12 cm long, 40 to 50 mm wide, thin, very glossy, glabrous, tapering at both ends, margins with irregularly spaced pointed teeth. In rainforest.	ESCALLONIACEAE (p.110)

Table 4.2

Trees. Leaves simple, alternate, margins entire.

			Family
Large trees in rain-forest	Leaf stalks longer than 15 mm	Leaves large, on long slender stalks, entire or 3, 5 or 7-lobed, deciduous. Flowers red. Fruit a leathery follicle.	STERCULIACEAE (p.78)
		Leaves round or nearly so, up to 30 cm diameter. Stinging hairs present on undersurface, stalks and branchlets.	URTICACEAE (p.63)
		Leaves 3 to 25 cm long, thick, shiny, dark green above, paler or rusty beneath. Milky sap in stems and stalks.	MORACEAE (p.61)
		Leaves 10 to 20 cm long, 6 to 12 cm wide, elliptic, on stalks 3 to 7 cm long. Fruit sticky.	NYCTAGINACEAE (p.66)
	Leaf stalks shorter than 15 mm	Leaves 6 to 15 cm long, elliptic, apex acutely pointed, aromatic, glossy green above, paler or glaucous beneath, midvein pale yellow. Fruit a black berry with 1 seed.	LAURACEAE (p.54)
		Leaves 4 to 12 cm long, ovate, apex acute, small depressions at junctions of mid and lateral veins. Fruit an ovoid black berry 12 to 20 mm long with 1 seed.	ICACINACEAE (p.261)
		Leaves 8 to 16 cm long, oblanceolate, bluntly pointed, exuding milky latex from broken stalk. Fruit a black plum about 50 mm diameter with 2 to 5 shiny brown seeds.	SAPOTACEAE (p.97)
Small to medium trees in or near rainforest	Leaves 2-ranked and/or stems zig-zag	Leaves 4 to 10 cm long, elliptic, blunt or rounded at apex, thick, not glossy, dark green above, pale and yellowish beneath.	EBENACEAE (p.96)
		Leaves 6 to 10 cm long, elliptic, shortly pointed, thin, shiny, unequal at base, 2 tiny stipules present in leaf axils.	EUPHORBIACEAE (p.98)
		Leaves 6 to 12 cm long, oblong-elliptic, soft, glossy both sides aromatic when crushed, oil glands minute.	EUPOMATIACEAE (p.53)
		Leaves 10 to 15 cm long, about 60 mm wide, ovate, shortly pointed at apex, small depressions in forks of lateral veins.	ICACINACEAE (p.261)
	Leaves alternate and clustered at ends of branches	Leaves 5 to 12 cm long, narrow ovate with undulate margins. Flowers white, fragrant, in terminal clusters. Fruit 10 mm diameter, orange. Widespread.	PITTOSPORACEAE (p.111)
		Leaves 10 to 15 cm long, widest above middle then tapering to a fine point, soft, thin. Flowers yellow, showy, fragrant. Northwards from Grose Vale.	PITTOSPORACEAE (p.111)
		Leaves 4 to 8 cm long, elliptic, with 5 to 7 veins running lengthwise, thin, glossy dark green above, pale green below.	EPACRIDACEAE (p.86)
	Leaf veins easily seen	Leaves 6 to 12 cm, elliptic, glossy dark green above, white beneath with dense hairs, young branches, inflorescence and leaf stalks with rusty brown hairs.	RHAMNACEAE (p.263)
		Leaves 10 to 12 cm long, elliptic, on stalks 10 to 20 mm long, often with a reddish tint.	ESCALLONIACEAE (p.110)
		Leaves 7 to 13 cm long, ovate, on stalks 10 to 20 mm long, numerous oil glands, aromatic smell. Illawarra.	RUTACEAE (p.231)
	Leaf veins obscure (except midvein)	Leaves 5 to 10 cm long, oblanceolate, soft, thin, dull green, tapering gradually to stalk. Shrub or small tree with corky bark.	SOLANACEAE (p.279)
		Leaves 5 to 10 cm long, glossy above, paler and duller beneath. Young shoots with short rusty hairs. Flowers and fruit in clusters on older stems.	MYRSINACEAE (p.97)
		Leaves 6 to 12 cm long, narrow lanceolate, tapering to both ends, apex acutely pointed. On rainforest margins near saltwater.	MYOPORACEAE (p.285)

Table 4.2

Trees. Leaves simple, alternate, margins entire.

		Family
Small to large trees in open forest and woodland	'Gum tree' with rough or smooth bark. Leaves with oil dots and eucalyptus smell when crushed.	MYRTACEAE (p.176)
	Leaves in vertical plane with a raised gland on the upper margin near the stalk. Veins 2 to 5 or penni-veined.	MIMOSACEAE (p.115)
	Leaves 10 to 20 cm long, leathery, veins strongly marked, tapering to a long pointed apex. Fruit a woody pear with 2 winged seeds.	PROTEACEAE (p.157)
	Leaves 5 to 10 cm long, obovate, very obtuse at apex, thick, leathery. Flowers white. Mangrove growing on mud flats.	MYRSINACEAE (p.97)
	Leaves reduced to minute scales. Fruit stalk enlarged and reddish. Small tree with drooping branchlets.	SANTALACEAE (p.255)

Table 4.3

Trees. Leaves simple, alternate, margins toothed.

			Family
Leaves thick, stiff or with a rough surface	Gnarled misshapen tree in dry forest. Leaves saw toothed, 8 to 16 cm long, 1 to 4 cm wide. Flowers in large erect grey-green spikes.		PROTEACEAE (p.157)
	Large shrubs or trees in moist forest or rain-forest	Medium to large tree along watercourses. Leaves approx 25 cm long, 80 mm wide, obovate, blade narrowing to stalk and ending abruptly, stalk swollen at top.	ELAEOCARPACEAE (p.77)
		Leaves rough on prickly stems, finely toothed, narrow elliptic or 3-lobed, approx 15 cm long. Flowers large, white-pink. Fruit a bristly capsule.	MALVACEAE (p.82)
		Leaves rough, finely toothed, ovate, 5 cm long. Flowers small. Fruit a small yellow berry.	MORACEAE (p.61)
		Leaves smooth, shiny on upper surface, strongly veined, margins with prickly teeth or entire.	MYRSINACEAE (p.97)
Leaves not noticeably thick or rough	Leaves large elliptic soft and thin	Leaf stalks grooved on upper surface with 2 to 4 raised glands at top. Leaves up to 18 cm long. Flowers small, yellowish, with numerous stamens.	EUPHORBIACEAE (p.98)
		Leaves up to 22 cm long, irregularly toothed, veins sunken above, raised below with short hairs. Fruit a purple-black berry.	ESCALLONIACEAE (p.110)
		Leaves up to 20 cm long, regularly and finely toothed. Flowers white, 5 mm diameter, scented, in narrow panicles. Fruit yellow-orange berry.	EHRETIACEAE (p.279)
	Leaves not as above	Leaves rhomboidal, 4 to 10 cm long, entire or with 1 to 3 angles on each margin, 3-veined from the base, the 2 lateral veins parallel with margins. Small tree.	FLACOURTIACEAE (p.83)
		Leaves elliptic, about 10 cm long, fine reticulate venation clearly visible. Flowers white, petals fringed. Fruit a blue berry. Small tree.	ELAEOCARPACEAE (p.77)
		Leaves narrow, elliptic, about 20 cm long on stalks 45 mm long. Flowers white, petals fringed. Fruit a blue berry. Tall tree in rainforest.	ELAEOCARPACEAE (p.77)

Table 4.4

Trees. Leaves compound (divided into 3 or more separate leaflets).

			Family
Leaflets toothed	Leaves pinnate	Leaflets 2 to 6 (mostly 4), about 10 cm long, sub-opposite, distantly and irregularly toothed. Small tree in rainforest. ...	SAPINDACEAE (p.226)
		Leaflets 11 or more, up to 20 cm long, opposite, a terminal leaflet present, finely toothed. Rainforest.	ARALIACEAE (p.248)
	Leaves bipinnate	Leaflets 5 to 9 per pinna, 3 to 6 cm long, thin, ovate-acuminate. Flowers purplish. Fruit a yellow berry.	MELIACEAE (p.230)
Leaflets with entire margins	Leaves 3-foliate	Leaves mostly opposite. Leaflets 8 to 10 cm long, 30 to 40 mm wide, on stalks 60 mm long, glossy above, paler and duller beneath.	RUTACEAE (p.231)
	Leaves pinnate	Leaflets 2 to 4, 5 to 7 cm long, alternate on rachis, rounded or notched at apex, glaucous or paler on undersurface.	SAPINDACEAE (p.226)
		Leaflets 4 to 10, 5 to 12 cm long, alternate, very oblique at base, exuding clear sap when cut. Uncommon tree.	ANACARDIACEAE (p.229)
		Leaflets 5 to 9, 7 to 9 cm long, opposite, symmetrical at base, a terminal leaflet present. Common shrub or small tree in rainforest and gullies.	MELIACEAE (p.230)
		Leaflets 5 to 11, 2 to 6 cm long, oblong, rounded at apex with small bristle, glossy dark green above, paler and finely hairy beneath. Flowers white, showy. Rainforests in south. ..	EUCRYPHIACEAE (p.107)
		Leaflets 5 to 11 (mostly 8), 6 to 12 cm long, alternate, broad, apex rounded or notched. Near the sea.	SAPINDACEAE (p.226)
		Leaflets 8 to 12, 15 to 30 cm long, 10 cm wide, oblong-elliptic. Branches, leaves and inflorescence rusty hairy.	SAPINDACEAE (p.226)
		Leaflets 10 to 20, about 10 cm long, in opposite pairs without a terminal leaflet, oblique at base. Tall tree in rainforest.	MELIACEAE (p.230)
		Leaflets many, on branches up to 3 m long. Tall palm in rainforest.	ARECACEAE (p.323)
	Leaves bipinnate	Pinnae few, leaflets 2 to 8 cm long, rhomboidal, asymmetric, distinctly alternate. Fruit a twisted pod.	CAESALPINIACEAE (p.130)
		Leaves variable on the one tree, both bipinnate and pinnate. Leaflets entire or toothed near apex, opposite with a terminal leaflet.	ARALIACEAE (p.248)
		Leaves with numerous leaflets (pinnules), 3 to 50 mm long. Prominent raised gland on rachis between some or all pinnae. Usually not in rainforest.	MIMOSACEAE (p.115)

Table 5.1

Aquatic plants.

			Family
Marine (saltwater) plants	Leaves strap-shaped, 3 to 4 mm wide, 5 to 50 cm long. Submerged plant with creeping rhizome, often exposed at low tide.		ZOSTERACEAE (p.323)
	Leaves strap-shaped, 10 to 11 mm wide, 5 to 70 cm long. Submerged plant with creeping rhizome, usually below low tide mark.		POSIDONIACEAE (p.322)
	Leaves opposite, oblong-elliptic 3 to 5 cm long, on stalks 2 to 8 cm long, in pairs at each node. Submerged plant with creeping stems.		HYDROCHARITACEAE (p.430)
Freshwater plants with leaves floating on surface	Leaves long, strap-shaped	Leaves 20 to 35 mm wide, up to 2 m long, rather thick, veins inconspicuous, upper part floating. Flowers small, in a spike 15 to 25 cm long on a stout, erect, emergent stem.	JUNCAGINACEAE (p.321)
		Leaves 10 to 20 mm wide with 5 to 9 conspicuous longitudinal veins and cross connections. Flowers floating, on long spiral stalks.	HYDROCHARITACEAE (p.430)
	Leaves elliptic to round or very small	Leaf blades ovate-elliptic, 4 to 12 cm long, with 5 to 7 main veins prominent on undersurface. Leaves arising from rootstock growing on bottom. Flowers white, 3 to 5 cm diameter.	HYDROCHARITACEAE (p.430)
		Leaf blades ovate-elliptic, 5 to 10 cm long, with 10 to 30 parallel veins, attached by long stalks to floating stems. Submerged leaves long, narrow. Flowers in a dense spike 2 to 5 cm long.	POTAMOGETONACEAE (p.322)
		Leaves 3 to 8 cm diameter arising on long stalks from the nodes of submerged stolons. Flowers yellow, 25 to 35 mm diameter, with fringed petals.	MENYANTHACEAE (p.274)
		Leaves oval, 5 to 10 cm long, 3 to 7 cm wide with centrally attached stalked. Submerged growth with thick gelatinous coating.	CABOMBACEAE (p.58)
		Small floating herb consisting of a flat or rounded disc 1 to 4 mm long with 1 or more roots attached below.	LEMNACEAE (p.325)
Freshwater plants with submerged leaves	Leaves long, strap-shaped, tufted, 10 to 20 mm wide. Flowers floating, attached to the bottom by long spiral stalks.		HYDROCHARITACEAE (p.430)
	Leaves linear lanceolate, 1 to 6 cm long, with flat or curled margins, or ovate and stem clasping, 1 to 4 cm long, attached to floating stems, alternate. Flower spikes 1 to 5 cm long.		POTAMOGETONACEAE (p.322)
	Leaves linear, 12 to 25 mm long, toothed, opposite or whorled, dilated at base into a short broad sheath. Flowers inconspicuous, submerged.		NAJADACEAE (p.322)
	Leaves 2 to 12 cm long, 1 mm wide, sheathed, attached alternately to slender stems 1 mm diameter. Flowers emergent, tiny, in a spikelet about 5 mm long. Sometimes forming dense submerged growth.		CYPERACEAE (p.334)
	Leaves divided into fine segments with attached bladder-traps. Flowers yellow, 2-lipped, 5 to 10 mm long, on erect emergent stalks about 10 cm high.		LENTIBULARIACEAE (p.285)
Freshwater plants with some leaves emergent	Leaves 1 to 5 cm long, finely divided, on long stalks. Herb with yellow buttercup flowers with usually 5 petals.		RANUNCULACEAE (p.58)
	Leaves 2 to 10 cm long, entire or deeply dissected, thick and fleshy. Flowers yellow, in slightly rounded heads 5 to 10 mm diameter. Small creeping herb.		ASTERACEAE (p.306)
	Leaves 3 to 6 cm long, narrow ovate, tapering to the stalk, attached alternately to floating stems. Flowers yellow, 10 mm long on stalks shorter than the leaves.		ONAGRACEAE (p.223)
	Emergent leaves 2 cm long, 1 to 2 mm wide, linear, entire or toothed, in whorls of 3 to 8 (mostly 6). Submerged leaves finely dissected. Flowers white, small, in upper axils.		HALORAGACEAE (p.224)

Table 5.1

Aquatic plants.

			Family
Tufted erect emergent plants	Leaf blades ovate	Blades 3 to 12 cm long with 3 to 7 longitudinal veins, on long erect stalks. Flowers white, 5 mm diameter, with 3 petals and 6 stamens, in a large open panicle.	ALISEMATACEAE (p.320)
		Blades 6 to 12 cm long with spreading pinnate veins, thick, matt, on long erect stalks. Flowers yellow, 20 to 25 mm diameter, with 5 lobes bearded at the base.	MENYANTHACEAE (p.274)
	Leaves narrow, at nodes on the stem	Leaves narrow lanceolate, 2 to 3 cm wide, tapering to a long point, with the bases encircling bamboo-like stems. Inflorescence a large plume-like whitish panicle.	POACEAE (p.349)
	Leaves narrow and sheathing on the lower part of the stem, or absent	Tall robust reed-like plant 1 to 3 m tall with leaves in 2 rows overlapping on the lower part of the stem. Flowers crowded into a dense velvety brown spike.	TYPHACEAE (p.356)
		Woolly-hairy perennial 0.5 to 1 m tall in ponds and slow flowing streams. Flowers yellow, 2 to 4 cm long enclosed in pointed brown bracts. Common.	PHILYDRACEAE (p.372)
		Perennial about 1 m tall with thick spongy sheathing leaves about 5 mm wide, midrib conspicuous on the underside. Flowers white, in globular clusters along the upper stem.	SPARGANIACEAE (p.355)
		Rush, in or near water. Leaves 1 to 2 mm diameter, or reduced to sheaths at the base of stems. Flowers tiny with 6 scaly segments, arranged in loose cymes or heads.	JUNCACEAE (p.332)
		Leafless spike rush with hollow stems 5 to 10 mm diameter with white transverse partitions, or narrower without partitions. Inflorescence a spikelet of overlapping glumes with a tiny flower in each axil. Spikelet 5 to 50 mm long. ...	CYPERACEAE (p.334)

Index

Common Name Index

Botanical Name Index